Illustrated Guide to the National Electrical Code®

2nd Edition

Charles R. Miller

DELMAR
THOMSON LEARNING™

Australia Canada Mexico Singapore Spain United Kingdom United States

Illustrated Guide to the National Electrical Code®, 2nd edition

Charles R. Miller

Business Unit Director:
Alar Elken

Executive Editor:
Sandy Clark

Acquisitions Editor:
Mark Huth

Editorial Assistant:
Dawn Daugherty

Developmental Editor:
Jennifer A. Thompson

Executive Marketing Manager:
Maura Theriault

Channel Manager:
Fair Huntoon

Marketing Coordinator:
Brian McGrath

Executive Production Manager:
Mary Ellen Black

Production Coordinator:
Toni Hansen

Project Editor:
Barbara L. Diaz

Art and Design Coordinator:
Cheri Plasse

Cover Design:
Charles Cummings, Advertising

Printed in the United States of America

4 5 XXX 05 04 03 02

For more information contact Delmar,

5 Maxwell Drive
Clifton Park, NY 12065

Or find us on the World Wide Web at

http://www.delmar.com

Library of Congress Cataloging-in-Publication Data

Miller, Charles R., 1955-
Illustrated guide to the National Electric Code / Charles R. Miller — 2nd. ed.
p. cm.
Includes index.
ISBN 0-7668-7334-X (alk. paper)
1. Electric engineering—Insurance requirements—United States. 2. Electric wiring—Insurance requirements—United States. 3. National Fire Protection Association.
National Electrical Code. I. Title
TK260.M52 2001
621.319'24'021873--dc21
2001028120

NOTICE TO THE READER

Publisher does not warrant or guarantee any of the products described herein or perform any independent analysis in connection with any of the product information contained herein. Publisher does not assume, and expressly disclaims, any obligation to obtain and include information other than that provided to it by the manufacturer.

The reader is expressly warned to consider and adopt all safety precautions that might be indicated by the activities herein and to avoid all potential hazards. By following the instructions contained herein, the reader willingly assumes all risks in connections with such instructions.

The publisher makes no representation or warranties of any kind, including but not limited to, the warranties of fitness for particular purpose or merchantability, nor are any such representations implied with respect to the material set forth herein, and the publisher takes no responsibility with respect to such material. The publisher shall not be liable for any special, consequential, or exemplary damages resulting, in whole or part, from the readers' use of, or reliance upon, this material.

Contents

SECTION 1 FOUNDATIONAL PROVISIONS

SECTION 2 ONE-FAMILY DWELLINGS

SECTION 3 MULTI-FAMILY DWELLINGS

SECTION 4 COMMERCIAL LOCATIONS

SECTION 5 SPECIAL OCCUPANCIES, AREAS, AND EQUIPMENT

Preface

Illustrated Guide to the NEC® offers an exciting new approach to understanding and applying the provisions of the *National Electrical Code®*. Unlike the *Code*, this text gathers and presents detailed information in a format, such as One-Family or Multi-Family Dwellings, based upon "type of occupancy." Code specifications applicable to a given type of occupancy are logically organized in easy-to-read units and graphically enhanced by numerous technical illustrations. Going an extra step, the occupancy-specific material is subdivided into specific rooms and areas. Information relevant to more than one type of occupancy is organized into independent units for easier reference. For instance, items such as raceways and conductors are covered in Unit 5 but are related to every type of occupancy.

Students who wish to acquire a comprehensive grasp of all electrical codes will want to study this text section by section and unit by unit. Practicing electricians who have specialized in one type of occupancy and who wish to understand an unfamiliar segment may want to focus on those new areas.

Example: An electrician who has been wiring commercial facilities for a number of years wants to wire a new house. Being unfamiliar with the codes concerning residential wiring, this individual can turn to Section 2, "One-Family Dwellings." Here, everything from receptacle placement to the placement of the service point is explained. Section 2 is made up of four units: Units 6 through 9. Unit 6, "General Provisions," contains general requirements for one-family dwellings, both interior and exterior. Unit 7, "Specific Provisions," addresses more complex issues, requiring additional provisions for specific areas such as kitchens, hallways, clothes closets, bathrooms, garages, basements, etc. Unit 8, "Load Calculations," simplifies the standard as well as optional load calculation methods for one-family dwellings. Unit 9, "Services and Electrical Equipment," is divided into five subheadings: Service-Entrance Wiring Methods, Service and Outside Wiring Clearances, Working Space Around Equipment, Service Equipment and Panelboards, and Grounding.

The "what," "when," "where" adoption of the provisions of the *NEC®* is under the discretionary control of state and local jurisdictions. State and local jurisdictions also have the liberty of appending additional codes, which in many cases may be more stringent than those outlined by the *NEC®*. The *Code* may be adopted in whole or in part. For example, while some local codes do not allow the use of nonmetallic-sheathed cable for residential or commercial wiring, others allow its use in residential, but not in commercial, wiring applications. To ensure compliance, obtain a copy of any additional rules and regulations for your area.

This guide's objective is to provide the information needed to complete your project—without the necessity of learning the *NEC®* from cover to cover. *Illustrated Guide to the National Electrical Code®* will bring your project to life as quickly and as accurately as any text on the market today. In the electrical field, as in any career, the learning experience never ends. Whether you are an electrician's apprentice, a master electrician, or an electrical inspector, *Illustrated Guide to the National Electrical Code®* has something for you. We believe you will find it to be a valuable addition to your reference library. In fact, you may want to include it in your toolbox or briefcase!

Please note that this guide book was completed after all the normal steps in the NFPA 70 review cycle—Proposals to Code-Making Panels, review by Technical Correlating Committee, Report on Proposals, Comments to Code-Making Panels, review by Technical Correlating Committee, Report on Comments, NFPA Annual Meeting, and ANSI Standards Council—and before the actual publication of the 2002 edition of the *NEC®*. Every effort has been made to be technically correct, but there is always the possibility of typographical errors or appeals made to the NFPA Board of Directors after the normal review cycle that could change the appearance or substance of the Code.

If changes do occur after the printing of this book, they will be included in the Instructors Guide and will be incorporated into the guidebook in its next printing.

Please note also that the Code has a standard method to introduce changes between review cycles, called "Tentative Interim Amendment," or TIA. These TIAs and typographical errors can be downloaded from the NFPA website, www.nfpa.org, to make your copy of the Code current.

ABOUT THE AUTHOR

For eighteen years, Charles R. Miller owned and operated a successful commercial electrical contracting company (Lighthouse Electric Co., Inc.) in Nashville, Tennessee. Throughout those years, he prided himself on solving problems abandoned by less-skilled or less-dedicated technicians. In 1988 he began operating a second company, dedicated to electrical-related training and known as Lighthouse Educational Services. Mr. Miller teaches custom-tailored classes and seminars covering various aspects of the electrical industry. Hundreds of students have taken advantage of his extensive experience in electrical contracting, regulatory exams (current electrical codes), and electrical-related business and law. Class and seminar attendees have included individuals employed by companies such as Ford, Textron, The Aerostructures Corporation, and Aladdin Industries; by academic institutions such as Tennessee State and Vanderbuilt Universities; and by governmental agencies, including NASA.

Charles Miller has dedicated over 5,000 hours to making *Illustrated Guide to the National Electrical Code®* a reality. His unsurpassed attention to detail is evident on every page. Since this book's conception, every day's waking hours have been consumed with careful planning and execution of content and design. His unwavering commitment to quality, from the first page in Unit 1 to the last page in Unit 19, has produced a technically superior, quintessentially user-friendly guide.

ACKNOWLEDGMENTS

I would like to say "thank you" to my children, Christin and Adam, for being patient and understanding during the extremely long hours and endless days working on this text. My mother, Evelyn Miller, gets a special "thank you" and "I love you" for a lifetime of support and encouragement. She called every day to check on me and quite often sent encouraging greeting cards that always came at just the right time.

Thank you to my Senior Editor at Delmar, Mark Huth, for the privilege of writing for such a professional publishing company. I also would like to thank the entire Delmar project team comprised of all of those listed on the copyright page at the front of this book.

Last, but not least, the author and Delmar would like to thank the following reviewers for their contributions:

Kermit Davis, Jr.
Spotsylvania Vocational Center

Greg Fletcher
Kennebec Valley Technical College

Lanny McMahill
City of Phoenix

Gary Reiman
Dunwoody Institute

Kevin Weigman
Northeast Wisconsin Technical College

Metrics (SI) and the *NEC* ®

The United States is the last major country in the world not using the metric system as the primary system. We have been very comfortable using English (United States Customary) values, however this is now changing. Manufacturers are now showing both inch-pound and metric dimensions in their catalogs. Plans and specifications for governmental new construction and renovation projects started after January 1, 1994 have been done using the metric system. You may not feel comfortable with metrics, but metrics are here to stay. You might just as well get familiar with the metric system.

The *NEC* ®and other *NFPA* Standards are becoming international standards. All measurements in the 2002 *NEC* ® are shown with metrics first, followed by the inch-pound value in parentheses. For example, 600 mm (24 in.).

In this guidebook, ease in understanding is of utmost importance. Therefore, inch-pound values are shown first, followed by metric values in parentheses. For example, 24 in. (600 mm).

A *soft metric conversion* is when the dimensions of a product already designed and manufactured to the inch-pound system have their dimensions converted to metric dimensions. The product does not change in size.

A *hard metric measurement* is where a product has been designed to SI metric dimensions. No conversion from inch-pound measurement units is involved. A *hard conversion* is where an existing product is redesigned into a new size.

In the 2002 edition of the *NEC* ®, existing inch-pound dimensions did not change. Metric conversions were made, then rounded off. Where rounding off would create a safety hazard, the metric conversions are mathematically identical.

For example, if a dimension is required to be six ft, it is shown in the *NEC* ® as 1.8 m (6 ft). Note that the 6 ft remains the same, and the metric value of 1.83 m has been rounded off to 1.8 m. This edition of the Illustrated Guide to the *NEC* ® reflects these rounded off changes, except that the inch-pound measurement is shown first, i.e., 6 ft (1.8 m).

Trade Sizes

A unique situation exists. Strange as it may seem, what electricians have been referring to for years has not been correct!

Raceway sizes have always been an approximation. For example, there has never been a ½-in. raceway! Measurements taken from the *NEC* ® for a few types of raceways show the following:

Trade Size	Inside Diameter (I.D.)
½ Electrical Metallic Tubing	0.622 in.
½ Electrical Nonmetallic Tubing	0.560 in.
½ Flexible Metal Conduit	0.635 in.
½ Rigid Metal Conduit	0.632 in.
½ Intermediate Metal Conduit	0.660 in.

You can readily see that the cross-sectional areas, critical when determining conductor fill, are different. It makes sense to refer to conduit, raceway, and tubing sizes as *trade sizes.* The *NEC* ® in *90.9(C)(1)* states that "where the actual measured size of a product is not the same as the nominal size, trade size designators shall be used rather than dimensions. Trade practices shall be followed in all cases." This edition of the Illustrated Guide to the *NEC* ®uses the term *trade size* when referring to conduits, raceways, and tubing. For example, instead of referring to a ½-in. EMT, it is referred to as trade size ½ EMT.

The *NEC* ® also uses the term *metric designator.* A ½-in. EMT is shown as *metric designator 16 (½).* A 1-in. EMT is shown as *metric designator 27 (1).* The numbers 16 and 27 are the *metric designator* values. The *(½)* and (1) are the *trade sizes.* The metric designator is the raceways' inside diameter—in rounded off millimeters (mm). Here are some of the more common sizes of conduit, raceways, and tubing. A complete table is found in the *NEC* ®, *Table 300.1(C).* Because of possible confusion, this text uses only the term *trade size* when referring to conduit and raceway sizes.

Metric Designator & Trade Size

Metric Designator	Trade Size
12	⅜
16	½
21	¾
27	1
35	1¼
41	1½
53	2
63	2½
78	3

Conduit knockouts in boxes do not measure up to what we call them. Here are some examples.

Trade Size Knockout	Actual Measurement
½	⅞ in.
¾	1³⁄₃₂ in.
1	1⅜ in.

Outlet boxes and device boxes use their outside measurements as their trade size. The volume allowed for conductors is based on the amount of free room inside the box, not outside the box. For example, a 4 in. x 4 in. x 1½ in. box has a volume of 21 cubic-in., not 24. Table 314.16(A) contains box-fill information for certain size metal boxes. Each box's *trade size* is shown in two columns—millimeters and inches. The volume allowed for conductors is also shown in two columns—cubic centimeters and cubic inches. In this text, the dimensions for device and junction boxes are shown in inches and cubic-inches. If metric dimensions are needed, simply reference Table 314.16(A).

For larger boxes in this text that are not shown in Table 314.16(A), metric values can be determined based on the English values that are given. The metric value (in millimeters) can be obtained by multiplying the size of the box (in inches) by 25.4. For example, what is the metric equivalency (in millimeters) of a pull box that measures 24 inches? Twenty-four inches is equal to 609.6 mm (24 x 25.4 = 609.6). Although the exact conversion is 25.4 millimeters per inch, most sections in the Code use the rounded number of 25 millimeters per inch. Therefore throughout most of the Code, 24 in. is equal to 600 mm.

Trade sizes for construction material will not change. A 2 x 4 is really a *name,* not an actual dimension. A 2 x 4 will still be referred to as a 2 x 4. This is its *trade size.*

In this text, measurements directly related to the *NEC* ® are given in both inch-pound and metric units. In many instances, only the inch-pound units are shown. This is particularly true for the examples of raceway calculations, box fill calculations, and load calculations for square foot areas.

Because the *NEC* ® rounded off most metric conversion values, a computation using metrics results in a different answer when compared to the same computation done using inch-pounds. For example, load calculations for a residence are based on 3 volt-amperes per square foot or 33 volt-amperes per square meter.

For a 40 ft x 50 ft dwelling: 3 VA x 40 ft x 50 ft = 6000 volt-amperes.

In metrics, using the rounded off values in the *NEC* ®: 33VA x 12 m x 15 m = 5940 volt-amperes.

The difference is small, but nevertheless there is a difference.

To show calculations in both units throughout this text would be very difficult to understand and would take up too much space. Calculations in either metrics or inch-pounds are in compliance with the *NEC* ®, 90.9(D). In 90.9(C)(3) we find that metric units are not required if the industry practice is to use inch-pound units.

It is interesting to note that the examples in *Chapter 9 of the NEC* ® use inch-pound units, not metrics.

FOUNDATIONAL PROVISIONS

section 1

UNIT 1 Introduction to the *National Electrical Code®*

Objectives

After studying this unit, the student should:

- be able to give a brief account of electricity in its infancy.
- be able to identify the catalyst which brought about the *National Electric Code* (*NEC*)®.
- understand how the *NEC®* began and its purpose.
- understand how changes to the *Code* evolve.
- be familiar with the terminology, presentation, and format of the *NEC®*.
- know what type of information is found in the *NEC®* (its layout).
- understand the *NEC®*'s concern with equipment and material standards.
- be able to recognize various trademark logos that denote listed and labeled products.
- comprehend the role of Nationally Recognized Testing Laboratories (NRTL) as well as the National Electrical Manufacturers Association (NEMA) and the expanded role of the National Fire Protection Association (NFPA).
- be familiar with this book's layout, text conventions, and illustration methods.
- be advised on how to study the *Illustrated Guide to the NEC®*.
- be aware that electrical requirements in addition to the *NEC®* may exist, and if so, that compliance is required.

THE *NATIONAL ELECTRICAL CODE®*

Just as an extensive education is required for doctors to perform the duties of their chosen field, a working knowledge of the *National Electrical Code* (*NEC*)® is a necessity for anyone practicing a profession in the electrical industry. The *NEC®* provides the standards by which all electrical installations are judged. Although other requirements, such as local ordinances and manufacturer instructions, must be applied, the *NEC®* is the foundation upon which successful installations are built. It is the most widely recognized and utilized compilation of technical rules for the installation and operation of electrical systems in the world today. Because of its widespread effect on the industry, it is important to understand the history of the *NEC®*.

The Beginning

In 1882, New York City was home to the first central-station electric generating plant developed by Thomas A. Edison. The Pearl Street Station began operation at 3:00 P.M. on Monday, September 4. Fifty-nine customers had reluctantly consented to have their houses wired on the promise of three free months of electric light. They were given the option of discarding the service if it proved to be unsatisfactory. But this new way of lighting was more than satisfactory . . . it was a sensation. The number of customers tripled in only four months. And, as they say, the rest is history. The new industry swept the nation: New construction included the installation of electricity, while property owners demanded that existing structures be updated, as well. New materials and equipment were developed and manufactured, and methods for installing and connecting these items to the electrical source were devised. For more than a decade, manufacturers, architects, engineers, inventors, electricians, and others worked independently to develop their contributions to the new technology. By 1895, there were as many as five different electric installation codes in use, and no single set of codes was accepted by all. To further complicate matters, there was an unexpected hazard darkening the prospects of this new industry.

Purpose and History of the *NEC*®

Electrically caused fires were becoming commonplace and, by 1897, the problem was reaching epidemic proportions. A diverse group of knowledgeable, concerned individuals assembled to address this critical issue. The need for standardization was apparent. The consensus of more than 1,200 individuals produced the first set of nationally-adopted rules to govern electrical installations and operations—the *National Electrical Code*®

The *NEC*® states its purpose as ". . . the practical safeguarding of persons and property from hazards arising from the use of electricity." This objective has remained constant throughout the *NEC*®'s existence, and the principles it contains continue to grow and change with the dynamic electrical industry.

Code Changes

The *NEC*® is regularly revised to reflect the evolution of products, materials, and installation techniques. Since 1911, the National Fire Protection Association (NFPA) of Quincy, Massachusetts, has been responsible for the maintenance and publication of the *NEC*®. The 2002 edition, which contains hundreds of reworded, as well as new, regulations, represents the diligent work of twenty code-making panels, composed strictly of volunteers from all professions within the electrical industry.

These panels are complemented by a host of private individuals who submit proposals (more than 4,200 for the 2002 edition), or comment on proposals already submitted, for changes to the *NEC*®. Anyone who wishes to participate can contact the National Fire Protection Association, 1 Batterymarch Park, Quincy, MA 02269, and request their free booklet, "The NFPA Standards-Making System." The current edition of the *NEC*® provides a form in the back of the book for submitting code change suggestions. (See the form which follows.)

Now that the background of the *NEC*® is an open book, let's take a closer look at what's inside it, and how we can go about understanding it.

FORM FOR PROPOSALS FOR 2005 *NATIONAL ELECTRICAL CODE*®

Mail to: Secretary, Standards Council
National Fire Protection Association
1 Batterymarch Park, P.O. Box 9101
Quincy, Massachusetts 02269-9101
Fax to: 617-770-3500

FOR OFFICE USE ONLY
Log # ____________
Date Rec'd ____________

Notes:
1. All proposals must be received by 5:00 p.m. EST on Friday, November 1, 2002
   Proposals received after 5:00 p.m. EST, Friday, November 1, 2002 will be returned to the submitter
2. Type or print legibly in black ink. Limit each proposal to a SINGLE section. Use a separate copy for each proposal.
3. If supplementary material (photographs, diagrams, reports, etc.) is included, you may be required to submit sufficient copies for all members and alternates of the technical committee.

Please indicate in which format you wish to receive your ROP/ROC: ☐ electronic or ☐ paper ☐ download

Date ____________ Name ____________ Tel. No. ____________

Company ____________

Street Address ____________

Please Indicate Organization Represented (if any) ____________

1. Section/Paragraph ____________

2. Proposal Recommends: (Check one) ☐ new text ☐ revised text ☐ deleted text

3. Proposal (include proposed new or revised wording, or identification of wording to be deleted): (Note: Proposed text should be in legislative format: i.e., use underscore to denote wording to be inserted (inserted wording) and strike-through to denote wording to be deleted

4. Statement of Problem and Substantiation for Proposal: (Note: State the problem that will be resolved by your recommendation; give the specific reason for your proposal including copies of tests, research papers, fire experience, etc If more than 200 words, it may be abstracted for publication)

5. ☐ This Proposal is original material. (Note Original material is considered to be the submitter's own idea based on or as a result of his/her own experience, thought, or research and, to the best of his/her knowledge, is not copied from another source)
☐ This Proposal is not original material; its source (if known) is as follows: ____________

If you need further information on the standards-making process, please contact the Standards Administration Department at 617-984-7249.
For technical assistance, please call NFPA at 617-770-3000.

I hereby grant the NFPA the non-exclusive, royalty-free rights, including non-exclusive, royalty-free rights in copyright, in this proposal and I understand that I acquire no rights in any publication of NFPA in which this proposal, in this or another similar or analogous form, is used.

____
Signature (Required)

PLEASE USE SEPARATE FORM FOR EACH PROPOSAL

NEC® Terminology, Presentation, and Format

Tables present a requirement's multiple application possibilities.

Table 210.21(B)(3) Receptacle Ratings for Various Size Circuits

Circuit Rating (Amperes)	Receptacle Rating (Amperes)
15	Not over 15
20	15 or 20
30	30
40	40 or 50
50	50

Diagrams, or figures, are used to further clarify *NEC*® applications.

Receptacles Caps

Figure 551.46(C) Configurations for grounding-type receptacles and attachment plug caps used for recreational vehicle supply cords and recreational vehicle lots.

NFPA document number followed by a page number.

Parts (subheadings) are used to break down Articles into simpler topics. (Not all Articles have subheadings.)

Sections are numerical listings where the Code requirements are located.

Vertical lines are placed in outside margins to identify material which has been added or altered since the last *NEC*® edition.

(FPN) stands for Fine Print Note and designates explanatory material. These are informational only and do not require compliance »90.5(C)«.

Exceptions appear in *italics* and explain when and where a specific rule does not apply.

70–62 ARTICLE 220 — BRANCH-CIRCUIT, FEEL

I. General

220.1 Scope. This article provides requirements for computing branch-circuit, feeder, and service loads.

Exception: Branch-circuit and feeder calculations for electrolytic cells as covered in 668.3(C)(1) and (4).

220.2 Computations.

(A) Voltages. Unless other voltages are specified, for purposes of computing branch-circuit and feeder loads, nominal system voltages of 120, 120/240, 208Y/120, 240, 347, 480Y/277, 480, 600Y/347, and 600 volts shall be used.

(B) Fractions of an Ampere. Where computations result in a fraction of an ampere that is less than 0.5, such fractions shall be permitted to be dropped.

220.3 Computation of Branch Circuit Loads. Branch-circuit loads shall be computed as shown in 220.3(A) through (C).

(A) Lighting Load for Specified Occupancies. A unit load of not less than that specified in Table 220.3(A) for occupancies specified therein shall constitute the minimum lighting load. The floor area for each floor shall be computed from the outside dimensions of the building, dwelling unit, or other area involved. For dwelling units, the computed floor area shall not include open porches, garages, or unused or unfinished spaces not adaptable for future use.

FPN: The unit values herein are based on minimum load conditions and 100 percent power factor and may not provide sufficient capacity for the installation contemplated.

(B) Other Loads — All Occupancies. In all occupancies, the minimum load for each outlet for general-use receptacles and outlets not used for general illumination shall not be less than that computed in 220.3(B)(1) through (11), the loads shown being based on nominal branch-circuit voltages.

Exception: The loads of outlets serving switchboards and switching frames in telephone exchanges shall be waived from the computations.

(1) Specific Appliances or Loads. An outlet for a specific appliance or other load not covered in (2) through (11) shall be computed based on the ampere rating of the appliance or load served.

(2) Electric Dryers and Household Electric Cooking Appliances. Load computations shall be permitted as specified in 220.18 for electric dryers and in 220.19 for electric ranges and other cooking appliances.

(3) Motor Loads. Outlets for motor loads shall be computed in accordance with the requirements in 430.22, 430.24, and 440.6.

(4) Recessed Luminaires (Lighting Fixtures). An outlet supplying recessed luminaire(s) [lighting fixture(s)] shall be computed based on the maximum volt-ampere rating of the equipment and lamps for which the luminaire(s) [fixture(s)] is rated.

(5) Heavy-Duty Lampholders. Outlets for heavy-duty lampholders shall be computed at a minimum of 600 volt-amperes.

CAUTION *Be advised that the local authority having jurisdiction has the ability to amend the Code requirements. Consult the proper authority to obtain applicable guidelines.*

Mandatory rules use the terms "shall" or "shall not" and require compliance »90.5(A)«.

Permissive rules contain the phrases "shall be permitted" or "shall not be required." These phrases normally describe options or alternative methods. Compliance is discretionary »90.5(B)«.

Formal Interpretations

Section 90-6 states: *To promote uniformity of interpretation and application of the provisions of this* Code, *formal interpretation procedures have been established and are found in the NFPA Regulations Governing Committee Projects.*

The *NEC*® Layout

The table of contents in the *NEC*® provides a breakdown of the information found in the book. Chapters 1 through 4 contain the most-often used Articles in the *Code*, because they include general, or basic, provisions. Chapter 1, while relatively brief, includes definitions essential to the proper application of the *NEC*®. It also includes an introduction and a variety of general requirements for electrical installations. More general requirements are found in Chapters 2, 3, and 4, addressing Wiring and Protection, Wiring Methods and Materials, and Equipment for General Use. Special issues are covered in Chapters 5 through 7. Chapter 5 contains information on Special Occupancies; Chapter 6, Special Equipment; and Chapter 7, Special Conditions. The contents of these chapters are applied in addition to the general rules given in earlier chapters. Chapter 8 covers Communications Systems and is basically independent of other chapters, except where cross-references are given. The final chapter, Chapter 9, contains Tables and Examples. Each chapter contains one or more Articles, while each Article contains Sections. Sections may be further subdivided by the use of lettered or numbered paragraphs. The *Code* is completed by Annexes A through F, an index; and a proposal form.

WIRING SYSTEM PRODUCT STANDARDS

In addition to installation rules, the *NEC*® is concerned with the type and quality of electrical wiring system materials. Two terms are synonymous with acceptability in this area: they are **Labeled** and **Listed**. These definitions, found in Article 100, are very similar. Similarities include: (1) an organization that is responsible for providing the listing or labeling; (2) these organizations must be acceptable to the authority having jurisdiction; (3) both are concerned with the evaluation of products; and (4) both maintain periodic inspection of the production (or manufacturing) of the equipment or materials which have been listed or labeled. A manufacturer of labeled equipment (or material) must continue to comply with the appropriate standards (or performance) under which the labeling was granted. Listed also means that the equipment, materials, or services meets appropriate designated standards or have been tested and found suitable for a specified purpose. This information is compiled and published by the organization. The Fine Print Note under **Listed** states that each organization may have different means for identifying listed equipment. In fact, some do not recognize equipment as listed unless it is also labeled. Listed or labeled equipment must be installed and used as instructed »110.3(B)«.

The organizations described below directly affect the *Code* as it relates to equipment and material acceptability and play a role in developing and maintaining the standards set forth in the *NEC*®.

Nationally Recognized Testing Laboratories (NRTLs)

Prior to 1989, there were only two organizations perceived as capable of providing safety certification of products that would be used nationwide. Because there were only two, innovative technology was slow to be tested and approved. When Congress created the Occupational Safety and Health Administration in the early 1970s, OSHA was directed to establish safety regulations for the workplace and for the monitoring of those regulations. OSHA adopted an explanation from the *National Electrical Code*® and included it in the Code of Federal Regulations. In part, it reads: "an installation or equipment is acceptable to the Assistant Secretary of Labor . . . if it is acceptable or certified, or listed, or labeled, or otherwise determined to be safe by a nationally recognized testing laboratory. . . ."

This was a start, but it did not identify the requirements for becoming a NRTL. Although OSHA introduced "Accreditation of Testing Laboratories" in 1973, the process through which a laboratory would receive accreditation was still missing. Vigorous cooperative efforts produced the OSHA regulation finalized in 1988 and called "OSHA Recognition Process for Nationally Recognized Testing Laboratories."

OSHA's NRTL program greatly benefits manufacturers by providing a system that certifies that a product meets national safety standards. Just as important, the door was opened for a greater number of laboratories to provide certification, and manufacturers are now better able to meet the demands of today's highly competitive market.

The aim of NRTLs is to ensure that electrical products properly safeguard against reasonable, identifiable risks. An extensive network of field personnel conduct unannounced inspections at manufacturing facilities which use the laboratory's "seal of approval." Some of the better-known trademarks of testing laboratories are shown below:

Some of the labels that appear on evaluated and certified electrical products, such as the ones below, will carry the trademarks of the testing laboratory, or the laboratory's standards being used for comparison.

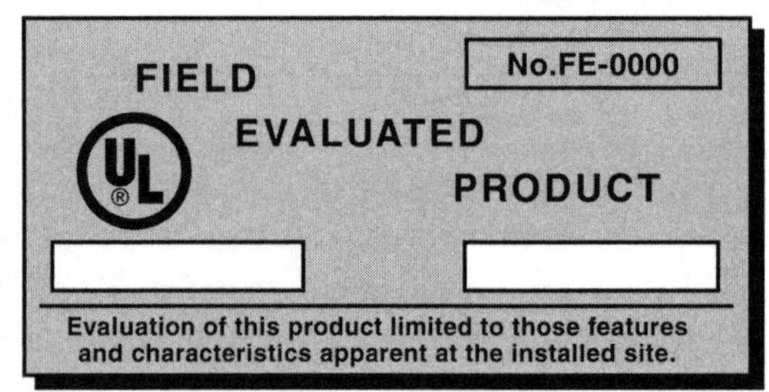

MET Laboratories

MET Laboratories, working with the Department of Labor as well as other agencies, served as a working example for the accreditation process for independent testing laboratories. In fact, MET became the first United States laboratory to successfully complete the process (1989) and thus became the first NRTL licensed by OSHA.

MET field inspectors interact with local electrical inspectors throughout the country to ensure product acceptance by all federal and state regulatory officials. The MET label is accepted by all fifty states, the federal government, and major retailers.

Underwriters Laboratories, Inc.

Prior to the formalization of NRTLs in 1989, electrical product standards were primarily written, and certification testing performed by, Underwriters Laboratories, Inc. (UL). Standards written by UL are still widely used. The appearance of the UL logo on a label indicates that the product complies with the UL standard. It does not mean, necessarily, that UL did the product testing. Although one of many NRTLs, Underwriters Laboratories is perhaps the most widely-recognized and respected testing laboratory in operation today. Founded in 1894, UL is a not-for-profit corporation whose mission is bring safer products to the marketplace and to serve the public through rigorous product safety testing. This organization offers a wide range of services, which include, but are not limited to: product listing, classification, component recognition, field certification, field engineering, facility registration, inspection, fact-finding, and research. As one can see from this list, UL plays a major role in guiding the safety of the electrical industry.

Intertek Testing Services

Select laboratories of Intertek Testing Services (ITS) have passed OSHA's stringent NRTL accreditation procedures and thereby have earned the right to issue product approvals and list products using the familiar ETL Listed and CE Marks. ITS has been conducting performance and reliability tests to nearly 200 safety standards applicable to workplace-related products since 1896. Intertek's comprehensive program includes testing, listing, labeling and quarterly follow-up inspections. While recognized internationally by its many Listed Marks, the ETL Listed Mark is accepted throughout the United States, by all jurisdictions for electrical products, when denoting compliance with nationally recognized standards such as ANSI, IEC, UL, CSA and CGA.

National Electrical Manufacturers Association (NEMA)

Founded in 1926, this organization is comprised of companies who manufacture equipment for all facets of electrical application, from generation through utilization. Their expansive objectives include product quality maintenance and improvement, safety standards for product manufacture and usage, and a variety of product standards, such as ratings and performance. NEMA contributes to the development of the *National Electrical Safety Code* as well as the *NEC*®.

National Fire Protection Association (NFPA)

The NFPA, more than a century old, dedicates itself to safety standards, gathering of statistical data, research, providing crucial information on fire protection, prevention and suppression methods, and much more. Boasting an internationally diverse membership of more than 65,000, this leading nonprofit organization publishes over 270 widely recognized codes and standards, including the *NEC*®. In addition, the NFPA is involved in training and education. Its primary pursuit is to protect lives and property from the often catastrophic hazards of fire.

THIS BOOK

The *Illustrated Guide to the NEC*® is designed to teach through visualization. If a picture is truly worth a thousand words, this book should provide a more in-depth look at the *National Electrical Code*® than can be found in any other single publication. These highly detailed illustrations are complemented with concise, easy to understand written information. Not intended as a how-to book, the *Illustrated Guide to the NEC*® instead strives to translate difficult material into simpler, straightforward principles. Once the reader understands how the *Code* translates in a specific area, the same techniques can be applied throughout.

Its Layout

Not only is the presentation of material different in this text from others on the market, but the organization of information also offers an new approach. After covering the fundamental provisions in the balance of Unit 1, this text proceeds to address code requirements by type of occupancy. Comprehensive information is given for one-family dwellings, multi-family dwellings, commercial locations, and special occupancies. To accomplish this task, information has been gathered logically from throughout the *Code* book and concentrated in one section, under the appropriate occupancy. Each occupancy type is broken down into its finite components, and each component thoroughly discussed and illustrated. (See Table of Contents.)

Text Conventions

General text will be grouped in small areas surrounding an illustration. **Notes** provide additional information, considered relevant to the point being discussed. **Cautions** indicate that particular care is needed during application. **Warnings** indicate impending danger, and are intended to prevent misunderstanding of a given rule.

Studying This Text

As the title implies, frequent references are made to the *National Electrical Code*®. Keep a copy of the latest edition close at hand. Any confusion about terminology not cleared up by the "Definitions" section of this text, may be explained by consulting the *Code's* Article 100: Definitions section. Whenever direct references are made to the *Code*, benefits will be gained by taking the time to read the suggested Article or Section. The *Illustrated Guide to the NEC*® is not intended, in any way, replace the *Code*. Each unit's "Competency Test" requires a thorough understanding of related *NEC*® subject matter. Use of this text alone is insufficient to successfully complete the test. It is, however, intended as an indispensable supplement to the *NEC*®.

Please note that when comparing computations made by both the English and Metric systems, slight differences will occur due to the conversion method used. These differences are not significant, and computations for both systems are therefore valid.

ADDITIONAL ELECTRICAL REQUIREMENTS

Local Ordinances

The importance of local (state, city, etc.) electrical codes cannot be over-emphasized. Local agencies can adopt the *NEC*® exactly as written or can amend the *Code* by incorporating more stringent regulations. While the *National Electrical Code*® represents the minimum standards for safety, some jurisdictions have additional restrictions. Obtain a copy of additional requirements (if any) for your area.

Engineers and/or architects who design electrical systems may also set requirements beyond the provisions of the *NEC*®. For example, an engineer might require the installation of 20 ampere circuits in areas where the *NEC*® allows 15-ampere circuits. Requirements from engineers and/or architects are found in additional documents, such as the following.

Plans and Specifications

If plans and specifications are provided for a project by knowledgeable engineers and/or architects, this information must be considered, and if need be, compared to the requirements set forth by the *NEC*®. It is unlikely that the plans or specifications provided by competent professionals will conflict with or contradict the *Code*. Nonetheless, it is best to be diligent in applying the governing principals of the *NEC*®.

Manufacturer Instructions

Some equipment or material may include instructions from the manufacturer. These instructions must be followed. For example, baseboard heaters generally include installation instructions. The *NEC*® does not prohibit the installation of receptacle outlets above baseboard heaters, but the manufacturer's instructions may prohibit the installation of their heater below receptacles.

SUMMARY

This introduction attempts to answer each question as it relates to the electrical industry in general, and to the *National Electrical Code*® in particular. It is not possible to do justice to the importance of the *Code* in a few short pages. With only a glimpse into its history and present-day supporting structure, this text moves on to the task of understanding the contents of the *Code*. The *Illustrated Guide to the NEC*® presents visually stimulating information in an occupancy organized, concise format. To begin the journey through the 2002 edition of the *National Electrical Code*®, simply turn the page, read, look, and understand.

UNIT 2

SECTION ONE: FOUNDATIONAL PROVISIONS

Definitions

Objectives

After studying this unit, the student should:

- understand the meaning of the term **accessible** (1) when applied to wiring methods, and (2) when applied to equipment.
- be able to identify accessible equipment that is not readily accessible.
- be able to accurately evaluate a location as accessible, readily accessible, or not readily accessible.
- be able to identify equipment classified as appliances.
- be familiar with the four categories of branch-circuits and be able to list their differences.
- be able to distinguish the difference between enclosed and guarded.
- be able to determine whether a load is continuous or noncontinuous.
- know the difference between branch-circuit conductors and feeder conductors.
- understand the terminology associated with grounded and grounding.
- know the maximum distance permitted for **within sight** situations.
- be able to give examples of damp, dry, and wet locations.
- comprehend the electrical vocabulary associated with the word **service**.
- be familiar with what constitutes a separately derived system.

Introduction

Understanding the terminology in the *National Electrical Code®* is of utmost importance. Knowing the correct definition of words and phrases as found in Article 100 is crucial to installing a hazard-free electrical system. Article 100 does not include commonly defined general or technical terms from related codes and standards. Normally, only terms used in two or more articles are defined in Article 100. Other terms are defined within the article in which they appear but may be referenced in Article 100. Part A of Article 100 contains definitions to be applied wherever the terms are used throughout the *NEC®*. Part B contains definitions applicable only to the parts of articles specifically covering installations and equipment operating at over 600 volts, nominal.

What is the difference between accessible and readily accessible? Which is appropriate in a given application? When sizing a branch-circuit or feeder, is the electrical load considered continuous or noncontinuous? What is the difference between a damp location and a wet location?

These and many other questions can be accurately answered only through a thorough understanding of *NEC®* terminology.

DEFINITIONS

Accessible (As Applied to Wiring Methods)

Wiring components are considered accessible when: (1) access can be gained without damaging the structure or finish of the building, or (2) they are not permanently closed in by the structure or finish of the building »Article 100«.

Ⓐ Conductors in junction boxes behind luminaires (lighting fixtures) are considered accessible if, by removing the luminaire (fixture), access to the conductors is available.

Ⓑ Conductors connected to switches and receptacles are accessible by removing the cover-plate and device.

Ⓒ Receptacles, behind furniture, are accessible because the furniture can be moved.

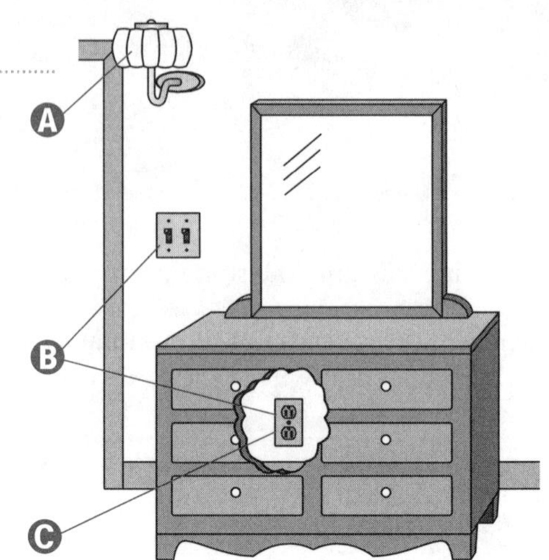

Accessible (As Applied to Equipment)

Ⓐ **Accessible** equipment is equipment not guarded by locked doors, elevators, or other effective means »Article 100«. Equipment installed in locations requiring the use of portable means, such as a ladder, is considered accessible, but not *readily accessible.*

Ⓑ Overcurrent devices do not have to be **readily accessible**, if located adjacent to the equipment, where access is achieved by the use of portable means »240.24(A)(4)«.

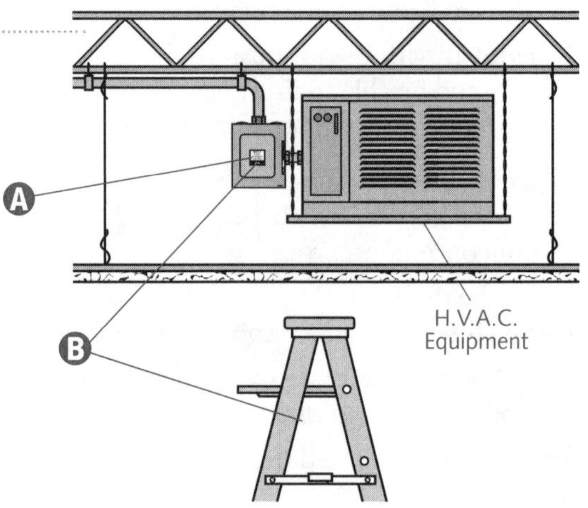

Accessible, Readily (Readily Accessible)

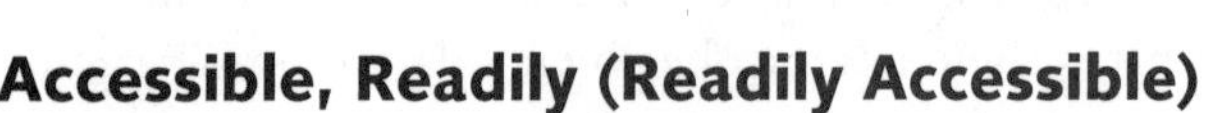

Readily accessible means capable of being reached quickly (for operation, renewal, or inspections) without having to climb over or remove obstacles, or resort to portable ladders, etc. »Article 100«.

Ⓐ The service disconnecting means must be readily accessible. It may be located either outside or inside, near the entry point of the service conductors »230.70(A)(1)«.

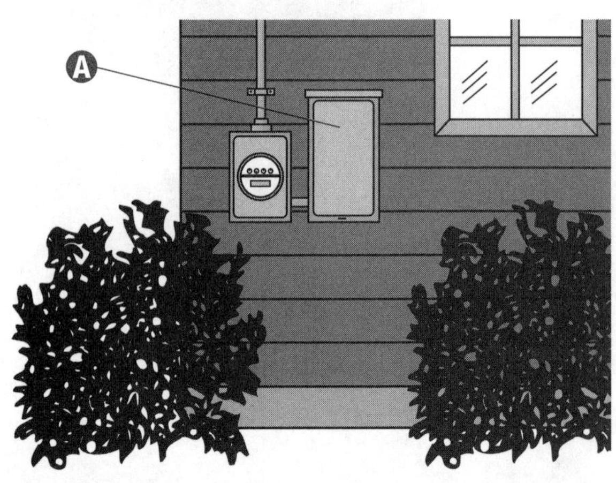

Accessible

Ⓐ Ready accessibility to wiring in luminaires (lighting fixtures) is not required. In most cases access can be gained through the use of a ladder, scaffolding, etc.

Ⓑ Conductors within junction boxes of recessed luminaires (fixtures) can be accessed by removing part of the luminaire (fixture), such as, trim, lamp, internal shell, etc.

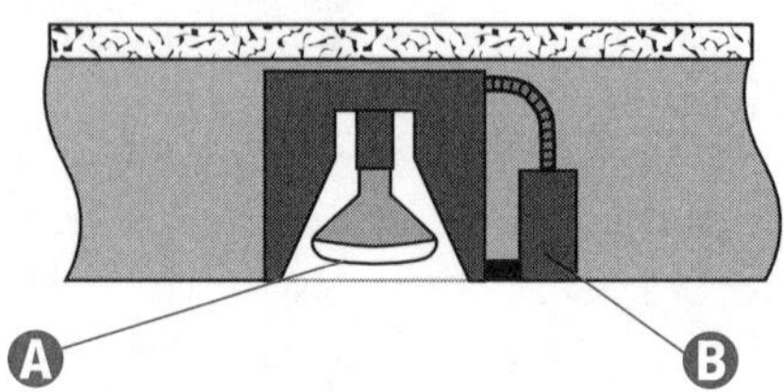

Appliance

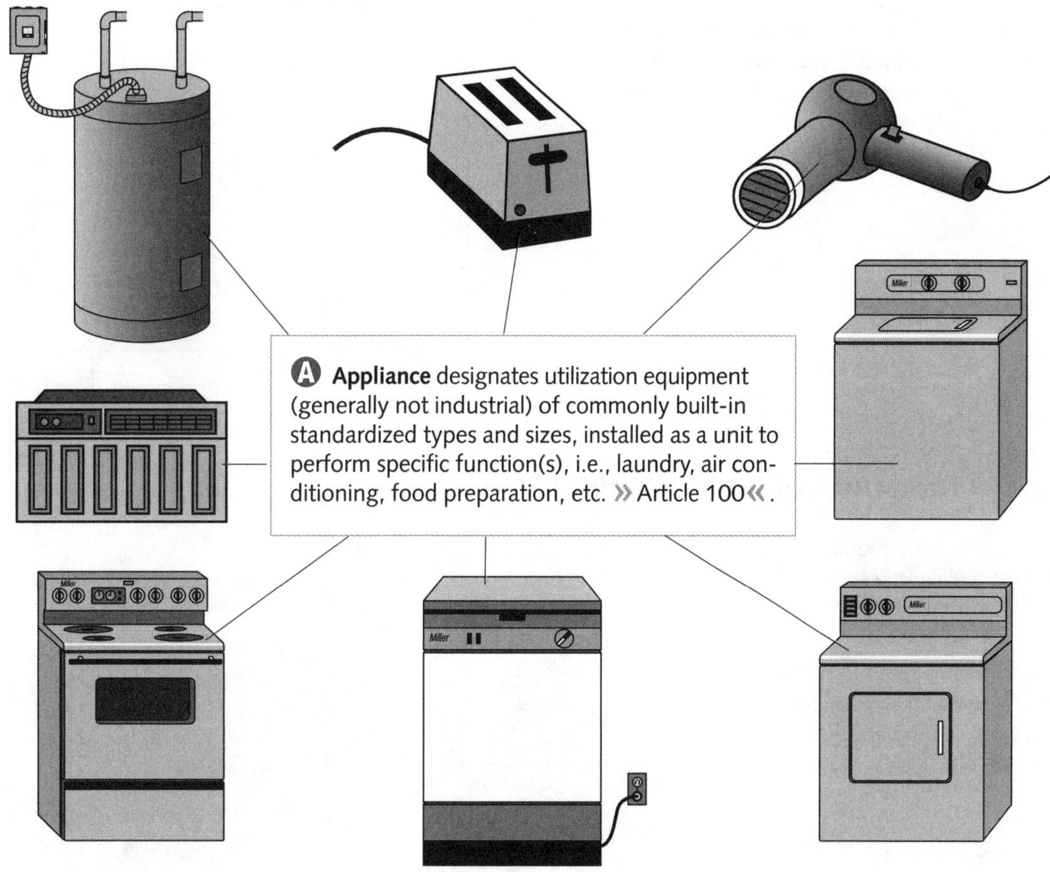

Not Readily Accessible

Ⓐ A receptacle installed in a ceiling, for a garage door opener, is not readily accessible. While receptacles installed in garages normally require ground-fault circuit-interrupter (GFCI) protection, receptacles that are not readily accessible do not » 210.8(A)(2) *Exception No. 1* «.

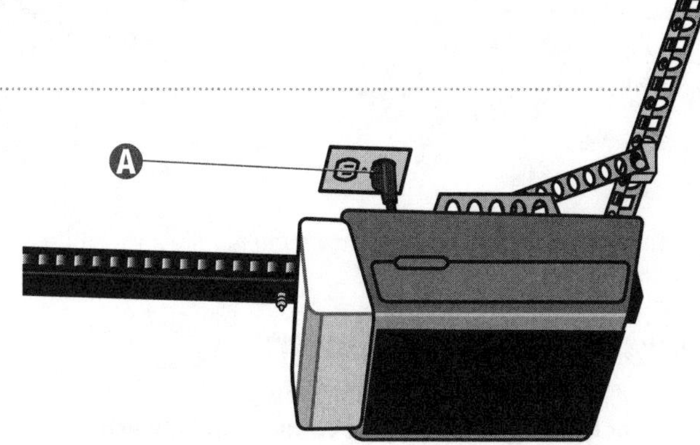

Attachment Plug (Plug Cap) (Plug)

Ⓐ An attachment plug (plug cap) (plug) is a device that, when inserted into a receptacle, establishes connection between the conductors of the attached flexible cord and the conductors permanently connected to the receptacle » Article 100 «.

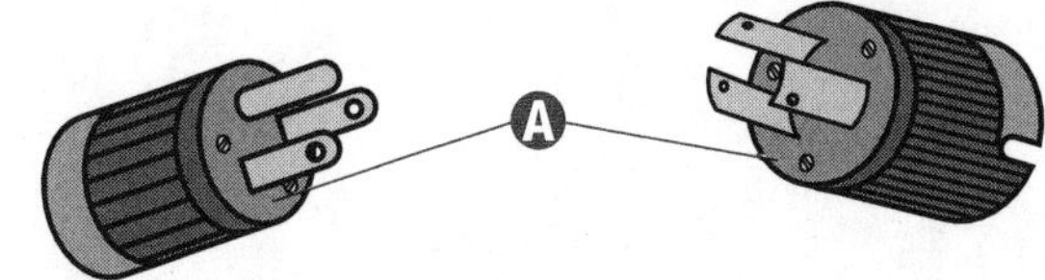

Bonding (Bonded)

Bonding is the permanent joining of metallic parts to form an electrically conductive path, ensuring continuity and the capacity to conduct safely any current likely to be imposed »Article 100«.

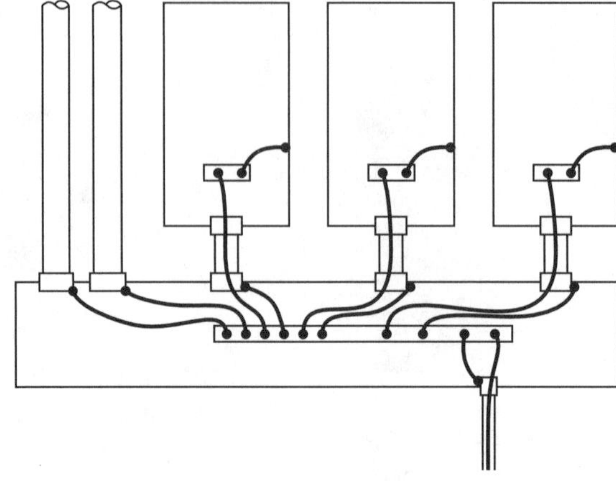

Bonding Jumper and Bonding Jumper, Main

A When metal parts are required to be electrically connected, a reliable conductor (bonding jumper) is installed, thereby guaranteeing the required electrical conductivity »Article 100«.

B The main bonding jumper is the connection at the service between the grounded circuit conductor and the equipment grounding conductor »Article 100«.

> NOTE
>
> *Main bonding jumpers must be made of copper or other corrosion-resistant material. A wire, bus, screw, or similar suitable conductor is acceptable as a main bonding jumper »250.28(A)«.*

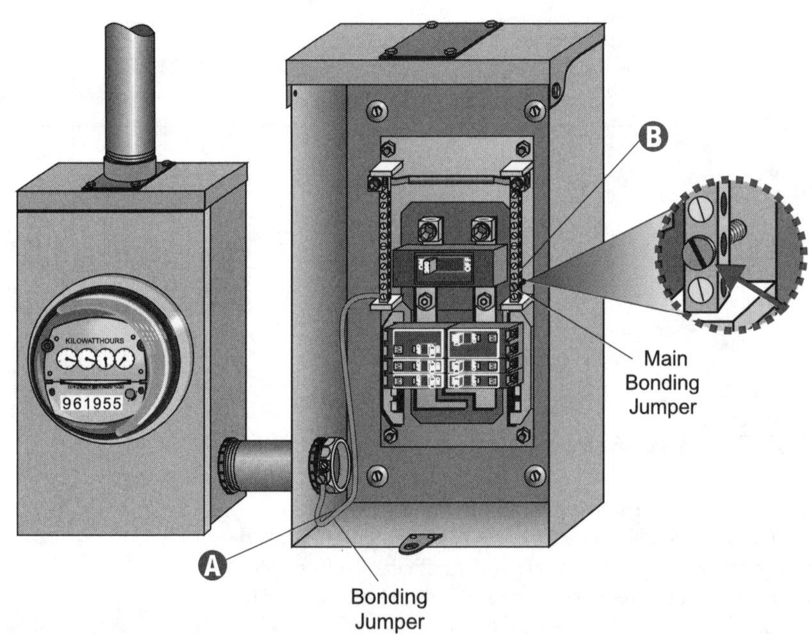

Bonding Jumper, Equipment

An equipment bonding jumper is the connection between two or more portions of the equipment grounding conductor »Article 100«.

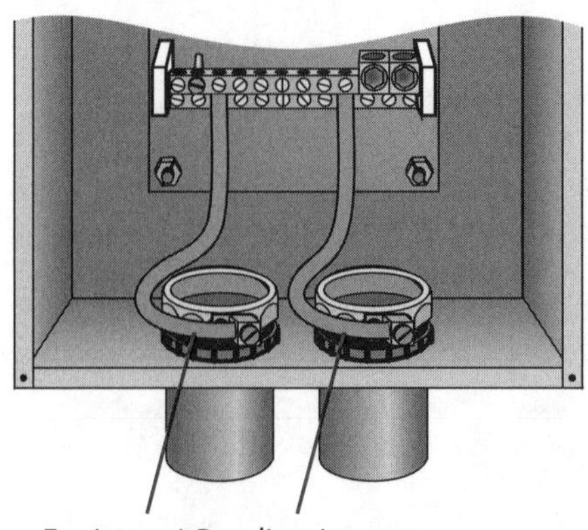

Branch-Circuit, Appliance

Ⓐ An appliance branch-circuit supplies energy to one or more outlets for the purpose of connecting appliance(s). These circuits exclude the connection of luminaires (lighting fixtures) unless they are part of the appliance being connected »Article 100«.

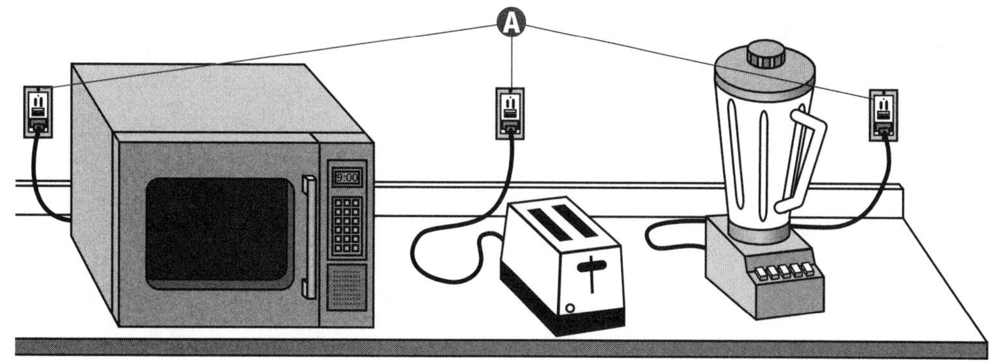

Branch-Circuit

Ⓐ The circuit conductors found between a circuit's final overcurrent protective device (such as the last fuse or breaker) and the circuit's outlet(s) is called a branch-circuit »Article 100«. Branch-circuits are divided into four categories: appliance, general purpose, individual, and multiwire.

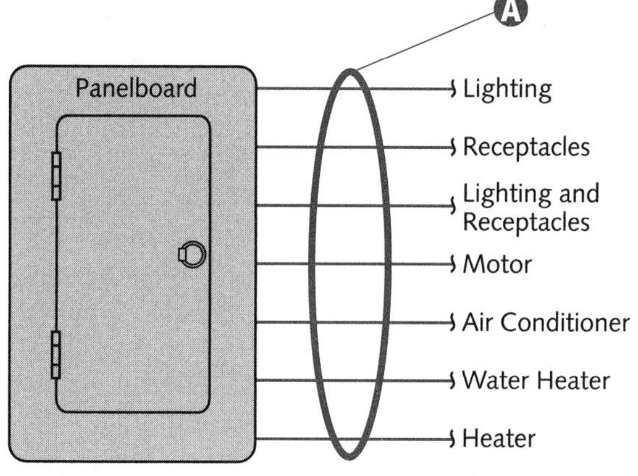

Branch-Circuit, Individual

Ⓐ An individual branch-circuit supplies only one piece of utilization equipment »Article 100«.

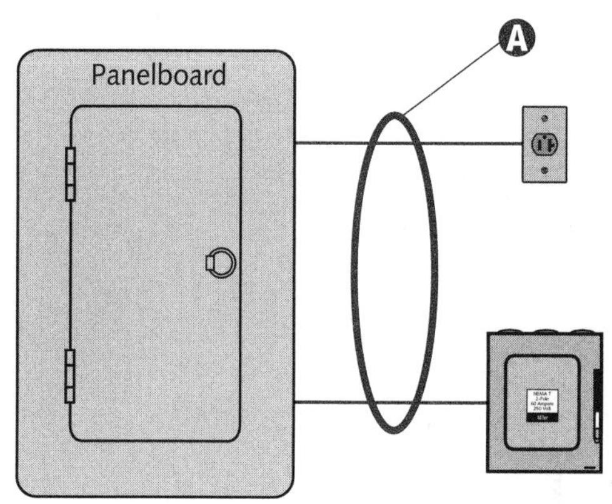

Branch-Circuit, General Purpose

Ⓐ A general purpose branch-circuit supplies two or more receptacles or outlets for lighting and appliances »Article 100«.

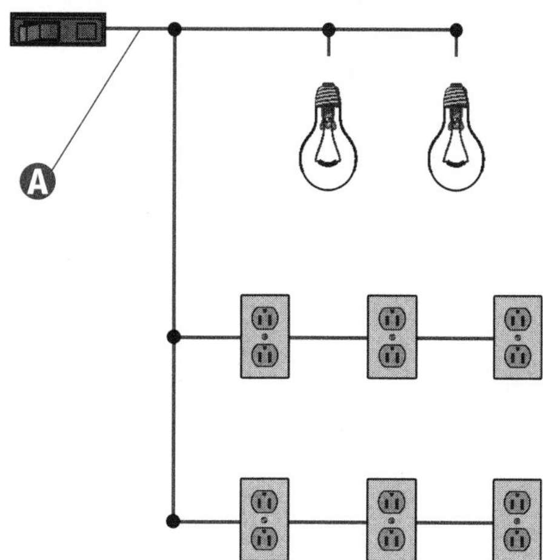

Concealed

A Concealed means rendered inaccessible by the structure or finish of the building »Article 100«.

B Conductors in concealed raceways, even though they may become accessible by withdrawing them, are still considered concealed »Article 100«.

C Raceway containing branch-circuit conductors in concrete slab.

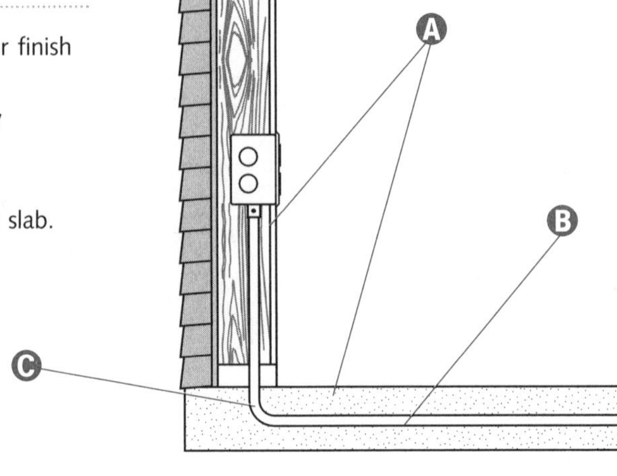

Branch-Circuit, Multiwire

A A voltmeter will not register a voltage (potential difference) when connected to the same ungrounded (hot) phase. Therefore, a multiwire circuit must not consist of conductors connected to the same phase.

B A multiwire branch-circuit consists of two or more ungrounded (hot) conductors that have a voltage between them »Article 100«.

C The one grounded (neutral) conductor of a multiwire circuit must be connected to the neutral or grounded conductor of the system »Article 100«.

D All conductors of a multiwire branch-circuit must originate from the same panelboard »210.4(A)«.

E There must be only one grounded (neutral) conductor, and there must be an equal voltage between it and each ungrounded conductor of the circuit »Article 100«.

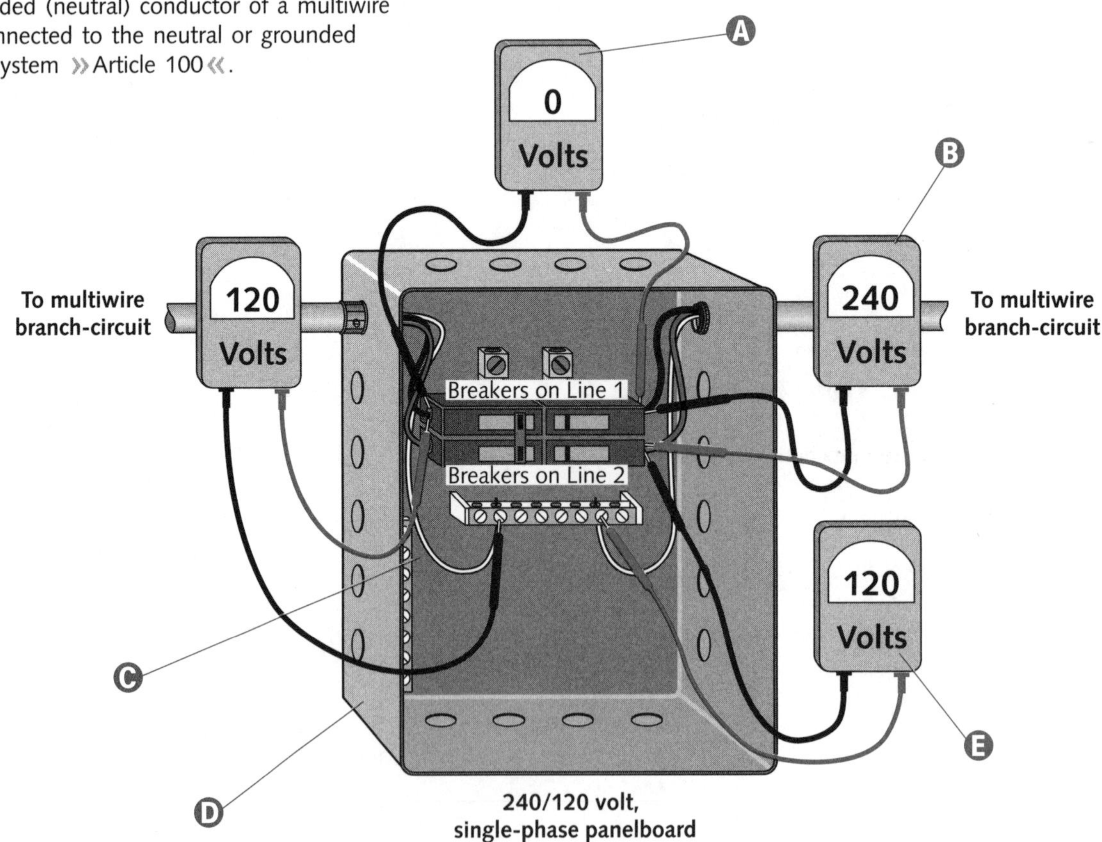

Receptacles on Multiwire Branch-Circuits

A multiwire receptacle circuit consists of one (or more) duplex receptacles, one (or more) multiple receptacles, two (or more) single receptacles, or combinations thereof.

A A duplex receptacle, with two circuits and one neutral on the same yoke, is also considered a multiwire circuit. Two circuits connected to one duplex receptacle, in dwelling units, must have a means to simultaneously disconnect all ungrounded (hot) conductors »210.4(B)«. This is accomplished through the use of either one double-pole breaker or two single-pole breakers with approved tie handles.

B As long as both circuits are not connected to the same receptacle, the two circuits can be thought of as separate circuits »210.4(A) and (B)«. The means of disconnect can be two single-pole breakers.

C Tab has been removed to allow separate feed of each outlet.

WARNING

In multiwire branch-circuits, the continuity of a grounded conductor shall not be dependent upon the device »300.13(B)«. If breaking the grounded conductor at the receptacle breaks the circuit down the line, then the grounded conductors must not be connected to the receptacle. Simply splice the grounded conductors and install a jumper wire to the receptacle.

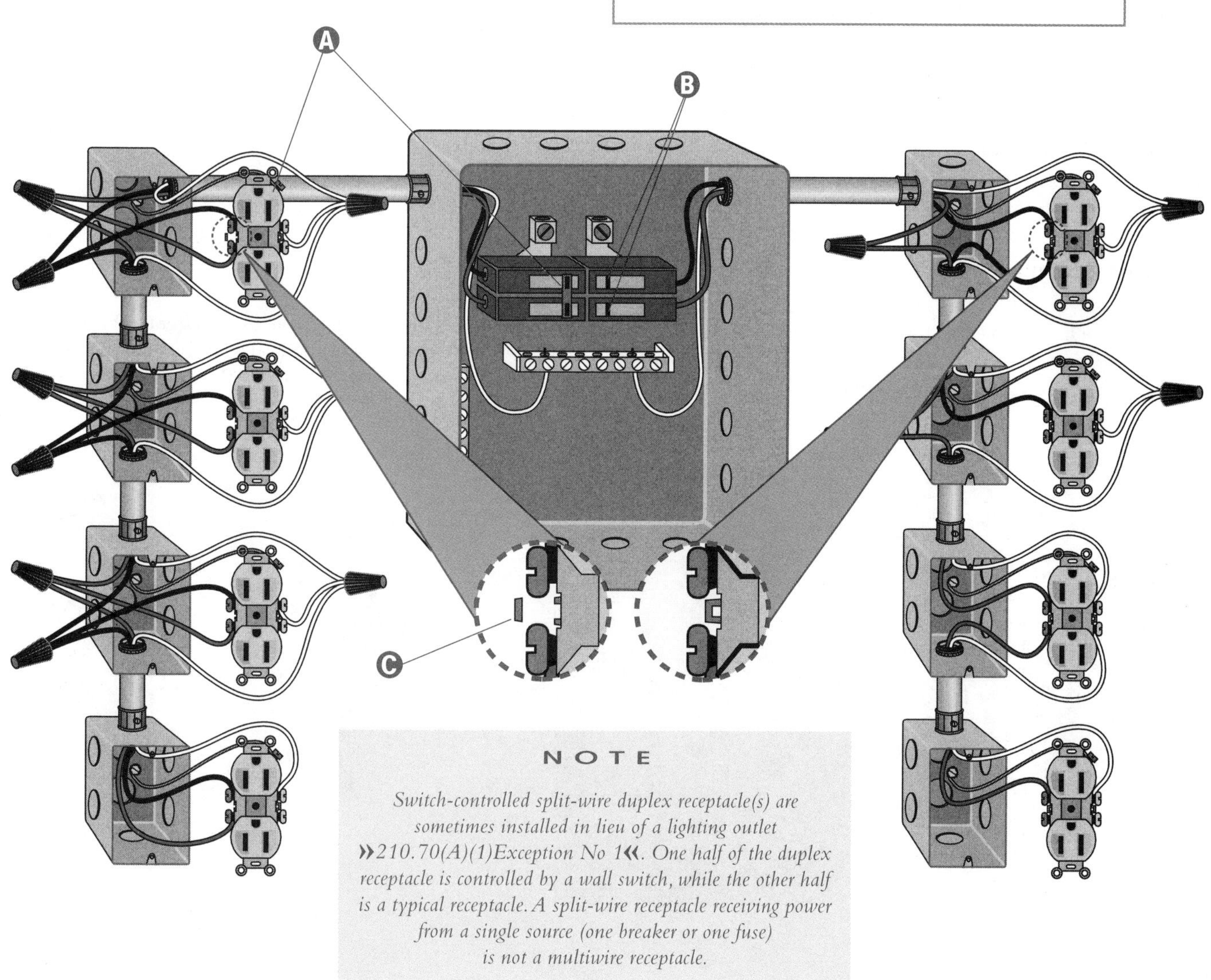

NOTE

Switch-controlled split-wire duplex receptacle(s) are sometimes installed in lieu of a lighting outlet »210.70(A)(1)Exception No 1«. One half of the duplex receptacle is controlled by a wall switch, while the other half is a typical receptacle. A split-wire receptacle receiving power from a single source (one breaker or one fuse) is not a multiwire receptacle.

Conduit Body (Condulet)

Ⓐ A conduit body is a separate portion of a conduit (or tubing) system providing access to the interior of the system through a removable cover(s) at a junction of multiple sections or at a terminal point of the system »Article 100«.

Ⓑ FS, FD, and larger boxes (cast or sheet metal) are not classified as conduit bodies »Article 100«.

Ⓒ A single conduit is not permitted as sole support for an FS-type or weatherproof junction box »370.23(E) and (F)«.

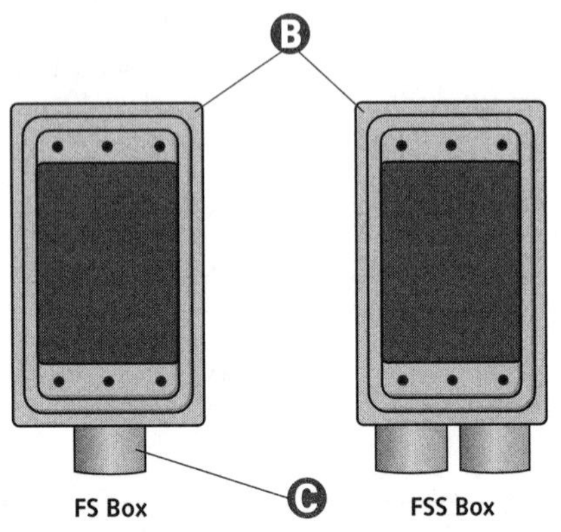

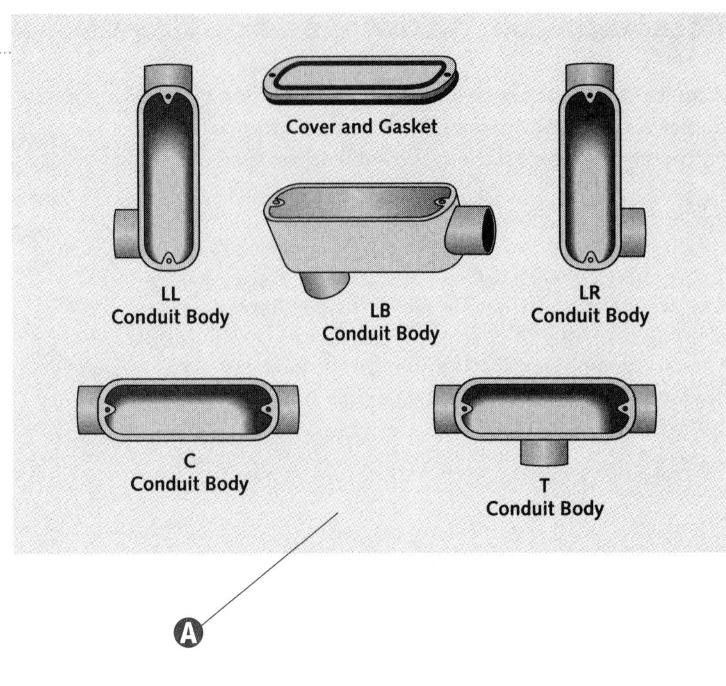

Continuous Load

A load having the maximum level of current sustained for three hours or more is referred to as a continuous load »Article 100«. Office lighting is an example of a continuous load.

Device

Ⓐ A unit of an electrical system, which carries but does not utilize electrical energy is known as a device »Article 100«.

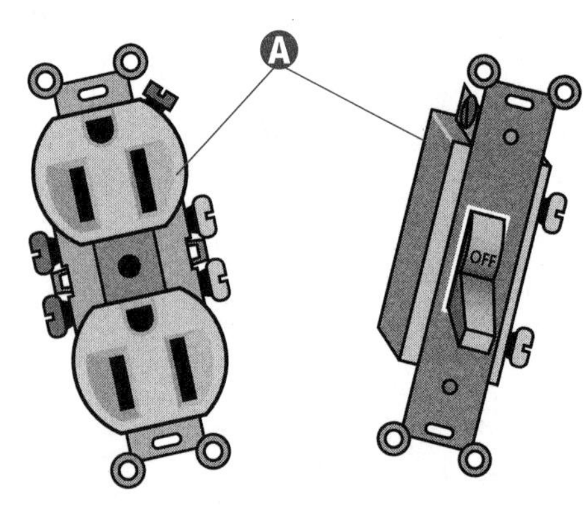

Enclosed

Equipment, conductors, etc., surrounded by a case, housing, fence or walls that prevent persons from accidentally contacting energized parts are referred to as enclosed »Article 100«.

Ⓐ Panelboards located within cabinets or cutout boxes are considered enclosed.

Ⓑ Electrical equipment installed within the perimeter of a fence, or similar area, qualifies as enclosed.

Ⓒ Conductors are also considered enclosed when installed in panels, wireways, raceways, etc.

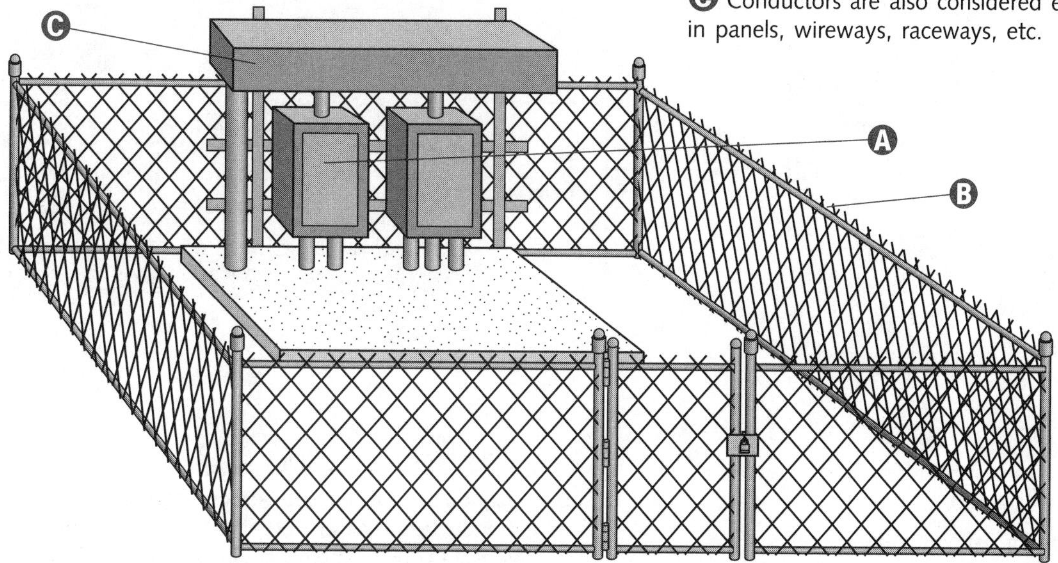

Enclosure

Ⓐ Any case, housing, apparatus, fence or walls surrounding an installation, designed to prevent personnel from accidentally contacting energized parts or to protect the equipment from physical damage, serves as an enclosure »Article 100«.

Equipment

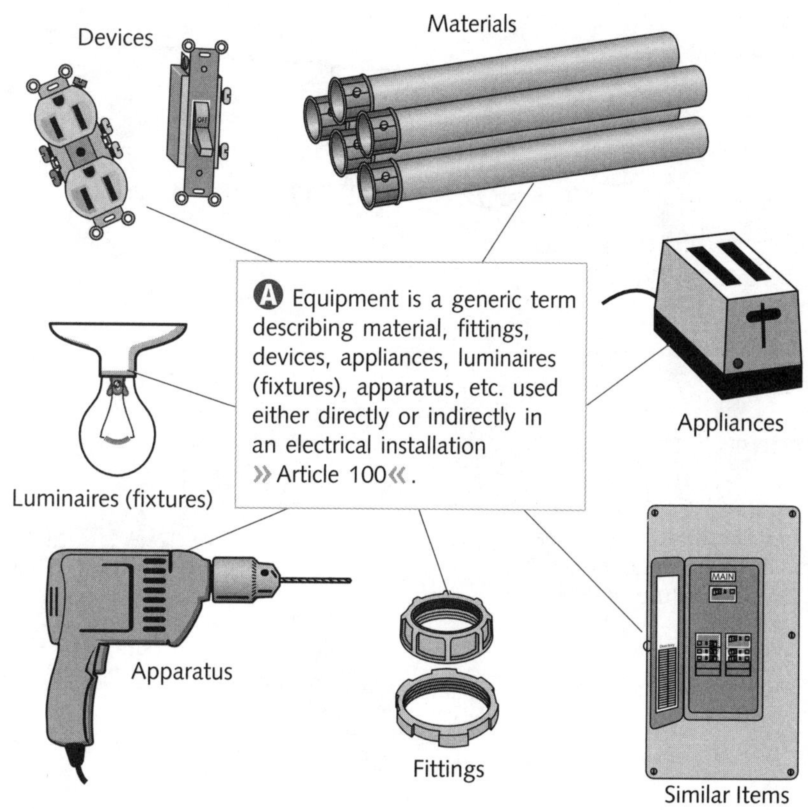

Feeder

A A feeder consists of all circuit conductors located between the service equipment, the source of a separately derived system, or other power supply source and the final branch-circuit overcurrent device »Article 100«.

B Branch-circuits (see definition).

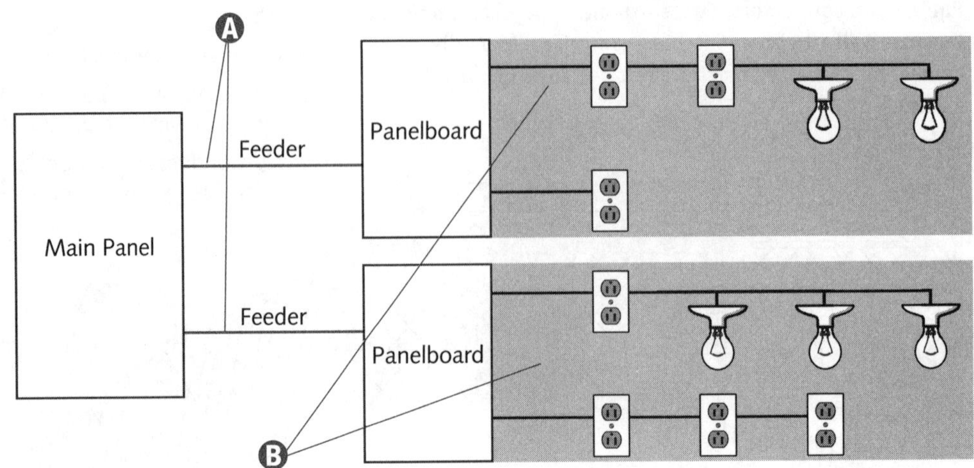

Festoon Lighting

A Festoon lighting is a string of outdoor lights suspended between two points »Article 100«.

B Overhead conductors for festoon lighting shall not be smaller than 12 AWG unless supported by messenger wires »225.6(B)«.

C Messenger wire, together with strain insulators, is used to support conductors in all spans exceeding 40 ft (12 m) in length. Conductors shall not be attached to any fire escape, downspout, or plumbing equipment »225.6(B)«.

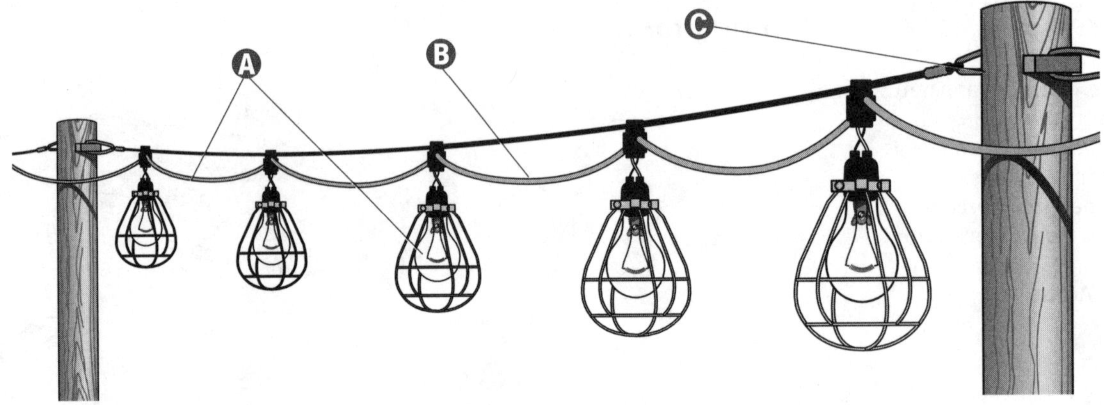

Fitting

A A fitting is an accessory such as a locknut, bushing, etc., whose function is primarily mechanical, rather than electrical, in nature »Article 100«.

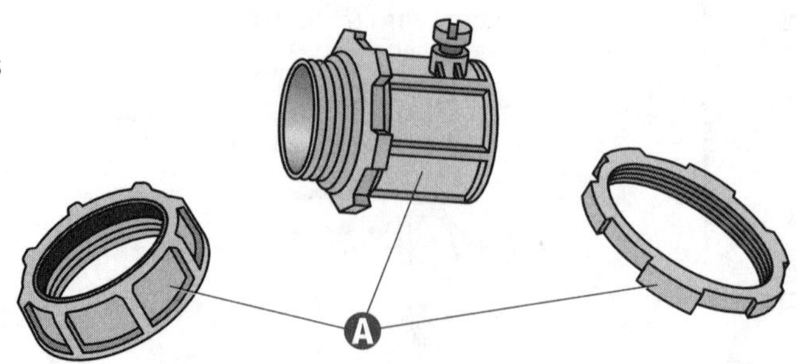

Grounded, Effectively Grounded, and Grounded Conductor

A A grounded conductor is a system or circuit conductor that is intentionally grounded »Article 100«.

B Grounded is defined as being connected to earth or to some conducting body serving in place of earth »Article 100«.

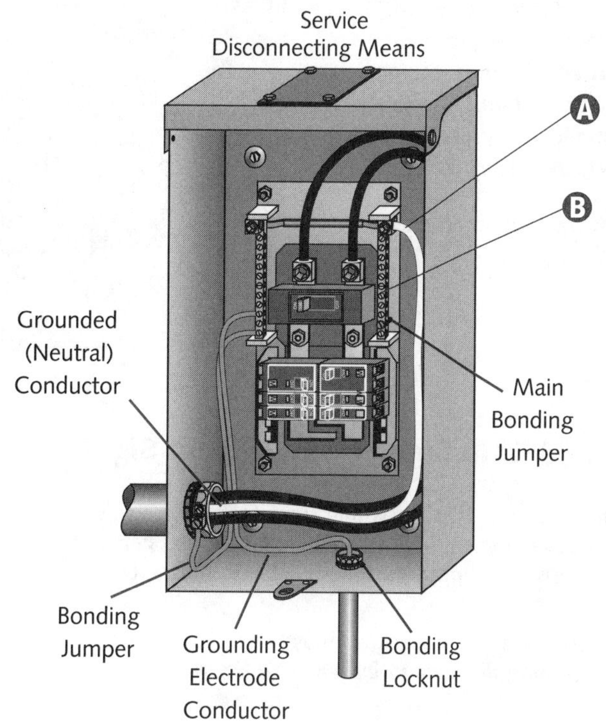

NOTE

A conductor is effectively grounded when an intentional connection is made to the earth through a ground connection or other connection having the same effect. Connections must have sufficiently low impedance and adequate current-carrying capacity to prevent hazardous voltage buildup »Article 100«.

Grounding Electrode Conductor and Equipment Grounding Conductor

A Main service equipment

B The conductor which connects the noncurrent-carrying metal parts of equipment, raceways, etc., to the system grounded conductor, the grounding electrode conductor (or both) at the service equipment (or source of a separately derived system) is called an equipment grounding conductor »Article 100«.

C System grounded (neutral) conductor

D Bonding jumpers

E Bonding jumper

F The grounding electrode conductor is the conductor connecting the grounding electrode to the equipment grounding conductor, the grounded conductor (or both) at the service equipment, at each building or structure where supplied from a common service, or at the source of a separately derived system »Article 100«.

G Grounding electrode

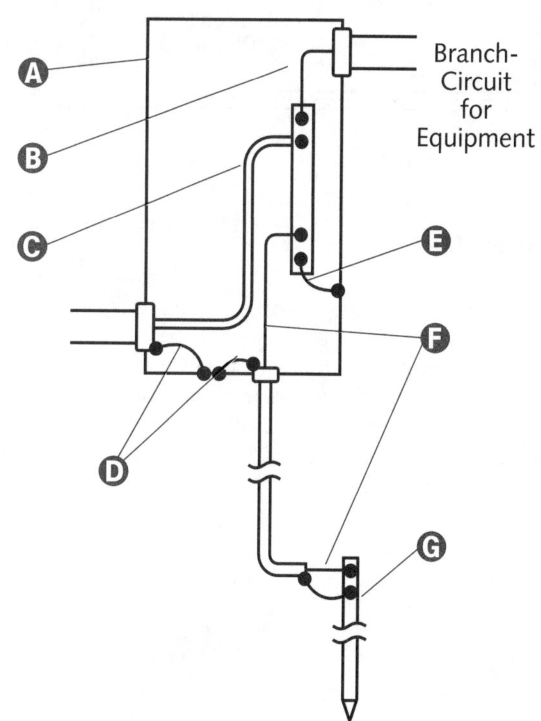

Guarded

Guarded is defined as covered, shielded, fenced, enclosed, or otherwise protected by means of suitable covers, casings, barriers, rails, screens, mats, or platforms effectively removing the likelihood of approach or contact by persons or objects »Article 100«.

Ⓐ Panelboards having doors or covers are considered guarded.

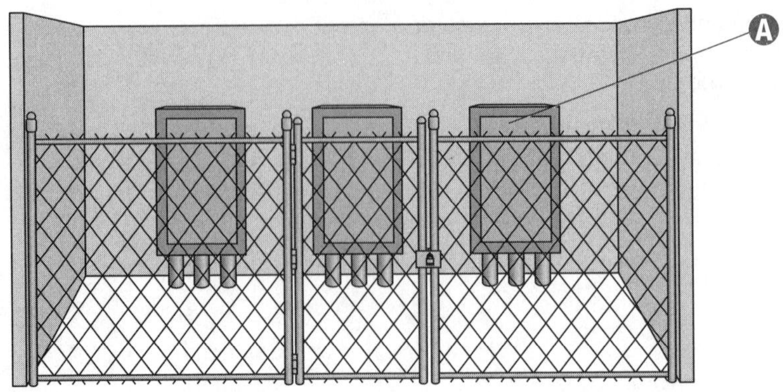

In Sight from (Within Sight from, Within Sight)

Ⓐ The *NEC*® terms "in sight from," "within sight from," or "within sight," etc., applied to equipment indicates that the specified items of equipment are visible and are not more than 50 ft (15 m) apart »Article 100«.

Ⓑ A motor controller disconnecting means must be located in sight from the controller location as required by 430.102(A).

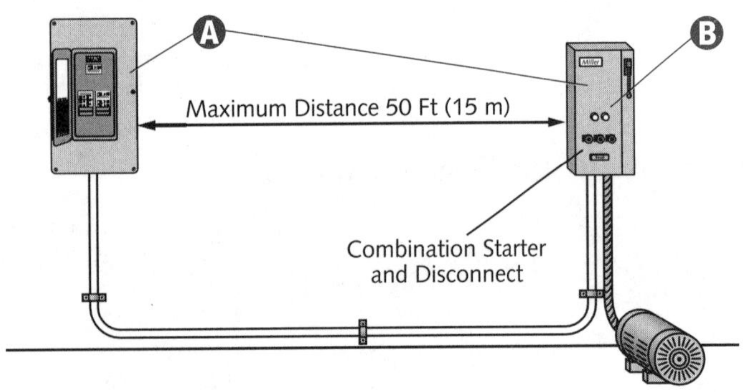

Lighting Outlet

Ⓐ An outlet intended for the direct connection of a lampholder, a luminaire (lighting fixture), or a pendant cord terminating in a lampholder is called a lighting outlet »Article 100«.

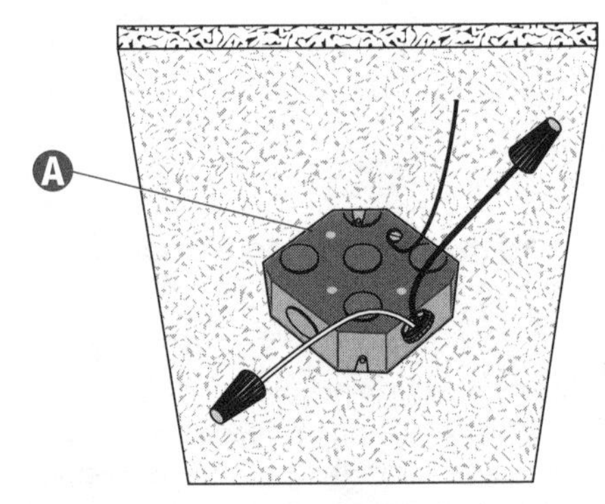

Location, Damp

Ⓐ Damp locations are those subject to moderate degrees of moisture. Exterior partially protected locations under canopies, marquees, roofed (open) porches, and similar sites are considered damp locations as well as interior locations such as some basements, barns, and cold-storage warehouses »Article 100«.

Location, Dry

A Dry locations are those not normally subject to moisture, except on a temporary basis such as a building under construction »Article 100«.

Location, Wet

A Installations in any of the following categories are wet locations: underground, within concrete slabs, in masonry (directly contacting the earth), areas subject to saturation (water or other liquids), and locations unprotected from weather »Article 100«.

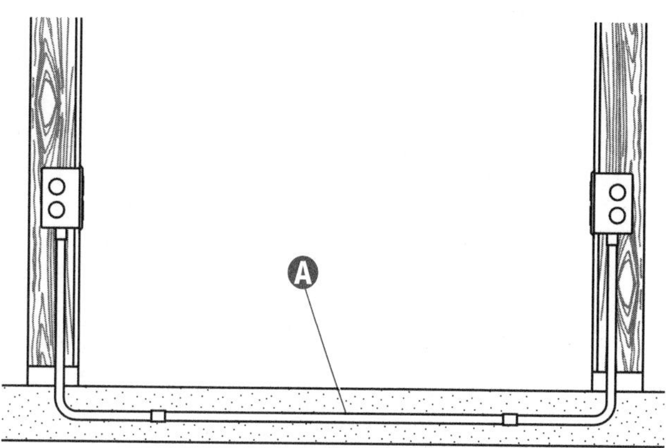

Multioutlet Assembly

A A surface, flush, or freestanding raceway designed to hold conductors and receptacles (assembled in the field or at the factory) is called a multi-outlet assembly »Article 100«.

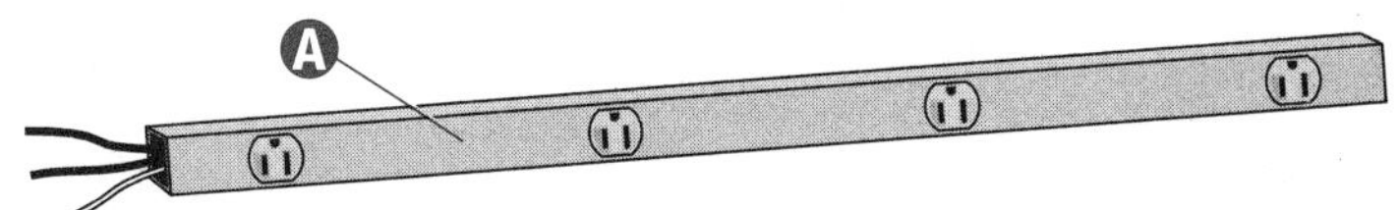

Outlet

A A point in a wiring system from which current is taken to supply utilization equipment is known as an outlet »Article 100«.

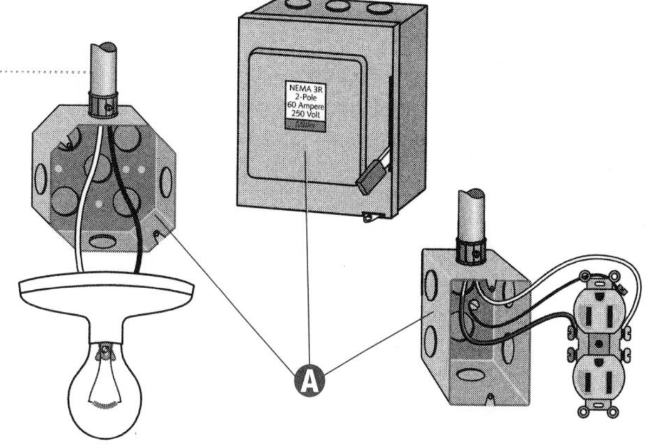

Plenum

Ⓐ The space above a suspended ceiling used for environmental air-handling purposes is an example of **other space used for environmental air** as described in 300.22(C).

Ⓑ A compartment or chamber having one or more attached air ducts and forming part of the air distribution system is known as a plenum »Article 100«.

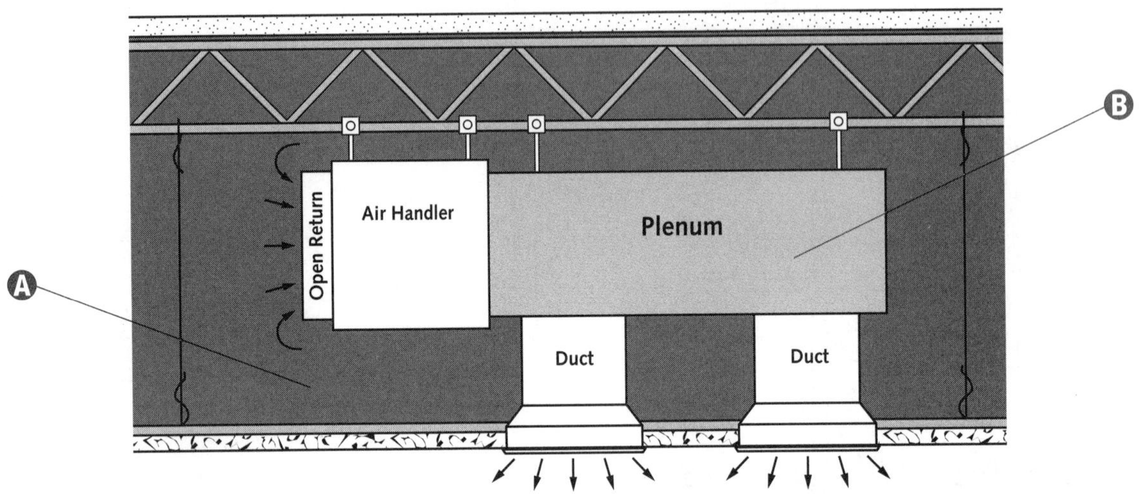

Receptacle

Ⓐ A contact device installed at an outlet for the connection of an attachment plug is a receptacle »Article 100«.

Ⓑ A single receptacle is a single contact device with no other contact device on the same yoke »Article 100«.

Ⓒ A multiple receptacle is a single device consisting of two or more receptacles »Article 100«.

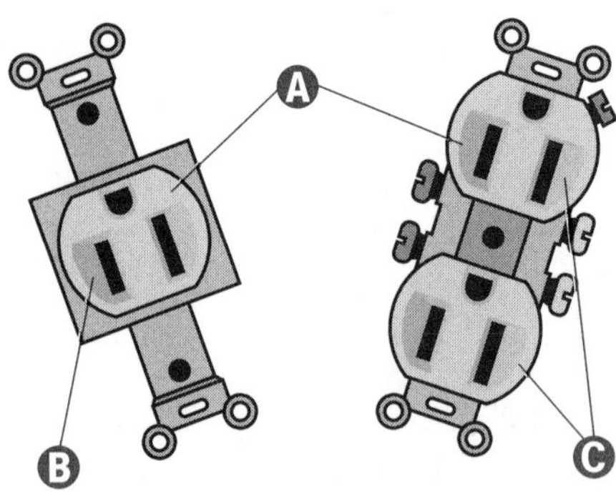

Separately Derived System

Ⓐ Article 450 contains provisions for transformers.

Ⓑ A separately derived system is a premises wiring system whose electric energy comes from a battery, solar photovoltaic system, generator, transformer, or converter windings to supply conductors originating in another system. Furthermore, a separately derived system has no direct electrical connection, including a solidly connected grounded circuit conductor »Article 100«.

NOTE

An alternate ac power source is not a separately derived system if the neutral is solidly interconnected to a service-supplied system neutral »250.20(D)FPN No. 1«.

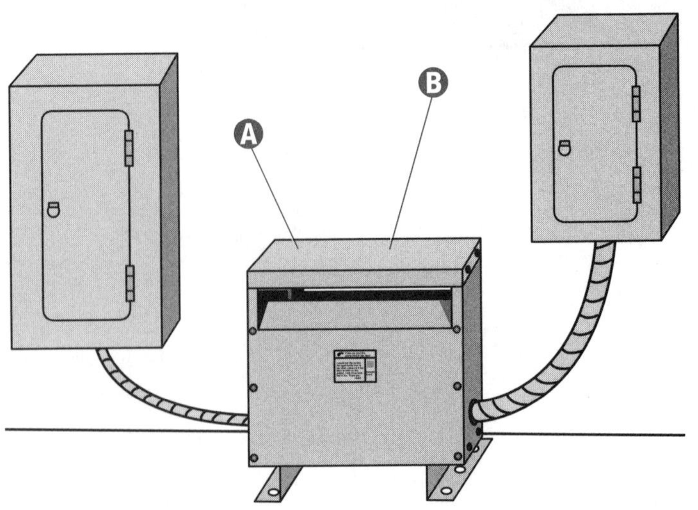

Service

Ⓐ The conductors and equipment which deliver energy from the serving utility to the wiring system of the premises are called the service »Article 100«.

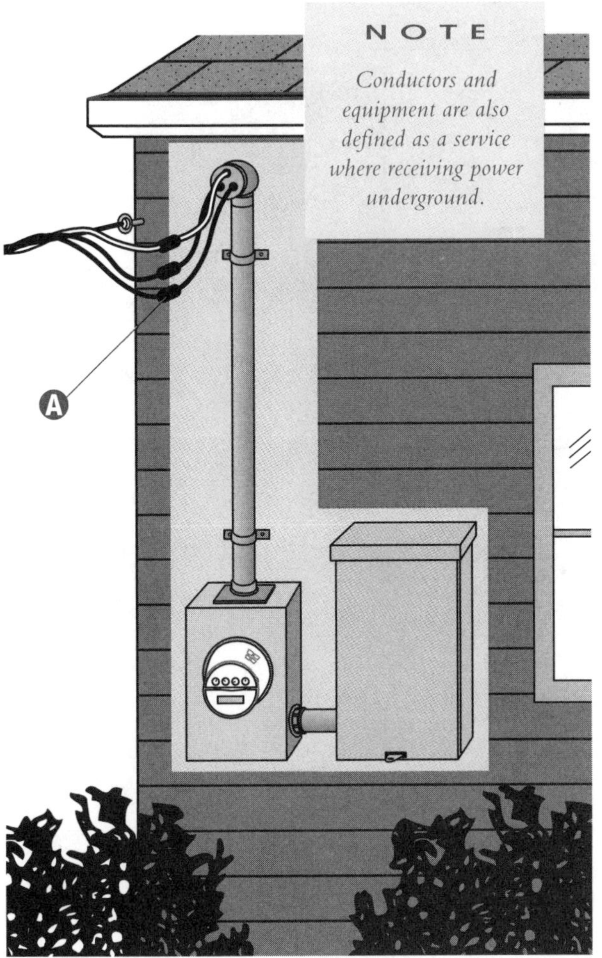

Service Conductors

Ⓐ The conductors from the service point to the service disconnecting means are known as service conductors »Article 100«.

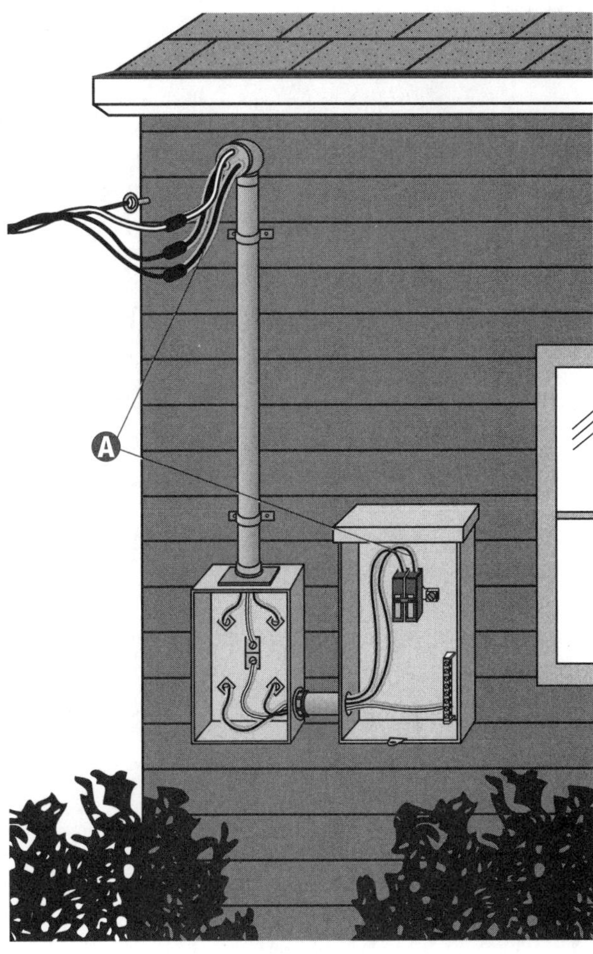

Service Drop

Ⓐ The overhead conductors from the last pole (or other aerial support), including splices (if any) to the service-entrance conductors at the building (or other structure), are the service drop »Article 100«.

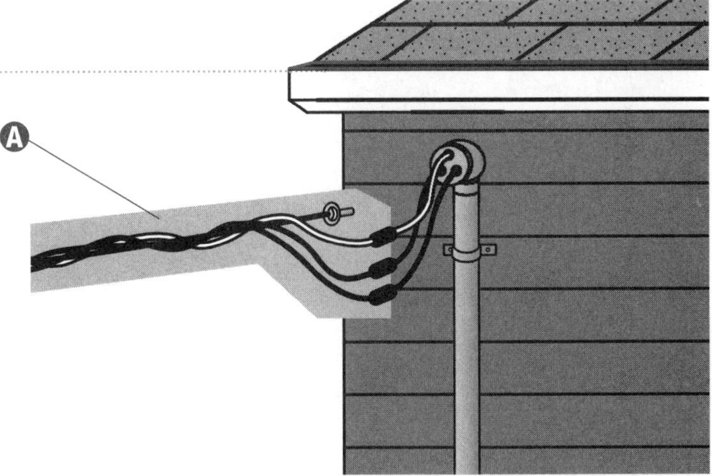

Service Equipment

A Service equipment is that equipment (usually circuit breaker(s), switch(es), fuse(s) and accessories) necessary to constitute the main control and cutoff which is connected to the load end of service conductors »Article 100«.

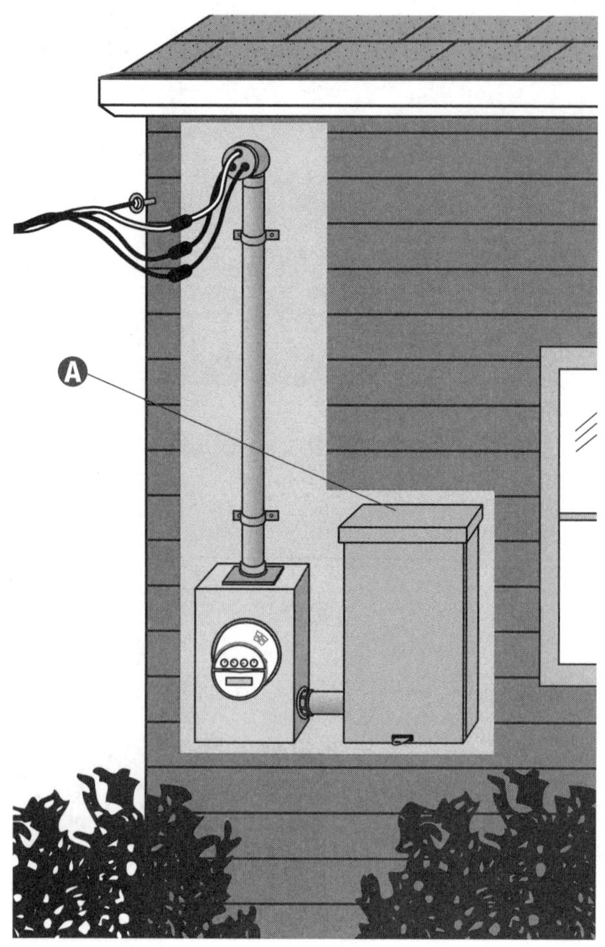

Service Lateral

A The underground service conductors between the main and the first point of connection to the service entrance conductors in a terminal box, meter base, etc. (inside or outside the building wall), is the service lateral. This includes any risers at a pole (or other structure) or from transformers »Article 100«.

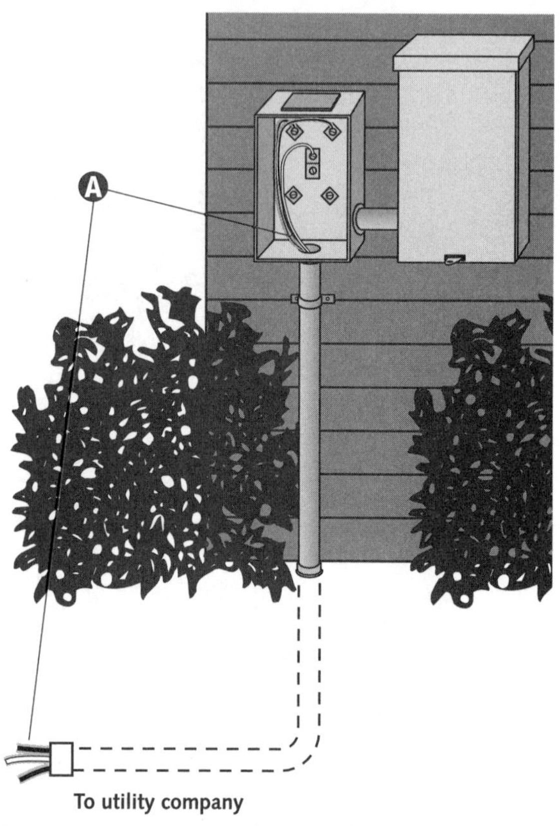

Service Point

A The underground service conductors between the main and the first point of connection to the service entrance conductors in a terminal box, meter base, etc. (inside or outside the building wall), is the service lateral. This includes any risers at a pole (or other structure) or from transformers »Article 100«.

Special Permission and Authority having Jurisdiction

A Special permission is the written consent of the authority having jurisdiction »Article 100«.

B The authority having jurisdiction (AHJ) for enforcing the *NEC*® may grant exception for the installation of conductors and equipment (not under the exclusive control of the electric utilities) used to connect the electric utility supply system to the service-entrance conductors of the premises served (provided such installations are outside a building or terminate immediately inside a building wall) »90.2(C)«.

C AHJ is the organization, office, or individual responsible for approving equipment, materials, an installation, or a procedure »Article 100«.

Summary

- Conductors within junction boxes must be accessible without damaging the construction or finish of the building or structure.
- Certain equipment, such as the service disconnecting means, must be readily accessible.
- The term appliance denotes more than just kitchen equipment.
- Branch-circuits are divided into four categories: appliance, general purpose, individual, and multiwire.
- General purpose branch-circuits may feed lights and receptacles or any combination thereof.
- An individual branch-circuit feeds only one piece of equipment.
- The terms bonded and grounded are not interchangeable.
- A dwelling multiwire circuit connected to one duplex receptacle must have a means to simultaneously disconnect all ungrounded (hot) conductors.
- A load having the maximum level of current sustained for three hours or more is a continuous load.
- Equipment is a general term encompassing a wide variety of items.
- A grounded conductor and a grounding conductor have different functions.
- One duplex receptacle is not defined as a single receptacle.
- Special permission is the written consent of the AHJ.

Unit 2 Competency Test

NEC® Reference	Answer	
________	________	1. A(n) _____ branch-circuit supplies two or more receptacles or outlets for lighting and appliances.
________	________	2. An electric circuit that controls another electric circuit through a relay is referred to as a(n) _____.
________	________	3. A(n) _____ branch-circuit consists of two or more ungrounded conductors having a potential difference between them, and a grounded conductor having equal potential difference between it and each ungrounded conductor of the circuit and that is connected to the neutral or grounded conductor of the system.
________	________	4. An intermittent operation in which load conditions are regularly recurrent is the definition of _____.
________	________	5. The _____ is the connection between the grounded circuit conductor and the equipment grounding conductor at the service.
________	________	6. A(n) _____ conductor is used to connect equipment or the grounded circuit of a wiring system to a grounding electrode or electrodes.
________	________	7. Rainproof, raintight, or watertight equipment can fulfill the requirements for _____ where varying weather conditions other than wetness, such as snow, ice, dust, or temperature extremes, are not a factor.

***NEC*® Reference**	**Answer**	
______	______	8. The purpose of the *NEC*® is the practical safeguarding of _____ from hazards arising from the use of electricity.
______	______	9. A(n) _____ is a manually-operated device used in conjunction with a transfer switch to provide a means of directly connecting load conductors to a power source and of disconnecting the transfer switch.
______	______	10. A type of surface, flush, or freestanding raceway, designed to hold conductors and receptacles, assembled in the field or at the factory is called a(n) _____.
______	______	11. A raceway encased in 4 in. (102 mm) of concrete on the fourth floor shall be considered a(n) _____ location.
______	______	12. A(n) _____ may consist of one or more sensing elements integral with the motor-compressor and an external control device.
______	______	13. An enclosure designed either for surface or flush mounting and provided with a frame, mat, or trim in which a swinging door or doors are or can be hung is called a(n) _____.
______	______	14. A(n) _____ is a shaftway, hatchway, well hole, or other vertical opening of space in which an elevator or dumbwaiter is designed to operate.
______	______	15. A multiwire branch circuit can supply: I. 120/240 volts to only one piece of utilization equipment II. 120/240 volts where all ungrounded conductors are opened simultaneously a) I only b) II only c) either I or II d) neither I nor II
______	______	16. An overcurrent protective device with a circuit opening fusible part that is heated and severed by the passage of overcurrent through it is the definition of a(n)_____.
______	______	17. _____ is a string of outdoor lights suspended between two points.
______	______	18. Name two items that must be present when defining an area as a bathroom.
______	______	19. A continuous load is where the _____ current is expected to continue for three hours or more. a) 80% b) average c) maximum d) 125%
______	______	20. A device is _____ if it is intentionally connected to earth through a ground connection and has sufficient current-carrying capacity to prevent the buildup of voltages.
______	______	21. A(n) _____ is a point on the wiring system at which current is taken to supply utilization equipment.
______	______	22. Circuit conductors between the final overcurrent device protecting the circuit and the outlet(s) is called a(n) _____.
______	______	23. _____ enclosures are constructed or protected so that exposure to a beating rain will not result in the entrance of water under specified test conditions.
______	______	24. Continuous duty is an operation at a substantially constant load for _____. a) 1 hour or more b) 1½ hours or more c) 3 hours or more d) an indefinitely long time
______	______	25. Ampacity is defined as the current in amperes that a conductor can carry continuously under the conditions of use without exceeding _____.
______	______	26. When a disconnecting means must be located within sight from a motor controller, the disconnect must be visible and not more than _____ ft from the controller.
______	______	27. A building containing three dwelling units is called a(n) _____.

UNIT

3

SECTION ONE: FOUNDATIONAL PROVISIONS

Boxes and Enclosures

Objectives

After studying this unit, the student should:

- be able to determine the cu.-in. capacity of boxes (metal and nonmetallic) when installing 6 AWG and smaller conductors.
- know which items are counted when calculating box fill and which are not.
- be aware that two identical switches, mounted side by side in a two-gang device box, could each have different cu.-in. volume allowances.
- be able to determine the minimum box size (including plaster ring, extension ring, etc.) that is needed if the number of conductors (6 AWG and smaller) is known.
- be familiar with box requirements when using nonmetallic-sheathed cable.
- know the minimum conductor length to be left inside boxes.
- understand that boxes and conduit bodies must remain accessible after installation.
- be familiar with mounting and supporting provisions for boxes and conduit bodies.
- be able to determine the type of box needed for various applications.
- understand calculation procedures for junction boxes containing 4 AWG and larger conductors.

Introduction

Choosing the right type and size of box (or enclosure) is very important to installing a system essentially free from hazard. Article 314 covers a variety of provisions concerning boxes (outlet, device, pull, and junction), conduit bodies, and fittings. Box selection must be based on requirements for a given location (such as dry, damp, wet, or hazardous). Boxes have particular requirements concerning the maximum number of conductors. Boxes containing 6 AWG and smaller conductors are required to have a minimum cu.-in. capacity which is determined by the size and number of conductors. Boxes containing 4 AWG and larger conductors are required to have a minimum height, width, and depth that is determined by the size and number of raceway entries. Article 314 contains provisions for installing, as well as supporting, boxes and conduit bodies. Boxes must be rigidly and securely fastened in place whether mounted *on* the surface, mounted to a framing member, or mounted *in* a finished surface. Under certain conditions, the only means of support a box needs is two threaded conduits. Access to conductors and devices located within boxes and conduit bodies must be available. Article 314 covers more information than the length of this book allows, such as manholes and boxes (pull and junction) for systems over 600 volts, nominal.

BOX FILL CALCULATIONS

Metal Boxes

The maximum number of conductors (for sizes 18 AWG through 6 AWG) permitted in various standard size metal boxes is listed in Table 314.16(A). Minimum cubic-in. capacities are also shown.

Ⓐ Volume does not have to be marked on metal boxes listed in Table 314.16(A) »314.16(A)(2)«.

Ⓑ A 3- by 2-in. (75- x 50-mm) device box, 2½ in. (65 mm) deep, has a volume of 12.5 in.^{3} (205 cm^3) »Table 314.16(A)«.

Ⓒ A 4-in. (100-mm) square box with a depth of 1½ in. (38 mm), has a volume of 21 in.^{3} (344 cm^3) »Table 314.16(A)«.

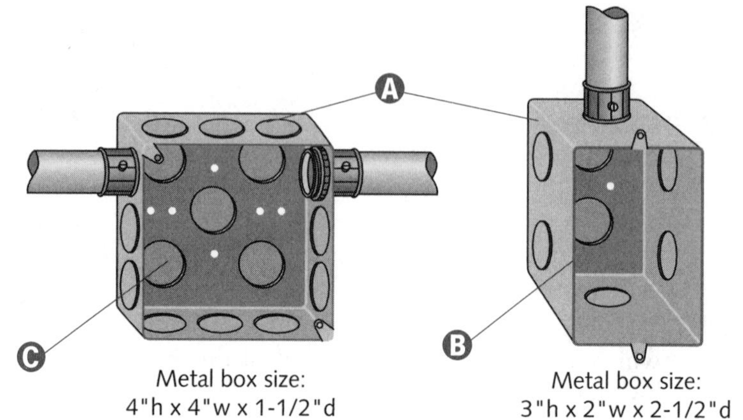

Metal box size:
4"h x 4"w x 1-1/2"d

Metal box size:
3"h x 2"w x 2-1/2"d

Nonmetallic Boxes

Ⓐ Volume shall be marked on all nonmetallic boxes and boxes with a volume of 100 in.^{3} (1650 cm^3) or less, except for boxes listed in Table 314.16(A) »314.16(A)(2)«.

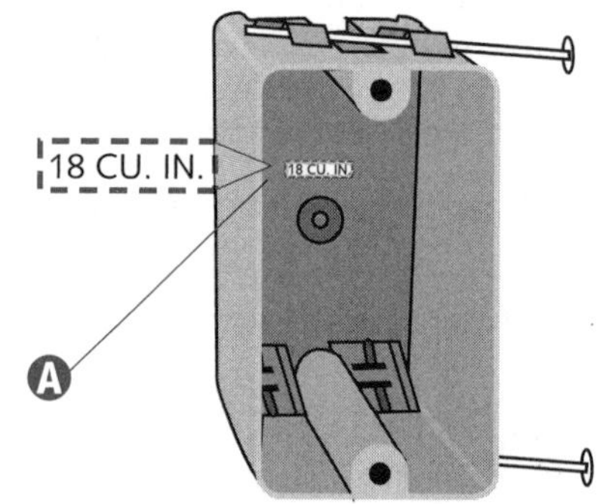

Additional Capacity

Ⓐ Additional capacity can only be calculated when the plaster ring, extension ring, etc., is clearly marked with a volume or when it corresponds in size to a box listed in Table 314.16(A).

Ⓑ Volume can be increased by using plaster (mud) rings, domed covers, extension rings, or similar items »314.16(A)«.

Ⓒ The combined volume of this box and plaster (mud) ring is 28.5 in.^{3} (467 cm^3) »Table 314.16(A)«.

Ⓓ The combined volume of the box and extension ring is 42 in.^{3} (688 cm^3) »Table 314.16(A)«.

Ⓔ A 4-in. (100-mm) square extension ring, having a depth of 1½ in. (38 mm), has a capacity of 21 in.^{3} (344 cm^3) because it has the same dimensions as a box listed in Table 314.16(A).

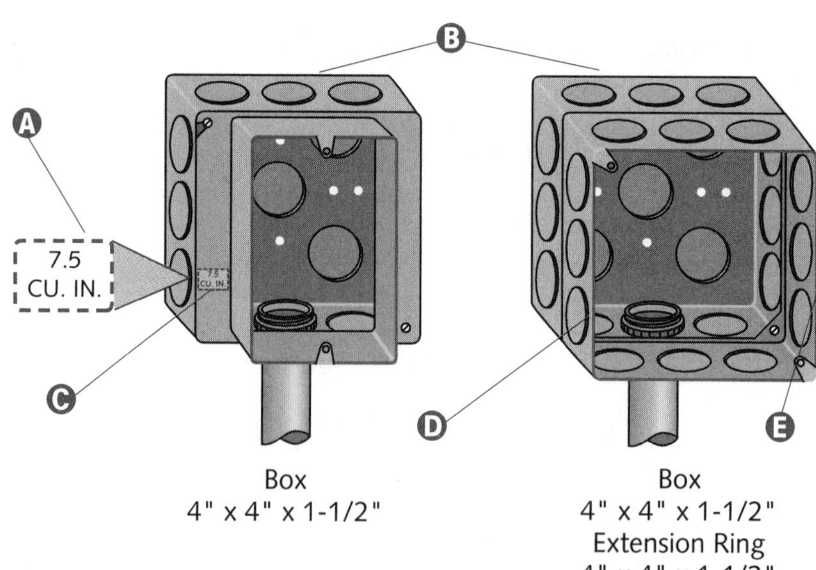

Box
4" x 4" x 1-1/2"

Box
4" x 4" x 1-1/2"
Extension Ring
4" x 4" x 1-1/2"

Equipment Grounding Conductor Fill

A Pigtails are not counted.

B Equipment grounding conductor(s) or equipment bonding jumpers entering a box count as one conductor »314.16(B)(5)«. Only the largest equipment grounding conductor shall be counted if multiple grounding conductors (of different sizes) enter the box.

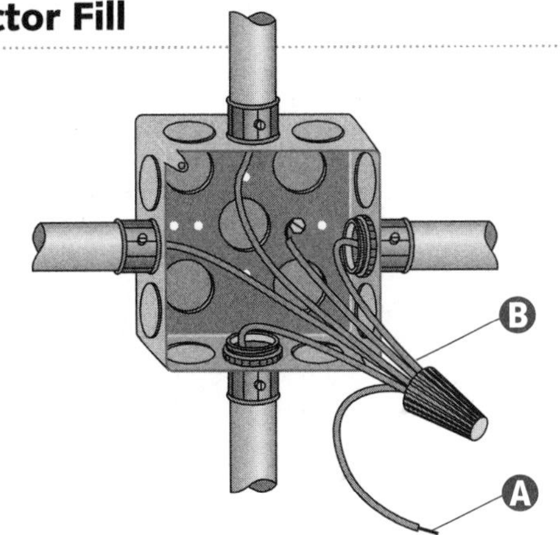

NOTE

One or more additional equipment grounding conductors (isolated equipment grounding conductors) as permitted in 250.146(D) shall be counted as one conductor »314.16(B)(5)«. When multiple isolated equipment grounding conductors of different sizes enter the box, only the largest is counted.

Determining the Number of Conductors

A Raceway fittings (connectors, hubs, etc.) are not counted.

B Each conductor originating outside the box which is terminated or spliced inside the box counts as one conductor »314.16(B)(1)«. These blue conductors are counted as two conductors.

C This box, as pictured, contains five conductors. If all of the conductors are 12 AWG, the box could hold a maximum of nine »Table 314.16(A)«.

D A conductor that does not leave the box, such as equipment bonding jumpers and pigtails, is not counted »314.16(B)(1)«.

E A conductor which passes through the box without splice or termination (unbroken) counts as one conductor »314.16(B)(1)«. This black unbroken conductor is counted as one conductor.

F Each conductor originating outside the box that is terminated or spliced inside the box counts as one conductor »314.16(B)(1)«. These white conductors count as two conductors.

G Wire connectors are not counted.

Metal box
4" x 4" x 1-1/2"

Cable Clamps and Connectors

A External cable connectors are not counted.

B Cable connector(s) with the clamping mechanism outside the box are not counted »314.16(B)(2)«.

C Two internal cable clamps count as one conductor.

D Internal cable clamp(s), whether factory or field supplied, shall be counted as one conductor »314.16(B)(2)«. Where more than one size conductor is present in the box, the clamp shall be counted as the largest conductor.

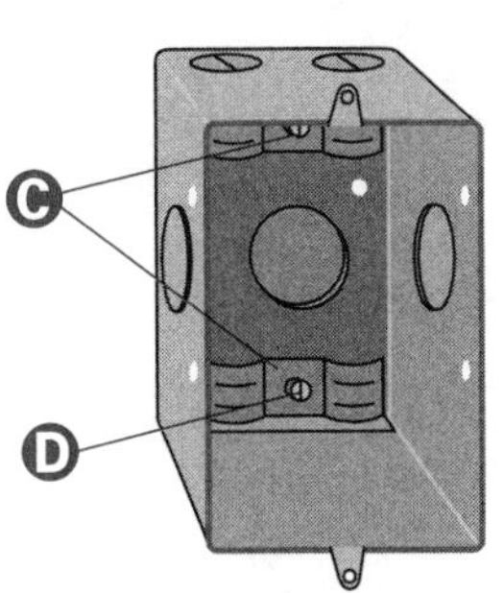

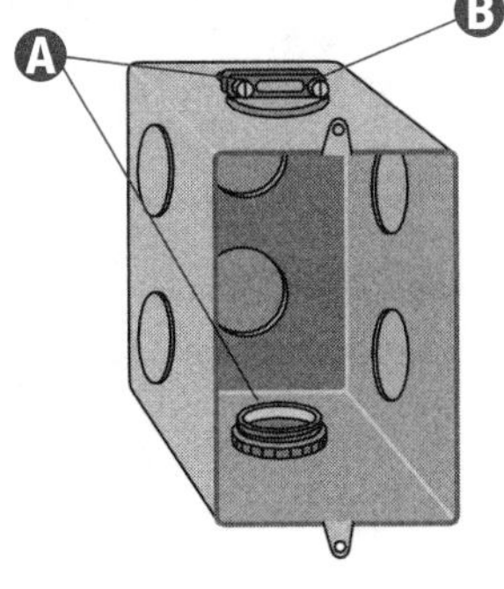

Luminaire (Fixture) Studs and Hickeys

A Luminaire (fixture) stud(s) shall be counted as one conductor »314.16(B)(3)«.

B One or more hickeys shall be counted as one conductor »314.16(B)(3)«.

C If conductors of different sizes are present in the box, each one shall be counted as the largest conductor »314.16(B)(3)«.

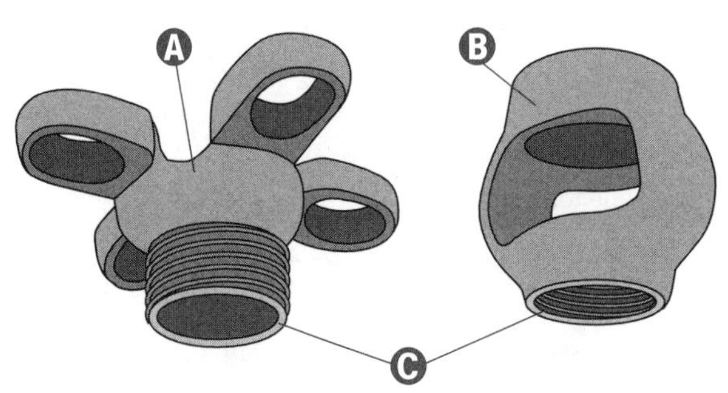

Devices or Equipment

A This duplex receptacle counts as two 12 AWG conductors.

B This single-pole switch counts as two 14 AWG conductors.

C Each mounting yoke or strap counts as two conductors. A mounting yoke or strap can contain one or more devices, such as a single receptacle, a duplex receptacle, a single switch, a double switch, a triple switch, or any combination. The size of the two conductors (when calculating box fill) shall be equal in size to the largest conductor connected to the device »314.16(B)(4)«.

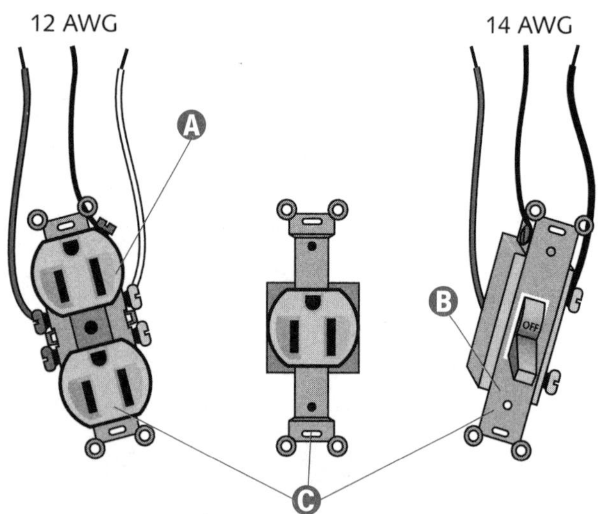

Volume Per Conductor

Table 314.16(B) lists the cu.-in. volume for conductors, sizes 18 AWG through 6 AWG.

Boxes, enclosures, and conduit bodies containing conductors, size 4 AWG or larger, must also comply with 314.28 provisions »314.16«.

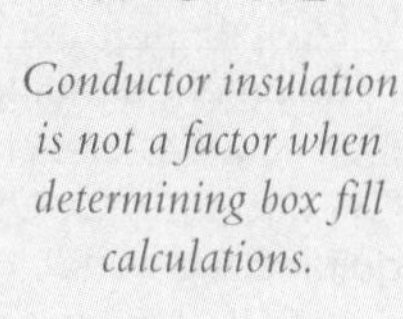

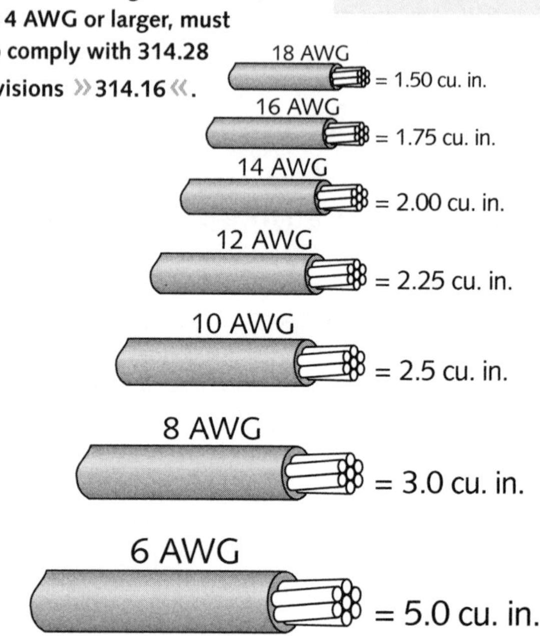

Box Fill Calculation

Wire connectors, pigtails, locknuts, bushings, raceway connectors, grounding screws, and equipment bonding jumpers are not factors when calculating box fill.

A These three conductors, although spliced, count as *three* conductors.

B Two conductors which terminate in the box count as *two* conductors.

C One receptacle counts as *two* 12 AWG conductors.

D An unbroken conductor counts as *one* conductor.

E These three equipment grounding conductors count as *one* conductor.

F These two conductors terminating in the box are counted as *two* conductors.

G A single unbroken conductor is counted as one conductor.

Total 12 AWG conductors	12
Volume per 12 AWG	2.25 in.^{3} (each) (36.9 cm^3 [each])
Minimum volume for conductors	$12 \times 2.25 = 27$ in.^{3} (12 x 36.9 = 442.8 cm^3)
Volume for box (including raised cover)	$21 + 6 = 27$ in.^{3} (344 + 98.4 = 442.4 cm^3)

This installation complies with 314.16 provisions.

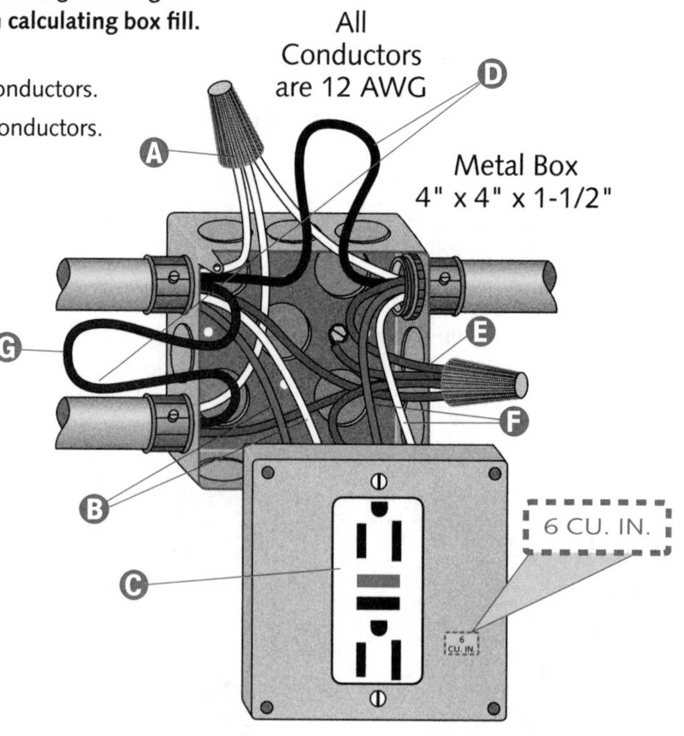

Domed Covers and Canopies

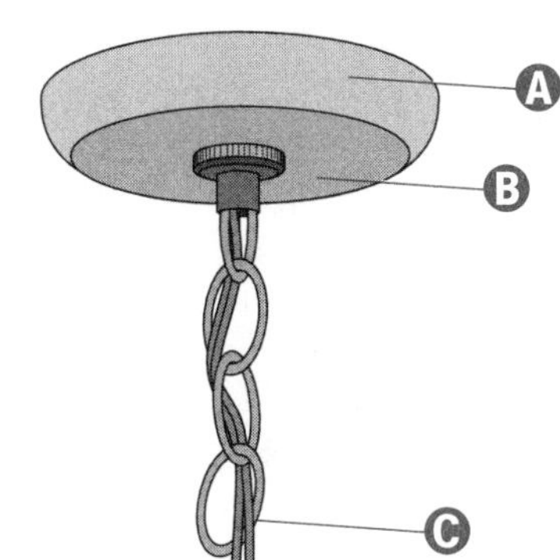

Ⓐ Four or fewer luminaire (fixture) wires (smaller than 14 AWG) and an equipment grounding conductor can be omitted from box fill calculations where they enter a box from a domed luminaire (fixture) (or similar canopy) and terminate within that box »314.16(B)(1) *Exception*«.

Ⓑ The size of the domed cover or canopy is not a factor.

Ⓒ Two luminaire (fixture) wires and one equipment ground are not counted.

Splices Inside Conduit Bodies

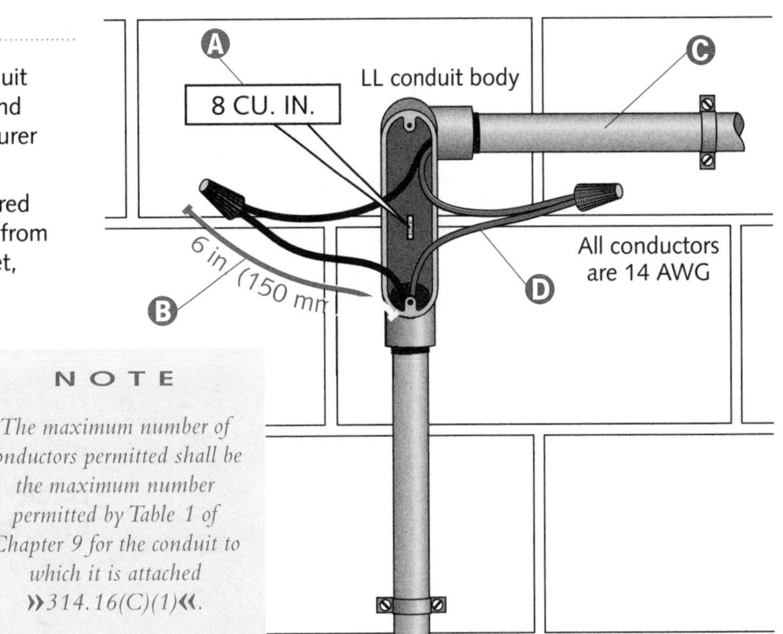

Ⓐ Splices, taps, or devices are permitted inside conduit bodies, provided that the cu. in. capacity is durably and legibly marked on the conduit body by the manufacturer »314.16(C)(2)«.

Ⓑ At least 6 in. (150 mm) of free conductor (measured from the point in the conduit body where it emerges from its raceway or cable sheath) shall be left at each outlet, junction, and switch point »300.14«.

Ⓒ A conduit body can be supported by rigid metal conduit, intermediate metal conduit, rigid nonmetallic conduit, or electrical metallic tubing »314.23(E) *Exception*«.

Ⓓ The maximum number of conductors shall be calculated in accordance with 314.16(B). Four 14 AWG conductors with a volume of 2.0 in.^{3} (32.8 cm^3) each require a total volume of 8 in.^{3} (131.2 cm^3).

NOTE

The maximum number of conductors permitted shall be the maximum number permitted by Table 1 of Chapter 9 for the conduit to which it is attached »314.16(C)(1)«.

GENERAL INSTALLATION

Securing Cables to Metal Boxes

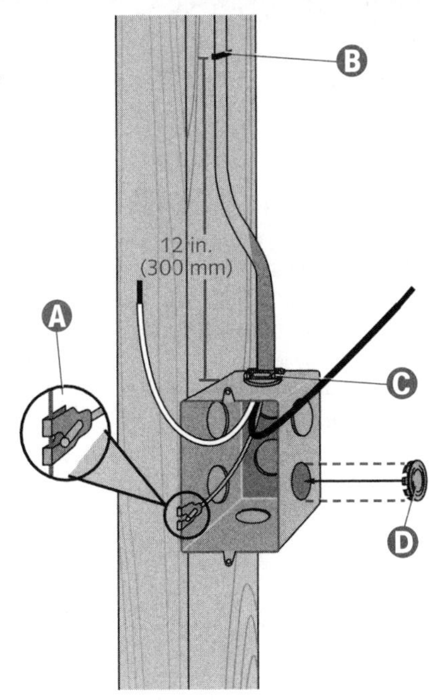

Conductors must be protected from abrasion where entering boxes, conduit bodies, or fittings »314.17«.

Ⓐ A connection must be made between the equipment grounding conductor(s) and a metal box by means of a listed grounding device or a grounding screw serving no other purpose »250.148(A)«. Grounding conductors *must not* be connected to enclosures with sheet metal screws »250.8«.

Ⓑ Nonmetallic-sheathed cable shall be secured within 12 in. (300 mm) of every cabinet, box, or fitting. The cable can be secured by using staples, cable ties, straps, or similar fittings which are designed and installed so that the cable remains undamaged »336.30«.

Ⓒ Cables must be securely fastened to any metal box or conduit body that they enter »314.17(B)«. The clamping mechanism can be internal or external to the box or conduit body.

Ⓓ Unused cable or raceway openings in boxes and conduit bodies must be closed so that the protection provided is at least equal to that provided by the wall of the box or conduit body »314.18«.

Single-Gang Nonmetallic Boxes

Ⓐ Nonmetallic-sheathed cable not fastened to the box must be secured within 8 in. (200 mm) of the box »314.17(C) *Exception*«.

Ⓑ At least 6 in. (150 mm) of free conductor (measured from the point in the box where it emerges from its raceway or cable sheath) shall be left at each outlet, junction, and switch point »300.14«.

Ⓒ The cable sheath shall extend into the box at least ¼ in. (6 mm) through the cable knockout or opening »314.17(C)«.

CAUTION *All permitted wiring methods must be secured to the box, unless the exception in 314.17(C) has been met.*

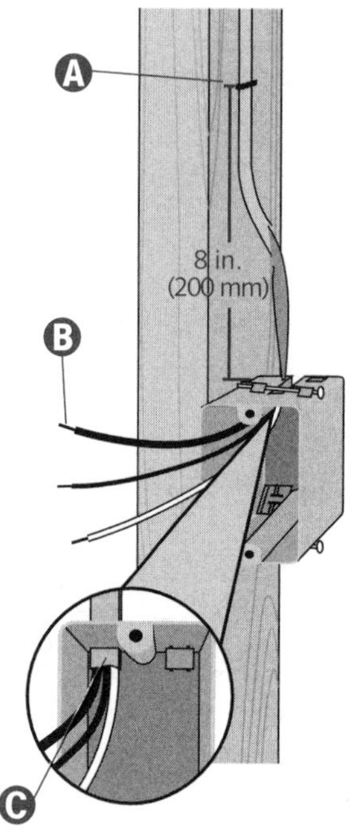

Securing Cables to Nonmetallic Boxes

Ⓐ Nonmetallic-sheathed cable shall be secured within 12 in. (300 mm) of every box »336.30«.

Ⓑ Cables entering a nonmetallic box must be secured to the box, unless it is a single-gang nonmetallic box »314.17(C)«.

Ⓒ The cable sheath shall extend into the box at least ¼ in. (6 mm) through the cable knockout or opening »314.17(C)«.

NOTE

*The exception pertaining to nonmetallic-sheathed cable entering a box which has no means of securing the cable to the box applies only to **single-gang** nonmetallic boxes »314.17(C) Exception«.*

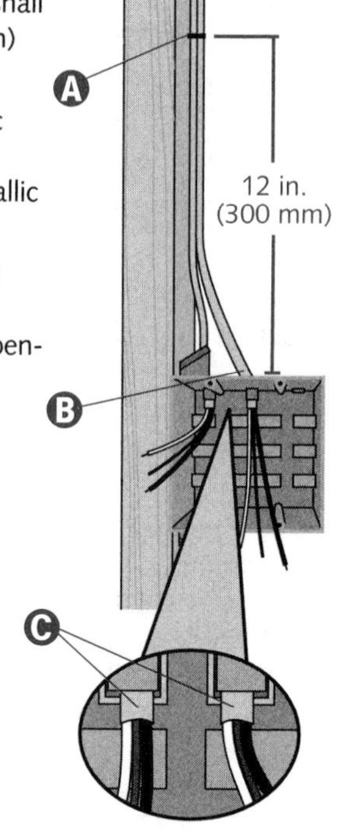

Boxes in Combustible and Noncombustible Materials

Ⓐ Boxes in walls or ceilings constructed of noncombustible material (concrete, tile, etc.) shall be installed so that the front edge of the box will be within ¼ in. (6 mm) of the finished surface »314.20«.

Ⓑ Boxes in walls or ceilings constructed of combustible material, such as wood, shall be flush with or extend beyond the finished surface »314.20«.

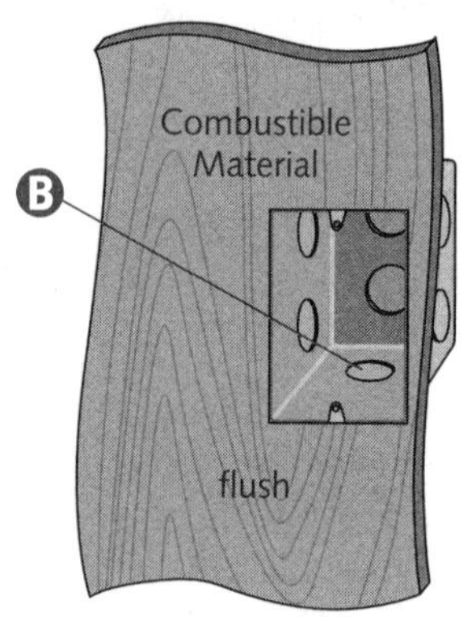

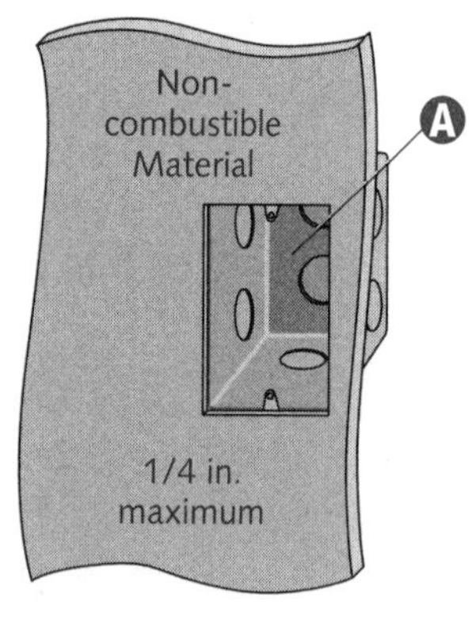

Back-to-Back Boxes in Fire-Resistant-Rated Wall

Ⓐ Qualified testing laboratories publish electrical construction material directories listing installation restrictions which apply to maintaining fire resistive ratings of assemblies involving penetrations, or openings. (An example is the minimum 24-in. [600-mm] horizontal separation usually required between boxes on opposite sides of the wall). These fire resistance directories, product listings, and building codes offer assistance in 300.21 compliance »300.21 (FPN)«.

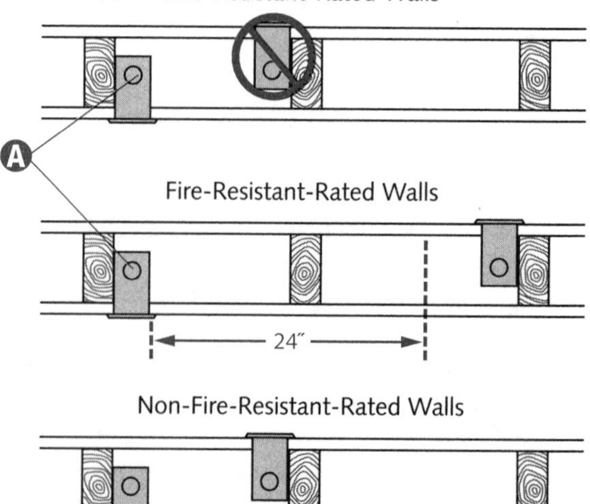

Gaps or Open Spaces

A Damaged or incomplete plaster, dry wall, or plasterboard surfaces must be repaired so that no gap or open space greater than ⅛ in. (3 mm) surrounds the box or fitting »314.21«.

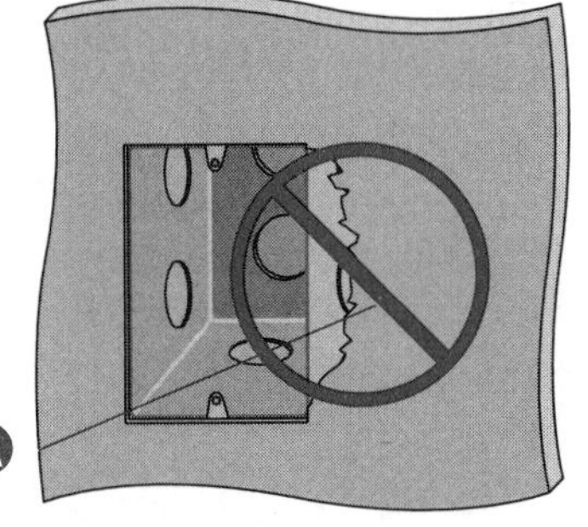

Minimum Internal Depth

A All metal boxes shall be grounded according to Article 250 »314.4«.

B All boxes must have an internal depth of at least ½ in. (12.7 mm). Boxes intended to enclose flush devices must be at least 15⁄16 in. (23.8 mm) internally deep »314.24«.

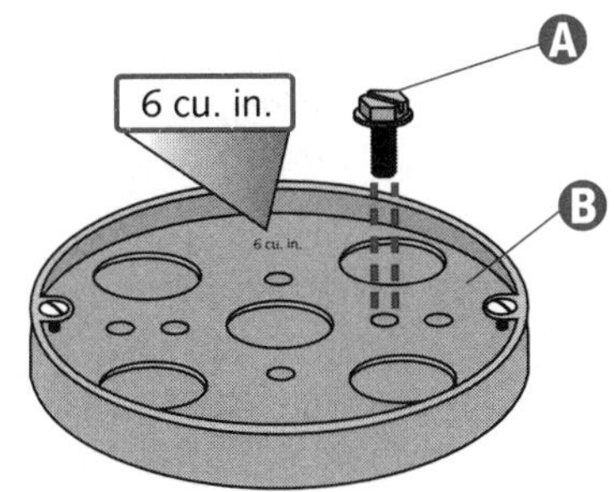

Exposed Surface Extensions

Equipment grounding, where required, must be in accordance with Article 250.

A Flush-mounted boxes requiring surface extensions, must have an extension ring mounted and mechanically secured to the flush box »314.22«.

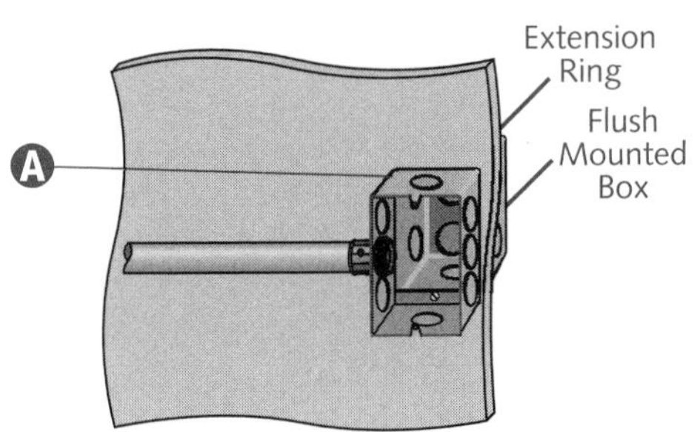

Surface Extensions Made from Covers

A The cover of a flush-mounted box can provide a surface extension where the cover is designed so that it is unlikely to fall off, or be removed if its securing means becomes loose. The wiring method must be flexible and be arranged so that any required grounding continuity is independent of the connection between the box and the cover »314.22 *Exception*«.

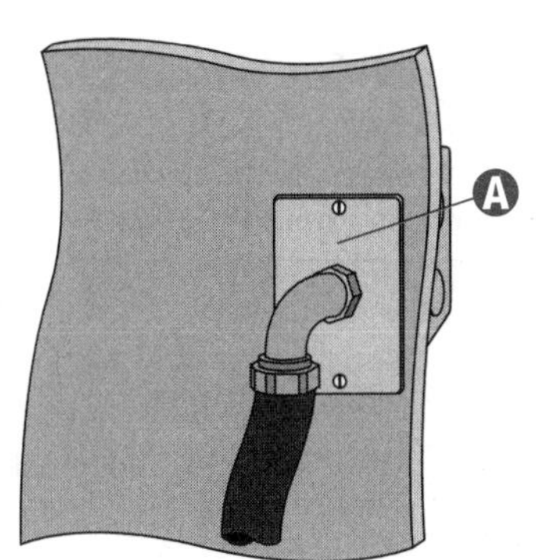

Metal Faceplates Covering Nonmetallic Boxes

A Snap switches (including dimmer switches) must be effectively grounded and must also provide a means to ground faceplates, even if a metal faceplate is not installed »404.9(B)«.

B Both metal and nonmetallic covers and plates shall be permitted. Metal covers or plates, when used, must comply with 250.110 grounding requirements »314.25(A)«.

C Should the snap switch enclosure, or the wiring method used, not have an equipment ground, a snap switch without a grounding connection can be used for replacement purposes only. If a snap switch is installed under this exception, which is located within reach of earth, grade conducting floors, or other conducting surfaces, a faceplate of nonconducting, noncombustible material must be installed »404.9(B) *Exception*«.

D There are two acceptable methods for grounding snap switches effectively: (1) The switch is mounted to a metal box using metal screws or to a nonmetallic box equipped with a means for grounding devices; (2) An equipment grounding conductor, or equipment bonding jumper, is connected to the equipment grounding termination on the snap switch »404.9(B)«.

CAUTION *Isolated ground receptacles, in nonmetallic boxes, must be covered with either a nonmetallic faceplate, or with an effectively grounded metal faceplate »406.2(D)(2)«.*

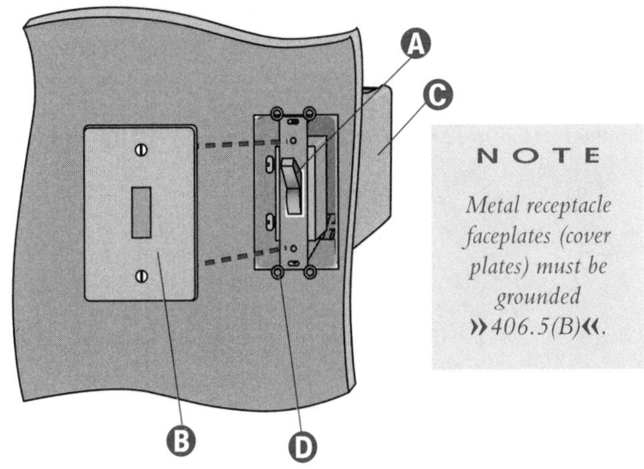

NOTE

Metal receptacle faceplates (cover plates) must be grounded »406.5(B)«.

Covers and Canopies

A To complete the installation, each box must have a cover, faceplate, lampholder, or luminaire (fixture) canopy »314.25«.

Receptacles Mounted on Covers

A Where receptacles are mounted, and supported by a cover, they must be secured by more than one screw, unless the box cover or device assembly is listed and identified as single-screw mounting »406.4(C)«.

Floor Boxes

A Boxes containing receptacles, located in the floor, must be listed for the specific application »314.27(C)«.

CAUTION *Only a limited number of boxes are listed specifically for wood floor construction.*

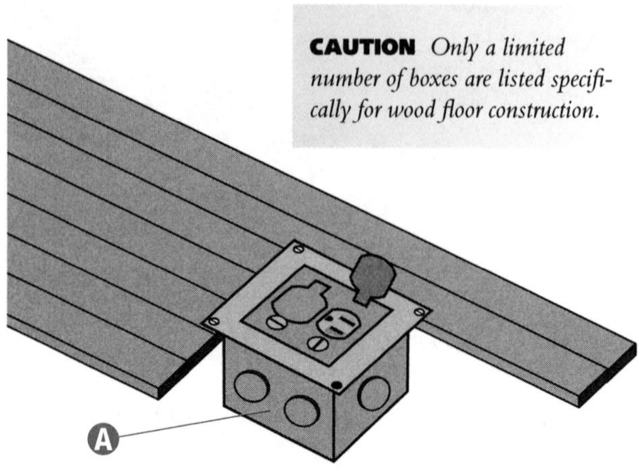

NOTE

Boxes located in elevated floors (such as show windows) do not have to be listed specifically as floor boxes, if the AHJ judges them free from likely exposure to physical damage, moisture, and dirt. Receptacles and covers shall be listed as an assembly for this type of location »314.27(C) Exception«.

Access to Outlet Boxes

A A means for connecting an equipment grounding conductor must be provided for luminaires (fixtures) with exposed metal parts »410.20«.

B Luminaires (fixtures) and equipment are considered grounded when mechanically connected to an equipment grounding conductor as specified in 250.118, and sized in accordance with 250.122 »410.21«.

C Electric-discharge luminaires (lighting fixtures), such as fluorescent luminaires, which are surface-mounted over concealed boxes (outlet, pull, or junction) shall have suitable openings in back of the luminaire (fixture) providing access to the box »410.14(B)«.

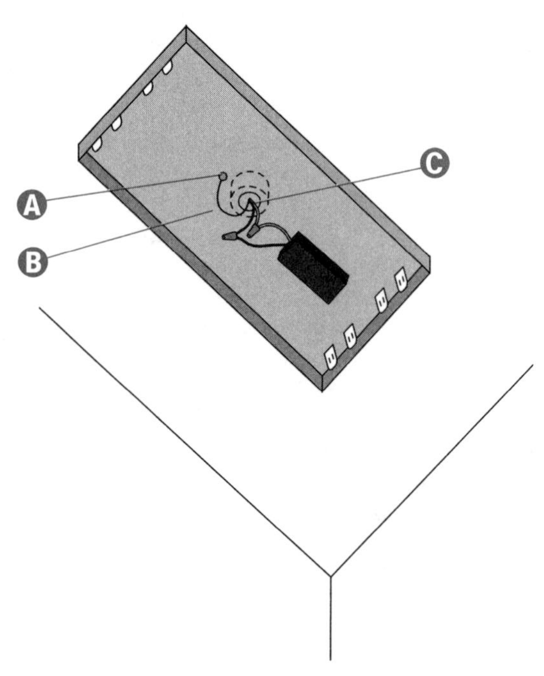

NOTE

Supplementary overcurrent protective devices, when used within luminaires (lighting fixtures), do not have to be readily accessible »240.10«.

BOX AND LUMINAIRE (FIXTURE) SUPPORT

Box and Enclosure Supports

Ⓐ Nails, where used, can attach brackets on the outside of the enclosure, or extend through the interior within ¼ in. (6 mm) of the back or ends of the enclosure »314.23(B)(1)«.

Ⓑ An enclosure, supported from a structural member of a building, shall be rigidly supported either directly, or by use of a brace (metal, polymeric, or wood) »314.23(B)«.

Ⓒ Wood braces must have a cross section of at least 1 in. by 2 in. (25 mm by 50 mm) »314.23(B)(2)«.

NOTE

An enclosure mounted on a building, or other surface, shall be rigidly and securely fastened in place. If the mounting surface does not provide rigid and secure support, additional support must be provided in accordance with 314.23 provisions »314.23(A)«.

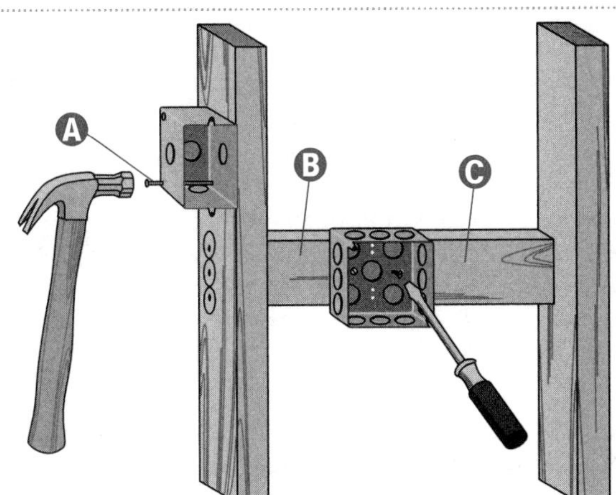

Mounting Enclosures in Finished Surfaces

Ⓐ No support is required where the cable is fished between access points, concealed in the finished surface, and where such supporting of the cable is impractical. Armored cable support provisions are found in 320.30(B)(1), for metal-clad cable in 330.30(B), and for nonmetallic-sheathed cable in 334.30 *Exception No. 1.*

Ⓑ An enclosure mounted in a finished surface must be rigidly secured to the surface by clamps, anchors, or fittings identified for the application »314.23(C)«.

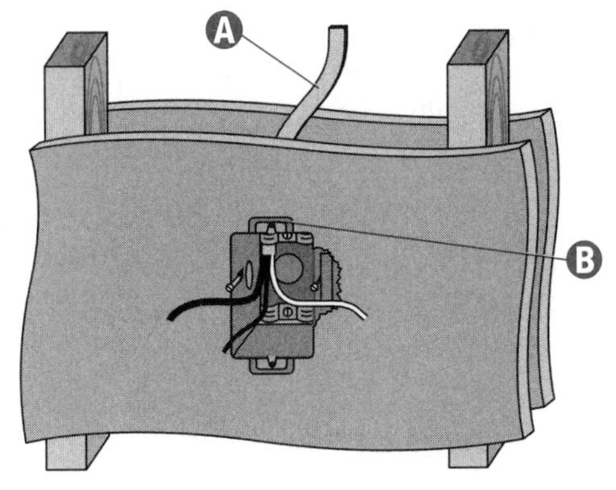

NOTE

No support is required for unbroken lengths (without coupling) of EMT that is fished between access points, concealed in the finished surface, and where such securing of the raceway is impractical »358.30(A) Exception No. 2«.

Enclosures in Suspended Ceilings

Ⓐ Raceways cannot be supported by ceiling grid support wires. Raceways can be secured to independent (additional) support wires that are secured at both ends »300.11(A)«.

Ⓑ An enclosure mounted in a suspended ceiling system shall be fastened to framing members by mechanical means (bolts, screws, rivets, clips, etc.) identified for use with the enclosure(s) and ceiling framing member(s) employed. The framing members must be adequately supported and securely fastened to one another as well as to the building structure »314.23(D)(1)«.

CAUTION *See 300.11(A)(1) and (A)(2) for wiring located within the cavity of a fire-rated and nonfire-rated floor-ceiling (or roof-ceiling) assembly.*

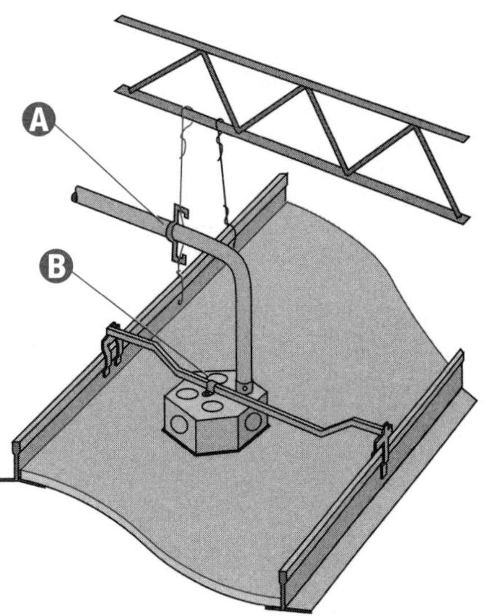

NOTE

Boxes can be secured to independent (additional) support wires that are attached at both ends »300.11(A)«.

Enclosures (No Device, Luminaires [Fixtures], or Lampholders) Supported by Raceways

A An enclosure which contains no devices, **or** which supports no luminaires (fixtures), can be supported by entering raceways when *all* of the following conditions are met: (1) the enclosure does not exceed 100 cu. in. (1650 cm^3) size; (2) the enclosure has threaded entries or hubs identified for the purpose; (3) the enclosure is supported by two or more conduits threaded wrenchtight into the enclosure or hubs; and (4) each conduit is secured within 36 in. (900 mm) of the enclosure, unless all entries are on the same side »314.23(E)«.

B Only rigid or intermediate metal conduit with threaded ends are permitted in 314.23(E).

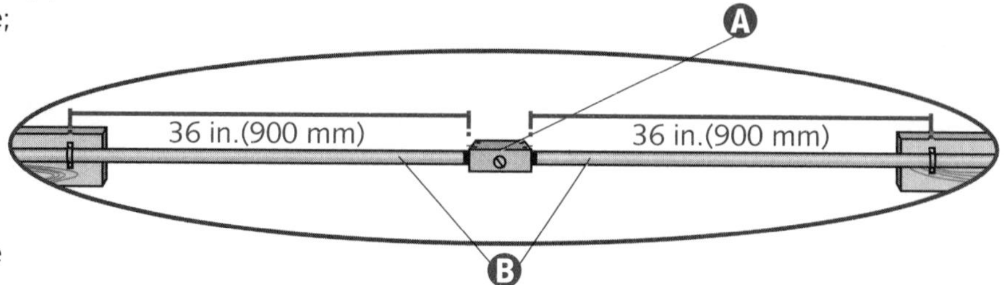

Enclosures (with Devices, Luminaires [Fixtures], or Lampholders) Supported by Raceways

A An enclosure that contains devices, supports luminaires (fixtures), lampholders, or other equipment can be supported by entering raceways when **all** of the following conditions are met: (1) the enclosure does not exceed 100 cu. in. (1650 cm^3) in size; (2) the enclosure has threaded entries or have hubs identified for the purpose; (3) the enclosure is supported by two or more conduits threaded wrenchtight into the enclosure or hubs; and (4) each conduit is secured within 18 in. (450 mm) of the enclosure »314.23(F)«.

B An outlet box can support a luminaire (light fixture) weighing 50 pounds (23 kg) or less, unless the box is listed for a weight greater than the luminaire (fixture) »314.27(B)«.

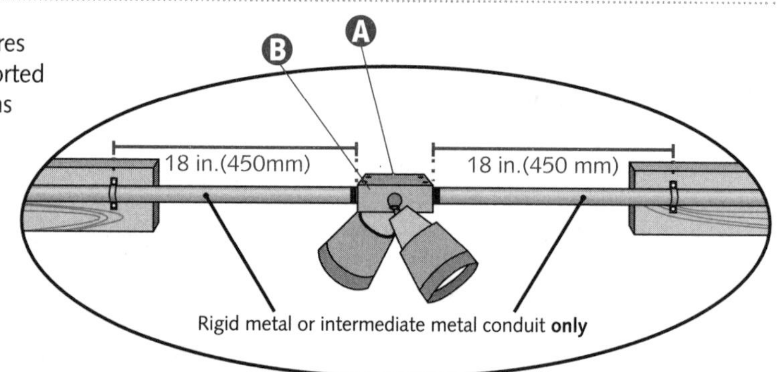

Enclosures with Conduit Entries on One Side

A An enclosure can be supported by conduits entering on the same side when *all* of the following conditions are met: (1) the enclosure does not exceed 100 cu. in. (1650 cm^3) in size; (2) the enclosure has threaded entries or hubs identified for the purpose; (3) the enclosure is supported by two, or more, conduits threaded wrenchtight into the enclosure or hubs; and (4) each conduit is secured within 18 in. (450 mm) of the enclosure »314.23(E)«.

B Unused cable or raceway openings in boxes and conduit bodies must be closed so that the protection provided is at least equal to that provided by the wall of the box or conduit body »314.18«.

C Boxes must be supported within 18 in. (450 mm), whether or not they contain devices or support fixtures, if all the conduits enter on the same side, unless the requirements of 314.23(F) *Exception No. 2* are met.

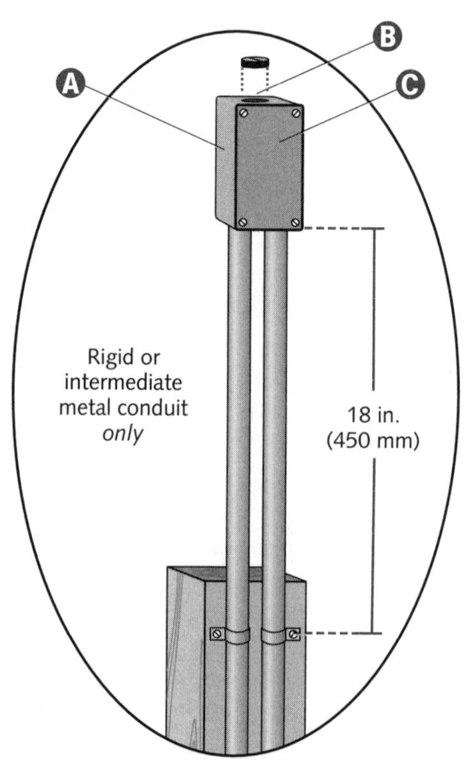

NOTE

Boxes cannot be supported by one conduit unless the requirements of 314.23(F) Exception No. 2 *are met.*

Support for Conduit Bodies (Condulets)

A Any size conduit body (condulet) not containing a device(s), luminaire(s) (fixtures[s]), lampholder(s), or other equipment can be supported by rigid metal, intermediate metal, or rigid nonmetallic conduit or electrical metallic tubing, provided the conduit body's trade size is no larger than the largest trade size of the supporting raceway »314.23(E) *Exception*«. Within this section, no consideration is given to splicing devices.

B Conduit body support must be rigid and secure »314.16(C)(2)«.

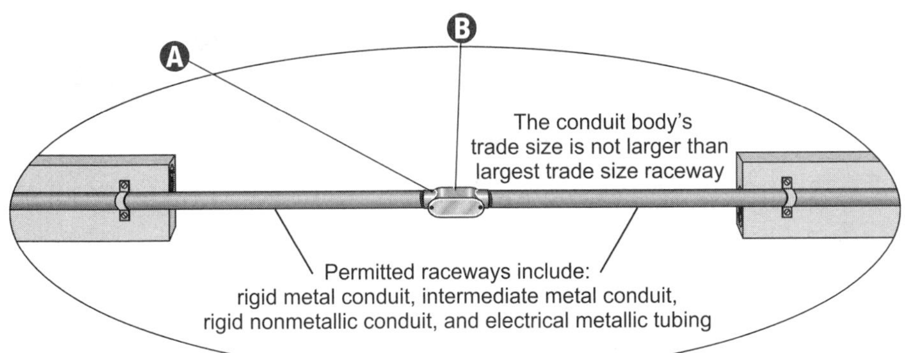

> **NOTE**
>
> *A raceway supported conduit body that contains a device(s), luminaire(s) (fixture[s]), lampholder(s), or other equipment must be supported by either rigid metal or intermediate metal conduit only. The trade size of the conduit body must not be larger than the largest trade size of the supporting raceway »314.23(F)* Exception«.
>
> *The only conduit body not requiring two support raceways is a conduit body constructed with only one conduit entry »314.23(E)* Exception«.

Supporting Luminaires (Fixtures) Using Lengths of Conduit Longer than 18 In.

A 314.23(F) *Exception No. 2* permits unbroken length(s) of rigid or intermediate metal conduit to support a box used for luminaire (fixture) or lampholder support, or to support a wiring enclosure within a luminaire (fixture) where all of the following conditions are met:

B The length of conduit extending beyond the last point of securely fastened support does not exceed 3 ft (900 mm).

C A luminaire (fixture) supported by a single conduit does not exceed 12 in. (300 mm), in any direction, from the point of conduit entry.

D The weight supported by any single conduit cannot exceed 20 pounds (9 kg).

E At the luminaire (fixture) end, each conduit (if more than one) is threaded wrenchtight into the box or wiring enclosure, or into hubs identified for the purpose.

F Where accessible to unqualified persons, the luminaire's (fixture's) lowest point is at least 8 ft (2.5 m) above grade (or standing area) and at least 3 ft (900 mm) [measured horizontally to the 8 ft (2.5 m) elevation] from windows, doors, porches, fire escapes, or similar locations.

G The unbroken conduit before the last point of support is 12 in. (300 mm) or greater, and that portion of the conduit is securely fastened not less than 12 in. (300 mm) from its last point of support.

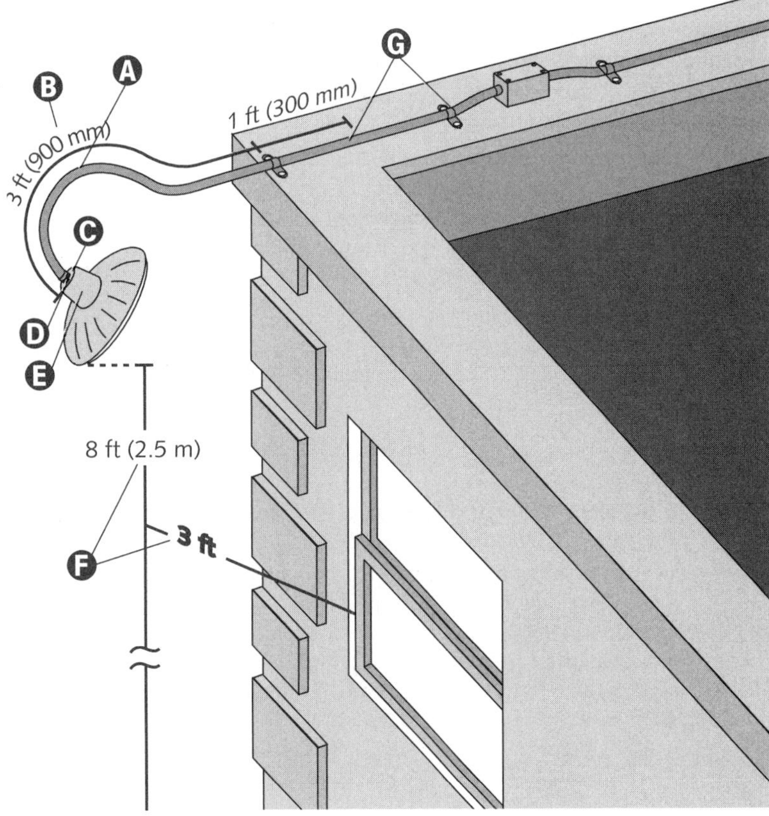

Strain Relief for Flexible Cords

Ⓐ Multiconductor cord or cable, used to support a box, must be protected in an approved manner so that the conductors are not subjected to strain. A strain-relief connector threaded into a box with a hub would be acceptable »314.23(H)(1)«.

Luminaire (Fixture) Hanger

Ⓐ 314.27(B) requires an independent support, such as a luminaire (fixture) hanger, for luminaires (fixtures) that weigh more than 50 pounds (23 kg), unless the box is listed.

Ⓑ When raceway fittings are used to support luminaire(s) (light fixture[s]), they must be capable of supporting the combined weight of the luminaire (fixture) assembly and lamp(s) »410.16(F)«.

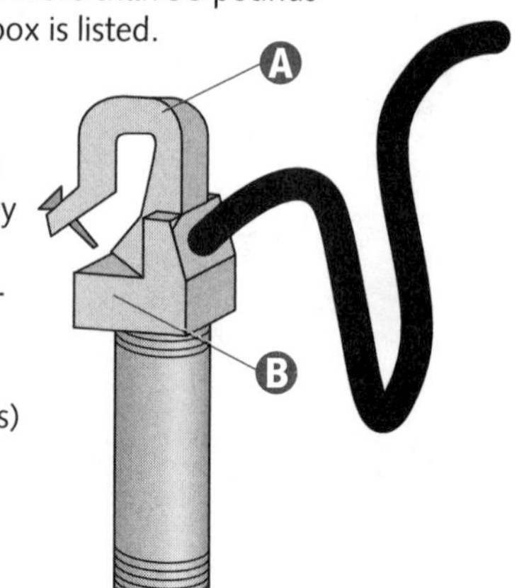

Conduit Stems Supporting Luminaires (Fixtures)

Ⓐ An outlet box can support luminaires (fixtures) weighing no more than 50 pounds (23 kg), unless the box is listed for a weight greater than the luminaire (fixture). A separate means for supporting a luminaire (light fixture) (independent of the outlet box) is also permitted »314.27(B)«.

Ⓑ Stems longer than 18 in. (450 mm) shall be connected to the wiring system with flexible fittings suitable for the location. At the luminaire (fixture) end, the conduit(s) shall be threaded wrenchtight into the box or wiring enclosure, or into hubs identified for the purpose »314.23(H)(2)«.

Ⓒ A box supporting lampholders, luminaires (lighting fixtures) or wiring enclosures within luminaires (fixtures) used in lieu of boxes in compliance with 300.15(B), must be supported by rigid or intermediate metal conduit stems »314.23(H)(2)«.

Ⓓ A luminaire (fixture) supported by a single conduit shall not exceed 12 in. (300 mm), in any horizontal direction, from the point of conduit entry »314.23(H)(2)«.

CAUTION *Any point of a luminaire (fixture) supported by a single conduit must be at least 8 ft (2.5 m) above grade (or standing area) and at least 3 ft (900 mm) (measured horizontally to the 8 ft [2.5 m] elevation) from windows, doors, porches, fire escapes, or similar locations, unless an effective means to prevent the threaded joint from loosening (such as a set screw) is used »314.23(H)(2)«.*

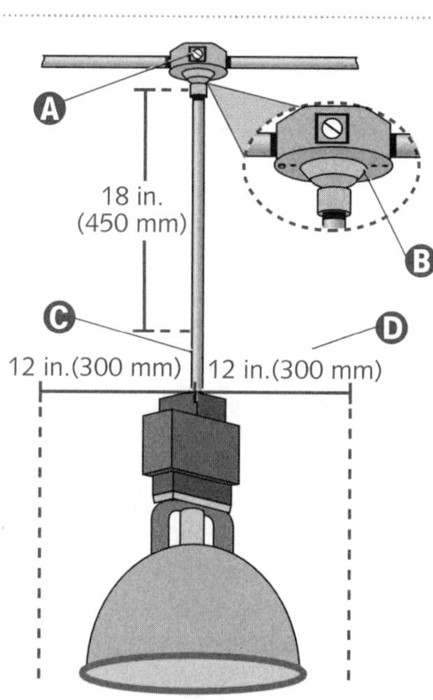

Device Boxes Supporting Luminaires (Fixtures)

Ⓐ A wall-mounted luminaire (fixture) weighing no more than 6 pounds (3 kg) can be supported by boxes (such as device boxes) not specifically designed to support luminaires (light fixtures), provided the luminaire (fixture) or its supporting yoke is secured to the box with at least two No. 6 or larger screws. Plaster rings, secured to other boxes, are also acceptable »314.27(A) *Exception*«.

Ⓑ Not over 6 pounds

Ⓒ Holes for two No. 6 (6/32) screws

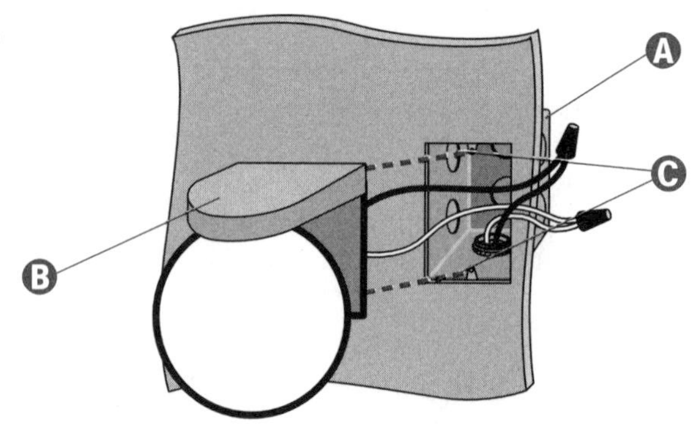

Mounting Nonmetallic Boxes

Ⓐ Supporting screws for nonmetallic boxes must be mounted outside of the box, unless the box is constructed in a manner which prevents contact between the conductors in the box and the supporting screws » 314.43 « .

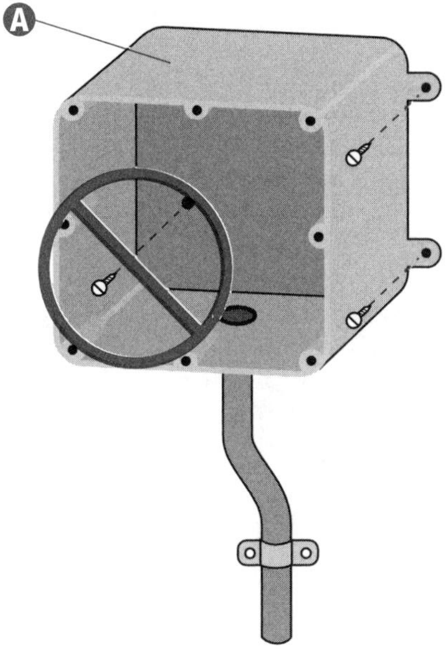

Luminaire (Lighting Fixture) Outlet Boxes

Ⓐ Boxes used at luminaire (lighting fixture) or lampholder outlets must be designed for the purpose. Every box used exclusively for lighting, must be designed and installed so that a luminaire (lighting fixture) may be attached » 314.27(A) « .

Ⓑ An outlet box can support a luminaire (light fixture) weighing no more than 50 pounds (23 kg), unless the outlet box is listed for the weight to be supported » 314.27(B) « .

Ⓒ Inspection of the connections between luminaire (fixture) conductors and circuit conductors must be possible without having to disconnect any part of the wiring (unless the luminaires [fixtures] are connected by attachment plugs and receptacles) » 410.16(B) « .

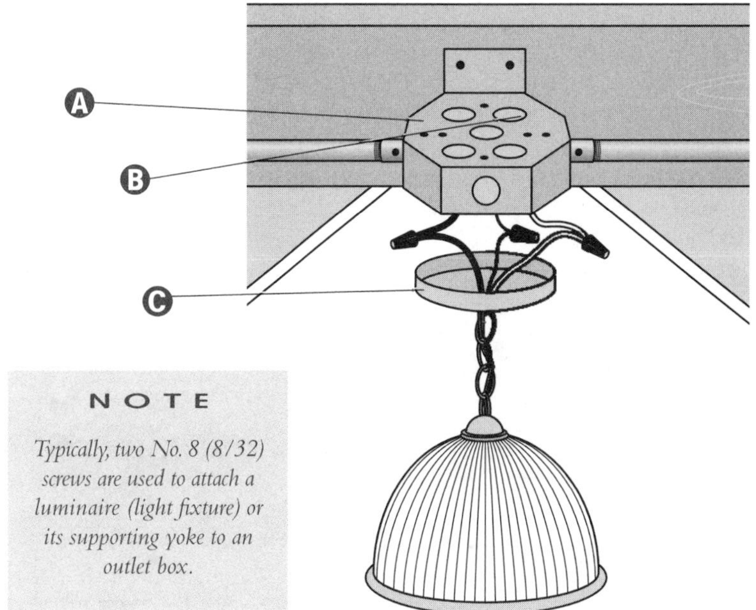

NOTE

Typically, two No. 8 (8/32) screws are used to attach a luminaire (light fixture) or its supporting yoke to an outlet box.

Ceiling-Suspended (Paddle) Fan Boxes

Ⓐ An outlet box cannot be used as the sole means of support for a ceiling-suspended (paddle) fan, unless the box is listed for the application and for the weight of the fan to be supported, such as, "ceiling fan outlet box" » 314.27(D) and 314.27(D) *Exception* « .

Ⓑ Ceiling-suspended (paddle) fans with a combined weight including accessories of not more than 35 pounds (16 kg) can be supported by outlet boxes identified for such use and installed in accordance with 314.23 and 314.27 » 422.18(A) « .

NOTE

Ceiling-suspended (paddle) fans, including accessories, if any, exceeding 35 pounds (16 kg), must be supported independently of the outlet box **»** *422.18(B)* **«**.

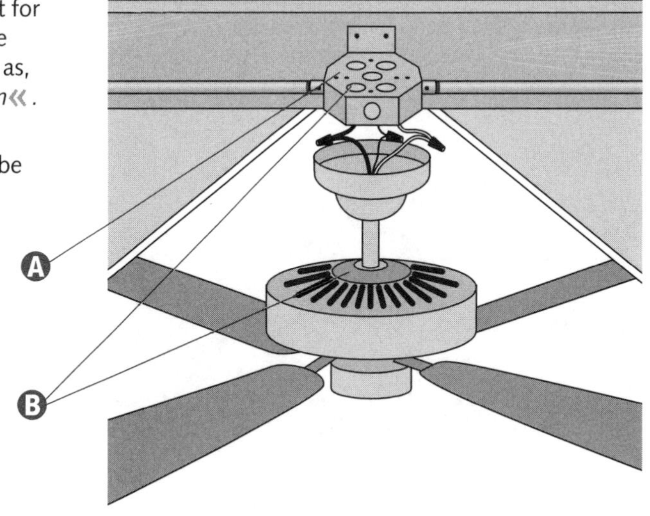

JUNCTION AND PULL BOX SIZING

Straight Pull—Two Raceways

314.16 is used to determine box size requirements for 6 AWG and smaller conductors. Calculations are based on the sizes and numbers of *conductors*. 314.28 is used to determine the box size requirements for 4 AWG and larger conductors (under 600 volts). Calculations here are based on the sizes and numbers of *raceways*.

A Box calculations for 4 AWG and larger conductors (under 600 volts) are performed based on the size and numbers of raceways »314.28«.

B Boxes containing straight pulls are sized according to the largest raceway entering the box. The length must be at least eight times the trade diameter of the largest raceway »314.28(A)(1)«.

C The box width must be large enough to provide proper installation of the raceway (or cable), including locknuts and bushings.

D The box depth must be large enough to provide proper installation of the raceway (or cable), including locknuts and bushings.

NOTE

Use the trade dimension that is applicable to the installation. For example, if millimeters or centimeters are needed to size the junction or pull box, use the metric designator in millimeters instead of the trade size in inches. A junction box is needed for a straight pull with two metric designator 53 raceways. (A 2-in. trade size raceway has a metric designator of 53 mm.) Calculate the minimum length by multiplying the metric designation by eight (53 by 8). The minimum size pull box required is 424 mm or 42.4 cm.

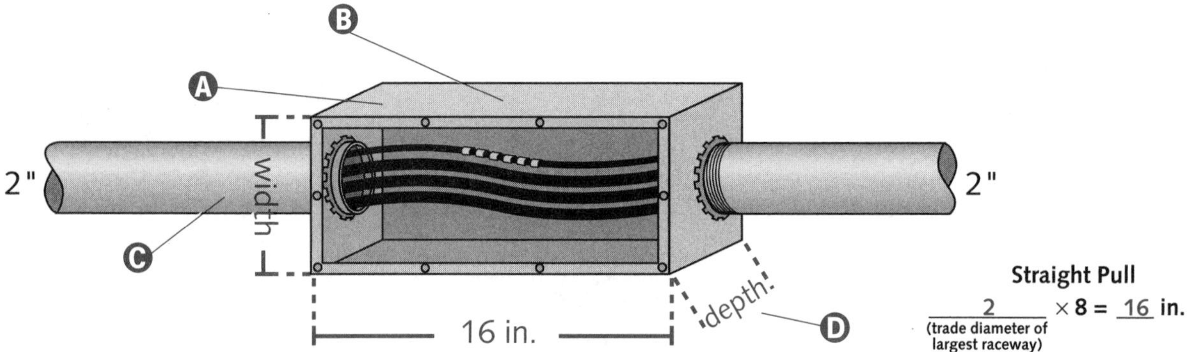

Straight Pull—Multiple Raceways

A The length must be at least eight times the trade diameter of the largest single raceway. No extra space is required for additional raceways when calculating the minimum length of straight pulls. However, additional space is needed for the width of additional raceways, including locknuts and bushings »314.28(A)(1)«.

B Conduit bodies and boxes (junction, pull, and outlet) must be installed so that the wiring they contain can be made accessible without removing any part of the building »314.29«.

CAUTION *All metal boxes shall be grounded in accordance with Article 250 provisions »314.4«.*

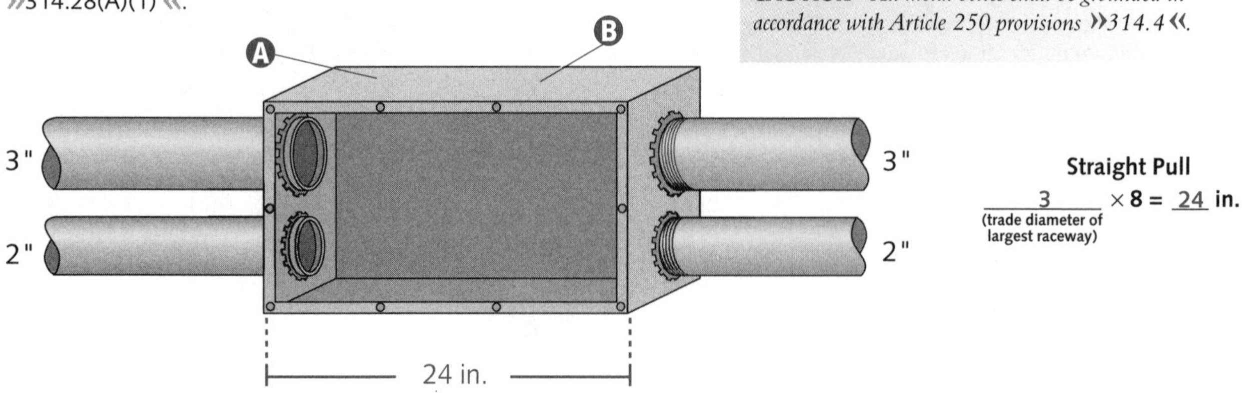

Angle Pull—Multiple Raceways

A Where splices, or where angle or U pulls are made, the distance between each raceway entry into the box and the opposite wall must be at least six times the trade diameter of the largest raceway in a row. This distance is increased for additional raceway entries (in the same row on the same wall of the box) by the sum of the diameters of all other raceway entries »314.28(A)(2)«.

B To calculate the dimension of a box with angle pulls, start with one wall where the raceways enter the box, and find the distance to the opposite wall of the box. The path of the conductors is irrelevant to this calculation.

C First, pick one wall and multiply the largest raceway (trade diameter) by 6. Add to that number the trade diameter of all other raceway(s) in the same row, on the same side of the box.

> **NOTE**
>
> *All pull boxes, junction boxes, and conduit bodies must have covers that are compatible with the box or conduit body construction and must be suitable for the conditions of use »314.28(C)«.*

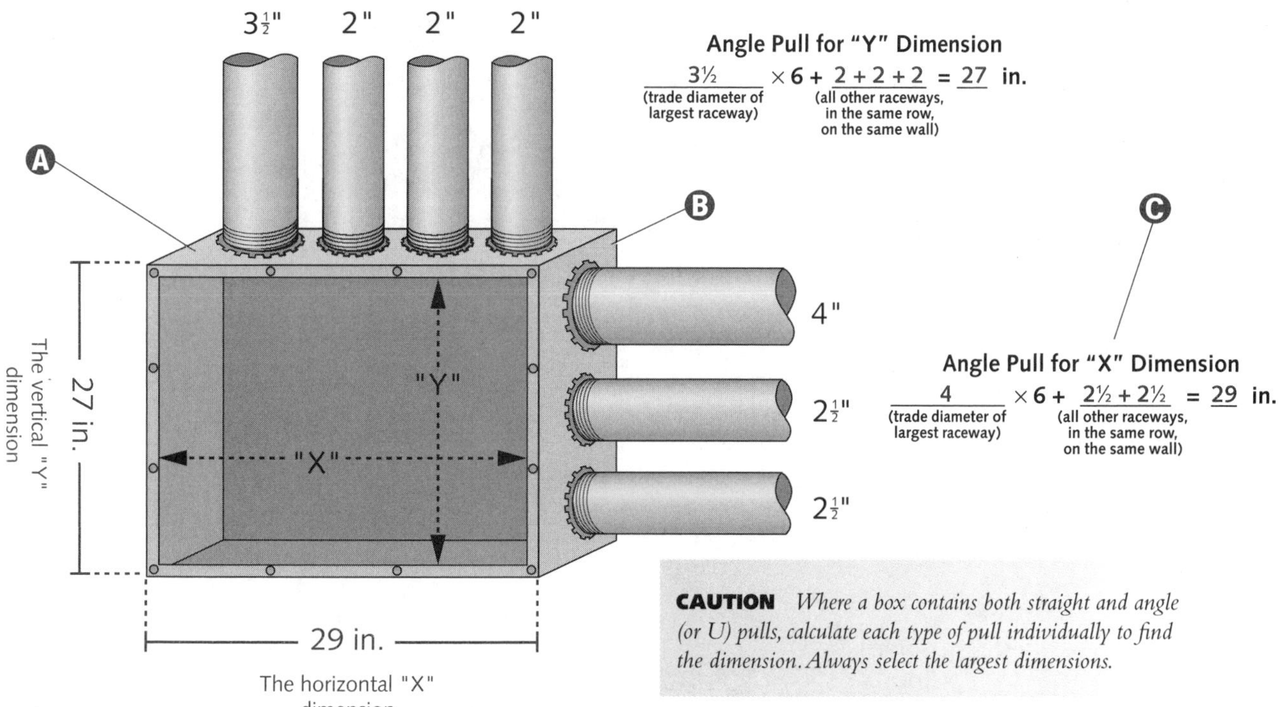

CAUTION *Where a box contains both straight and angle (or U) pulls, calculate each type of pull individually to find the dimension. Always select the largest dimensions.*

Raceways Enclosing the Same Conductors

A Raceways containing the same conductors cannot be closer than six times the larger raceway, even if they are on different walls »314.28(A)(2)«.

B Since no other raceways enter on the same wall of the box, no additional raceway diameters are added.

2 (trade diameter of largest raceway) × 6 + 0 (all other raceways, in the same row, on the same wall) = 12 in.

2 (trade diameter of largest raceway) × 6 + 0 (all other raceways, in the same row, on the same wall) = 12 in.

B

Raceways Enclosing the Same Conductors

2 (trade diameter of largest raceway) × 6 = 12 in.

12 in.

2"

A

12 in.

12 in.

2"

Angle Pull—Multiple Rows

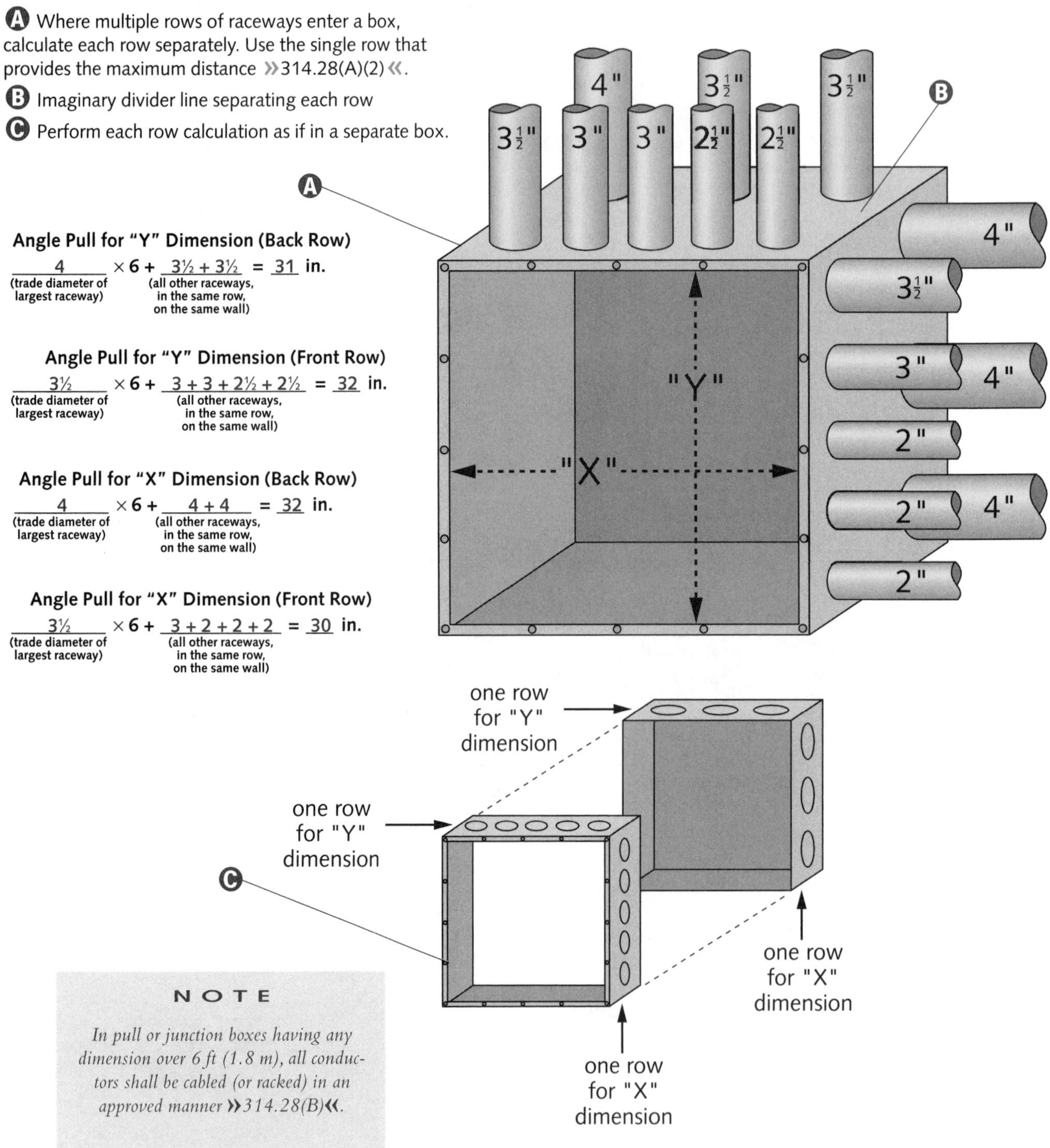

NOTE

In pull or junction boxes having any dimension over 6 ft (1.8 m), all conductors shall be cabled (or racked) in an approved manner »314.28(B)«.

U Pull—Two Raceways

A Use the angle pull method to calculate U pulls. A box with conduit entries only on one wall has a minimum distance to the opposite wall. Multiply the largest raceway by six and add the sum of the trade diameters of the other raceway(s) entering the same wall »314.28(A)(2)«.

B The minimum box width must include:12 in. between raceways plus the thickness of the two raceways (including enough area to provide proper installation of locknuts and bushings).

C The distance between raceways enclosing the same conductor(s) must be at least 6 times the trade diameter of the largest raceway »314.28(A)(2)«.

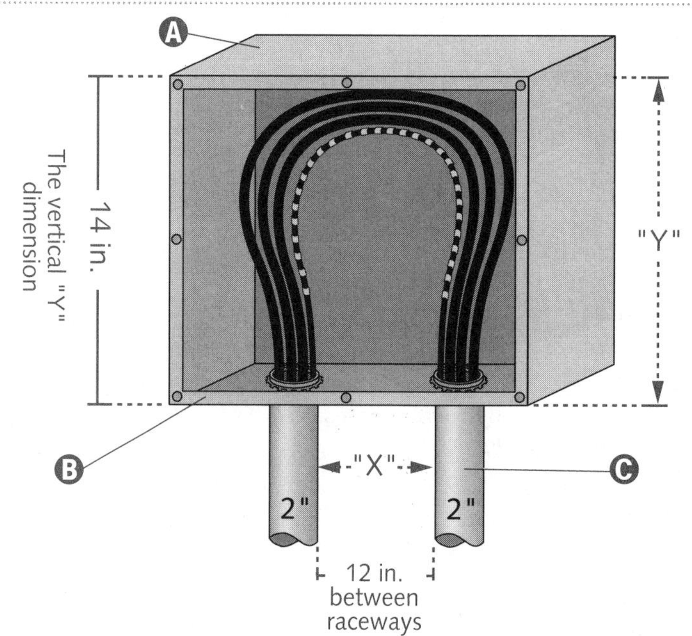

NOTE

Ungrounded conductors (4 AWG and larger) entering a cabinet, box, enclosure, or raceway from another raceway must be fitted with a substantial, smoothly rounded insulating surface, except where the conductors are insulated from the fitting or raceway by a securely fastened-in-place material (such as a plastic bushing) »*300.4(F)*«.

U Pull for "Y" Dimension

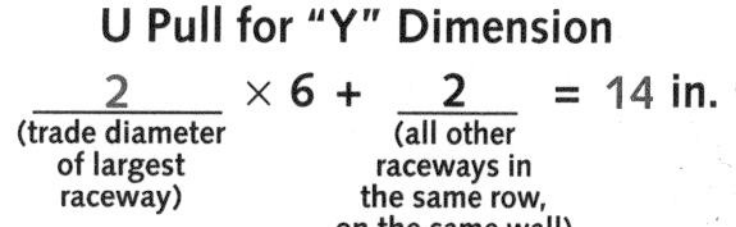

Raceways Enclosing the Same Conductors

2 (trade diameter of largest raceway) × 6 = 12 in.

Raceways Entering Opposite Removable Covers

A Where angle pulls are made, the distance between each raceway entry into the box and the opposite wall of the box must be at least six times the trade diameter of the largest raceway »314.28(A)(2)«. Where no other raceways enter the same wall of the box, no additional raceway diameters are added.

B Where a raceway or cable enters the wall of a box (or conduit body) opposite a removable cover, the distance from the entry wall to the cover can be determined by the distance requirements for one wire per terminal found in Table 312.6(A) »314.28(A)(2) *Exception*«.

C The minimum distance between raceways, enclosing the same conductor(s), is six times the trade diameter of the largest raceway »314.28(A)(2)«.

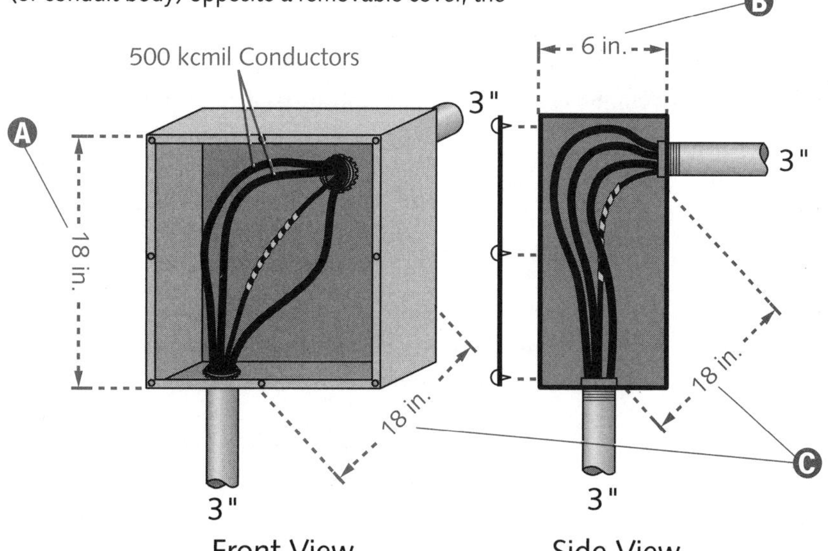

Raceways or Cables Entering Boxes Opposite from Removable Covers

Wire Size (AWG or kcmil)	Minimum Distance from Wall to Cover (in.)	mm
4–3	2	50.8
2	2½	63.5
1	3	76.2
1/0–2/0	3½	88.9
3/0–4/0	4	102
250	4½	114
300–350	5	127
400–500	6	152
600–700	8	203
750–900	8	203
1000–1250	10	254
1500–2000	12	305

Boxes and Conduit Bodies Not Meeting 314.28 and Chapter 9

Ⓐ Smaller-size conduit bodies can generally hold the same number and sizes of conductors (6 AWG and smaller) as the raceway entering the conduit body. Conduit bodies (and boxes) are manufactured which are smaller than those required by 314.28(A)(1) and (2). These may be used if approved and permanently marked showing the maximum number and size conductors allowed »314.28(A)(3)«.

Ⓑ Three 250 kcmil conductors are the maximum number and size permitted in this conduit body.

Ⓒ According to Chapter 9 and Table C8, four 500 kcmil THHN conductors can be installed in a 3-in. rigid metal conduit.

Ⓓ If a conduit body is durably and legibly marked by the manufacturer with the volume, it can contain splices, taps, or devices »314.16(C)(2)«.

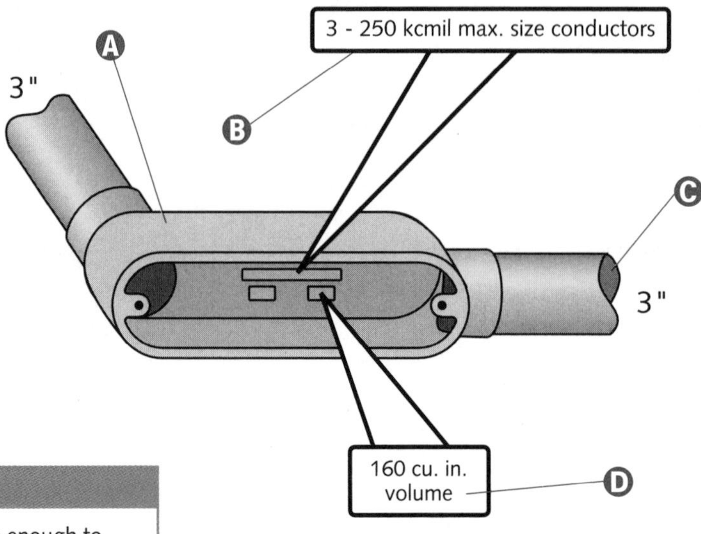

WARNING

Conduit bodies do not have to be large enough to contain the same size and number of conductors permitted in the raceway which enters the conduit body. Some conduit bodies are marked with the maximum size and number of conductors.

Summary

- Limitations apply to the size and number of conductors permitted in boxes (junction, pull, etc.) and conduit bodies.
- One strap (yoke) counts as two of the largest conductors connected to the device.
- Nonmetallic boxes may list the maximum number of conductors permitted in the box, but that number must be reduced if switches, receptacles, etc., are installed in the box.
- Plaster rings, raised covers, extension rings, etc., can provide additional cu.-in. capacity to boxes.
- Conduit bodies can contain splices, taps, or devices, if the conduit body is durably and legibly marked by the manufacturer with the volume.
- The minimum length of free conductor (6 in. [150 mm]) is measured from the point in the box where it emerges from its raceway or cable sheath.
- All metal boxes must be grounded in accordance with Article 250 provisions.
- Boxes and conduit bodies must remain accessible.
- In suspended ceilings, boxes can be secured by support wires that are installed in addition to the ceiling grid support wires.
- Where boxes enclose 4 AWG and larger conductors, different calculation methods apply to straight pulls than to angle pulls.
- There is a minimum distance required between raceways enclosing the same conductors (4 AWG and larger) within a pull or junction box.
- Conduit bodies may not allow the same size and number of conductors as is permitted in the conduit (or tubing) which enters the conduit body.

Unit 3 Competency Test

***NEC*® Reference** **Answer**

__________ __________ 1. In pull boxes or junction boxes having any dimension over _____ in., all conductors shall be cabled or racked up in an approved manner.

a) 6 b) 12 c) 36 d) 72

__________ __________ 2. A nonmetallic surface extension can be run in any direction from an existing outlet, but cannot be run on the floor or within _____ in(es). from the floor.

__________ __________ 3. What are the minimum dimensions for "X" and "Y" in the drawing below?

__________ __________ 4. No box shall have an internal depth of less than _____in(es)., unless the box is intended to enclose flush devices.

__________ __________ 5. A $4\frac{11}{16}$" square box that is $2\frac{1}{8}$" deep already contains two receptacles, two internal cable clamps, four 12 AWG THHN copper conductors (2 black and 2 white), two grounding conductors, three wirenuts, two pigtails, and one equipment bonding jumper. (A flat plaster ring was used to secure the receptacles to the box.) How many more 12 AWG THHN copper conductors can be added to this box?

__________ __________ 6. In damp or wet locations, surface type enclosures (within the scope of Article 312) shall be so placed or equipped as to prevent moisture or water from entering and accumulating within the cabinet or cutout box, and shall be mounted so there is at least _____ in(es). airspace between the enclosure and the wall or other supporting surface.

__________ __________ 7. The minimum thickness in in(es). for a steel box measuring 6" × 4" × $3\frac{1}{2}$" is _____ in(es). thick.

__________ __________ 8. Covers of outlet boxes and conduit bodies having holes through which flexible cord pendants may pass shall be provided with approved _____ or shall have smooth, well-rounded surfaces upon which the cord may bear.

__________ __________ 9. A weatherproof metal junction box measuring 4" h × 4" w × 6" d has been installed to support a luminaire (light fixture). Two rigid metal conduits have been threaded wrenchtight into the threaded entries of the enclosure. The box has not yet been secured to the brick wall, but both conduits have been strapped to the wall 16 in. from the box. What additional support (from the list below) is required for this box?

I. Two metal screws with plastic anchors

II. Two toggle bolts

III. Two drive pins

a) none b) I or II only c) I or III only d) I, II, or III

NEC® Reference	Answer	
________	________	10. Where permanent barriers are installed in a junction box, each section shall be considered as a(n) _____.
________	________	11. What is the maximum number of 12 AWG THW conductors permitted in a 4" octagon box that is 1½" deep? (The box contains a fixture stud and a hickey.)
________	________	12. A means must be provided in each metal box for the connection of a(n) _____.
________	________	13. In walls or ceiling of noncombustible material, boxes shall be installed so that the front edge of the box will not be set back of the finished surface more than _____in(es).
________	________	14. Nonmetallic boxes shall be suitable for the lowest _____ conductor entering the box.
________	________	15. An electrician needs to install a surfacemounted, nonmetallic, weatherproof cabinet on the outside of a concrete block building. Since the cabinet must be mounted in a wet location, it must be mounted so there is at least _____ in(es). of airspace between the cabinet and the wall.
________	________	16. How many 14 AWG THW conductors are permitted in a 3" × 2" × 2½" device box?
________	________	17. Nonmetallic surface extensions must be secured in place by approved means at intervals not exceeding _____in(es).
________	________	18. A 2½" rigid metal conduit enters the back of a pull box with a removable cover on the opposite side. The raceway encloses four 4/0 AWG THHN copper conductors and one 4 AWG equipment grounding conductor. The minimum depth (distance from the entry wall to the cover) for this pull box is _____in(es).
________	________	19. Where nonmetallic boxes are used with open wiring or concealed knob-and-tube wiring, the conductors must enter the box through _____.
________	________	20. Luminaires (light fixtures) weighing no more than _____ pounds can be supported by boxes (such as device boxes) not specifically designed to support luminaires (light fixtures), provided the luminaire (fixture) (or its supporting yoke) is secured to the box with at least two No. 6 or larger screws.
________	________	21. For straight pulls, the length of the box shall not be less than _____ the outside diameter, over sheath, of the largest shielded conductor or cable entering the box on systems over 600 volts.

UNIT

4 Cables

SECTION ONE: FOUNDATIONAL PROVISIONS

Objectives

After studying this unit, the student should:

- know that cables must be installed at least 1¼ in. (32 mm) from the nearest edge of wood framing members, unless a steel plate (or bushing) has been installed.
- understand that nonmetallic-sheathed cable passing through metal framing must be protected by bushings (or grommets) covering all metal edges.
- be aware that openings around electrical penetrations through fire-resistant-rated construction must be sealed using approved methods to maintain the fire resistant rating.
- be able to determine what cables are permitted in spaces used for environmental air-handling purposes.
- know the support requirements for MC, AC, and nonmetallic-sheathed cable.
- be aware that a minimum bending radius must be maintained with cables.
- understand that cables must be protected from physical damage.
- be familiar with general, as well as specific, installation provisions for MC, AC, and nonmetallic-sheathed cable.
- understand conductor identification and the permissible re-identification of certain conductors.
- have a good grasp of underground installation provisions.
- be introduced to flat conductor, integrated gas spacer, mineral-insulated, and medium voltage cables.

Introduction

Unit 4 contains regulations for cable systems that are used often (such as metal-clad cable and nonmetallic-sheathed cable), as well as cable systems that are used rarely (such as integrated gas spacer cable and flat conductor cable. A key provision found in most of the cable articles (and some of the raceway articles) states that the cable (or raceway) must not be installed where subject to physical damage. Cable protection can be achieved by several methods, including: installing the cable a minimum distance from the outside edge of a framing member; installing a steel plate over the cable; or installing the cable in an approved raceway system. Cables installed in attics have different provisions dependent upon the type of attic entrance. Minimum support distances (from box or enclosure to the first support, or between supports) are included in this unit. Conductor identification (and permissible re-identification) for cables is also presented in this unit.

Most of the cable Articles, found in Chapter 3 of the *NEC®*, have sections titled **Uses Permitted** and **Uses Not Permitted**. These sections give locations (some specific) where the cable system can (or cannot) be installed. Since most of these provisions are clearly understandable, this book will not repeat much of this information. Also, be aware that state and local jurisdictions restrict the usage of certain cables in some (if not all) types of occupancies.

GENERAL INSTALLATION

Setback for Bored Holes

A Holes must be bored so that the edge of the hole is at least 1¼ in. (32 mm) from the nearest edge of the wood member »300.4(A)(1)«.

B The 1¼-in. (32-mm) setback applies to all locations (concealed and exposed).

C The 1¼-in. (32-mm) setback applies to studs, joists, rafters, etc.

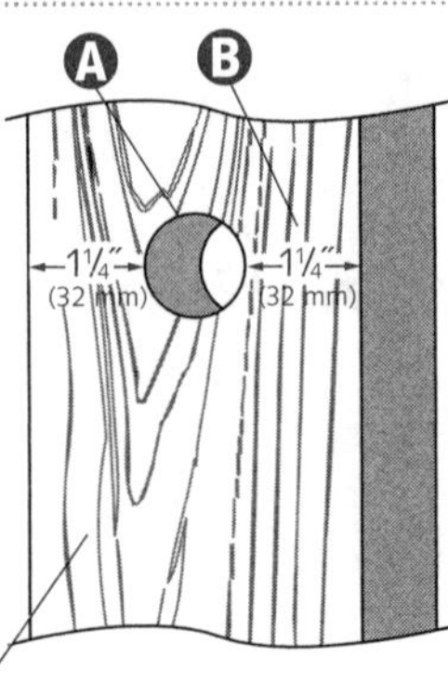

NOTE

The largest hole that can be bored in a 3½-in. wood stud without using a steel plate (clip or sleeve) is 1 in. (provided the drill bit cuts straight through the center of the stud). The 1¼-in. (32-mm) setback must be maintained from both edges (front and back). 3½ in. minus 2½ in. (1¼ setback, two edges) leaves a remainder of 1 in.

Metal Framing Members

A Nonmetallic-sheathed cable passing through holes (or slots) in metal framing must be protected by listed bushings (or grommets) covering all metal edges and securely fastened in the opening prior to cable installation »300.4(B)(1)«.

B 300.4 applies to all locations (concealed and exposed).

C Openings (slots or holes) can be punched (either factory or field), cut, or drilled into the metal members »300.4(B)(1)«.

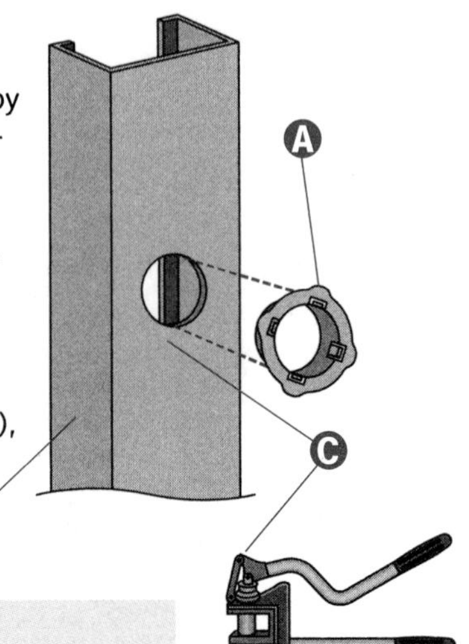

NOTE

Where nails (or screws) are likely to penetrate nonmetallic-sheathed cable (or electrical nonmetallic tubing), a steel plate (clip or sleeve) at least ¹⁄₁₆-in. (1.6-mm) thick shall be used to protect the cable or tubing »300.4(B)(2)«.

Steel Plates

A Where the bored hole is less than 1¼ in. (32 mm) from the edge, the cable (or raceway-type wiring method) shall be protected from penetration, by screws or nails, by a steel plate (or bushing) »300.4(A)(1)«.

B Cables, installed in a groove, covered by wallboard, paneling, carpeting, etc., shall be protected by ¹⁄₁₆-in. (1.6-mm) thick steel plate (sleeve or equivalent) or by at least 1-¼ in. (32-mm) free space for the full length of the groove in which the cable is installed »300.4(E)«.

C Where there is no potential of weakening building structure, cables (or raceways) can be laid within notches in wood studs, joists, rafters, etc. »300.4(A)(2)«.

D The steel plate (or bushing) must be at least ¹⁄₁₆-in. (1.6-mm) thick, and of appropriate length and width to adequately cover the wiring »300.4(A)(1)«.

E The steel plate [with a minimum thickness of ¹⁄₁₆-in. (1.6-mm)] must be installed before the building finish is completed »300.4(A)(2)«.

F Requirements for notched wood members apply to all locations (concealed and exposed).

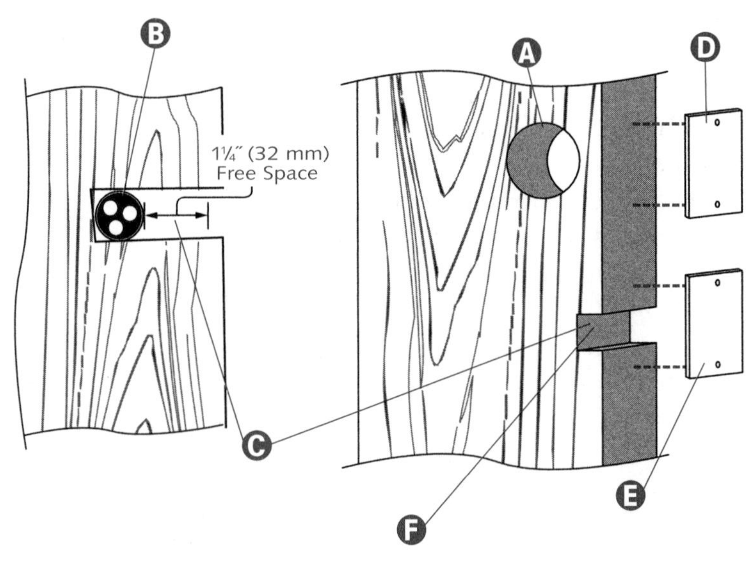

Parallel to Framing Members

A Cables installed parallel to framing members (studs, joists, etc.,) shall be installed and supported so that the nearest outside surface of the cable is at least 1¼ in. (32 mm) from the nearest edge of the framing member where nails (or screws) are likely to penetrate »300.4(D)«.

B If a 1-¼ in. (32-mm) setback is not possible, the cable must be protected from penetration by nails (or screws) by a minimum 1⁄16-in. (1.6-mm) thick steel plate (sleeve, etc.) »300.4(D)«.

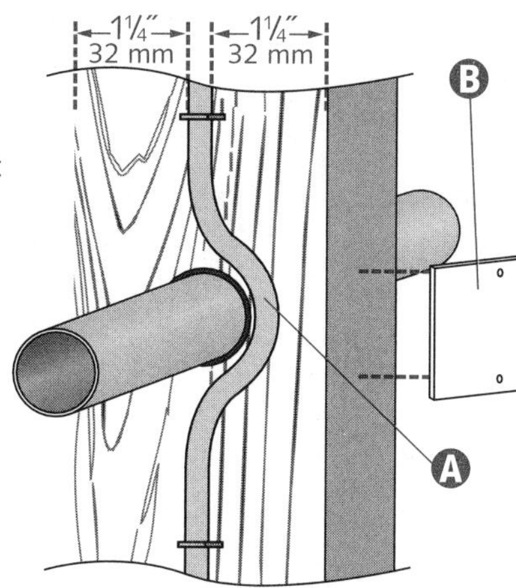

NOTE

For concealed work in finished buildings or finished panels for prefabricated buildings, where such support is impracticable, cables can be fished between access points »300.4(D) Exception No. 2«.

Maintaining the Integrity of Fire-Resistant-Rated Construction

A Electrical installation openings which penetrate fire-resistant-rated structures (walls, partitions, floors, ceilings, etc.) must be firestopped using approved methods that maintain the fire resistance rating »300.21«.

CAUTION *Some state and local jurisdictions may require that all penetrations (fire rated and non-fire rated) be sealed.*

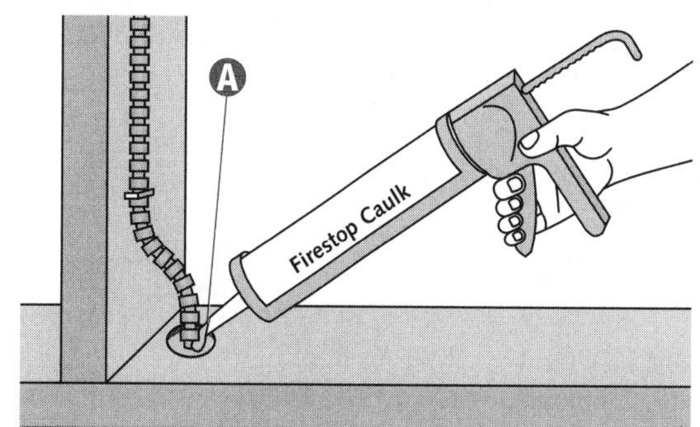

NOTE

In hollow spaces, vertical shafts, and ventilation (or air-handling) ducts, electrical installations must not substantially increase the possible spread of fire or products of combustion »300.21«.

Wiring in Ducts, Plenums, and Other Air-Handling Spaces

A Only Type MI and MC cables, employing a smooth or corrugated impervious metal sheath without an overall nonmetallic covering, can be installed in ducts (or plenums) specifically fabricated to transport environmental air »300.22(B)«. 300.22(B) also lists, other raceways permitted in ducts and plenums of this type.

B 300.22(C) applies to spaces used for environmental air-handling purposes other than ducts and plenums specified in 300.22(A) and (B). Types AC, MI, MC (without an overall nonmetallic covering), or other factory-assembled multiconductor cables (control or power) listed specifically for the use must be installed in these spaces. All other types of cables and conductors must be installed in one of the raceways (or wireways) named in 300.22(C)(1).

CAUTION *Ducts used to transport dust, loose stock, or flammable vapors cannot contain wiring systems of any type. No wiring system (of any type) shall be installed within a duct or shaft containing only such ducts used for vapor removal or for ventilation of commercial-grade cooking equipment »300.22(A)«.*

NOTE

Metallic manufactured wiring systems (without nonmetallic sheath) having listed prefabricated cable assemblies is permitted in this type of installation »300.22(C)(1)«.

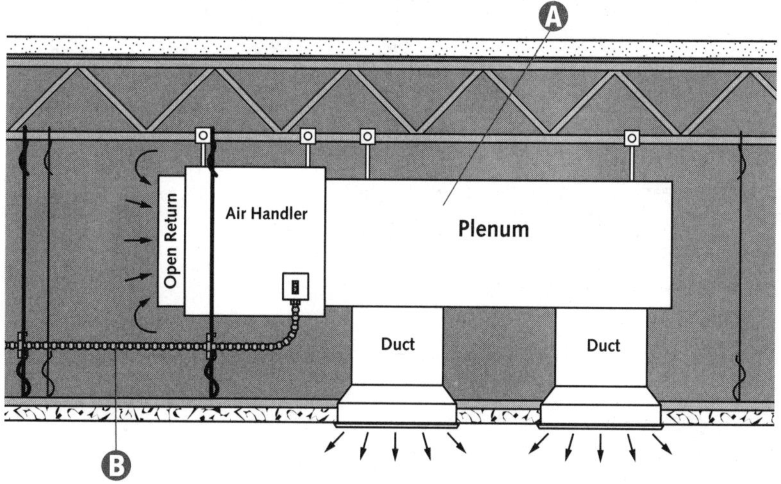

Wiring Within Air-Handling Spaces in Dwelling Units

A 300.22 does not apply to **dwelling unit** joist or stud spaces where wiring passes perpendicular to the long dimension of such spaces »300.22(C) *Exception*«.

B Conductors must remain at least 1¼ in. (32 mm) from the nearest edge of a wood member »300.4(A)(1)«.

C Joist or stud spaces, used for environmental air-handling purposes, cannot contain nonmetallic-sheathed cable which runs along the long dimension.

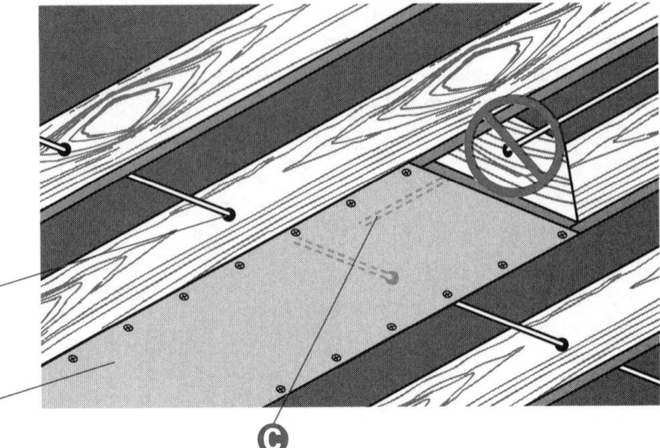

Securing Cables

Generally, all cables must be secured to the box, cabinet, etc. »314.17(B) and (C)«.

Staples, cable ties, straps, etc. must be designed to secure cables and shall be installed so that cables are not damaged.

A Type MC cable must be supported and secured at intervals not exceeding 6 ft (1.8 m) »330.30(A)«.

B Type AC cable must be supported and secured at intervals not exceeding 4½ ft (1.4 m) »320.30«.

C Nonmetallic-sheathed cable must be supported and secured at intervals not exceeding 4½ ft (1.4 m) »334.30«.

D Type MC cables containing no more than four conductors, sized 10 AWG or smaller, must be secured within 12 in. (300 mm) of every box, cabinet, fitting, etc. »330.30(A)«.

E Type AC cable must be secured within 12 in. (300 mm) of every box, cabinet, or fitting »320.30«.

F Nonmetallic-sheathed cable (secured to the box) must be secured within 12 in. (300 mm) of every box, cabinet, or fitting »334.30«.

G Nonmetallic-sheathed cable **not** secured to a single-gang box must be secured within 8 in. (200 mm) »314.17(C) *Exception*«.

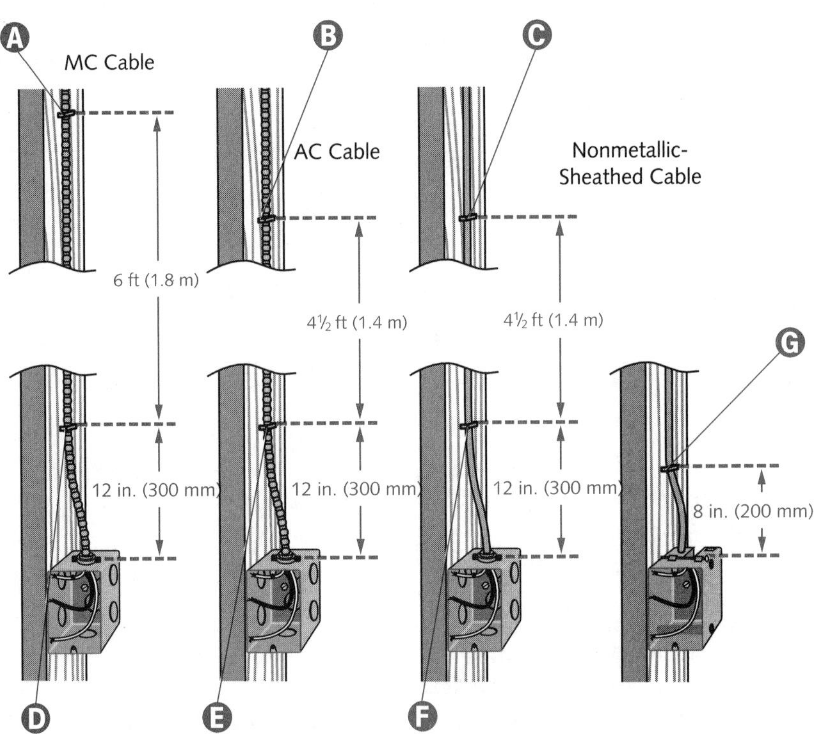

NOTE

Cables fished between access points, where concealed in finished buildings (or structures) and where support is impracticable, can remain unsupported and unsecured »320.30(B)(1), 330.30(B), and 334.30 Exception No. 1«.

Cables Passing Through Framing Members

A Cables running horizontally (or diagonally) are considered supported and secured when passing through a framing member (wood, metal, etc.), unless the support intervals exceed those listed for the specified cable. Cables installed in notches and protected by a ⅟₁₆-in. (1.6-mm) steel plate are also considered secured »320.30(A), 330.30(A)(1), and 334.30«.

B Cables must be secured within 12 in. (300 mm) of every box, cabinet, fitting, etc.

C This cable is not "secured" because the support interval exceeds the maximum distance permitted.

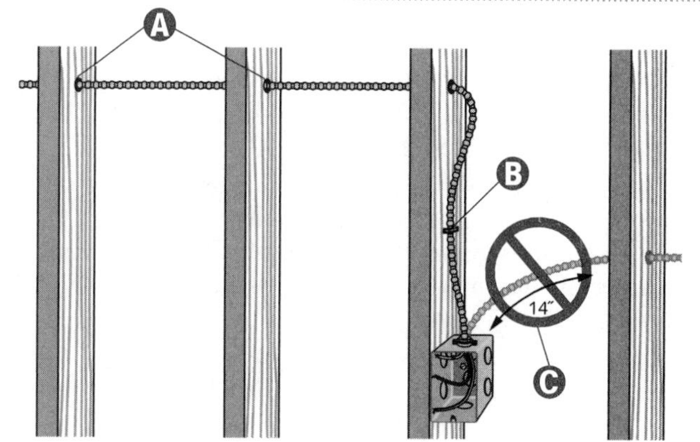

Attics Without Permanent Stairs or ladders

A Attics and roof spaces not accessible by permanent stairs or ladders require protection only within 6 ft (1.8 m) of the nearest edge of the scuttle hole or attic entrance »320.23(A), 330.23, and 334.23«.

B Protection is required for cables located within 6 ft (1.8 m) (measured both vertically and horizontally) of the attic entrance.

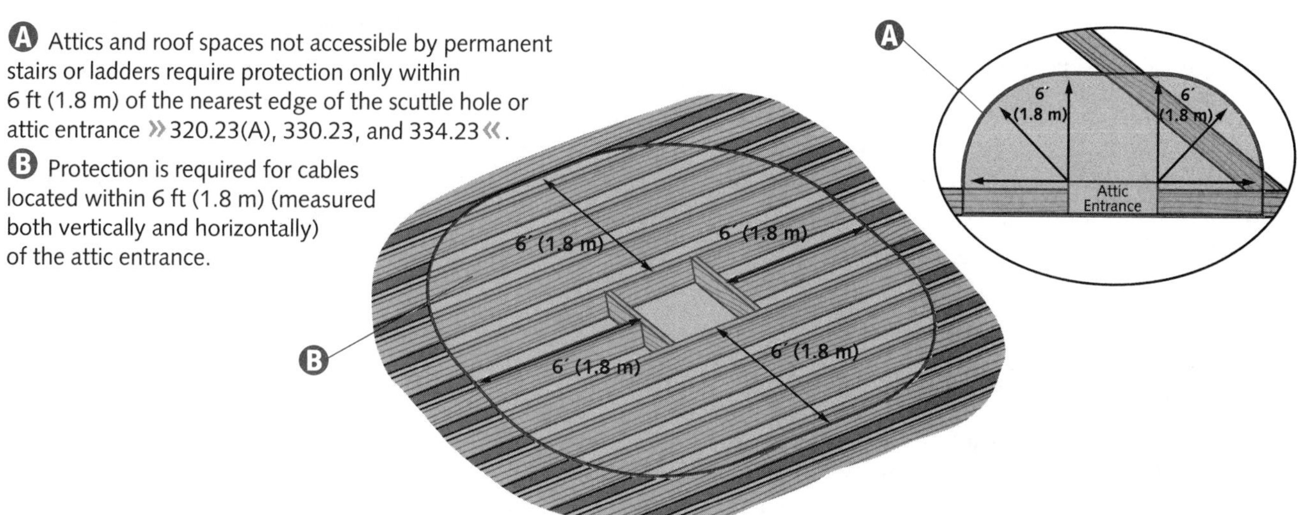

Bending Radius

Type AC cable must have a bending radius at least five times the diameter of the cable »320.24«.

Type MC cable (interlocked-type or corrugated sheath) has a bending radius of seven times the external diameter of the metallic sheath »330.24(B)«.

Nonmetallic-sheathed cable has a minimum bending radius of five times the cable diameter »334.24«.

A A curve radius is measured from the inner edge of the bend.

Example: If the outside diameter of Type MC cable (interlocked-type) is ½ (.5) in. (13 mm), then the radius will need a measurement of 3½ in. (89 mm) (.5 x 7 = 3.5 in.).

CAUTION *All bends shall be made so that the cable will not be damaged.*

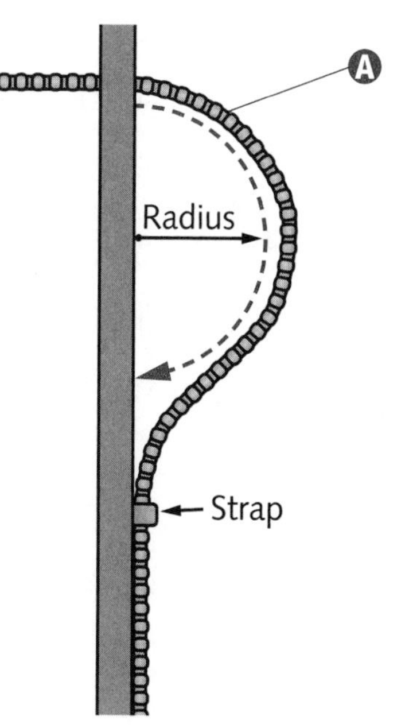

Attics With Permanent Stairs or Ladders

Ⓐ Cables within 7 ft (2.1 m) of the floor (or floor joist) can run through the studs if meeting the requirements of 300.4(D). Cables more than 7 ft (2.1 m) above the floor (or floor joist) can run across the face of framing members.

Ⓑ If cables are installed across the face of rafters or studding and they are located within 7 ft (2.1 m) of the floor (or floor joist) in attics or roof spaces that are accessible, the cable must be protected by substantial guard strips that are at least as high as the cable »320.23(A)«.

Ⓒ Cables running across the top of floor joists must be protected by substantial guard strips that are of sufficient height to adequately protect the cable »320.23(A)«.

Ⓓ Cable installed parallel to the sides of rafters, studs, or floor joists does not require guard strips or running boards; however, the installation must also comply with 300.4(D) »320.23(B)«.

Ⓔ Cables not installed on the face (or surface) do not require guard strips or running boards.

Ⓕ Cables installed in attics (and roof spaces) accessible by permanent stairs (or ladders) must meet 320.23 provisions.

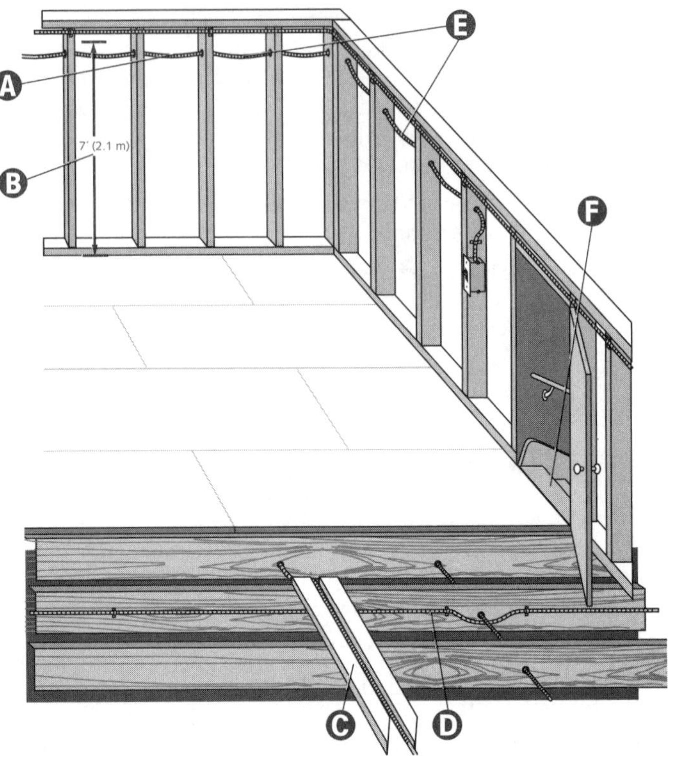

NOTE

Type MC cable and nonmetallic-sheathed cable must also comply with 320.23 »330.23 and 334.23«.

Insulating (Anti-Short) Bushings

Ⓐ All armor termination points of AC cable shall have a fitting to protect the wires from abrasion, unless the outlet boxes (or fittings) are designed to afford equivalent protection. In addition, an insulating bushing (or its equivalent protection) shall be provided between the conductors and the armor »320.40«.

Ⓑ Type AC cable shall have a flexible metal tape armor »320.100«.

Ⓒ The connector or clamp which fastens Type AC cable to boxes or cabinets must be designed so that the insulating bushing (or its equivalent) is visible for inspections »320.40«.

Ⓓ Type AC cables shall have an internal bonding strip (copper or aluminum) in intimate contact with the armor for its entire length »320.100«.

Ⓔ Some insulating (anti-short) bushings are manufactured so that part of the bushing extends past the connector or clamp (once installed), effectively acting as a flag. This flag increases the visibility of the bushing after installation.

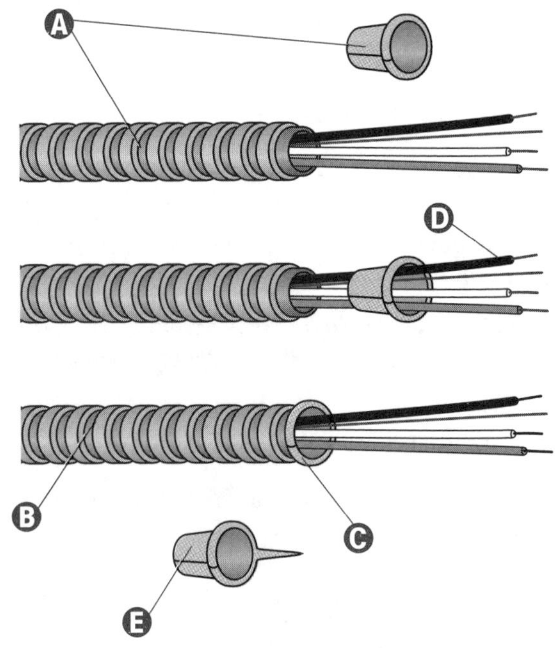

NOTE

Insulating bushings in Type MC cable are optional rather than required. Although Article 330 does not require an insulating bushing between the conductors and the armor, it is a good practice to use one.

Exposed Runs of Type AC and MC Cables Under Joists

A Exposed runs of Type AC cable installed on the underside of joists must be supported (and secured) at *every* joist and must be located so that physical damage is avoided »320.15«.

B Type MC cable must be supported (and secured) at intervals of 6 ft (1.8 m) or less »330.30(A)«.

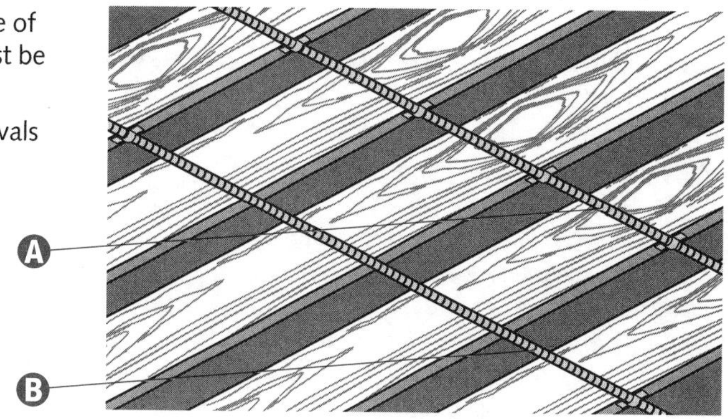

Exposed Nonmetallic-Sheathed Cable in Unfinished Basements

A Exposed nonmetallic-sheathed cable (smaller than 8 AWG if three-conductor, or 6 AWG if two-conductor), running at angles with joists in unfinished basements, must either be run through bored holes in the joists or on running boards »334.15(C)«.

B The nearest outside surface of installed and supported cable must be at least 1¼ in. (32 mm) from the nearest edge of the framing member where nails or screws are likely to penetrate »300.4(A)(1)«.

C Three-conductor 8 AWG, or two-conductor 6 AWG, or larger, can be secured directly to the lower edges of the joists »334.15(C)«.

D Where the number of current carrying conductors in multiconductor cables are bundled or stacked longer than 24 in. (600 mm) without maintaining spacing, reduce the allowable ampacity of each conductor as shown in Table 310.15(B)(2)(a) »310.15(B)(2)(a)«.

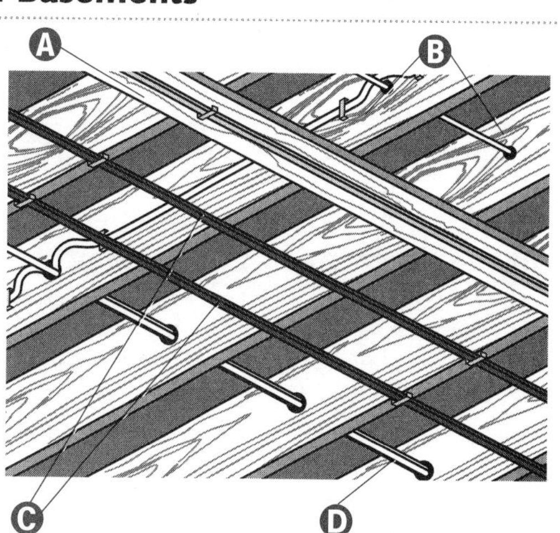

Exposed Nonmetallic-Sheathed Cable Passing Through a Floor

Cables entering or exiting conduit (or tubing) used for support or protection against physical damage require fittings on the conduit (or tubing) ends to prevent cable abrasion »300.15(C)«.

A Nonmetallic-sheathed cable passing through a floor, must be enclosed in rigid (or intermediate) metal conduit, electrical metallic tubing, Schedule 80 PVC rigid nonmetallic conduit, listed surface metal (or nonmetallic) raceway, or other metal pipe that extends at least 6 in. (150 mm) above the floor »334.15(B)«.

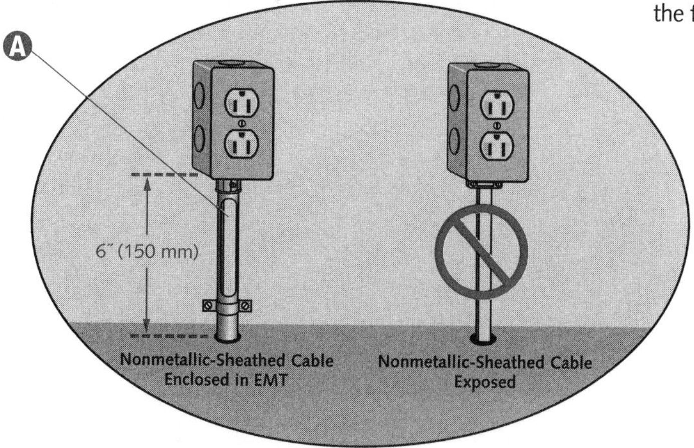

NOTE

Raceway (or cable) openings in a fire-resistant-related floor must be firestopped using approved methods to maintain the fire resistance rating »300.21«.

If subject to physical damage, conductors must be adequately protected »300.4«. Nonmetallic-sheathed cable can be protected by conduit, electrical metallic tubing, Schedule 80 PVC rigid non-metallic conduit, pipe, guard strips, listed surface metal (or nonmetallic) raceway, or other means »334.15(B)«.

CONDUCTOR IDENTIFICATION

General Conductor Identification Provisions

A Generally, a conductor with a continuous white (or gray) covering shall be used only as a grounded circuit conductor »200.7«.

B Equipment grounding conductors can be bare, covered, or insulated. Individually covered (or insulated) equipment grounding conductors must have a continuous outer finish that is either green or green with yellow stripe(s) »250.119«.

C Conductors used as ungrounded (hot) conductors, whether single conductors or in multiconductor cables, must be finished in a way that clearly distinguishes them from grounded and grounding conductors »310.12(C)«. Ungrounded (hot) conductors (except for a **high leg** conductor) can be any color other than white, gray, green, or green with yellow stripe(s).

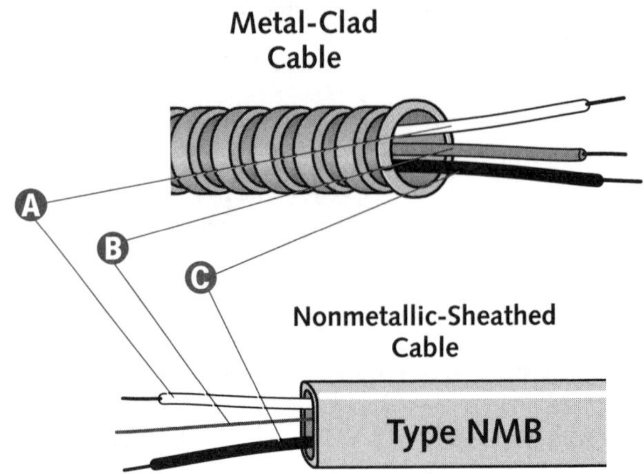

Three-Phase Conductor Identification

A Generally, ungrounded (hot) conductors can be any color except white, gray, green, or green with yellow stripe(s). A widely accepted practice is to identify three-phase ungrounded (hot) conductors as black, red, and blue in 208/120 volt, four-wire, wye-connected systems; and brown, orange, and yellow in 480/277 volt, four-wire, wye-connected systems.

B Grounded (or neutral) conductor

C Grounding conductor

D Interlocking-armor Type MC cable requires an equipment grounding conductor.

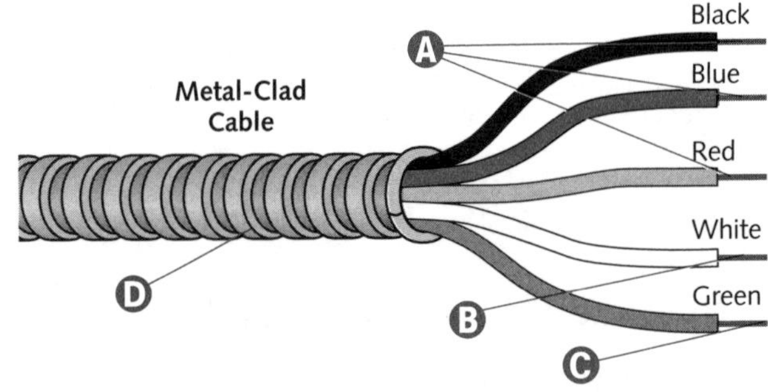

White Conductor Used as Ungrounded Switch Leg (Loop)

A A cable's white (or gray) conductor can be used as a switch loop (leg) in single-pole, 3-way, and 4-way switch installations, even though it is an ungrounded (hot) conductor »200.7(C)(2)«.

B The white (or gray) conductor's new use must be permanently re-identified by painting, or other effective means, at its terminations, and at each location where the conductor is visible and accessible »200.7(C)(2)«.

C Blue tape has been wrapped around the white conductor, re-identifying it as an ungrounded (hot) conductor. White, gray, green, or green with yellow strip(s) tape (or paint) cannot be used.

D The white (or gray) conductor can be used to feed a switch or can be a traveler in 3-way (or 4-way) switch installations. The white (or gray) conductor cannot be used as a return conductor from the switch to the switched outlet »200.7(C)(2)«.

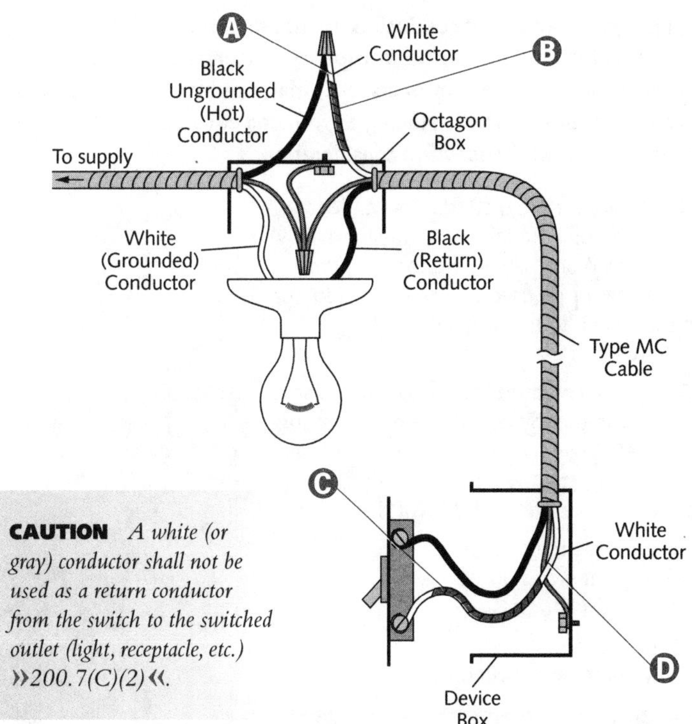

CAUTION *A white (or gray) conductor shall not be used as a return conductor from the switch to the switched outlet (light, receptacle, etc.) »200.7(C)(2)«.*

White Conductor Used as an Ungrounded (Hot) Conductor

A The white (or gray) conductor can be used as an ungrounded (hot) conductor if it is part of a cable assembly. It must be permanently re-identified as an ungrounded conductor, by painting (or other effective means) at its termination, and at each location where the conductor is visible and accessible »200.7(C)(1)«.

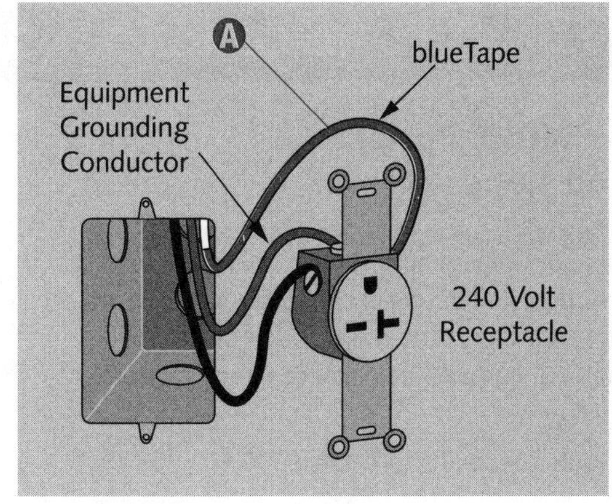

WARNING

A white (or gray) conductor in a conduit (or raceway) cannot be used (or re-identified) as an ungrounded conductor.

White Conductor Used as a Traveler

Although grounded circuit conductors are not required at switch locations, new product technology (such as occupancy sensors, home automation switches, programmable controllers, etc.) may require a grounded circuit conductor. For this reason, installing a grounded circuit conductor at switch locations is a design issue worth consideration.

Ⓐ The white (or gray) conductor can feed a switch or can be a traveler in 3-way (or 4-way) switch installations. The white (or gray) conductor must not be used as a return conductor from the switch to the switched outlet »200.7(C)(2)«.

Ⓑ The white (or gray) conductor's new use must be permanently re-identified by painting (or other effective means), at its terminations, and at each location where the conductor is visible and accessible »200.7(C)(1)«.

Ⓒ Grounded circuit conductors are not required at switch locations »404.2(A) *Exception*«.

Ⓓ Different colors can be used for travelers in 3-way and 4-way switches.

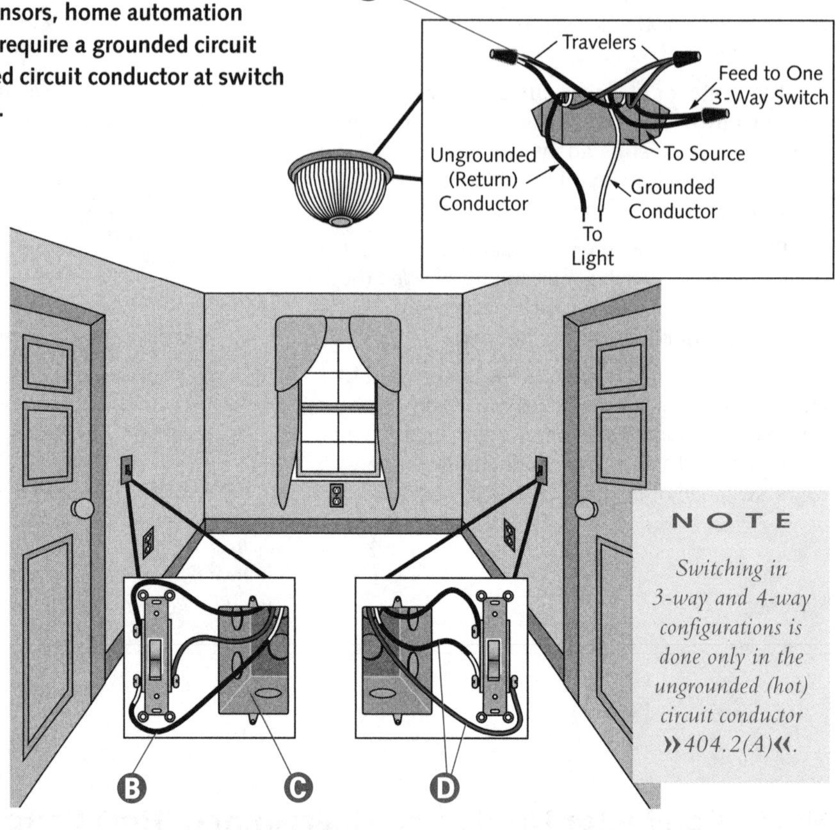

NOTE

Switching in 3-way and 4-way configurations is done only in the ungrounded (hot) circuit conductor »404.2(A)«.

WARNING

A grounding conductor (green, green with yellow strip(s) or bare) cannot be used (or reidentified) as an underground (hot) or a grounded (white) conductor.

UNDERGROUND INSTALLATIONS

Cover for Direct Burial Cables and Conductors

Cover measurement commences at the top surface of any direct burial wiring method (conductor, cable, conduit or other raceway) and ends at the closest point of the finished grade's top surface (including concrete, etc.) »Table 300.5«.

Direct burial cables or conductors. Table 300.5

NOTE

If the cover depth requirements of Table 300.5 cannot be met due to the presence of solid rock, the wiring must be installed in a raceway (metal or nonmetallic) approved for direct burial. The raceways must then be covered by a minimum of 2 in. (50 mm) of concrete that extends down to the solid rock surface »Table 300.5, Note 4«.

Raceways must enclose cables and conductors located under buildings »Table 300.5(C) and Table 300.5«.

All locations **NOT** specified in Table 300.5

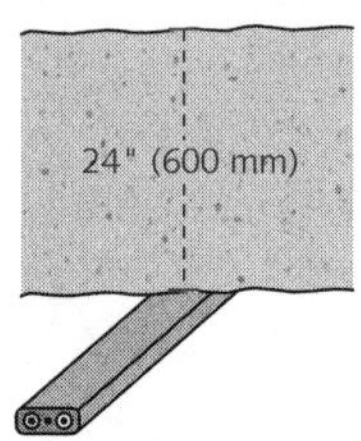

Streets, highways, roads, alleys, driveways, and parking lots

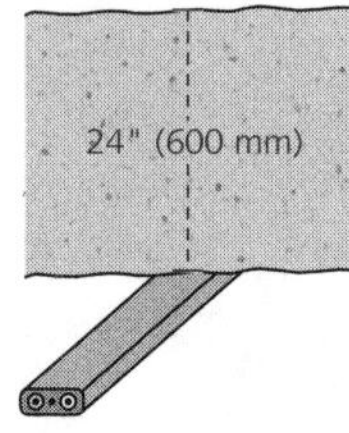

Under minimum of 4-in. (100-mm) thick concrete exterior slab with no vehicular traffic [the slab must extend at least 6 in. (150 mm) beyond the underground installation]

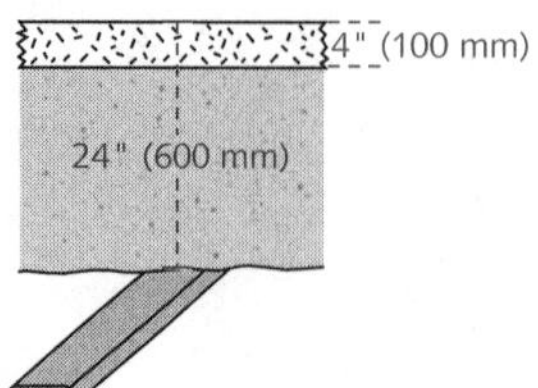

In trench below 2-in. (50-mm) thick concrete or equivalent

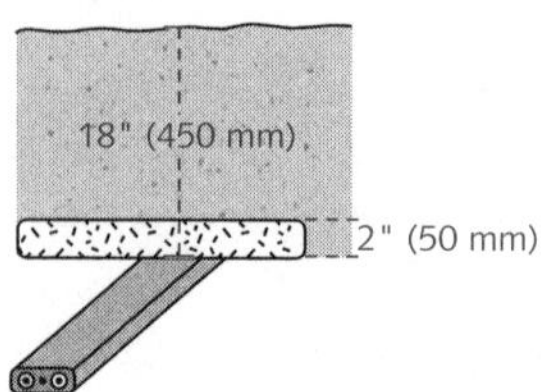

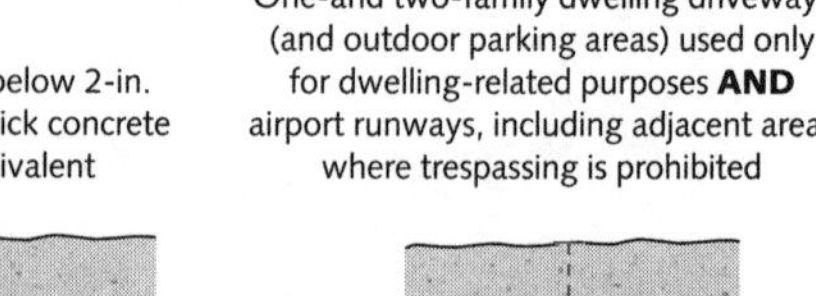

One-and two-family dwelling driveways (and outdoor parking areas) used only for dwelling-related purposes **AND** airport runways, including adjacent areas where trespassing is prohibited

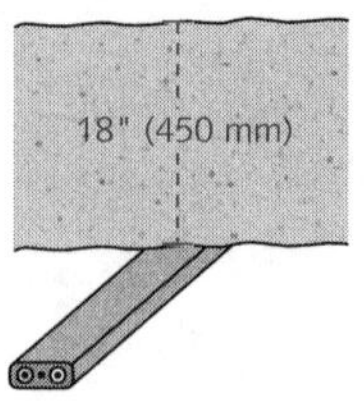

Protection of Conductors and Cables

Cable installed under a building must be in a raceway which extends beyond the building's outside walls »300.5(C)«.

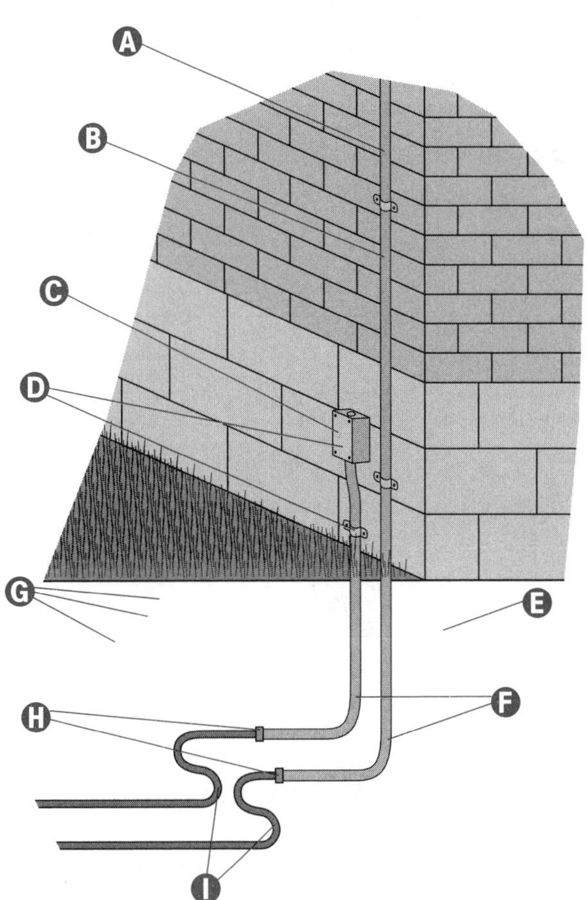

A Protection must extend at least 8 ft (2.5 m) above finished grade »300.5(D)«.

B Direct buried conductors and cables emerging from the ground must be protected by enclosures or raceways »300.5(D)«.

C Conductors entering a building shall be protected to the point of entry »300.5(D)«.

D Raceways and boxes shall be fastened in place securely »300.11(A)«.

E Protection must extend below grade to a depth matching Table 300.5 cover requirements, up to a maximum depth of 18 in. (450 mm) »300.5(D)«.

F Direct buried conductors, raceways, or cables subject to movement by settlement (or frost) must be arranged to prevent damage to both the enclosed conductors and the equipment connected to the raceways »300.5(J)«.

G Excavation backfill cannot contain large rock, paving materials, cinders, large or sharp substances, or corrosive materials where damage to raceways or cables may occur. Provide protection in the form of granular or selected material, suitable running boards, suitable sleeves, or other approved means to prevent physical damage to the raceway or cable »300.5(F)«.

H Where the conductors or cables emerge as a direct burial wiring method, install a bushing or fitting to protect conductors from abrasion on the end of the conduit (or tubing) that terminates underground. A seal incorporating the same protective characteristics can be used in lieu of a bushing »300.5(H) and 300.15(C)«.

I 300.5 recognizes "S" loops in underground direct burial to raceway transitions, expansion joints in raceway risers to fixed equipment, and the provision of flexible connections to equipment subject to settlement or frost heaves »300.5(J)(FPN)«.

Cables Permitted Underground

Cable Types	Article
IGS Integrated Gas Spacer Cable	326
MV Medium Voltage Cable	328
MI Mineral-Insulated, Metal-Sheathed Cable	332
MC Metal-Clad Cable (where identified for such use)	330
USE Underground Service-Entrance Cable	338
UF Underground Feeder (and Branch-Circuit) Cable	340

Cover for Residential Branch-Circuits

Residential branch-circuits rated 120 volts (or less) with GFCI protection and maximum overcurrent protection of 20 amperes. Table 300.5

CAUTION

Raceways must enclose cables and conductors located under buildings. These raceways must extend beyond the building's outside walls ››*300.5(C)*‹‹.

NOTE

The minimum cover for residential branch-circuits, in or under airport runways (including adjacent areas where trespassing is prohibited) is 18 in. (450 mm) ››*Table 300.5*‹‹.

Lesser depths are permitted where cables and conductors rise for terminations (or splices) or where access is otherwise required ››*Table 300.5, Note 2*‹‹.

All locations **NOT** specified in Table 300.5

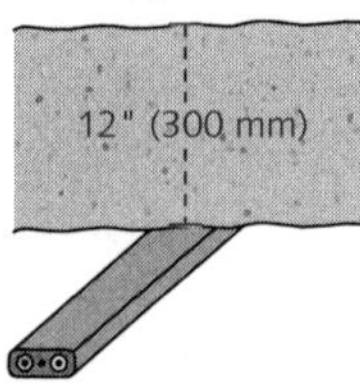

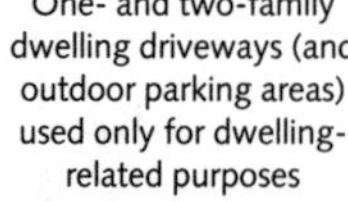

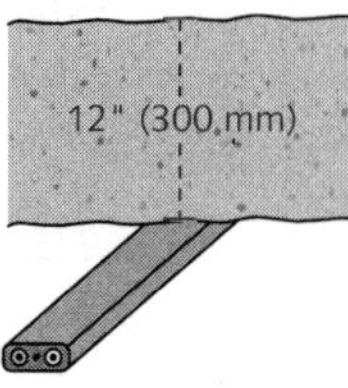

Under minimum of 4-in. (100-mm) thick concrete exterior slab with no vehicular traffic [the slab must extend at least 6 in. (150 mm) beyond the underground installation]

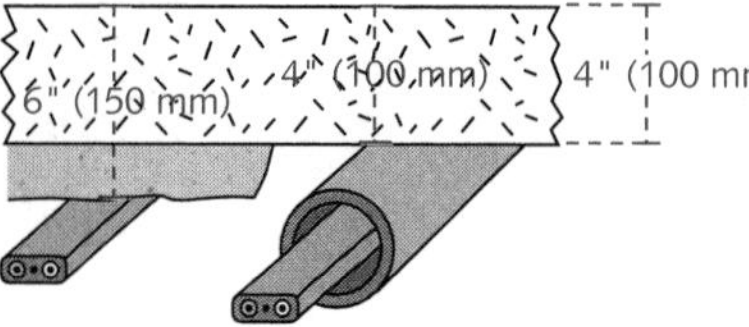

In trench below 2-in. (50-mm) thick concrete or equivalent

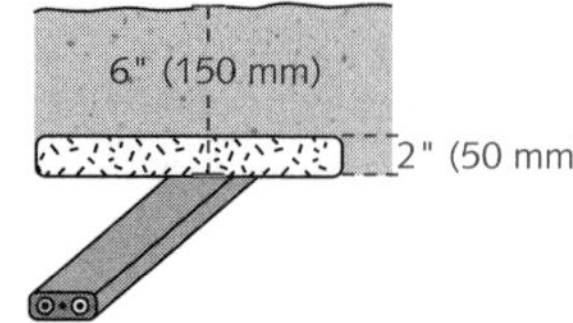

Streets, highways, roads, alleys, driveways, and parking lots

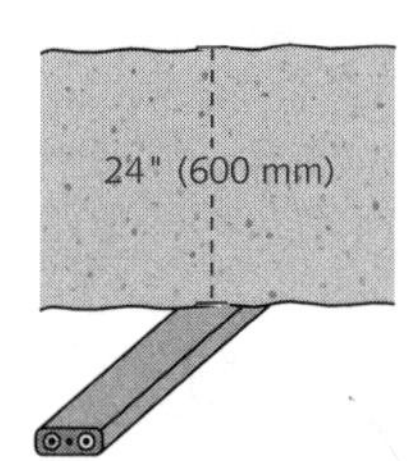

Cover for Low Voltage Circuits

Circuits for control of irrigation and landscape lighting limited to not more than 30 volts and installed with Type UF or in other identified cable (or raceway). Table 300.5

NOTE

Where a wiring method listed in Table 300.5, columns 1–3, is used for a circuit type in columns 4 and 5, the shallower depth of burial is allowed ››*Table 300.5, Note 3*‹‹.

All locations **NOT** specified in Table 300.5

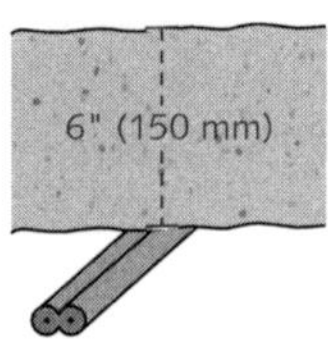

Under minimum of 4-in. (100-mm) thick concrete exterior slab with no vehicular traffic (the slab must extend at least 6 in. (150 mm) beyond the underground installation)

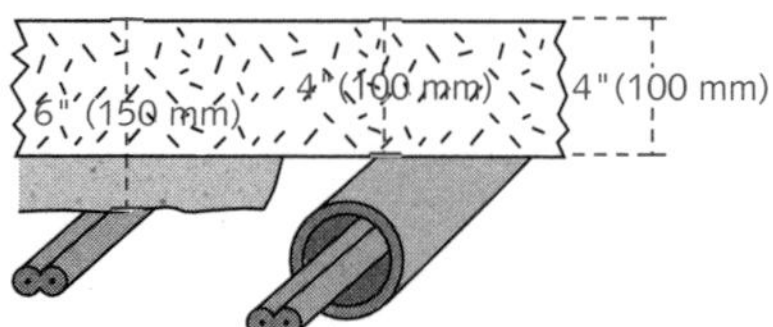

In trench below 2-in. (50-mm) thick concrete or equivalent

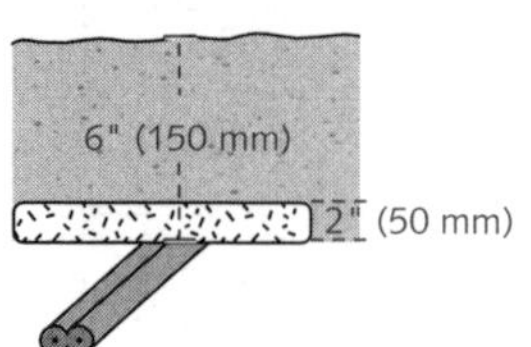

Streets, highways, roads, alleys, driveways, and parking lots

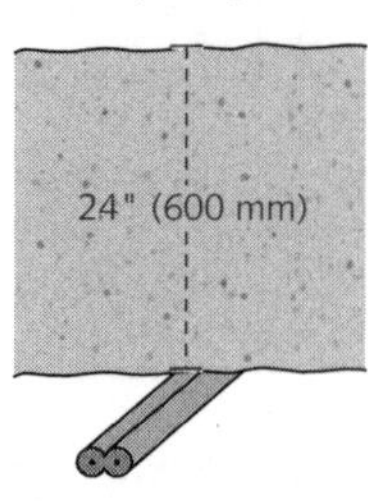

One- and two-family dwelling driveways (and outdoor parking areas) used only for dwelling-related purposes **AND** airport runways, including adjacent areas where trespassing is prohibited

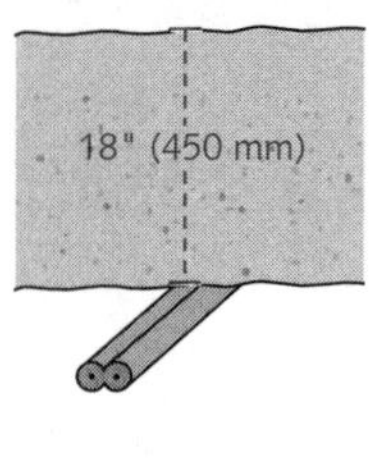

SPECIAL APPLICATION CABLES

Flat Conductor Cable (Under-Carpet Wiring)

A Power feed, grounding connection, and shield system connection between the FCC system and other wiring systems shall be accomplished through a transition assembly identified for this use »324.40(D)«.

B Carpet squares are required to cover floor-mounted Type FCC cable, cable connectors, and insulating ends. They cannot be larger than 36 in. (914 mm) square. Release-type adhesives must be used when adhering carpet squares to the floor »324.10(H)«.

C All receptacles, receptacle housings, and self-contained devices used with the FCC system shall be identified for this use and shall be connected to the Type FCC cable as well as the metal shields. Connection from any grounding conductor of the Type FCC cable to the shield system must be made at *each* receptacle »324.42(A)«.

D All FCC system components must be securely anchored to the floor (or wall) using an adhesive or mechanical system identified for this use. Floors must be prepared to ensure attachment of the FCC system to the floor until the carpet squares are placed »324.30«.

E Type FCC cable can be installed on hard, solid, smooth, continuous floor surfaces made of concrete, ceramic, composition flooring, wood, or similar materials »324.10(C)«.

F A metal top shield shall be installed over all floor-mounted Type FCC cable, connectors, and insulating ends. The top shield must completely cover all cable runs, corners, connectors, and ends »324.30«.

G Type FCC cable consists of three or more flat copper conductors placed edge-to-edge and separated by an enclosing insulation assembly »324.100(A)«. While the maximum current rating for general-purpose and appliance branch-circuits is 20 amperes, individual branch-circuits can have a current rating up to 30 amperes »324.10(B)(2)«.

H All Type FCC cable, connectors, and insulating ends require a bottom shield, either metallic or nonmetallic »324.40(C)(2) and 324.100(B)(1)«.

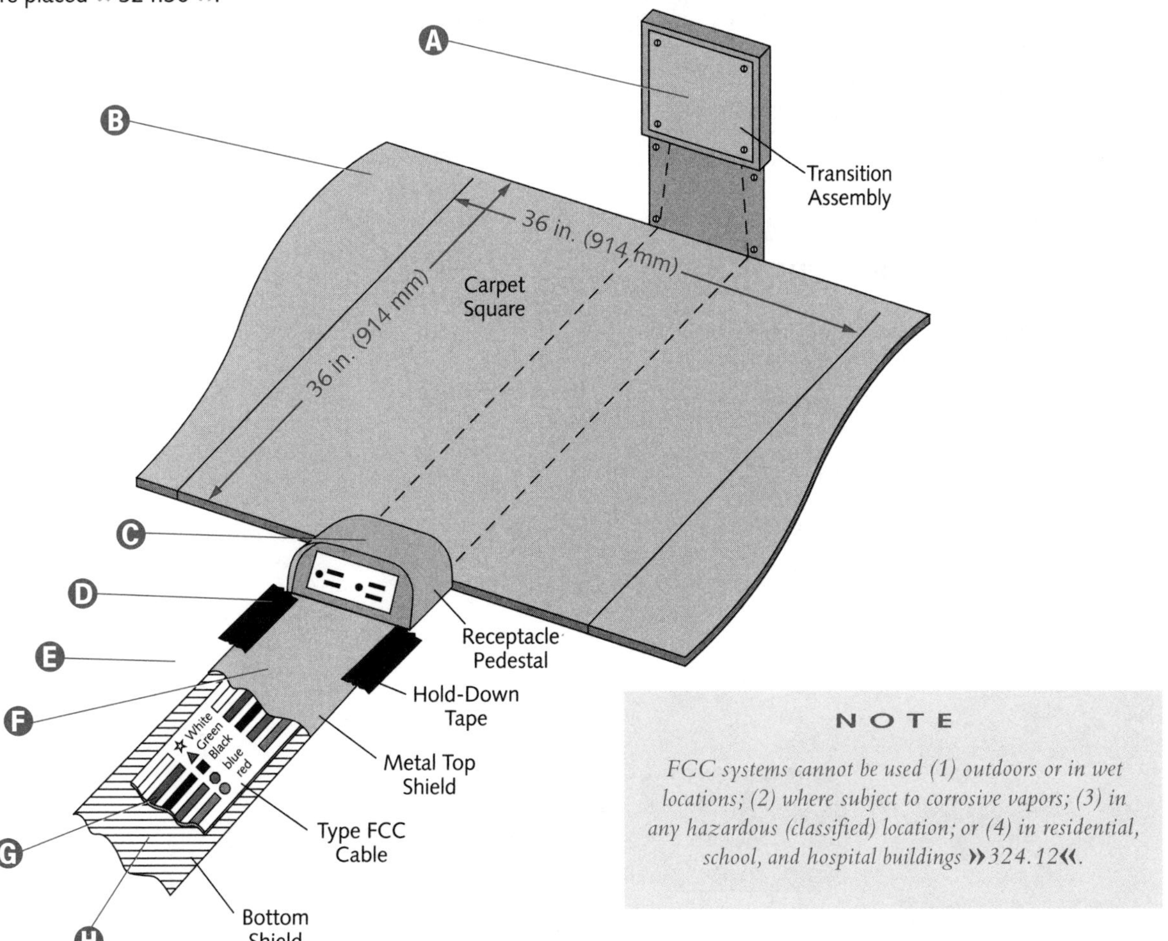

NOTE

FCC systems cannot be used (1) outdoors or in wet locations; (2) where subject to corrosive vapors; (3) in any hazardous (classified) location; or (4) in residential, school, and hospital buildings »324.12«.

Integrated Gas Spacer Cable (Type IGS)

The maximum bending radius is found in Table 326.24.

A run of Type IGS cable between pull boxes (or terminations) cannot contain more than the equivalent of four quarter bends (360° total), including those bends located immediately adjacent to the pull box (or terminations) »326.26«.

Terminations and splices must be identified as suitable for maintaining the gas pressure within the conduit. A valve and cap must be provided for each length of the cable and conduit for monitoring and maintenance of gas pressure in the conduit »326.40«.

Each conductor contains ½-in. (12.7-mm) diameter solid aluminum rod(s). A conductor assembly contains from one to nineteen of these rods, which are laid parallel. Conductor size ranges from a 250-kcmil minimum to a 4750-kcmil maximum »326.104«.

Ⓐ Type IGS cable is a manufactured assembly of conductor(s), each individually insulated and enclosed in a loose-fit nonmetallic flexible conduit as an integrated gas spacer cable rated 0-600 volts »326.2«. It is used in underground (direct burial) applications as service-entrance, feeder, or branch-circuit conductors »326.10«.

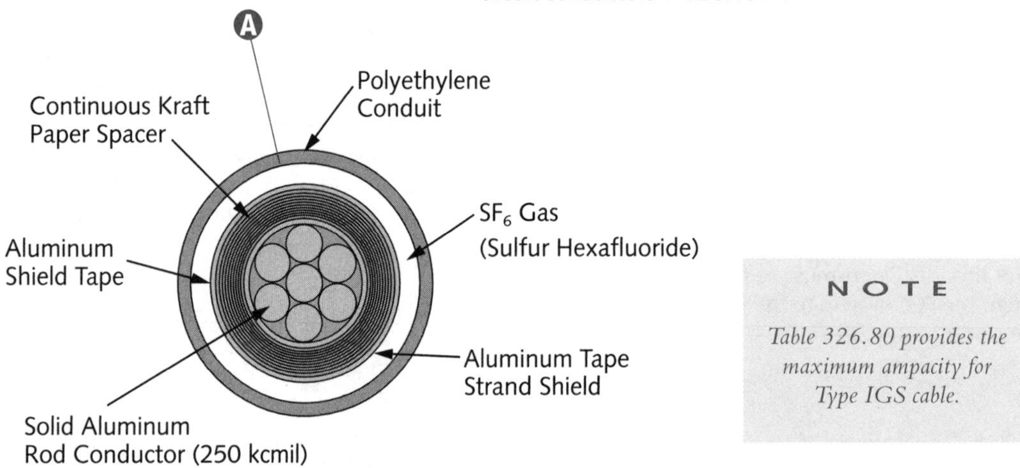

NOTE

Table 326.80 provides the maximum ampacity for Type IGS cable.

Medium Voltage Cable (Type MV)

Ⓐ Type MV is a solid dielectric insulated cable (single or multiconductor) rated 2001 volts or higher »328.2«.

Ⓑ Type MV cables can be used on power systems rated up to 35,000 volts, nominal »328.10«.

Ⓒ Type MV cables must have copper, aluminum, or copper-clad aluminum conductors and be constructed in accordance with Article 310 »328.100«.

Ⓓ Conductor ampacity is found in 310.60, unless the cable is installed in a cable tray, in which case the ampacity shall be determined in accordance with 392.13 »328.80«.

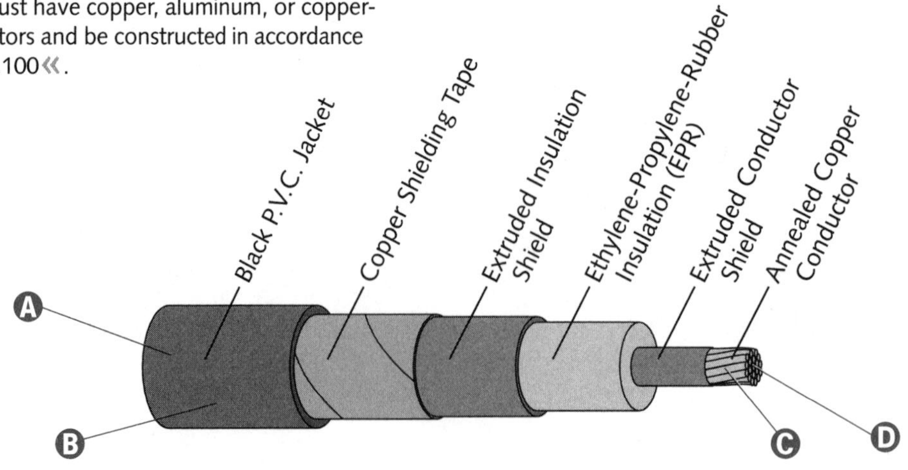

Mineral-Insulated, Metal-Sheathed Cable (Type MI)

Type MI cable must be securely supported at intervals no greater than 6 ft (1.8 m) except where fished »332.30(1)«.

A Type MI (mineral-insulated, metal-sheathed) cable is a manufactured assembly of conductor(s) insulated with a highly compressed refractory mineral insulation and enclosed in a liquidtight (and gastight) continuous copper (or alloy steel) sheath »332.2«.

B The cable must not be damaged when bent. The radius of the inner edge of any bend cannot be less than five times the external diameter of the metallic sheath for cable that is not more than ¾-in. (19-mm) in external diameter. Cable diameters greater than ¾-in. (19-mm) must have a bend radius not less than ten but not more than 1 in. (25 mm) times the diameter »332.24«.

C When using single-conductor cables, all phase conductors and the neutral conductor (if any) must be grouped together to minimize induced voltages on the sheath »300.20(A)«.

D Only fittings identified for this use can connect Type MI cable to boxes, cabinets, or other equipment »332.40(A)«.

E If these single-conductor cables had entered a ferrous enclosure, compliance with 300.20(B) would be required to prevent heating from induction.

F Where Type MI cable terminates, a seal shall be provided *immediately* after stripping to prevent the entrance of moisture into the insulation. Each conductor extending beyond the sheath must be provided with an insulating material »332.40(B)«.

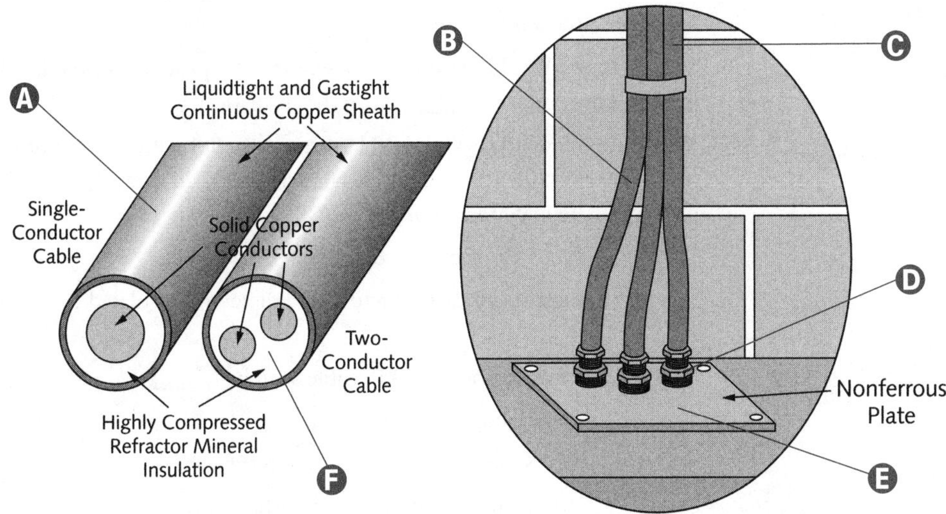

Summary

- Cables, when installed, must be protected from the possibility of physical damage.
- Conductors must be installed at least 1¼ in. (32 mm) from the nearest edge of the framing member, unless a steel plate [at least 1/16-in. (1.6-mm) thick] has been installed to protect the conductor(s).
- Conductors passing through metal studs (or framing members) may require protection in the form of bushings or steel plates.
- Fire-resistant-rated construction must remain intact once the electrical system has been installed.
- In dwellings, wiring can pass through the short dimension of a joist (or stud) space being used for environmental air-handling purposes.
- Cables must be secured within a certain distance from boxes (junction, device, etc.), unless the cable has been fished between access points.
- The maximum distance between supports depends upon the type of cable.
- The minimum bending radius must be observed when installing a cable.
- The installation of nonmetallic-sheathed cable is prohibited in structures exceeding three floors above grade, except for one- and two-family dwellings.
- Certain sizes of nonmetallic-sheathed cable can be attached directly to the underside of exposed joists in unfinished basements.
- In a cable system, a white conductor used as a feed (or traveler) with switches (single-pole, 3-way, and 4-way) must be re-identified.
- Table 300.5 contains minimum cover requirements for direct buried cables and conductors.
- Chapter 3 of the *NEC*® contains provisions for certain specialized cables.

Unit 4 Competency Test

NEC® Reference	Answer	
_______	_______	1. Nonmetallic-sheathed cable must be secured within _____ in. from a two-gang plastic device box mounted in a wall.
_______	_______	2. Floor-mounted Type FCC cable, cable connectors, and insulating ends shall be covered with carpet squares no larger than _____ in. square.
_______	_______	3. Type AC cable shall be secured by staples, cable ties, straps, hangers, or similar fittings at intervals not exceeding _____ ft.
_______	_______	4. The minimum bending radius for three-in. diameter integrated gas spacer cable is _____ in.
_______	_______	5. _____ is a factory assembly of one or more insulated circuit conductors with or without optical-fiber members enclosed in an armor of interlocking metal tape or a smooth or corrugated metallic sheath.
_______	_______	6. Where Type MI cable terminates, a seal shall be provided immediately after stripping to prevent the entrance of _____ into the insulation.
_______	_______	7. Where nonmetallic-sheathed cable is run at angles with joists in unfinished basements, it shall be permissible to secure cables not smaller than three _____ AWG conductors directly to the lower edges of the joists. a) 12 b) 10 c) 8 d) 6
_______	_______	8. _____ is a single or multiconductor solid dielectric insulated cable rated 2001 volts or higher.
_______	_______	9. The overall covering of Type UF cable shall be _____. I. flame retardant II. suitable for direct burial in the earth III. moisture, fungus, and corrosion resistant a) II only b) I and II only c) II and III only d) I, II and III
_______	_______	10. Power feed, grounding connection, and shield system connection between the FCC system and other wiring systems shall be accomplished in a(n) _____ identified for this use.
_______	_______	11. Type AC cable shall be permitted to be unsupported where the cable is not more than _____ in. in length at terminals where flexibility is necessary. a) 2 b) 6 c) 24 d) 72
_______	_______	12. The temperature limitations on which the ampacities of mineral-insulated, metal-sheathed, cable are based shall be determined by the _____. a) insulation of the cable b) composition of the outer sheath c) insulating materials used in the end seal d) highest ambient temperature surrounding the cable
_______	_______	13. The minimum bending radius for metal-clad (corrugated sheath) cable is _____ times the external diameter of the metallic sheath.
_______	_______	14. In both exposed and concealed locations, where a cable or raceway-type wiring method is installed through bored holes in joists, rafters, or wood members, holes shall be bored so that the edge of the hole is not less than _____ in(es). from the nearest edge of the wood member.

NEC® Reference	Answer	
______	______	15. Where run across the top of floor joists, or within _____ ft of the floor or floor joists across the face of rafters or studding, in an attic that has permanent stairs, nonmetallic-sheathed cable must be protected by substantial guard strips that are at least as high as the cable.
______	______	16. Type FCC cable shall be clearly and durably marked on both sides at intervals of not more than 24 in. with the information required by 310.11(A) and with what additional information? I. material of conductors and maximum ampacity II. rated frequency and number of phases III. maximum temperature rating a) I only b) I and II only c) I and III only d) I, II, and III
______	______	17. _____ is a factory assembly of one or more conductors insulated with a highly compressed refractory mineral insulation and enclosed in a liquidtight and gastight continuous copper or alloy steel sheath.
______	______	18. Armored cable installed in thermal insulation shall have conductors rated at _____. The ampacity of the cable installed in these applications shall be that of 60°C conductors. a) 60°C b) 75°C c) 90°C d) 90°F
______	______	19. Type MC cable must be supported (and secured) at intervals not exceeding _____ in. a) 4½ b) 6 c) 54 d) 72
______	______	20. The maximum ampacity of a 500 kcmil, integrated gas spacer cable (Type IGS) is _____ amperes.
______	______	21. Type UF cable, buried underground, feeding a 15-ampere, ground-fault circuit-interrupter in a one-family dwelling must have a minimum cover of _____ in. a) 6 b) 12 c) 18 d) 24
______	______	22. Splices, in knob-and-tube wiring, shall be _____ unless approved splicing devices are used.
______	______	23. A nonmetallic-sheathed cable, passing through a bored hole less than 1¼ in. from the edge of a wood framing member, must be protected by a steel plate (or bushing) at least _____ in(es) thick.
______	______	24. In FCC systems, general-purpose and appliance branch-circuits have a maximum current rating of _____ amperes, while individual branch-circuits have a maximum current rating of _____ amperes.
______	______	25. Type MI cable shall be supported securely at intervals not exceeding _____ ft by straps, staples, hangers, or similar fittings designed and installed so as not to damage the cable.
______	______	26. Metal-clad cable, installed outside of buildings or as aerial cable, shall comply with what additional Article(s)?
______	______	27. Type UF cable, buried under a one-family dwelling driveway, must have a minimum cover of _____ in.

SECTION ONE: FOUNDATIONAL PROVISIONS

Raceways and Conductors

Objectives

After studying this unit, the student should:

- know the maximum number of bends for raceways installed between pull points.
- have a good understanding of rigid and intermediate metal-conduit provisions.
- be able to determine the type of raceway best suited for each installation.
- be familiar with IMC, RMC, and EMT support requirements.
- understand the unique characteristics of the four types of rigid nonmetallic conduits.
- thoroughly understand rigid nonmetallic conduit and electrical nonmetallic tubing provisions, both general and specific.
- be familiar with the different types of flexible conduit, and be able to locate each related Article.
- be aware of raceway types other than conduit and tubing.
- successfully calculate the electrical trade-size conduit required for each project.
- have a basic grasp of conductor properties.
- be familiar with conductor temperature limitations.
- know and understand the provisions pertaining to conductors connected in parallel.
- understand ampacity correction factors, including continuous loads, and how to apply them.

Introduction

Raceways are the conduits, or similar housing, whose dual function is to facilitate the installation (or removal) of electrical conductors and to provide protection from mechanical damage. Article 100 defines raceway as an enclosed channel of metal (or nonmetallic) materials designed expressly for holding wires, cables, or busbars, with additional functions as permitted in the *Code*. The term **raceway** applies to more than just tubular conduit.

This unit contains regulations for raceway systems that are frequently used (such as electrical metallic tubing and wireways), as well as raceway systems that are rarely used (such as, cellular concrete floor raceways.) In addition to defining different raceways, other important information is covered, such as minimum and maximum size, installation provisions, fittings, and support requirements. Instructions are presented on how to perform raceway fill calculations, a prerequisite in determining the size of raceway needed for particular circuit(s). A brief explanation of conductors, including ampacity correction factors and conductor temperature limitations, can be found near the conclusion of this unit. Understanding this unit will make finding the best raceway system for a particular project much easier.

Most raceway Articles have sections entitled **Uses Permitted** and **Uses Not Permitted** (*NEC*® Chapter 3). These sections give locations (some specific) where the raceway system can (or cannot) be installed. Because most of these provisions are clearly understandable, this book will not repeat much of this information. Although not addressed in this unit, another important area is underground raceway installation (300.5). Underground installations were covered in Unit 4, which explained minimum cover requirements for cables. Use Table 300.5 (Minimum Cover Requirements) to find the minimum burial depth for different types of raceways installed in a variety of locations.

GENERAL PROVISIONS

Suspended (Drop) Ceilings

A Raceways can be supported by independent (additional) support wires, secured at both ends. Support wires and associated fittings must provide adequate support »300.11(A)«.

B Boxes can be secured by attaching independent (additional) support wires at both ends. Secure support must be provided by the support wires and associated fittings »300.11(A) and 314.23(D)(2)«.

C Ceiling grid support wires cannot be used to support raceways.

D Certain boxes can also be mounted to suspended ceiling framing members »314.23(D)(2)«.

E EMT conduit must be fastened securely within 3 ft (900 mm) of all boxes (outlet, junction, device, etc.) »358.30(B)«.

F Suspended ceiling system framing members, used to support luminaires (fixtures), must be securely fastened to one another as well as to the building structure at appropriate intervals. Luminaires (fixtures) must be securely fastened to the ceiling framing member by mechanical means (bolts, screws, or rivets, etc.). Listed clips, identified for use with the type of ceiling framing member(s) and luminaire(s) (fixture(s)), are also permitted »410.16(C)«.

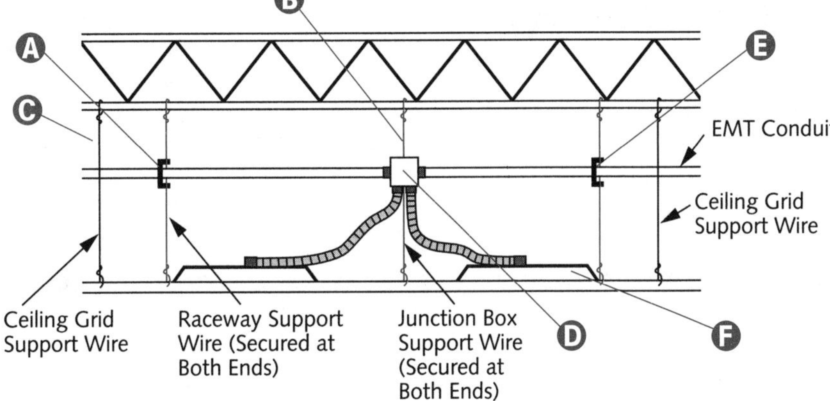

Maximum Bends in One Run

A The equivalent of four quarter bends (360° total) is the maximum allowed between pull points, e.g., conduit bodies and boxes. Because the total bends in this conduit run is 340°, this installation falls within *NEC*® specifications.

B The bend maximum of 360° applies to the following raceways: IGS (326.26), RMC (344.26), IMC (346.26), FMC (348.26), LFMC (350.26), RNC (352.26), NUC (354.26), LFNC (356.26), EMT (358.26), and ENT (362.26).

C Generally, raceway installation must be complete between outlet, junction, or splicing points prior to the installation of conductors »300.18(A)«.

D All bends are counted, even those located immediately adjacent to the pull box (or termination). A box offset with two 10° bends counts as 20°.

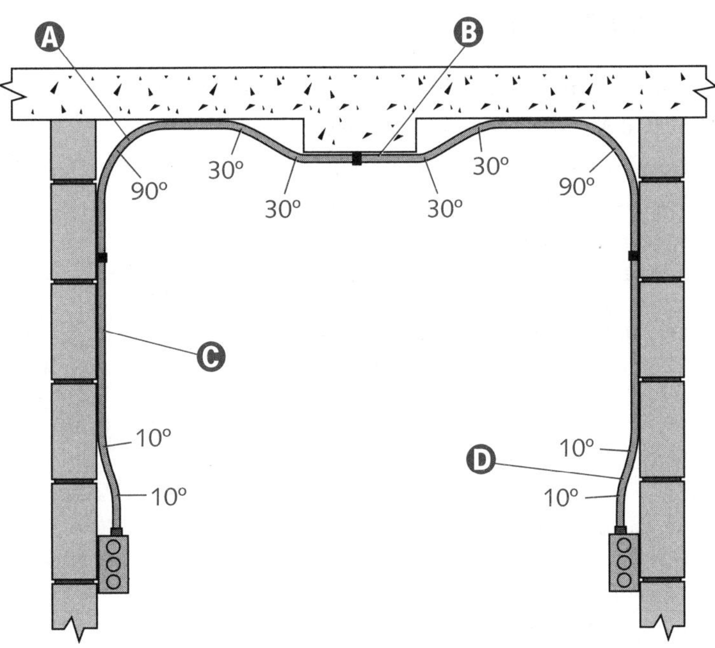

NOTE

Metal raceways must not be supported, terminated, or connected by welding unless specifically permitted by design or Code specifications »300.18(B)«.

Insulated Fittings

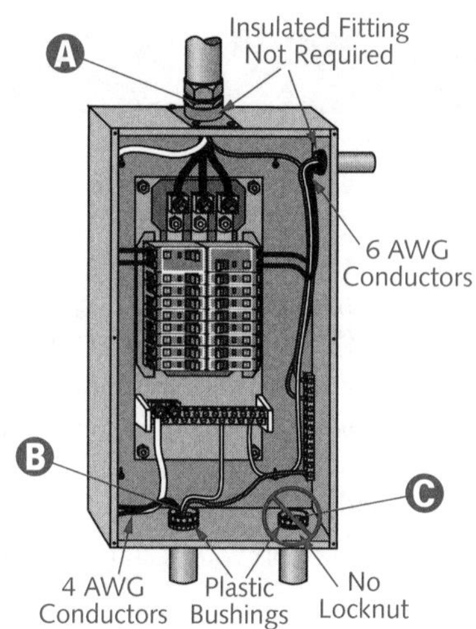

Ⓐ An insulated fitting is not required if a smoothly rounded (or flared) entry for conductors is provided by threaded hubs (or bosses) that are an integral part of a cabinet, box enclosure, or raceway »300.4(F) *Exception*«.

Ⓑ Where raceways containing 4 AWG or larger ungrounded (hot) conductors enter a cabinet (box, enclosure, or raceway), the conductors must be protected by a substantial fitting that provides a smoothly rounded insulating surface, unless the conductors are separated from the fitting (or raceway) by other securely attached substantial insulating material »300.4(F)«.

Ⓒ Conduit bushings, constructed completely of insulating material, cannot be used to secure a fitting or raceway »300.4(F)«. An insulating (plastic, thermoplastic, etc.) bushing cannot replace a locknut.

Raceways Supporting Raceways

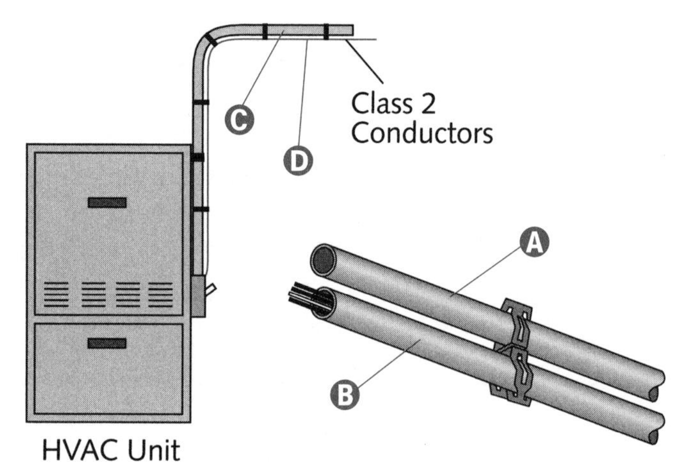

Ⓐ Conduits (not used as raceways) may support raceways, if fastened securely in place with approved fittings »300.11(B)«.

Ⓑ Raceways *cannot* be used to support other raceways, cables, conductors, or non-electric equipment, unless identified for that purpose »300.11(B)(1)«.

Ⓒ This conduit contains power supply conductors for HVAC unit.

Ⓓ Raceways, containing power supply conductors for electrically controlled equipment, can support Class 2 circuit conductors (or cables) used solely for connection to the equipment control circuits »300.11(B)(2)«.

Conduit Thickness

Ⓐ Internal diameters for each electrical trade size conduit and tubing are listed in Table 4 of Chapter 9.

Ⓑ Ten 10 AWG THHN conductors are permitted in ¾" EMT.

Ⓒ Eleven 10 AWG THHN conductors are permitted in ¾" IMC.

Ⓓ Ten 10 AWG THHN conductors are permitted in ¾" RMC.

Ⓔ Annex C (preceding the index in the back of the *Code* book) can be used to find the maximum number of conductors permitted in each electrical trade size conduit or tubing. The conductors must be the same size (total cross-sectional area including insulation) when using Annex C.

Ⓕ Nine 10 AWG THHN conductors are permitted in ¾" schedule 40 PVC.

Ⓖ Seven 10 AWG THHN conductors are permitted in ¾" schedule 80 PVC.

Ⓗ The internal diameters of different types of conduit vary. Therefore a group of conductors that fit into one conduit may not fit into another type, even though it is the same electrical trade size.

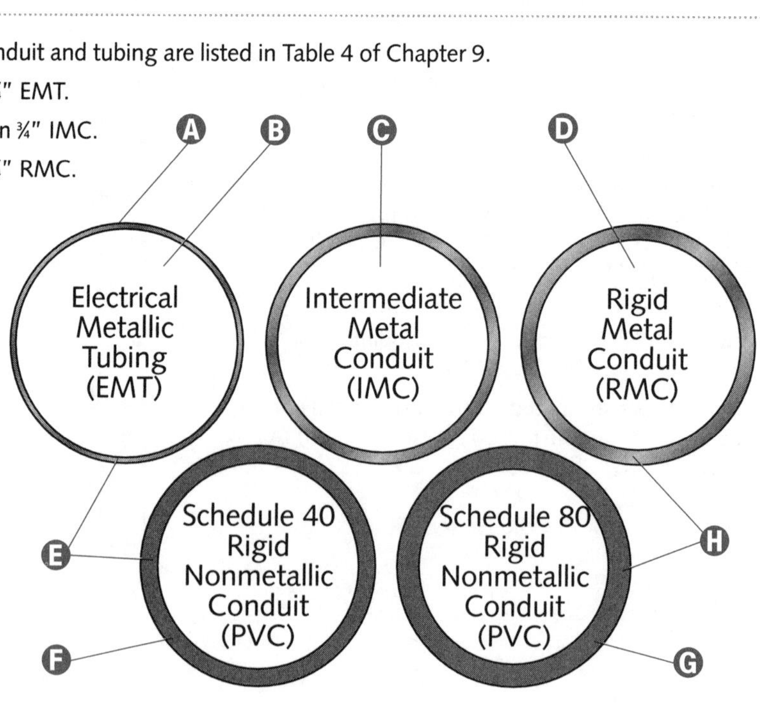

Bending Radius

Ⓐ Bends must be made so that the conduit or tubing remains undamaged with its internal diameter basically undiminished. For any field bend, the radius of the curve to the centerline of the conduit must not be less than indicated in Table 346.24. Raceways permitted to use the "One Shot and Full Shoe Benders" column includes: RMC »344.24«, IMC »346.24«, and EMT »358.24«.

Ⓑ The bending radius of certain other raceways must not be less than shown in the column titled "Other Bends." They include: FMC »348.26«, LFMC »350.26«, RNC »352.24«, LFNC »356.26«, and ENT »362.24«.

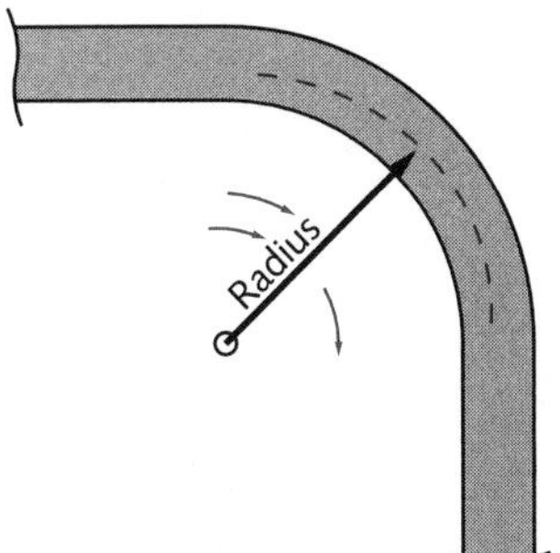

Table 344.24 Radius of Conduit Bends

Conduit Size	One Shot and Full Shoe Benders	Other Bends
Trade Size	Bending Radius (in in.)	Bending Radius (in in.)
½	4	4
¾	4½	5
1	5¾	6
1¼	7¼	8
1½	8¼	10
2	9½	12
2½	10½	15
3	13	18
3½	15	21
4	16	24
5	24	30
6	30	36

Ⓐ Ⓑ

NON-FLEXIBLE CONDUIT (AND TUBING)

Rigid Metal Conduit

Ⓐ Rigid Metal Conduit (RMC) ia a raceway of circular cross section. RMC is manufactured in both ferrous (such as steel) and nonferrous (such as aluminum) metal. Other special use types include: silicone bronze and stainless steel »344.2«.

Ⓑ RMC is the heaviest (thickest walled) classification of metal conduit.

Ⓒ Threadless couplings and connectors used with conduit must be made tight. Those buried in masonry (or concrete) must be the concrete-tight type. If installed in wet locations, they must be the raintight type »344.42(A)«.

Ⓓ The minimum approved electrical trade size for RMC is ½ in. »344.20(A)«.

Ⓔ The maximum approved electrical trade size for RMC is 6 in. »344.20(B)«.

Ⓕ RMC can be used under all atmospheric conditions, within all types of occupancies »344.10(A)«.

Ⓖ Contact of dissimilar metals (except for the combination of aluminum and steel) should be avoided wherever possible to lessen the potential for galvanic action »344.14«.

Ⓗ RMC usually ships in standard lengths of 10 ft (3 m), including the coupling. Normally one coupling is furnished with each length »344.130«.

Ⓘ Coupled with listed fittings, RMC provides electrical continuity; and therefore, can serve as an equipment grounding conductor »344.2«.

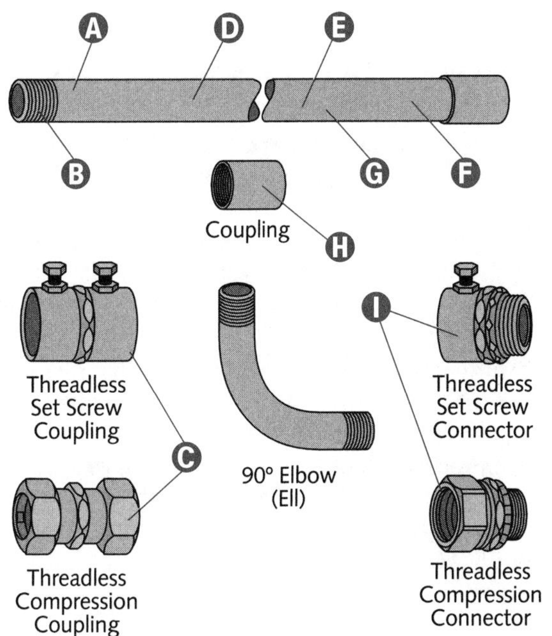

NOTE

RMC, elbows, couplings, and fittings can be installed in concrete, in direct contact with the earth, or in areas subject to severe corrosive influences where corrosion protection is provided and where judged suitable for the condition »344.10(B)«.

Intermediate Metal Conduit

A The minimum approved electrical trade size for IMC is ½ in. »346.20(A)«.

B The maximum approved electrical trade size for IMC is 4 in. »346.20(B)«.

C The definition for Intermediate Metal Conduit (IMC) and RMC is essentially the same »346.2«. However, as the name implies, IMC is lighter in weight and is constructed with thinner walls than RMC. Unlike RMC, IMC is only manufactured of steel which provides protective strength equivalent to thicker-walled conduits.

D Coupled with listed fittings, IMC provides electrical continuity; therefore, it can serve as an equipment grounding conductor »346.2«.

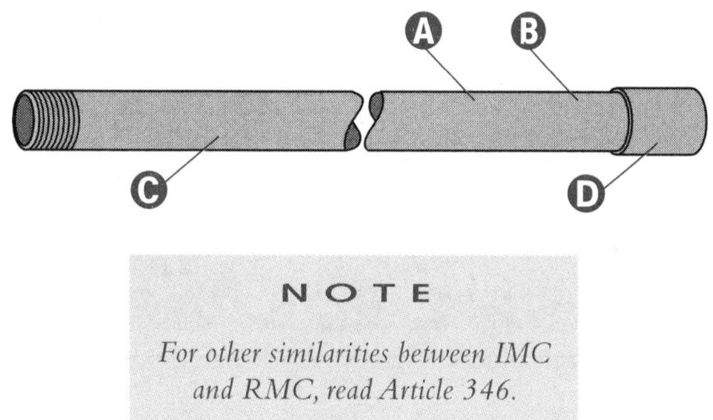

NOTE

For other similarities between IMC and RMC, read Article 346.

Cutting, Reaming, and Threading

A Electrical metallic tubing must not be threaded. Factory threaded integral couplings can be used »358.28(B)«.

B All EMT cut ends must be reamed (or otherwise finished) to remove rough edges »358.28(A)«.

C All cut ends of IMC and RMC must be reamed (or otherwise finished) to remove rough edges »344.28 and 346.28«.

D Running threads cannot be used on conduit for coupling connections »344.42(B) and 346.42 (B)«.

E Threading conduit in the field requires a standard cutting die with a ¾-in. taper per ft (1 in 16) »344.28 and 346.28«.

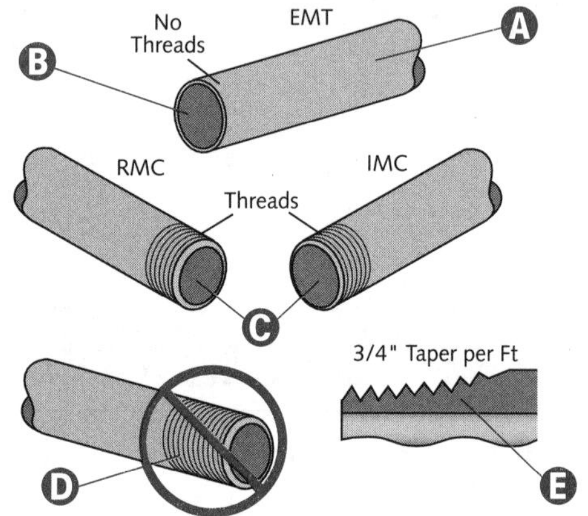

RMC, IMC, and EMT Support Requirements

A The maximum support interval for RMC, IMC, and EMT is 10 ft (3 m) »344.30(B)(1), 346.30(B)(1), and 358.30(A)«.

B In addition, each conduit (or tube) must be securely fastened within 3 ft (900 mm) of each conduit termination (outlet box, junction box, device box, cabinet, conduit body, etc.) »344.30(A), 346.30(A), and 358.30(A)«.

C If structural members are not available within 3 ft (900 mm), a distance of 5 ft (1.5 m) is acceptable for RMC and IMC »344.30(A) and 346.30(A)«.

D Unbroken lengths of EMT, i.e. without coupling, can be fastened within 5 ft (1.5 m) where structural members do not readily permit fastening within 3 ft (900 mm) »358.30(A) *Exception No. 1*«.

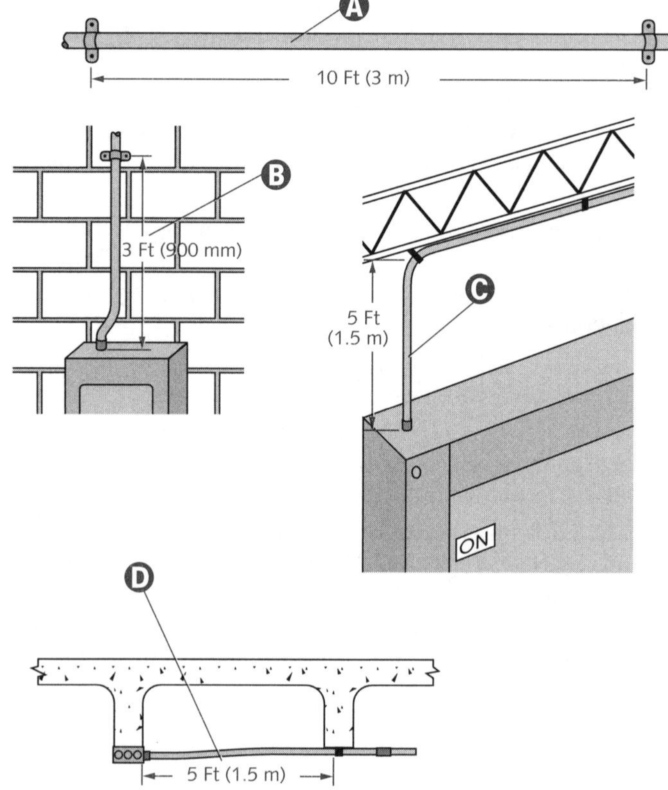

NOTE

RMC, IMC, and EMT must be installed as a complete system before installing conductors »344.30, 346.30, and 358.30«.

It is not required (where approved) that RMC and IMC be securely fastened within 3 ft (900 mm) of the service head for above-the-roof mast termination »344.30(A) and 346.30(A)«.

RMC and IMC With Threaded Couplings

A Straight runs of RMC and IMC, made up with threaded couplings, can be supported in accordance with Table 344.30(B)(2), provided such supports prevent transmission of stresses to termination where conduit is deflected between supports »344.30(B)(2)«.

B The distance between supports increases as the conduit size increases.

C The distance between supports can be increased to 20 ft (6 m) for exposed vertical risers from industrial machinery or fixed equipment, provided: (1) the conduit is made up with threaded couplings; (2) the conduit is firmly supported at the top and bottom of the riser; and (3) no other means of intermediate support is readily available »344.30(B)(3)«.

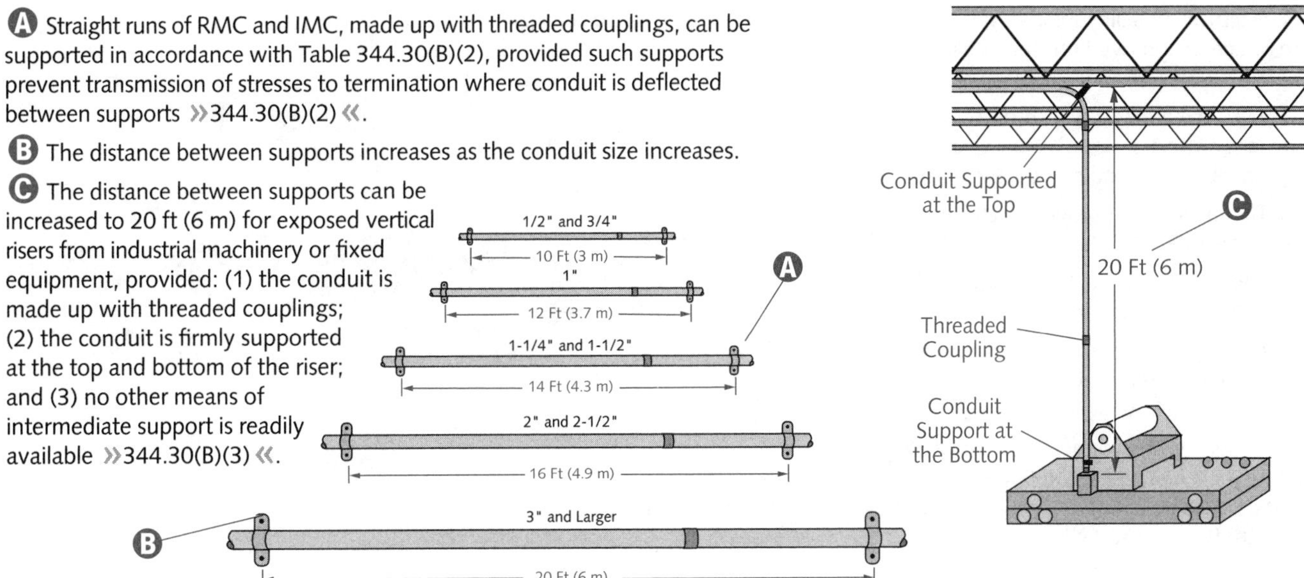

Horizontal Runs Through Framing Members

A Horizontal runs of RMC, IMC, and EMT that are supported by openings through framing members [at intervals not greater than 10 ft (3 m)] and are fastened securely within 3 ft (900 mm) of termination points, are permitted »344.30(B)(4), 346.30(B)(4), and 358.30(B)«.

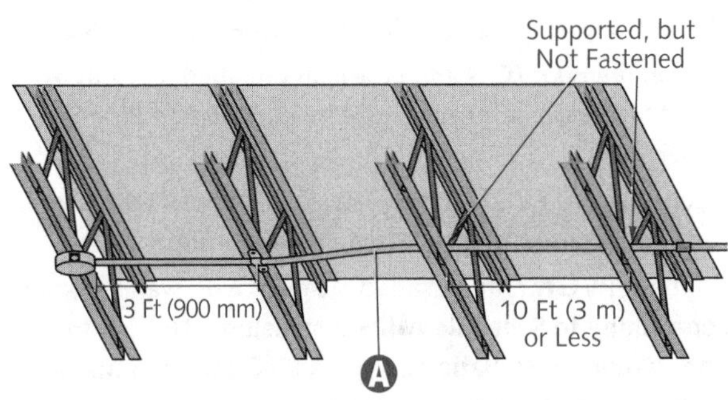

EMT Fished in Walls

A For concealed work in finished buildings (or prefinished wall panels) where standard securing is impracticable, unbroken lengths (without coupling) of electrical metallic tubing can be fished »358.30(A) *Exception No. 2*«.

B No coupling in fished portion of wall.

C Normally, EMT must be fastened within 3 ft (900 mm) of device boxes.

D Finished wall provides no access to secure tubing.

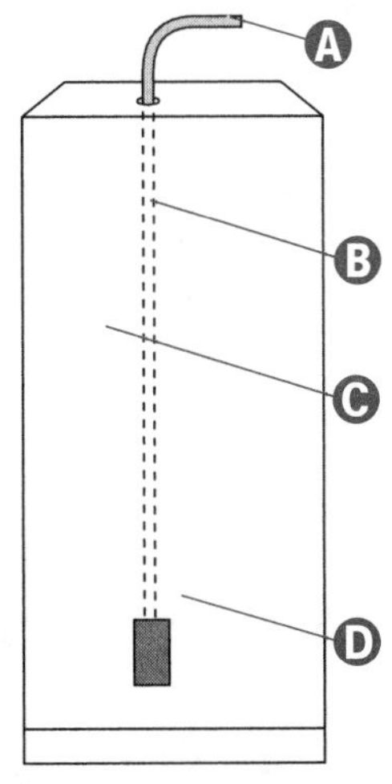

EMT Used as an Equipment Grounding Conductor

A EMT, can serve as an equipment grounding conductor »250.118(4)«.

B A raceway used as the equipment grounding conductor, as provided in 250.118 and 250.134(A), must comply with 250.2(D) »250.122(A)«.

C EMT must be supported according to 358.30 provisions.

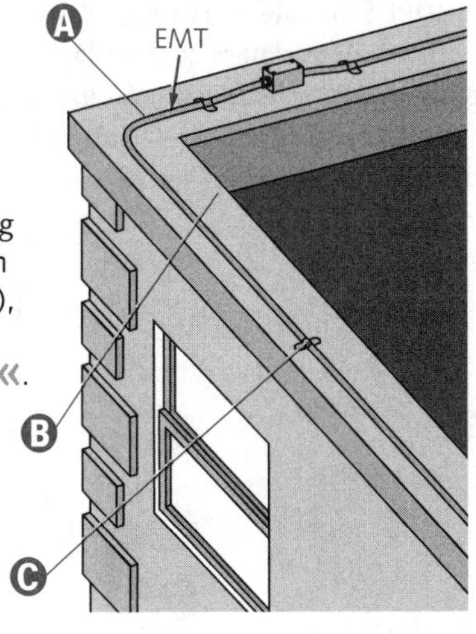

Electrical Metallic Tubing

A Electrical metallic tubing (EMT) is an unthreaded thinwall raceway of circular cross section approved for the installation of electrical conductors when joined together with appropriate fittings »358.2«. EMT is also referred to as thinwall.

B EMT, the thinnest walled classification of metal non-flexible raceways, provides protection from all but severe physical damage. Read 358.10 and 12 for additional information on uses (permitted and not permitted).

C The minimum approved electrical trade size for EMT is ½ in. »358.20(A)«.

D The maximum approved electrical trade size for EMT is 4 in. »358.20(B)«.

E All EMT cut ends must be reamed (or otherwise finished) to remove rough edges »358.28(A)«.

F Couplings and connectors used with tubing must be made tight. Buried in masonry (or concrete), they must be the concrete-tight type. Installed in wet locations, they must be the raintight type »358.42«.

> **NOTE**
>
> *The number of conductors permitted in EMT must not exceed the percentage fill specified in Table 1, Chapter 9 »358.22«.*

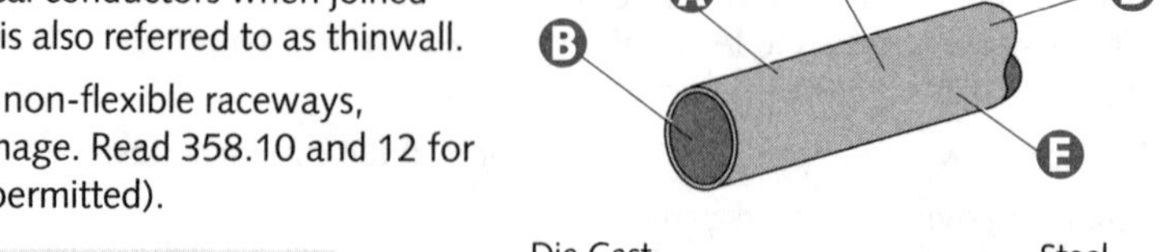

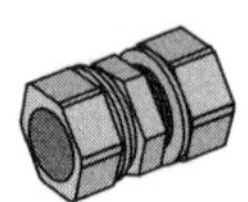

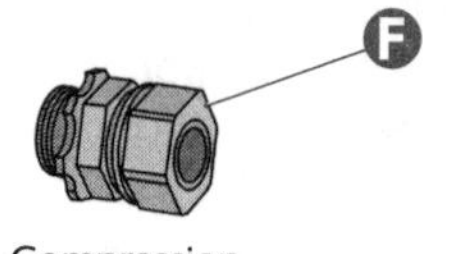

Rigid Nonmetallic Conduit

NEC® Chapter 9, Table 4, identifies five types of RNC:

- **Schedule 80 PVC is an extra-heavy-walled raceway with a wall thickness conforming to Schedule 80-Iron Pipe Size (IPS) dimensions.**
- **Schedule 40 PVC is a heavy-walled raceway with a wall thickness conforming to Schedule 40-IPS dimensions.**
- **Type A PVC is a thin-walled raceway with wall thickness conforming to Schedule A-IPS dimensions. Limited to underground installations, Type A PVC conduit must be laid with its entire length in concrete.**
- **Type EB PVC is a thin walled raceway with wall thickness designed to achieve a duct stiffness of 20 lbs./in./in. Type EB PVC conduit, limited to underground installations, must be laid with its entire length in concrete in outdoor trenches.**
- **HDPE Schedule 40 is a high density polyethylene raceway with a wall thickness conforming to Schedule 40-IPS dimensions. HDPE Schedule 40 conduit, also limited to underground installations, can be direct buried with or without being encased in concrete.**

A Rigid Nonmetallic Conduit (RNC), commonly referred to as PVC (polyvinyl chloride), is a nonmetallic raceway of circular cross section, with or without integral couplings.

B All joints between lengths of conduit, and between conduit and couplings, fittings, and boxes, must be made by an approved method »352.48«.

C All cut ends must be trimmed inside and out to remove rough edges »352.28«.

D Expansion fittings must be provided for RNC to compensate for thermal expansion and contraction where the length change is expected to be 0.25 in. (6 mm) or greater, in accordance with Table 352.44(A) and (B), in a straight run between securely mounted items such as boxes, cabinets, elbows, or other conduit terminations »352.44 and 300.7(B)«.

E The minimum approved electrical trade size for RNC is ½ in. »352.20(A)«.

F The maximum approved electrical trade size for RNC is 6 in. »352.20(B)«.

G Expansion and contraction problems generally do not arise in underground RNC applications.

H Only listed fittings can be used with RNC »352.6«.

> **NOTE**
>
> *Read 352.10 and 12 for a description of uses (permitted and not permitted) of RNC.*

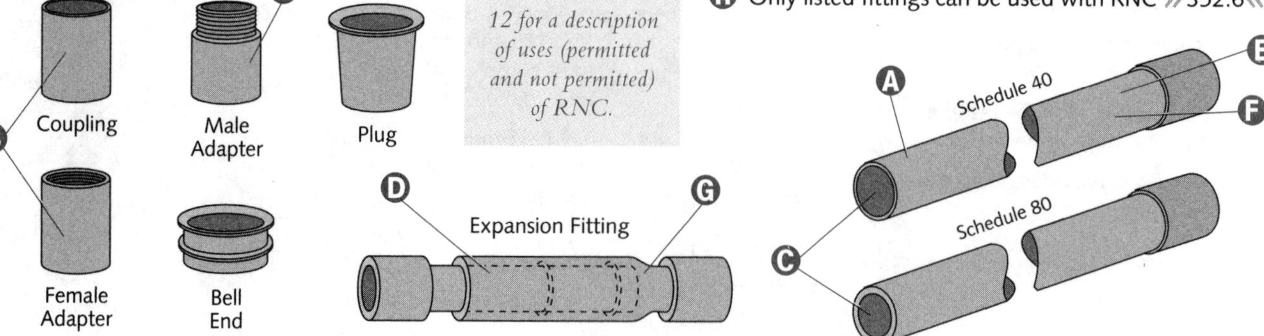

Securing RNC

A Each conduit must be fastened securely within 3 ft (900 mm) of all termination points »352.30(A)«.

B Table 352.30 support provisions must be followed for RNC »352.30(B)«.

C RNC must be fastened so that movement from thermal expansion or contraction is permitted »352.30«.

> **NOTE**
>
> *Horizontal runs of RNC that are supported by openings through framing members (at intervals not greater than those in Table 352.30) and are securely fastened within 3 ft (900 mm) of termination points, are permitted »352.30(B)«.*
>
> *RNC, listed for securing at other than 3 ft (900 mm), can be installed in accordance with the listing »352.30(A)«.*

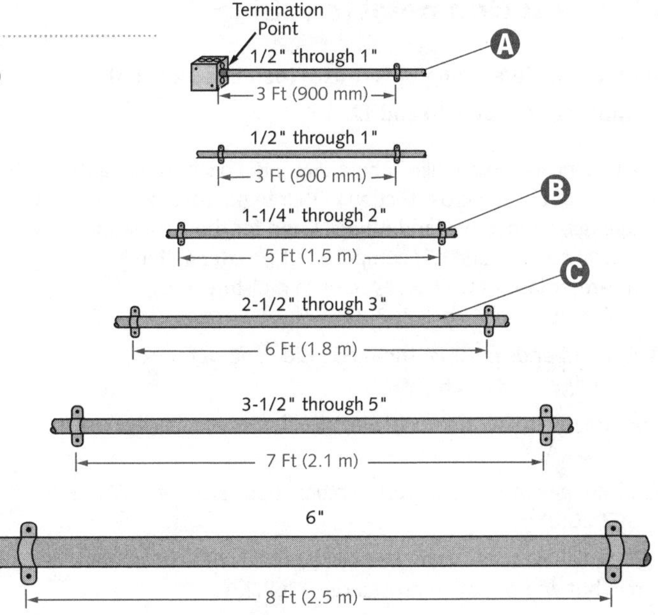

Bending RNC

A RNC bends must be made so that the conduit remains undamaged and the internal diameter of the conduit is not reduced »352.24«.

B The conduit must not be damaged.

C Field bends must be made only with bending equipment identified for the purpose. The radius of the curve to the centerline of such bends must not be less than shown in Table 344.24 »352.24«.

D Installing PVC plugs in the ends of larger conduits (2 to 6 in.) can prevent conduit collapses or other deformities.

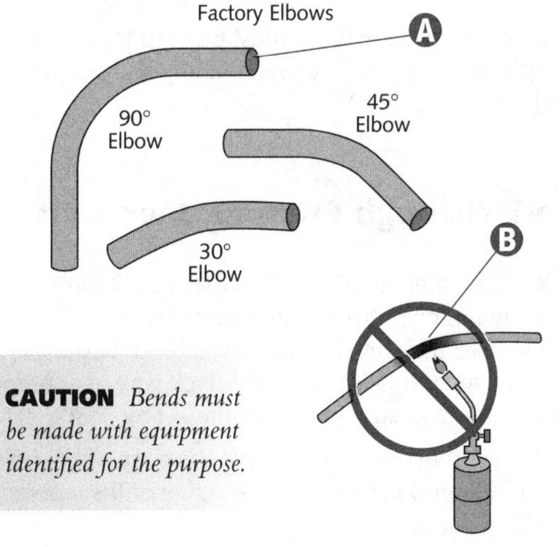

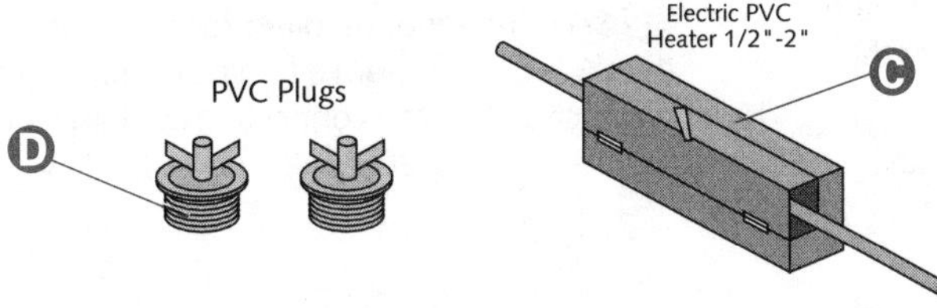

CAUTION *Bends must be made with equipment identified for the purpose.*

Securing ENT

A The equivalent of four quarter bends (360° total) is the maximum between pull points, e.g., conduit bodies and boxes »362.26«.

B Electrical Nonmetallic Tubing must be secured every 3 ft (900 mm) or less »362.30(A)«.

C Bends must be made so that the tubing is undamaged and the tubing's internal diameter is not reduced. While bends can be made manually (without auxiliary equipment) the radius of the curve of the inner edge of each bend must not be less than shown in Table 344.24 »362.24«.

D ENT must be fastened securely within 3 ft (900 mm) of each outlet box, device box, junction box, cabinet, or other termination point »362.30(A)«.

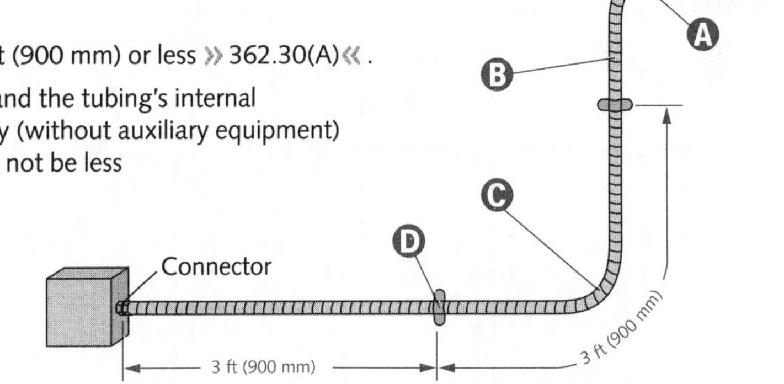

Electrical Nonmetallic Tubing

For the complete listing of ENT uses (permitted and not permitted), read 362.10 and 12.

Ⓐ Electrical Nonmetallic Tubing (ENT) is a pliable corrugated raceway of circular cross section with integral (or associated) couplings, connectors, and fittings listed for the installation of electrical conductors. It is composed of a material that is flame-retardant as well as resistant to moisture and chemical atmospheres »362.2«.

Ⓑ All cut ends must be trimmed inside and out to remove rough edges »362.26«.

Ⓒ The minimum approved electrical trade size for ENT is ½ in. »362.20(A)«.

Ⓓ The maximum approved electrical trade size for ENT is 2 in. »362.20(B)«.

Ⓔ A pliable raceway can be bent by hand with reasonable force, but without other assistance »362.2«.

Ⓕ Outside diameters are such that standard rigid PVC conduit couplings and connectors can be used on ENT of PVC construction. ENT installation instructions outline the procedure to follow when installing PVC conduit fittings that are cemented in place. Specific cement requirements and application methods are provided.

Ⓖ An approved method must be used for all joints between lengths of tubing and between tubing and couplings, fittings, and boxes »362.48«.

CAUTION

ENT must not be stored or installed where exposed to direct sunlight, unless identified as "sunlight resistant" »362.12(9)«.

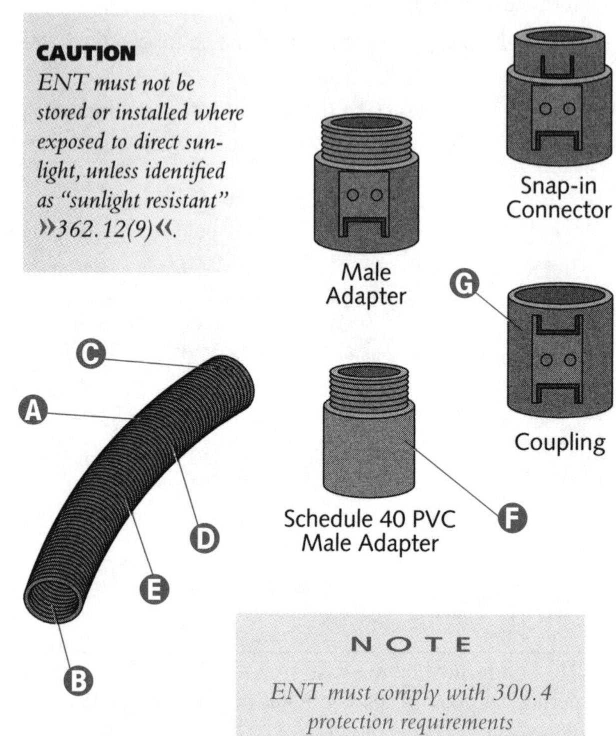

NOTE

ENT must comply with 300.4 protection requirements

ENT Through Framing Members

Ⓐ If the bored hole is less than 1¼ in. (32 mm) from the edge of the framing member, the raceway must be protected from penetration (by screws, nails, etc.) by a steel plate (or bushing). The steel plate must be at least 1⁄16 in. (1.6 mm) thick, and of appropriate length and width to adequately cover the wiring »300.4(A)(1)«.

Ⓑ Holes must be bored so that the edge of the hole is at least 1¼ in. (32 mm) from the nearest edge of the wood member »300.4(A)(1)«.

Ⓒ ENT must be securely fastened in place within 3 ft (900 mm) of all termination points »362.30(A)«.

Ⓓ Horizontal runs of ENT can be supported by openings through framing members at intervals not greater than 3 ft (900 mm) »362.30(B)«.

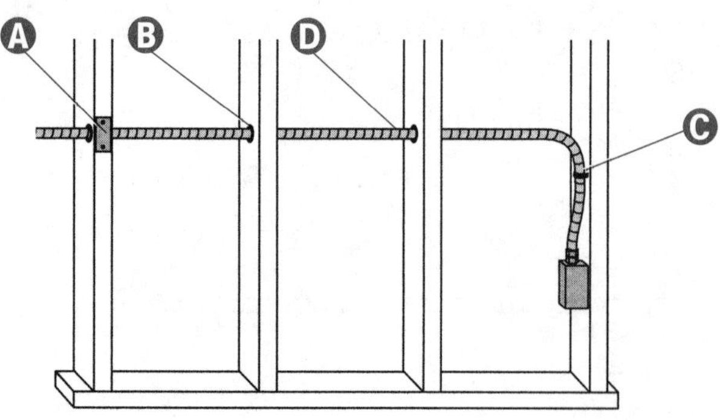

ENT Luminaire (Fixture) Whips

Ⓐ ENT must be secured within 3 ft (900 mm) of all termination points »362.30(A)«.

Ⓑ Securing is not required for ENT in lengths not more than 6 ft (1.8 m) from a luminaire (fixture) terminal connection [for tap connections to luminaires (lighting fixtures)] »362.30(A) *Exception*«.

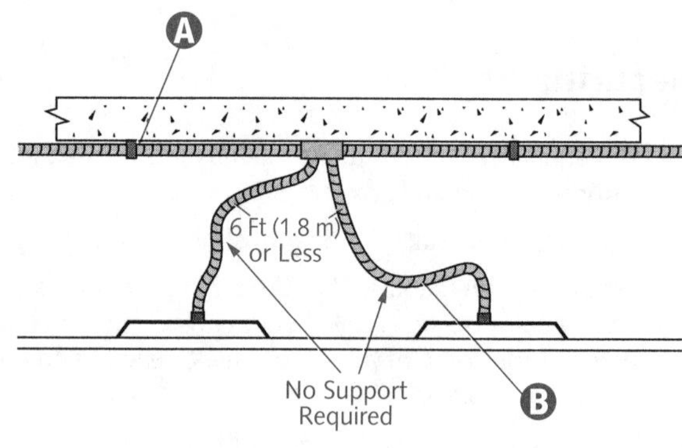

FLEXIBLE CONDUIT

Flexible Metal Conduit

A The minimum electrical trade size for FMC is ½ in., unless a 348.20(A) provision is met.

B Flexible Metal Conduit (FMC) is a raceway of circular cross section made of helically wound, formed, and interlocked metal strips »348.2«. FMC must be listed, and can be used in both exposed and concealed locations »348.6« and »348.10«.

C Flexible metal conduit is often referred to as *Greenfield*, or simply *Flex*.

D Fittings used with FMC must be listed »348.6«.

E The maximum approved electrical trade size for FMC is 4 in. »348.20(B)«.

F The equivalent of four quarter bends (360° total) is the maximum between pull points, e.g., conduit bodies and boxes »348.26«.

G Bends must be made so that the conduit is not damaged and the conduit's internal diameter is not effectively reduced. The radius of the curve of the inner edge of any field bend must not be less than shown in Table 344.24 »348.24«.

H All cut ends must be trimmed (or otherwise finished) to remove rough edges, except where fittings are used which thread into the convolutions »348.28«.

I Angle connectors cannot be used for concealed raceway installations »348.42«.

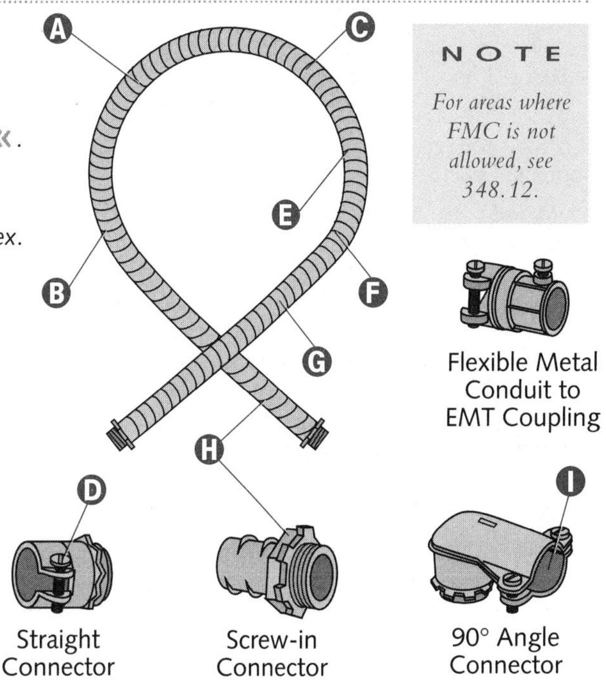

NOTE

For areas where FMC is not allowed, see 348.12.

FMC Support

Support is not required for fished FMC »348.30(A) *Exception No. 1*«.

A FMC must be fastened securely in place (by an approved means) at intervals of 4½ ft (1.4 m), or less »348.30(A)«.

B FMC must be secured within 12 in. (300 mm) of each box, cabinet, conduit body, or other conduit termination »348.30(A)«.

C Refer to Table 348.22 for the maximum number of conductors permitted in ⅜ in. flex.

D If the length of FMC from a luminaire (fixture) terminal for tap connections, as permitted in 410.67(C), to luminaires (light fixtures) is 6 ft (1.8 m), or less, strapping is not required »348.30(A) *Exception No. 3*«.

NOTE

Horizontal runs of FMC can be supported by openings through framing members [at intervals not greater than 4½ ft (1.4 m)], if securely fastened within 12 in. (300 mm) of each termination point »348.30(B)«.

E FMC of ⅜ in. electrical trade size can be used: (1) for enclosing the leads of motors as permitted in 430.145(B); (2) in lengths not in excess of 6 ft (1.8 m) as part of a listed assembly, for tap connections to luminaires (lighting fixtures) as permitted in 410.67(C), or for utilization equipment; (3) for manufactured wiring systems as permitted in 604.6(A); (4) in hoistways, as permitted in 620.21(A)(1); (5) as part of a listed assembly to connect wired luminaire (fixture) sections as permitted in 410.77(C) »348.20(A)«.

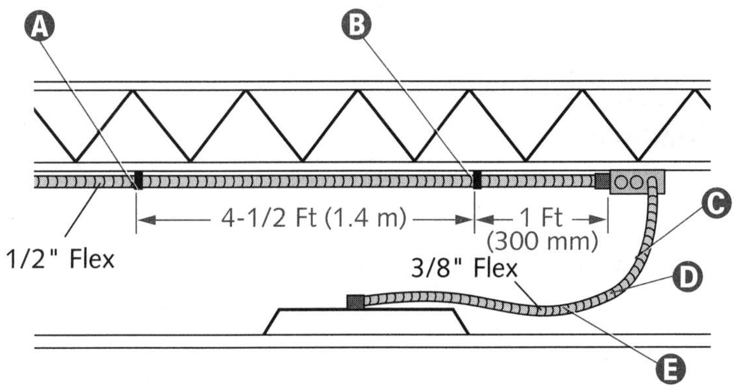

FMC Installation

Ⓐ Grounding conductors are required for circuits over 20 amperes.

Ⓑ For lengths of 3 ft (900 mm), or less, requiring flexibility, the 12-in. (300-mm) securing distance can be waived »348.30(A) *Exception No. 2*«.

Ⓒ If an equipment bonding jumper is required around FMC, it must be installed in accordance with 250.102 »348.60«.

Ⓓ Listed FMC can serve as a grounding means if: (1) the conduit is terminated in fittings listed for grounding; (2) the circuit conductors contained therein are protected by overcurrent devices rated at 20 amperes or less; (3) the total length in any ground return path is 6 ft (1.8 m) or less; and (4) the conduit is not installed for flexibility »250.118(6)«.

Ⓔ FMC must be secured within 12 in. (300 mm) of each box, cabinet, conduit body, or other conduit termination »348.30(A)«.

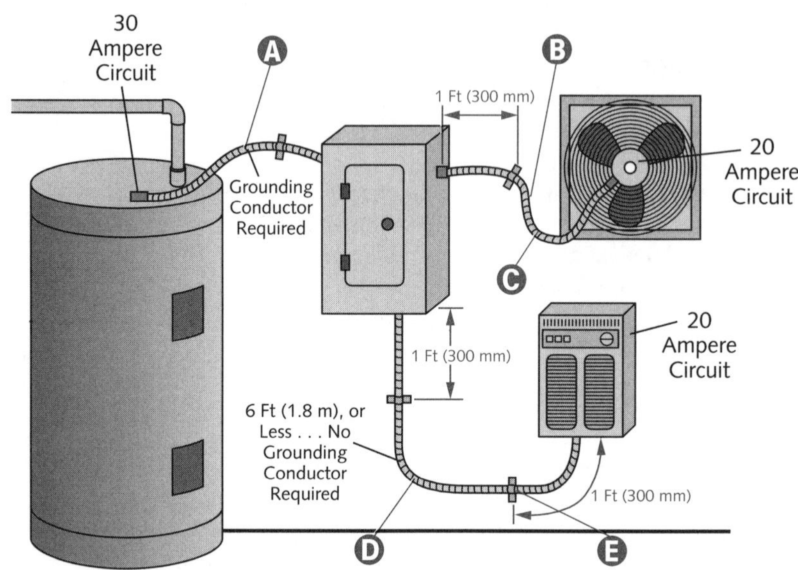

CAUTION *An equipment grounding conductor must be installed in FMC used to connect equipment where flexibility must be maintained »348.60«.*

Liquidtight Flexible Metal Conduit

The equivalent of four quarter-bends (360° total) is the maximum allowed between pull points, e.g., conduit bodies and boxes »350.26«.

Ⓐ The minimum electrical trade size for LFMC is ½ in., except as provided in 348.20(A) »350.20(A)«. The number of conductors allowed in ⅜ in. conduit must not exceed the limit set in Table 348.22 »350.22(B)«.

Ⓑ Liquidtight Flexible Metal Conduit (LFMC) is a raceway of circular cross section having an outer liquidtight, nonmetallic, sunlight-resistant jacket over an inner flexible metal core with associated couplings, connectors, and fittings and approved for the installation »350.2«.

Ⓒ LFMC and associated fittings must be listed »350.6«.

Ⓓ Angle connectors cannot be used for concealed raceway installations »350.42«.

Ⓔ The maximum approved electrical trade size for LFMC is 4 in. »350.20(B)«.

Ⓕ LFMC is often referred to as *Sealtite* (a registered trademark).

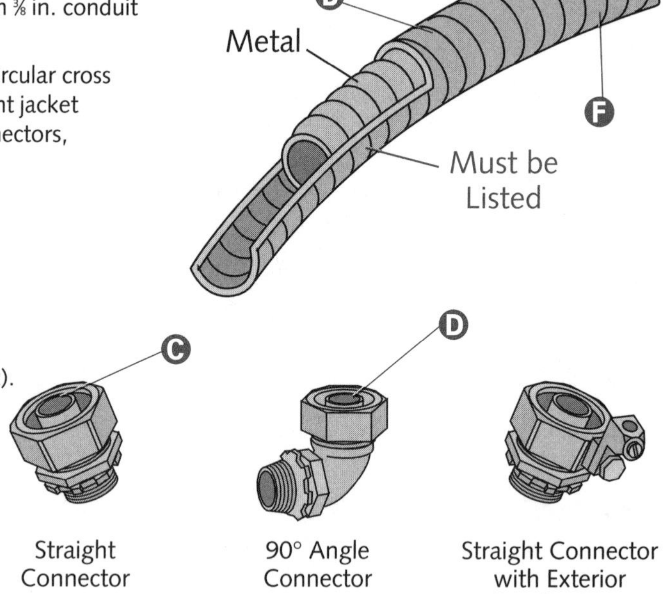

NOTE

Support requirements for LFMC are listed in 350.30.

Grounding Liquidtight Flexible Metal Conduit

A Listed LFMC, ⅜" through 1¼", can be used as a grounding means if: (1) the total length of flexible conduit, in any ground return path, is not more than 6 ft (1.8 m); (2) the conduit is terminated in fittings listed for grounding; and, (3) the circuit conductors contained therein are protected by overcurrent devices rated at 20 amperes or less for ⅜" and ½" electrical trade sizes and 60 amperes or less for ¾" through 1¼" electrical trade sizes »250.118(7)«.

B LFMC can serve as a grounding means as covered in 250.118. If an equipment bonding jumper is required around LFMC, it must be installed in accordance with 250.102 »350.60«.

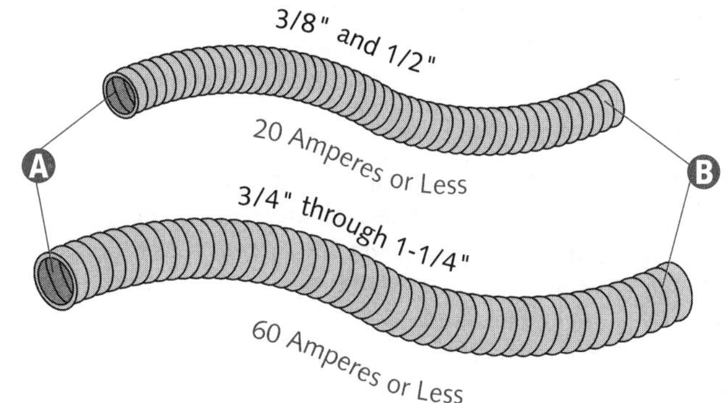

CAUTION *LFMC used to connect equipment, where flexibility is required, must have an equipment grounding conductor »350.60«.*

NOTE

An equipment bonding jumper can be installed either inside or outside of a raceway. If installed on the raceway's exterior, the jumper cannot exceed 6 ft (1.8 m) in length and must be routed with the raceway »250.102(E)«.

Liquidtight Flexible Nonmetallic Conduit

The equivalent of four quarter-bends (360° total) is the maximum allowed between pull points, e.g., conduit bodies and boxes »356.26«.

A Type LFNC-A has a smooth seamless inner core surrounded by reinforcement layer(s), and bonded to a smooth seamless cover »356.2(1)«.

B Type LFNC-B has a smooth inner surface having integral reinforcement within the conduit wall »356.2(2)«. (Outside appearance is similar to LFMC).

C Type LFNC-C has a corrugated internal and external surface having no integral reinforcement within the conduit wall »356.2(3)«. (Similar to ENT, except more flexible.)

D LFNC and associated fittings must be listed »356.6«.

E Angle connectors cannot be used for concealed raceway installations »356.42«.

F Liquidtight Flexible Nonmetallic Conduit (LFNC) is a raceway of circular cross section of various types, including: (1) FNMC-A; (2) FNMC-B; and (3) FNMC-C »356.2«.

G The maximum approved electrical trade size for LFNC is 4 in. »356.20(B)«.

H The minimum electrical trade size for LFNC is ½ in., unless a 356.20(A) provision applies.

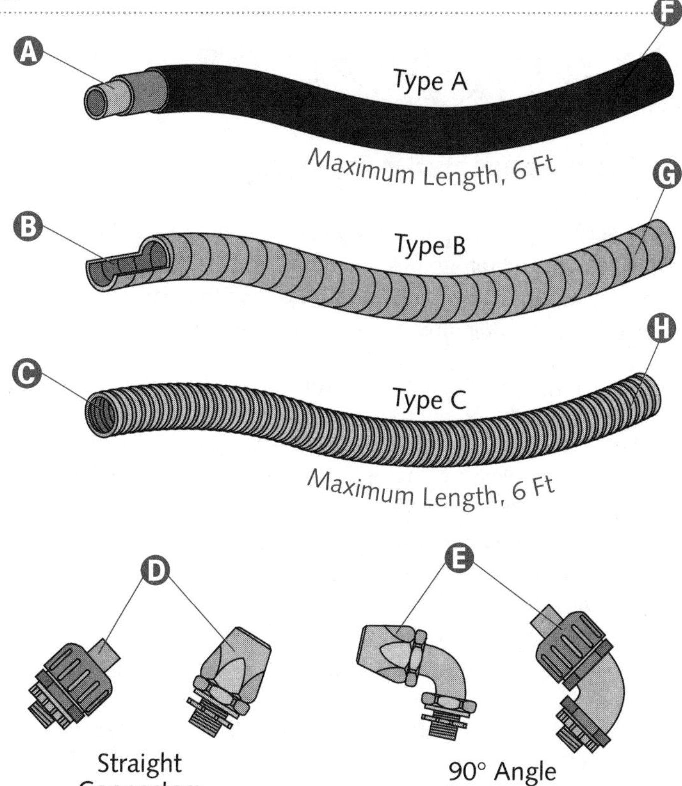

Liquidtight Flexible Nonmetallic Conduit Installation

Strapping is not required for LFNC lengths of 6 ft (1.8 m), or less, from a luminaire (fixture) terminal connection for tap conductors to luminaires (light fixtures) as permitted in 410.67(C) »356.30(2)«.

Support is not required for fished LFNC »356.30(2)«.

Horizontal runs of LFNC can be supported by openings through framing members [at intervals not greater than 3 ft (900 mm)], if securely fastened within 12 in. (300 mm) of each termination point »356.30(3)«.

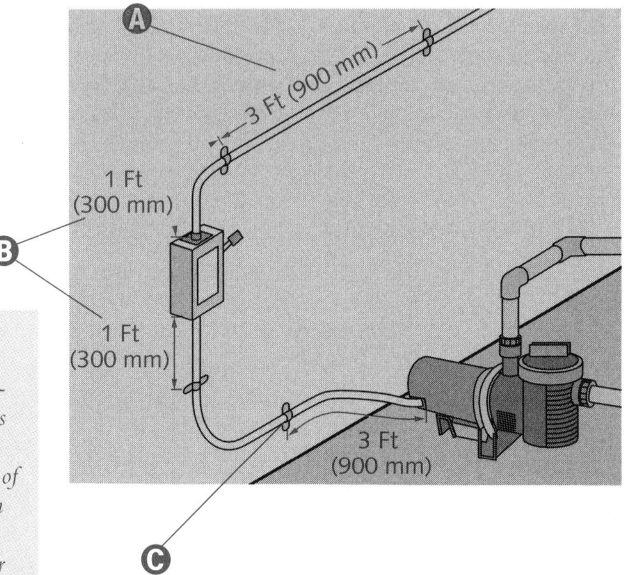

A LFNC must be securely fastened at intervals not greater than 3 ft (900 mm) »356.30(1)«.

B LFNC must be secured within 12 in. (300 mm) of each connection to every outlet box, junction box, cabinet, or fitting »356.30(1)«.

C Where flexibility is necessary, LFNC can be secured within 3 ft (900 mm) of the termination »356.30(2)«.

If an equipment grounding conductor is required for the circuits installed in LFNC, it can be installed either inside or outside of the conduit. Where installed on the outside, the length of the equipment grounding conductor must not exceed 6 ft (1.8 m) and must be routed with the raceway »250.102(E)«.

OTHER RACEWAYS

Surface Nonmetallic Raceways

Where surface nonmetallic raceways are used in combination for both signaling and for lighting and power circuits, the different wiring systems must be run in separate compartments, identified by sharply contrasting interior finish colors »388.70«.

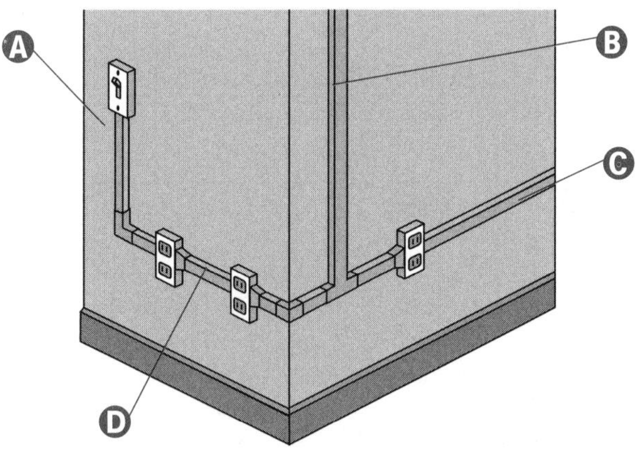

A Surface nonmetallic raceway construction must be visibly distinguishable from other raceways. Surface nonmetallic raceways and their fittings must be designed so that sections can be mechanically coupled together, and installed without subjecting the wires to abrasion »388.100«.

B The size and number of conductors installed in any raceway must not exceed the raceway's design limitations »388.21 and 388.22«.

C Splices and taps are permitted within surface nonmetallic raceway having a removable cover which remains accessible once installed. The conductors, at the point of splices and taps, must not fill the raceway to more than 75% of its area. Splices and taps, in surface nonmetallic raceways without removable covers, must be made only in junction boxes. All splices and taps must be made using approved methods »388.56«.

D Surface nonmetallic raceways can be installed in dry locations »388.10(1)«. NEC® 388.12 lists the locations where these raceways cannot be used.

NOTE

Unbroken lengths of surface nonmetallic raceways can pass transversely through dry walls, partitions, and floors. Conductors must be accessible on both sides of the wall, partition, or floor »388.10(2)«.

Where Article 250 requires equipment grounding, a separate equipment grounding conductor must be installed within the raceway »388.60«.

Surface Metal Raceways

A Surface metal raceway construction must be visibly distinguishable from other raceways. Surface metal raceways and their fittings must be designed so that the sections can be electrically and mechanically coupled together, and installed without subjecting the wires to abrasion. Nonmetallic covers and accessories can be used on surface metal raceways only if identified for such use »386.100«.

B Uses (permitted and not permitted) are located in 386.10 and 386.12.

C The number and size of conductors installed in any raceway must not exceed the raceway's design limitations »386.21 and 386.22«.

D 310.15(B)(2)(a) derating factors do not apply to conductors installed in surface metal raceways where all of the following conditions are met: (1) the cross-sectional area of the raceway exceeds 4 square in. (2500 mm^2); (2) the current-carrying conductors do not exceed 30 in number; (3) the sum of the cross-sectional areas of all contained conductors does not exceed 20% of the interior cross-sectional area of the surface metal raceway »386.22«.

E Unbroken lengths of surface metal raceway can pass transversely through dry walls, dry partitions, and dry floors. Access to the conductors must be maintained on both sides of the wall, partition, or floor »386.10(4)«.

F Where surface metal raceways are used in combination for both signaling and for lighting and power circuits, the different wiring systems must be run in separate compartments identified by sharply contrasting interior finish colors, while maintaining the same relative position of compartments throughout the premises »386.70«.

G Splices and taps are permitted within surface metal raceway having a removable cover that remains accessible after installation. The conductors, at the point of splices and taps, must not fill the raceway to more than 75% of its area. Splices and taps in surface metal raceways without removable covers can be made only in junction boxes. Use only approved methods for splices and taps »386.56«.

H Multioutlet assemblies are covered in Article 380.

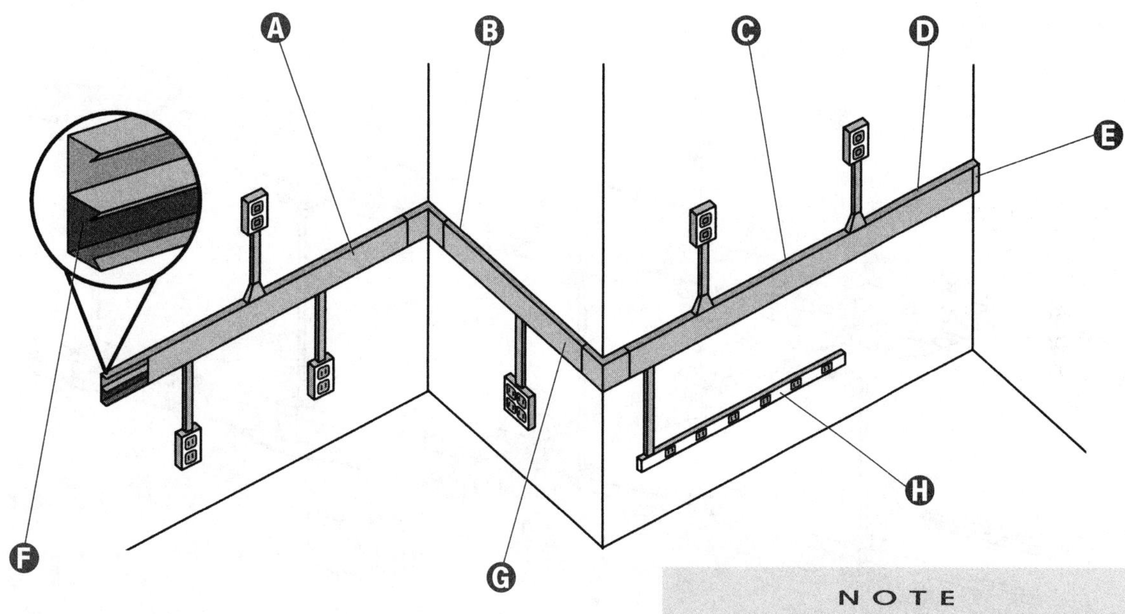

NOTE

If a surface metal raceway enclosure provides a transition from another wiring method, it must have a means to connect an equipment grounding conductor »386.60«.

Strut-Type Channel Raceways

Use Table 384.22 to determine the maximum number of conductors permitted in strut-type channel raceway. Apply the appropriate Chapter 9 Tables for the outside diameter (O.D.) dimensions of the type and size of wire used »384.22«.

310.15(B)(2)(a) derating factors do not apply to conductors installed in strut-type channel raceways, where all of the following conditions are met: (1) the cross-sectional area of the raceway exceeds 4 square in. (2500 mm^2); (2) the current-carrying conductors do not exceed 30 in number; and (3) the sum of the cross-sectional areas of all contained conductors does not exceed 20% of the interior cross-sectional area of the strut-type channel raceway »384.22«.

Conductors larger than that for which the strut-type channel raceway is listed must not be used »384.21«.

A surface mount strut-type channel raceway must be secured to the mounting surface with external retention straps at intervals not exceeding 10 ft (3 m), and within 3 ft (900 mm) of each raceway termination (outlet box, cabinet, junction box, etc.) »384.30(A)«.

Ⓐ Strut-type channel raceways, closure strips, and accessories must be listed and identified for such use »384.6«.

Ⓑ Splices and taps are permitted in raceways that provide access after installation via a removable cover. The conductors, at any point of splices or taps, must not fill the raceway to more than 75%. All splices and taps must be made by approved methods »384.56«.

Ⓒ Uses (permitted and not permitted), are covered in 384.10 and 384.12.

Ⓓ Strut-type channel raceway enclosures providing a transition to (or from) other wiring methods must accommodate the connection of an equipment grounding conductor. A strut-type channel raceway can be used as an equipment grounding conductor »384.60«.

Ⓔ Strut-type channel raceways can be suspended in air provided an approved method (designed for the purpose) is applied at intervals of not more than 10 ft (3 m) »384.30(B)«

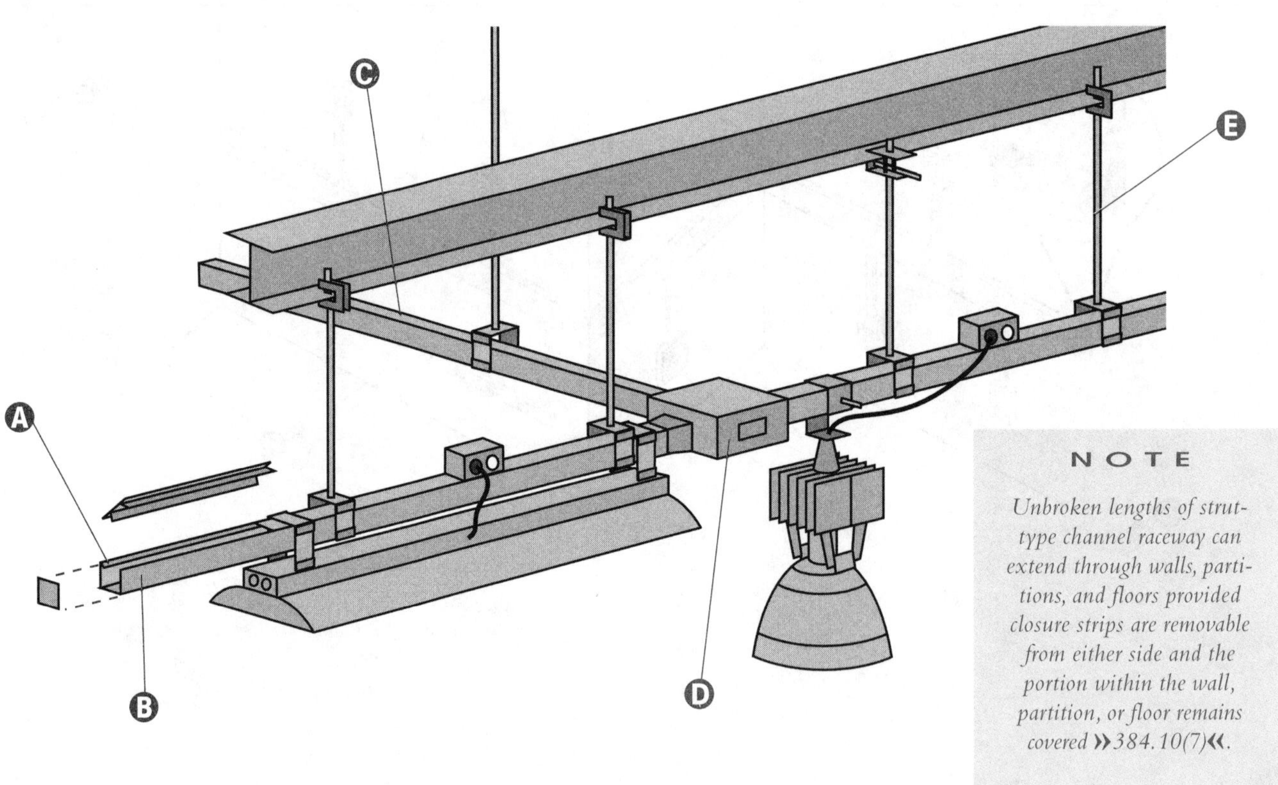

NOTE

Unbroken lengths of strut-type channel raceway can extend through walls, partitions, and floors provided closure strips are removable from either side and the portion within the wall, partition, or floor remains covered »384.10(7)«.

Underfloor Raceways

Ⓐ Connections between raceways and distribution centers and wall outlets must be made by approved fittings or by any of the wiring methods in Chapter 3, where installed according to the provisions of the respective articles »390.15«.

Ⓑ Underfloor raceways can be installed beneath a concrete (or other flooring material) surface. In office occupancies, installations flush with the concrete floor and covered with linoleum (or equivalent covering) are acceptable »390.2(A)«.

Ⓒ The combined cross-sectional area of all conductors or cables must not exceed 40% of the interior cross-sectional area of the raceway »390.5«.

Ⓓ The size of conductors installed must not be larger than that for which the underfloor raceway is designed »390.4«.

Ⓔ Inserts must be leveled and sealed to prevent the entrance of concrete. Metal raceway inserts must also be metal and must be electrically continuous with the raceway »390.14«.

Ⓕ Seal and level junction boxes to the floor grade to prevent entrance of water or concrete. Underfloor metal raceway junction boxes must also be metal and must be electrically continuous with the raceway »390.13«.

Ⓖ Splices and taps are acceptable only within junction boxes. Continuous, unbroken conductor connecting the individual outlets (so-called loop wiring) is not considered a splice or tap »390.6«.

Ⓗ Underfloor raceways must be laid so that a straight line from the center of one junction box, to the center of the next junction box, will coincide with the centerline of the raceway system. Raceways must be held firmly in place to prevent misalignment during construction »390.8«.

Ⓘ Raceway dead ends must be closed »390.10«.

Ⓙ Install a suitable marker at, or near, each end of straight raceway runs to locate the last insert »390.9«.

CAUTION *When an outlet is abandoned, discontinued, or removed, the supplying circuit conductors must be removed from the raceway. Splices or reinsulated conductors, such as would occur in the case of abandoned outlets on loop wiring, are not allowed in raceways »390.7«.*

NOTE

390.2(B) lists areas where underfloor raceways are not permitted.

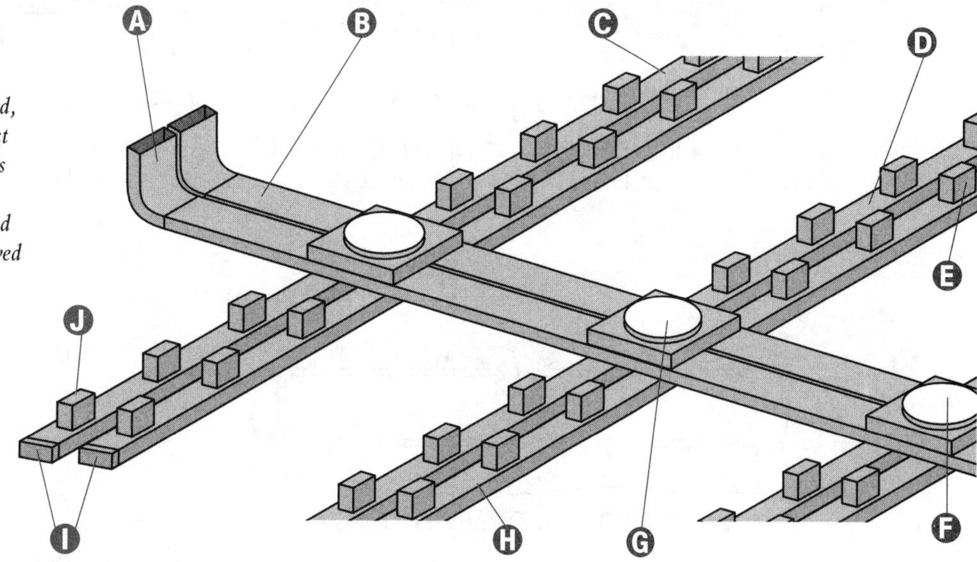

Underfloor Raceway Covering

A Half-round and flat-top raceways not over 4 in. (100 mm) in width must be covered by ¾ in. (20 mm) or more of concrete (or wood) »390.3(A)«.

B Flat-top raceways, greater than 4 in. (100 mm) but not more than 8 in. (200 mm) wide, with a minimum of 1-in. (25-mm) spacing between the raceways, must be covered with at least 1 in. (25 mm) of concrete »390.3(B)«.

C Trench-type flush raceways having removable covers can be laid flush with the floor's surface. Such approved raceways shall be designed so that the cover plates provide adequate mechanical protection and rigidity equivalent to junction box covers »390.3(C)«.

D In office occupancies, approved flat-top metal raceways [of 4 in. (100 mm) or less in width], can be flush with the concrete floor surface, provided they are covered with a minimum 1⁄16-in. (1.6-mm) floor covering, such as linoleum. Where more than one (but fewer than four) single raceways are installed flush with the concrete, they must be contiguous with one another and be joined forming a rigid assembly »390.3(D)«.

E Flat-top raceways, greater than 4 in. (100 mm) but not more than 8 in. (200 mm) wide, that are spaced less than 1 in. (25 mm) apart must be covered with at least 1½ in. (38 mm) of concrete »390.3(B)«.

F Splices and taps are permitted in trench-type flush raceways whose removable covers are accessible after installation. The conductors, at the point of any splices or taps, must not fill more than 75% of the raceway's area »390.6«.

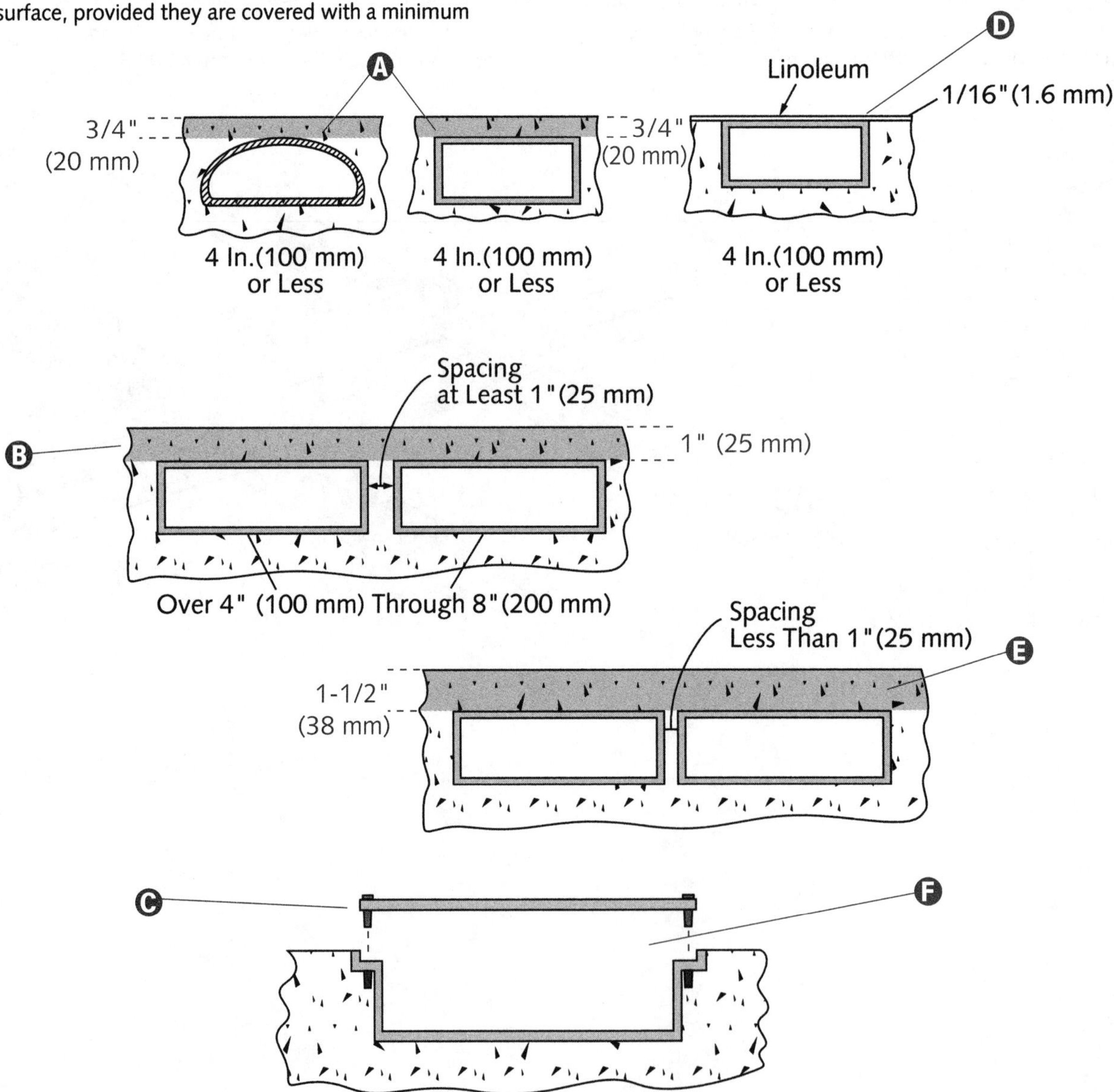

Cellular Metal Floor Raceways

A The combined cross-sectional area of all conductors or cables must not exceed 40% of the cell or header's interior cross-sectional area »374.5«.

B Junction boxes must be leveled to the floor grade and sealed against the entrance of water or concrete. Junction boxes used with these raceways must be metal and must be electrically continuous with the raceway »374.9«.

C For Article 374 purposes, a **cellular metal floor raceway** consists of the hollow spaces of cellular metal floors, together with suitable fittings, that may be approved as enclosures for electric conductors »374.2«.

D Conductors larger than 1/0 AWG can be installed only by special permission »374.4«.

E Inserts must be leveled to the floor grade and sealed against the entrance of concrete. Only metal inserts can be used with metal raceways and each must be electrically continuous with the raceway »374.10«.

F Connections between raceways and distribution centers and wall outlets must be made of FMC (where not installed in concrete), RMC, IMC, EMT, or approved fittings »374.11«.

G Splices and taps are made only in header access units or junction boxes. Continuous, unbroken conductor connecting the individual outlets (so-called loop wiring) is not considered a splice or tap »374.6«.

H A "header" is a transverse raceway for electric conductors, providing access to predetermined cells of a cellular metal floor, accommodating the installation of electric conductors from a distribution center to the cells »374.2«.

I A "cell" is a single, enclosed tubular space in a cellular metal floor member, whose axis is parallel to the axis of the metal floor member »374.2«.

CAUTION *When an outlet is removed, abandoned, or discontinued, remove the section of circuit conductors supplying the outlet from the raceway. Raceways must not contain splices or reinsulated conductors as would occur with abandoned outlet loop wiring »374.7«.*

NOTE

374.3 lists areas where cellular metal floor raceways cannot be used.

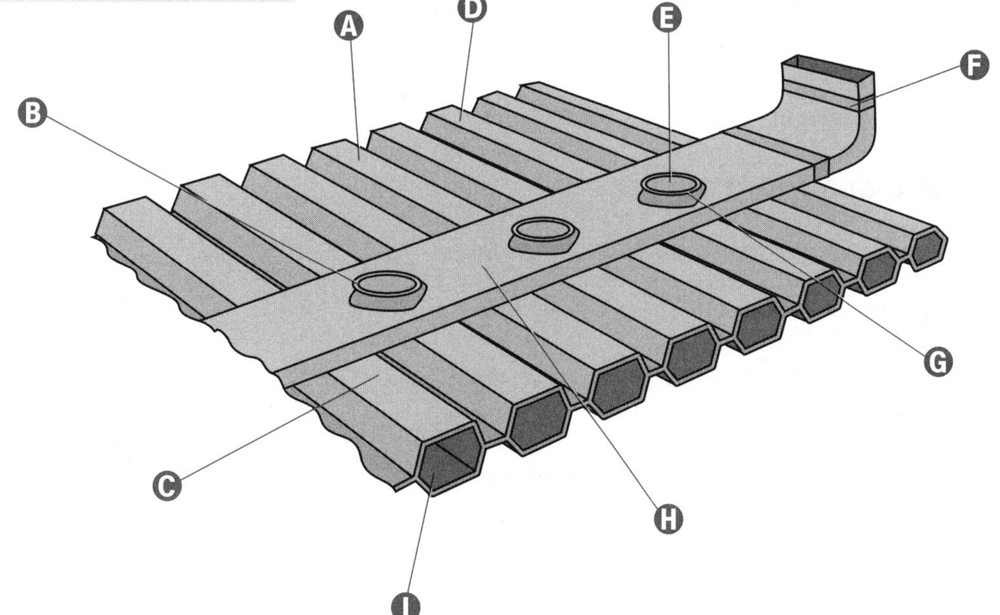

Cellular Concrete Floor Raceways

A Splices and taps are allowed only in header access units or junction boxes. Continuous, unbroken conductor connecting the individual outlets (so-called loop wiring) is not considered a splice or tap »372.12«.

B A "header" is a transverse metal raceway for electric conductors that provides access to predetermined cells of a precast cellular concrete floor and permits the installation of conductors from a distribution center to the floor cells »372.2«.

C Inserts must be leveled and sealed against the entrance of concrete. Inserts must be metal and must be fitted with grounded-type receptacles. A grounding conductor must connect the insert receptacles to a positive ground connection on the header »372.9«.

D The header must be installed in a straight line and at right angles to the cells »372.5«.

E Mechanically secure the header to the top of the precast cellular concrete floor. The header must be electrically continuous throughout its entire length and must be electrically bonded to the distribution center enclosure »372.5«.

F Conductor larger than 1/0 AWG can be installed only by special permission »372.10«.

G The combined cross-sectional area of all conductors or cables must not exceed 40% of the cell or header's cross-sectional area »372.11«.

H A "cell" is a single, enclosed tubular space in a floor made of precast cellular concrete slabs, whose direction is parallel to the direction of the floor member »372.2«.

I Cellular concrete floor raceways are hollow spaces in floors constructed of precast cellular concrete slabs (together with suitable metal fittings) designed to provide access to the floor cells »372.1«.

NOTE

372.4 lists areas where cellular concrete floor raceways cannot be used.

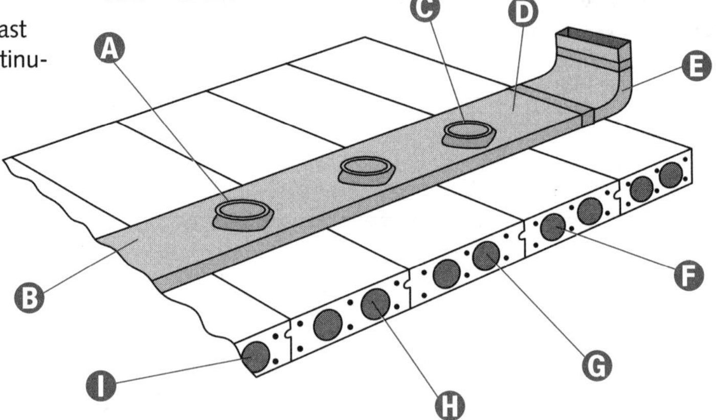

CAUTION *When an outlet is removed, abandoned, or discontinued, remove the section of circuit conductors supplying the outlet from the raceway. Raceways must not contain splices or reinsulated conductors, such as loop wiring of abandoned outlets »372.13«.*

Wireways (Metal and Nonmetallic)

A Wireways (sheet-metal and flame-retardant, nonmetallic) are troughs with hinged or removable covers for housing and protecting electric wires (and cables). The complete wireway system is installed before conductors are laid in place »376.2 and 378.2«.

B The sum of the cross-sectional area of all contained conductors must not exceed 20% of the wireway's interior cross-sectional area »376.22 and 378.22«.

C No conductor larger than that for which the wireway is designed shall be installed »376.21 and 378.21«.

D Within **metal wireways**, the derating factors in 310.15(B)(2)(a) are only applicable where the number of current-carrying conductors, including neutral conductors classified as current-carrying under the provisions of 310.15(B)(4), exceeds 30 »376.22«.

E Derating factors, specified in 310.15(B)(2)(a), are applicable to **nonmetallic wireways** containing more than three current-carrying conductors »378.22«.

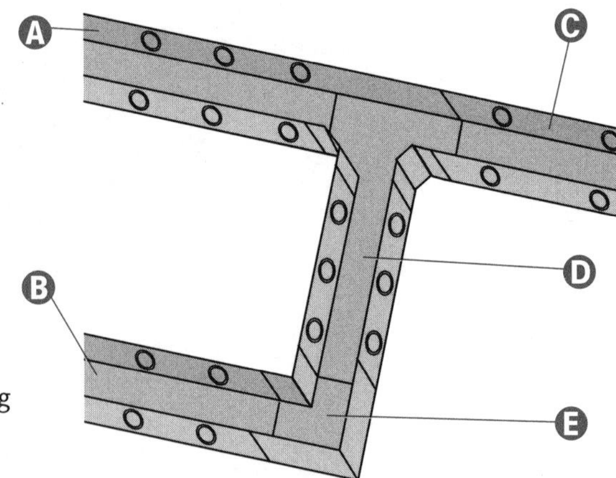

NOTE

Conductors for signaling circuits or controller conductors between a motor and its starter (used only for starting duty) are not considered as current-carrying conductors »376.22 and 378.22«.

The manufacturer's name (or trademark) on all wireways must remain visible after installation. In addition, the interior cross-sectional area (in square in.) must be marked on all nonmetallic wireways »376.120 and 378.120«.

CAUTION *Wireways containing 4 AWG or larger conductors and used as junction or pull boxes must be sized according to 314.28 or 314.71.*

Wireway Installation

Listed nonmetallic wireways are permitted: (1) only for exposed work, except as permitted in 378.10(4); (2) where subject to corrosive environments and (3) in wet locations where listed for the purpose »378.10«.

Nonmetallic wireways shall not be used: (1) where subject to physical damage; (2) in any hazardous (classified) location, except as permitted in 504.20; (3) where exposed to sunlight unless listed and marked as suitable for the purpose; (4) where subject to ambient temperatures other than those for which nonmetallic wireway is listed; and, (5) for conductors whose insulation temperature limitations would exceed those for which the raceway is listed »378.12«.

A Accessible splices and taps are permitted within a wireway. The conductors, at the point of any splice or tap, must not fill the wireway to more than 75% of its area »376.56 and 378.56«.

B Close all dead ends with listed fittings »376.58 and 378.58«.

C Where insulated conductors are deflected within a wireway, either at the ends or where conduits, fittings, or other raceways (or cables) enter (or leave) the wireway, or where the direction of the wireway is deflected greater than 30°, 312.6(A) dimensions apply »376.23(A) and 378.23(A)«.

D Where 4 AWG or larger insulated conductors enter a wireway through a raceway or cable, the distance between raceway and cable entries (enclosing the same conductor) must not be less than that required in 314.28(A)(1) for straight pulls, and 314.28(A)(2) for angle pulls »376.23(B) and 378.23(B)«.

E Unbroken wireway lengths can pass transversely through walls. Access to the conductors must be maintained on both sides of the wall »376.10(4) and 378.10(4)«.

F **Metal** wireways can be used: (1) for exposed work except as permitted in 376.10(4); (2) where subject to corrosive environments where identified for the use; or (3) in hazardous (classified) locations as permitted by 501.4(B) for Class 1, Division 2 locations; 502.4(B) for Class II, Division 2 locations; and 504.20 for intrinsically safe wiring »376.10«.

G **Metal** wireways are not permitted where subject to severe physical damage or severe corrosive environments »376.12«.

H Extensions from wireways can be made with cord pendants (installed in accordance with 400.10) or with any Chapter 3 wiring method that includes an equipment grounding means. Where a separate equipment grounding conductor is employed, connection of the wiring method's equipment grounding conductors to the wireway must comply with 250.8 and 250.12 »376.70 and 378.70«.

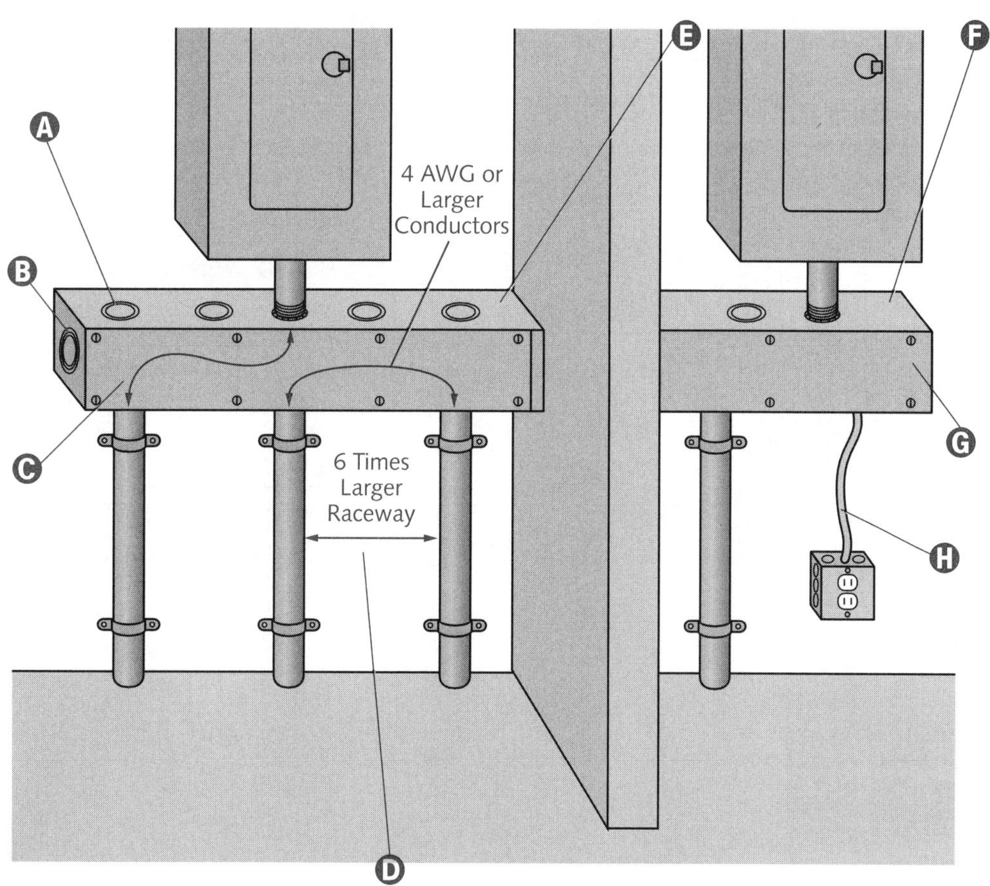

CAUTION *Extreme cold may cause nonmetallic wireways to become brittle and, therefore, more susceptible to damage from physical contact »378.10(3) FPN«.*

NOTE

Wireways installed in wet locations must be listed for the purpose »376.10(3) and 378.10(3)«.

Wireway Supports

A Horizontally run wireways must be supported at each end »376.30(A)«.

B The maximum distance between supports for individual lengths of metal wireway must not exceed 10 ft (3 m) »376.30(A)«.

C Nonmetallic wireway support intervals must not exceed 3 ft (900 mm) »378.30(A)«.

D Horizontally run individual, nonmetallic wireway lengths of more than 3 ft (900 mm) must be supported at each end or joint »378.30(A)«.

E Individual, nonmetallic wireways must be supported as listed, and in no case shall the distance between supports exceed 10 ft (3 m) »378.30(A)«.

F Metal wireway support intervals must not exceed 5 ft (1.5 m) »376.30(A)«.

G Individual, metal wireway lengths [longer than 5 ft (1.5 m)], run horizontally, must be supported at each end or joint, unless listed otherwise »376.30(A)«.

H Horizontally run wireways must be supported at each end »378.30(A)«.

I Nonmetallic wireway expansion fittings must be provided to compensate for thermal expansion and contraction where the change in length is expected to be 0.25 in. (6 mm) or more in a straight run »378.44«. Expansion characteristics of PVC nonmetallic wireway (which is the same as PVC rigid nonmetallic conduit) are listed in Table 352.44(A).

NOTE

Nonmetallic wireway vertical runs must be securely supported at intervals of 4 ft (1.2 m) or less (except as otherwise listed), and can have only one joint between supports. Adjoining nonmetallic wireway sections must be securely fastened together to form a rigid joint »378.30(B)«.

Metal wireway vertical runs must be securely supported at intervals not exceeding 15 ft (4.5 m), and are limited to one joint between supports. Adjacent wireway sections must be securely fastened together to form a rigid joint »376.30(B)«.

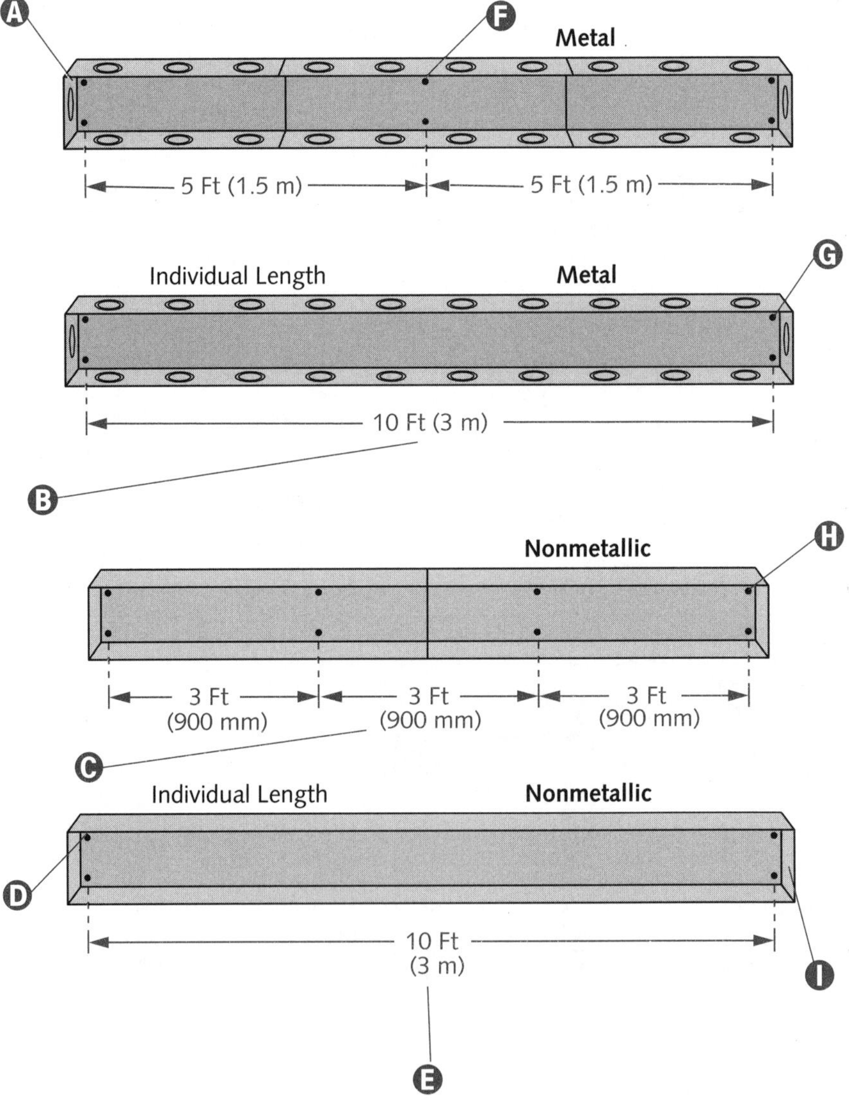

RACEWAY FILL

Same Size Conductors

A Annex C (located in the back of the *Code* book) can be used to find the maximum number of conductors permitted in a particular raceway (conduit or tubing). The conductor maximum is based on all conductors (in the raceway) being the same size (total cross-sectional area including insulation) » Chapter 9, Note1 «.

B Tables C1 through C12 (in Annex C) are based on the 40% fill for three or more conductors as permitted in Table 1 of Chapter 9.

C Annex C also lists the maximum number of fixture wires permitted in each type of conduit (or tubing).

Raceways Containing Multiconductor Cables

A A multiconductor cable of two or more conductors is treated as a single conductor for percentage conduit fill area calculations. To calculate the cross-sectional area of cables with elliptical cross sections, use the major diameter of the ellipse as a circle diameter » Chapter 9, Note 9 «.

NOTE

While Table 1 applies only to complete conduit or tubing systems, it does not apply to sections of conduit used to protect exposed wiring from physical damage » Chapter 9, Note 2 «.

Different Size Conductors

A Tables 4 and 5 (of Chapter 9) are used when a combination of different size conductors is installed in a single raceway » Chapter 9, Note 6 «.

B Use Table 4's 40% column when there are three or more conductors.

C Use Tables 5 and 5A for dimensions of insulated conductors.

D Include equipment grounding or bonding conductors, if any, when calculating raceway fill » Chapter 9, Note 3 «.

E Table 4 contains trade size information for twelve different types of conduits and tubing. Columns include internal diameters and total area, also referred to as the cross-sectional area. Table 4 also shows the maximum percent fill for tubing or conduits containing different numbers of conductors.

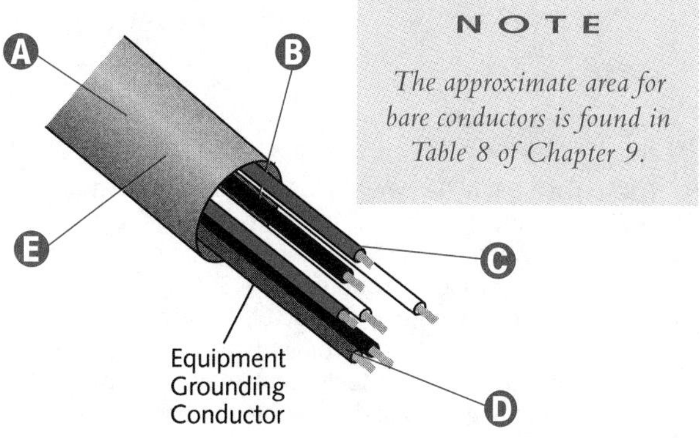

NOTE

The approximate area for bare conductors is found in Table 8 of Chapter 9.

Nipple Fill

A A length of conduit (or tubing) measuring 24 in. (600 mm) or less is considered a nipple » Table 310.15(B)(2)(a) *Exception No. 3* and Chapter 9, Note 4 «.

B For nipples, disregard 310.15(B)(2)(a) adjustment factors » Chapter 9, Note 4 «.

C Installed between boxes, cabinets, and similar enclosures, conduit or tubing nipples can be filled to 60% of their total cross-sectional area » Chapter 9, Note 4 «.

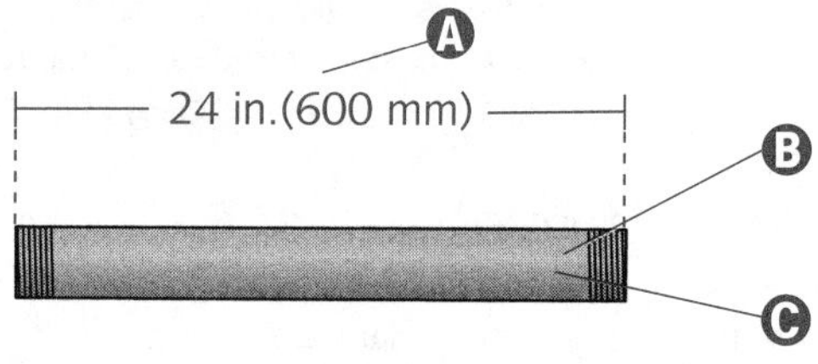

NOTE

When calculating the maximum number of same size conductors (total cross-sectional area including insulation) in a conduit or tubing, the next-higher whole number is used when the calculation results in a decimal of 0.8 or larger » Chapter 9, Note 7 «.

Raceway Fill Percentage

A A conduit having a single conductor can be filled to 53% of the conduit's cross-sectional area.

B A conduit containing exactly two conductors can only be filled to 31% of the cross-sectional area of the conduit or tubing.

C A conduit containing three or more conductors can be filled to 40% of its cross-sectional area.

D When pulling three conductors or cables into a raceway, if the ratio of the raceway (inside diameter) to the conductor or cable (outside diameter) is between 2.8 and 3.2, jamming can occur. Jamming is less likely to occur when pulling four or more conductors or cables into a raceway »Chapter 9, Table 1, FPN No. 2 «.

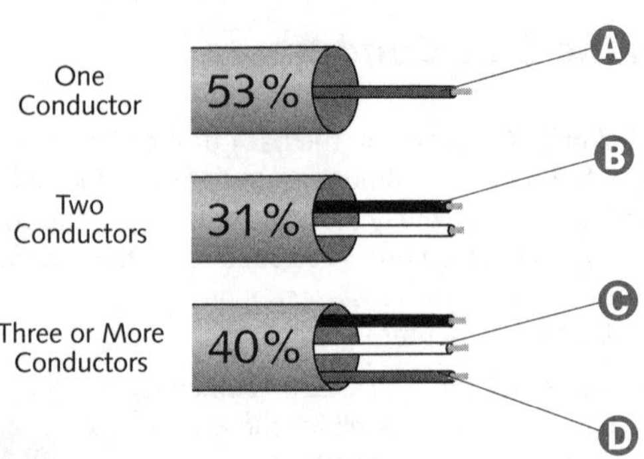

NOTE

Tables 1 and 4 (in Chapter 9) list maximum fill percentages for conduits and tubing.

CAUTION *Table 1 is based on common conditions of proper cabling and alignment of conductors where the length of the pull and the number of bends are within reasonable limits. Be advised that, for certain conditions, a larger size conduit (or fewer conductors) should be considered »Chapter 9, Table 1, FPN No. 1«.*

CONDUCTORS

Conductors in Parallel

A Conductors of one phase, neutral, or grounded circuit conductors can have different physical characteristics from those of another phase, neutral, or grounded circuit conductors and still achieve balance. For example: neutral conductors do not have to be the same length as Phase A conductors, and Phase A conductors do not have to be the same length as Phase B conductors, etc. »310.4 «.

B As a general rule, 1/0 AWG and larger size conductors can be connected in parallel (electrically joined at both ends to form a single conductor) »310.4 «.

C The paralleled conductors in each phase, neutral, or grounded circuit conductor must have the same characteristics »310.4 «. For example . . . all paralleled, Phase C conductors must:

(1) be the same length,
(2) have the same conductor material,
(3) be the same size in circular mil area,
(4) have the same insulation type, and
(5) be terminated in the same manner.

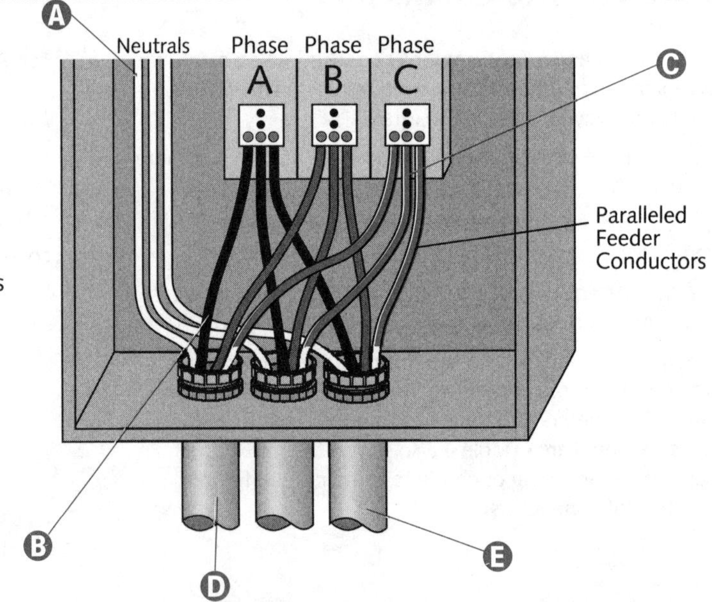

D Conductors carrying alternating current, installed in metal enclosures or **metal raceways**, must be so arranged as to avoid heating the surrounding metal by induction »300.20(A) «.

E If run in separate raceways (or cables), the raceways (or cables), must have the same physical characteristics »310.4 «.

NOTE

Equipment grounding conductors, used with conductors in parallel, must comply with 310.4 requirements, applying the sizing requirements of 250.122.

WARNING

Each metal raceway (housing paralleled conductors) must contain all phase conductors, the grounded conductor (where used) and all equipment grounding conductors »300.20(A) «. For example: a Phase A conductor, a Phase B conductor, a Phase C conductor, and, if used, a neutral (or grounded) conductor and an equipment grounding conductor must be in each raceway.

Maximum Ampacities

A Tables 310.16 through 310.22 list certain conductor maximum temperature ratings.

B Table 310.13 contains specific conductor information, such as trade name, type letter, maximum operating temperature, application provisions, insulation, size, and outer covering.

C Tables 310.16 through 310.22 list aluminum (or copper-clad aluminum) as well as copper conductors.

D Maximum ampacities are listed in Tables 310.16 through 310.22. Other factors must be considered before using these ampacities: temperature limitations »110.14(C)«, continuous loads »210.19(A) and 215.2(A)«, ambient temperature »Tables 310.16 through 310.22«, and the number of current-carrying conductors »310.15(B)(2)(a)«, to name a few.

60°C (140°F)	75°C (167°F)	90°C (194°F)
70 amperes	85 amperes	95 amperes
TW	THW	THHN
4 AWG Copper	4 AWG Copper	4 AWG Copper

Conductor Temperature Limitations—100 Amperes or Less

A Equipment termination provisions for circuits rated 100 amperes or less, or marked for 14 AWG through 1 AWG conductors, are used only for conductors rated 60°C »110.14(C)(1)(a)«. Because the lowest temperature (weakest link) is 60°C, the ampacity of this conductor cannot exceed 70 amperes.

B A 4 AWG copper conductor with a temperature rating of 60°C (140°F) has an ampacity of 70 amperes »Table 310.16«.

C A temperature rating of 60/75°C means that the termination is suitable for 60° or 75°C conductors.

D Conductor temperature limitations can be compared to the strength of a chain. A chain is only as strong as its weakest link. One potential weak link for conductors is the termination (connection) point. The conductor ampacity cannot be higher than the lowest temperature rating of any connected termination, conductor, or device »110.14(C)«.

E The ampacity of a 4 AWG THHN copper conductor is 95 amperes »Table 310.16«.

F A 4 AWG copper conductor with a temperature rating of 75°C (167°F) has an ampacity of 85 amperes »Table 310.16«.

G Equipment termination provisions for circuits rated 100 amperes or less, or marked for 14 AWG through 1 AWG conductors, can be used for conductors up to their maximum ampacities if the equipment is listed and identified for use with such conductors »110.14(C)(1)(a)(3)«. The ampacity of this conductor now has a rating of 85 amperes because the lowest temperature (weakest link) is 75°C.

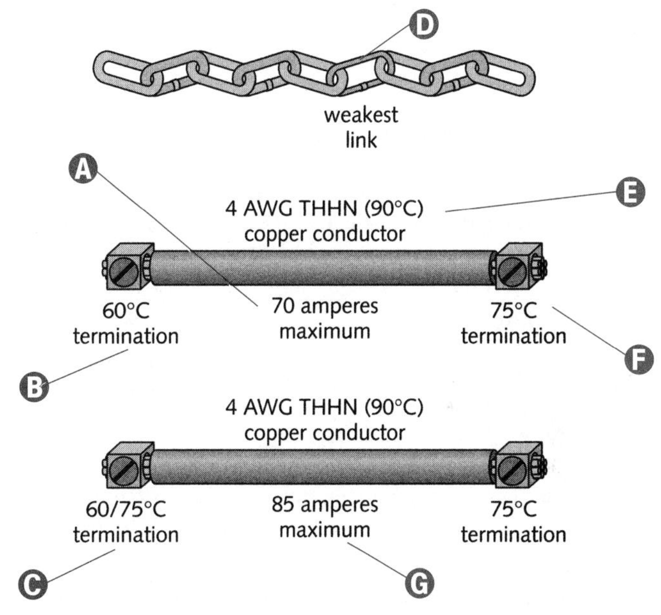

NOTE

When using conductors with temperature ratings higher than those specified for terminations, the higher ampacity can be used for ampacity adjustment, correction, or both »110.14(C)«. For example, an adjustment factor of 80% must be applied to 4 AWG THHN copper conductors because there will be more than three current-carrying conductors in the raceway. The connected load requires an overcurrent protection and conductor ampacity rating of at least 70 amperes, but the lowest temperature rated termination is 60°C. A 70-ampere overcurrent protective device can be installed with 4 AWG THHN (90°C) copper conductors because the allowable ampacity (after the adjustment factor has been applied) is 76 amperes (95 x 80% = 76 amperes). Had derating not been possible (using the higher ampacity), the maximum overcurrent protective device would be 60 amperes (70 x 80% = 56 amperes).

Conductor Properties

A A 6 AWG (or smaller) insulated grounded conductor must be identified by a continuous white or gray outer finish or by three continuous white stripes on other than green insulation along its entire length »200.6(A)«.

B Table 310.13 lists conductor insulations and applications.

C An insulated grounded conductor larger than 6 AWG shall be identified by either a continuous white or gray outer finish or by three continuous white stripes on other than green insulation along its entire length or, at the time of installation, by a distinctive white marking at its terminations. This marking must encircle the conductor (or insulation) »200.6(B)«.

D Ampacities of insulated conductors rated 0 through 2000 volts, 60° through 90°C (140° through 194°F), with three or fewer current-carrying conductors installed in raceway, cable, or earth (directly buried) are found in Table 310.16. The ampacities listed there are based on an ambient temperature of 30°C (86°F). A variety of insulations are listed for both copper and aluminum (or copper-clad aluminum) conductors.

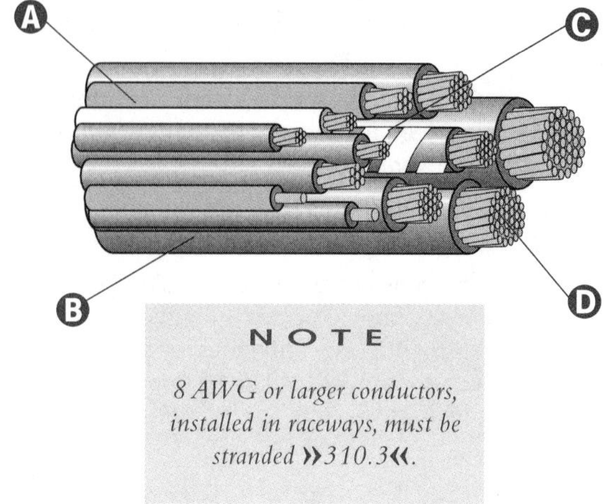

NOTE

8 AWG or larger conductors, installed in raceways, must be stranded »310.3«.

Conductor Temperature Limitations—Over 100 Amperes

Equipment termination provisions for circuits rated over 100 amperes, or marked for conductors larger than 1 AWG, are used only for conductors with higher temperature ratings provided the conductor ampacity is based on the 75°C (167°F) ampacity of the conductor size used »110.14(C)(1)(b)«.

Where the overcurrent device is rated over 800 amperes, the ampacity of the conductors it protects must be equal to, or greater than, the overcurrent device rating »240.4(C)«.

When running three sets of paralleled copper conductors (for a 1200 ampere overcurrent device), the minimum size is 600 kcmil, unless the equipment is listed and identified for use with 90°C (194°F) conductors.

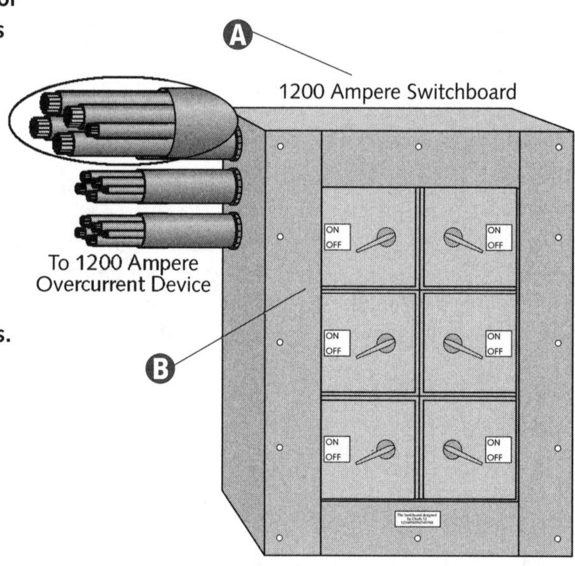

A Switchboard and panelboard provisions are found in Article 408.

B When running *four* sets of paralleled copper conductors (for a 1200 ampere overcurrent device), the minimum size is 350 kcmil, unless the equipment is listed and identified for use with 90°C (194°F) conductors.

CAUTION *Three paralleled sets of 500 kcmil THHN copper conductors do not meet NEC® provisions if the overcurrent device is 1200 amperes, unless the equipment is listed and identified for use with 90°C (194°F) conductors. Because of 110.14(C)(1)(b), the maximum ampacity (at 75°C) for these conductors is only 380 amperes. The total combined ampacity for each phase is, therefore, only 1140 amperes. 240.4(C) requires a minimum conductor ampacity of 1200 amperes.*

NOTE

The minimum size conductors (aluminum, copper-clad aluminum, or copper) that can be connected in parallel (electrically joined at both ends to form a single conductor) is size 1/0 AWG, unless meeting an exception »310.4«.

Ampacity Correction and Adjustment Factors

The following are not counted as current-carrying conductors when using Table 310.15(B)(2)(a):

- Conductors of different systems, as provided in 300.3, installed in a common raceway or cable, unless the number of power and lighting conductors (current-carrying) exceeds three »310.15(B)(2)(a) *Exception No. 1*«.
- Conductors in cable trays, unless required by 392.11 »310.15(B)(2)(a) *Exception No. 2*«.
- Conductors in nipples of 24 in. (600 mm) or less »310.15(B)(2)(a) *Exception No. 3*«.
- Neutral conductors of normally balanced circuits containing three or more conductors »310.15(B)(4)(a)«.
- Grounding and bonding conductors »310.15(B)(5)«.

The following shall be counted as current-carrying conductors when using Table 310.15(B)(2)(a):

- The neutral conductor in a 3-wire circuit consisting of two phase (hot) wires and the neutral that is fed from a 4-wire, 3-phase wye-connected system »310.15(B)(4)(b)«.
- The neutral conductor of a 4-wire, 3-phase wye circuit where the major portion of the load consists of **nonlinear loads**, because harmonic currents are present in the neutral conductor »310.15(B)(4)(c)«. Examples: Electronic equipment (such as computers), electronic/electric-discharge lighting (such as fluorescent lights and lights with ballast), adjustable speed drive systems, and similar equipment »Article 100—Definitions«.
- A grounded conductor of any 2-wire circuit.

A When there are more than three current-carrying conductors in a raceway or cable, the allowable ampacity of each conductor must be reduced (derated) as shown in Table 310.15(B)(2)(a) »310.15(B)(2)(a)«.

B Ambient temperature is the temperature of the conductor's surrounding environment. It is the temperature of the environment in which the conductor, raceway, or cable is installed. This environment may include air, water, earth, or a combination thereof.

C A grounding (or bonding) conductor does not count as a current-carrying conductor »310.15(B)(5)«.

D When the ambient temperature exceeds 30°C (86°F), the table ampacity has to be multiplied by the correction factor.

E Ampacity correction factors for ambient temperatures are found at the bottom of Tables 310.16 through 310.22. These tables are used to correct the ampacity of conductors, by multiplying the ampacity of the conductor by the appropriate factor.

NOTE

*The maximum ampacities given in Tables 310.16 through 310.22 are based on a maximum of three current-carrying conductors and a maximum ambient temperature of 30°C (86°F). When these limitations are exceeded, the ampacities listed must be reduced, or **derated**.*

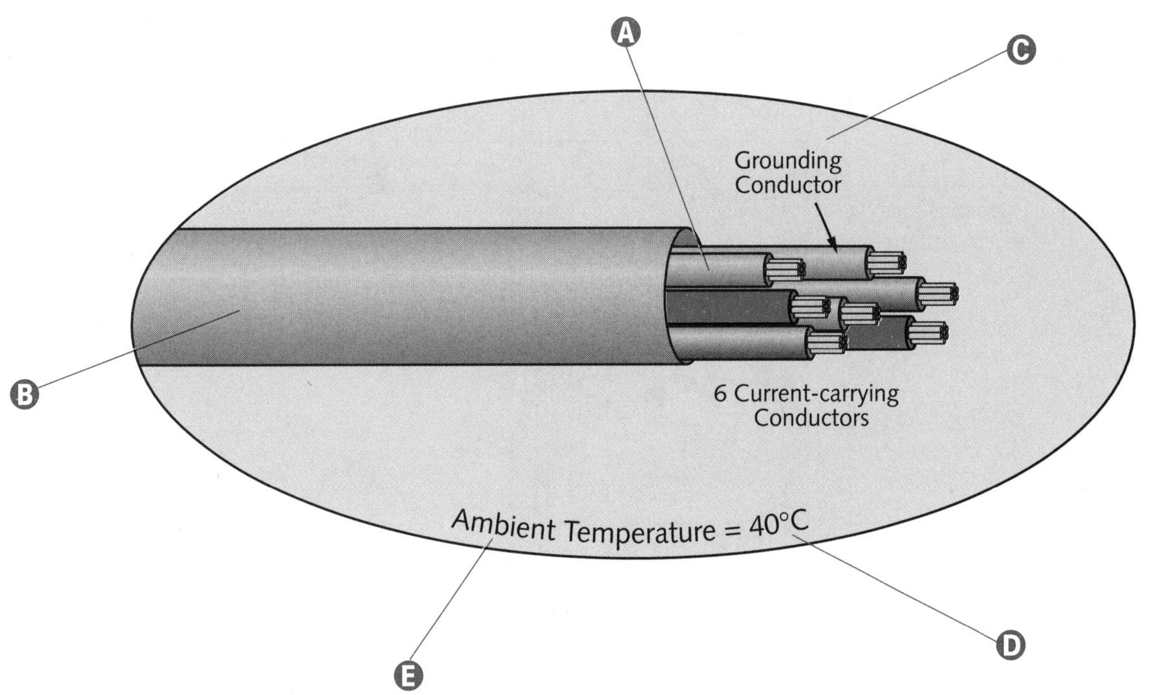

Continuous Loads

To determine the maximum conductor ampacity and overcurrent protection, start with the actual Table ampacity for the type and size conductor, then derate as follows:

- **12 AWG THHN copper conductors have an ampacity of 30 amperes »Table 310.16«.**
- **Instead of multiplying the load by 125%, the conductor ampacity can be multiplied by 80% (the reciprocal of 125% is 80%). This calculation must be performed before applying any adjustment or correction factors »210.19(A)«.**

30 amperes x 80% = 24 amperes

- **Next, multiply the ampacity by 80%, because there are six current-carrying conductors in the conduit »Table310.15(B)(2)(a)«.**

24 amperes x 80% = 19.2 amperes

- **Finally, multiply the ampacity by .87, because of the 110°F ambient temperature. This correction factor is found near the bottom of Table 310.16, in the 90°C column, in the row across from 105–113°F.**

19.2 amperes x .87 = 16.7 amperes

- **The maximum ampacity for these 12 AWG conductors is 16.7 amperes. The next-higher standard-size circuit breaker above 16.7 is 20 amperes.**
- **One last precaution: the load on the breaker cannot exceed 16 amperes because of the continuous load »210.20(A)«.**

A The maximum continuous load current permitted on a 20-ampere overcurrent device is 16 amperes (20 ÷ 125% = 16).

B A continuous load is expected to maintain maximum current for three hours or more »Article 100—Definitions«. Fluorescent office lighting represents a continuous load.

C A continuous load must be multiplied by 125% *before* applying any adjustment or correction factors to conductors (branch-circuit or feeder) »210.19(A) and 215.2(A)«.

D Where a branch-circuit (or feeder) supplies any combination of continuous and noncontinuous loads, the rating of the overcurrent device shall not be less than the total of the noncontinuous load plus 125% of the continuous load »210.20(A) and 215.3«.

E After conductor adjustment or correction factors have been applied, overcurrent protection can be determined. Where the adjusted ampacity does not correspond to a standard size fuse or circuit breaker, the next-higher standard size can be used only: (1) if it does not exceed 800 amperes; and (2) the conductors being protected are not part of a multioutlet branch-circuit supplying receptacles for cord- and plug-connected portable loads »240.3(B)«.

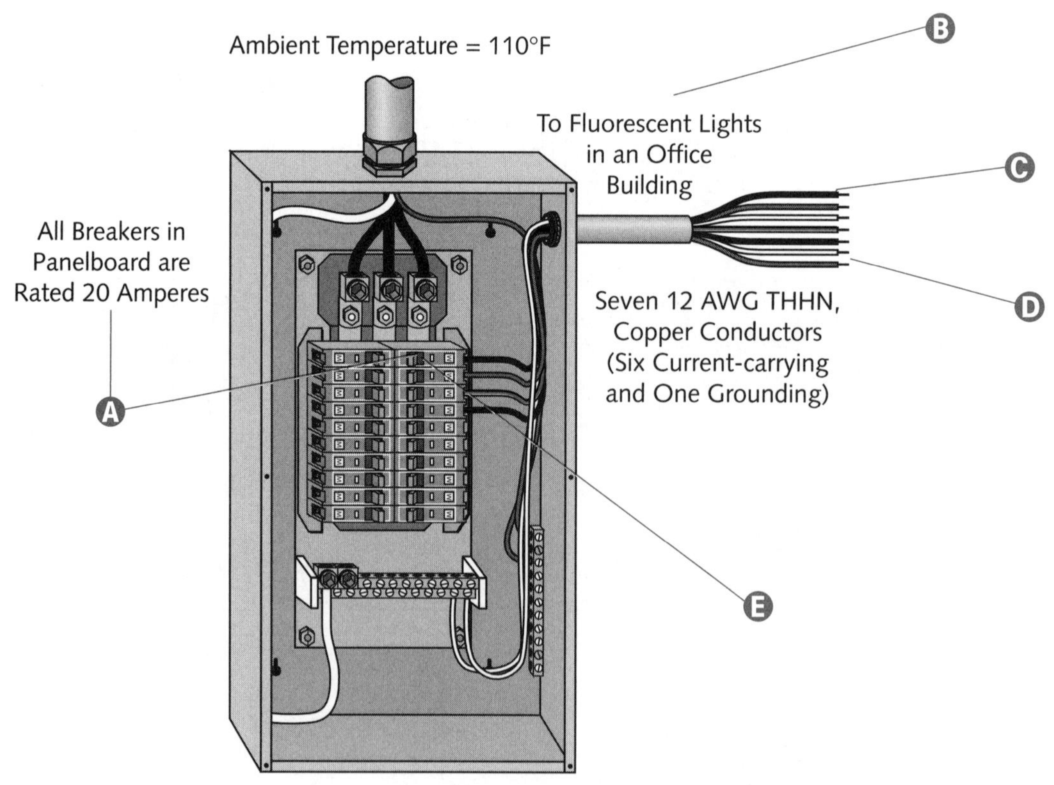

Support Requirements for Conductors in Vertical Raceways

A Conductors in vertical raceways must be supported if the vertical rise exceeds Table 300.19(A) values »300.19(A)«. While three specific methods for supporting cables are listed in 300.19(B)(1), (2), and (3), other equally effective methods are also permitted.

B Clamping devices constructed of, or employing, insulating wedges inserted in raceway ends are one type of conductor support. If clamping the insulation does not adequately support the cable, the conductor must also be clamped »300.19(B)(1)«.

C One cable support must be provided at the top of the vertical raceway, or as close to the top as practical. Intermediate supports must be provided so that supported conductor lengths do not exceed those specified In Table 300.19(A) »300.19(A)«.

D Covered boxes can be inserted at the required intervals in which insulating supports are installed and secured in a satisfactory manner to withstand the weight of the conductors attached thereto »300.19(B)(2)«.

E Cables can be supported in junction boxes, by deflecting the cables at least 90° and carrying them horizontally to a distance not less than twice the diameter of the cable, provided the cables are carried on at least two insulating support and are secured thereto by tie wires (if desired) »300.19(B)(3)«.

F In using this, cables must be supported at intervals not greater than 20% of the values in Table 300.19(A) »300.19(B)(3)«.

Table 300.19(A)
Maximum Distance between Conductor Supports in Vertical Raceways (in ft)

AWG or Circular-Mil Size of Conductor	*Aluminum or Copper-Clad Aluminum*	*Copper*
18 AWG through 8 AWG	100	100
6 AWG through 1/0 AWG	200	100
2/0 AWG through 4/0 AWG	180	80
250 through 350 kcmil	135	60
400 through 500 kcmil	120	50
600 through 750 kcmil	95	40
Over 750 kcmil	85	35

C

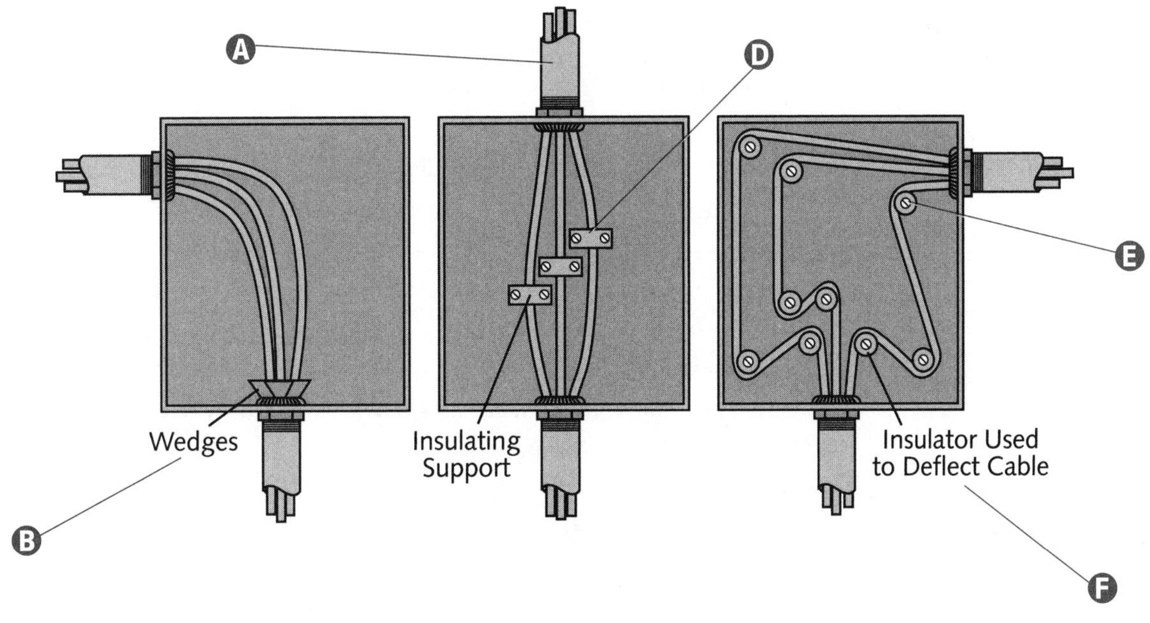

Summary

- Raceways cannot be supported by ceiling grid support wires, but can be supported by independent (additional) support wires.
- Raceways cannot support other raceways, cables, conductors, or non-electric equipment.
- The maximum total of conduit bends, between pull points, is four quarter-bends (360°).
- RMC is a heavy-walled metal raceway that can be threaded.
- IMC has thinner walls than does rigid, and it can also be threaded.
- Unless made up with threaded couplings, RMC and IMC support requirements are the same as EMT.
- EMT, the thinnest walled classification of metal non-flexible raceways, provides protection from all but severe physical damage.
- RNC is commonly referred to as PVC.
- The distance between RNC supports increases as conduit size increases.
- Certain raceways, permitted as luminaire (fixture) whips, do not require support if installed in lengths of six ft or less.
- Wireways are troughs with hinged or removable covers for housing and protecting conductors and cables.
- Wireways can pass transversely through walls if the length passing through the wall is unbroken and if conductors can be accessed on both sides of the wall.
- Raceway fill can be calculated from Tables 4 and 5, located in Chapter 9.
- Conduit (and tubing) nipples can be filled to 60% of their cross-sectional area.
- Certain conductor properties are listed in Table 310.13.
- Conductor temperature limitations must be considered when determining overcurrent protection.
- Ambient temperature, the number of current-carrying conductors, and continuous loads can alter maximum conductor ampacity or overcurrent protection.

Unit 5 Competency Test

NEC® **Reference** **Answer**

1. Where exposed to the weather, raceways shall be:
   a) rainproof and arranged to drain.
   b) watertight and arranged to drain.
   c) weatherproof and arranged to drain.
   d) raintight and arranged to drain.
2. Most conduit and tubing type raceways have a maximum equivalent of four _____ bends between pull points, e.g., conduit bodies and boxes.
3. The cross-sectional area of 2" EMT is _____ square in.
4. All cut ends of conduits shall be reamed or otherwise finished to remove rough edges. Where conduit is threaded in the field, a standard cutting die with a _____ -in. taper per ft (1 in 16) shall be used.
5. Multioutlet assemblies shall not be installed:
   I. in hoistways
   II. within dry partitions
   III. within dry locations
   a) I only b) III only c) I and II only d) I, II, and III
6. Splices and taps shall be permitted within a wireway provided they are accessible. The conductors, including splices and taps, shall not fill the wireway to more than _____ of its area at that point.

***NEC*® Reference**	**Answer**

7. A _____ shall be defined as a transverse raceway for electric conductors, providing access to predetermined cells of a cellular metal floor, thereby permitting the installation of electric conductors from a distribution center to the cells.
8. In both exposed and concealed locations, where a cable or raceway-type wiring method is installed through bored holes in joist, rafters, or wood members, holes shall be bored so that the edge of the hole is not less than _____ in(es). from the nearest edge of the wood member.
9. How many 14 AWG THHN conductors, including an equipment grounding conductor, can be installed in a ⅜" flexible metal conduit using inside fittings?
10. What is the maximum distance (in ft) between supports for a straight run of 2" IMC made up with threaded couplings?
11. The sum of cross-sectional areas of all contained conductors at any cross section of a wireway shall not exceed _____ fill.
12. Where conductors carrying alternating current are installed in metal raceways, they shall be so arranged as to avoid heating the surrounding metal by _____.
13. A _____ is a type of surface, flush, or freestanding raceway designed to hold conductors and receptacles, assembled in the field or at the factory.
14. Where practicable, contact between dissimilar metals anywhere in the system shall be avoided to eliminate the possibility of _____.
15. What is the maximum number of 4 AWG THHN conductors permitted in 1½" rigid metal conduit?
16. Where the same 4 AWG conductors enter and then exit a wireway through conduit or tubing, the distance between those raceway entries shall not be less than _____ times the trade diameter of the larger raceway.
17. All components of an exposed wiring system, installed where walls are frequently washed, must be mounted with a minimum of _____ in(es). airspace between components (boxes, fittings, conduits, etc.) and the wall (supporting surface).
18. Where an equipment bonding jumper is installed on the outside of a raceway, the length shall not exceed _____ ft and shall be routed with the raceway or enclosure.
19. Connections from headers to cabinets and other enclosures, in cellular concrete floor raceways, shall be made by means of:

    I. listed metal raceways

    II. listed nonmetallic raceways

    III. listed fittings

    a) I only b) I and II only c) I and III only d) I, II, and III
20. ENT shall be secured at least every _____ ft.
21. IMC encased in a concrete slab, on the third floor of an office building, is considered a _____ location.
22. When an outlet from an underfloor raceway is discontinued, the sections of circuit conductors supplying the outlet:
    a) shall be permitted to be spliced
    b) shall be removed from the raceway.
    c) shall be permitted to be reinsulated.
    d) shall be handled like abandoned outlets on loop wiring.

***NEC*® Reference**	**Answer**	
______	______	23. RMC, elbows, couplings, and fittings shall be permitted to be installed in concrete, in direct contact with the earth, or in areas subject to severe corrosive influences where protected by ______ and judged suitable for the condition.
______	______	24. Where installed in raceways, conductors of size ______ and larger shall be stranded.
______	______	25. 3/0 AWG aluminum conductors in vertical raceways must be supported if the vertical rise is greater than ______ ft.
______	______	26. Where raceways containing ungrounded conductors ______ or larger enter a cabinet, box enclosure, or raceway, the conductors shall be protected by a substantial fitting providing a smoothly rounded insulating surface, unless the conductors are separated from the fitting or raceway by substantial insulating materials securely fastened in place.
______	______	27. Nonmetallic wireways shall be supported where run horizontally at intervals not to exceed ______ ft, and at each end or joint, unless listed for other support intervals.
______	______	28. Luminaires (fixtures) shall not be used as a raceway for circuit conductors unless ______ for use as a raceway.
______	______	29. If a 1¼" EMT raceway containing three conductors is already filled to 20%, what is the raceway's cross-sectional area?
______	______	30. Expansion fittings and telescoping sections of metal raceways shall be made electrically continuous by ______ or other means.
______	______	31. RMC shall be securely fastened within ______ ft of each outlet box, junction box, device box, cabinet, conduit body, or other conduit termination.
______	______	32. The maximum distance between supports for 4" RNC is ______ ft.
______	______	33. When calculating the maximum number of conductors permitted in a conduit or tubing, all of the same size (total cross-sectional area including insulation), the next-higher whole number shall be used to determine the maximum number of conductors permitted when the calculation results in a decimal of ______ or larger.
______	______	34. Where insulated conductors are deflected within a wireway, either at the ends or where conduits, fittings, or other raceways or cables enter or leave the wireway, or where the direction of the wireway is deflected greater than ______, dimensions corresponding to 312.6(A) shall apply.

section 2

ONE-FAMILY DWELLINGS

UNIT 6

General Provisions

Objectives

After studying this unit, the student should:

- be able to calculate the minimum number of 15- and 20-ampere branch-circuits in a one-family dwelling.
- know the requirements for single receptacles on individual branch-circuits.
- understand the branch-circuit ratings allowed for general-purpose receptacles.
- know how to layout general-purpose receptacles in a dwelling.
- know the receptacle ratings allowed on various size branch-circuits.
- understand the requirements for receptacle boxes.
- have a good understanding of split-wire receptacles.
- know the requirements for wet bar receptacles.
- be familiar with the provisions concerning the minimum length of free conductors inside boxes.
- know the receptacle replacement requirements.
- understand the general requirements for lighting and switches.
- be familiar with the use of a white conductor as an ungrounded conductor.
- understand general requirements for devices and luminaire (fixture) boxes.
- know the provisions for outdoor receptacles.
- be familiar with the provisions concerning illuminating outdoor entrances and exits.
- understand the requirements for attaching receptacles and/or lighting to vegetation (such as trees).

Introduction

Unit 6 contains provisions concerning general areas, both inside and outside, of one-family dwellings. A one-family dwelling is a building consisting solely of one dwelling unit » Article 100«. Learning codes that pertain to one-family dwellings is of great importance, because these codes are the foundation for all residential wiring. One-family dwellings are built in all shapes and sizes. Some general codes apply to every dwelling, such as the placement of receptacles in habitable rooms. Other codes permit the installation of nonmetallic-sheathed cable in any level one-family (or two-family) dwelling, but prohibit its installation in multifamily dwellings exceeding three floors above grade. For this reason, wiring methods are explained in separate units. For information on codes pertaining to nonmetallic-sheathed cable, refer to Unit 4.

ELECTRICAL FLOOR PLAN (BLUEPRINT)

Electrical blueprints (floor plans) may be optional on a one-family dwelling. Sometimes it is the responsibility of the electrician to design and install the electrical system. An electrical floor plan is shown below. The floor plan shows receptacles, lighting, switches, and other items which might be found on a residential plan. The NEC® references (shown here in blue), are, of course, not present on general plans.

This chapter covers everything in a one-family dwelling from a single receptacle in an individual branch-circuit to illumination of entrances and exits. Branch-circuits will be explained briefly in this unit and in greater detail in later units. Receptacles have both general and specific codes. General codes are explained in this unit, while specific codes for rooms and areas with special receptacle requirements are explained in Unit 7. This unit also includes general requirements for lighting and switches. Specific requirements can be found in Unit 7.

> **CAUTION** *Some state and local jurisdictions may not allow the use of nonmetallic-sheathed cable in any dwelling. Obtain a copy of any additional rules and regulations for your area.*

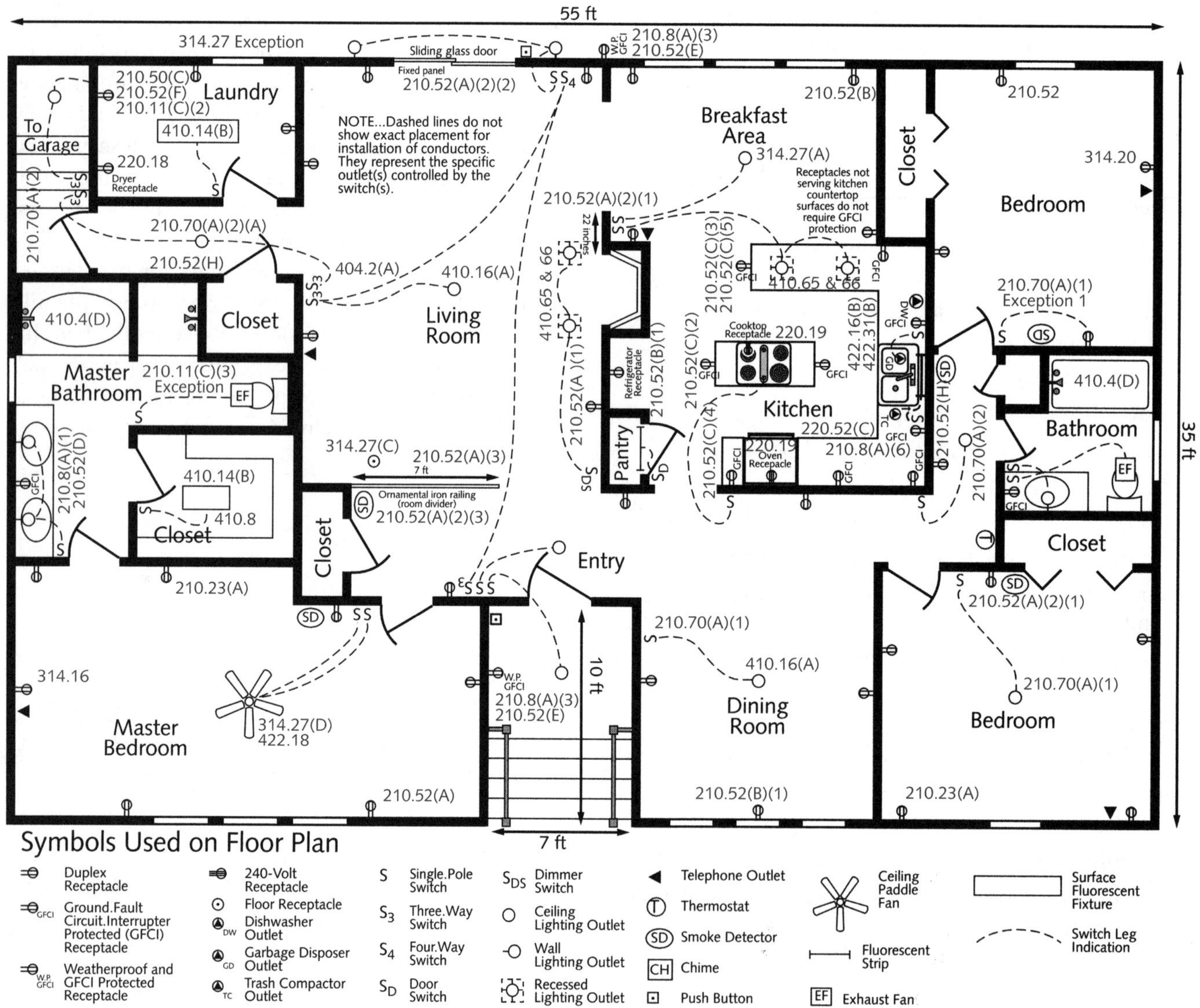

BRANCH-CIRCUITS

General Purpose Branch-Circuit

A general purpose branch-circuit may feed lights, receptacles, or both.

Ⓐ General purpose branch-circuit overcurrent protection shall not exceed 15 amperes for 14 AWG copper; 20 amperes for 12 AWG copper; and 30 amperes for 10 AWG copper »240.4(D)«.

Ⓑ Branch-circuit conductors shall have an ampacity not less than the maximum load served »210.19(A)(1)«.

Required Branch-Circuits

Ⓐ A minimum of two 20-ampere small appliance branch-circuits are required »210.11(C)(1) and 210.52(B)(1)«.

Ⓑ At least one 20-ampere laundry branch-circuit is required »210.11(C)(2) and 210.52(F)«.

Ⓒ A minimum of one 20-ampere bathroom branch-circuit is required »210.11(C)(3) and 210.52(D)«.

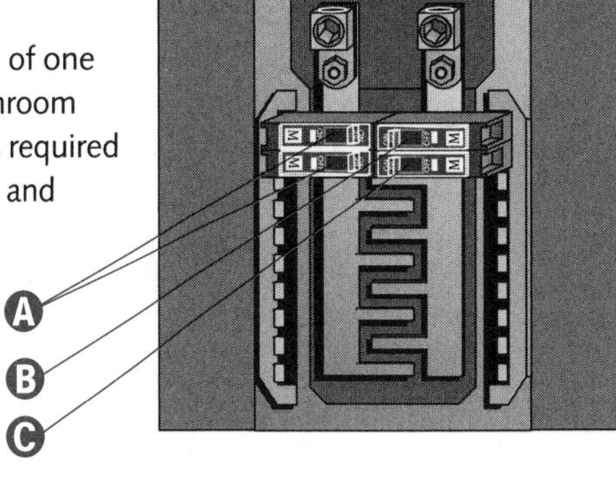

Calculating 15-Ampere Branch-Circuits

Ⓐ Calculation to find the minimum number of 15-ampere branch-circuits:

- Calculate the general lighting and receptacle load by using Table 220.3(A).
- First, multiply 1400 sq ft by 3 volt-amperes.
  1400 × 3 = 4200 volt-amperes
- Next, divide the total volt-amperes by 120 volts to find the total amperes.
  4200 ÷ 120 = 35 amperes
- Finally, divide 35 amperes by 15 (for a 15-ampere circuit) to find the minimum number of circuits. 35 ÷ 15 = 2.33 (Round up to the next whole number if the result of the calculation is not a whole number.)
- If 15-ampere circuits are installed for general lighting and receptacles, then at least three circuits are required.

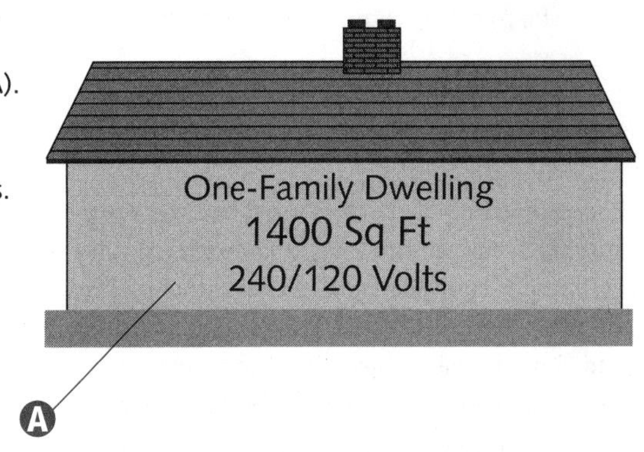

Calculating 20-Ampere Branch-Circuits

Ⓐ Calculation to find the minimum number of 20-ampere branch-circuits:

- Calculate the general lighting and receptacle load by using Table 220.3(A).
- First, multiply 1400 sq ft by 3 volt-amperes.
  1400 × 3 = 4200 volt-amperes
- Next, divide the total volt-amperes by 120 volts to find the total amperes. 4200 ÷ 120 = 35 amperes
- Finally, divide 35 amperes by 20 (for a 20-ampere circuit) to find the minimum number of circuits. 35 ÷ 20 = 1.75 (Round up to the next whole number if the result of the calculation is not a whole number.)
- If 20-ampere circuits are installed for general lighting and receptacles, then at least two circuits are required.

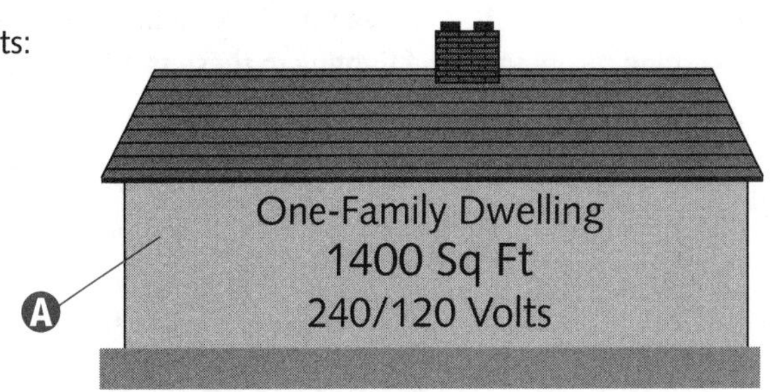

NOTE

At least four 20-ampere branch-circuits are required (in addition to any installed for general lighting and receptacles): two small appliance branch-circuits, one laundry branch circuit, and one bathroom branch-circuit.

Individual Branch-Circuit for a Water Heater

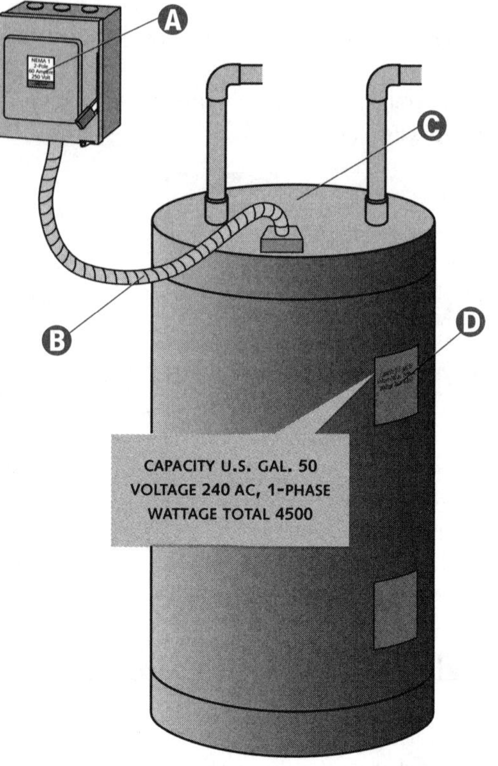

A Calculation to find the maximum overcurrent protective device:

- The overcurrent protection shall not exceed the rating marked on the appliance. If the rating is not marked, the overcurrent protection shall not exceed 150% of the rated current »422.11(E)(3)«. 18.75 × 150% = 28.13
- If the calculated rating does not correspond to a standard size fuse or breaker, as found in 240.6(A), the next higher standard rating shall be permitted »422.11(E)(3)«. The next standard size fuse or breaker higher than the calculated rating of 28.13 amperes is 30 amperes.
- The maximum overcurrent protective device (fuse or breaker) is 30 amperes.

B The minimum conductor size is 10 AWG copper »Table 310.16«. 12 AWG conductors are not permitted, because the overcurrent protection shall not exceed 20 amperes for 12 AWG copper, which is insufficient for this application »240.4(D)«.

C A branch-circuit supplying a fixed storage-type water heater (120-gallon capacity or less) shall have a rating not less than 125% of the water heater nameplate rating »422.13«.

D Calculation to find the minimum circuit ampacity:

- First, divide the wattage by the voltage to find amperage. 4500 ÷ 240 = 18.75
- Next, multiply the amperage by 125% as required by 422.13. 18.75 × 125% = 23.44
- The minimum ampacity required for this water heater is 23.44 amperes.

Ampere Rating of Receptacles

Receptacle branch-circuits must have a rating of 20 amperes for laundry areas, bathrooms, kitchens (except for refrigeration equipment supplied from an individual branch-circuit rated 15 amperes or greater), dining rooms, pantries, breakfast rooms, and similar areas. Throughout the remainder of a one-family dwelling, 15-ampere receptacle branch-circuits are allowed.

Lighting and receptacles can share the same branch-circuit, except for small-appliance receptacles, bathroom receptacles (unless the circuit supplies a single bathroom), and laundry receptacles »210.23(A)«. See Unit 7 for additional information concerning receptacles and lighting in these specific areas.

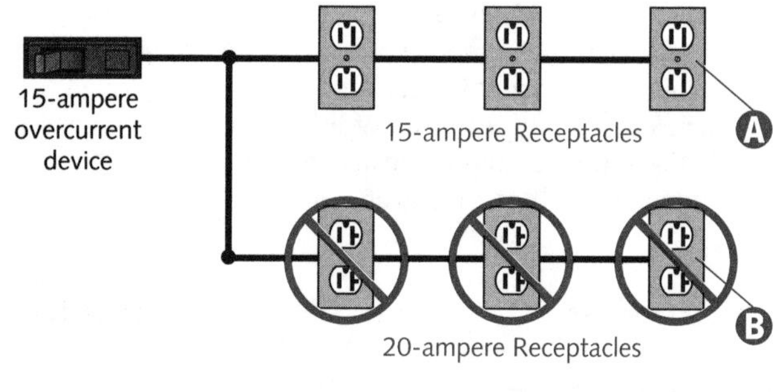

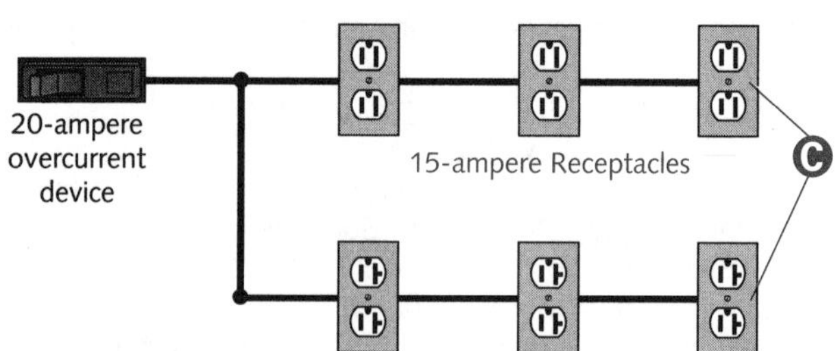

A Receptacles installed on 15-ampere circuits must have an ampere rating of not more than 15 amperes »Table 210.21(B)(3)«.

B A 20-ampere receptacle cannot be installed on a 15-ampere circuit.

C Receptacles installed on 20-ampere circuits can have an ampere rating of either 15 or 20 »Table 210.21(B)(3)«.

NOTE

Branch-circuits for receptacles cannot be smaller than 14 AWG »210.19(D)«. Generally, 15-ampere circuits require 14 AWG, and 20-ampere circuits require 12 AWG, copper conductors. Larger conductors may be needed to compensate for voltage drop, ambient temperature, or where more than three current-carrying conductors exist in a raceway or cable.

A Single Receptacle on an Individual Branch-Circuit

A A single receptacle installed on a branch-circuit with no other device or outlet, shall have an ampere rating equal to the rating of that circuit »210.21(B)(1)«.

B A duplex receptacle is not a single receptacle. As defined in Article 100, a single receptacle is a single contact device with no other contact device on the some yoke.

C Receptacles on 20-ampere circuits can have a rating of either 15 or 20 amperes »Table 210.21(B)(3)«.

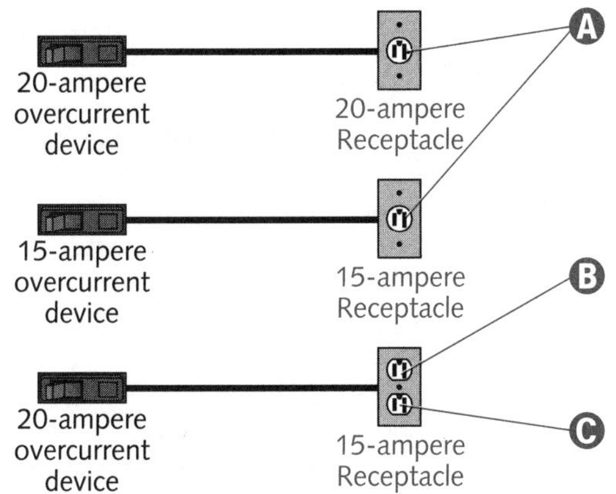

RECEPTACLES

General Receptacle Placement

Wall space determines the minimum number of receptacles, in a given dwelling. Receptacle outlets shall be installed in kitchens, family rooms, dining rooms, living rooms, parlors, libraries, dens, sun rooms, bedrooms, recreation rooms, or similar rooms or areas in accordance with the provisions specified in 210.52(A).

A Receptacles shall be installed so that no point measured horizontally along the floor line in any wall space is more than 6 ft (1.8 m), from a receptacle outlet »210.52(A)(1)«.

B The maximum distance between receptacles is 12 ft (3.6 m).

NOTE

Dwelling unit receptacle outlet general provisions (210.52) apply to any part of a basement containing habitable rooms, such as a den, recreational room, etc. »210.52(G)«.

Wall Spaces 2 Ft in Width

A A receptacle is required for any wall space 2 ft (600 mm) or more in width (including space measured around corners), unbroken along the floor line by doorways, fireplaces, and similar openings »210.52(A)(2)(1)«.

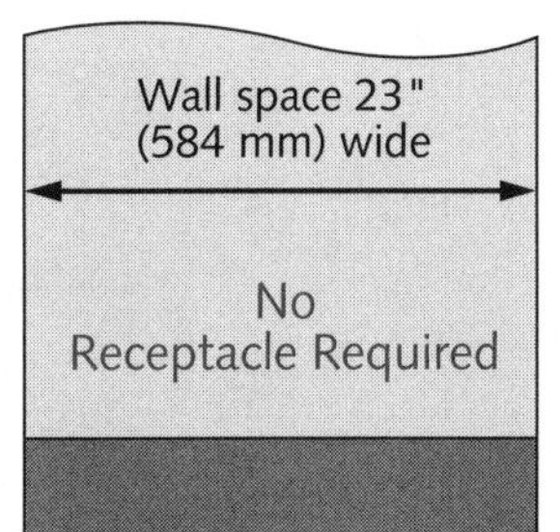

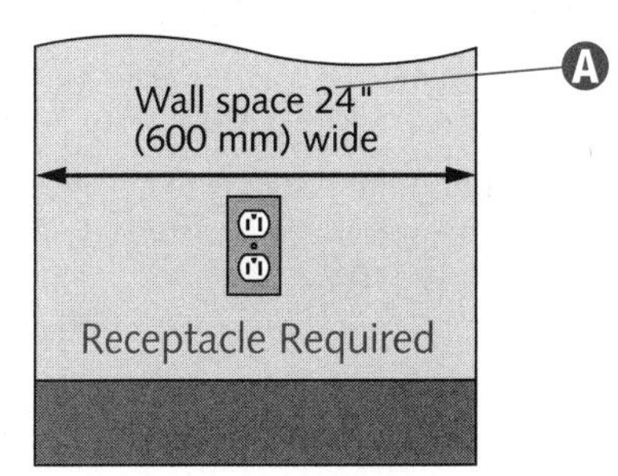

Maximum Distance to a Receptacle

A An easy way to understand the placement of dwelling receptacles is to imagine having a floor lamp with a 6-ft cord. Anywhere this lamp is placed around the wall, there should be a receptacle within reach, without using as extension cord. Even when placed beside a door opening, an outlet should be within reach. If the lamp is placed next to a wall that is at least 24 in. wide, an outlet should be available within that wall space.

B The maximum distance to any receptacle, along the floor line, measured horizontally, shall be 6 ft (1.8 m) »210.52(A)(1)«.

WARNING

All branch-circuits that supply 125-volt, single-phase, 15- and 20-ampere outlets located in dwelling-unit bedrooms must be protected by an arc-fault circuit interrupter listed to provide protection of the entire branch-circuit. »210.12(B)«.

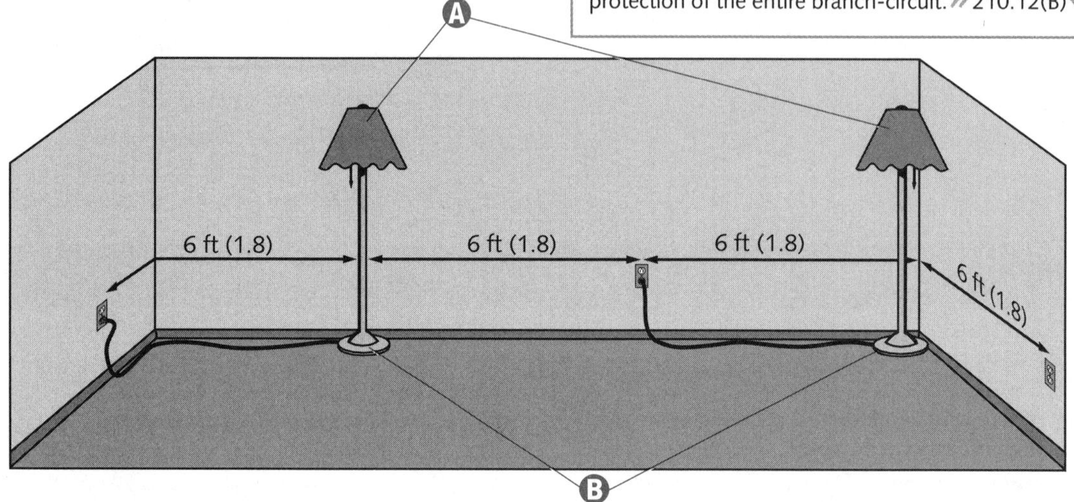

Space Measured Around Corners

A The maximum distance **between receptacles**, along the floor line, measured horizontally, shall be 12 ft (3.6 m) »210.52(A)(1)«.

B Wall space includes space measured around corners »210.52(A)(2)(1)«.

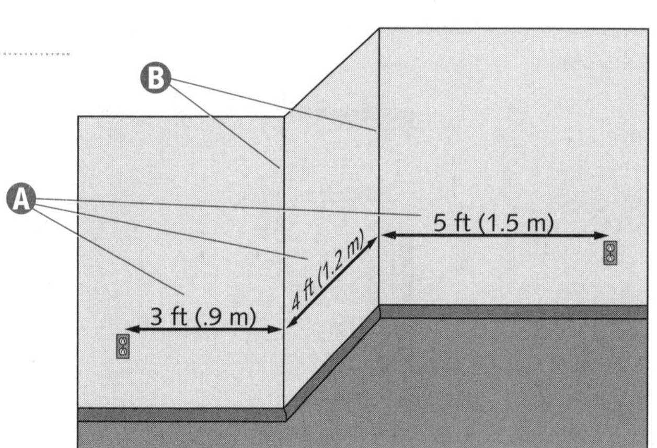

Fixed Panels

A Glass door fixed panels in exterior walls are counted as wall space »210.52(A)(2)(2)«.

B Sliding panels in exterior walls are not counted as wall space »210.52(A)(2)(2)«.

C The maximum distance to any receptacle, along the floor line, measured horizontally, shall be 6 ft (1.8 m) »210.52(A)(1)«.

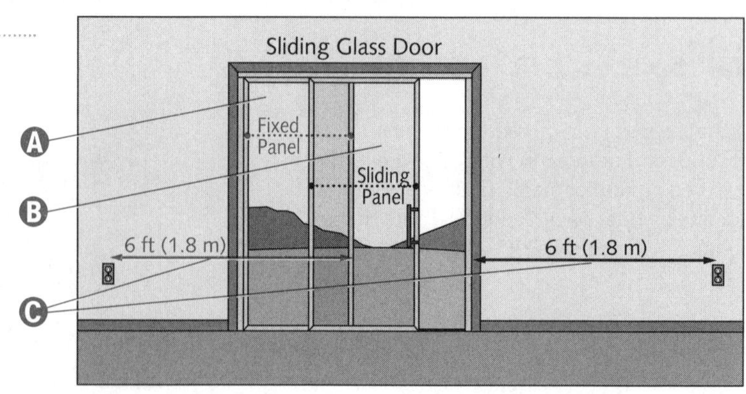

LIGHTING AND SWITCHING

Outdoor Entrances and Exits

A At least one wall-switch-controlled lighting outlet shall be installed in every habitable room of a dwelling »210.70(A)(1)«.

B At least one wall-switch-controlled lighting outlet is required to provide exterior side illumination of outdoor entrances or exits having grade level access »210.70(A)(2)(b)«.

NOTE

Remote, central, or automatic lighting control is permitted for outdoor entrances »210.70(A)(2) Exception«.

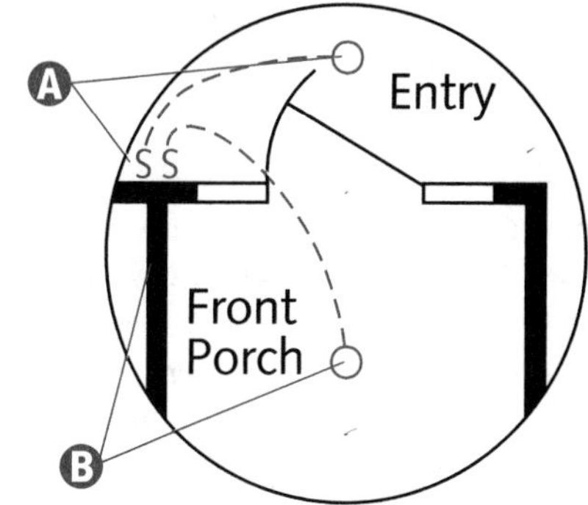

General Lighting

All switches shall be located so that they may be operated from a readily accessible place, not more than 6 ft, 7 in. (2.0 m), from the floor to the center of the switch operating handle »404.80(A)«. No other height requirements are stipulated for wall switches.

A Dwelling hallways require a minimum of one wall switch-controlled lighting outlet »210.70(A)(2)(a)«.

B At least one wall-switch-controlled lighting outlet shall be installed in every habitable room of a dwelling »210.70(A)(1)«.

C Except for kitchens and bathrooms, every habitable room is permitted one or more wall-switch-controlled receptacle(s) in lieu of a lighting outlet »210.70(A)(1) *Exception No.1*«. Switch-controlled receptacles allow for cord-connected lighting.

D At least one wall switch-controlled lighting outlet shall be installed in every bathroom of a dwelling »210.70(A)(1)«.

NOTE

Occupancy sensors are permitted to control lighting outlets in lieu of wall switches »210.70(A)(1) Exception No. 2«. One of two requirements must be met in order to use occupancy sensors: either a regular wall switch must be present in addition to the sensor, or the sensor must be equipped with a manual override that allows the sensor to function as a switch. Either type of wall switch shall be located at a customary location.

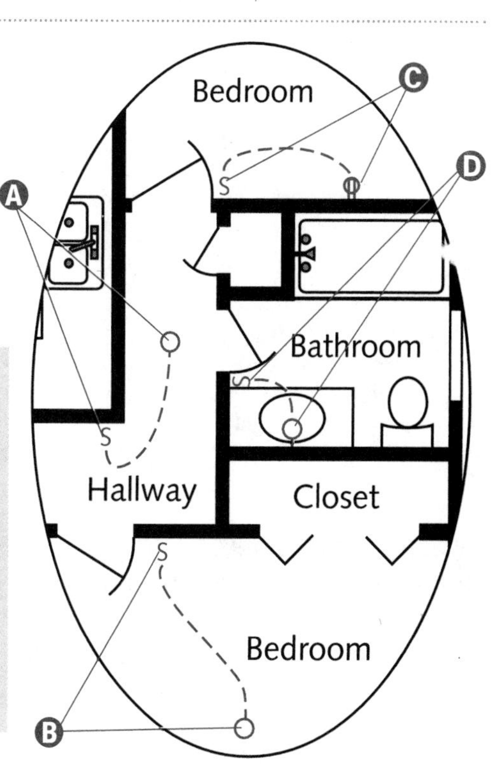

Ceiling-Suspended (Paddle) Fans

A A listed box is required for a ceiling-suspended (paddle) fan where the box is the only support for the fan »314.27(D)«. Those boxes are permitted to support ceiling fans weighing no more than 35 pounds (16 kg) »422.18(A)«. Include any light kit or accessory when determining the total weight of the fan.

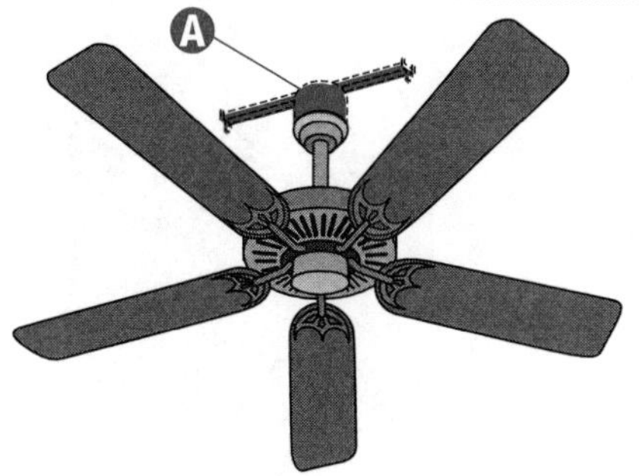

NOTE

Ceiling-suspended (paddle) fans exceeding 35 pounds (16 kg) shall be supported independently of the outlet box »422.18(B)«.

Cables In talled to Feed Switches

A Metal switch bo es shall be effectively grounded »404.12«.

B A white conductor, in cable, can be used as an ungrounded conductor whe supplying power to a switch, but not as a return conductor f m the switch to the switched outlet. The conductor s l be permanently re-identified to indicate its use by pain or other effective means, at its terminations, and at eac cation where the conductor is visible and accessible »200. C)(2)«. The re-identified conductor can be any color, exc t white, gray, or green.

C Switches, including dimmer switches, shall be effectively grounded by one of the two methods fou n 404.9(B): (1) the switch is mounted with metal screws t metal box or to a nonmetallic box with integral means for grounding devices, or (2) an equipment grounding conductor or equipment bonding jumper is connected to the snap switch's equipment grounding termination »404.9(B)«.

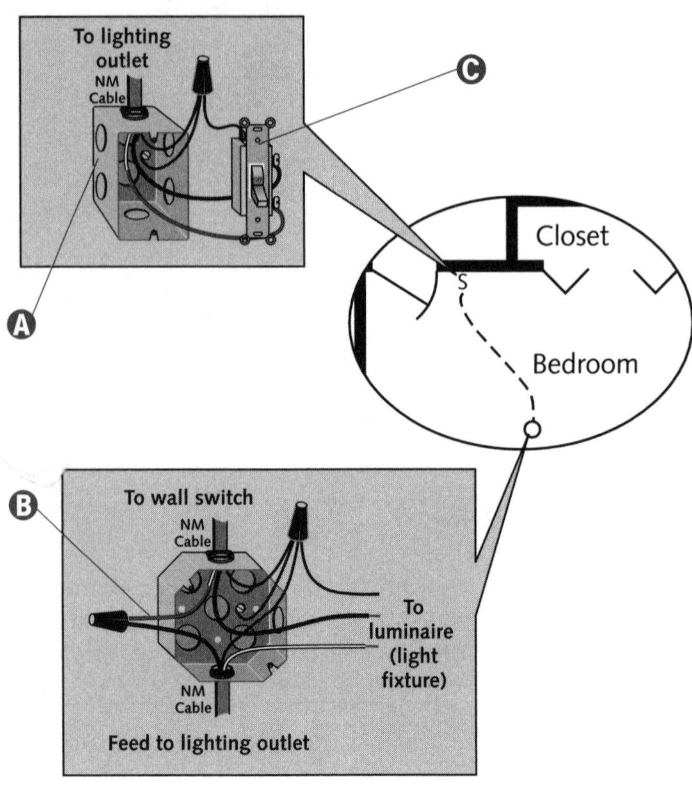

NOTE

An outlet box can support a luminaire (light fixture) weighing no more than 50 pounds (22.7 kg), unless the box is listed for a weight greater than the luminaire (fixture). A separate means for supporting a luminaire (light fixture) (independent of the outlet box) is also permitted »410.16(A)«.

No provision permits the installation of a smaller conductor as a switch loop. Switch loops (legs), that are part of a 20-ampere branch-circuit, cannot be smaller than 12 AWG copper conductors. Table 210.24 requires 12 AWG conductors throughout a 20-ampere branch-circuit, except for taps. Taps and switch loops are not the same.

OUTDOOR RECEPTACLES AND LIGHTING

Required Outdoor Receptacles

A One-family dwellings require at least one receptacle outlet (accessible at grade level and not more than 6 ft, 6 in. (2.0 m) above grade) at the front and back of the dwelling »210.52(E)«.

B *Each* grade level unit of a two-family dwelling requires at least one receptacle outlet (accessible at grade level and not more than 6 ft, 6 in. (2.0 m) above grade) at the front and back of the dwelling »210.52(E)«.

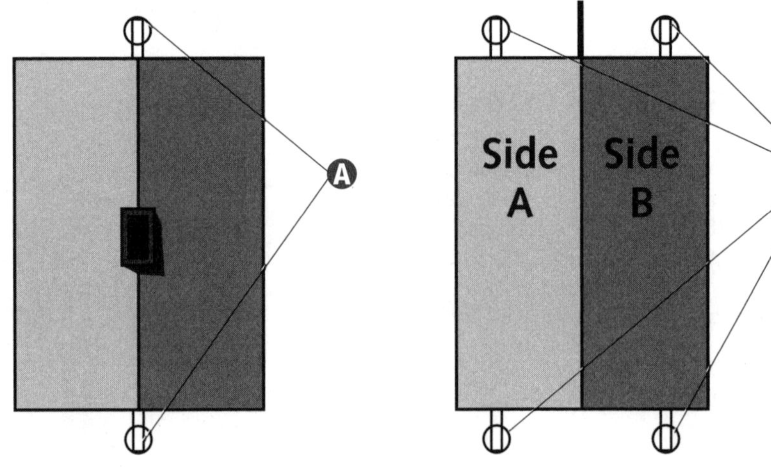

Receptacle Enclosures and Covers

A An outdoor receptacle, whether in a location protected from the weather or in other damp locations, requires a weatherproof receptacle enclosure when the receptacle is covered (attachment plug not inserted and receptacle cover closed). A receptacle location protected from the weather includes areas such as: under roofed, open porches, canopies, marquees, etc., which are not subjected to beating rain or water run-off »406.8(A)«.

B The enclosure for a receptacle installed in a wall flush-mounted outlet box requires the use of a weatherproof faceplate assembly that provides a watertight seal between the plate and the wall »406.8(E)«.

C An installation suitable for wet locations shall also be considered suitable for damp locations »406.8(A)«.

D All 15- and 20-ampere, 125- and 250-volt receptacles installed outdoors in a wet location must have an enclosure that is weatherproof whether the attachment plug is inserted or removed »406.8(B)(1)«.

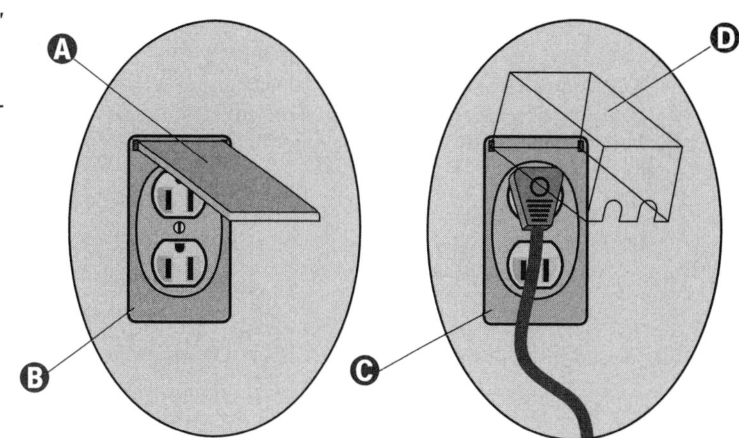

Receptacles More Than 6 Ft 6 In. Above Grade

A At least one receptacle outlet, accessible at grade level and not more than 6 ft, 6 in. (2.0 m) above grade, shall be installed at the front *and* back of the dwelling »210.52(E)«.

B While receptacles are permitted more than 6 ft, 6 in. (2.0 m) above grade, they do not count as required outdoor receptacles.

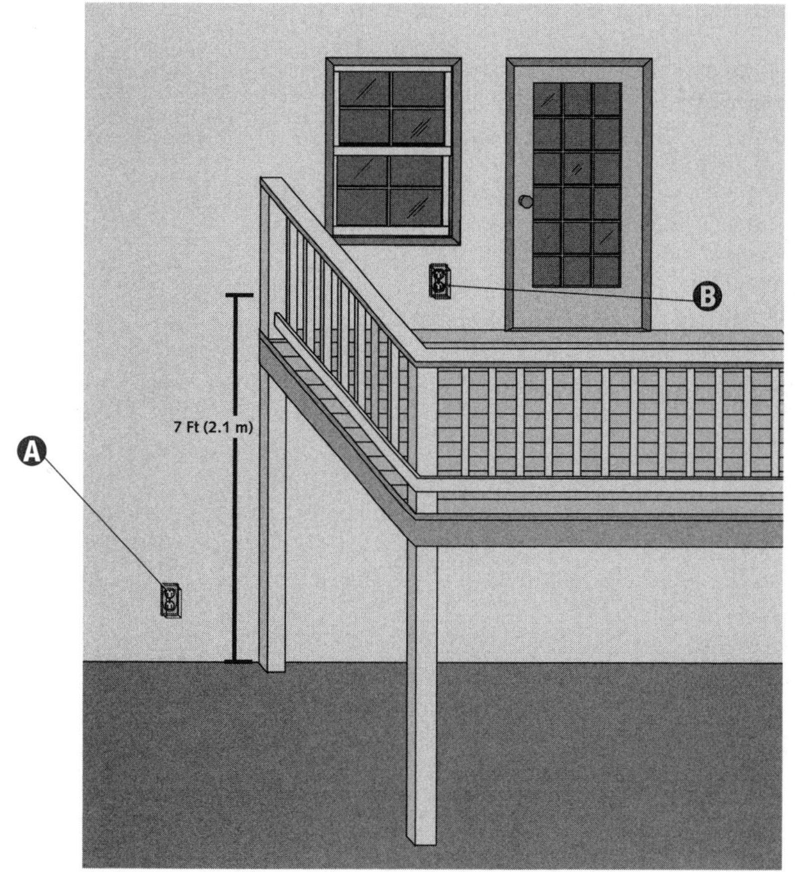

NOTE

GFCI protection is not required for receptacles that are not readily accessible and are supplied by a dedicated branch-circuit for electric snow-melting or de-icing equipment. The installation must comply with Article 426 »210.8(A)(3) Exception«. *(See 426.28 for ground-fault protection of equipment.) Outdoor receptacles that are not readily accessible (such as under an eave) require GFCI protection.*

CAUTION *Outdoor receptacles require GFCI protection »210.8(A)(3)«.*

Support for Receptacle Enclosures

A As a general rule, a single conduit cannot support an enclosure that contains a device »314.23(F)«.

B Rigidly and securely fasten in place all boxes and raceways »300.11(A) and 314.23(A)«.

C Supporting an enclosure from grade requires rigid support in the form of conduit or a brace (metal, polymeric, or wood) »314.23(B)«. Wood brace cross sections must be at least 1 in. by 2 in. (25 mm x 50 mm). Wood braces must be treated to withstand the conditions when used in wet locations »314.23(B)(2)«.

D An enclosure containing devices can be supported by two conduits under the conditions found in »314.23(F)«: (1) The box shall not exceed 100 cubic in.; (2) The box has either threaded entries or hubs identified for the purpose; (3) The box shall be supported by two or more conduits threaded wrenchtight into the enclosure or hubs; (4) Each conduit shall be secured within 18 in. (450 mm) of the enclosure.

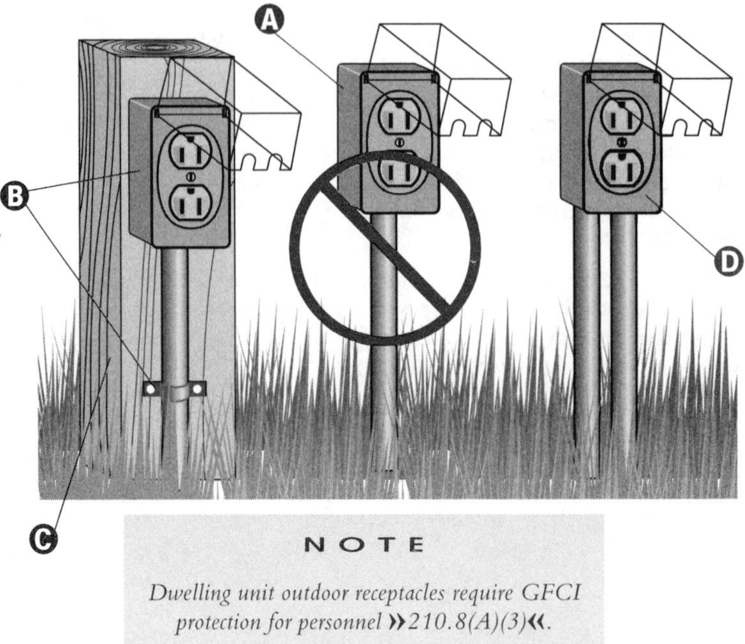

NOTE

Dwelling unit outdoor receptacles require GFCI protection for personnel »210.8(A)(3)«.

Illumination of Entrances and Exits

A Install at least one wall-switch-controlled lighting outlet to provide illumination on the exterior side of outdoor entrances and exits (doorways) »210.70(A)(2)(b)«. It is not required that the luminaire (light fixture) be located adjacent to the doorway, as long as illumination around the doorway is provided.

B Install at least one receptacle outlet at the front and back of the dwelling »210.52(E)«. Outdoor receptacles require GFCI protection »210.8(A)(3)«.

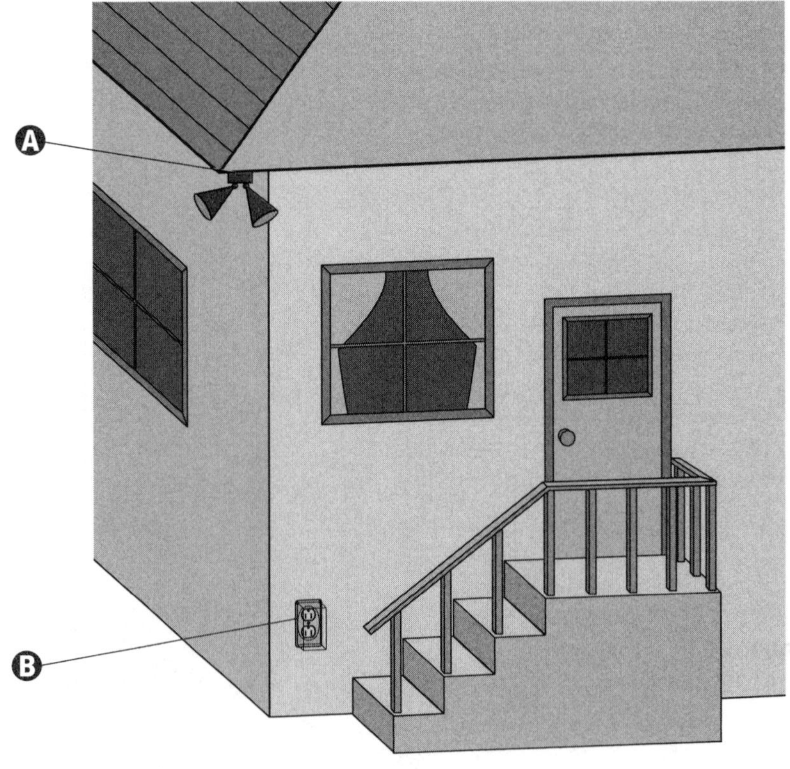

Wiring in Trees

A Outdoor luminaires (lighting fixtures) and associated equipment can be supported by trees »410.16(H)«.

B Vegetation (such as trees) cannot be used to support overhead conductor spans »225.26«.

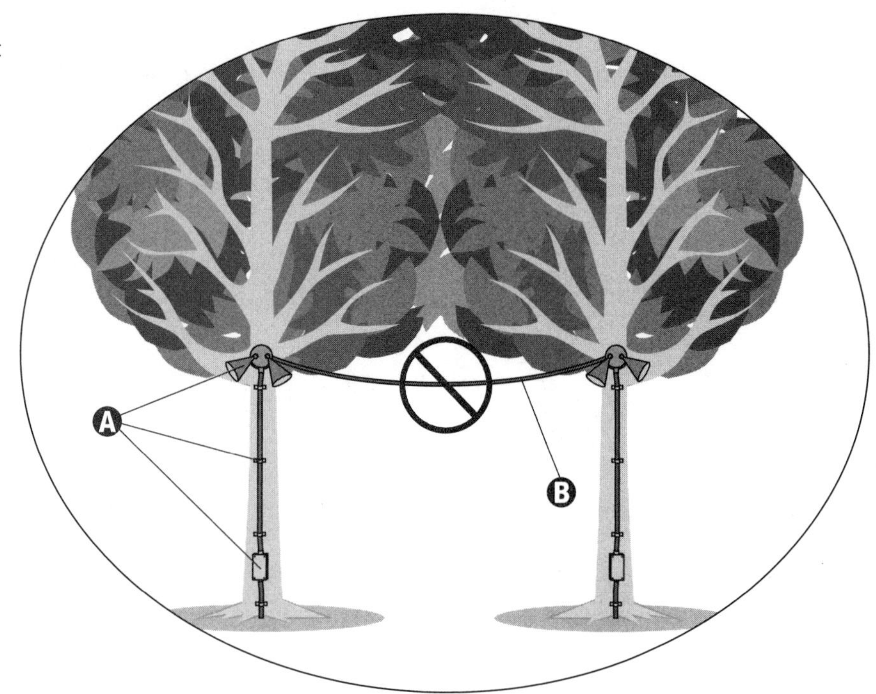

> **NOTE**
>
> *Direct buried conductors and cables, emerging from the ground, shall be protected by enclosures or raceways that extend from the minimum cover distance required by 300.5(A) below grade to a point 8 ft (2.5 m) or more above finished grade »300.5(D)«.*

Summary

- General-purpose branch-circuits may feed lights, receptacles, or any combination thereof.
- An individual branch-circuit feeds only one piece of equipment.
- The maximum distance to any receptacle, measured horizontally along the floor line, is 6 ft (1.8 m).
- A receptacle is required for any wall space 2 ft (600 mm) or wider.
- Twenty-ampere receptacles are not permitted on 15-ampere branch-circuits.
- Both 15- and 20-ampere receptacles are permitted on 20-ampere branch-circuits.
- GFCI-protected receptacles may replace nongrounded-type receptacles where the receptacle enclosure is without a grounding means.
- GFCI-protected receptacles, on nongrounded systems, must be marked in accordance with 406.3(D)(3) provisions.
- A means to simultaneously disconnect all ungrounded conductors must be provided where there are two circuits on the same yoke.
- A switch-controlled lighting outlet shall be installed in every habitable room of a dwelling, including bathrooms, hallways, and stairways.
- Boxes are permitted to support luminaires (light fixtures) weighing 50 pounds (22.7 kg) or less.
- Listed boxes are permitted to support ceiling fans weighing no more than 35 pounds (16 kg).
- One-family dwellings require at least one exterior GFCI protected receptacle outlet at the front and back of the dwelling.
- Outdoor entrances and exits must have at least one wall-switch-controlled lighting outlet.
- Although lighting and receptacles can be supported by trees, trees cannot support overhead conductor spans.

Unit 6 Competency Test

NEC® Reference	Answer	
______	______	1. A 20-ampere rated duplex receptacle may be installed on a _____ branch-circuit. a) 15-ampere b) 20-ampere c) 15- or 20-ampere d) 20- or 25-ampere
______	______	2. The continuity of a grounded conductor shall not be dependent upon the device in _____. a) branch-circuits not having a grounded conductor b) multiwire branch-circuits c) individual branch-circuits d) branch-circuits
______	______	3. Receptacles located _____ ft above the floor are not counted in the required number of receptacles along the wall.
______	______	4. At least _____ in. of free conductor, measured from the point in the box where it emerges from its raceway or cable sheath, shall be left at each outlet, junction, and switch point for splices or the connection of luminaires (fixtures) or devices.
______	______	5. In nonmetallic-sheathed cable, the equipment grounding conductor for 15-, 20-, and 30-ampere branch-circuits _____. a) is required only with aluminum or copper-clad aluminum cable b) must be the same size as the insulated circuit conductors c) may be one size smaller than the insulated circuit conductors d) may be two sizes smaller than the insulated circuit conductors
______	______	6. Receptacles installed for the attachment of portable cords shall be rated at not less than 15 amperes, 125 volts, or _____ amperes, 250 volts, and shall be of a type not suitable for use as lampholders.
______	______	7. A luminaire (fixture) that weighs more than 6 pounds or exceeds _____ in. in any dimension shall not be supported by the screw shell of a lampholder.
______	______	8. Which of the following is not a standard classification for a branch-circuit supplying several loads? a) 20 amperes b) 25 amperes c) 30 amperes d) 50 amperes
______	______	9. A cord connector that is supported by a permanently-installed cord pendant shall be considered a(n) _____ outlet.
______	______	10. Grounding-type receptacles shall be installed only on circuits of the _____ for which they are rated, except as provided in Tables 210.21(B)(2) and (B)(3). I. voltage class II. wattage III. current a) I only b) I and III only c) I and II only d) I, II, and III
______	______	11. Ceiling suspended fans that do not exceed _____ pounds in weight, with or without accessories, shall be permitted to be supported by outlet boxes identified for such use and supported in accordance with 314.23 and 314.27.
______	______	12. Receptacle outlets in floors shall not be counted as part of the required number of receptacle outlets unless located within _____ in. of the wall.

NEC® Reference **Answer**

__________ __________ 13. Luminaires (fixtures) shall be wired with conductors having insulation suitable for the environmental conditions, _____, _____, and _____ to which the conductors will be subjected.

__________ __________ 14. A (luminaire) light fixture that weighs more than 50 pounds shall be supported _____.

__________ __________ 15. One 15-ampere rated duplex receptacle may be installed on a _____ ampere individual branch-circuit.

__________ __________ 16. A branch-circuit that supplies only one utilization equipment is known as a(n) _____ branch-circuit.

__________ __________ 17. The maximum distance between receptacles in a one-family dwelling is _____ ft.

__________ __________ 18. A nongrounding-type receptacle has been replaced with a new grounding-type duplex receptacle. The 15-ampere circuit breaker feeding the circuit has been replaced with a 15-ampere ground-fault circuit-breaker. The new receptacle must be marked _____.

I. "Two-Wire Circuit"

II. "No Equipment Ground"

III. "GFCI Protected"

a) III only b) I and III only c) II and III only d) I, II, and III

__________ __________ 19. An outlet where one or more receptacles are installed is called a(n) _____.

__________ __________ 20. What is the maximum distance that a receptacle wall box can be set back, from the finished surface of a 1/4-in. wood paneling wall?

a) 0.0625 in. b) 0.125 in. c) 0.25 in. d) 0.0 in.

__________ __________ 21. Switch-controlled receptacles (in lieu of a lighting outlet) are permitted in all but which of the following?

a) bedroom b) library c) dining room d) hallway

__________ __________ 22. A 50-ampere rated single receptacle may be installed on a _____ -ampere individual branch-circuit.

__________ __________ 23. A(n) _____ is a system or circuit conductor that is intentionally grounded.

__________ __________ 24. The definition of a branch-circuit is _____.

a) the circuit conductors between the service and the sub-panel

b) the circuit conductors prior to the final overcurrent device protecting the circuit

c) the circuit conductors between the final overcurrent device protecting the circuit and the outlet(s)

d) the circuit conductors between the final overload device protecting the circuit and the outlet(s)

__________ __________ 25. In dwelling units, a duplex receptacle with two circuits on the same yoke must have _____ as a means of disconnect.

I. two single-pole circuit breakers without tie handles

II. one double-pole circuit breaker

III. two single-pole circuit breakers with tie handles

a) I only b) I or II only c) II only d) II or III only

UNIT 7

SECTION TWO: ONE-FAMILY DWELLINGS

Specific Provisions

Objectives

After studying this unit, the student should:

- know the required ampere rating for receptacles and branch-circuits in kitchens, pantries, dining rooms, breakfast areas, and similar areas.
- be familiar with the requirements for kitchen countertop receptacle placement, including islands and peninsulas.
- know the minimum number of circuits required for kitchens, pantries, dining rooms, breakfast areas, and similar areas.
- understand requirements pertaining to permanently connected, as well as cord- and plug-connected, appliances.
- be able to perform load calculations for appliance branch-circuits.
- know the specific provisions for placement of GFCI-protected receptacles.
- understand the requirements for hallway and stairway receptacles and lighting.
- be able to identify which luminaires (fixtures) are, and which are not, permitted in closets, and the placement thereof.
- understand the comprehensive definition of a bathroom and know the requirements for receptacles, lighting, and fans.
- know the requirements for receptacles and lighting in attached garages, detached garages, and basements.
- understand the requirements for laundry (including clothes dryers) receptacles and branch-circuits.
- be familiar with the requirements for attic and crawl space lighting and receptacles.
- understand lighting and receptacle requirements around HVAC equipment.

Introduction

Unit 6 contains provisions concerning general areas, both inside and outside, of one-family dwellings. The material in Unit 7 addresses more complex issues, requiring additional provisions for specific areas (such as kitchens, hallways, clothes closets, bathrooms, garages, basements, etc.). For instance, a kitchen may require the general provisions found in Unit 6, as well as specific provisions found in this unit. A dining room having no countertop surface requires receptacle placement by the general rule, but it must be fed from a 20-ampere branch-circuit. Some areas require only one receptacle outlet, while bathroom receptacle requirements depend on the number and location of sinks. This unit also includes specific requirements for lighting outlets and switches. The pages that follow provide significant information on specific provisions for areas such as kitchens, dining rooms, breakfast areas, hallways, stairways, clothes closets, bathrooms, attached as well as detached garages, basements, attics, and crawl spaces.

KITCHENS, DINING ROOMS, AND BREAKFAST AREAS

Receptacle and Branch-Circuit Rating

Receptacle placement in a kitchen is determined by 210.52(A) and (C). All of the general receptacle placement provisions found in 210.52(A) apply, as well as additional requirements found in 210.52(C) for countertops.

Ⓐ Receptacles installed on the 20-ampere small appliance circuits may be rated either 15 or 20 amperes »Table 210.2(B)(3)«.

All receptacles installed to serve kitchen countertops shall have GFCI protection »210.8(A)(6)«.

Ⓑ Twenty-four in. (600 mm) is the maximum distance to any receptacle, measured horizontally along the wall line »210.52(C)(1)«. An easy way to understand countertop receptacle placement is to imagine a toaster with a 24-in. cord. Anywhere this toaster is placed around the countertop wall, there should be a receptacle within reach.

> **NOTE**
>
> *In kitchens, pantries, breakfast rooms, and dining areas, circuits that serve countertop surface receptacles and general purpose receptacles shall be 20-ampere small appliance branch-circuits »210.52(B)(1)«.*

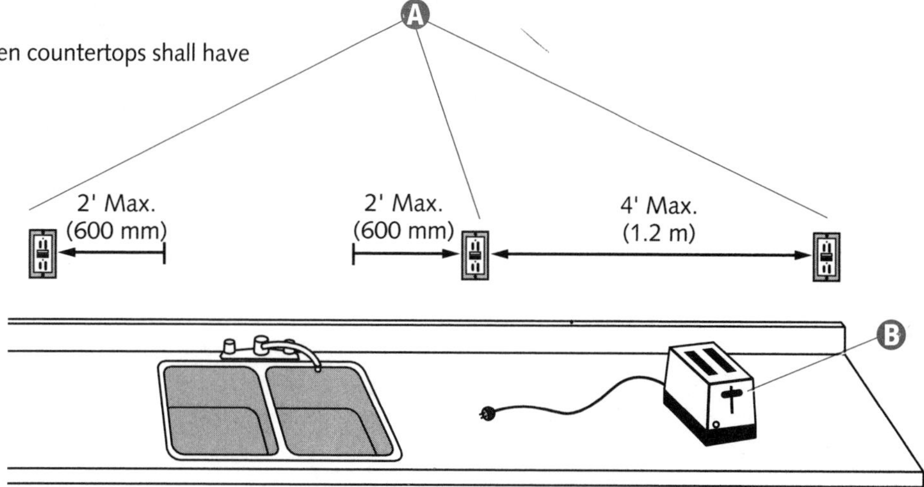

Other Outlets Fed from Small Appliance Branch-Circuits

Ⓐ A dedicated clock outlet, installed to provide power and support, is permitted on a small appliance branch-circuit «210.52(B)(2) *Exception No.1*«.

Ⓑ A hood fan is not permitted on a small appliance branch-circuit »210.52(B)(2)«.

Ⓒ Outdoor receptacles are not permitted on small appliance branch-circuits »210.52(B)(2)«.

Ⓓ A lighting outlet is not permitted on a small appliance branch-circuit »210.52(B)(2)«.

Ⓔ Circuits feeding receptacles in kitchens, pantries, breakfast rooms, dining rooms, or similar areas shall not feed receptacles outside of these areas »210.52(B)(2)«.

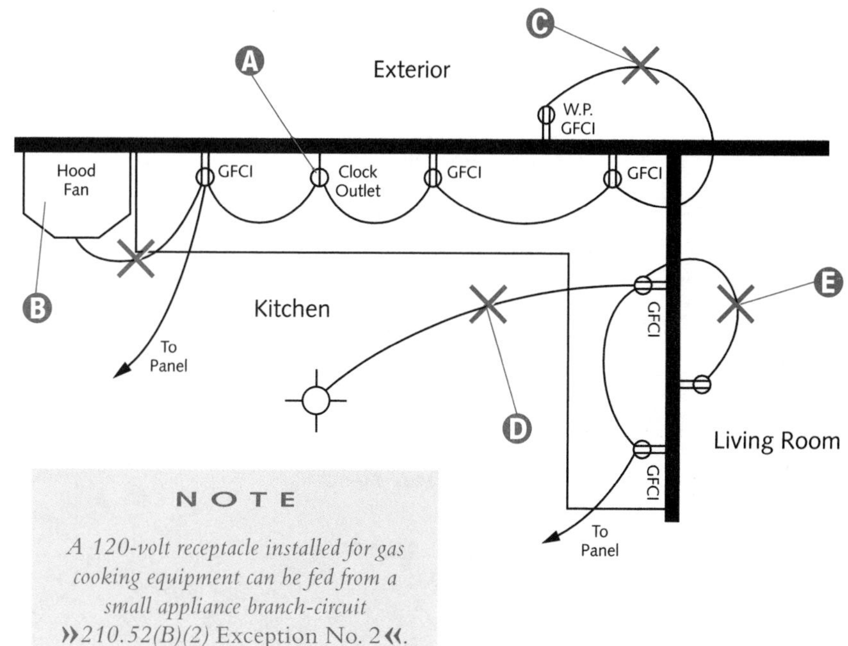

> **NOTE**
>
> *A 120-volt receptacle installed for gas cooking equipment can be fed from a small appliance branch-circuit »210.52(B)(2)* Exception No. 2«.

Kitchen Countertop Receptacle Placement

All **receptacles installed to serve kitchen countertop surfaces shall have GFCI protection »210.8(A)(6)«.**

Branch-circuits in kitchens shall be 20-ampere circuits »210.52(B)(1)«.

A minimum of two small appliance branch-circuits are required for receptacles serving kitchen countertops »210.52(B)(3)«. Either, or both, of these small appliance branch-circuits may feed other receptacles in the kitchen, pantry, breakfast room, and/or dining area.

Receptacles located more than 20 in. (500 mm) above countertops are allowed but do not count as required receptacles »210.52(C)(5)«.

Receptacles inside appliance garages are permitted, but do not count as required countertop receptacles »210.52(C)(5)«.

Ⓐ Twenty-four in. (600 mm) is the maximum distance to any receptacle, along the wall line, measured horizontally »210.52(C)(1)«.

Ⓑ A receptacle installed for refrigeration equipment may be fed from one of the 20-ampere small appliance branch-circuits, or from an individual branch-circuit rated 15 or 20 amperes »210.52(B)(1) *Exception No 2*«.

Ⓒ Receptacles not serving kitchen countertops do not require GFCI protection.

Ⓓ A receptacle installed behind a refrigerator does not count as a required countertop receptacle »210.52(C)(5)«.

Ⓔ Each wall counter space 12 in. (300 mm) or wider requires a receptacle »210.52(C)(1)«.

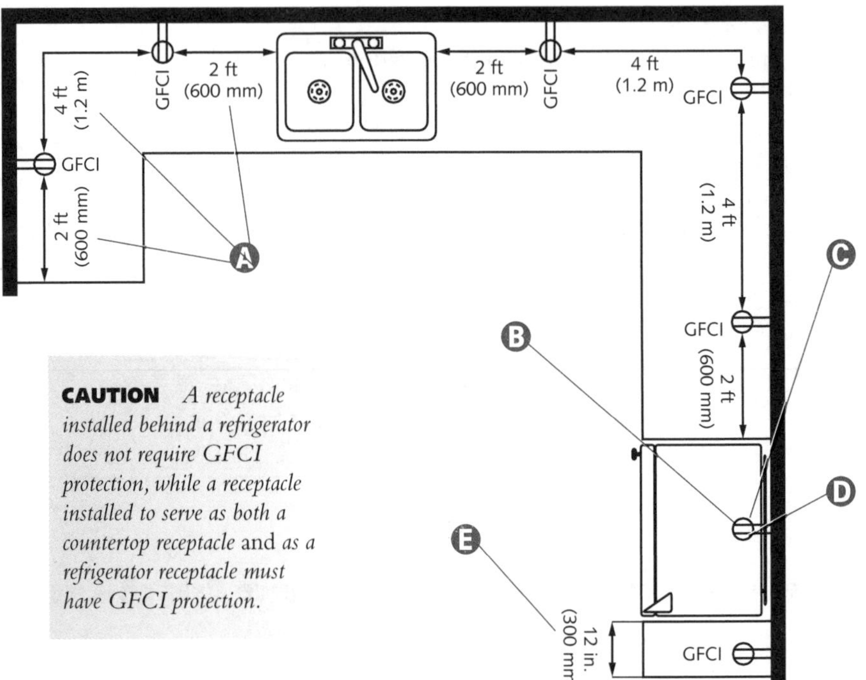

CAUTION *A receptacle installed behind a refrigerator does not require GFCI protection, while a receptacle installed to serve as both a countertop receptacle and as a refrigerator receptacle must have GFCI protection.*

Kitchen Countertop Receptacles

Ⓐ As a general rule, receptacle outlets shall be located above the countertop »210.52(C)(5)«. Installation of receptacles below the countertop are allowed by some exceptions (see 210.52(C)(5) *Exception*).

Ⓑ GFCI protection is not needed for receptacles installed along the wall to meet 210.52(A) requirements.

Ⓒ GFCI protection is required for receptacles installed to serve kitchen countertops »210.8(A)(6)«.

Ⓓ Receptacles located inside cabinets or cupboards, which do not serve kitchen countertop surfaces, do not require GFCI protection »210.8(A)(6)«.

Ⓔ Receptacles located inside cabinets or cupboards do not count as required kitchen countertop receptacles »210.52«.

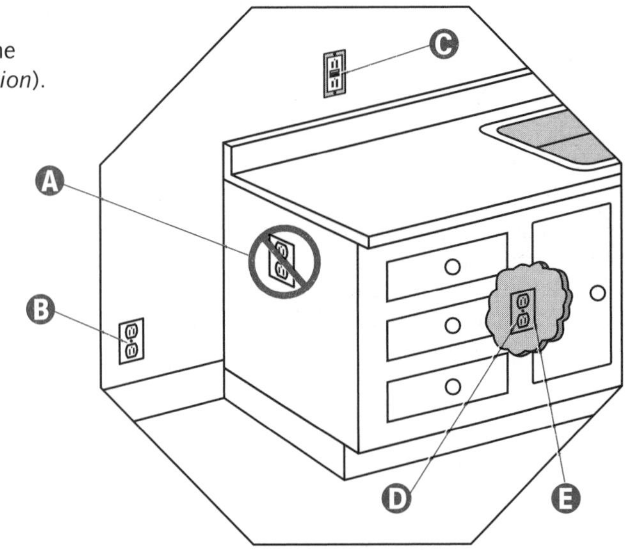

Island Receptacle Placement

At least one receptacle outlet shall be installed at each island counter space with a long dimension of 24 in. (600 mm), or more, and a short dimension of 12 in. (300 mm) or more »210.52(C)(2)«.

A receptacle is required for each island counter space with at least a 12- by 24-in. (300- x 600-mm) area that is separated from other counter space because of range tops, refrigerators, or sinks »210.52(C)(4)«.

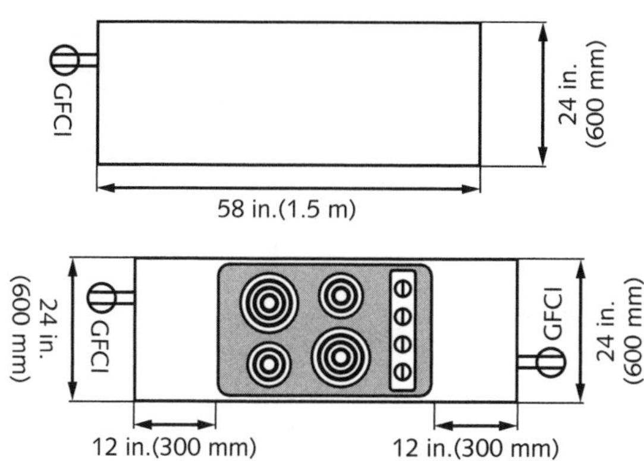

Peninsular Receptacle Placement

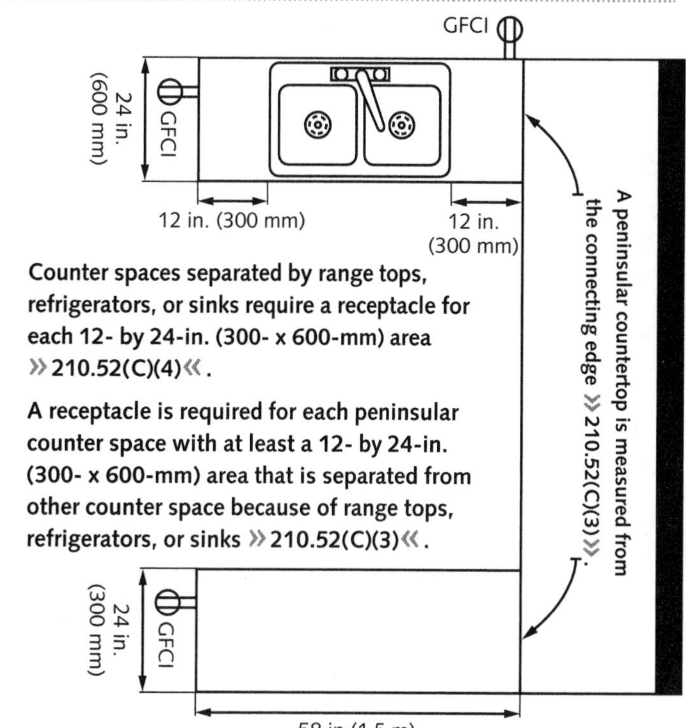

Counter spaces separated by range tops, refrigerators, or sinks require a receptacle for each 12- by 24-in. (300- x 600-mm) area »210.52(C)(4)«.

A receptacle is required for each peninsular counter space with at least a 12- by 24-in. (300- x 600-mm) area that is separated from other counter space because of range tops, refrigerators, or sinks »210.52(C)(3)«.

Receptacles Mounted Below Countertops

Where an island or peninsular countertop is flat across its entire surface, receptacle installation below the countertop is permitted under certain conditions. If a wall, backsplash, overhead cabinet, or similar area is available above the counter, the receptacle must be mounted above the countertop. Receptacles installed more than 20 in. (500 mm) above the countertop, such as in an overhead cabinet, are permitted but are not included as required countertop receptacles »210.52(C)(5) *Exception*«.

CAUTION *A peninsular counter side without doors or drawers might be considered wall space by the AHJ. If so, and if longer than 6 ft (1.8 m), a receptacle is required.*

A Receptacles installed face-up in the work surface or countertop are not allowed »210.52(C)(5)«.

B GFCI protection is required for all receptacles installed to serve kitchen countertop surfaces »210.8(A)(6)«.

C While receptacles located more than 12 in. (300 mm) below countertops are permitted, they are not to be included as required countertop receptacles »210.52(C)(5) *Exception*«.

D Receptacles mounted in the cabinet are permitted under an overhanging countertop but are not counted as required countertop receptacles where the countertop extends more than 6 in. (150 mm) beyond its support base »210.52(C)(5) *Exception*«.

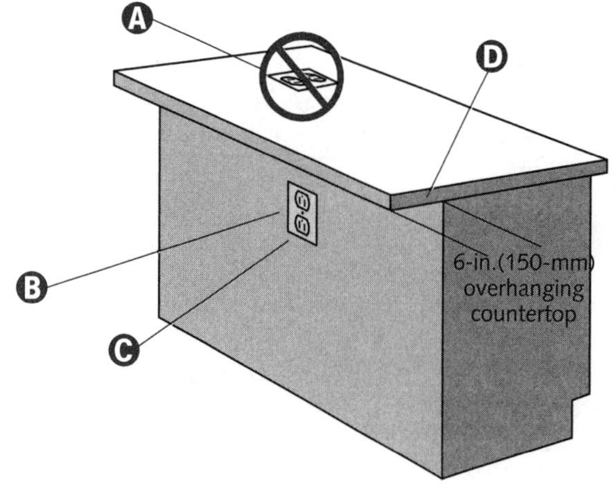

Permanently Connected Appliances

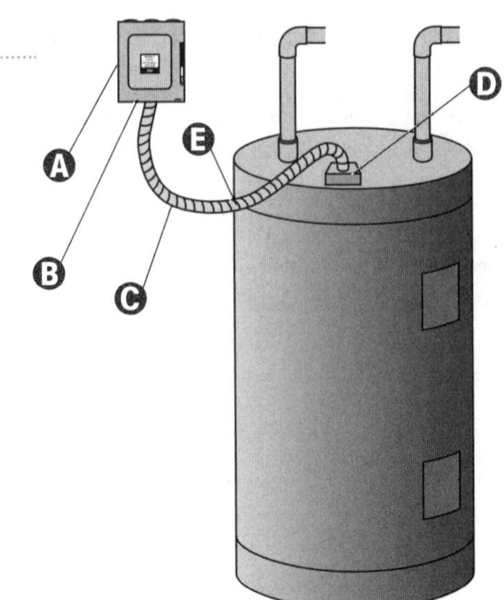

A A disconnecting means is not required if the branch-circuit switch or circuit breaker can be seen while working on the appliance »422.31(B) «.

B A disconnecting means is not required if the branch-circuit switch or circuit breaker can be locked in the open (off) position »422.31(B) «.

C Water heaters shall not be connected with flexible cords »422.16(A) «.

D Grounding must meet 250.134 requirements for equipment fastened-in-place or connected by permanent wiring methods.

E Adequately protect conductors where subject to physical damage »300.4 «.

> **NOTE**
>
> *A means must be provided to disconnect all ungrounded (hot) conductors from the appliance »422.30«.*

A fastened-in-place appliance (dishwasher, garbage disposer, etc.) shall not be connected to a small appliance branch-circuit »210.52(B)(2) «.

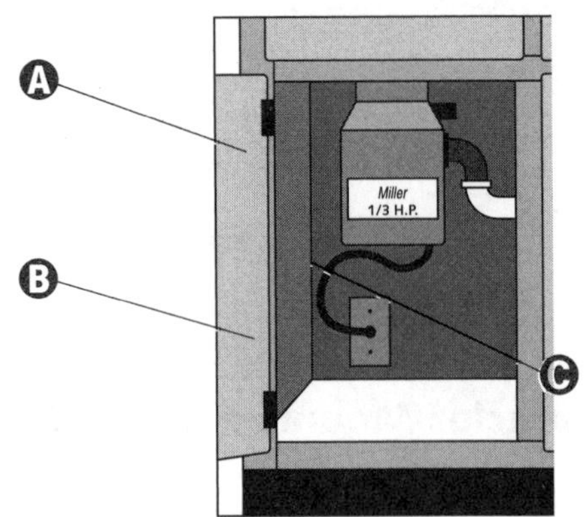

A A permanently connected appliance, rated over 300 volt-amperes or ⅛ horsepower, must have a means to disconnect all ungrounded (hot) conductors from the appliance »422.30 «.

B A disconnecting means is not required if the branch-circuit switch or circuit breaker can be locked in the open (off) position »422.31(B) «.

C Grounding must meet 250.134 requirements for equipment fastened-in-place or connected by permanent wiring methods.

Kitchen Waste (Garbage) Disposers

Some appliances (dishwashers, trash compactors, and kitchen garbage disposers) can be cord and plug connected. This arrangement provides a means of disconnect where accessible »422.33(A) «.

Kitchen waste disposers can be cord and plug connected under certain conditions »422.16(B)(1) «. The flexible cord must be identified as suitable for the purpose by the appliance manufacturer's installation instructions, and must meet *all* of the following conditions:

- The flexible cord shall be terminated with a grounding-type attachment plug. A listed kitchen disposer, distinctly marked as protected by a system of double insulation or its equivalent, does not require termination with a grounding-type attachment plug.
- The cord must be between 18 and 36 in. (450 and 900 mm) in length.
- The receptacle must be located so the flexible cord will not become damaged.
- The receptacle shall be accessible.

A The switch controlling a kitchen waste (garbage) disposer serves as a disconnecting means. Also see 430.109(C).

B Appliance receptacles installed inside cabinets, as permitted by 422.16, do not require GFCI protection.

C Grounding must meet 250.138 requirements for cord- and plug-connected equipment.

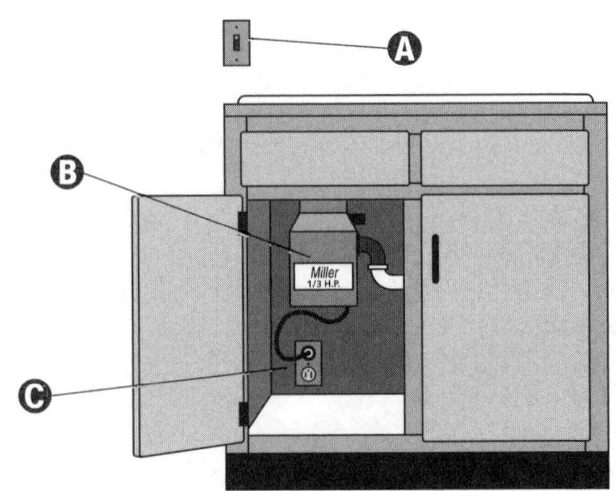

Built-In Dishwashers and Trash Compactors

Built-in dishwashers and trash compactors can be cord and plug connected under certain conditions »422.16(B)(2)«. The flexible cord must be identified as suitable for the purpose by the appliance manufacturer's installation instructions, and must meet all of the following conditions:

- The flexible cord shall be terminated with a grounding-type attachment plug. A listed dishwasher or trash compactor, distinctly marked as protected by a system of double insulation, or its equivalent, does not require termination with a grounding-type attachment plug.
- The cord must be between 3 and 4 ft (0.9 and 1.2 m) in length when measured from the face of the attachment plug to the plane of the rear of the appliance.
- The receptacle must be located so the flexible cord will not become damaged.
- The receptacle shall be accessible.

Ⓐ Where accessible, cord and plug arrangements may serve as a disconnecting means »422.33(A)«. A receptacle installed under the sink (inside the cabinet) may serve as a means of disconnect.

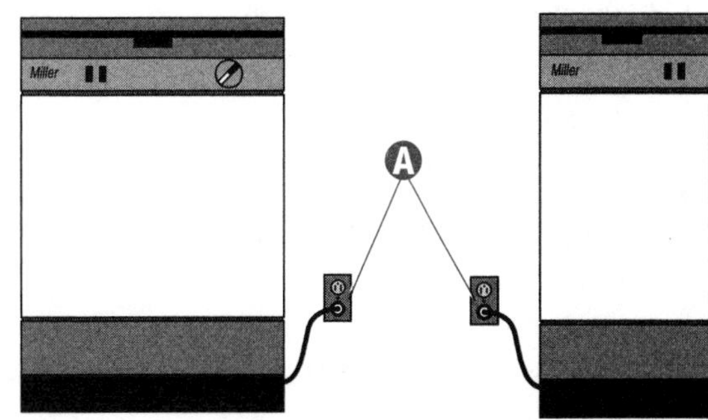

Appliance Branch-Circuit Rating

Branch-circuits, rated 15 or 20 amperes, may feed lighting units, utilization equipment, or a combination of both. The rating of any one cord- and plug-connected utilization equipment shall not exceed 80% of the branch-circuit ampere rating »210.23(A)«. Cord- and plug-connected equipment may be portable, or may be fastened-in-place equipment. Fastened-in-place utilization equipment shall not exceed 50% of the branch-circuit ampere rating where the circuit also feeds lighting units, and/or non-fastened-in-place utilization equipment »210.23(A)(2)«.

NOTE

Local codes may require a separate branch-circuit for the dishwasher and garbage disposer, i.e., one branch-circuit for each.

Ⓐ The full load current for a single-phase motor is found in Table 430.148.

Ⓑ This receptacle could be split-wired to accommodate two individual branch-circuits.

Ⓒ The total load shall not exceed the branch-circuit rating »220.4«. For motor-operated, fastened-in-place appliances with a motor larger than ⅛ horsepower, multiply the largest motor by 125% and add 100% of the other loads »220.4(A)«.

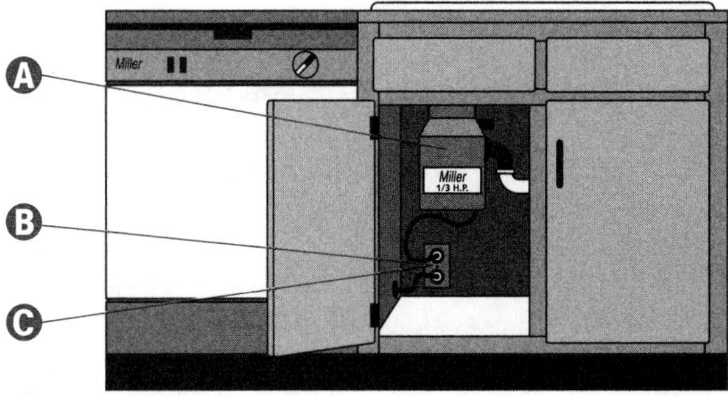

Calculation for One Counter-Mounted Cooking Unit and One Wall-Mounted Oven on the Same Circuit

Calculation for one cooktop with a nameplate rating of 6.5 kW at 240 volts and one oven with a nameplate rating of 4 kW at 240 volts supplied from one branch-circuit:

- Add the rating of the cooktop and oven and compute the branch-circuit in accordance with Table 220.19.
- 6.5 kW + 4 kW = 10.5 kW
- Table 220.19, Column C for one unit less than 12 kW = 8 kW
- 8 kW = 8000 watts
  8000 watts ÷ 240 volts = 33.3 amperes
- Overcurrent protective device = 35 amperes »240.4(B)«.
- Conductor size = 8 AWG »Table 310.16«.

Calculation for One Counter-Mounted Cooking Unit (Cooktop)

Branch-circuit calculation for one cooktop with a nameplate rating of 6.5 kW at 240 volts:

- Use the nameplate rating to determine the branch-circuit conductor size and the overcurrent protection for one counter-mounted cooking unit »Table 220.19, Note 4«.
- 6.5 kW = 6500 watts
  6500 watts ÷ 240 = 27.1 amperes
- Overcurrent protective device = 30 amperes »240.4(B)«.
- Conductor size = 10 AWG »Table 310.16«.

Calculation for One Wall-Mounted Oven

Branch-circuit calculation for one oven with a nameplate rating of 4 kW at 240 volts:

- Use the nameplate rating to determine the branch-circuit conductor size and the overcurrent protection for one wall-mounted oven »Table 220.19, Note 4«.
- 4 kW = 4000 watts
  4000 watts ÷ 240 = 16.7 amperes
- Overcurrent protective device = 20 amperes »240.4(B)«.
- Conductor size = 12 AWG »Table 310.16«.

Counter-Mounted Cooking Units (Cooktops) and Wall-Mounted Ovens

A The nameplate rating of a counter-mounted cooking unit or a wall oven shall be the minimum branch-circuit load »Table 220.19, Note 4«.

B Wall-mounted ovens and counter-mounted cooking units may be connected permanently or by cord and plug »422.16(A)(3)«.

C A disconnecting means is required »422.30«.

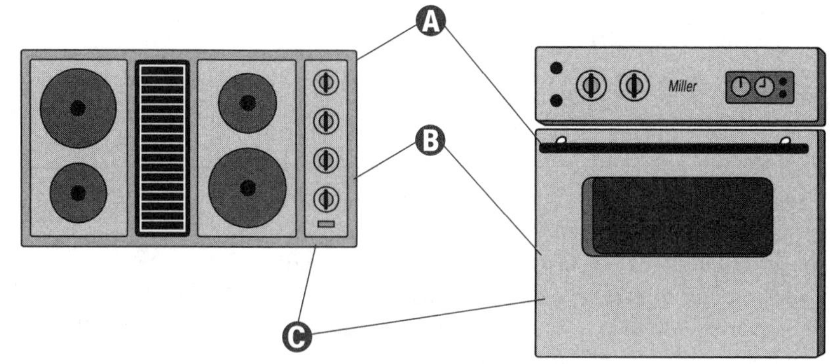

Cooktop and Wall-Mounted Oven(s) on Same Branch-Circuit

A Tap conductors from a 50-ampere branch-circuit must have an ampacity of at least 20 and not less than the ampacity of the individual unit served »210.19(C) *Exception No. 1*«.

B Where a single branch-circuit feeds one cooktop and one or two ovens (in the same room), add the nameplate ratings of the appliances, treating the total kW as one range to be found in Table 220.19 »Table 220.19, Note 4«.

C No disconnecting means is required if the branch-circuit switch or circuit breaker can be locked in the open (off) position »422.31(B)«.

D It is permissible to feed one cooktop and one wall-mounted oven, or one cooktop and two wall-mounted ovens, with one branch-circuit where all of the units are in the same room »Table 220.19 Note 4«.

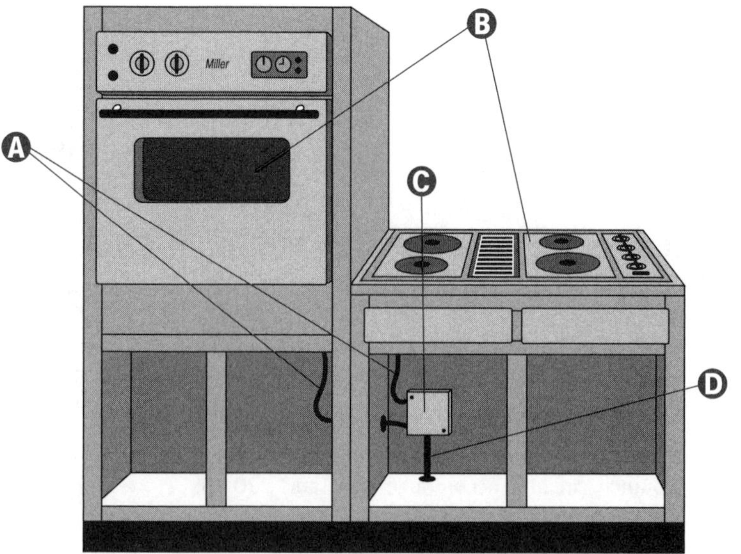

Electric Ranges

Ⓐ Grounding must meet the 250.138 requirements for cord- and plug- connected equipment.

Ⓑ It is permissible to compute the branch-circuit load for one range in accordance with Table 220.19 »Table 220.19 Note 4«.

Ⓒ A disconnecting means is required »422.30«. An accessible separable connector or an accessible plug and receptacle can serve as the disconnecting means »422.33(A)«.

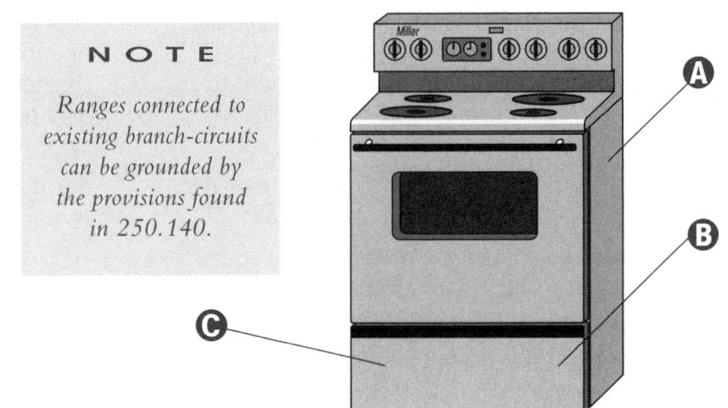

NOTE

Ranges connected to existing branch-circuits can be grounded by the provisions found in 250.140.

Ⓐ Most household electric ranges are cord- and plug- connected. The attachment plug and receptacle can serve as the disconnecting means if they are accessible by removing a drawer on the front of the range »422.33(B)«.

Ⓑ New branch-circuit conductors installed for a range must have an insulated grounded (neutral) conductor and a grounding means. Grounding must meet 250.138 requirements for cord-and plug-connected equipment.

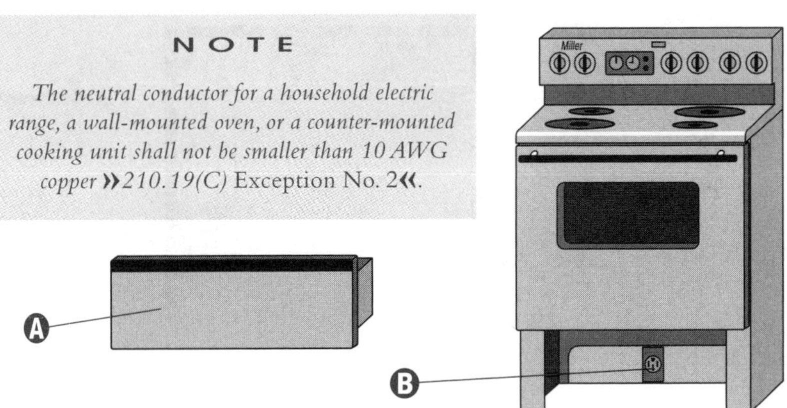

NOTE

The neutral conductor for a household electric range, a wall-mounted oven, or a counter-mounted cooking unit shall not be smaller than 10 AWG copper »210.19(C) Exception No. 2«.

Calculation for Ranges Rated 8¾ kW Through 12 kW

Branch-circuit calculation for one range with a nameplate rating of 12 kW at 240 volts:

- Compute the branch-circuit load for one range in accordance with Table 220.19
- 12 kW = 8 kW (Column C)
  8 kW = 8000 watts
  8000 watts ÷ 240 = 33.3 amperes
- Overcurrent protective device = 40 amperes »240.4(B)«.
- Conductor size = 8 AWG »Table 310.16«.

NOTE

The minimum branch-circuit rating for a range rated 8¾ kW, or more, is 40 amperes »210.19(C)«.

Calculation for Ranges Rated More Than 12 kW

Branch-circuit calculation for one range with a nameplate rating of 15 kW at 240 volts:

- Compute the branch-circuit load for one range in accordance with Table 220.19.
- Column C is for 12-kW and under ranges.
  Note 1 applies to ranges over 12 kW.
- First, subtract 12-kW from the 15-kW range.
  15 kW – 12 kW = 3 kW
- Next, multiply 3 kW by 5% = 15%
- Then, multiply 8 kW (Column A) by 15%.
  8 kW x 15% = 1.2 kW
- Finally, add 1.2 kW to 8 kW.
  1.2 kW + 8 kW = 9.2 kW = 9200 watts
- 9200 watts + 240 volts = 38.3 amperes
- Overcurrent protective device = 40 amperes »240.4(B)«.
- Conductor size = 8 AWG »Table 310.16«.

Kitchen, Dining Room, and Breakfast Area

A A minimum of two small appliance branch-circuits are required for receptacles serving kitchen countertops »210.52(B)(3)«. Either, or both, of these small appliance branch-circuits may feed other receptacles in the kitchen, pantry, breakfast room, and/or dining area.

B Non-countertop receptacle placement is determined by wall space »210.52(A)«.

C The receptacle outlet installed for refrigeration equipment can be supplied from an individual branch-circuit rated 15 amperes, or greater »210.52(B)(1) *Exception No. 2*«.

D Range hoods and lights are not permitted on small appliance branch-circuits »210.52(B)(2)«.

E All receptacles in the kitchen, pantry, breakfast room, dining room, or similar area shall be supplied from 20-ampere branch-circuits (except for refrigeration equipment) »210.52(B)(1)«.

F A minimum of two circuits are required for receptacles located in the kitchen, pantry, breakfast room, dining room, or similar areas »210.52(B)(1)«.

G At least one wall switch-controlled lighting outlet shall be installed in dining rooms, kitchens, and breakfast areas »210.70(A)(1)«. One or more receptacles controlled by a wall switch are permitted in lieu of a lighting outlet in dining rooms and breakfast areas, but not in kitchens »210.70(A)(1) *Exception No. 1*«.

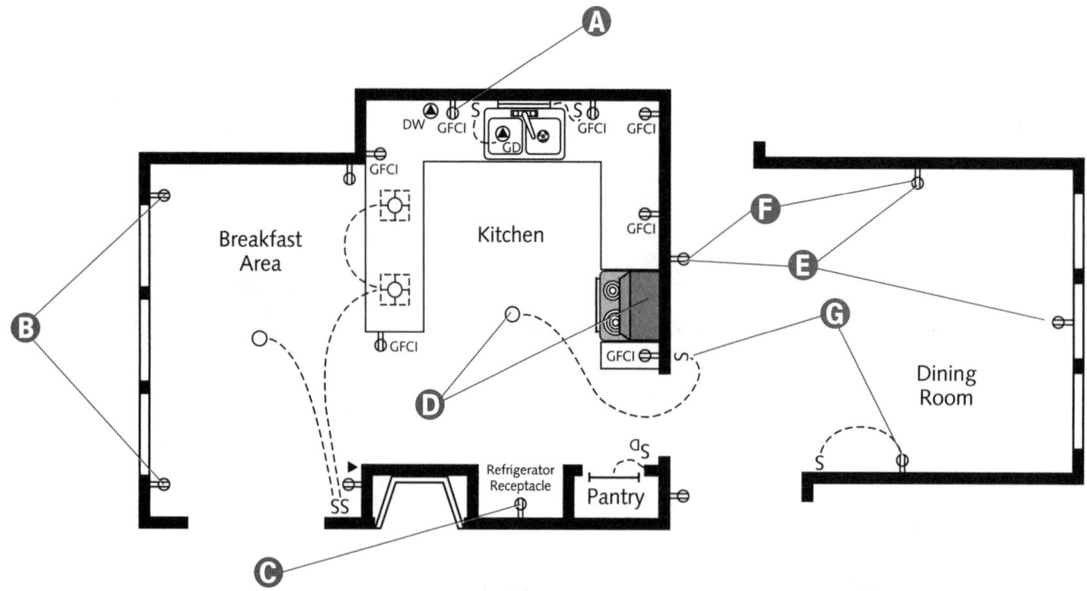

BEDROOMS

Arc-Fault Circuit-Interrupter Protection

An *arc-fault circuit interrupter* is a device intended to provide protection from the effects of arc faults by recognizing characteristics unique to arcing and by functioning to de-energize the circuit when an arc fault is detected » 210.12(A)«.

A All branch-circuits that supply 125-volt, single-phase, 15- and 20-ampere outlets, installed in dwelling unit bedrooms, must be protected by an arc-fault circuit interrupter(s) »210.12(B)«.

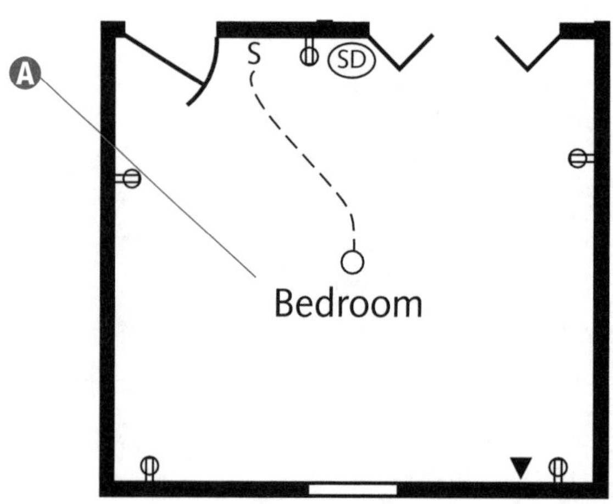

HALLWAYS AND STAIRWAYS

Hallway and Stairway Lighting

A minimum of one lighting outlet is required in each hallway and stairway »210.70(A)(2)(a)«. The light must be wall-switch-controlled, except where controlled by remote, central, or automatic means »210.70(A)(2) *Exception*«. A receptacle controlled by a wall switch is not permitted.

Ⓐ Interior stairways of six or more risers (steps) shall have a wall switch located at each floor level and landing level that includes an entryway »210.70(A)(2)(c)«.

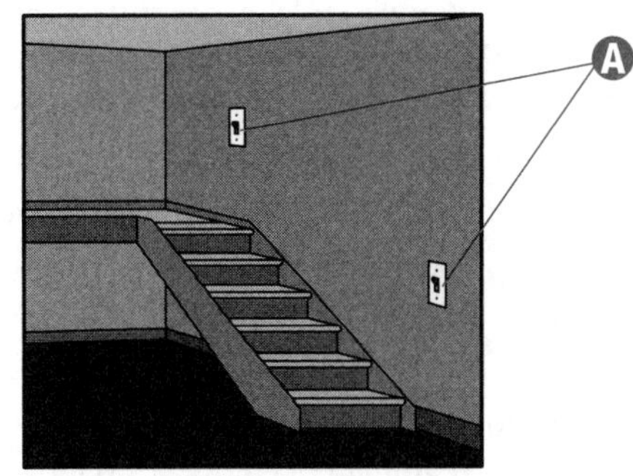

Hallway Receptacle Placement

Receptacles in hallways are not subject to the general provisions for placement determined by wall space.

Ⓐ If length is less than 10 ft (3.0 m), no receptacle is required.

Ⓑ Hallways measuring 10 ft (3.0 m) or more in length must have at least one receptacle outlet »210.52(H)«. Length is determined by measuring along the hallway center line, without passing through a doorway.

Ⓒ If length is 10 ft (3.0 m) or more, a receptacle is required.

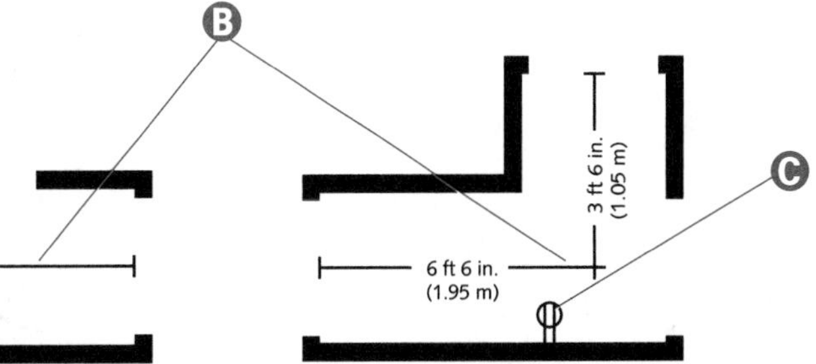

CLOTHES CLOSETS

Luminaire (Fixture) Types *Not* Permitted in Clothes Closets »410.8(C)«

Ⓐ Pendant luminaires (fixtures) or lampholders

Ⓑ Incandescent luminaires (fixtures) with open or partially enclosed lamps

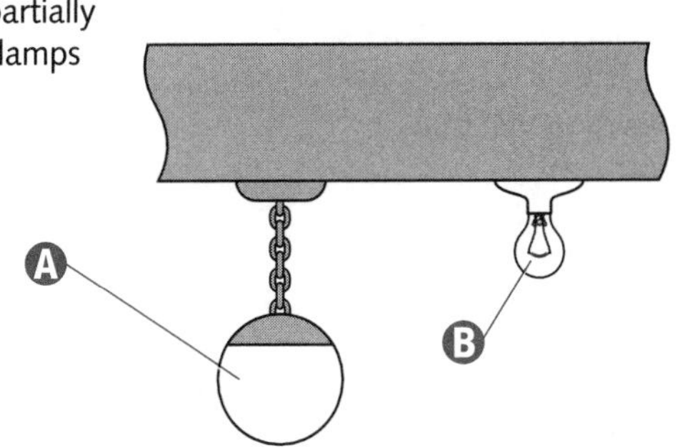

Luminaire (Fixture) Types Permitted in Clothes Closets »410.8(B)«

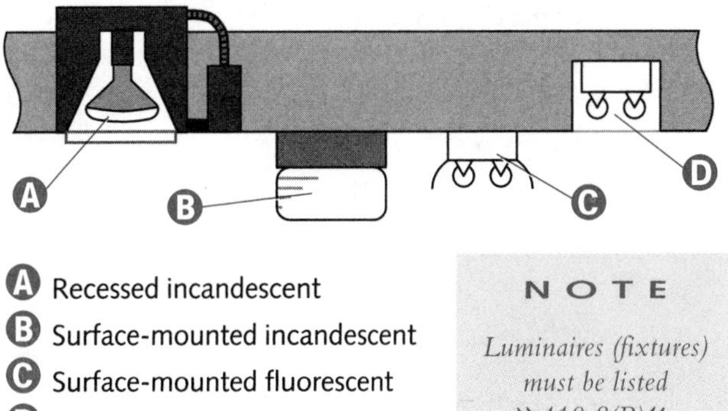

Ⓐ Recessed incandescent

Ⓑ Surface-mounted incandescent

Ⓒ Surface-mounted fluorescent

Ⓓ Recessed fluorescent

NOTE

Luminaires (fixtures) must be listed »410.8(B)«.

Dedicated Space in Clothes Closets

Receptacles in clothes closets are permitted, but are not required.

Lighting outlets are permitted in closets, but are not required.

Ⓐ Surface-mounted incandescent luminaires (fixtures) shall have a minimum 12-in. (300-mm) clearance between the luminaire (fixture) and the nearest point of a storage space »410.8(D)(1)«.

Ⓑ Storage space

Ⓒ 12 in. (300 mm) or the width of the shelf, whichever is greater »410.8(A)«.

Ⓓ 6 ft (1.8 m) or the highest clothes-hanging rod »410.8(A)«.

Ⓔ 24 in. (600 mm) »410.8(A)«.

Ⓕ Surface-mounted fluorescent luminaires (fixtures) shall have a minimum 6-in. (150-mm) clearance between the luminaire (fixture) and the nearest point of a storage space »410.8(D)(2)«.

Ⓖ Storage space

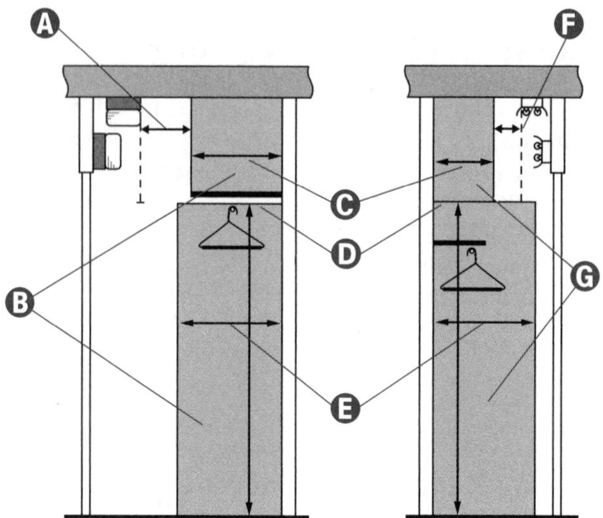

Ⓐ 6 ft (1.8 m) or the highest clothes-hanging rod »410.8(A)«.

Ⓑ 24 in. (600 mm) »410.8(A)«.

Ⓒ Recessed incandescent, 6 in. (150 mm) »410.8(D)(3)«.

Ⓓ Surface-mounted incandescent, 12 in.(300 mm) »410.8(D)(1)«.

Ⓔ Surface-mounted fluorescent, 6 in. (150 mm) »410.8(D)(2)«.

Ⓕ Recessed fluorescent, 6 in. (150 mm) »410.8(D)(4)«.

Ⓖ 12 in. (300 mm) or the width of the shelf, whichever is greater »410.8(A)«.

Ⓗ For a closet with access to both sides of a hanging rod, the storage space shall include the volume below the highest rod extending 12 in. (300 mm) on either side of the rod on a plane horizontal to the floor and extending the entire length of the rod »410.8(A)«.

Ⓘ 6 ft (1.8 m) or the highest clothes-hanging rod »410.8(A)«.

Ⓙ 24 in.(600 mm) »410.8(A)«.

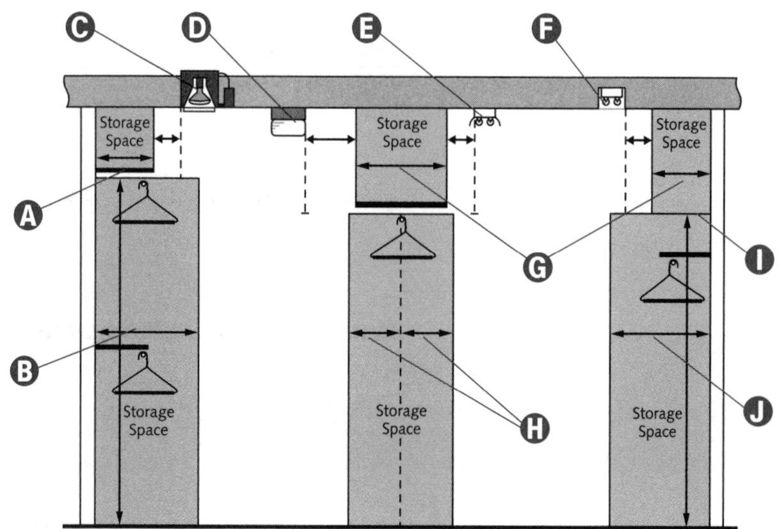

NOTE

Overcurrent devices shall not be located in the vicinity of easily ignitible material and, therefore, may not be installed in clothes closets »240.24(D)«.

BATHROOMS

Bathroom "Area"

The general provisions for receptacle placement by wall space do not apply to bathrooms.

A GFCI protection for personnel is required for every 125-volt receptacle located in the bathroom area »210.8(A)(1)«.

B **Bathroom Definition:** An area including a basin (lavatory or sink) with one or more of the following: a toilet, a tub, or a shower »Article 100—Definitions«. A bathroom is an **area**—not necessarily a single room.

C One or more lighting outlets, controlled by a wall switch, are required in bathrooms »210.70(A)(1)«.

NOTE

Service disconnecting means shall not be located in bathrooms »230.70(A)(2)«. Overcurrent devices, other than supplementary overcurrent protection, shall not be installed in bathrooms »240.24(E)«. Supplementary overcurrent protection is not branch-circuit overcurrent protection. It is an additional overcurrent protection usually installed within luminaires (light fixtures), appliances, and other equipment. It is not required that supplementary overcurrent protection be readily accessible »240.10«.

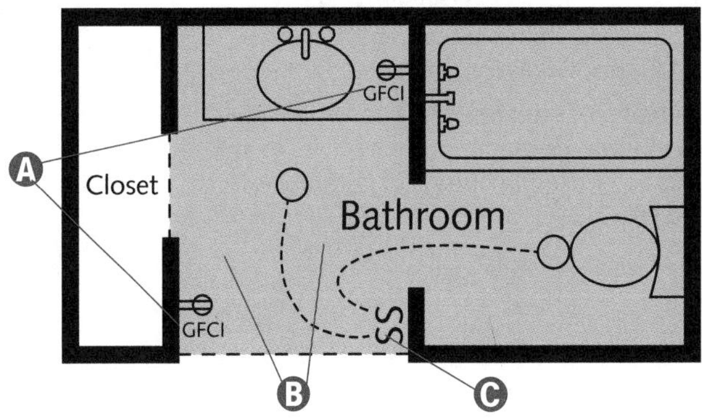

Receptacle Within 36 In. of Sink

A At least one wall receptacle shall be located within 36 in. (900 mm) of the outside edge of each basin (lavatory or sink) »210.52(D)«.

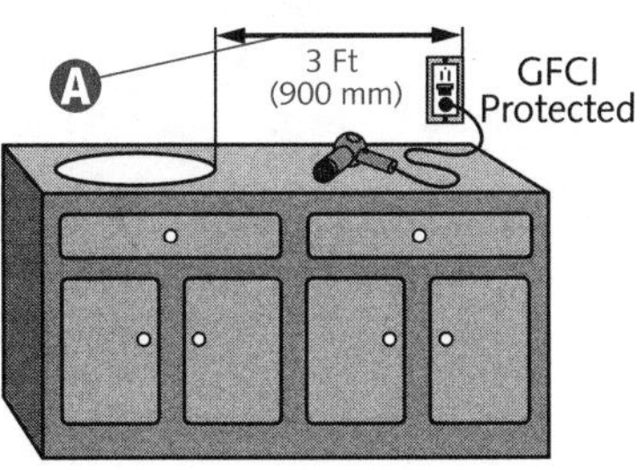

Bathroom Branch-Circuit Rating

A Bathroom receptacle outlets shall be supplied by at least one 20-ampere branch-circuit. »210.11(C)(3)«.

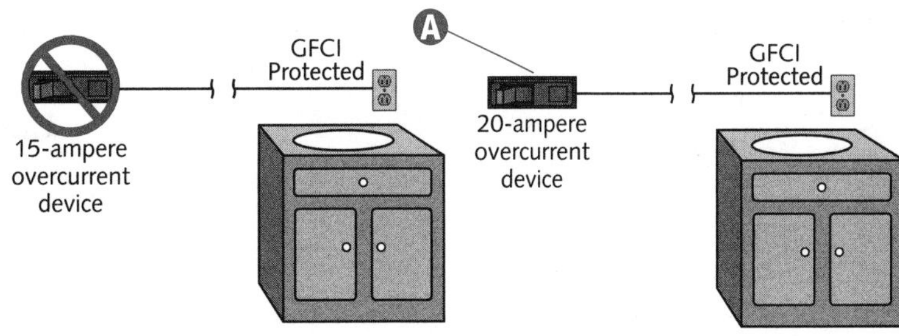

Bathroom Receptacles

Receptacles are not permitted in a face-up position in countertops or similar work surfaces »406.4(E)«.

A At least one wall receptacle shall be located within 36 in. (900 mm) of the outside edge of each basin (lavatory or sink) »210.52(D)«.

B Receptacles may have an individual rating of either 15 or 20 amperes, but must be supplied from a 20-ampere branch-circuit »210.11(C)(3)«. GFCI protection for personnel is required for all bathroom receptacles »210.8(A)(1)«.

CAUTION *The receptacle outlet must be located on a wall that is adjacent to the basin location »210.52(D)«.*

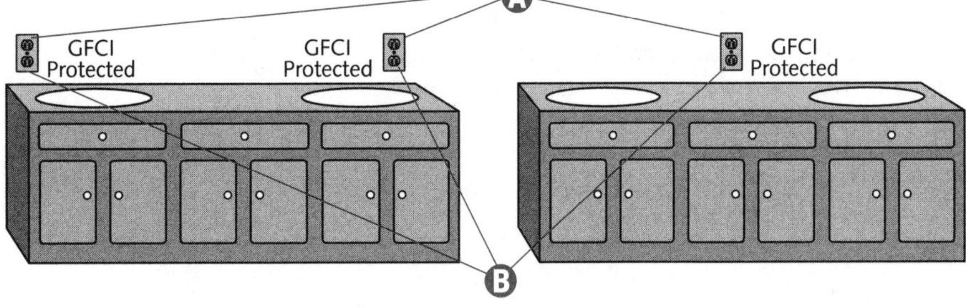

Bathroom

A A branch-circuit providing power to a bathroom receptacle may also provide power to other bathroom receptacles, whether in the same bathroom, or in different bathrooms »210.11(C)(3)«.

B A branch-circuit providing power to a bathroom receptacle may also provide power to other equipment, such as lighting and exhaust fans, but only within the same bathroom »210.11(C)(3) *Exception*«. If this is done, however, that branch-circuit cannot be used to provide power to any other bathrooms.

C A branch-circuit providing power to bathroom receptacles cannot provide power to any receptacle or lighting outside of bathrooms »210.11(C)(3)«.

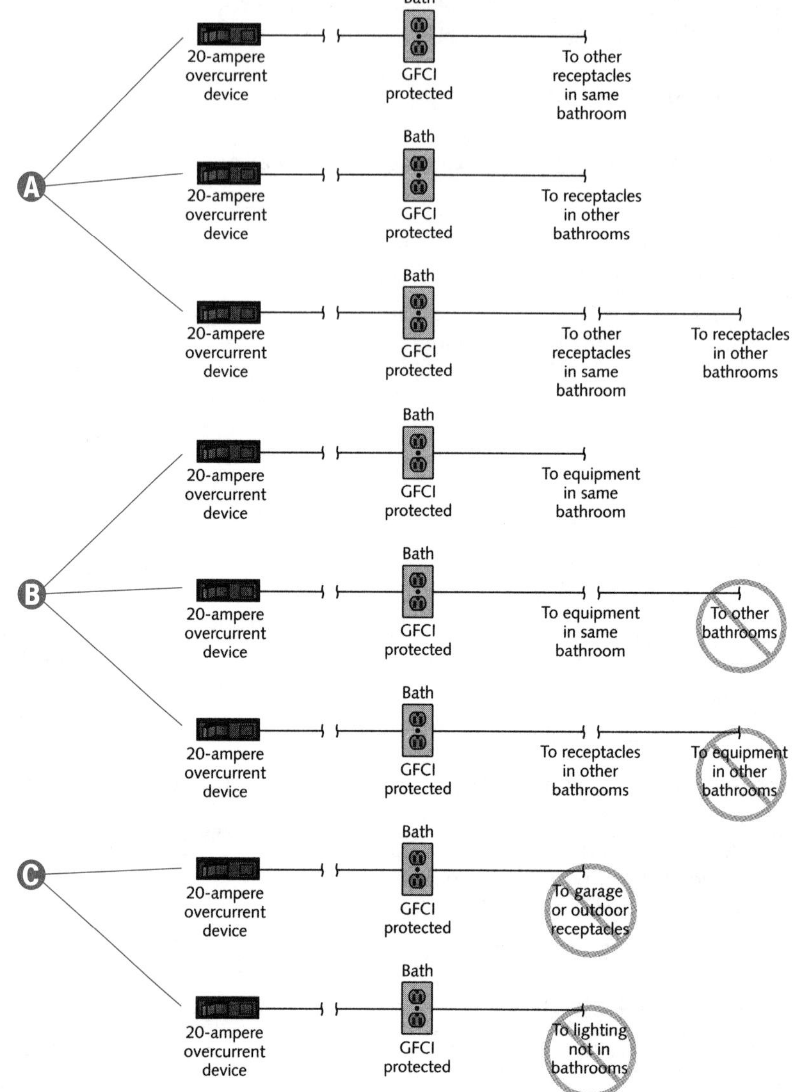

Bathtub and Shower Zone

A Luminaire (fixture) permitted (not in zone)

B Luminaire (fixture) not permitted

C Hanging luminaires (fixtures), cord-connected luminaires (fixtures), lighting track, pendants, and ceiling-suspended (paddle) fans are *not* permitted within a certain zone of a bathtub »410.4(D)«. The bathtub zone measures 3 ft (900 mm) horizontally and 8 ft (2.5 m) vertically from the top of the bathtub rim or shower stall threshold, and includes the area directly over the tub.

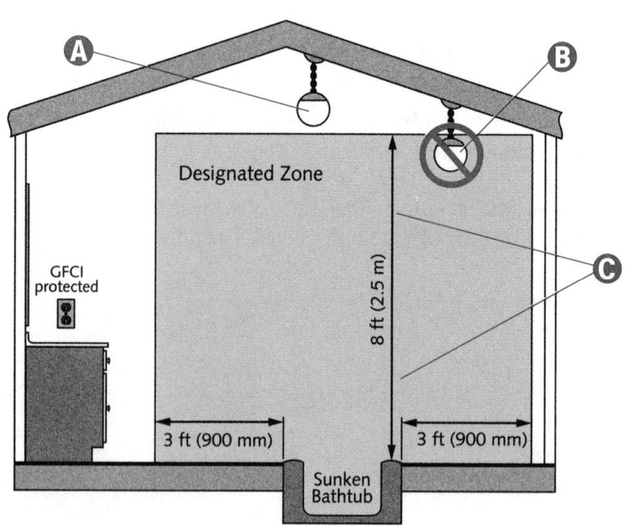

Bathtub and Shower Zone (continued)

Ⓐ Recessed lights, surface-mounted lights, and exhaust fans are permitted within the bathtub zone. Exposed metal parts must be grounded in accordance with 250.110.

Ⓑ Hanging luminaires (fixtures), cord-connected luminaires (fixtures), lighting track, pendants, and ceiling-mounted (paddle) fans are permitted only *outside* the zone.

CAUTION *A receptacle shall not be installed within a bathtub or shower space* ›› *406.8(C)* ‹‹.

Ⓒ The bathtub zone measures 3 ft (900 mm) horizontally and 8 ft (2.5 m) vertically from the top of the bathtub rim or shower stall threshold. The zone is all encompassing and includes the area directly over the tub or shower stall ›› 410.4(D) ‹‹.

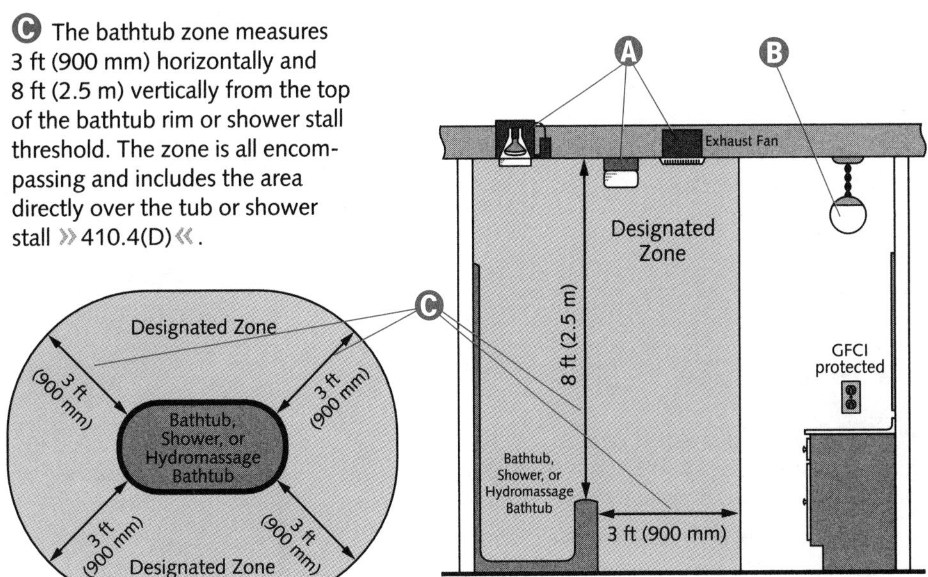

BASEMENTS AND GARAGES

Garage Receptacles

Ⓐ If a duplex receptacle is installed, GFCI protection is required. If a single receptacle is installed for an appliance, GFCI protection is not required ›› 210.8(A)(2) *Exception No. 2* ‹‹.

Ⓑ Laundry equipment receptacle(s) are not counted as required garage receptacle(s) ›› 210.52(G) ‹‹.

Ⓒ At least one receptacle is required in an attached garage ›› 210.52(G) ‹‹. (Receptacle placement is not determined by wall space.)

Ⓓ Receptacles not readily accessible do not require GFCI protection ›› 210.8(A)(2) *Exception No.1* ‹‹.

Ⓔ All receptacles installed in garages must have GFCI protection, unless installed under one of the exceptions found in 210.8(A)(2).

Ⓕ If a duplex receptacle is installed for two appliances, GFCI protection is not required. The two appliances must be cord and plug connected and must be located within a dedicated space where in normal use they are not easily moved from one place to another ›› 210.8(A)(2) *Exception No. 2* ‹‹.

Ⓖ Non-GFCI protected receptacles installed by an exception found in 210.8(A)(2) do not count toward the one garage receptacle requirement ›› 210.8(A)(2) ‹‹.

Ⓗ A single receptacle installed on a 20-ampere branch-circuit must have a rating of 20 amperes ›› 210.21(B)(1) ‹‹.

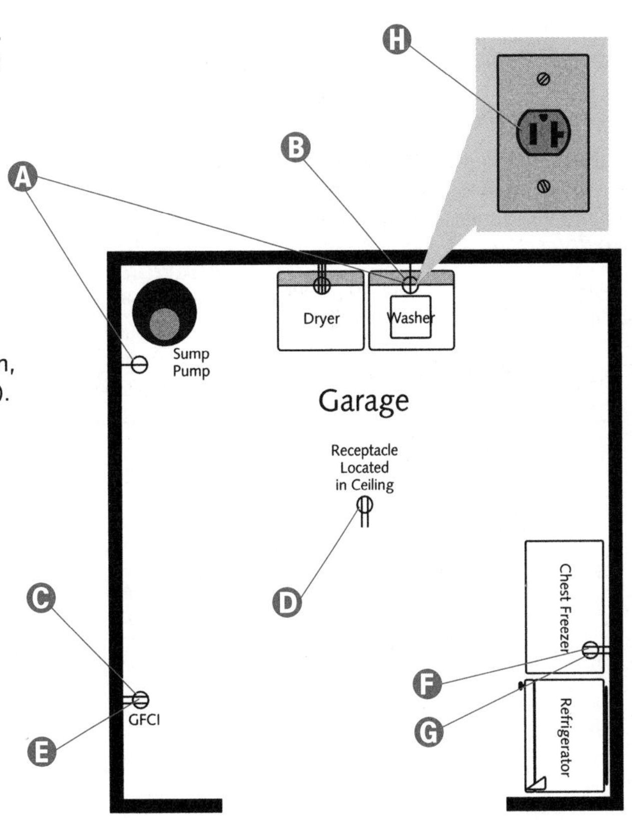

GFCI Receptacles in Garage

A If a duplex receptacle is installed, GFCI protection is required »210.8(A)(2)«.

B If a single receptacle is installed, GFCI protection is *not* required »210.8(A)(2) *Exception No. 2*«.

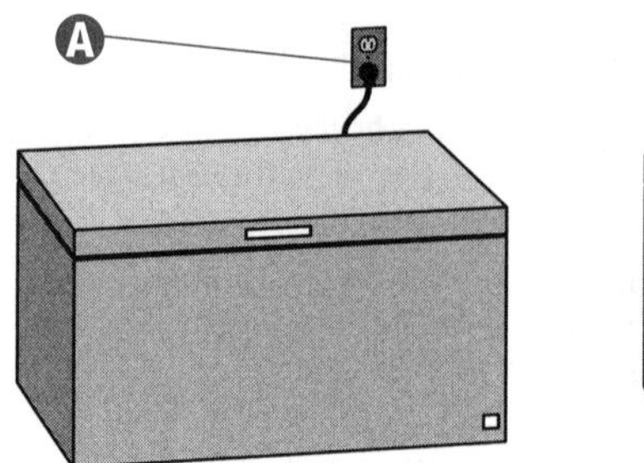

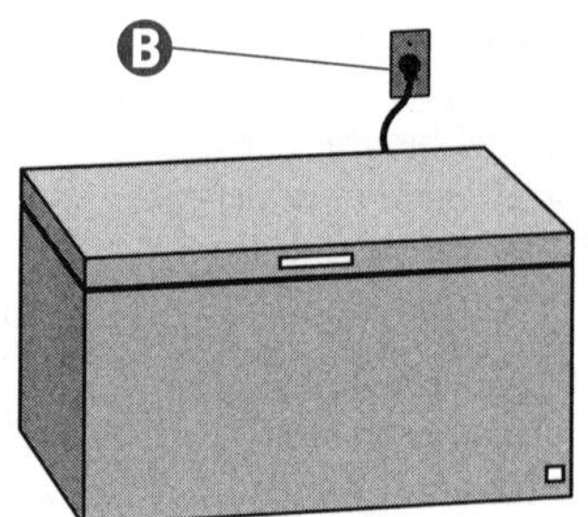

Lighting

A Basements and attached garages require at least one receptacle each »210.52(G)«. (Receptacle placement is not determined by wall space.)

B At least one wall-switch-controlled lighting outlet is required to provide illumination on the exterior side of an outside entrance or exit with grade level access »210.70(A)(2)(b)«.

C A vehicle garage door is not considered an outside entrance or exit and therefore does not require a wall-switch-controlled lighting outlet »210.70(A)(2)«.

D A lighting outlet shall be installed at, or near, equipment that may require servicing »210.70(A)(3)«.

E At least one wall-switch-controlled lighting outlet is required for basements and attached garages »210.70(A)(2) and (3)«.

F Interior stairways with 6 steps, or more, shall have a wall switch located at each floor level and landing that includes an entryway »210.70(A)(2)(c)«.

G A lighting outlet installed to provide illumination for the general area (basement, garage, etc.) may also serve as the equipment lighting, provided it is located at, or near, the equipment.

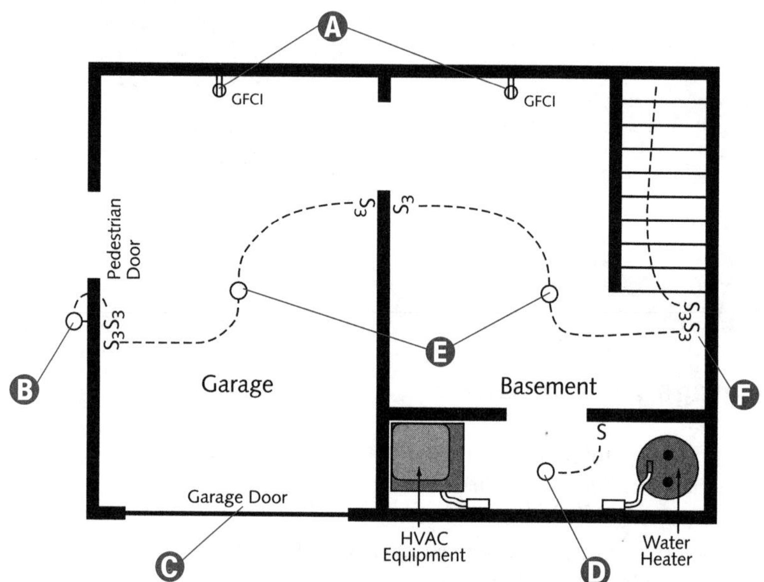

CAUTION *Where part of the basement is finished into one or more habitable rooms, each separate unfurnished portion requires at least one receptacle »210.52(G)«.*

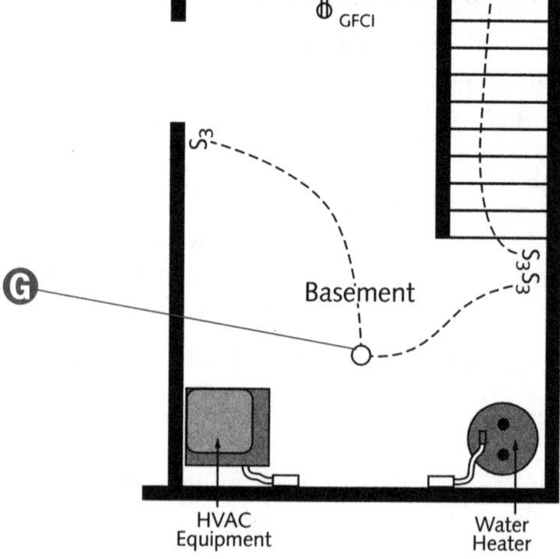

NOTE

Unfinished basements are defined as portions or areas of the basement not intended as habitable rooms and limited to storage areas, work areas, and the like »210.8(A)(5)«.

Disconnecting Means

A A disconnecting means is not required if the branch-circuit switch or circuit breaker is within 50 ft (15 m) and can be seen while working on the appliance »422.31(B)«.

B The branch-circuit switch or circuit breaker cannot be seen while working on the appliance. A disconnecting means is required unless the branch-circuit switch or circuit breaker can be locked in the open (off) position »422.31(B)«.

C A permanently connected appliance rated over 300 volt-amperes or ⅛ horsepower requires a means to disconnect all ungrounded (hot) conductors from the appliance »422.30«.

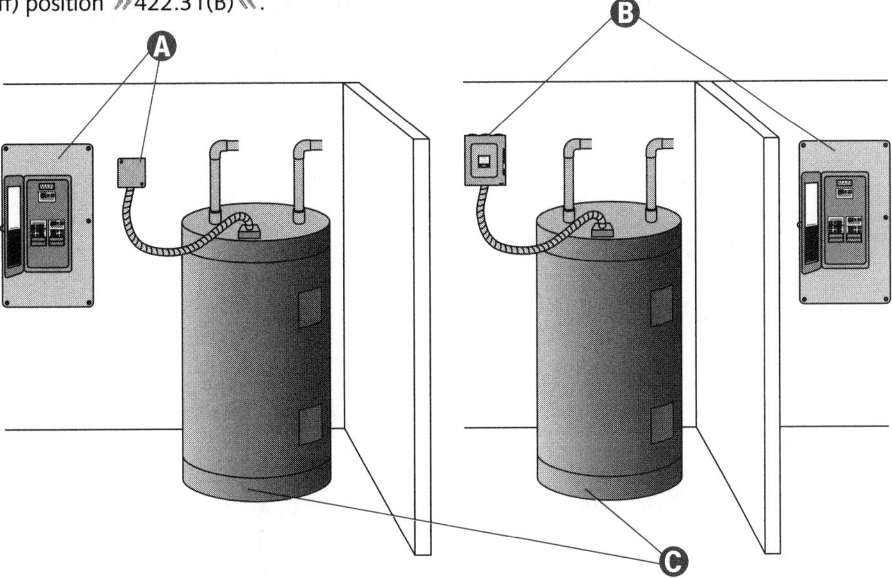

Detached Garages

A At least one receptacle is required in each detached garage with electric power »210.52(G)«.

B All receptacles installed in detached garages must have GFCI protection, unless installed under one of the exceptions found in 210.8(A)(2).

C At least one wall-switch-controlled lighting outlet is required for detached garages with electric power »210.70(A)(2)(a)«.

D A detached garage, with electric power, requires at least one wall-switch-controlled lighting outlet to provide illumination on the exterior side of an outside entrance or exit with grade-level access »210.70(A)(2)(b)«.

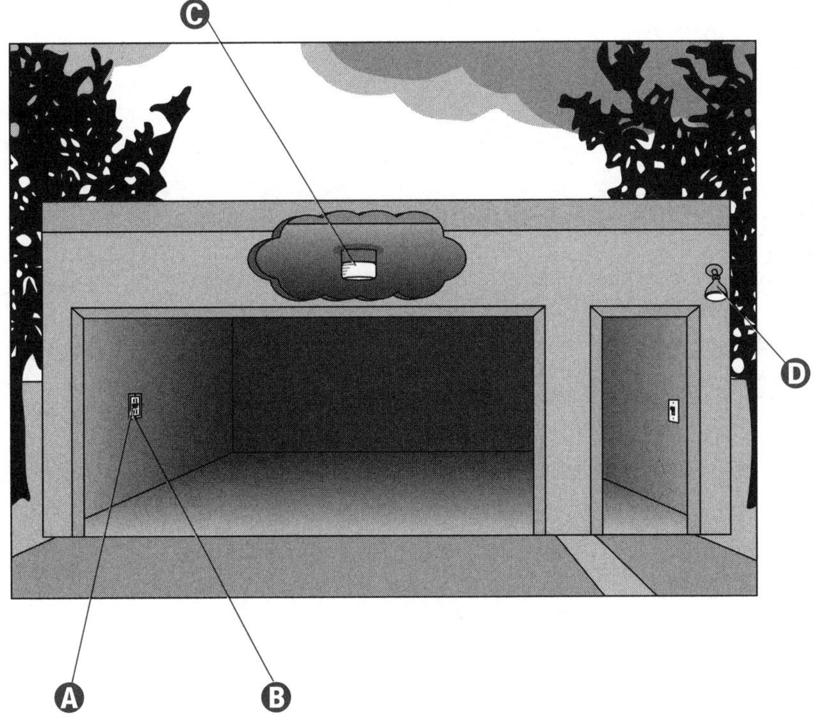

NOTE

Lighting and receptacles are not required in detached garages. There are no provisions to provide electric power to detached garages. However, when electric power is provided, the provisions for attached garages must be met »210.52(G)«.

Accessory Buildings

Ⓐ All receptacles installed in accessory buildings having a floor located at, or below, grade level not intended as habitable rooms and limited to storage areas, work areas, and areas of similar use must have GFCI protection, unless installed under one of the exceptions found in 210.8(A)(2).

Ⓑ Lighting and receptacles are not required in accessory buildings. There are no provisions to provide electric power to detached garages.

LAUNDRY AREAS

Receptacles and Lighting

Ⓐ The clothes dryer outlet must have an insulated grounded (neutral) conductor *and* an equipment grounding conductor »250.138«.

Ⓑ A lighting outlet is required for spaces containing equipment that require servicing »210.70(A)(3)«. A lighted area immediately adjacent to a laundry area (such as a hallway) may provide sufficient light for both areas.

Ⓒ Laundry receptacle outlet(s) must be fed from a 20-ampere branch-circuit »210.11(C)(2)«.

Ⓓ At least one receptacle is required for the laundry area »210.52(F)«.

NOTE

Electric clothes dryers connected to existing branch-circuits are permitted to follow 250.140.

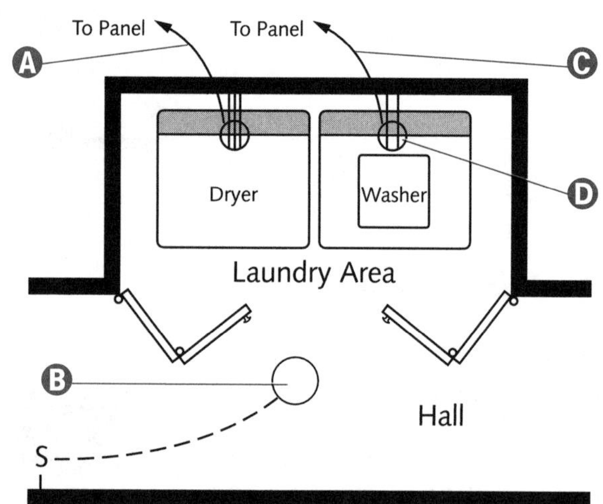

Receptacles and Lighting

Ⓐ Laundry receptacle outlet(s) must be fed from a 20-ampere branch-circuit »210.11(C)(2)«.

Ⓑ Multiple receptacles are permitted on the laundry circuit as long as all of the outlet(s) are within the laundry area »210.11(C)(2)«.

Ⓒ A 240-volt receptacle outlet for a clothes dryer is not a requirement (the dryer could be a gas dryer).

Ⓓ Utility rooms require a lighting outlet »210.70(A)(3)«.

Ⓔ Receptacle outlet(s) outside of the laundry area are not permitted on a laundry circuit »210.11(C)(2)«.

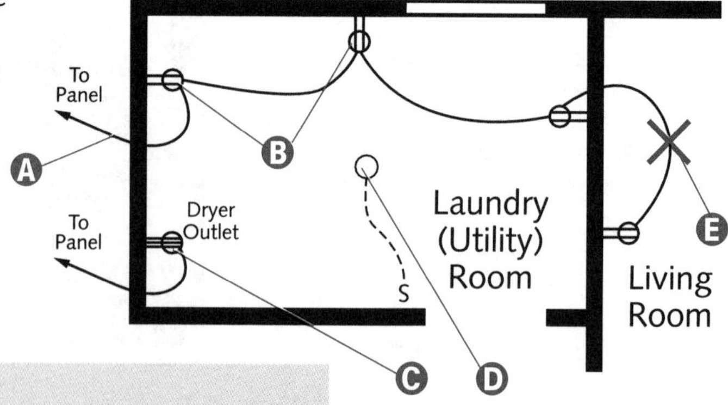

NOTE

A laundry room/area is not subject to the general provisions for receptacle placement found under 210.52(A). Although additional laundry area receptacles are permitted, only one receptacle outlet is required »210.52(F)«.

Receptacles and Lighting (continued)

A receptacle outlet installed for a specific appliance, such as laundry equipment, shall be installed within 6 ft (1.8 m) of the intended location of the appliance »210.50(C)«.

Ⓐ The clothes dryer outlet must have an insulated grounded (neutral) conductor *and* an equipment grounding conductor »250.138«.

Ⓑ Laundry receptacle outlet(s) must be fed from a 20-ampere branch-circuit »210.11(C)(2)«. A duplex receptacle installed on an individual 20-ampere branch-circuit may have either a 15- or 20-ampere rating »210.21(B)(2)«.

Ⓒ Height requirements for laundry receptacle outlet(s) are not specified, except they cannot be located more than 5½ ft (1.7 m) above the floor »210.52«.

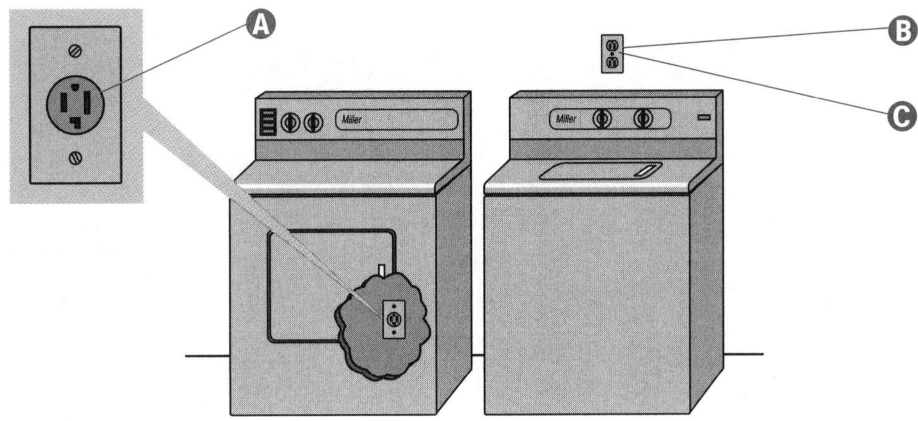

ATTIC AND CRAWL SPACES

Attic Spaces

A receptacle is not required for attic spaces that do not contain equipment that may require servicing.

Attic spaces not used for storage or attics that do not contain equipment that may require servicing, do not require lighting outlets.

Ⓐ Attic spaces used for storage or containing equipment that may require servicing must have at least one lighting outlet »210.70(A)(3)«.

Ⓑ At least one switch shall be located at the usual entry and exit to the attic space »210.70(A)(3)«. If the luminaire (light fixture) contains a switch and is located at the usual point of entry, a separate switch may not be required.

Ⓒ Cables not installed within framing members may need protection up to a height of at least 7 ft (2.1 m) »320.23(A), 330.23, and 334.23«.

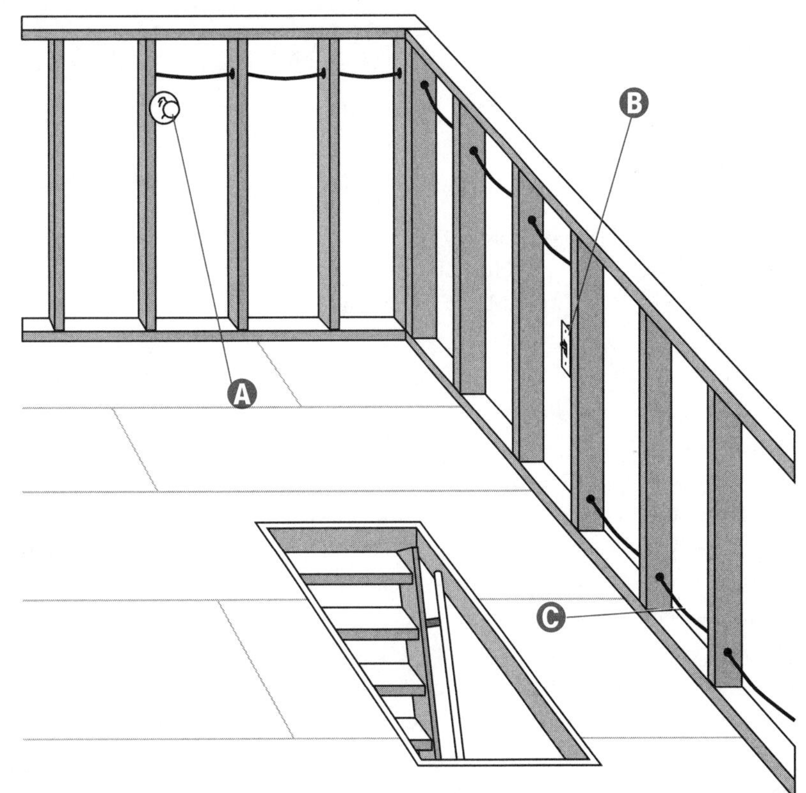

Crawl and Underfloor Spaces

A Crawl spaces with lighting outlet(s) require at least one switch located at the usual entry and exit to the crawl space »210.70(A)(3)«. If the luminaire (light fixture) contains a switch and is located at the usual point of entry, a separate switch may not be required.

B A 125-volt, single-phase, 15- or 20-ampere rated receptacle outlet shall be installed at an accessible location for the servicing of equipment. The receptacle shall be located on the same level and within 25 ft (7.5 m) of the equipment. The receptacle shall not be connected to the load side of the equipment disconnecting means »210.63«.

C Receptacle outlet(s) require GFCI protection where installed in a crawl space that is not above grade level »210.8(A)(4)«.

D Underfloor spaces used for storage containing equipment that may require servicing must have at least one lighting outlet. A lighting outlet shall be located at or near equipment that may require servicing »210.70(A)(3)«.

E A disconnecting means must be provided for HVAC equipment »440.3(B), 422.30, and 424.19«.

A GFCI protected receptacle located outside of the crawl space may meet the requirements of 210.63 if the receptacle is on the same level and within 25 ft (7.5 m) of the equipment.

No lighting outlet(s) are required for underfloor spaces that are not used for storage or contain no equipment which may require servicing.

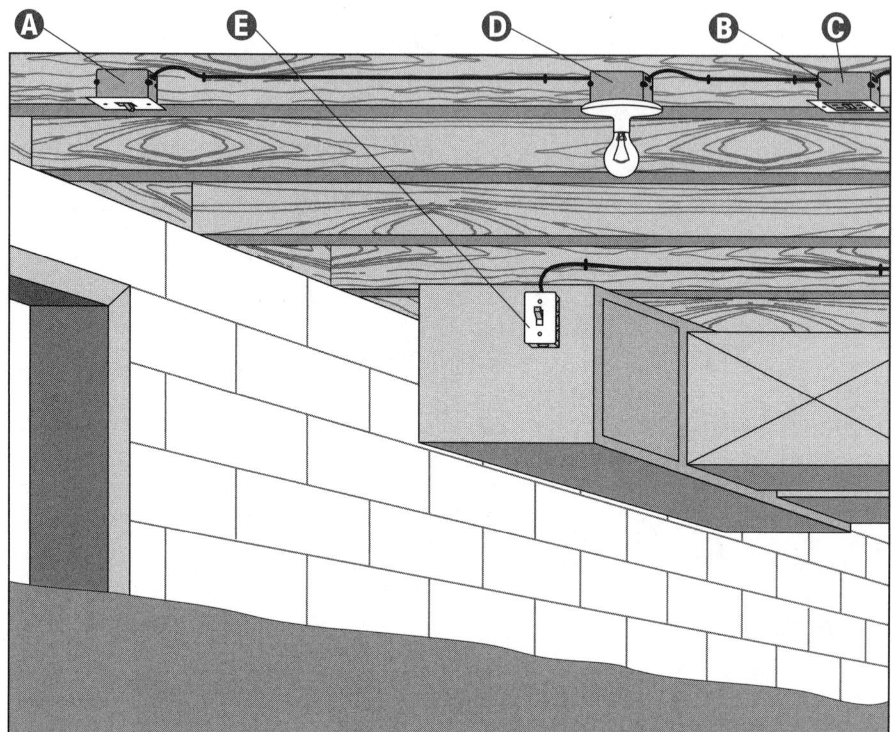

Summary

- Receptacle branch-circuits rated 20 amperes are required for small appliance branch-circuits in kitchens, pantries, dining rooms, breakfast areas, and similar areas.
- Twenty-four in. (600 mm) is the maximum distance to any countertop receptacle, along the wall line, measured horizontally.
- General provisions for countertop receptacle placement do not apply to islands and peninsulas.
- A minimum of two circuits is required for receptacles serving kitchen countertops.
- Receptacles located in kitchens, pantries, dining rooms, breakfast areas, and similar areas, also require a minimum of two circuits.
- All receptacles installed to serve kitchen countertops shall have GFCI protection.
- Some appliances can be cord and plug connected.
- Permanently connected as well as cord and plug connected appliances require a disconnecting means.
- Branch-circuit conductors installed for ranges and clothes dryers must have an insulated grounded (neutral) conductor and a grounding means.
- Hallways measuring 10 ft (3.0 m) or more in length require only one receptacle.
- Clothes closets have dedicated space where luminaires (light fixtures) are not permitted.
- Only certain types of luminaires (light fixtures) are permitted in the dedicated space within bathrooms.
- Receptacles in bathrooms must be GFCI protected and be supplied by 20-ampere branch-circuit(s).
- As a general rule, garage and basement receptacles require GFCI protection.

- Lighting is required at or near equipment that may require servicing.
- At least one laundry receptacle outlet is required and must be fed from a 20-ampere branch-circuit.
- Under certain conditions, attics and crawl spaces may require a lighting outlet and receptacle.
- A receptacle is required on the same level and within 25 ft (7.5 m) of HVAC equipment that may require servicing.

Unit 7 Competency Test

NEC® Reference	Answer	
______	______	1. The maximum distance between kitchen countertop receptacles is _____ ft.
______	______	2. Hanging luminaires (fixtures), located directly above any part of the bathtub, shall be installed so that the luminaire (fixture) is not less than _____ ft above the top of the bathtub rim.
______	______	3. A 15-ampere rated duplex receptacle may be installed on a _____ branch-circuit. a) 15-ampere b) 20-ampere c) 15- or 20-ampere d) 15-, 20-, or 25-ampere
______	______	4. The minimum-size THHN copper neutral conductor permitted for an 8¾-kW, 240-volt, 3-wire household electric range is _____. a) 12 AWG b) 10 AWG c) 8 AWG d) 6 AWG
______	______	5. Luminaires (fixtures) shall be constructed, or installed, so that adjacent combustible material will not be subjected to temperatures in excess of _____ degrees Fahrenheit.
______	______	6. A one-family dwelling has a front entrance on the east side of the house, an attached garage which has two 10 ft wide doors on the south side, and a back entrance door on the west side of the house. This one-family dwelling requires _____ lighting outlet(s) for these outdoor entrances.
______	______	7. Overcurrent devices shall not be located in the vicinity of easily ignitible material such as in _____.
______	______	8. A single receptacle installed on an individual branch-circuit shall have an ampere rating not less than _____% of the rating of the branch-circuit. a) 125 b) 100 c) 80 d) 50
______	______	9. Appliance receptacle outlets installed for specific appliances, such as laundry equipment, shall be placed within _____ ft of the intended location of the appliance.
______	______	10. Tap conductors for household cooking equipment supplied from a 50-ampere branch-circuit shall have an ampacity of not less than _____ amperes.
______	______	11. One cooktop and two ovens are connected to one branch-circuit in a kitchen of a one-family dwelling. The cooktop has a nameplate rating of 4 kW at 240 volts, while each oven has a nameplate rating of 5.5 kW at 240 volts. The minimum load calculation for this branch-circuit is _____ watts.
______	______	12. The rating of a single cord- and plug-connected utilization equipment shall not exceed _____% of the branch-circuit ampere rating.
______	______	13. A recessed incandescent luminaire (fixture) with a completely enclosed lamp installed in the ceiling of a clothes closet must have a minimum clearance of _____ in. between the luminaire (fixture) and the nearest point of the storage space.
______	______	14. Where the *NEC*® specifies that one equipment shall be "within sight" of another equipment, the specified equipment shall be visible and not more than _____ ft distant from the other.

***NEC*® Reference** **Answer**

__________ __________ 15. In a dwelling unit, a hallway measuring 21½ ft in length requires a minimum of _____ receptacle outlet(s).

__________ __________ 16. A permanently connected dishwasher has a nameplate rating of 780 volt-amperes. The circuit breaker is permitted to serve as the disconnecting means where it is:

I. within sight from the dishwasher.

II. capable of being locked in the open position.

III. capable of being locked in the closed position.

a) I only b) I or III only c) I or II only d) I, II, or III

__________ __________ 17. Interior stairway lighting outlets shall be controlled by a wall switch at each floor level, where the difference between levels is _____ steps or more.

__________ __________ 18. Supplementary overcurrent devices shall not be required to be _____.

__________ __________ 19. The maximum-length cord permitted on an electrically operated trash compactor is _____ in.

__________ __________ 20. For a one-family dwelling, at least one receptacle outlet shall be installed in each:

I. basement

II. attached garage

III. detached garage

a) I only b) I and II only c) II and III only d) I, II and III

__________ __________ 21. A refrigerator receptacle can be fed from a:

I. 15-ampere individual branch-circuit.

II. 20-ampere individual branch-circuit.

III. 20-ampere small appliance branch-circuit.

a) II only b) III only c) II or III d) I, II, or III

__________ __________ 22. A receptacle installed on the laundry branch-circuit, having no other receptacles, can be a:

I. 15-ampere single receptacle.

II. 15-ampere duplex receptacle.

III. 20-ampere duplex receptacle

a) III only b) I or III c) II or III d) I, II, or III

__________ __________ 23. At least one GFCI-protected receptacle shall be located within _____ ft from the outside edge of each bathroom sink.

__________ __________ 24. Receptacles are permitted under an overhanging countertop, but are not counted as required countertop receptacles, where the countertop extends more than _____ in. beyond its support base.

__________ __________ 25. A branch-circuit providing power to a bathroom receptacle may also provide power to:

I. two other receptacles, two luminaires (light fixtures), and an exhaust fan in the same bathroom.

II. two other receptacles in the same bathroom and receptacles in two additional bathrooms.

III. one luminaire (light fixture) in the same bathroom and one luminaire (light fixture) in another bathroom.

a) I only b) I or II only c) II or III only d) I, II, or III

SECTION TWO: ONE-FAMILY DWELLINGS

Load Calculations

Objectives

After studying this unit, the student should:

- be able to calculate the general lighting load in a one-family dwelling.
- know the minimum volt-ampere requirements for small appliance and laundry branch-circuits.
- know how to apply demand factors to the general lighting load.
- be able to apply demand factors to fastened-in-place appliances.
- be able to calculate feeder-demand loads for household clothes dryers.
- know how to calculate feeder-demand loads for household cooking equipment
- be able to calculate heating and air-conditioning feeder-demand loads.
- be able to calculate a one-family dwelling service and/or feeder (using the standard method.)
- be able to calculate a one-family dwelling service and/or feeder (using the optional method.)
- know how to size service and feeder conductors.
- be able to calculate and choose the appropriate size neutral conductor.
- understand how the grounding electrode conductor is selected.

Introduction

Load calculations must be performed in order to determine the size of services and feeders. Service conductors are the conductors between the power provider and the disconnecting means, whether a main breaker or main lug panel. A feeder is defined as all circuit conductors between the service disconnecting means, and the panelboard containing fuses or breakers which feed branch-circuits. Additional load calculations will be needed for dwellings with panelboard(s) that are not part of the service. For example, a separate panelboard might be installed near the middle of the house to help eliminate voltage drop in branch-circuit conductors. Since the remote panelboard (sub-panel) does not carry the total load, an additional calculation is needed to size the feeder and panelboard.

This Unit simplifies the standard, as well as optional, load calculation methods for one-family dwellings. A blank form is provided at the beginning of each method. The floor plan of the one-family dwelling being calculated is found on page 96 while the specifications are on page 138. Both calculation methods will use the same specifications. Each line on each form is explained in detail using that information. The completed form follows the explanations, showing all of the information, together in one place.

COMPILING INFORMATION ESSENTIAL TO LOAD CALCULATIONS

Total Floor Area Square Footage

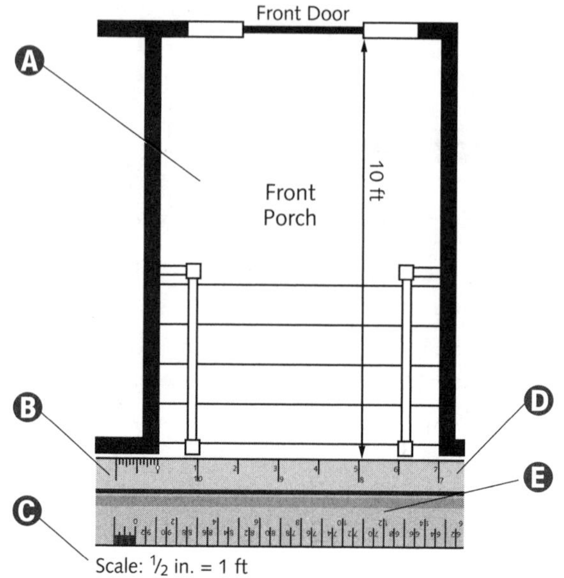

A Open porches are not included in the calculated floor area of dwelling units »220.3(A)«. Although this porch has a roof, it is considered open, and therefore, is not included in the calculation.

B Scale rulers are also useful in determining the location of outlets (receptacles, lights, switches, etc.).

C Before a scale ruler can be used to determine dimensions, the scale used to produce the blueprint (floor plan) must be known. (Then the area of an exterior porch, for example, is simple to calculate.)

D Scale rulers are useful in determining the total square footage of a structure (house, building, etc.) that has been drawn on a set of blueprints (floor plans).

E A regular tape measure (ruler, yard stick, etc.) can be used, but the use of a scale ruler makes the job much easier.

Gathering Information

Gathering certain information is necessary to accurately perform load calculations. All the essential data can be identified by simply answering the questions found on the load calculation form, either standard or optional. Enter the information on the appropriate line of the load calculation forms provided with this text book.

A What is the dwelling's total square-foot area, using the outside dimensions?

B What fastened-in-place appliances will be installed and what is the volt-ampere load of each?

C How many small appliance branch-circuits will be installed?

D How many laundry circuit(s) will be installed?

E If an electric clothes dryer will be installed, what is the volt-ampere load?

F What household cooking equipment will be installed and what is the volt-ampere load of each?

G Will there be an air-conditioning system? If so, what is the total volt-ampere load, including the compressor and fan motors? If an electric heating system will be installed, what is the total volt-ampere load, including the strip heat and blower motor?

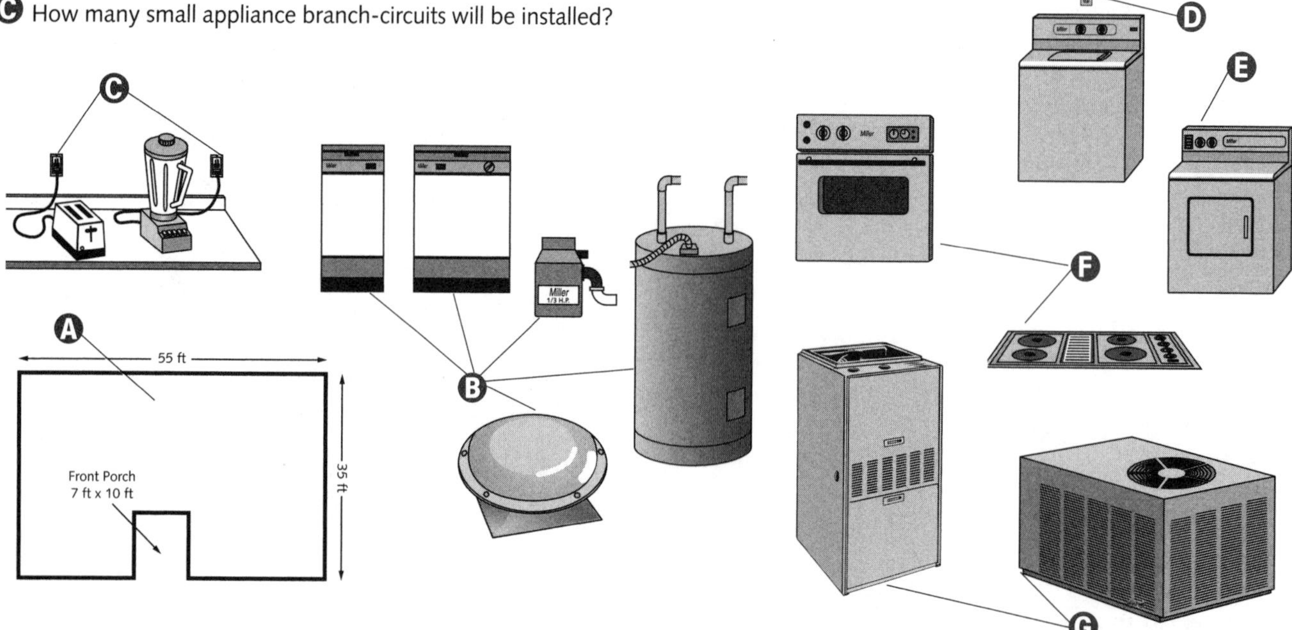

STANDARD METHOD: ONE-FAMILY DWELLINGS

Standard Method Load Calculation for One-Family Dwellings

1 General Lighting and Receptacle Loads 220.3(A)
Do not include open porches, garages, or unused or unfinished spaces not adaptable for future use.
3 × ______ (sq ft using outside dimensions) = **1**

2 Small Appliance Branch-Circuits 220.16(A)
*At least **two** small appliance branch-circuits must be included. 210.11(C)(1)*
1500 × ______ (minimum of two) = **2**

3 Laundry Branch Circuit(s) 220.16(B)
*At least **one** laundry branch-circuit must be included. 210.11(C)(2)*
1500 × ______ (minimum of one) = **3**

4 Add lines 1, 2, and 3 **4**

Lines 5 through 8 utilize the demand factors found in Table 220.11.

5 ______ (line 4) – 3000 = **5** (if 117,000 or less, skip to line 8)

6 ______ (line 5, if more than 117,000) – 117,000 = **6**

7 ______ (line 6) × 25% = **7**

8 ______ (smaller of line 5 or 117,000) × 35% = **8**

9 Total General Lighting and Receptacle Load 3000 + ______ (line 7) + ______ (line 8) = **9**

10 Fastened-In-Place Appliances 220.17
Use the nameplate rating.
Do not include electric ranges, clothes dryers, space-heating equipment, or air-conditioning equipment.

water heater /	dishwasher /	disposer /
compactor /	/	/
/	/	/

If fewer than four units, put total volt-amperes on line 10.
If four or more units, multiply total volt-amperes by 75%. ______ (volt-amps of four or more) × 75% = **10**

11 Clothes Dryers 220.18
(If present, otherwise skip to line 12.) Use 5000 watts or the nameplate rating, whichever is larger.
(The neutral demand load is 70% for feeders. 220.22) **11**

12 Ranges, Ovens, Cooktops, and Other Household Cooking Appliances Over 1750 Watts 220.19
(If present, otherwise skip to line 13.) Use Table 220.19 and all of the applicable Notes.
(The neutral demand load is 70% for feeders. 220.22) **12**

13 Heating or Air-Conditioning System (Compare the heat and A/C, and omit the smaller.) 220.21
Include the air handler when using either one. For heat pumps, include the compressor and the maximum amount of electric heat that can be energized while the compressor is running. **13**

14 Largest Motor (one motor only) 220.14 and 430.24
Multiply the volt-amperes of the largest motor by 25%. ______ (volt-amps of largest motor) × 25% = **14**

15 Total Volt-Ampere Demand Load: *Add lines 9 through 14 to find the minimum required volt-amperes.* **15**

16 Minimum Amperes
Divide the total volt-amperes by the voltage
______ (line 15) ÷ ______ (voltage) = **16** ______ (minimum amperes)

17 **Minimum Size Service and/or Feeder 240.6(A)** **17**

18 Size the Service and/or Feeder Conductors.
Use 310.15(B)(6) to find the service conductors up to 400 amperes.
Ratings in excess of 400 amperes shall comply with Table 310.16.
310.15(B)(6) also applies to feeder conductors serving as the main power feeder.
Minimum Size Conductors **18**

19 Size the Neutral Conductor 220.22
310.15(B)(6) states that the neutral service and/or feeder conductor can be smaller than the ungrounded (hot) conductors, provided the requirements of 215.2, 220.22, and 230.42 are met.
250.24(B)(1) states that the neutral cannot be smaller than the required grounding electrode conductor specified in Table 250.66.
Minimum Size Neutral Conductor **19**

20 Size the Grounding Electrode Conductor (for Service) 250.66
Using line 18 to find the grounding electrode conductor in Table 250.66.
Size the Equipment Grounding Conductor (for Feeder) 250.122
Use line 17 to find the equipment grounding conductor in Table 250.122.
Equipment grounding conductor types are listed in 250.118.
Minimum Size Grounding Electrode Conductor **20**

Sample One-Family Dwelling Load Calculation

Assume water heater, clothes dryer, counter-mounted cooking unit, wall-mounted oven, and electric heat kW ratings equivalent to kVA.

> NOTE
>
> *The information to the right will be used throughout the remainder of this unit to demonstrate the use of the load calculation forms, both standard and optional.*

Dwelling outside dimensions	35 ft x 55 ft
Front porch (included within outside dimensions)	7 ft x 10 ft
Small appliance branch-circuits	four
Laundry branch-circuits	two
Water heater	4.5 kW, 240 volt
Dishwasher	10 amperes, 120 volt
Waste (garbage) disposer	½HP, 115 volt
Trash compactor	7.5 amperes, 120 volt
Three attic fans (4.2 amperes each)	12.6 amperes, 120 volt
Clothes dryer	5.5 kW, 240 volt
Counter-mounted cooking unit (cook top)	7 kW, 240 volt
Wall-mounted oven	6 kW, 240 volt
Electric heat (3 banks @ 5 kW each)	15 kW, 240 volt
Air handler (blower motor)	3.2 amperes, 115 volt
Air-conditioner compressor	16.6 amperes, 230 volt
Condenser fan motor	2 amperes, 115 volt

Line 1—General Lighting and Receptacle Loads

A In Table 220.3(A) the general lighting loads are listed according to the type of occupancy and are determined by requiring a minimum lighting load for each square foot of floor area.

B The lighting load for residential dwelling units is 3 volt-amperes (VA) per sq ft.

C The dwelling floor plan on page 96 has an outside dimension of 35' × 55' (1925 sq ft), but there is an open porch with a dimension of 7 ft × 10 ft (70 sq ft). Since open porches are not included in the total area, 70 is deducted from 1925, leaving 1855 square ft.

D The floor area for each floor shall be computed using the outside dimensions. The total floor area does not include open porches, garages, or unused (including unfinished) spaces that are not adaptable for future use »220.3(A)«.

> NOTE
>
> *All 15 or 20 ampere general-use receptacle outlets (except for small appliance and laundry), are considered part of the general lighting load. The single asterisk (*) footnote at the bottom of Table 220.3(A) refers to 220.3(B)(10), which states that no additional load calculations are required for these outlets.*

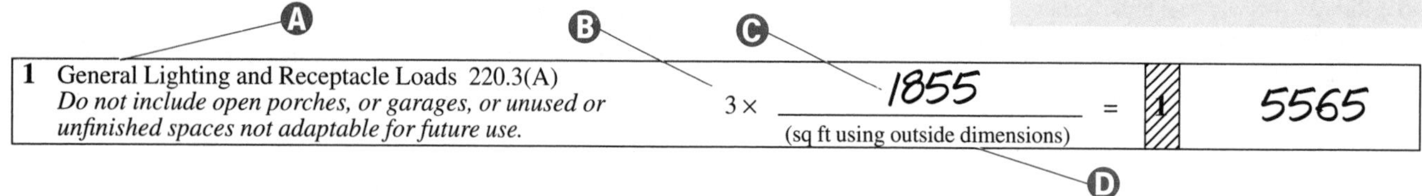

1 General Lighting and Receptacle Loads 220.3(A) *Do not include open porches, or garages, or unused or unfinished spaces not adaptable for future use.*	3 × 1855 (sq ft using outside dimensions)	=	1	5565

Line 2—Small Appliance Branch-Circuits

A Small appliance branch-circuits must be included when calculating a dwelling unit service »220.16(A)«.

B Each small appliance branch-circuit is calculated at no less than 1500 volt-amperes »220.16(A)«.

C These loads can be included with the general lighting load and subjected to the demand factors provided in Table 220.11 »220.16(A)«.

D A dwelling unit must have at least two small appliance branch-circuits »210.11(C)(1)«.

> NOTE
>
> *An individual branch-circuit for refrigeration equipment (permitted by 210.52(B)(1)* Exception No. 2*) can be excluded from this calculation »220.16(A)* Exception«.
>
> *If a dwelling has more than one feeder, a separate load calculation is needed for each. It is not necessary to include this step if a feeder does not supply small appliance branch-circuit(s). A load calculation for a feeder supplying small appliance branch-circuit(s) must include at least 1500 volt-amperes for each 2-wire circuit »220.16(A)«.*

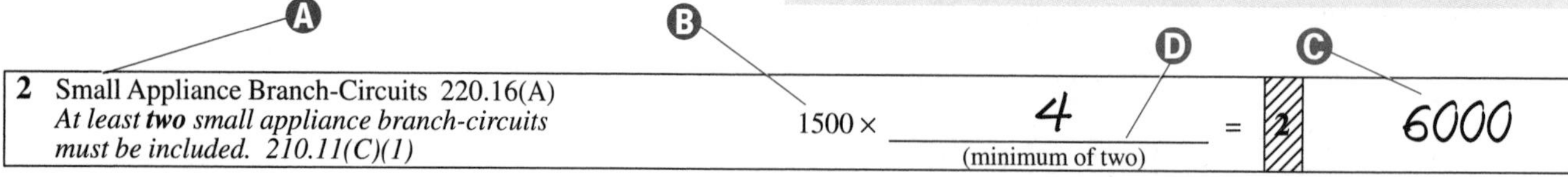

2 Small Appliance Branch-Circuits 220.16(A) *At least **two** small appliance branch-circuits must be included. 210.11(C)(1)*	1500 × 4 (minimum of two)	=	2	6000

Line 3—Laundry Branch Circuit(s)

A Laundry branch-circuit(s) must be included when calculating a dwelling unit service »220.16(B)«.

B Each laundry branch-circuit is calculated at no less than 1500 volt-amperes »220.16(B)«.

C At least one laundry branch-circuit is required per dwelling unit »210.11(C)(2)«.

D These loads can be included with the general lighting load and subjected to the demand factors provided in Table 220.11 »220.16(B)«.

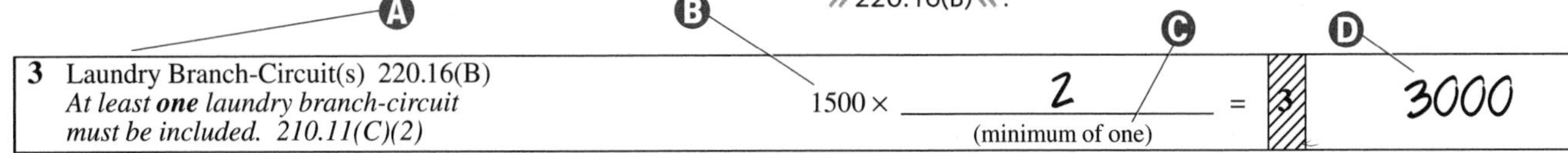

3 Laundry Branch-Circuit(s) 220.16(B) *At least **one** laundry branch-circuit must be included. 210.11(C)(2)*	1500 × 2 (minimum of one) =	3	3000

Lines 4 Through 8—Applying Demand Factors Found in Table 220.11

A Line 4 is simply the total of lines 1 through 3.

B Lines 5 through 8 outline the procedure that derates by using demand factors provided in Table 220.11.

C Line 5 is the result of subtracting 3000 from the number in Line 4.

D The first 3000 (or less) volt-amperes must remain at 100% »Table 220.11«.

E If Line 5 is 117,000 or less, then Lines 6 and 7 can be skipped.

F If Line 4 is 120,000 or less, then Lines 6 and 7 will remain empty.

G Insert the smaller of Line 5 or 117,000.

H The next 117,000 (from 3001 to 120,000) is multiplied by 35% »Table 220.11«.

I The result found on line 8 has been rounded up from 4047.75 to the next whole number, thus eliminating the decimal.

NOTE

Generally, one-family dwelling load calculations will not use Lines 6 and 7. (For example: a one-family dwelling with a total floor area of 25,000 sq ft, 20 small appliance branch-circuits, and 10 laundry circuits will not need Lines 6 and 7.)

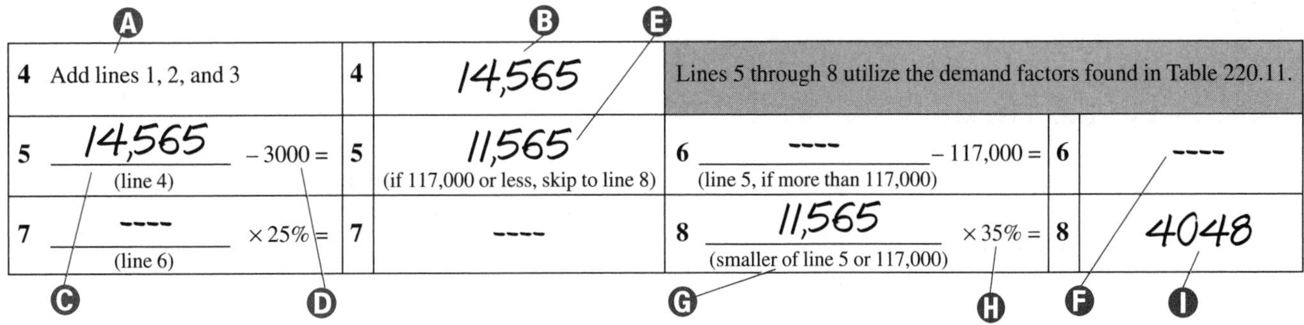

4 Add lines 1, 2, and 3	4	14,565	Lines 5 through 8 utilize the demand factors found in Table 220.11.		
5 14,565 (line 4) − 3000 =	5	11,565 (if 117,000 or less, skip to line 8)	6 ---- (line 5, if more than 117,000) − 117,000 =	6	----
7 ---- (line 6) × 25% =	7	----	8 11,565 (smaller of line 5 or 117,000) × 35% =	8	4048

Line 9—Total General Lighting (and Receptacle) Load

A Line 9 is the total of general lighting, and general-use receptacles, small appliance branch-circuits, and laundry branch circuit(s) after derating.

B If Line 7 was skipped (left blank), leave this blank also.

C This shaded block draws attention to the first of six lines (boxes) that will be added together to give the total volt-ampere demand load for the one-family dwelling.

D If performing a load calculation for a feeder (and panel) that supplies less than 1000 sq ft of floor area and does not supply small appliance or laundry branch-circuits, the result could be less than 3000. For example, if a feeder is installed to supply only 800 sq ft of floor area, and will not supply any small appliance or laundry branch-circuits, then Line 9 will be 2400 (3 VA per sq ft × 800 sq ft = 2400).

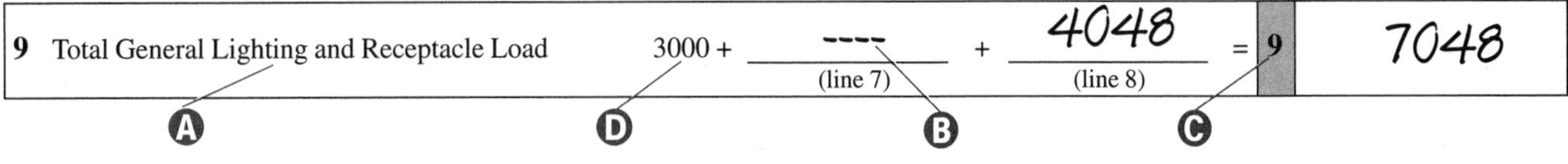

9 Total General Lighting and Receptacle Load	3000 + ---- (line 7) + 4048 (line 8) =	9	7048

Line 10—Fastened-In-Place Appliances

A The term **appliance** designates utilization equipment commonly built in standardized types and sizes, installed as a unit to perform specific function(s) »Article 100«.

B For the purpose of this load calculation, a kW rating will be the same as a kVA rating.

C Multiplying voltage by amperage produces volt-amperes (VA).

D Horsepower ratings must be converted to volt-amperes (VA) before they can be added to this form. The full load current ratings for single-phase motors are found in Table 430.148. Multiply the amperes (9.8) by the rated voltage (115) to find the volt-amperes (1127).

E No derating is allowed when there are only 1, 2, or 3 fastened-in-place appliances.

F If the feeder being calculated contains at least 4 fastened-in-place appliances, the combined volt-ampere rating of those appliances is multiplied by 75% and placed on Line 10 »220.17«.

G Space is provided for additional fastened-in-place appliances.

> **NOTE**
>
> *Electric ranges, dryers, space-heating equipment, and air-conditioning equipment are not included as fastened-in-place appliances »220.17«.*
>
> *Household cooking appliances, individually rated in excess of 1750 watts, are derated under 220.19 provisions.*

10 Fastened-In-Place Appliances 220.17 *Use the nameplate rating.* *Do not include electric ranges, clothes dryers, space-heating equipment, or air-conditioning equipment.*	water heater / 4500	dishwasher / 1200	disposer / 1127	
	compactor / 900	3 attic fans / 1512	/	
	/	/	/	
If fewer than four units, put total volt-amperes on line 10. If four or more units, multiply total volt-amperes by 75%.		9239 (volt-amps of four or more)	× 75% = **10**	6929

Line 11—Clothes Dryers

A A clothes dryer is not a requirement for a load calculation. Skip this line if there is no clothes dryer.

B If one of the circuits on the feeder being calculated is a clothes dryer, the rating for the dryer must be at least 5000 watts (volt-amperes) »220.18«.

C A clothes dryer neutral load on a feeder is calculated at 70% »220.22«.

D If the nameplate rating is more than 5000 watts (volt-amperes), the larger number is used »220.18«.

E 5.5 kW is equal to 5500 watts (volt-amperes).

11 Clothes Dryers 220.18 *(If present, otherwise skip to line 12.) Use 5000 watts or the nameplate rating, whichever is larger.* (The neutral demand load is 70% for feeders. 220.22)	**11**	5500

Line 12—Household Cooking Appliances

A Household cooking appliances (ranges, wall-mounted ovens, counter-mounted cooking units, etc.) are not required in a load calculation. Skip this line if there are no cooking appliances rated over 1¾ kW.

B Individual household cooking appliances rated more than 1750 watts (volt-amperes) can be derated by using Table 220.19 demand factors (and all of the applicable notes).

C The neutral load on a feeder is calculated at 70% for household electric ranges, wall-mounted ovens, and counter-mounted cooking units »220.22«.

D A 7 kW counter-mounted cooking unit (cook top) and a 6 kW wall-mounted oven are included in this calculation. Both units fall within the parameters of Column B. The demand factor percentage found in that column for 2 units is 65%. Multiply the total kW (13) by the demand factor percentage (65%) to find the derated load (8.45 kW). 8.45 kW is equal to 8450 watts (volt-amperes).

12 Ranges, Ovens, Cooktops, and Other Household Cooking Appliances Over 1750 Watts 220.19 *(If present, otherwise skip to line 13.) Use Table 220.19 and all of the applicable Notes.* (The neutral demand load is 70% for feeders. 220.22)	12	8450

Line 13—Heating or Air-Conditioning System

A The smaller of two (or more) **noncoincident loads** can be omitted, as long as they will never be energized simultaneously »220.21«. By definition, noncoincident means not occurring at the same time. A dwelling's electric heating system and air-conditioning system may be noncoincident loads.

B The air handler (blower motor) has a rating of 368 volt-amperes (3.2 × 115 = 368). Since both loads will be energized simultaneously, add the air handler to the 15 kW of electric heat.

C Fixed electric space heating loads are computed at 100% of the total connected load »220.15«.

D The air handler (blower) works with either the heating or the air conditioning. Add it to both calculations.

E The air-conditioner compressor (16.6 × 230 = 3818 VA), condenser fan motor (2 × 115 = 230 VA), and blower motor (3.2 × 115 = 368 VA) have a combined load of 4416 volt-amperes. Because this is less than the heating load, omit the A/C compressor and the condenser fan motor.

NOTE

A heat pump with supplementary heat is not considered a noncoincident load. Add the compressor full-load current to the maximum amount of heat that can be energized while the compressor is running. Remember to include all associated motor(s)—blower, condenser, etc.

13 Heating or Air-Conditioning System (Compare the heat and A/C, and omit the smaller.) 220.21 *Include the air handler when using either one. For heat pumps, include the compressor and the maximum amount of electric heat that can be energized while the compressor is running.*	13	15,368

Line 14—Largest Motor

A Multiply the largest motor in the dwelling calculation by 25% »220.14 and 430.24«.

B The largest motor is usually the air-conditioner compressor. If the air conditioner load has been omitted, use the second largest motor. In this case, use the ½-HP-garbage disposer motor.

C The result of 281.75 has been rounded up, thus eliminating the decimal.

14 Largest Motor (one motor only) 220.14 and 430.24 *Multiply the volt-amperes of the largest motor by 25%.*	1127 (volt-amps of largest motor) × 25% =	14	282

Line 15—Total Demand Load

A Add lines 9 through 14 to find the service's minimum volt-ampere load.

15 Total Volt-Ampere Demand Load: *Add lines 9 through 14 to find the minimum required volt-amperes.*	**15**	43,577

Lines 16 and 17—Minimum Service and/or Feeder

A Place the total volt-ampere amount found in Line 15 here.

B Write down the source voltage that will supply the feeder.

C Fractions of an ampere 0.5 and higher are rounded up, while fractions less than 0.5 are dropped »220.2(B)«.

D The service (or feeder) overcurrent protection must be higher than the number found in Line 16. 240.6(A) provides a list of standard ampere ratings for fuses and circuit breakers.

E Find the amperage by dividing the volt-amperes by the voltage.

F This is the minimum amperage rating required for the service and/or feeder, being calculated.

G The next standard size fuse or breaker above 182 amperes is 200 »240.6(A)«.

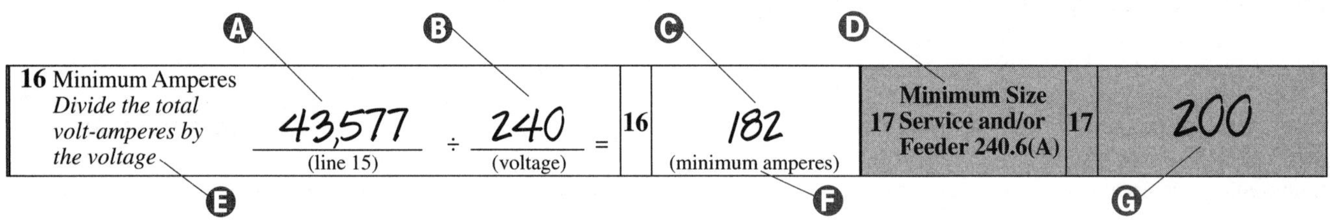

16 Minimum Amperes *Divide the total volt-amperes by the voltage*	43,577 (line 15) ÷ 240 (voltage) =	**16**	182 (minimum amperes)	**Minimum Size 17 Service and/or Feeder 240.6(A)**	**17**	200

Line 18—Minimum Size Conductors

A Use 310.15(B)(6) to find service entrance and service lateral conductors.

B 310.15(B)(6) also applies to feeder conductors that serve as the main power feeder to a dwelling unit and are installed in a raceway or cable (with or without an equipment grounding conductor). It is not required that the feeder conductors be larger than the service-entrance conductors »310.15(B)(6)«.

C The overcurrent device rating must be 400 amperes, or less.

D Table 310.16 requires 3/0 AWG (75°C) copper conductors for 200 amperes. However, since this is a dwelling service, use Table 310.15(B)(6), which allows 2/0 AWG copper conductors.

E The main power feeder is the feeder(s) that supplies the lighting and appliance branch-circuit panelboard(s) »310.15(B)(6)«.

F Table 310.15(B)(6) provides sizes for both copper and aluminum conductors. Copper-clad aluminum conductors are listed in the column with aluminum.

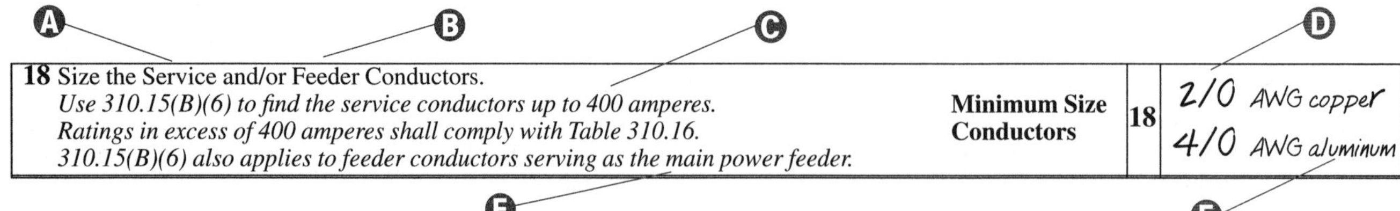

18 Size the Service and/or Feeder Conductors. *Use 310.15(B)(6) to find the service conductors up to 400 amperes. Ratings in excess of 400 amperes shall comply with Table 310.16. 310.15(B)(6) also applies to feeder conductors serving as the main power feeder.*	**Minimum Size Conductors**	**18**	2/0 AWG copper 4/0 AWG aluminum

Line 19—Neutral Conductor

A No neutral conductor is connected to the water heater in this calculation.

B All fastened-in-place appliances utilizing a grounded conductor must be included.

C The electric heat and A/C compressor are 240 volt loads (no neutral).

CAUTION *The neutral must not be smaller than the required grounding electrode conductor specified in Table 250.66 »250.24(B)(1)«.*

D If the largest motor is 120 (115) volts, use the motor load in Line 14.

E For the purpose of this load calculation, the terms **Neutral** and **Grounded** are synonymous.

F The feeder (or service) neutral load is the maximum unbalance of the load determined by Article 220 »220.22«.

G The neutral (grounded) conductor can be smaller than the ungrounded (hot) conductors provided certain requirements in 215.2, 220.22, and 230.42 are met »310.15(B)(6)«.

H All numbers will be shown as volt-amperes.

I Since all of the general lighting and receptacle loads have a rating of 120 volts, the total volt-ampere rating (from Line 9) must be included in the neutral calculation.

J A demand factor of 75% can be applied to the nameplate rating load of 4 or more fastened-in-place appliances »220.17«.

K Multiply the clothes dryer load (Line 11) by 70% to determine the neutral load »220.21«.

L Multiplying Line 12 by 70% provides the neutral cooking appliances load »220.21«.

M Because both of these motors will be energized simultaneously, thereby contributing to the neutral load, they must be included in the neutral calculation.

N The total neutral volt-ampere load (21,247) divided by the voltage (240) renders the minimum neutral amperes (89).

O Size the neutral by finding the maximum unbalanced load.

P A 3 AWG (75°C) copper conductor is required for 89 amperes »Table 310.16«.

Q A 2 AWG (75°C) aluminum conductor is required for 89 amperes »Table 310.16«.

General lighting and receptacles (Line 9) .... 7048 (H) (I)
Fastened-in-place appliances
(A) **Water heater 0**
(B) **Dishwasher 1200**
Waste (garbage) disposer........ 1127
Trash compactor 900
Three attic fans .... 1512 (J)
Total........................ 4739

Total fastened-in-place appliances (4739 × 75%)... 3554 (K)
Clothes dryer (Line 11 × 70%).................. 3850
Cooking appliances (Line 12 × 70%)............. 5915 (L)
(C) **Electric heat.................................. 0**
Air-conditioner compressor 0
Air handler (blower motor)...................... 368 (M)
Condenser fan motor 230
(D) **Largest motor 282**
TOTAL .... 21,247 (N)

(E) (F) (G) (D) (P)

19 Size the Neutral Conductor 220.22 *310.15(B)(6) states that the neutral service and/or feeder conductor can be smaller than the ungrounded (hot) conductors, provided the requirements of 215.2, 220.22, and 230.42 are met.* *250.24(B)(1) states that the neutral cannot be smaller than the required grounding electrode conductor specified in Table 250.66.*	**Minimum Size Neutral Conductor**	**19**	3 copper 2 aluminum

(Q)

Line 20—Grounding Electrode Conductor

The following page shows the complete one-family dwelling load calculation according to the standard method.

A Use 310.15(B)(6) to find service entrance and service lateral conductors.

B The minimum-size grounding electrode conductor is 4 AWG copper (or 2 AWG aluminum) »Table 250.66«.

C In outdoor applications, aluminum (or copper-clad aluminum) grounding conductors must not be installed within 18 in. (450 mm) of the earth »250.64(A)«.

(A) (B) (C)

20 Size the Grounding Electrode Conductor (for Service) 250.66 *Using line 18 to find the grounding electrode conductor in Table 250.66.* Size the Equipment Grounding Conductor (for Feeder) 250.122 *Use line 17 to find the equipment grounding conductor in Table 250.122.* *Equipment grounding conductor types are listed in 250.118.*	**Minimum Size Grounding Electrode Conductor**	**20**	4 copper 2 aluminum

Standard Method Load Calculation for One-Family Dwellings

1 General Lighting and Receptacle Loads 220.3(A)
Do not include open porches, or garages, or unused or unfinished spaces not adaptable for future use.
3 × 1855 (sq ft using outside dimensions) = **1** 5565

2 Small Appliance Branch-Circuits 220.16(A)
*At least **two** small appliance branch-circuits must be included. 210.11(C)(1)*
1500 × 4 (minimum of two) = **2** 6000

3 Laundry Branch-Circuit(s) 220.16(B)
*At least **one** laundry branch-circuit must be included. 210.11(C)(2)*
1500 × 2 (minimum of one) = **3** 3000

4 Add lines 1, 2, and 3 — **4** 14,565

Lines 5 through 8 utilize the demand factors found in Table 220.11.

5 14,565 (line 4) − 3000 = **5** 11,565 (if 117,000 or less, skip to line 8)

6 ---- (line 5, if more than 117,000) − 117,000 = **6** ----

7 ---- (line 6) × 25% = **7** ----

8 11,565 (smaller of line 5 or 117,000) × 35% = **8** 4048

9 Total General Lighting and Receptacle Load 3000 + ---- (line 7) + 4048 (line 8) = **9** 7048

10 Fastened-In-Place Appliances 220.17
Use the nameplate rating. Do not include electric ranges, clothes dryers, space-heating equipment, or air-conditioning equipment.

water heater / 4500	dishwasher / 1200	disposer / 1127
compactor / 900	3 attic fans / 1512	/
/	/	/

If fewer than four units, put total volt-amperes on line 10.
If four or more units, multiply total volt-amperes by 75%.
9239 (volt-amps of four or more) × 75% = **10** 6929

11 Clothes Dryers 220.18
(If present, otherwise skip to line 12.) Use 5000 watts or the nameplate rating, whichever is larger.
(The neutral demand load is 70% for feeders. 220.22) — **11** 5500

12 Ranges, Ovens, Cooktops, and Other Household Cooking Appliances Over 1750 Watts 220.19
(If present, otherwise skip to line 13.) Use Table 220.19 and all of the applicable Notes.
(The neutral demand load is 70% for feeders. 220.22) — **12** 8450

13 Heating or Air-Conditioning System (Compare the heat and A/C, and omit the smaller.) 220.21
Include the air handler when using either one. For heat pumps, include the compressor and the maximum amount of electric heat that can be energized while the compressor is running. — **13** 15,368

14 Largest Motor (one motor only) 220.14 and 430.24
Multiply the volt-amperes of the largest motor by 25%.
1127 (volt-amps of largest motor) × 25% = **14** 282

15 Total Volt-Ampere Demand Load: *Add lines 9 through 14 to find the minimum required volt-amperes.* — **15** 43,577

16 Minimum Amperes
Divide the total volt-amperes by the voltage
43,577 (line 15) ÷ 240 (voltage) = **16** 182 (minimum amperes)

17 Minimum Size Service and/or Feeder 240.6(A) — **17** 200

18 Size the Service and/or Feeder Conductors.
Use 310.15(B)(6) to find the service conductors up to 400 amperes. Ratings in excess of 400 amperes shall comply with Table 310.16. 310.15(B)(6) also applies to feeder conductors serving as the main power feeder.
Minimum Size Conductors — **18** 2/0 AWG copper; 4/0 AWG aluminum

19 Size the Neutral Conductor 220.22
310.15(B)(6) states that the neutral service and/or feeder conductor can be smaller than the ungrounded (hot) conductors, provided the requirements of 215.2, 220.22, and 230.42 are met. 250.24(B)(1) states that the neutral cannot be smaller than the required grounding electrode conductor specified in Table 250.66.
Minimum Size Neutral Conductor — **19** 3 copper; 2 aluminum

20 Size the Grounding Electrode Conductor (for Service) 250.66
Using line 18 to find the grounding electrode conductor in Table 250.66.
Size the Equipment Grounding Conductor (for Feeder) 250.122
Use line 17 to find the equipment grounding conductor in Table 250.122. Equipment grounding conductor types are listed in 250.118.
Minimum Size Grounding Electrode Conductor — **20** 4 copper; 2 aluminum

OPTIONAL METHOD: ONE-FAMILY DWELLINGS

Optional Method Load Calculation for One-Family Dwellings

1 General Lighting and Receptacle Loads 220.30(B)(2)
Do not include open porches, garages, or unused or unfinished spaces not adaptable for future use.
3 × ________ (sq ft using outside dimensions) = **1** ________

2 Small Appliance Branch-Circuits 220.30(B)(1)
At least ***two*** *small appliance branch-circuits must be included. 210.11(C)(1)*
1500 × ________ (minimum of two) = **2** ________

3 Laundry Branch-Circuit(s) 220.30(B)(1)
At least ***one*** *laundry branch-circuit must be included. 210.11(C)(2)*
1500 × ________ (minimum of one) = **3** ________

4 Appliances 220.30(B)(3) and (4)
Use the nameplate rating of ***ALL*** *appliances (fastened-in-place, permanently connected, or connected to a specific circuit), ranges, ovens, cook tops, motors, and clothes dryers.*
Convert any nameplate rating given in amperes to volt-amperes by multiplying the amperes by the rated voltage.

Do not include any heating or air-conditioning equipment in this section.

Total volt-amperes of all appliances LISTED BELOW **4** ________

water heater /	clothes dryer /	range /
dishwasher /	disposer /	/
/	/	/
/	/	/
/	/	/

5 Apply 220.30(B) demand factor to the total of lines 1 through 4.

________ (total of lines 1 through 4) – 10,000 = ________ × 40% = ________ + 10,000 = **5** ________

6 Heating and/or Air-Conditioning System 220.30(C)
Use the nameplate rating(s) in volt-amperes for all applicable systems in lines ***a*** *through* ***e***.

a) Air-conditioning and cooling system(s), including heat pump compressors and supplemental heating, unless the controller prevents the compressor and supplemental heating from operating at the same time.
________ × 100% = **a)** ________

b) Electric thermal storage and other heating systems where the usual load is expected to be continuous at full nameplate value: *Systems qualifying under this selection shall not be figured under any other selection in 220.30(C).*
________ × 100% = **b)** ________

c) Central electric space-heating equipment, including integral supplemental heating in heat pumps where the controller prevents the compressor and supplemental heating from operating at the same time.
________ × 65% = **c)** ________

d) Electric space-heating equipment, if less than four separately controlled units:
________ × 65% = **d)** ________

e) Electric space-heating equipment, if four or more separately controlled units:
________ × 40% = **e)** ________

7 Total Volt-Ampere Demand Load: ________ (largest VA rating from line 6a through 6e) + ________ (line 5) = **7** ________

8 Minimum Amperes
Divide the total volt-amperes by the voltage
________ (line 7) ÷ ________ (voltage) = **8** ________ (minimum amperes)

9 **Minimum Size Service and/or Feeder 240.6(A)** **9** ________ (minimum is 100 amperes)

10 Size the Service and/or Feeder Conductors.
Use 310.15(B)(6) to find the service conductors up to 400 amperes.
Ratings in excess of 400 amperes shall comply with Table 310.16.
310.15(B)(6) also applies to feeder conductors serving as the main power feeder.
Minimum Size Conductors **10** ________

11 Size the Neutral Conductor 220.22
Note: There is no optional method for calculating the neutral conductor.
310.15(B)(6) states that the neutral service and/or feeder conductor can be smaller than the ungrounded (hot) conductors, provided the requirements of 215.2, 220.22, and 230.42 are met.
250.24(B)(1) states that the neutral cannot be smaller than the required grounding electrode conductor specified in Table 250.66.
Minimum Size Neutral Conductor **11** ________

12 Size the Grounding Electrode Conductor Table 250.66
Using line 10 to find the grounding electrode conductor in Table 250.66.
Size the Equipment Grounding Conductor (for Feeder) 250.122
Use line 9 to find the equipment grounding conductor in Table 250.122.
Equipment grounding conductor types are listed in 250.118.
Minimum Size Grounding Electrode Conductor **12** ________

Lines 1 Through 3—General Lighting and Receptacle Loads

A In the **Optional Method**, Lines 1, 2 and 3 are calculated exactly as described in the **Standard Method**. (See explanations on pages 138 and 139).

A

1	General Lighting and Receptacle Loads 220.30(B)(2) *Do not include open porches, garages, or unused or unfinished spaces not adaptable for future use.*	3 ×	1855 (sq ft using outside dimensions)	= 1	5565
2	Small Appliance Branch-Circuits 220.30(B)(1) *At least* ***two*** *small appliance branch-circuits must be included. 210.11(C)(1)*	1500 ×	4 (minimum of two)	= 2	6000
3	Laundry Branch-Circuit(s) 220.30(B)(1) *At least* ***one*** *laundry branch-circuit must be included. 210.11(C)(2)*	1500 ×	2 (minimum of one)	= 3	3000

Line 4—Appliances

A This Section includes appliances (fastened-in-place, permanently connected, or connected to a specific circuit), ranges, wall-mounted ovens, counter-mounted cooking units, clothes dryers, and water heaters »220.30(B)(3)«.

B Heating and air-conditioning systems are *not* included in this computation.

C Use the exact nameplate rating of the clothes dryer, even if less than 5000 volt-amperes.

D Do not apply the demand factors in Table 220.19 to household cooking equipment (ranges, cook tops, and ovens). List the nameplate ratings as they appear.

A B C

4 Appliances 220.30(B)(3) and (4)	*Do not include any heating or air-conditioning equipment in this section.*		Total volt-amperes of all appliances LISTED BELOW	4	27,739
Use the nameplate rating of ***ALL*** *appliances (fastened-in-place, permanently connected, or connected to a specific circuit), ranges, ovens, cook tops, motors, and clothes dryers. Convert any nameplate rating given in amperes to volt-amperes by multiplying the amperes by the rated voltage.*	water heater / 4500	clothes dryer / 5500	range /		
	dishwasher / 1200	disposer / 1127	cook top / 7000		
	compactor / 900	3 attic fans / 1512	oven / 6000		
	/	/	/		
	/	/	/		

D

Line 5—Applying Demand Factors to Lines 1 Through 4

A This demand factor is found in 220.30(B).

B This is the newly calculated load which includes everything except the heating and air-conditioning systems.

C (5565 + 6000 + 3000 + 27,739)

D Because the first 10,000 volt-amperes are calculated at 100%, subtract 10,000 here.

E The remainder of the load is calculated at 40%.

F The result on Line 5 has been rounded up from 12,921.6.

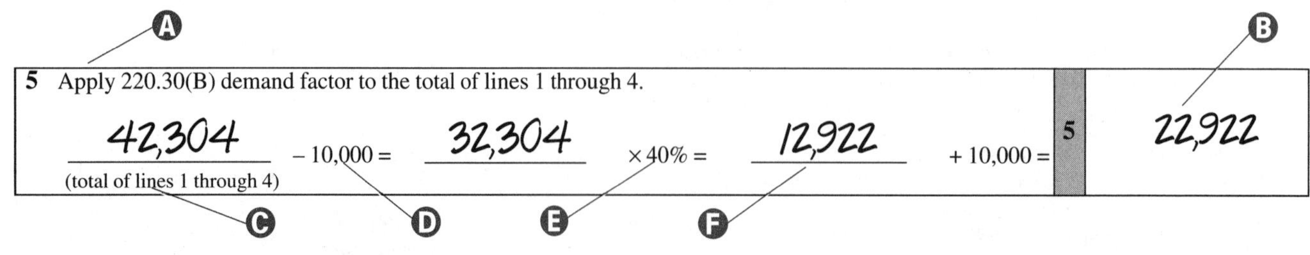

A B

5 Apply 220.30(B) demand factor to the total of lines 1 through 4.

42,304 (total of lines 1 through 4) – 10,000 = 32,304 × 40% = 12,922 + 10,000 = 5 22,922

C D E F

Line 6—Heating and/or Air-Conditioning Systems

Ⓐ The air-conditioner compressor (16.6 × 230 = 3818 VA), condenser fan motor (2 × 115 = 230 VA), and blower motor (3.2 × 115 = 368 VA) have a combined load of 4416 volt-amperes.

Ⓑ The electric heat (3 × 5000 = 15,000) added to the blower motor (3.2 × 115 = 368) is 15,368 volt-amperes.

> **NOTE**
>
> *The total volt-ampere rating of one, two, or three separately controlled heating units is multiplied by 65%. The total volt-ampere rating of four (or more) separately controlled heating units is multiplied by 40%.*

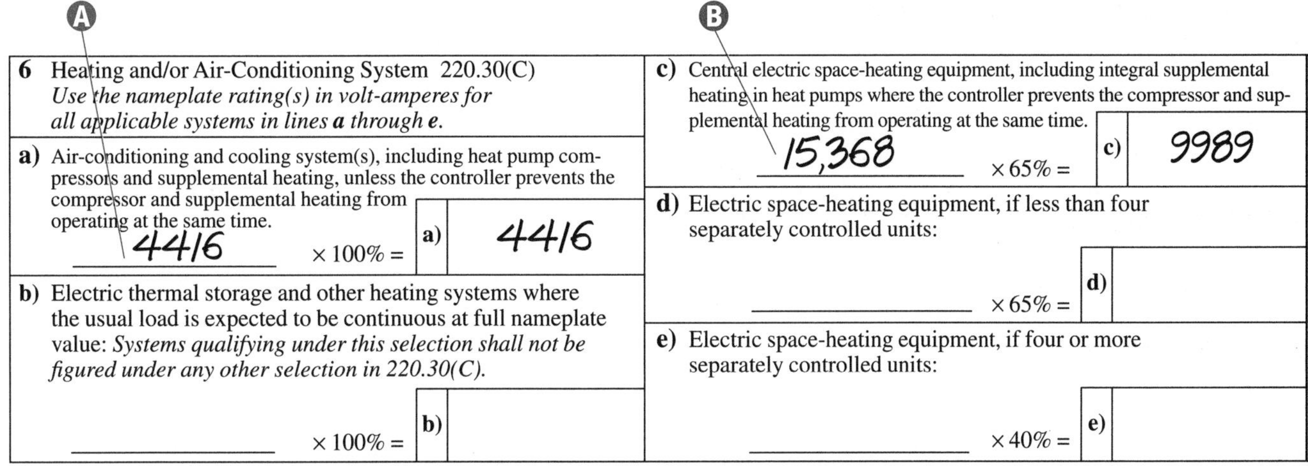

6 Heating and/or Air-Conditioning System 220.30(C) *Use the nameplate rating(s) in volt-amperes for all applicable systems in lines **a** through **e**.*	**c)** Central electric space-heating equipment, including integral supplemental heating in heat pumps where the controller prevents the compressor and supplemental heating from operating at the same time. 15,368 × 65% = **c)** 9989
a) Air-conditioning and cooling system(s), including heat pump compressors and supplemental heating, unless the controller prevents the compressor and supplemental heating from operating at the same time. 4416 × 100% = **a)** 4416	**d)** Electric space-heating equipment, if less than four separately controlled units: ______ × 65% = **d)**
b) Electric thermal storage and other heating systems where the usual load is expected to be continuous at full nameplate value: *Systems qualifying under this selection shall not be figured under any other selection in 220.30(C).* ______ × 100% = **b)**	**e)** Electric space-heating equipment, if four or more separately controlled units: ______ × 40% = **e)**

Line 7—Total Volt-Ampere Demand Load

Ⓐ The largest volt-ampere rating of the heating and/or air-conditioning system(s), after application of demand factors.

Ⓑ The volt-ampere demand load from Line 5.

Ⓒ The total volt-ampere load calculated by the optional method. (For comparison, this one- family dwelling, calculated by the standard method, is 43,577 volt-amperes.)

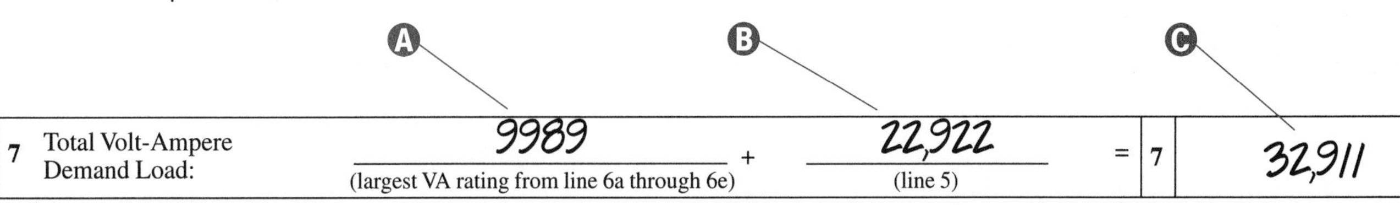

7 Total Volt-Ampere Demand Load:	9989 (largest VA rating from line 6a through 6e)	+	22,922 (line 5)	= **7**	32,911

Lines 8 and 9—Minimum Service and/or Feeder

Ⓐ The volt-ampere load from Line 7.

Ⓑ The source voltage.

Ⓒ The fraction of an ampere (.13) has been dropped.

Ⓓ The overcurrent protection chosen for the service (or feeder) must be higher than the number found on Line 8.

240.6(A) lists standard ampere ratings for fuses and circuit breakers.

Ⓔ The next standard size fuse or breaker above 137 amperes is 150. (By the standard method, this same dwelling requires a 200-ampere service.)

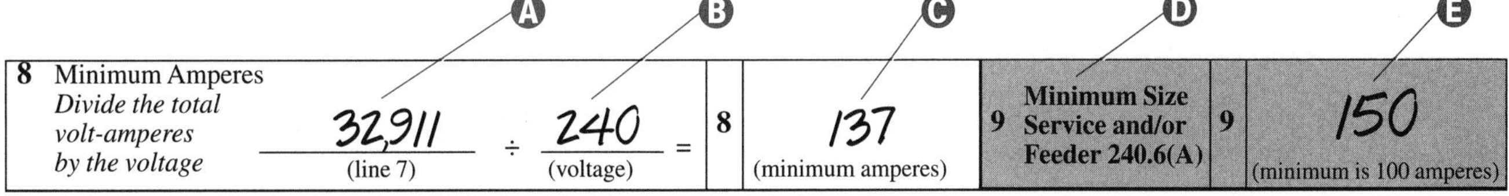

8 Minimum Amperes *Divide the total volt-amperes by the voltage*	32,911 (line 7)	÷	240 (voltage)	= **8**	137 (minimum amperes)	**9** Minimum Size Service and/or Feeder 240.6(A)	**9**	150 (minimum is 100 amperes)

Line 10—Minimum Size Conductors

A Use 310.15(B)(6) to find service conductors.

B 310.15(B)(6) also applies to feeder conductors that serve as the main power feeder to a dwelling unit and are installed in a raceway or cable (with or without an equipment grounding conductor). It is not required that the feeder conductors be larger than the service-entrance conductors »310.15(B)(6)«.

C The overcurrent device rating must be 400 amperes or less.

D Table 310.16 requires 1/0 AWG (75°C) copper conductors for 200 amperes. However, since this is a dwelling service, use Table 310.15(B)(6), which allows 1 AWG copper conductors.

E The main power feeder is the feeder(s) that supplies the lighting and appliance branch-circuit panelboard(s) »310.15(B)(6)«.

F Table 310.15(B)(6) provides copper and aluminum (and copper-clad aluminum) conductor sizes.

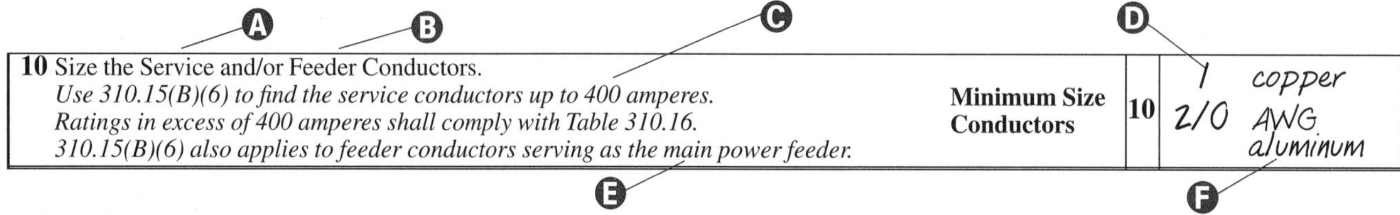

10 Size the Service and/or Feeder Conductors. *Use 310.15(B)(6) to find the service conductors up to 400 amperes.* *Ratings in excess of 400 amperes shall comply with Table 310.16.* *310.15(B)(6) also applies to feeder conductors serving as the main power feeder.*	**Minimum Size Conductors**	**10**	1 copper 2/0 AWG aluminum

Line 11—Neutral Conductor

A The neutral conductor must be calculated by the standard method.

B Review the complete neutral conductor explanation on page 143, since the calculation method is the same.

> NOTE
>
> *There is no optional method for calculating the neutral conductor.*

11 Size the Neutral Conductor 220.22 **Note: There is no optional method for calculating the neutral conductor.** *310.15(B)(6) states that the neutral service and/or feeder conductor can be smaller than the ungrounded (hot) conductors, provided the requirements of 215.2, 220.22, and 230.42 are met.* *250.24(B)(1) states that the neutral cannot be smaller than the required grounding electrode conductor specified in Table 250.66.*	**Minimum Size Neutral Conductor**	**11**	3 copper 2 aluminum

Line 12—Grounding Electrode Conductor

On the following page is the complete one-family dwelling load calculation according to the optional method.

A The minimum size grounding electrode conductor is 6 AWG copper (or 4 AWG aluminum) »Table 250.66«.

12 Size the Grounding Electrode Conductor Table 250.66 *Using line 10 to find the grounding electrode conductor in Table 250.66.* Size the Equipment Grounding Conductor (for Feeder) 250.122 *Use line 9 to find the equipment grounding conductor in Table 250.122.* *Equipment grounding conductor types are listed in 250.118.*	**Minimum Size Grounding Electrode Conductor**	**12**	6 copper 4 aluminum

Optional Method Load Calculation for One-Family Dwellings

1 General Lighting and Receptacle Loads 220.30(B)(2)
Do not include open porches, garages, or unused or unfinished spaces not adaptable for future use.
3 × 1855 (sq ft using outside dimensions) = **1** 5565

2 Small Appliance Branch-Circuits 220.30(B)(1)
*At least **two** small appliance branch-circuits must be included. 210.11(C)(1)*
1500 × 4 (minimum of two) = **2** 6000

3 Laundry Branch-Circuit(s) 220.30(B)(1)
*At least **one** laundry branch-circuit must be included. 210.11(C)(2)*
1500 × 2 (minimum of one) = **3** 3000

4 Appliances 220.30(B)(3) and (4)
*Use the nameplate rating of **ALL** appliances (fastened-in-place, permanently connected, or connected to a specific circuit), ranges, ovens, cook tops, motors, and clothes dryers.*
Convert any nameplate rating given in amperes to volt-amperes by multiplying the amperes by the rated voltage.

Do not include any heating or air-conditioning equipment in this section.

Total volt-amperes of all appliances LISTED BELOW **4** 27,739

water heater / 4500	clothes dryer / 5500	range /
dishwasher / 1200	disposer / 1127	cook top / 7000
compactor / 900	3 attic fans / 1512	oven / 6000
/	/	/
/	/	/

5 Apply 220.30(B) demand factor to the total of lines 1 through 4.
42,304 (total of lines 1 through 4) – 10,000 = 32,304 × 40% = 12,922 + 10,000 = **5** 22,922

6 Heating and/or Air-Conditioning System 220.30(C)
*Use the nameplate rating(s) in volt-amperes for all applicable systems in lines **a** through **e**.*

a) Air-conditioning and cooling system(s), including heat pump compressors and supplemental heating, unless the controller prevents the compressor and supplemental heating from operating at the same time.
4416 × 100% = **a)** 4416

b) Electric thermal storage and other heating systems where the usual load is expected to be continuous at full nameplate value: *Systems qualifying under this selection shall not be figured under any other selection in 220.30(C).*
× 100% = **b)**

c) Central electric space-heating equipment, including integral supplemental heating in heat pumps where the controller prevents the compressor and supplemental heating from operating at the same time.
15,368 × 65% = **c)** 9989

d) Electric space-heating equipment, if less than four separately controlled units:
× 65% = **d)**

e) Electric space-heating equipment, if four or more separately controlled units:
× 40% = **e)**

7 Total Volt-Ampere Demand Load: 9989 (largest VA rating from line 6a through 6e) + 22,922 (line 5) = **7** 32,911

8 Minimum Amperes
Divide the total volt-amperes by the voltage
32,911 (line 7) ÷ 240 (voltage) = **8** 137 (minimum amperes)

9 Minimum Size Service and/or Feeder 240.6(A) **9** 150 (minimum is 100 amperes)

10 Size the Service and/or Feeder Conductors.
Use 310.15(B)(6) to find the service conductors up to 400 amperes.
Ratings in excess of 400 amperes shall comply with Table 310.16.
310.15(B)(6) also applies to feeder conductors serving as the main power feeder.
Minimum Size Conductors **10** 1 copper; 2/0 AWG aluminum

11 Size the Neutral Conductor 220.22
Note: There is no optional method for calculating the neutral conductor.
310.15(B)(6) states that the neutral service and/or feeder conductor can be smaller than the ungrounded (hot) conductors, provided the requirements of 215.2, 220.22, and 230.42 are met.
250.24(B)(1) states that the neutral cannot be smaller than the required grounding electrode conductor specified in Table 250.66.
Minimum Size Neutral Conductor **11** 3 copper; 2 aluminum

12 Size the Grounding Electrode Conductor Table 250.66
Using line 10 to find the grounding electrode conductor in Table 250.66.
Size the Equipment Grounding Conductor (for Feeder) 250.122
Use line 9 to find the equipment grounding conductor in Table 250.122.
Equipment grounding conductor types are listed in 250.118.
Minimum Size Grounding Electrode Conductor **12** 6 copper; 4 aluminum

Summary

- A dwelling unit load calculation, for general lighting and receptacles, is determined by total sq ft.
- For this purpose, Table 220.3(A) requires a unit load of 3 volt-amperes per sq ft.
- At least 1 laundry and 2 small appliance branch-circuits are required when performing a service (or main power feeder) calculation.
- Laundry and small appliance branch-circuits are calculated using a rating of at least 1500 volt-amperes each.
- Table 220.11 demand factors are applied to the general lighting, small appliance, and laundry loads.
- The total volt-ampere rating of four or more fastened-in-place appliances is multiplied by 75%.
- Demand loads for household cooking appliances (over 1¾ kW rating) are found in Table 220.19.
- The heating and air-conditioning loads are compared, and the larger of the two is used.
- For heat pumps, include the compressor and the maximum amount of electric heat that can be energized while the compressor is running.
- Two load calculation methods are provided in Article 220, Standard and Optional.
- The neutral conductor can be smaller than the ungrounded (hot) conductors, provided the requirements of 215.2, 220.22, and 230.42 are met.
- Table 250.66 is used to size grounding electrode conductors.

Unit 8 Competency Test

***NEC®* Reference** **Answer**

______ ______ 1. It shall be permissible to apply a demand factor of _____ % to the nameplate rating load of _____ or more appliances fastened in place, other than electric ranges, clothes dryers, space-heating equipment, or air-conditioning equipment, that are served by the same feeder in a one-family dwelling.

______ ______ 2. What is the lighting demand load (before derating) for a house with outside dimensions of 30 ft × 48 ft on the first floor and 22 ft × 42 ft on the second floor?

______ ______ 3. A homeowner wants to add twelve general-purpose receptacle outlets to the electrical drawings, before the house is built. What additional load, in volt-amperes, will this add to the service-load calculation?

______ ______ 4. In each dwelling unit, the feeder load shall be computed at _____ volt-amperes for each 2-wire small appliance branch-circuit required by 210.11(C)(1).

______ ______ 5. What is the demand load for the service for two 3½-kW wall-mounted ovens and one 5-kW counter-mounted cooking unit? All of the appliances are supplied from a single branch-circuit and are located in the same room of a one-family dwelling.

a) 12 kW b) 9.6 kW c) 8 kW d) 6.6 kW

______ ______ 6. A 4.5-kVA, 240-volt clothes dryer will contribute _____ amperes to the neutral load, when calculating the service by the standard method.

______ ______ 7. A one-family dwelling contains the following appliances: a 1-kVA, 115-volt dishwasher; a ¼-HP, 115-volt waste (garbage) disposer, and a ⅓-HP, 115-volt trash compactor. Using the standard method, what is the load contribution, in volt-amperes, for these appliances?

______ ______ 8. For one-family dwellings, the service disconnecting means must have a rating of at least _____ amperes, 3-wire.

______ ______ 9. A one-family dwelling has a 175-ampere, 240-volt service that is fed with 75°C conductors. What is the minimum size copper grounding electrode conductor required?

NEC® Reference	Answer	
________	________	10. A one-family dwelling has six 3-kW electric wall heaters (with individual thermostats) and five room air-conditioners. Two air-conditioners are rated 10.5 amperes at 230 volts, and the others are rated 7.2 amperes at 115 volts. A service calculation is being performed, using the optional method. How many volt-amperes should be included (after demand factors) for the heating and air-conditioning? (Assume heater kW ratings equivalent to kVA.)
________	________	11. In each dwelling unit, a feeder load of at least _____ volt-amperes must be included for each 2-wire laundry branch-circuit installed as required by 210.11(C)(2).
________	________	12. A 12-kW, 240-volt range will contribute _____ watts to the load, when calculating the service by the standard method.

Questions 13 through 20 are based on a 120/240 volt, 3-wire, single-phase one-family dwelling containing the following:

Floor area	2500 sq ft	Clothes dryer	4 kW, 240 volt
Range	12 kW, 240 volt	Electric heat (2 banks @ 3 kW each)	6 kW, 240 volt
Water heater	4 kW, 240 volt	Air handler (blower motor)	¼ HP, 115 volt
Dishwasher	1.5 kW, 120 volt	Air-conditioner compressor	5 HP, 230 volt
Waste (garbage) disposer	⅓ HP, 115 volt	Condenser fan motor	⅙ HP, 115 volt

Assume water heater, clothes dryer, range, and electric heat kW ratings equivalent to kVA.

Using Part B (the standard method), of Article 220:

________	________	13. What is the minimum ampere rating for the service?
________	________	14. What is the minimum size service?
________	________	15. What is the minimum size THWN copper conductor that can be installed as the service (ungrounded) conductor?
________	________	16. What is the minimum size grounding electrode conductor?

Using the Optional Method:

________	________	17. What is the minimum ampere rating for the service?
________	________	18. What is the minimum size service?
________	________	19. What is the minimum size THWN copper conductor that can be installed as the service (ungrounded) conductor?
________	________	20. What is the minimum size grounding electrode conductor?
________	________	21. What is the minimum size THWN copper conductor that can be installed as the neutral conductor?

UNIT 9

SECTION TWO: ONE-FAMILY DWELLINGS

Services and Electrical Equipment

Objectives

After studying this unit, the student should:

- be able to determine adequate strength for a mast supporting service-drop conductors.
- understand service-entrance cable provisions.
- know the definition of a service lateral and understand the applicable provisions.
- be familiar with minimum wiring clearance requirements (vertical, horizontal, etc.) for service and outside wiring.
- have a good understanding of the amount of three-dimensional working space required around service equipment, panelboards, electric equipment, etc.
- thoroughly understand panelboards, cabinets, and cutout boxes.
- be familiar with the multitude of provisions relating to circuit breakers and fuses.
- know and understand different aspects of service equipment and panelboard installations.
- be able to identify the appropriate Table—310.15(B)(6) or 310.16—for conductor sizing.
- understand that electric equipment must be installed in a neat and workmanlike manner.
- have a detailed understanding of grounding and bonding procedures.
- be able to properly position grounded and grounding conductors in remote panelboards (subpanels).
- have an extensive understanding of the grounding system, as a whole.
- understand how to ground and bond panelboards on the supply side, as well as the load side, of the service disconnecting means.
- be able to prevent objectionable current flow in grounding conductors.
- be familiar with the different grounding possibilities for panelboards installed in separate buildings or structures.

Introduction

Because the service is the heart of the electrical system, the importance of understanding this unit is enormous. Many of the regulations learned in this unit will be applicable in all types of occupancies, not merely single-family dwellings. Unit 9 is divided into five sections. The first, **Service-entrance Wiring Methods**, addresses both overhead and underground service entrance provisions. The section entitled **Service and Outside Wiring Clearances** covers the clearances required above roofs and final grade. Other clearances (horizontal, vertical, and diagonal) are shown for windows, doors, balconies, platforms, etc. Next, the **Working Space Around Equipment** section illustrates common working zone requirements. **Service Equipment and Panelboards** explains panelboards, cutout boxes, and cabinets. Specific provisions pertaining to circuit breakers, fuses, and circuit breaker panels are also covered. **Grounding**, the last section, explains in detail Article 250 provisions relating to service equipment and panelboards. Refer to Unit 9 often when studying (or installing) services, whether one-family, multi-family dwellings, commercial, or even industrial buildings.

Since many local jurisdictions and electric utility companies have supplemental requirements, exercise caution in compiling a complete set of service installation requirements. Examples of these requirements include, but are not limited to: restrictions on the type of conduit used for service mast; minimum conduit trade size; specific strapping methods; restrictions on the use of service-entrance cables; special underground service provisions; limitations on the maximum distance service-entrance conductors can run within a building or structure; and various grounding provisions. Obtain a copy of these additional rules (if any) for your area.

SERVICE-ENTRANCE WIRING METHODS

Service Mast as Support

A Service-drop and service-entrance conductors must be arranged so that water cannot enter the service raceway or equipment »230.54(G)«.

B It is not required (where approved) that rigid (and intermediate) metal conduit be securely fastened within 3 ft (900 mm) of the service head for above-the-roof termination of a mast »344.30(A) and 346.30(A)«.

C Open conductors must be attached to fittings identified as suitable for use with service conductors or to noncombustible, nonabsorbent insulators securely attached to the building or other structure »230.27«.

D A service mast used to support service-drop conductors must be of adequate strength or be supported by braces (or guys) to safely withstand the strain imposed by the service drop »230.28«.

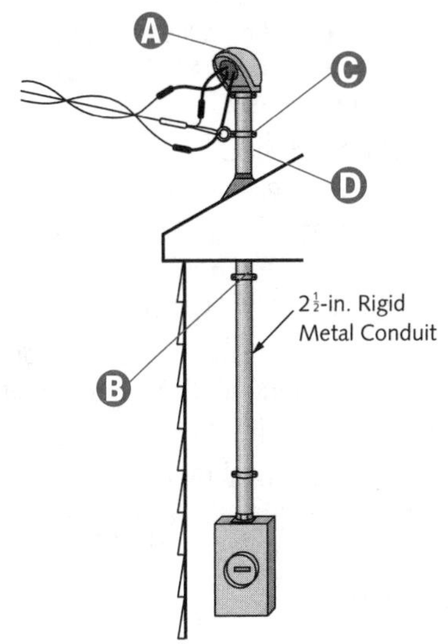

Service Heads (Weather Heads)

A Service raceways must be equipped with a raintight service head at the point of connection to service-drop conductors »230.54(A)«.

B Raceway-type service masts require raceway fittings that are identified for use with service masts »230.28«.

C Conductors of different potential must be brought out of service heads through separately bushed openings »230.54(E)«.

D An insulated grounded conductor, larger than 6 AWG, must be identified either by a continuous white or gray outer finish or by three continuous white stripes on other than green insulation along its entire length or by a distinctive white marking at its terminations during installation. This marking must completely encircle the conductor or insulation »200.6(B)«.

E Conductors must have adequate mechanical strength and sufficient ampacity to carry the current for the load (as computed in accordance with Article 220) »230.23(A)«.

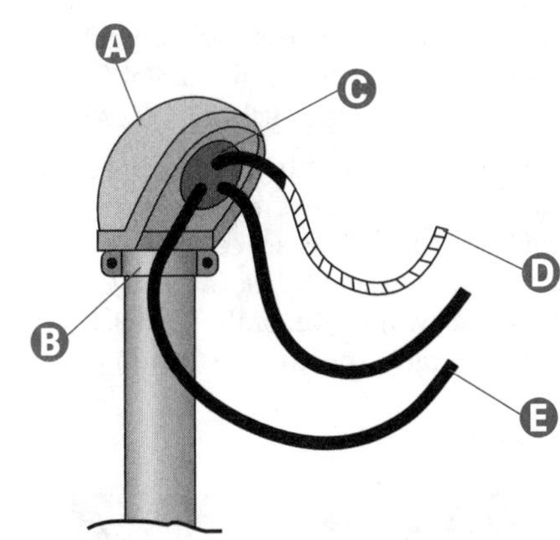

Service Mast Attachments

A The service drop consists of the overhead conductors, inclusive of any splices, extending from the last pole (or other aerial support) to the service-entrance conductors at the building (or similar structure) »Article 100«.

B Nothing, except electrical service-drop conductors, can be attached to the service mast »230.28«.

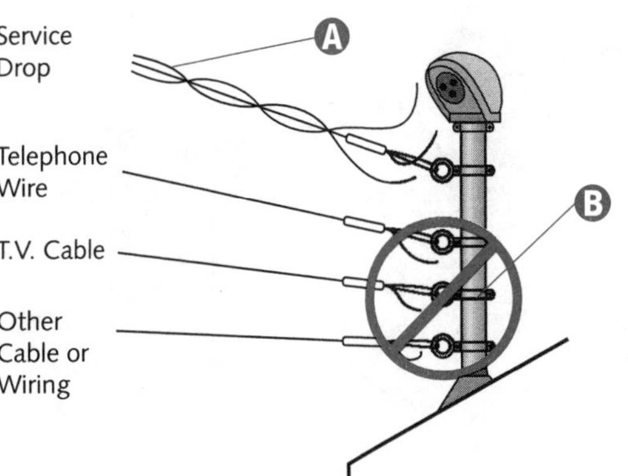

Service Drops Attached to Buildings

A Open conductors must be attached to fittings identified specifically for use with service conductors, or to noncombustible, nonabsorbent insulators that are securely attached to the building (or other structure) »230.27«.

B EMT (electrical metallic tubing) couplings and connectors shall be made up tight. If installed in wet locations, they must be of the raintight type »358.42«.

C Service heads must be located above the point where service-drop conductors are attached to the building or other structure »230.54(C)«.

D Where it is impracticable to locate the service head above the point of attachment, the service head must be located within 24 in. (600 mm) of that point »230.54(C) *Exception*«.

E Since this service mast does not support service-drop conductors, another wiring method (EMT) is used. Other allowable service-entrance methods are listed in 230.43.

F Electric equipment must be firmly secured to the surface on which it is mounted. Wooden plugs driven into holes in masonry, concrete, plaster, etc., are not acceptable »110.13(A)«.

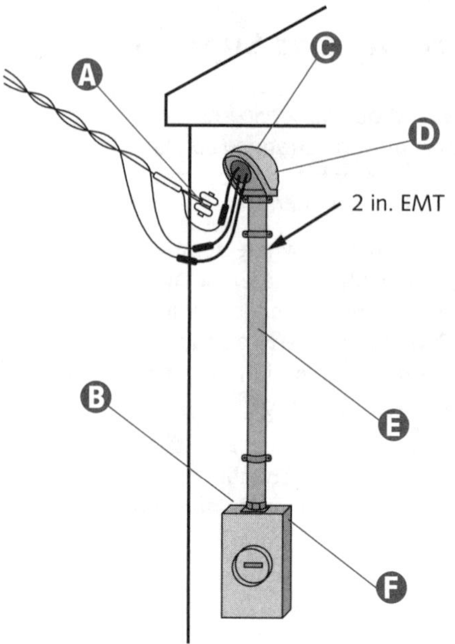

Support of Overhead Conductors

A Vegetation (such as trees) cannot be used to support overhead conductor spans, except for temporary wiring in accordance with Article 527 »225.26«.

B Drip loops must be formed on each individual conductor. To prevent moisture penetration, service-entrance conductors are connected to the service-drop conductors either (1) below the level of the service head, or (2) below the level of the termination of the service-entrance cable sheath »230.54(F)«.

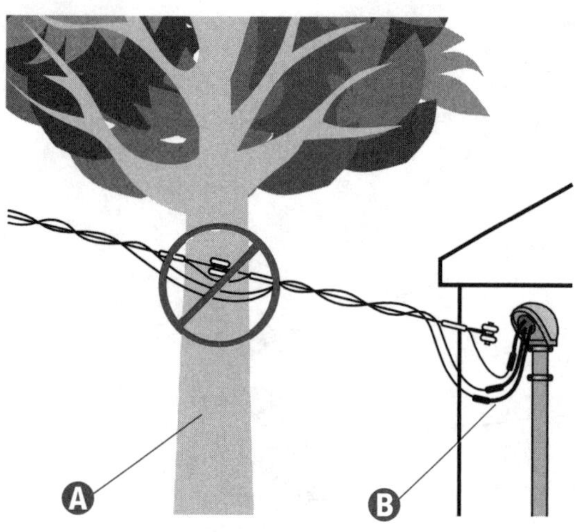

Goosenecks in Service-Entrance Cables

A Bends in service-entrance cable must be made so that the cable will not be damaged. In addition, the radius of the curve of the inner edge of any bend must be at least five times the diameter of the cable »338.24«.

B Goosenecks in service-entrance cables must be located above the point of attachment of the service-drop conductors to the building (or other structure) »230.54(C)«.

C Type SE cable can be formed in a gooseneck and taped with a self-sealing weather resistant thermoplastic »230.54(B) *Exception*«.

D Drip loops must be formed on all individual conductors »230.54(F)«.

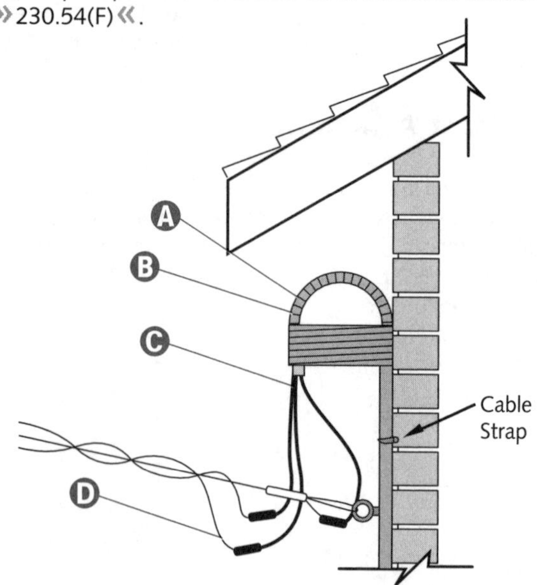

Service-Entrance Cable

A Service cables must be fitted with a raintight service head, except for Type SE cable that has been formed in a gooseneck »230.54(B)«.

B If Type SE cable consists of at least two conductors, one can be uninsulated »338.2«.

C Service-entrance cable is a single conductor or multiconductor assembly (with or without an overall covering), primarily used for services. Type SE has a flame-retardant, moisture-resistant covering »338.2«.

D Service cables must be securely fastened in place »230.54(D)«.

E Service head conductors of different potential must be brought out through separately bushed openings »230.54(E)«.

F Entrance cables must be supported by straps (or other approved means) within 12 in. (300 mm) of every service head »230.51(A)«.

G Entrance cables must be supported at intervals not to exceed 30 in. (750 mm) »230.51(A)«.

H Entrance cables must be supported within 12 in. (300 mm) of every raceway or enclosure connection »230.51(A)«.

I Service-entrance conductors must be arranged so that water cannot enter service raceway or equipment. »230.54(G)«.

CAUTION *Service cables, where subject to physical damage, must be appropriately protected »230.50«.*

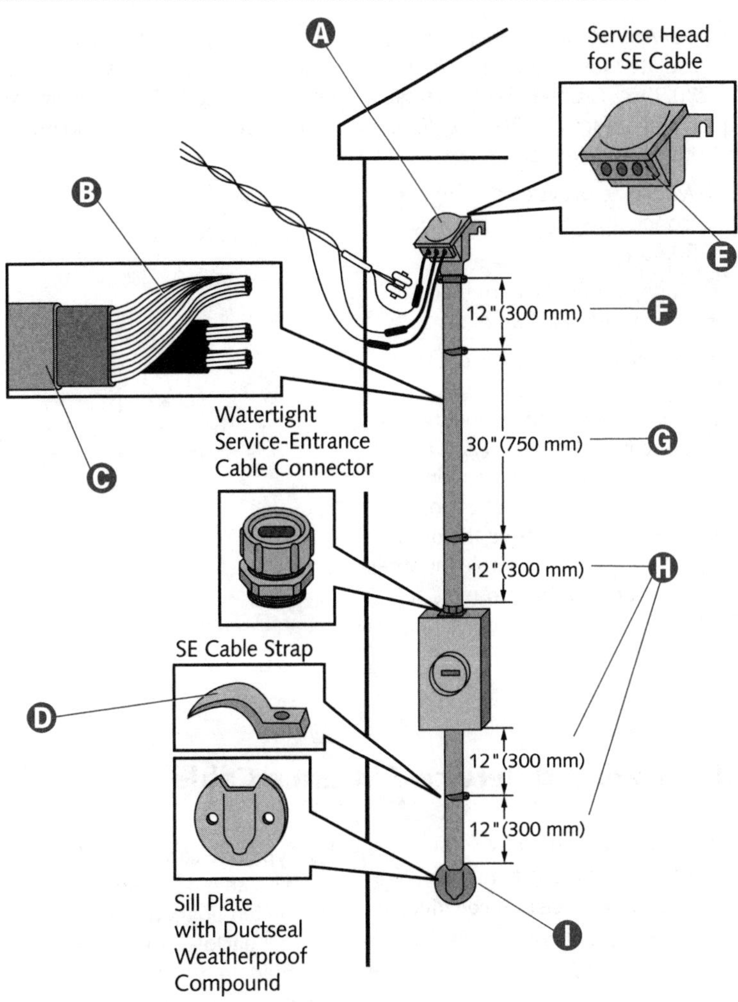

Conductors Considered Outside of a Building

A Building or structure interior wall

B An interior service disconnecting means must be installed as close as possible to the point of entrance of the service conductors and must be readily accessible. A raceway, beneath a 2-in. (50-mm) concrete slab, lies outside the building. Under this condition, the service disconnecting means could be installed on (or in) an interior wall of the building. (See Caution.)

C Conductors installed under 2 in. (50 mm) or more of concrete beneath a building (or other structure) are considered outside »230.6(1)«.

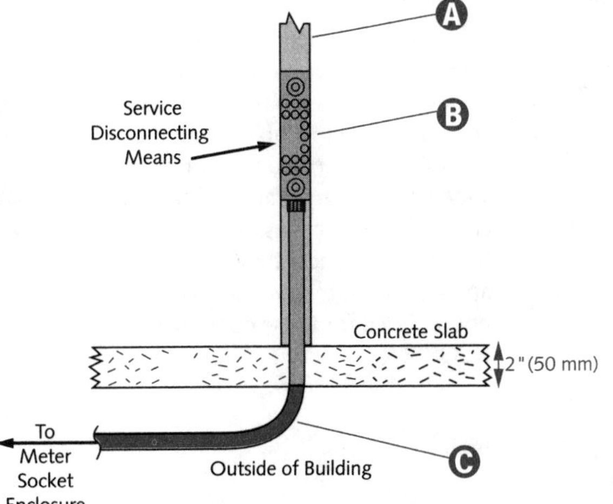

CAUTION *Although not stipulated in the NEC®, some local jurisdictions have adopted specific distance requirements for service conductors. [For example, the length for each service conductor between the meter socket enclosure and the service disconnecting means could be limited to only six ft (1.8 m)].*

Underground Services

A Service-lateral conductors must be insulated for the applied voltage »230.30«. Grounded conductors, qualifying under 230.30 *Exception*, can be uninsulated (bare).

B Metal raceways containing service conductors must be grounded »250.80«.

C Wiring methods, permitted for services, are listed in 230.43.

D Service lateral consists of the underground service conductors between the street main and the first point of connection to the service-entrance conductors in a terminal box, meter base, or other enclosure whether inside or outside the building wall. Any risers at a pole (or other structure) or from transformers are included »Article 100«.

E Direct buried conduit (or other raceways) must be installed in adherence with Table 300.5 minimum cover requirements »300.5(A)«.

NOTE

Service-lateral conductors must have adequate mechanical strength and sufficient ampacity to carry the current for the load (as computed in accordance with Article 220) »230.31(A)«.

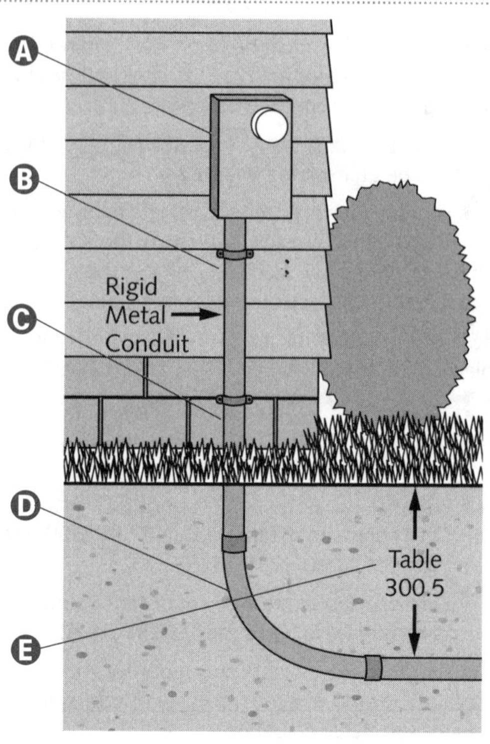

Underground Service-Entrance Cable

Although Type USE cable identified for underground use has a moisture-resistant covering, a flame-retardant covering is not required »338.2«.

If Type USE cable contains at least two conductors, one can be uninsulated »338.2«.

A Underground service-lateral conductors must be protected against damage in accordance with 300.5. Service-lateral conductors entering a building must be installed in accordance with 230.6 or must be protected by an approved raceway wiring method identified in 230.43 »230.32«.

B Where the conductors or cables emerge as a direct burial wiring method, a bushing (or fitting) must be installed to protect conductors from abrasion on the end of the conduit (or tubing) that terminates underground. A seal providing the same level of protection can be used instead of a bushing »300.5(H) and 300.15(C)«.

C Type USE cable, used for service laterals, can emerge aboveground at outside terminations (in meter bases or other enclosures) provided 300.5(D) protection requirements are met »338.10(A)«.

D When installed, direct buried cable must meet Table 300.5 minimum cover requirements »300.5(A)«.

E Cabled, single-conductor, Type USE constructions, recognized for underground use, may have a bare copper conductor cabled with the assembly. Underground approved, Type USE single, parallel, or cabled conductor assemblies, may have a bare copper concentric conductor applied. An outer overall covering is not required for these constructions »338.2«.

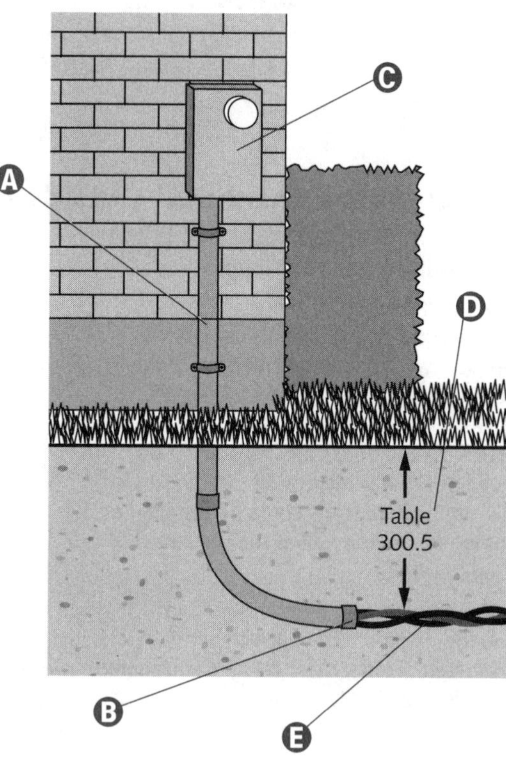

Wiring Methods

A Service-entrance conductors must be installed according to applicable *Code* requirements for the type of wiring method used and limited to the following methods: (1) open wiring on insulators; (2) Type IGS cable; (3) RMC; (4) IMC; (5) EMT; (6) ENT; (7) service-entrance cables; (8) wireways; (9) busways; (10) auxiliary gutters; (11) RNC; (12) cablebus; (13) Type MC cable; (14) mineral-insulated, metal-sheathed cable; (15) FMC (or LFMC) not longer than 6 ft (1.8 m) between raceways, or between raceway and service equipment, having equipment bonding jumper routed with the FMC (or LFMC) according to 250.102(A), (B), (C), and (E) provisions; or (16) LFNC »230.43«.

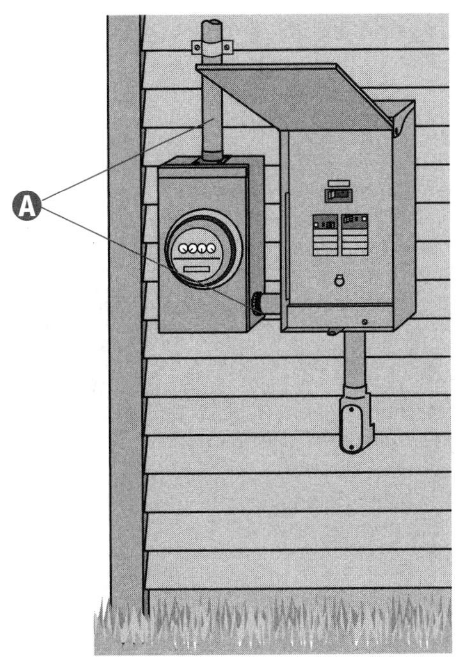

NOTE

The wiring method of choice must be installed according to provisions for that particular method. For example, RMC must be installed according to Article 344 provisions.

SERVICE AND OUTSIDE WIRING CLEARANCES

Accessibility to Pedestrians

A The point of attachment of the service-drop conductors to a building (or other structure) must meet 230.24 minimum clearances. In all cases this point of attachment must be at least 10 ft (3.0 m) above finished grade »230.26«.

B Overhead conductors, accessible only to pedestrian traffic, must have a minimum 10-ft (3.0-m) vertical clearance above finished grade, sidewalks, platforms, or projections from which they might be reached. The voltage must not exceed 150 to ground »230.24(B)(1) and 225.18(1)«.

C The 3-ft (900-mm) minimum requirement does not apply to conductors run above the top level of a window »230.9(A) *Exception* and 225.19(D)(1) *Exception*«.

D The 3-ft (900-mm) clearance from windows, found in 230.9, does not apply to raceways and service equipment.

E A drip loop's lowest point has a minimum vertical clearance of 10 ft (3.0 m) »230.24(B)(1)«.

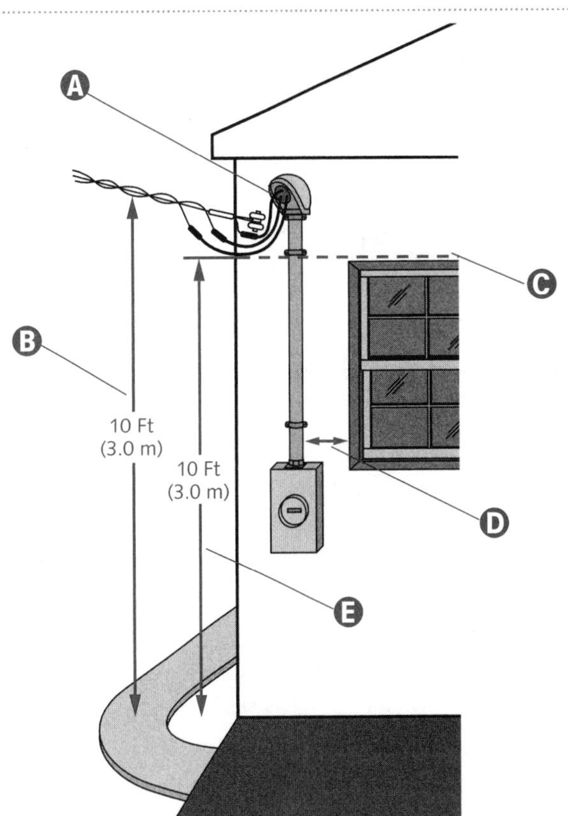

Vertical Clearances

A A 10-ft (3.0-m) minimum vertical clearance is required from any platform or projection, from which overhead conductors (of 150 volts, or less, to ground) might be reached »230.24(B)(1) and 225.18(1)«.

B A minimum 12-ft (3.7-m) vertical clearance is required over residential property and driveways, and over commercial areas (not subject to truck traffic) provided conductor voltage does not exceed 300 volts to ground »230.24(B)(2) and 225.18(2)«.

C Conductor voltage exceeding 300 volts to ground has a minimum vertical clearance of 15 ft (4.5 m) »230.24(B)(3) and 225.18(3)«.

D Public streets, alleys, roads, parking areas subject to truck traffic, driveways on non-residential property, and other land traversed by vehicles (such as cultivated, grazing, forest, and orchard) require a minimum 18-ft (5.5-m) vertical clearance »230.24(B)(4) and 225.18(4)«.

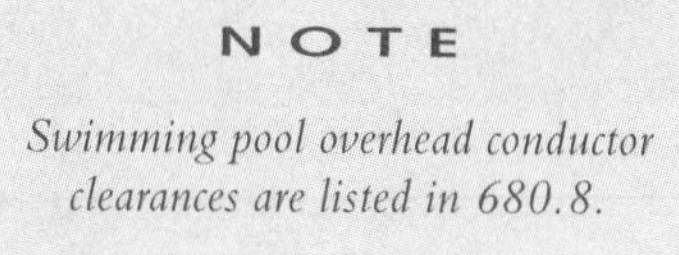
NOTE

Swimming pool overhead conductor clearances are listed in 680.8.

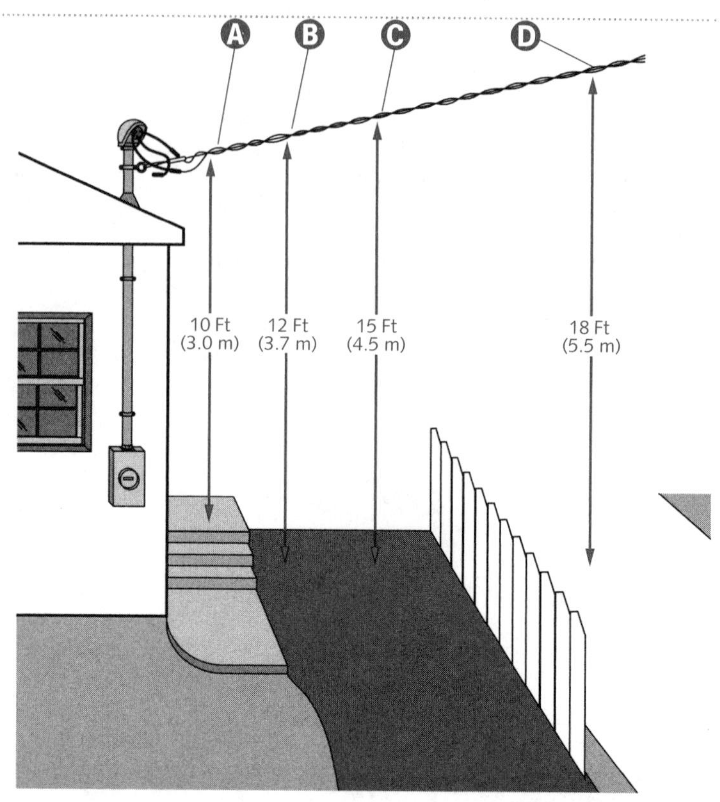

Clearance from Windows and Accessible Areas

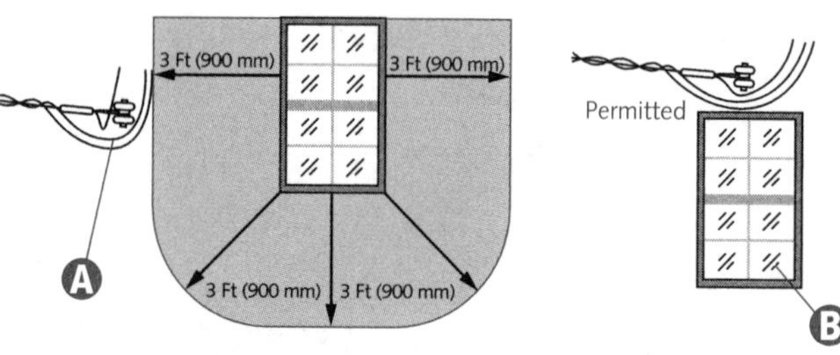

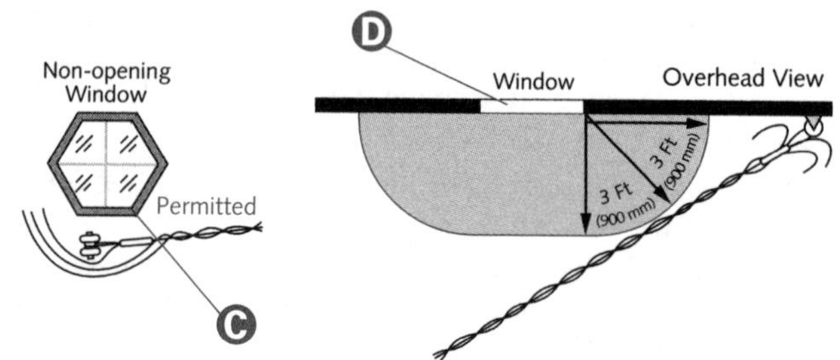

A Service conductors, installed as open conductors or multi-conductor cable (having no overall outer jacket), must have a clearance of at least 3 ft (900 mm) from windows (designed to open), doors, porches, balconies, ladders, stairs, fire escapes, and similar locations »230.9(A) and 225.19(D)(1)«.

B The 3-ft (900-mm) clearance requirement does not apply to conductors run above the top level of a window »230.9(A) *Exception* and 225.19(D)(1) *Exception*«.

C The 3-ft (900-mm) clearance is not required for windows that are not designed to open.

D The 3-ft (900-mm) clearance applies to the entire perimeter of windows, doors, etc.

Conductors Obstructing Openings

A Overhead conductors must not obstruct building entrances and must not be installed beneath openings designed for passage of materials, such as openings in farm and commercial buildings »230.9(C) and 225.19(D)(3)«.

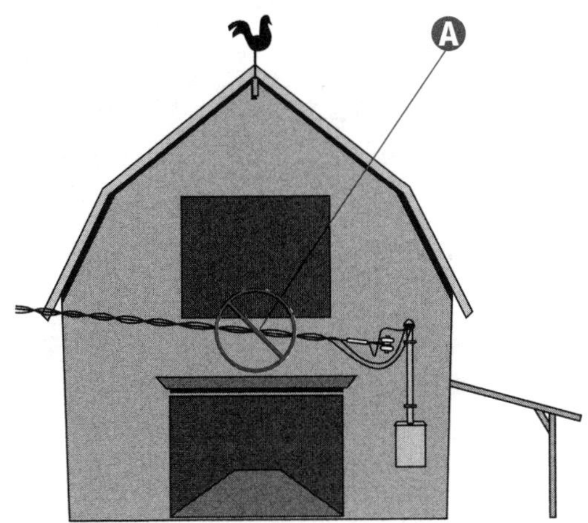

NOTE

A minimum 18-ft (5.5-m) vertical clearance is required over land routinely traversed by vehicles (such as cultivated, grazing, forest, and orchard) »230.24(B)(4) and 225.18(4)«.

Conductors Above Overhanging Portions of Roofs

A A maximum of 6 ft (1.8 m) of service-drop conductors pass over the roof.

B Conductors terminate in a through-the-roof raceway (or other approved support).

C The vertical clearance for service-drop conductors can be reduced to 18 in. (450 mm) provided all 230.24(A) *Exception No. 3* requirements are met.

D The voltage between conductors does not exceed 300 volts.

E The horizontal service-drop conductor measurement cannot exceed 4 ft (1.2 m).

F This reduction in clearance applies to conductors passing above roof overhang only.

NOTE

If all the requirements of 225.19(A) Exception No. 3 are met, the vertical clearance for the overhead spans of open conductors, including multiconductor cables, can be reduced to 18 in. (450 mm).

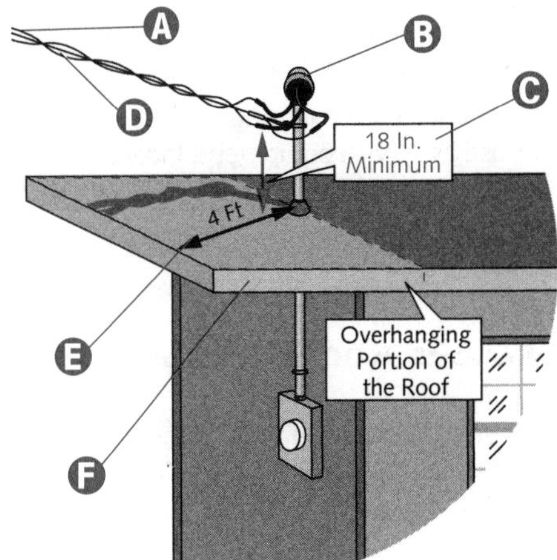

Conductor Clearance Above Flat Roofs

A Conductors must have a vertical clearance of at least 8 ft (2.4 m) above the roof surface »230.24(A) and 225.19(A)«.

B While service-drop conductors must not be readily accessible, they must comply with 230.24(A) through (D) for services not over 600 volts, nominal »230.24«.

C A service mast supporting service-drop conductors must be of adequate strength or be supported by braces (or guys) to safely withstand the strain imposed by the service drop »230.28«.

D The vertical clearance above roof level must be maintained for a distance of at least 3 ft (900 mm) from all edges of the roof »230.24(A) and 225.19(A)«.

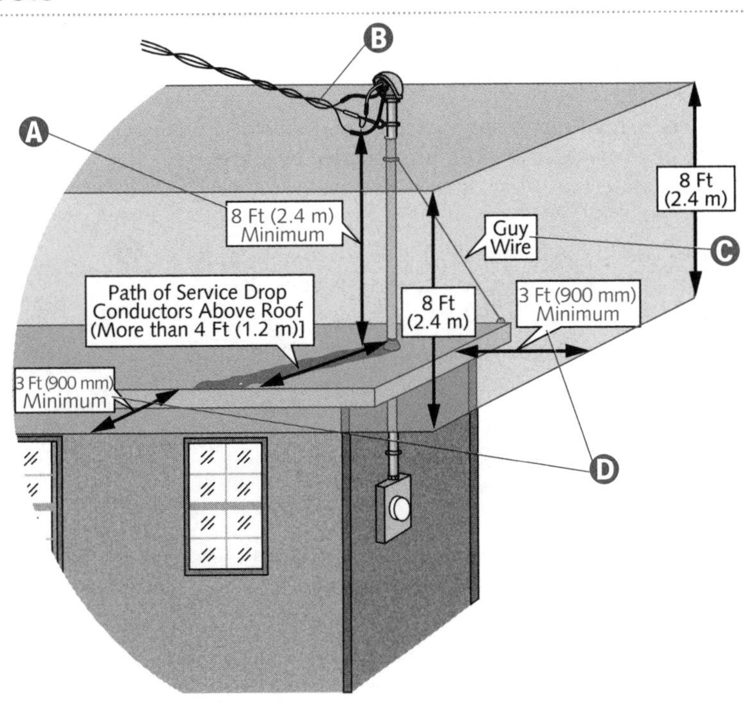

Conductors Above Sloped Roofs

A The voltage between the ungrounded (hot) conductors is 240 volts.

B Where the voltage between conductors does not exceed 300, and the roof has a slope of not less than 4 in. in 12 (100 mm in 300), a 3-ft (900-mm) clearance is permitted »230.24(A) *Exception No. 2* and 225.19(A) *Exception No. 2*«.

C This roof, as denoted by this symbol, has a slope (or slant) of 6 in. in 12.

D Unlike 230.24(B), which stipulates that "vertical clearance" also includes the drip loop, 230.24 provisions mention only the service-drop conductor. Because the end of the service-drop conductors could be part (or all) of the drip loop, caution is advised.

E The horizontal service-drop measurement is less than 4 ft (1.2 m). However, the conductors pass over more than just the overhanging portion of this roof. Therefore, the clearance cannot be reduced to 18 in. (450 mm) »230.24(A) *Exception No. 3* and 225.19(A) *Exception No. 3*«.

F Service equipment rated at 600 volts, or less, must be marked as suitable for use as service equipment. Individual meter socket enclosures are not considered service equipment »230.66«.

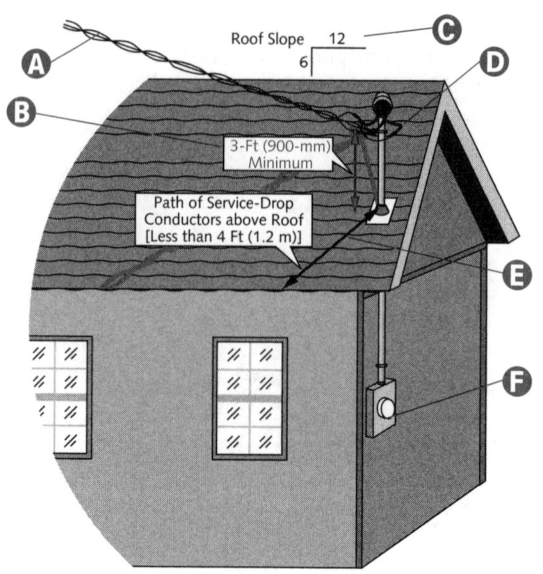

Various Roof Designs

A Service-drop conductors above a roof with a slope of at least 4 in. in 12 (100 mm in 300) has a minimum clearance of 3 ft (900 mm) »230.24(A) *Exception No. 2*«.

B This roof has a slope of 4 in. in 12.

C The point of attachment is usually the lowest point of the service-drop. If, however the conductors run above the roof's highest point (ridge or peak), a taller service mast may be required.

D This roof has a slope of 6 in. in 12 (more than 4 in 12).

E The supply voltage for all three service-drop conductor illustrations is 240 volts.

F A service mast supporting service-drop conductors must be of adequate strength or be supported by braces (or guys) to safely withstand the strain imposed by the service drop »230.28«.

G Service-drop conductors above a roof with a slope less than 4 in. in 12 (100 mm in 300) have a minimum clearance of 8 ft (2.5 m) »230.24(A)«.

H This roof has a slope of 3 in. in 12 (less than 4 in 12).

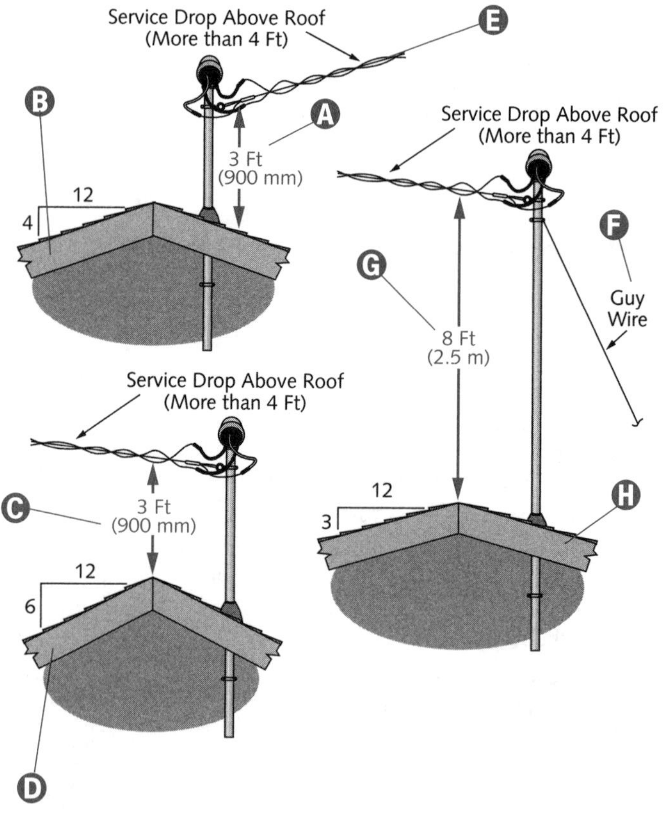

NOTE

225.19 lists the vertical clearance for the overhead span of open conductors and open multiconductor cables.

Because each of the illustrated service-drops has a horizontal measurement of more than 4 ft, the amount of slope determines the minimum clearance requirement [either 3 ft (900 mm) or 8 (2.5 m)].

Clearance Reduction

A No more than 6 linear ft (1.8 m) of service-drop conductors pass over the roof.

B Voltage between conductors does not exceed 300 volts.

C The horizontal service-drop conductor measurement does not exceed 4 ft (1.2 m).

D Conductors terminate in a through-the-roof raceway (or other approved support.)

E The service-drop's minimum vertical clearance is 18 in. (450 mm) where all 230.24(A) *Exception No. 3* requirements have been met.

F The reduction in clearance [to 18 in. (450 mm)] applies to sloped roofs as well as flat roofs where the service-drop passes above the overhang portion of the roof only.

> **NOTE**
>
> *If all the requirements of 225.19(A)* Exception No. 3 *are met, the vertical clearance for the overhead spans of open conductors, including multiconductor cables, can be reduced to 18 in. (450 mm).*

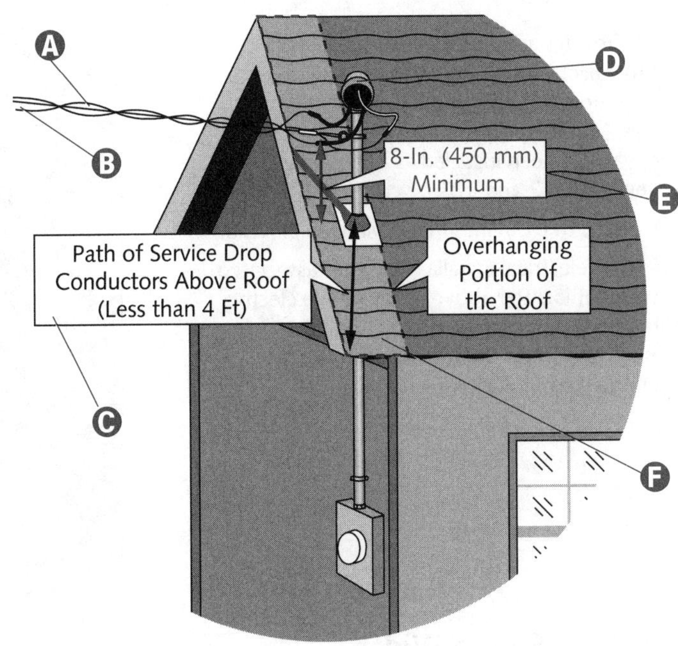

WORKING SPACE AROUND EQUIPMENT

Headroom and Height

A Overcurrent devices cannot be located in the vicinity of easily ignitible material (such as clothes closets) »240.24(D)«. Although mentioned specifically, clothes closets are only one type of area that may contain easily ignitible materials.

B A minimum of 6½ ft (2.0 m) of working space headroom is required around service equipment, switchboards, panelboards, or motor control centers »110.26(E)«.

C A clear work space must extend from the grade, floor, or platform to the height required by 110.26(E) »110.26(A)(3)«.

D Work space illumination must be provided for all indoor installations of service equipment, panelboards, etc. Supplemental luminaires (light fixtures) are not required where the work space is illuminated by an adjacent light source »110.26(D)«.

E Sufficient working space access must be provided and maintained about all electric equipment to permit ready and safe equipment operation and maintenance »110.26«.

F Working space required by 110.26 must not be used for storage »110.26(B)«.

> **NOTE**
>
> *Compliance with the minimum headroom of working space is optional for 200 amperes, or less, service equipment or panelboards, in existing dwellings »110.26(E)* Exception«.
>
> *Electric equipment work space must have at least one entrance of sufficient area to provide access »110.26(C)«.*

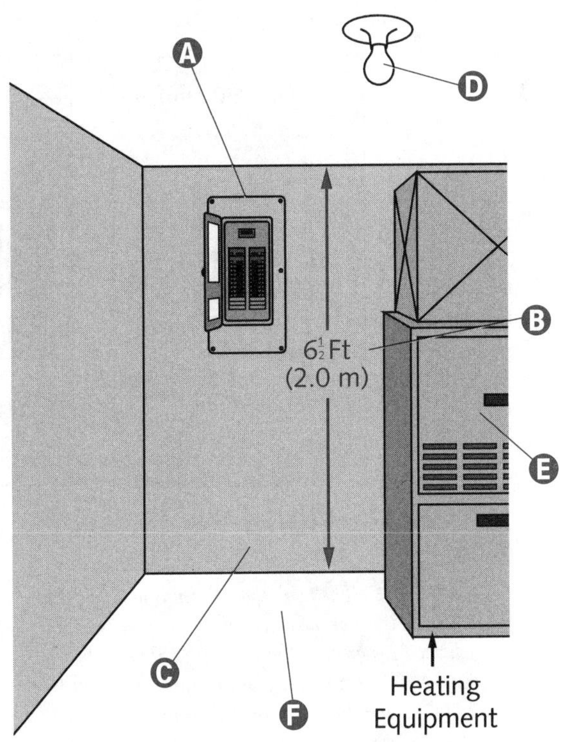

Maximum Depth of Associated Equipment

A Wireways are troughs (sheet-metal or nonmetallic) with hinged or removable covers used to house and protect electric wires and cables. Conductors are laid in place after the wireway has been installed as a complete system »376.2 and 378.2«.

B Article 376 outlines metal wireway provisions.

C In an electrical installation, other associated equipment located above or below the electric equipment cannot extend more than 6 in. (150 mm) beyond the front of that electrical equipment »110.26(A)(3)«.

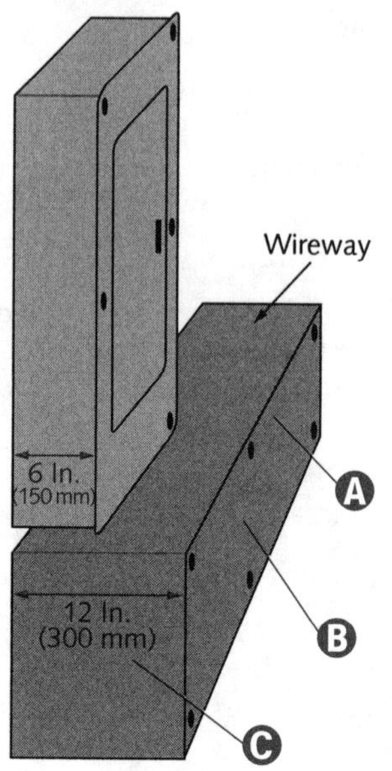

CAUTION *Surrounding space (equal to the width and depth of the equipment) must be clear of foreign systems, unless protection is provided against damage from condensation, leaks, or breaks in such foreign systems. This zone extends from the top of the equipment to the structural ceiling. A dropped, suspended, or similar ceiling that does not strengthen the building is not considered a structural ceiling. Sprinkler protection is permitted within the dedicated space, where the piping complies with 110.26 »110.26(F)(1)(b)-(d)«.*

Working Space Width

A The frontal work space width of electric equipment must be the greater of the width of the equipment or 30 in. (750 mm) »110.26(A)(2)«.

B 110.26 does not stipulate that the panel (or equipment) be located in the center of the work space. It can be located anywhere within the required space.

C A panel can be located in the middle, with equal amounts of space on each side, or . . .

D A panel can be located on the far left side, with the additional space on the right, or . . .

E A panel can be located on the far right, with the additional space on the left side.

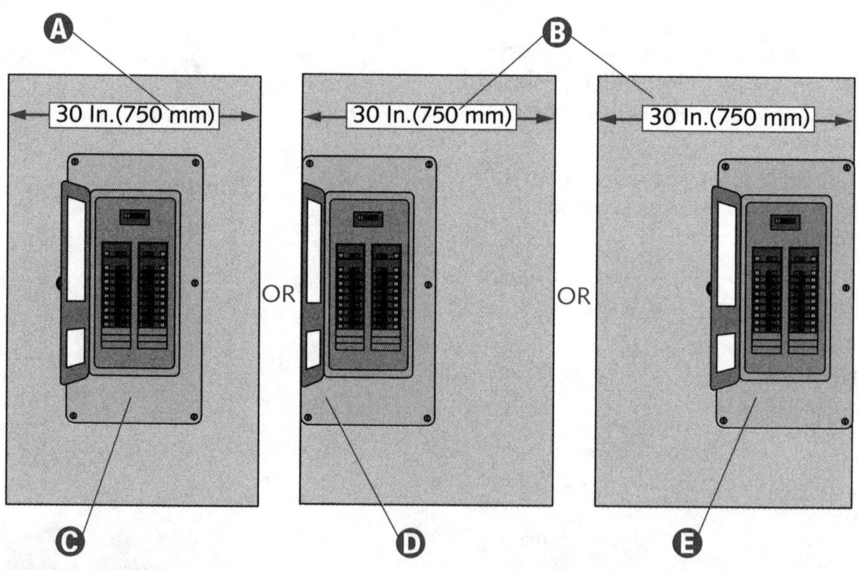

CAUTION *The work space extends to a height of 6½ ft (2.0 m) or the height of the equipment, whichever is greater. Above 6½ ft (2.0 m) or the height of the equipment, the dedicated space must equal the width and depth of the equipment »110.26(E) and (F)«.*

NOTE

The working space must always permit at least a 90° opening of equipment doors or hinged panels »110.26(A)(2)«.

Overlapping Work Space

A The work space width required in front of electric equipment is the greater of the width of the equipment or 30 in. (750 mm) »110.26(A)(2)«.

B In this configuration, it is not necessary that each panel have its own dedicated working space.

C A panel (or equipment) can share the dedicated work space of other equipment associated with the electrical installation.

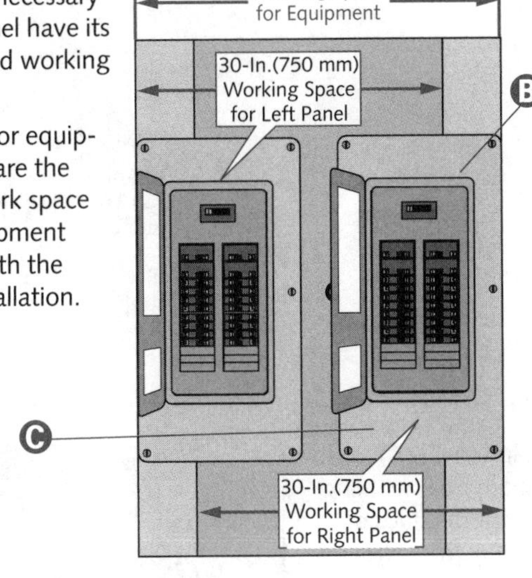

Equipment Behind Doors

A Equipment can be located behind a door provided a 3-ft (900-mm) working space is achieved by closing the door.

B Overcurrent devices must be readily accessible unless one of the conditions in 240.24(A)(1) through (4) apply »240.24(A)«. (The term "readily accessible" is defined in Unit 2 of this book.)

Working Space Depth

A Concrete, brick, or tile walls are recognized as grounded »Table 110.26(A) Note 2«.

B Table 110.26(A) specifies working space depth requirements in the direction of access to live parts. Distances are measured from the live parts (if exposed) or from the front or opening (if enclosed) »110.26(A)(1)«.

C The minimum work space depth for equipment operating at 150 volts, nominal (or less) to ground is 3 ft (900 mm), regardless of the conditions found in Table 110.26(A). For equipment operating between 151 and 600 volts, to ground, refer to Table 110.26(A).

D Electric equipment must be firmly secured to the surface on which it is mounted. Wooden plugs driven into holes in masonry, concrete, plaster, or similar materials are not acceptable »110.13(A)«.

E Except as elsewhere required or permitted, the work space area for equipment operating at 600 volts, nominal (or less) to ground, which is likely to require examination, adjustment, servicing, or maintenance while energized, must comply with (1), (2), and (3) of 110.26(A) »110.26(A)«.

NOTE

Installation and working space requirements for equipment operating over 600 volts, nominal are found in 110.30 through 110.40.

F The minimum headroom is 6½ ft (2.0 m), except for existing dwellings whose service does not exceed 200 amperes »110.26(E)«.

G The minimum acceptable width of working space is 30 in. (750 mm) »110.26(A)(2)«.

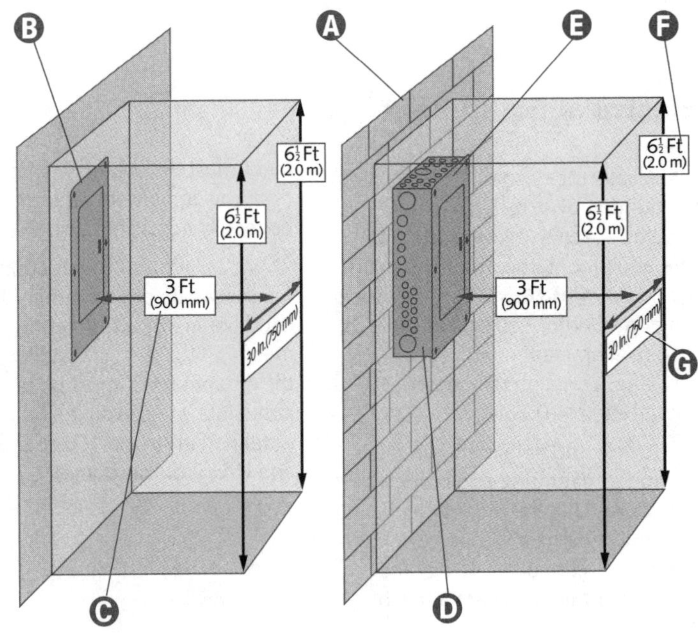

Dedicated Space

A The space equal to the width and depth of the equipment and extending from the floor to a height of 6 ft (1.8 m) above the equipment or to the structural ceiling, whichever is lower, must be dedicated to the electrical system. A dropped, suspended, or similar ceiling, that does not strengthen the building is not considered a structural ceiling »110.26(F)(1)(A) and (D)«.

B All switchboards, panelboards, distribution boards, and motor control centers must be located in dedicated spaces and must be protected from damage »110.26(F)«.

NOTE

Outdoor clear work space includes the zone described in 110.26(A). No architectural appurtenances or other equipment can occupy in this zone »110.26(F)(2)«.

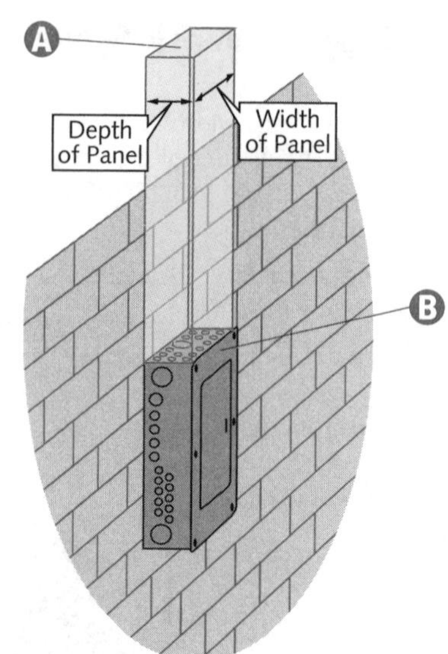

Outdoor Service Equipment

A Outdoor electrical equipment must be installed in suitable enclosures and must be protected from accidental piping system spills and leaks, and from accidental contact by unauthorized personnel, or vehicular traffic »110.26(F)(2)«.

B Panelboards must be mounted in dead front enclosures (cabinets, cutout boxes, etc.) specifically designed for that purpose »408.18«. Dead front means having no exposed live parts on the operating side of the equipment »Article 100«.

C The zone described in 110.26(A) must be included in the minimum clear working space. No architectural appurtenance or other equipment can occupy this zone »110.26(F)(2)«.

NOTE

Panelboards in damp or wet locations must comply with 312.2(A) »408.17«.

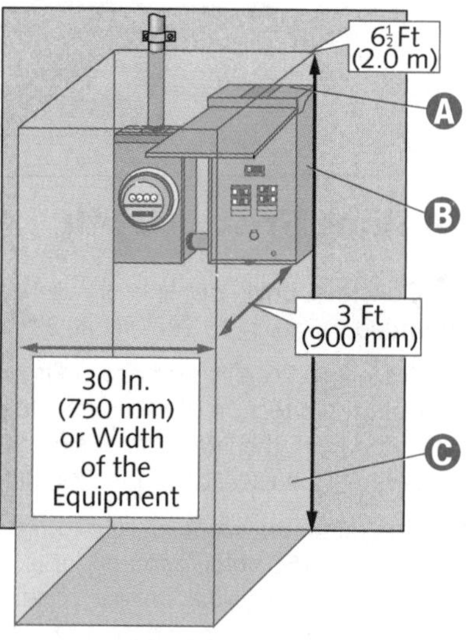

Electrical Equipment

A All electric equipment must be surrounded by sufficient, unobstructed space to allow ready and safe equipment operation and maintenance »110.26«. Working space is mandatory for equipment likely to require examination, adjustment, servicing, or maintenance while energized »110.26(A)«.

B Article 440 details air-conditioning and refrigerating equipment provisions. General requirements, disconnecting means, overcurrent protection, conductor sizing, and room air conditioners are a few of the covered topics.

C The disconnecting means for air-conditioning or refrigerating equipment must be located within sight from the equipment and must also be readily accessible »440.14«.

D Installation of the disconnecting means can be on, or within, the air-conditioning or refrigerating equipment »440.14«. If installed within the equipment, it must be readily accessible. ("Accessible, Readily" is defined in Article 100 of the *NEC*® and Unit 2 of this book.)

NOTE

Equipment grounding conductors are found in Article 250, Part VI.

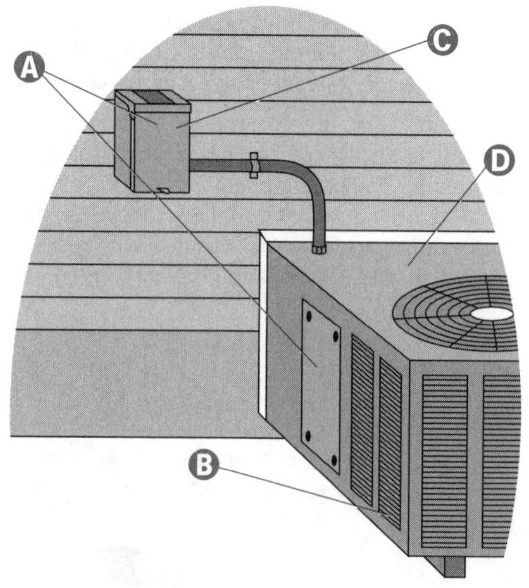

SERVICE EQUIPMENT AND PANELBOARDS

Panelboards

A A panelboard (1) consists of one, or more, panel units designed for assembly, thereby forming a single panel; (2) includes buses and automatic overcurrent devices; (3) may contain switches for controlling circuits (light, heat, or power); (4) are designed for mounting in a cabinet or cutout box; (5) are designed for placement in or against a wall, partition, or other support; and (6) are accessible only from the front »Article 100«. (Panelboards include, but are not limited to, circuit breaker panels and fused or non-fused switches (disconnects).

B Panelboards include overcurrent devices.

C Panelboards include buses.

D Panelboards are mounted in cabinets, or cutout boxes, which are accessible only from the front.

E Panelboards can be single or multi-panel units.

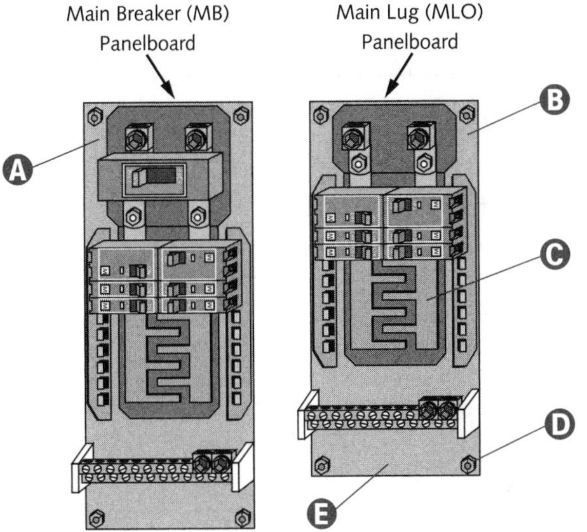

Enclosures

A A cabinet is an enclosure that by design may be surface or flush mounted, having a frame, mat, or trim attached to which are (or can be) swinging door(s) »Article 100«.

B Cabinets can be either flush or surface mounted.

C Cutout boxes have swinging doors or covers attached directly to the enclosure.

D Cabinets include trims with swinging doors.

E **Type 1** enclosures, intended for indoor use, primarily provide protection against contact with the enclosed equipment in locations not subject to unusual service conditions.

F Another type of enclosure, a cutout box, is designed for surface mounting and has swinging doors (or covers) secured directly to and extending from one of the enclosure walls »Article 100«.

G Cutout boxes are surface mounted.

H Dead front is defined as having no exposed live parts on the operating side of the equipment »Article 100«.

I **Type 3R** enclosures, intended for outdoor use, primarily provide protection against falling precipitation and must remain undamaged by external formation of ice. They are not intended to provide protection against conditions such as dust, internal condensation, or internal icing.

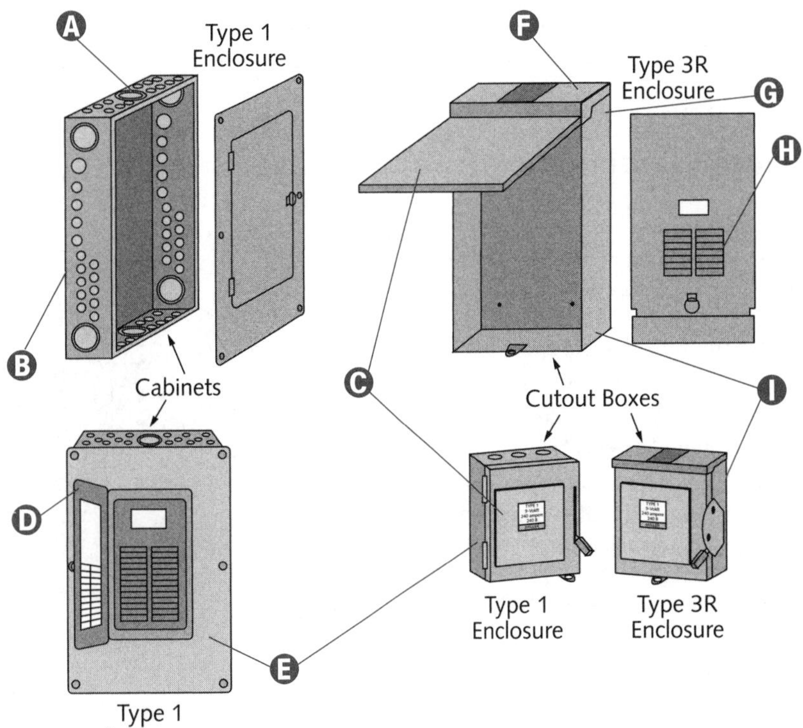

Maximum Number of Overcurrent Devices

Ⓐ When counting the number of overcurrent devices within a panelboard, the main breaker is not added.

Ⓑ A hand-operable circuit breaker (equipped with a lever or handle), or a power-operated circuit breaker (capable of being opened by hand in the event of a power failure), can serve as a switch depending on the number of poles. »380.11«.

Ⓒ A 2-pole circuit breaker counts as two overcurrent devices for the purpose of 408.15 »408.15«.

Ⓓ Twin (dual) circuit breakers count as two overcurrent devices. Similarly, quad circuit breakers count as four.

Ⓔ A maximum of 42 overcurrent devices (exclusive of those provided for in the mains) can be installed in any one cabinet or cutout box of a lighting and appliance branch-circuit panelboard »408.15«.

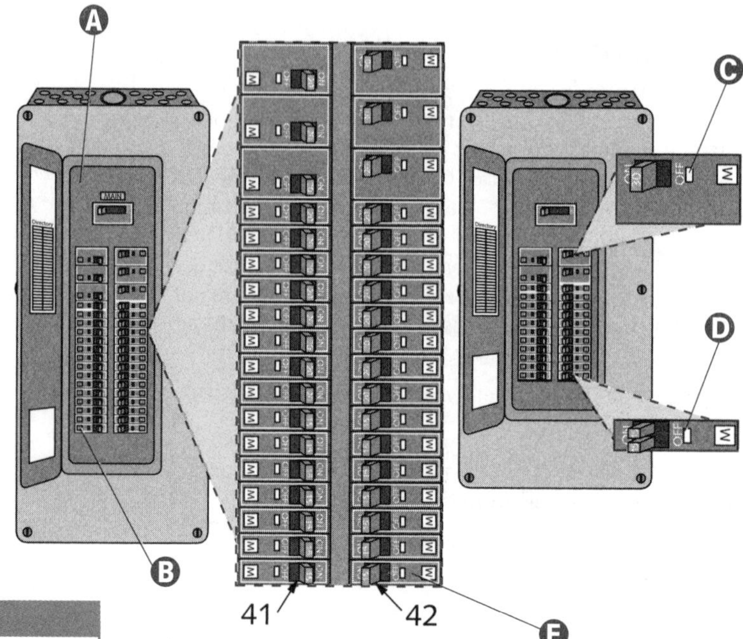

WARNING

Lighting and appliance branch-circuit panelboards physically inhibit the installation of excess overcurrent devices. These panelboards accommodate only that number for which the panelboard was designed, rated, and approved »408.15«.

Unused Openings

Ⓐ The purpose of all panelboard circuits (and circuit modifications) must be legibly identified on a circuit directory affixed to the face or inside of the panel door(s) »408.4«. The marking must be durable enough to withstand the surrounding environment »110.22«.

Ⓑ Busbars, both insulated and bare, must be rigidly mounted »408.31«.

Ⓒ Within the scope of Article 312 (Cabinets, Cutout Boxes, and Meter Socket Enclosures), unused openings must be closed in a manner that affords protection equivalent to that of the enclosure »312.4 and 110.12(A)«.

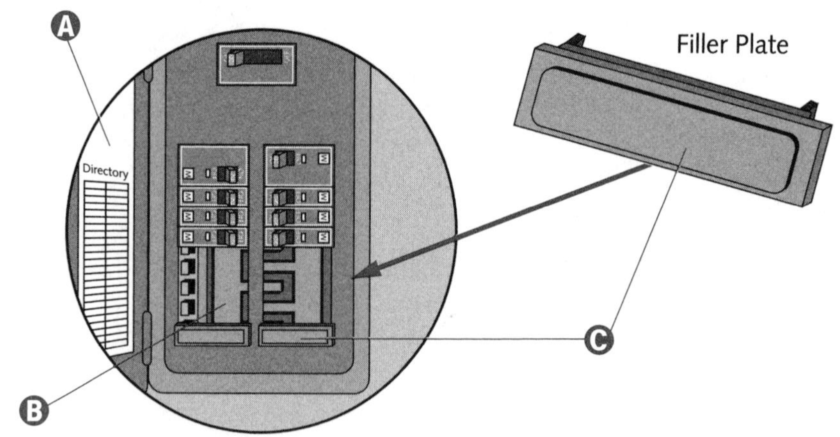

Circuit Breakers

The standard ampere ratings for fuses and inverse time circuit breakers are: 15, 20, 25, 30, 35, 40, 45, 50, 60, 70, 80, 90, 100, 110, 125, 150, 175, 200, 225, 250, 300, 350, 400, 450, 500, 600, 700, 800, 1000, 1200, 1600, 2000, 2500, 3000, 4000, 5000, and 6000 amperes. Fuses and inverse time circuit breakers with nonstandard ampere ratings can be used. Additional standard ampere ratings for fuses are: 1, 3, 6, 10, and 601 ›› 240.6(A)‹‹.

Ⓐ A **circuit breaker** is a device designed to both open and close a circuit by nonautomatic means, and to open the circuit automatically when subjected to a predetermined overcurrent without being damaged (when properly applied within its rating) ›› Article 100 ‹‹.

Ⓑ All circuit breakers having an interrupting rating other than 5000 amperes must be appropriately marked. The interrupting rating is not required on circuit breakers used for supplementary protection ›› 240.83(C) ‹‹. (Note: The ampere interrupting rating (AIR) is not the current rating of the circuit breaker.)

Ⓒ Circuit breakers must have a voltage rating (clearly marked) no less than the nominal system voltage that indicates their ability to interrupt fault current (between phases or phase to ground) ›› 240.83(E) ‹‹.

Ⓓ Circuit breakers and switches (general-use and motor-circuit), enclosure mounted according to 404.3, must clearly identify the open (off) or closed (on) positions ›› 404.7 ‹‹.

Ⓔ For vertically-operated circuit breakers, the "up" position of the handle must be the "on" position ›› 240.81 and 404.7 ‹‹.

Ⓕ A circuit breaker's ampere rating must be durably marked and remain visible after installation. When necessary to ensure visibility, removal of a trim or cover is permitted ›› 240.83(A) ‹‹.

Ⓖ A circuit breaker's **Inverse time** is a qualifying term indicating an intentional delay in the tripping action of the breaker. The length of the delay decreases as the magnitude of the current increases ›› Article 100 ‹‹.

Ⓗ A **nonadjustable** circuit breaker does not allow the value of current that triggers the tripping action, or the time required for its operation, to be altered ›› Article 100 ‹‹.

Ⓘ A circuit breaker's open (off) or closed (on) positions must be clearly identified ›› 240.81 and 404.7 ‹‹.

Ⓙ The ampere rating of any circuit breaker rated at 100 amperes or less, and 600 volts or less, must be molded, stamped, etched, or similarly marked into the handle or escutcheon area ›› 240.83(B) ‹‹.

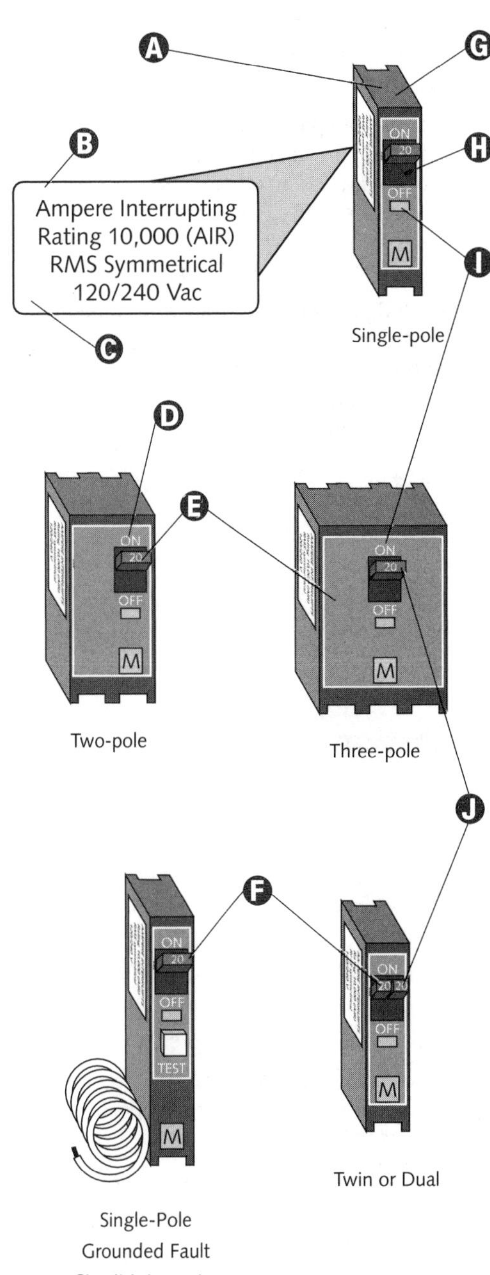

NOTE

Circuit breakers must be trip free as well as manually operable in both the closed and opened positions. While a circuit breaker may normally operate by electrical or pneumatic means, a means of manual operation must also be provided ››240.80 ‹‹.

Circuit breakers used as switches in 120-volt and 277-volt fluorescent lighting circuits must be listed and be marked "SWD" or "HID" ››240.83(D)‹‹.

CAUTION *Circuit breakers and fuses can be connected in parallel provided they are factory assembled and listed as a unit. Individual fuses, circuit breakers, or combinations cannot otherwise be connected in parallel ›› 240.8 ‹‹.*

Back-Fed Devices

A Since a bolt-on breaker is not a plug-in type overcurrent protection device, no additional fastener (screw, clip, etc.) is needed.

B These remote panelboard (subpanel) feeder conductors originated from a main panelboard. The conductors are connected to the load side terminals of the circuit breaker, instead of connecting to the main lugs. In this configuration (the reverse of normal operation), the conductor feeds the breaker, thus the term back-fed device.

C Back-fed plug-in-type overcurrent protection devices or plug-in-type main lug assemblies, used to terminate field installed ungrounded supply conductors, must be secured to the panelboard by an additional fastener mounting means »408.16(F)«.

A back-fed device functioning as a service disconnect must be permanently marked to identify it as a service disconnect »230.70(B)«.

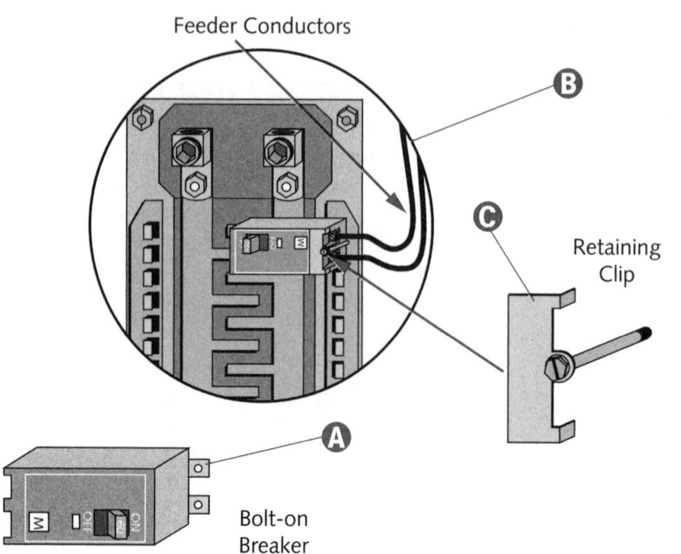

Installation

A Enclosures installed in wet locations must be weatherproof »312.2(A)«.

B Cabinets and cutout boxes must have sufficient space to prevent crowding of installed conductors »312.7«.

C Panelboards in damp or wet locations must be installed in compliance with 312.2(A) »408.17«.

D Concrete, brick, and tile walls are considered to be grounded »Table 110.26(A) Note 2«.

E The service conductors must be attached to the service disconnecting means by pressure connectors, clamps, or other approved means »230.81«.

F In damp or wet locations, surface-type enclosures (within the scope of Article 312) must be located or equipped so that moisture or water cannot enter or accumulate within the cabinet or cutout box. A minimum of ¼-in. (6-mm) airspace between the enclosure and the wall or other supporting surface is required »312.2(A)«.

G Electric equipment must be firmly secured to the surface on which it is mounted. Wooden plugs driven into holes in masonry, concrete, plaster, or similar materials are not acceptable »110.13(A)«.

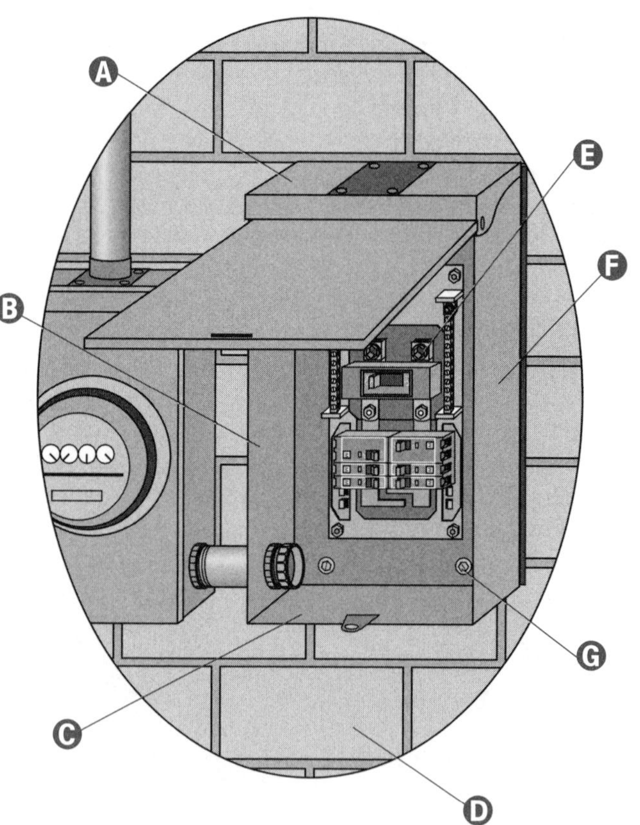

Grouping of Service Disconnects

Ⓐ Energized parts of service equipment must be enclosed as specified in 230.62(A), or guarded as specified in 230.62(B) »230.62«.

Ⓑ Two to six circuit breakers, or sets of fuses, are acceptable as the overcurrent device to provide overload protection »230.90(A) *Exception No. 3*«.

Ⓒ A service disconnecting means consisting of more than one switch or circuit breaker, as permitted by 230.71, must have a combined rating of all switches or circuit breakers used, not less than the rating required by 230.79 »230.80«.

Ⓓ Multiple service disconnects (from two to six) as permitted in 230.71, must be grouped »230.72(A)«.

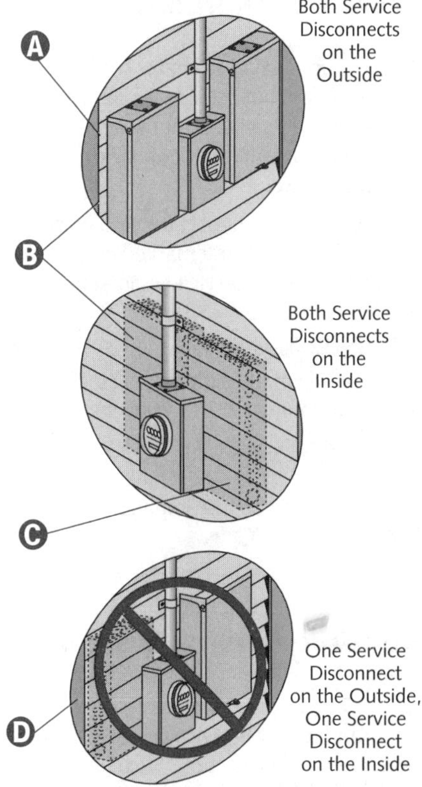

NOTE

Equipment whose purpose is to interrupt current at fault levels must have an interrupting rating sufficient for the nominal circuit voltage, as well as the current available at the equipment's line terminals. Equipment intended to interrupt current at other than fault levels must have an interrupting rating at nominal circuit voltage sufficient for current interruption »*110.9*«.

Location of Service Disconnecting Means

Ⓐ The service disconnecting means shall be installed at a readily accessible location either outside of a building or structure, or inside as near as possible to the point of entrance of the service conductors »230.70(A)(1)«.

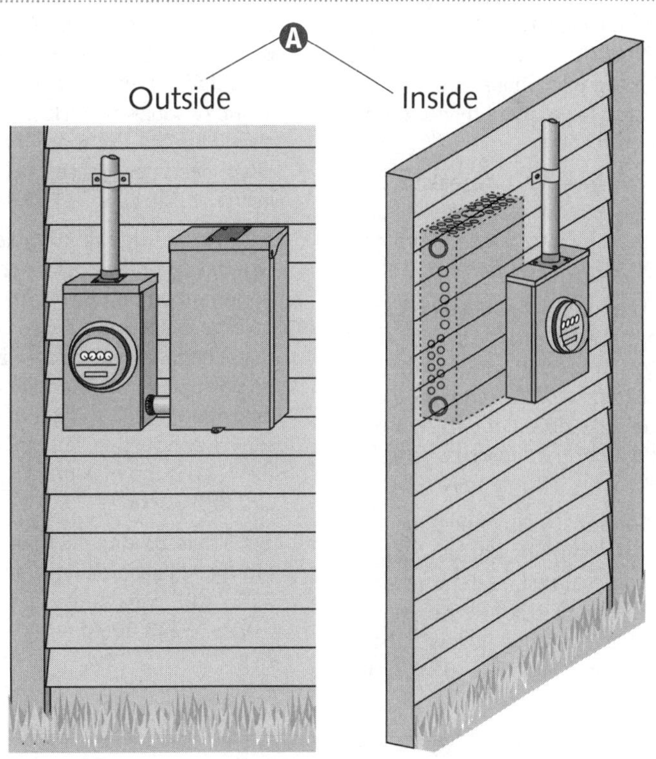

NOTE

Maximum service conductor lengths are not specified within a building or structure. The provision merely states that the disconnecting means must be located nearest the point of service conductor entrance. Since the length of service conductors can alter available short-circuit current, the length should be kept to a minimum. Some local jurisdictions define a maximum acceptable length for service conductors.

CAUTION *Neither service disconnecting means nor overcurrent devices (other than supplementary overcurrent protection) can be located in bathrooms as defined in Article 100* »*230.70(A)(2) and 240.24(E)*«.

Service Equipment

A Because individual meter socket enclosures are not considered service equipment, a "suitable for use as service equipment" marking is not required »230.66«.

B Service equipment rated 600 volts, or less, must be clearly marked as suitable for use as service equipment »230.66«.

C Each service disconnect must be permanently marked identifying it as a service disconnect »230.70(B)«.

D A one-family dwelling service disconnecting means must have a minimum rating of 100 amperes, 3-wire »230.79(C)«.

E A means to disconnect all conductors in a building (or other structure) from the service-entrance conductors must be provided »230.70«.

WARNING

Local utility companies often set parameters (height, grounding, etc.) for metering equipment installation. Local jurisdictions may also have special regulations concerning service equipment. Obtain copies of all applicable provisions for the area in which the equipment will be installed.

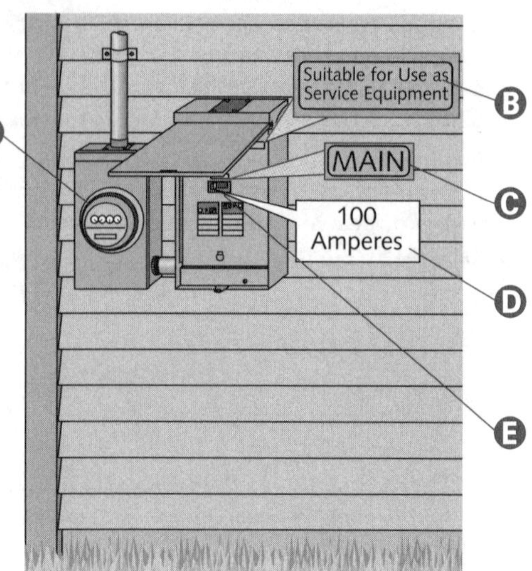

Multiple Service Disconnects

A 230.90 requires that each ungrounded (hot) service conductor have overload protection. Such protection shall be provided by an overcurrent device in series with each ungrounded service conductor having a rating (or setting) not higher than the allowable conductor ampacity »230.90(A)«. Under 230.90(A), one of the five exceptions, allow more than one overcurrent device for overload protection.

B Two to six circuit breakers, or sets of fuses, can serve as the overcurrent device, providing overload protection »230.90(A) *Exception No. 3*«.

C The total circuit breaker (or fuse) ratings can exceed the service conductors ampacity, provided the calculated load (in accordance with Article 220) does not exceed the ampacity of the service conductors »230.90(A) *Exception No. 3*«.

D The service disconnecting means must have a rating not less than the load to be carried, as determined in accordance with Article 220. In no case shall the rating be lower than specified in 230.79(A), (B), (C), or (D) »230.79«.

E Each service disconnect must be permanently marked as a service disconnect »230.70(B)«. Each individual service disconnect must be marked. A single marking for multiple-service disconnects is not permitted.

F The service disconnecting means for each service (permitted by 230.2), or for each set of service-entrance conductors (permitted by 230.40, *Exception Nos. 1, 3, 4, or 5*) shall consist of not more than six switches or six circuit breakers whether mounted in a single enclosure, in a group of separate enclosures, in or on a switchboard. No more than six disconnects per service can be grouped in any one location »230.71(A)«.

G Two or three single-pole switches or breakers, individually operable, are permitted on multiwire circuits (one pole for each ungrounded conductor). The multipole disconnect must be equipped with "handle ties" or a "master handle" to disconnect all service conductors. Six operations of the hand is the maximum allowed »230.71(B)«.

H Single-pole circuit breakers, grouped in accordance with 230.71(B), constitute one protective device »230.90(A)«.

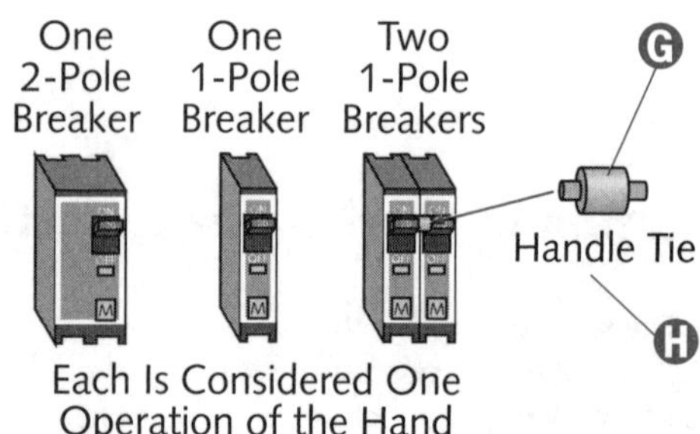

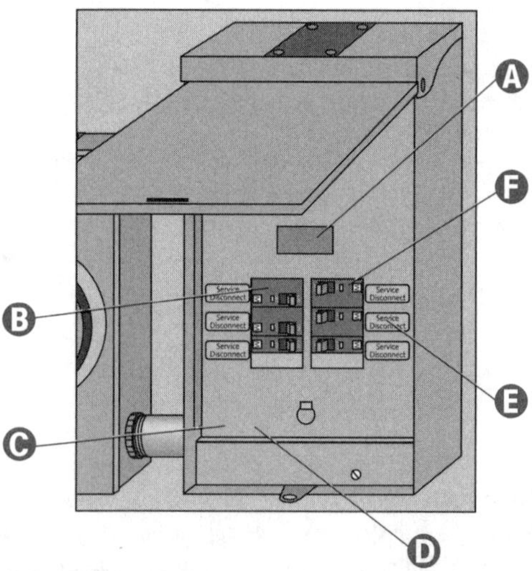

NOTE

A set of fuses shall be considered all the fused required to protect all of a circuit's ungrounded conductors »230.90(A)«.

Power Panelboards

A According to Article 408, panelboards are classified as (1) lighting and appliance branch-circuit or (2) power panelboards »408.14«.

B A power panelboard has no more than 10% of its overcurrent devices protecting lighting and appliance branch-circuits »408.14(B)«.

C An individual overcurrent protective device is not required for a power panelboard, used as service equipment, with multiple disconnecting means in accordance with 230.71 »408.16(B) *Exception*«.

Power Panelboard

NOTE

In addition to the requirements of 408.13, a power panelboard with supply conductors that include a neutral and that has more than 10% of its overcurrent devices protecting branch-circuits rated 30 amperes or less must be protected by an overcurrent protective device having a rating not greater than that of the panelboard. The overcurrent protective device must be located within, or at any point on the panelboard's supply side »408.16(B)«.

Lighting and Appliance Branch-Circuit Panelboards

A Raceway containing feeder conductors

B Panelboards are classified as (1) lighting and appliance branch-circuit or (2) power panelboards »408.14«.

C The front edge of cabinets (panelboards) situated in walls constructed of noncombustible material (concrete, tile, etc.) must be within ¼ in. (6 mm) of the finished surface »312.3«.

D Individual protection for a lighting and appliance panelboard is not required if the feeder has overcurrent protection not greater than the panelboard's rating »408.16(A) *Exception No. 1*«.

E A lighting and appliance branch-circuit panelboard has more than 10% of its overcurrent devices protecting lighting and appliance branch-circuits »408.14(A)«.

F A lighting and appliance branch-circuit has a connection to the panelboard neutral, and has overcurrent protection of 30 amperes or less in one or more conductors »408.14«.

G A grounding connection must not be made to any grounded circuit conductor on the load side of the service disconnecting means, except as otherwise permitted in Article 250 »250.24(A)(5)«.

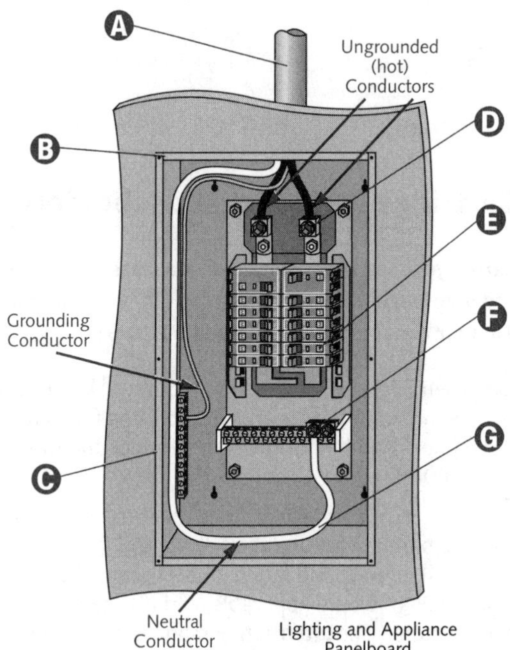

Lighting and Appliance Panelboard

NOTE

Each lighting and appliance branch-circuit panelboard must be individually protected on the supply side by no more than two main circuit breakers (or two sets of fuses) having a combined rating not greater than the panelboard's rating »408.16(A)«.

Panelboards used as service equipment in pre-existing individual residential occupancy installations, do not require individual protection for lighting and appliance branch-circuit panelboards »408.16(A) Exception No. 2«.

CAUTION *Cabinets in walls constructed of combustible material (wood, etc.) must be flush with or extend beyond the finished surface »312.3«.*

Common-Trip and Single-Pole Circuit Breakers

A Circuit breakers must open all ungrounded conductors of the circuit unless permitted in 240.20(B)(1), (B)(2), and (B)(3) »240.20(B)«.

B Individual single-pole circuit breakers, with or without approved handle ties, are permitted as protection for each ungrounded conductor in multiwire branch-circuits serving only single-phase, line-to-neutral loads, except where limited by 210.4(B) »240.20(B)(1)«.

C In grounded systems, individual single-pole circuit breakers with approved handle ties are permitted as protection for each ungrounded conductor for line-to-line connected loads for single-phase circuits or 3-wire direct-current circuits »240.20(B)(2)«.

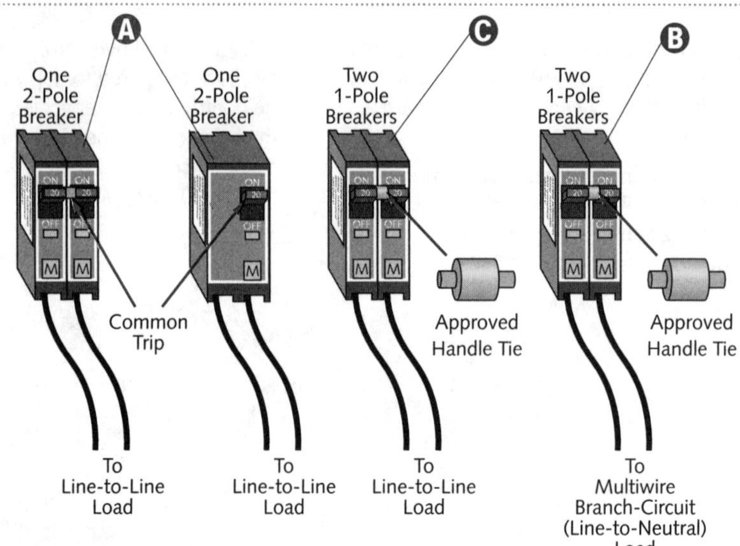

NOTE

For line-to-line loads in 4-wire, 3-phase systems or 5-wire, 2-phase systems having a grounded neutral and no conductor operating at a voltage greater than permitted in 210.6, individual single-pole circuit breakers with approved handle ties can protect each ungrounded (hot) conductor »*240.20(B)(3)*«.

Cartridge Fuses and Fuseholders

Cartridge fuses and fuseholders are classified according to voltage and amperage ranges. Fuses rated 600 volts, nominal or less, can be used for equal or lesser voltages »240.61«.

A Fuseholders are designed so that it will be difficult to put a fuse of any given class into a fuseholder intended for a current lower, or voltage higher, than that of the class to which the fuse belongs »240.60(B)«.

B Fuses must be plainly marked (by printing on or a label attached to the fuse barrel), showing the following: (1) ampere rating, (2) voltage rating, (3) interrupting rating if other than 10,000 amperes, (4) "current-limiting" where applicable, and (5) the name or trademark of the manufacturer. The interrupting rating is not required on fuses used for supplementary protection »240.60(C)«.

C Fuseholders for current-limiting fuses must not permit insertion of noncurrent-limiting fuses »240.60(B)«.

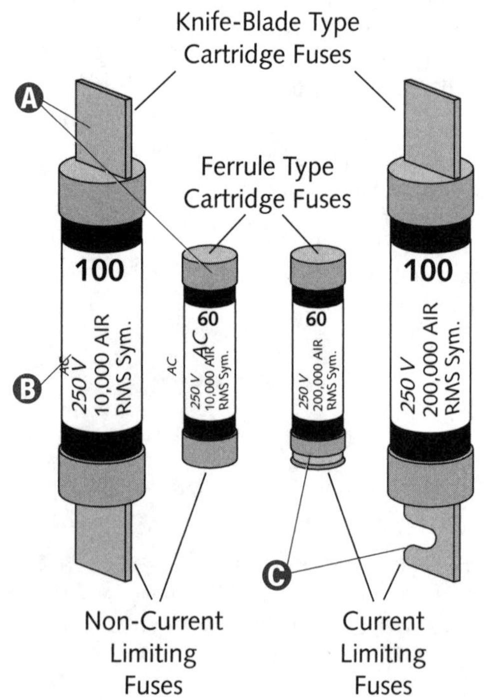

NOTE

Cartridge fuses, and fuseholders with a 300-volt rating, can be used in the following: (1) circuits not exceeding 300 volts between conductors, and (2) single-phase, line-to-neutral circuits supplied from a 3-phase, 4-wire solidly grounded neutral source with line-to-neutral voltage not exceeding 300 volts »*240.60(A)*«.

Conductor Sizing

A Table 310.15(B)(6) lists conductors that can be used as service-entrance, service-lateral, and feeder conductors serving as the main power feeder to a dwelling unit and being installed in raceway or cable (with or without an equipment grounding conductor) »310.15(B)(6)«.

B The minimum size service-entrance conductors for this 200-ampere disconnect are 2/0 AWG copper or 4/0 AWG aluminum (or copper-clad aluminum) »Table 310.15(B)(6)«.

C The minimum size feeder conductors for the 100-ampere remote panelboard are 4 AWG copper or 2 AWG aluminum (or copper-clad aluminum). As 310.15(B)(6) specifies, the main power feeder is that feeder(s) supplying the lighting and appliance branch-circuit panelboard(s) »310.15(B)(6)«.

D Article 440 outlines provisions relating to disconnecting means, branch-circuit overload protection, and branch-circuit conductors for air-conditioning and refrigeration equipment.

E Feeder conductors (not used as the main power feeder) are found in Table 310.16.

> **NOTE**
>
> *The grounded (neutral) conductor can be smaller than the ungrounded (hot) conductors, provided 215.2, 220.22, and 230.42 requirements are met »310.15(B)(6)«. The grounded conductor must not be smaller than the required grounding electrode conductor specified in Table 250.66 »250.24(B)(1)«.*
>
> *Dwelling unit feeder conductors are not required to be larger than the service-entrance conductors »310.15(B)(6)«.*

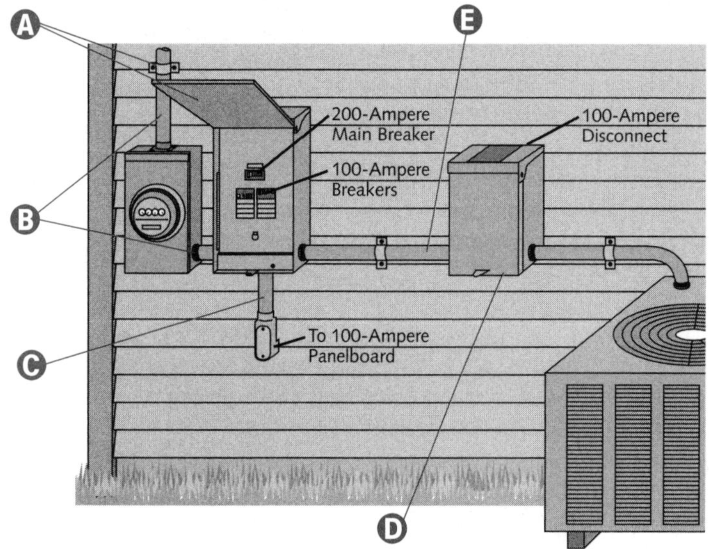

Installation

A Electric equipment must be installed in a neat and workmanlike manner »110.12«. "Workmanlike," by definition, implies skillful and characteristic of a good workman.

B All electric equipment internal parts (busbars, wiring terminals, insulators, etc.) must remain undamaged and uncontaminated by foreign materials (paint, plaster, cleaners, abrasives, corrosive residues, etc.). Equipment parts that are broken, bent, cut, corroded, impaired due to chemical action, heat, or otherwise damaged in such a way as to adversely affect safe operation or mechanical strength must not be used »110.12(C)«.

C 312.10 and 312.11 itemize construction specifications for cabinets, cutout boxes, and meter socket enclosures.

D Cabinets and cutout boxes must have space to sufficiently accommodate (without crowding) all installed conductors »312.7«.

E Conductor sizes are expressed in American Wire Gage (AWG) or in circular mils »110.6«.

> **NOTE**
>
> *Unless adequate space is available, enclosures for switches or overcurrent devices must not be used as junction boxes, auxiliary gutters, or raceways for conductors feeding through or (tapping off) to other switches or overcurrent devices. Conductors must not fill more than 40% of a cross-sectional area of the space, and conductors, splices, and taps must not fill the wiring space at any cross section to more than 75% of the cross-sectional area »312.8«.*

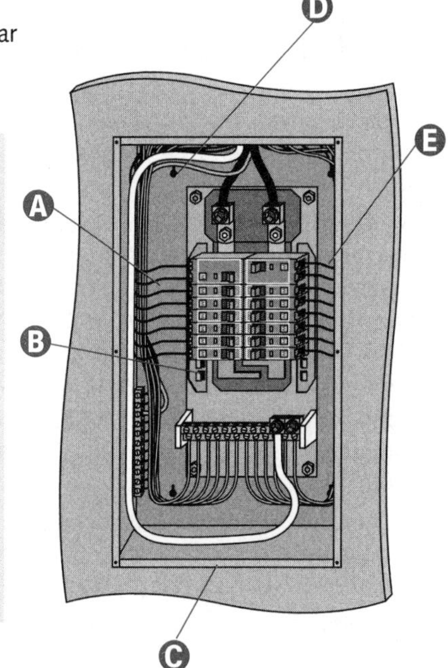

Plug Fuses, Fuseholders, and Adapters

A Plug fuses, of 15-ampere and lower rating, are identified by a hexagonal configuration in a prominent location (window cap, cap, etc.) to distinguish them from fuses of higher ampere ratings »240.50(C)«.

B Type S adapters must fit Edison-base fuseholders »240.54(A)«.

C Once inserted into a fuseholder, Type S adapters cannot be removed »240.54(C)«.

D Type S fuses are classified at not over 125 volts, and 0 to 15 amperes, 16 to 20 amperes; or 21 to 30 amperes »240.53(A)«.

E Edison-base type fuseholders must be installed only where they are to accept Type S fuses by the use of adapters »240.52«.

F Plug fuses rated 16 to 30 amperes are identified by means other than a hexagonal configuration.

G Type S fuseholders and adapters are designed so that either the fuseholder itself, or the fuseholder with a Type S adapter inserted, can be used only for a Type S fuse »240.54(B)«.

H Type S fuses of an ampere classification, specified in 240.53(A), are not interchangeable with a lower-ampere classification. They can be used only in a Type S fuseholder or a fuseholder with a Type S adapter inserted »240.53(B)«.

I Type S fuses, fuseholders, and adapters are designed to make tampering or shunting (bridging) difficult »240.54(D)«.

NOTE

The screw shell of a plug-type fuseholder must be connected to the circuit's load side »240.50(E)«.

WARNING

Edison-base type plug fuses can be used only as replacements in existing installations exhibiting no evidence of overfusing or tampering »240.51(B)«.

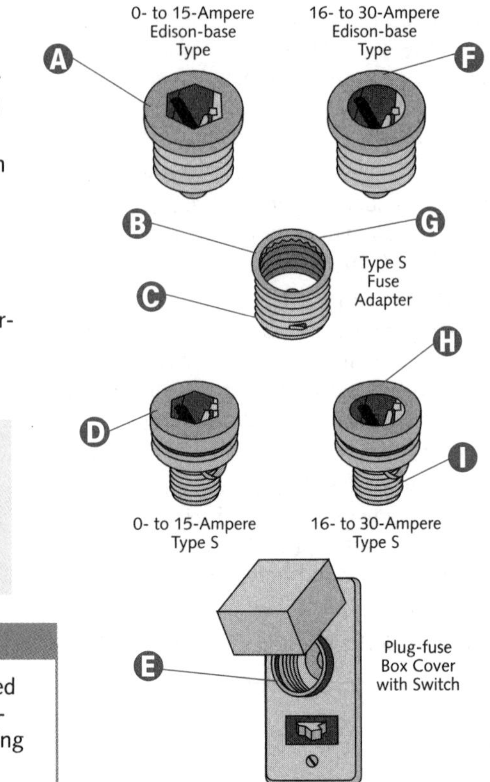

GROUNDING

Grounded (Neutral) Conductor in the Service Disconnecting Means

A Grounding and bonding jumpers must be connected by exothermic welding, or by listed means, such as pressure connectors and clamps »250.8«.

B An ac system grounded at any point and operating at less than 1000 volts must have a grounded conductor that is run to each service disconnecting means and bonded to each service disconnecting means enclosure »250.24(B)«.

C A grounded system must have an unspliced main bonding jumper connecting the equipment grounding conductor(s) and the service-disconnect enclosure to the grounded conductor of the system (within each service disconnect enclosure) »250.28«.

D A single bar in the service-disconnect enclosure may serve as the neutral bar (or bus) and the equipment grounding terminal bar (or bus).

E The grounded conductor must be routed with the phase conductors; cannot be smaller than the required grounding electrode conductor specified in Table 250.66; and is not required to be larger than the largest ungrounded service-entrance conductor »250.24(B)(1)«.

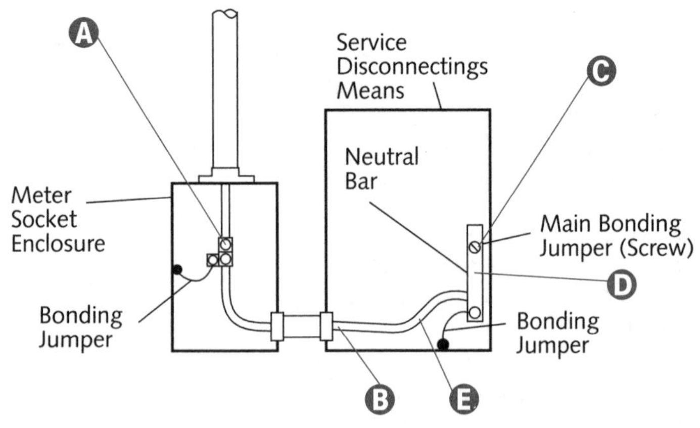

NOTE

The bonding jumper (screw, strap, etc.) is installed only in service equipment, and not in subpanels.

Main Bonding Jumper

A A main bonding jumper is the connection between the grounded circuit conductor (neutral) and the equipment grounding conductor at the service »Article100«.

B Main bonding jumpers must be attached by exothermic welding, or other listed means such as pressure connectors or clamps. Sheet-metal screws are not acceptable »250.28(C)«.

C A grounded system must have an unspliced main bonding jumper connecting the equipment grounding conductor(s) and the service-disconnect enclosure to the grounded conductor of the system (within each service disconnect enclosure) »250.28«.

D In cases where the main bonding jumper is a screw only, it must be identified with a green finish that remains visible after installation »250.28(B)«.

E Main bonding jumpers must be made of copper or other corrosion-resistant material. A wire, bus, screw, or similar suitable conductor is acceptable as a main bonding jumper »250.28(A)«.

NOTE

The main bonding jumper cannot be smaller than the sizes shown in Table 250.66 for grounding electrode conductors »250.28(D)«. (Listed main bonding jumpers furnished with panelboards meet the provisions of 250.28.)

CAUTION *Panelboards, listed as suitable for use as service equipment, are furnished with a main bonding jumper that has not yet been installed. When using the panelboard as service equipment, install the main bonding jumper (screw, strap, etc.). When the panelboard is used for other purposes, such as a subpanel, do not install the main bonding jumper.*

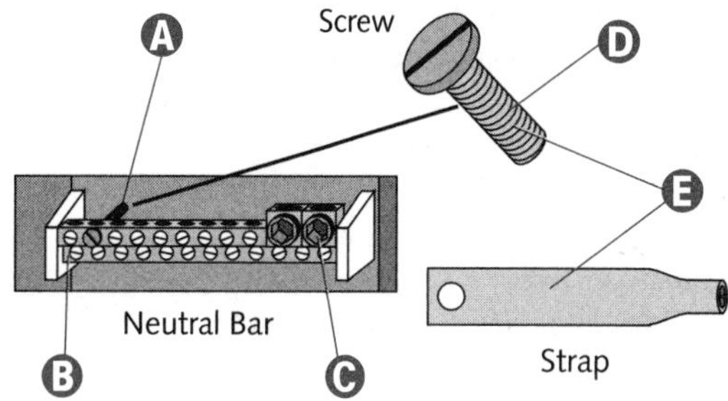

Remote Panelboards (Subpanels)

A Unless grounded by connection to the grounded circuit conductor (permitted by 250.32, 250.140, and 250.142), noncurrent-carrying metal parts of equipment, raceways, and other enclosures must be grounded (where required) by one of the following methods: (A) by any of the equipment grounding conductors permitted by 250.118, or (B) by an equipment grounding conductor contained within the same raceway, cable, or otherwise run with the circuit conductors »250.134«.

B Equipment grounding conductors can be bare, covered, or insulated. Individually covered or insulated equipment grounding conductors must have a continuous outer finish, either green or green with yellow stripe(s), except as otherwise permitted by Article 250, Part VI »250.119«.

C Only in service disconnecting means can the neutral terminal bar be connected to the enclosure or equipment grounding terminal bar. At this point, the neutral bar is isolated from the equipment grounding system.

D A grounded circuit conductor cannot be used for grounding noncurrent-carrying metal parts of equipment on the load side of the service disconnecting means, unless an exception has been met »250.142(B)«.

E Panelboard cabinets and panelboard frames, if metal, must be in physical contact with one another and must be grounded. Grounding conductors must not be connected to a terminal bar provided for grounded conductors (may be a neutral) unless the bar is identified for that purpose and is located where interconnection between equipment grounding conductors and grounded circuit conductors comply with Article 250 »408.20«.

F A grounding connection must *not* be made to any grounded circuit conductor on the load side of the service disconnecting means except as otherwise allowed in Article 250 »250.24(A)(5)«.

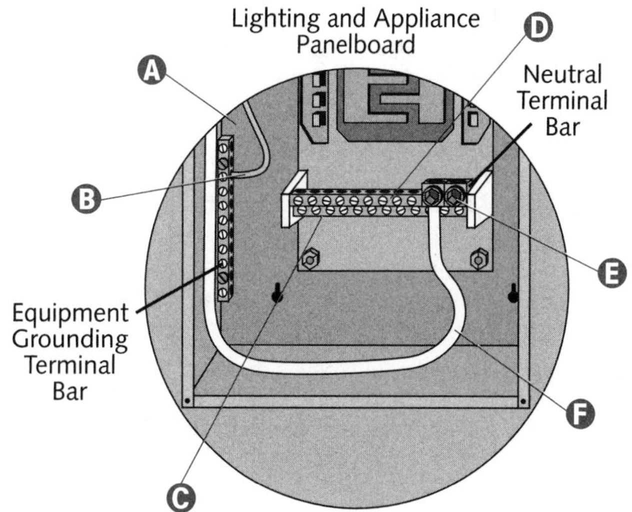

NOTE

The bonding jumper (screw, strap, etc.) is installed only in the service disconnecting means and not in subpanels.

Grounding Electrode Conductor Connections

A The size of the grounding electrode conductor on an ac system (grounded or ungrounded) must meet Table 250.66 specifications, except as permitted in 250.66(A) through (C).

B A premises wiring system, supplied by a grounded ac service, must have a grounding electrode conductor connected to the grounded service conductor, at each service, in accordance with 250.24(A)(1) through (5) »250.24(A)«.

C A grounding electrode conductor must connect the equipment grounding conductors, the service-equipment enclosures, and, where the system is grounded, the grounded service conductor to the grounding electrode(s) »250.24(C)«.

D Splices in grounding electrode conductors are acceptable only if by means of irreversible compression-type connectors (listed for the purpose) or by exothermic welding »250.53(D)(3)«.

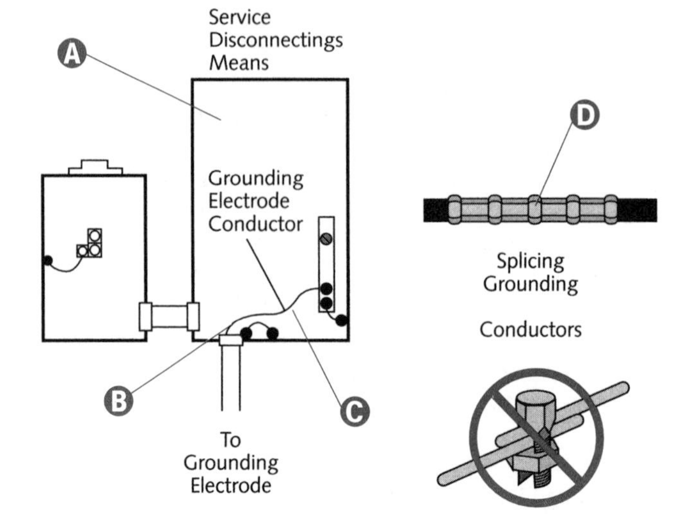

NOTE

The grounding electrode conductor can be connected to the grounded service conductor at any accessible point from the load end of the service drop (or service lateral), to and including the terminal (or bus) to which the grounded service conductor is connected at the service disconnecting means »250.24(A)(1)«.

Methods of Bonding at the Service

A One method is to bond equipment to the grounded service conductor in a 250.8 approved manner »250.94(1)«.

B Connections utilizing threaded couplings or threaded bosses on enclosures must be wrench-tight »250.94(2)«.

C Approved devices, such as bonding-type locknuts and bushings are also permitted to ensure electrical continuity at service equipment »250.94(4)«.

D Standard locknuts or bushings cannot be the sole means for bonding as required by 250.94 »250.94«.

E Threadless couplings and connectors are approved where made up tight for metal raceways and metal-clad cables. Standard locknuts and bushings cannot be used to bond these items »250.94(3)«.

F Bonding jumpers, which meet other Article 250 requirements, must be used around concentric (or eccentric) knockouts that are punched (or otherwise formed) in a way that impairs the electrical connection to ground »250.94«.

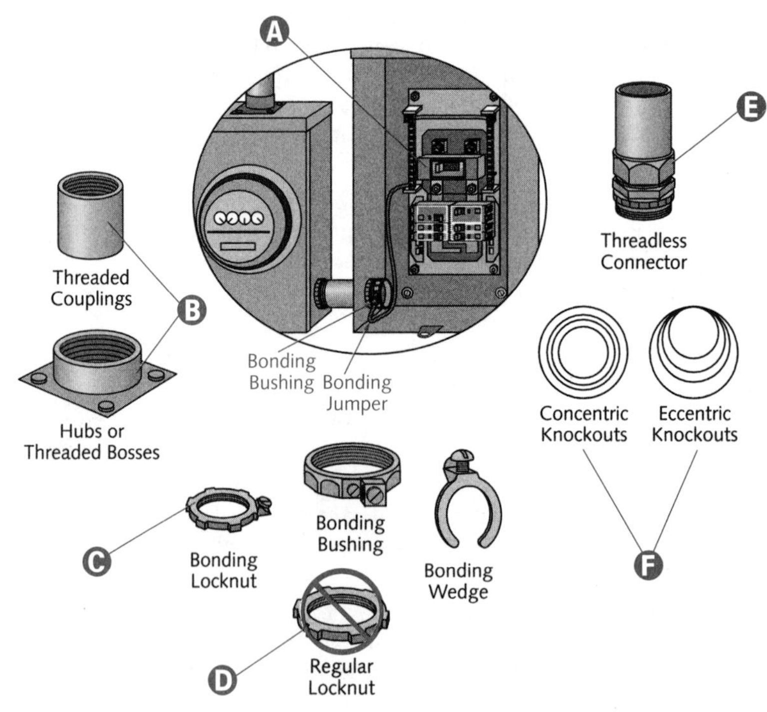

Bonding at the Service

A Bonding is the permanent joining of metallic parts to form an electrically conductive path, thereby ensuring continuity and capacity to safely conduct any current likely to be imposed »Article 100«.

B Hubs (or threaded bosses) are acceptable when bonding service equipment enclosures to threaded conduits or threadless connectors »250.94(2) and (3)«.

C All service equipment noncurrent-carrying metal parts (raceways, enclosures containing service conductors, meter fittings, metal raceway, or armor enclosing a grounding electrode conductor, etc.) must be effectively bonded together. »250.92(A)«.

D Bonding, where necessary, must be provided not only to ensure electrical continuity but also ample capacity to safely conduct any likely fault current »250.90«.

E Electrical continuity at service equipment, service raceways, and service conductor enclosures must be accomplished by one of the approved methods in 250.94.

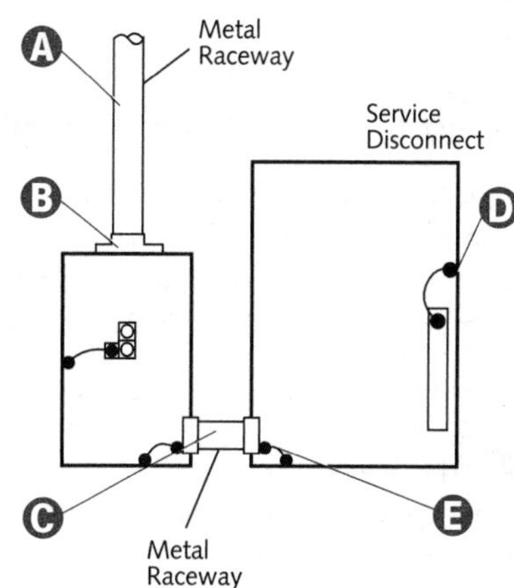

Grounding Electrode System

A The metal frame of the building or structure, where effectively grounded, must be bonded »250.52(A)(2)«.

B A ground ring, which encircles the building or structure in direct contact with the earth at a depth of at least 2½ ft (750 mm)below the surface, consisting of at least 20 ft (6.0 m) of bare copper conductor not smaller than 2 AWG is allowed »250.52(A)(4) and 250.53(G)«.

C A metal underground water pipe must be in direct contact with the earth for a minimum of 10 ft (3.0 m) (including any metal well casing effectively bonded to the pipe) and electrically continuous (or made so by bonding around insulating joints or sections or insulating pipe) to the points where the grounding electrode conductor and the bonding conductors are connected »250.52(A)(1)«.

D An electrode encased by at least 2 in. (50 mm) of concrete, located within and near the bottom of a concrete foundation or footing in direct contact with the earth, consisting of one or more bare or zinc galvanized or other electrically conductive coated steel reinforcing bars each at least ½ in. (12.7 mm) in diameter and 20 ft (6.0 m) continuous length, or consisting of at least 20 ft (6.0 m) of bare copper conductor not smaller than 4 AWG, is acceptable »250.52(A)(3)«.

E Grounding path continuity, or the bonding connection to interior piping, must not rely on water meters, filtering devices, or similar equipment »250.53(E)(1)«.

F Bonding jumper(s) shall be installed in accordance with 250.64(A),(B), and (E), shall be sized in accordance with 250.66, and shall be connected as specified in 250.70 »250.53(C)«.

G Interior metal water piping located more than 5 ft (1.5 m) from the point of entrance to the building cannot be used as part of the grounding electrode system, nor as a conductor to interconnect electrodes that are part of the grounding electrode system, unless the exception for industrial and commercial buildings is met »250.52(A)(1)«.

> **NOTE**
>
> *An unspliced grounding electrode conductor can be run to any convenient grounding electrode available in the system, or to grounding electrode(s) individually. It must be sized for the largest grounding electrode conductor required among all the electrodes being connected »250.53(D)(1) and (2)«.*

> **CAUTION** *An underground metal water pipe requires an additional electrode of a type specified in 250.52(A) through (7). The supplemental electrode can be bonded to the grounding electrode conductor, the grounded service-entrance conductor, the nonflexible grounded service raceway, or to any grounded service enclosure »250.53(E)(2)«.*

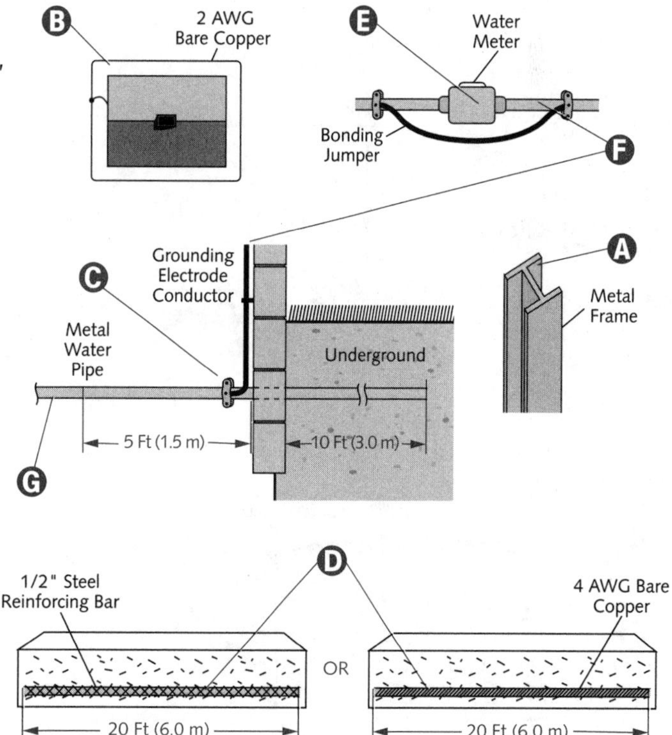

Rod, Pipe, and Plate Electrodes

If none of the electrodes specified in 250.52(A)(1) through (A)(6) is available, electrode(s) specified in 250.52(A)(4) through (A)(7) must be used »250.50«.

Pipe or conduit electrodes cannot be smaller than ¾-in. trade size; and, if iron or steel, shall have a galvanized outer surface (or shall be otherwise metal-coated or corrosion protected) »250.52(A)(5)(a)«.

Aluminum electrodes are not permitted »250.52(B)(2)«.

A The electrode's upper end must be flush with, or below, ground level unless the aboveground end and the grounding electrode conductor attachment are adequately protected against physical damage as specified in 250.10 »250.53(H)«.

B Electrodes of iron or steel rods must be at least ⅝ in. (15.87 mm) in diameter. Stainless steel rods less than ⅝ in. (15.87 mm) in diameter, nonferrous rods, or their equivalent must be listed, and be ½ in. (12.7 mm) or more in diameter »250.52(A)(5)(6)«.

C In multiple electrode configurations, each electrode of one grounding system (including that used for air terminals) must be at least 6 ft (1.8 m) from any other electrode of another grounding system. Two or more grounding electrodes, effectively bonded together, are considered a single grounding electrode system »250.53(B)«.

D A single electrode consisting of a rod, pipe, or plate that does not have a resistance to ground of 25 ohms (or less) shall be augmented by an additional electrode of any type specified in 250.52(A)(2) through (A)(7). Multiple electrodes installed to meet the requirements of 250.56 must be at least 6 ft (1.83 m) apart »250.56«.

E A metal underground gas piping system *cannot* be used as a grounding electrode »250.52(B)«.

F At least 8 ft (2.5 m) of electrode length must be in contact with the soil.

G Where rock bottom is encountered, the electrode can be driven at an oblique angle not to exceed 45° from vertical »250.53(H)«.

H Where rock bottom is encountered, the electrode can be buried in a trench that is at least 2½ ft (750 mm) deep »250.53(H)«.

I Each plate electrode must expose at least 2 sq ft (0.186 m²) of surface to exterior soil »250.52(A)(6)«. (One sq ft of surface area on each side of a 1 ft x 1 ft plate equals 2 sq ft.)

J Spacing rods [longer than 8 ft (2.5 m)] farther than 6 ft (1.8 m) apart improves paralleling efficiency »250.56 FPN«.

K Iron or steel plate electrodes must be at least ¼ in. (6.35 mm) thick, while nonferrous metal plates must be at least 0.06 in. (1.52 mm) thick »250.52(A)(6)«.

L Plate electrodes must be installed at least 2½ ft (750 mm) below the earth's surface »250.53(I)«.

NOTE

An underground metal water pipe must have an additional electrode of a type specified in 250.52(A)(2) through (A)(7) »250.53(E)(2)«.

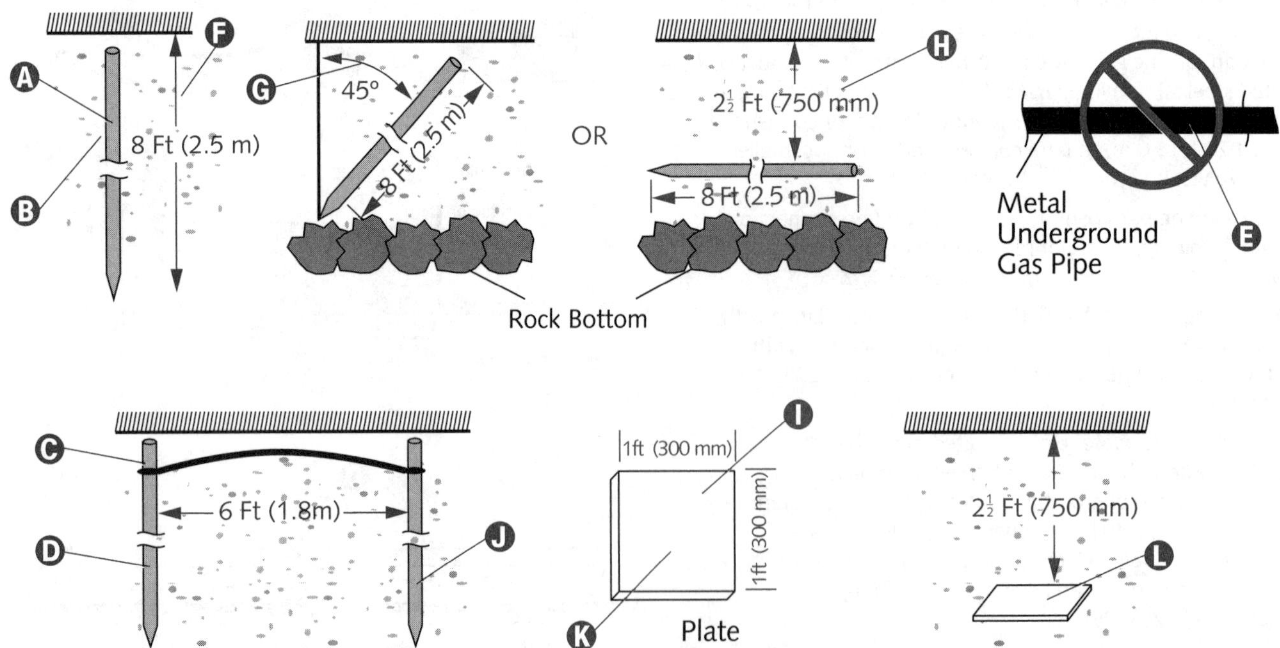

Grounding Electrode Conductor

The grounding electrode conductor must be made of copper, aluminum, or copper-clad aluminum. The material selected shall be inherently resistant to corrosive conditions existing at the site of installation, or shall be suitably protected against corrosion. The conductor can be solid (or stranded), insulated, covered, or bare »250.62«.

Aluminum or copper-clad aluminum grounding conductors (insulated or bare) must not be used in direct contact with masonry, or earth, or where subject to corrosive conditions. If used outside, aluminum or copper-clad aluminum grounding conductors must not be installed within 18 in. (450 mm) of the earth »250.64(A)«.

A A grounding electrode conductor must be used to connect the equipment grounding conductors, the service-equipment enclosures, and, in a grounded system, grounded service conductors to the grounding electrode »250.24(C)«.

B The grounding electrode conductor must be installed in one continuous length (without a splice or joint), unless spliced by an irreversible means »250.53(D)(3) and 250.64(C)«.

C A grounding electrode conductor (or its enclosure) must be securely fastened to the surface on which it is carried »250.64(B)«.

D Where a raceway is used as protection for a grounding conductor, the installation must comply with appropriate raceway article requirements »250.64(E)«.

E Metal raceways which enclose grounding electrode conductors must be bonded on each end »250.64(E)«.

F Metal enclosures, enclosing grounding electrode conductors, must be electrically continuous from the point of cabinet (or equipment) attachment to the grounding electrode, and must be fastened securely to the ground clamp (or fitting). Metal enclosures, not physically continuous from cabinet (or equipment) to the grounding electrode, must be made electrically continuous by bonding each end to the grounding conductor »250.64(E)«.

G The grounding conductor must be connected to the grounding electrode by exothermic welding, or by listed means, such as lugs, pressure connectors, clamps, etc. Connections depending on solder cannot be used »250.70«.

H Only one conductor can be connected to the grounding electrode by a single clamp or fitting, unless it is listed for multiple conductors »250.70«.

I One variety of ground-rod clamp listed for direct burial

J Ground clamps must be listed for the grounding electrode and grounding electrode conductor materials; and, if used on pipe (rod or other buried electrodes) must also be listed for direct soil burial »250.70«.

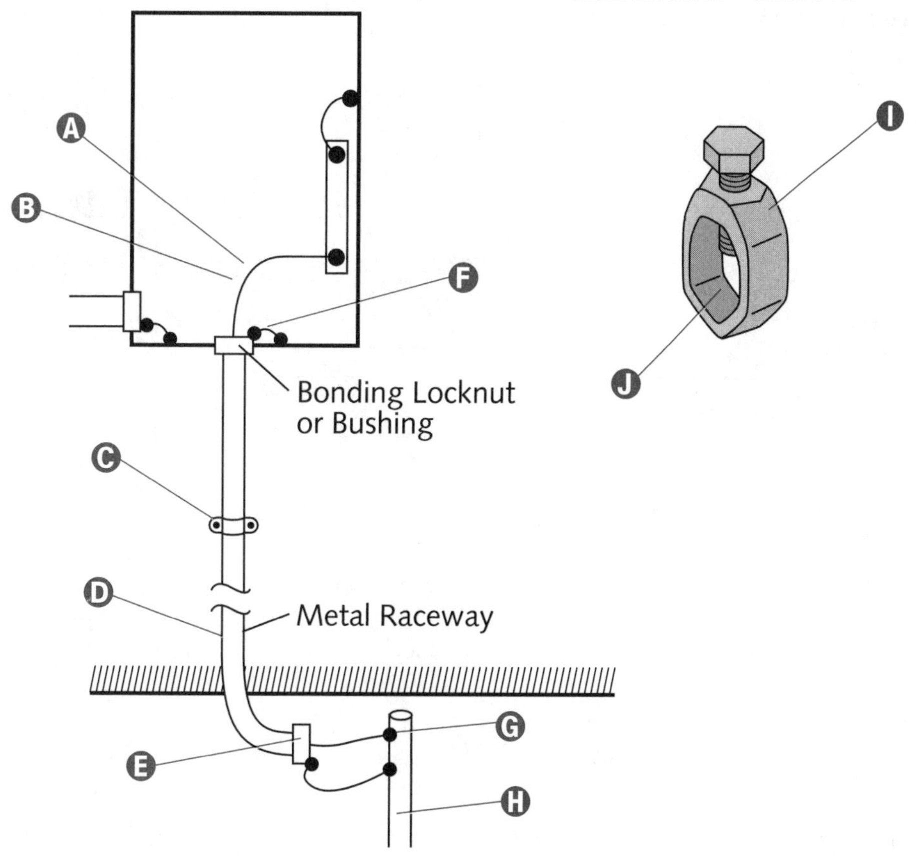

Grounding Electrode Conductor Sizing

A 4 AWG or larger conductor (copper or aluminum) must be protected if exposed to severe physical damage. A 6 AWG grounding conductor, not exposed to physical damage, can run along the surface of the building construction without metal covering or protection provided it is securely fastened to the construction. Otherwise, it must be housed in RMC, IMC, RNC, EMT, or cable armor »250.64(B)«.

Size 8 AWG and smaller grounding electrode conductors must be installed in RMC, IMC, RNC, EMT, or cable armor »250.64(B)«.

A While the grounding electrode conductor must be sized in accordance with 250.66, the tap conductors can be sized in accordance with the grounding electrode conductors specified in 250.66 for the largest conductor serving respective enclosures. The tap conductors must be connected to the grounding electrode conductor so that the grounding electrode conductor remains without a splice or joint »250.64(D)«.

B A 4 AWG copper (or 2 AWG aluminum or copper-clad aluminum) grounding electrode conductor is required if the service-entrance conductors are 2/0 AWG or 3/0 AWG copper »Table 250.66«.

C Interior metal water piping, located more than 5 ft (1.5 m) from the point of entrance to the building, cannot be used as part of the grounding electrode system nor as a conductor to interconnect electrodes that are part of the system »250.52(A)(1)«.

D To ensure a permanent and effective ground for a metal piping system used as a grounding electrode, effective bonding must be provided as necessary around insulated joints (and sections) as well as around any equipment likely to be disconnected for repairs or replacement »250.68(B)«.

E If the grounding electrode conductor is connected to rod, pipe, or plate electrodes as permitted in 250.52(A)(5) or (A)(6), that portion of the conductor serving as the sole connection to the grounding electrode is not required to be larger than 6 AWG copper wire, or 4 AWG aluminum wire »250.66(A)«.

F Where the grounding electrode conductor is connected to a concrete encased electrode as permitted in 250.52(A)(3), that portion of the conductor serving as the sole connection to the grounding electrode is not required to be larger than 4 AWG copper wire »250.66(B)«.

G If, as 250.52(A)(4) permits, the grounding electrode conductor is connected to a ground ring, that portion of the conductor serving as the sole connection to the grounding electrode does not have to be larger than the ground ring conductor »250.66(C)«.

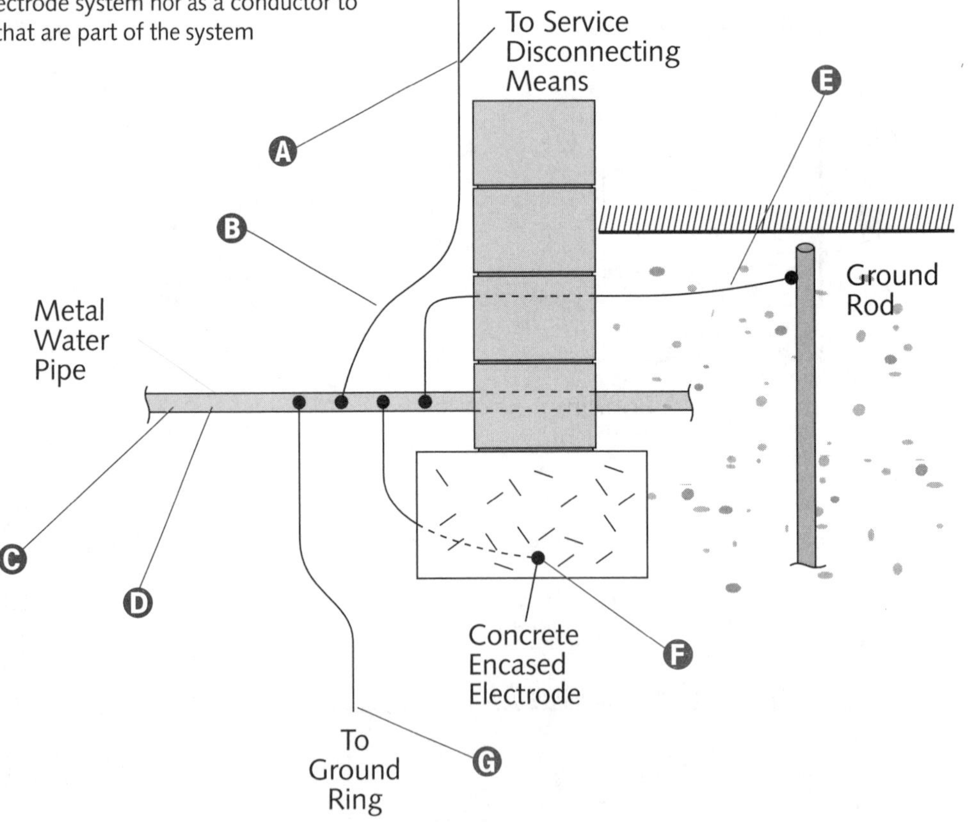

Equipment Bonding Jumper on the Supply Side of the Service

A Grounding electrode conductor and equipment bonding jumper sizes, on the supply side of the service, are determined by service-entrance conductor sizes.

B Unit 8 explains grounded (neutral) conductor sizing.

C The grounding electrode conductor must be sized in accordance with 250.66 »250.64(D)«.

D Bonding jumpers, on the supply side of the service, cannot be smaller than the sizes shown in Table 250.66 for grounding electrode conductors »250.102(C)«.

CAUTION *Metal piping system(s) with the potential to become energized must be bonded to the service equipment enclosure, the grounded conductor at the service, the grounding electrode conductor (if of sufficient size), or to the grounding electrode(s) used. The bonding jumper must be sized in accordance with Table 250.122 by using the rating of the circuit that may energize the piping system(s). The equipment grounding conductor of this circuit can serve as the bonding means »250.104(B)«.*

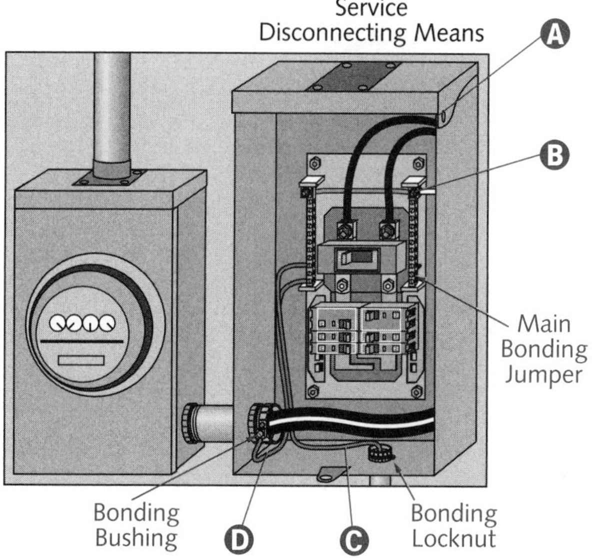

NOTE

The interior metal water piping system must be bonded either to the service equipment enclosure, the grounded conductor at the service, the grounding electrode conductor (if of sufficient size), or to the grounding electrode(s) used. The bonding jumper must be sized in accordance with Table 250.66, and installed in accordance with 250.64(A), (B), and (E). The bonding jumper's point of attachment must be accessible »250.104(A) and (A)(1)«.

Equipment Bonding Jumper on the Load Side of the Service

Equipment bonding jumpers (on the load side of the service overcurrent devices) must be: (1) sized, as a minimum, in accordance with Table 250.122; but (2) are not required to be larger than the circuit conductors supplying the equipment; and (3) cannot be smaller than 14 AWG copper »250.102(D)«.

A A panelboard (on the load side of the service overcurrent devices) protected by a 100 ampere overcurrent protection device (circuit breaker or fuse), requires an 8 AWG copper equipment grounding conductor »Table 250.122«.

B Wire-type copper, aluminum, or copper-clad aluminum equipment grounding conductors must not be less than shown in Table 250.122, but are not required to be larger than the circuit conductors supplying the equipment. When a raceway, cable armor, or sheath is used as the equipment grounding conductor (as provided in 134(A) and 250.118), it must comply with 250.2(D) »250.122(A)«.

C Metal panelboard cabinets and frames must be grounded and in physical contact with each other. Where the panelboard is used with nonmetallic raceway (or cable), or where separate grounding conductors are provided, a terminal bar for the grounding conductors must be secured within the cabinet. The terminal bar must be bonded to the metal cabinet and panelboard frame. Otherwise it shall be connected to the grounding conductor that is run with the conductors feeding the panelboard »408.20«.

D Equipment grounding conductor and equipment bonding jumper sizes (on the load side of the service) are determined by the size of the overcurrent device which protects the equipment.

E Unit 8 explains grounded (neutral) conductor sizing.

NOTE

Where conductor sizes are adjusted to compensate for voltage drop, equipment grounding conductors, if required, must be adjusted proportionately according to circular mil area »250.122(B)«.

An equipment grounding conductor run with (or enclosing) the circuit conductors must be one of those listed in 250.118.

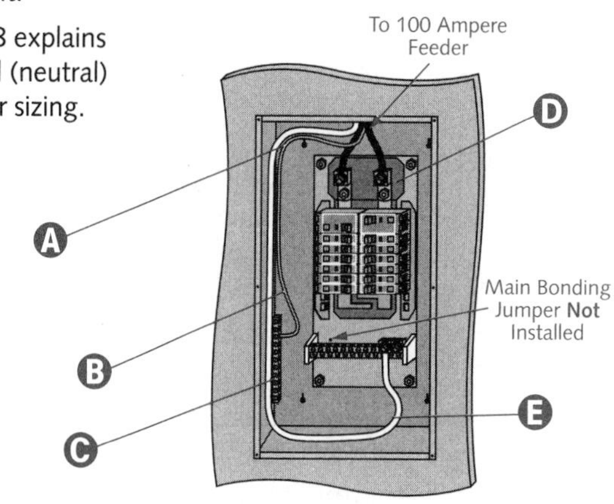

Lighting and Appliance Panelboard

Separate Buildings or Structures

A grounding electrode is not required at separate buildings (or structures) where only one branch-circuit supplies the building (or structure), and the branch-circuit includes an equipment grounding conductor for grounding the conductive noncurrent-carrying parts of all equipment »250.32(A) *Exception*«.

A Where multiple buildings (or structures) are supplied from a common ac service by a feeder(s) (or branch-circuit[s]), each building must have a grounding electrode (as described in Part III) connected to the metal enclosure of the building disconnecting means. If there are no existing grounding electrodes, a grounding electrode must be installed that meets Part III requirements »250.32(A)«.

B A remote panelboard (subpanel) installed in a detached garage (or structure) is not considered a service disconnecting means.

C An equipment grounding conductor (described in 250.118) must be run with the supply conductors and connected to the building (or structure) disconnecting means and to the grounding electrode(s). The equipment grounding conductor must be used for grounding or bonding or equipment, structures, or frames required to be grounded or bonded. Any installed grounded conductor must not be connected to the equipment grounding conductor or the grounding electrode(s) »250.32(B)(1)«.

D Where (1) an equipment grounding conductor is *not* run with the supply to the building or structure; and (2) there are no continuous metallic paths bonded to the grounding system in both buildings or structures involved; and (3) ground-fault protection of equipment has not been installed on the common ac service, the grounded circuit conductor run with the supply to the building or structure must be connected to the building or structure disconnecting means and to the grounding electrode(s) and must be used for grounding or bonding equipment, structures, or frames requiring bonding or grounding »250.32(B)(2)«.

E The size of the grounding electrode conductor to the grounding electrode(s) must not be less than given in Table 250.66, and does not have to be larger than the largest ungrounded supply conductor »250.32(E)«.

F The grounding electrode conductor installation must comply with Article 250, Part III »250.32(E)«.

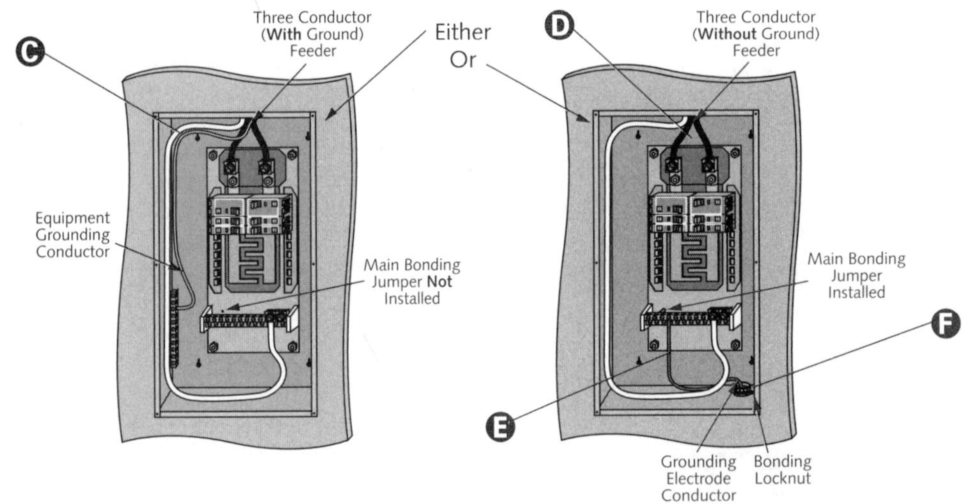

NOTE

Where one or more disconnecting means supply one or more additional buildings (or structures) under single management, and where these disconnecting means are located remote from those buildings (or structures) according to 225.32, Exception No. 1 *and 2 provisions, all of 230.32(D)(1) through (3) conditions must be met »250.32(D)«.*

Preventing Objectionable Current

Electrically conductive materials, such as metal water piping, metal gas piping, and structural steel members, that are likely to become energized, must be bonded to the supply system grounded conductor »250.2(C)«.

Temporary currents resulting from accidental conditions, such as ground-fault currents, that occur only while the grounding conductors are performing their intended protective functions, are not classified as objectionable current for the purpose specified in 250.6(A) and (B) »250.6(C)«.

Ⓐ Noncurrent-carrying conductive materials enclosing electrical conductors (or equipment), or forming part of such equipment, must be connected to earth so as to limit the voltage to ground on these materials. Where the electrical system is grounded, these materials must be connected together and to the supply system grounded conductor as specified by Article 250 »250.2(B)«.

Ⓑ If multiple grounding connections result in an objectionable current, one or more of the following alterations in 250.6(B)(1) through (4) are permitted, provided that the requirements of 250.2(D) are met: (1) discontinue one or more, but not all, of such grounding connections; (2) change the location of the grounding connections; (3) interrupt the continuity of the conductor (or conductive path) linking the grounding connections; and (4) take other suitable remedial action approved by the AHJ »250.6(B)«.

Ⓒ The grounding of electric systems, circuit conductors, surge arresters, and conductive noncurrent-carrying materials (and equipment) must be installed and arranged in a manner that will prevent objectionable current over the grounding conductors or grounding paths »250.6(A)«.

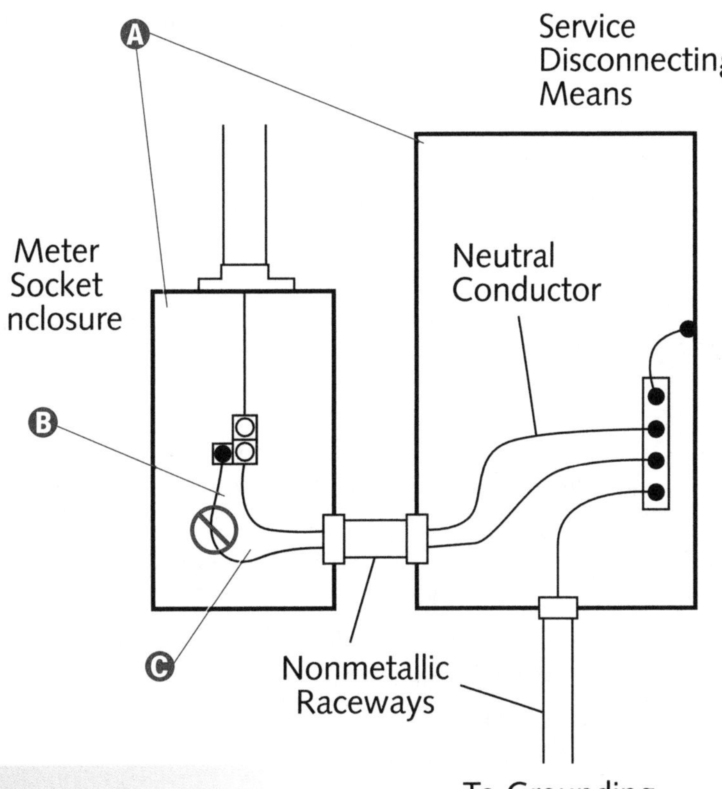

CAUTION *Connecting a grounded (neutral) and grounding conductor in parallel creates objectionable current on the grounding conductor, because the unbalanced current flows through both the grounded and grounding conductor. Should the grounded conductor lose its continuity (become disconnected), the grounding conductor would then carry all of the unbalanced (neutral) current. A hazard would exist if the grounding conductor's ampacity rating is less than the current flowing through it.*

Summary

- A service mast must safely withstand the strain imposed by service-drop conductors.
- Nothing, except electrical service-drop conductors, can be attached to the service mast.
- Wiring methods, permitted for services, are listed in 230.43.
- Service-drop conductor vertical ground clearances are found in 230.24(B)
- Service-drop conductor roof clearances are in 230.24(A).
- A minimum of 6½ ft (2.0 m) of work-space headroom is required around service equipment and panelboards.
- A minimum frontal work space width is 30 in. (750 mm) for service equipment and panelboards.
- The depth of this work space must be at least 3 ft (900 mm).
- Panelboards are mounted in cabinets or cutout boxes.
- A maximum of 42 overcurrent devices can be installed in any one panelboard.
- Each service disconnect must be permanently marked identifying it as a service disconnect.
- The service disconnecting means must be installed at a readily accessible location whether inside or outside.
- Two to six circuit breakers, or sets of fuses, can serve as the overcurrent device, providing overload protection.
- Table 310.15(B)(6) can be used to size conductors that serve as the main power feeder to a dwelling.
- Conductors not serving as the main power feeder are sized from Table 310.16.
- Electric equipment must be installed in a neat and workmanlike manner.
- Grounding and bonding conductors must be connected by listed means.
- The grounded (neutral) conductor is isolated from the equipment grounding conductor in subpanels.
- Grounding electrode conductors and bonding jumpers on the service supply side are sized from Table 250.66.
- Equipment grounding conductors and bonding jumpers, on the load side of the service, are sized according to Table 250.122.
- All noncurrent-carrying metal parts of service equipment must be effectively bonded together.
- Grounding electrode systems (and conductors) are located in Part III of Article 250.
- Panelboards, in separate buildings or structures, must be installed according to 250.32 provisions.

Unit 9 Competency Test

***NEC*® Reference** **Answer**

______ ______ 1. The point of attachment of the service-drop conductors to a building must not be less than _____ ft above finish grade.

______ ______ 2. An equipment bonding jumper on the supply side of the service must not be smaller than the sizes shown in Table _____.

______ ______ 3. Service conductors run above the top level of a window shall be:

a) 8 ft above window.

b) accessible.

c) 3 ft above window.

d) considered out of reach.

______ ______ 4. Every circuit breaker having an interrupting rating other than _____ amperes shall have its interrupting rating shown on the circuit breaker.

______ ______ 5. Energized parts of service equipment that are not enclosed shall be installed on a _____ and guarded in accordance with 110.18 and 110.27.

I. switchboard

II. panelboard

III. control board

a) II only b) I or II only c) II or III only d) I, II, or III

NEC® Reference	Answer	
______	______	6. A lighting and appliance branch-circuit panelboard is one having more than _____ % of its overcurrent devices protecting lighting and appliance branch-circuits.
______	______	7. Service-entrance cables shall be supported by straps or other approved means within _____ of every service head, gooseneck, or connection to a raceway or enclosure. a) 1 ft b) 6 ft c) 12 ft d) 30 ft
______	______	8. Bonding jumpers meeting the other requirements of this article shall be used around _____ knockouts that are punched or otherwise formed so as to impair the electrical connection to ground.
______	______	9. Residential 240-volt, service-drop conductors must have a vertical clearance above a driveway of at least _____ ft.
______	______	10. A grounding ring encircling the building, in direct contact with the earth, must consist of at least _____ ft of bare copper conductor not smaller than 2 AWG.
______	______	11. Plug fuses of 15-ampere and lower rating shall be identified by a(n) _____ configuration of the window, cap, or other prominent part to distinguish them from fuses of higher ampere ratings.
______	______	12. Where more than one grounding electrode is used, each electrode of one grounding system (including that used for air terminals) shall not be less than _____ ft from any other electrode of another grounding system.
______	______	13. Not more than _____ overcurrent devices (other than those provided for in the mains) of a lighting and appliance branch-circuit panelboard shall be installed in any one cabinet or cutout box.
______	______	14. Service heads and _____ in service-entrance cables shall be located above the point of attachment of the service-drop conductors to the building or other structure.
______	______	15. The width of the working space in front of the electric equipment shall be the width of the equipment or _____ in., whichever is greater.
______	______	16. Which of the following is not a true statement concerning an equipment grounding conductor? a) Under certain conditions, equipment grounding conductors shall be required to be larger than circuit conductors supplying the equipment. b) One equipment grounding conductor may serve multiple circuits. c) Under certain conditions, equipment grounding conductors may be run in parallel. d) The size of equipment grounding conductors shall be increased where ungrounded conductors are increased in size.
______	______	17. The minimum headroom of working spaces about service equipment, switchboards, panelboards, or motor control centers shall be _____ ft.
______	______	18. Service-drop conductors shall have a vertical clearance of not less than _____ ft above the roof surface.
______	______	19. In damp or wet locations, surface-type cabinets, cutout boxes, and meter socket enclosures shall be so placed or equipped as to prevent moisture or water from entering and accumulating within them, and mounted so there is at least _____ in(es). airspace between the enclosure and the wall or other supporting surface.

***NEC*® Reference** **Answer**

_______ _______ 20. Other equipment associated with the electrical installation located above or below the electrical installation shall be permitted to extend not more than _____ beyond the front of the electrical equipment.

a) 0 in. b) 6 in. c) 12 in. d) 24 in.

_______ _______ 21. What size copper equipment bonding jumper is required for a 200-ampere panelboard, located on the load side of the service overcurrent device? (The conductors feeding the panelboard are 2/0 AWG copper conductors.)

_______ _______ 22. Plug fuses of the Edison-base type shall be classified at not over 125 volts and _____ amperes.

_______ _______ 23. A _____ is used to connect the grounding electrode to the equipment grounding conductor, to the grounded conductor, or to both, at the service equipment.

_______ _______ 24. A single electrode consisting of a rod, pipe, or plate that does not have a resistance to ground of _____ or less shall be augmented by one additional electrode.

_______ _______ 25. A disconnecting means serving a hermetic refrigerant motor-compressor shall be selected on the basis of the nameplate rated-load current or branch-circuit selection current, whichever is greater, and locked-rotor current, respectively, of the motor-compressor. The ampere rating shall be at least _____ of the nameplate rated-load current or branch-circuit selection current, whichever is greater.

_______ _______ 26. Circuit breakers rated _____ or less and 600 volts or less shall have the ampere rating molded, stamped, etched, or similarly marked into their handles or escutcheon areas.

_______ _______ 27. A dwelling has a 175-ampere service that is fed with THW copper conductors. What is the minimum size copper grounding electrode conductor?

_______ _______ 28. The _____ is the connection between the grounded circuit conductor and the equipment grounding conductor at the service.

_______ _______ 29. Which of the following is not a standard ampere rating for fuses?

a) 25 amperes b) 50 amperes c) 75 amperes d) 601 amperes

_______ _______ 30. The paralleling efficiency of ground rods longer than _____ ft is improved by spacing greater than 6 feet.

_______ _______ 31. Where the service-entrance phase conductors are larger than 1100 kcmil copper, the bonding jumper shall have an area not less than _____ % of the area of the largest phase conductor.

_______ _______ 32. Service conductors are the conductors from the service point to the _____.

_______ _______ 33. The minimum depth of working space for a 120/240 volt panelboard, with exposed live parts on one side and grounded parts on the other, shall be at least _____ ft.

_______ _______ 34. Each plate electrode shall expose not less than _____ sq ft of surface to exterior soil and shall be installed not less than _____ ft below the surface of the earth.

_______ _______ 35. Where installed on the outside of a raceway, the length of the equipment bonding jumper shall not exceed _____ ft and shall be routed with the raceway.

MULTI-FAMILY DWELLINGS

section 3

UNIT 10

Comprehensive Provisions

Objectives

After studying this unit, the student should:

- be aware that general receptacle and lighting installation provisions for one-family dwellings are almost identical to those for multi-family structures.
- know that certain exceptions permit the installation of more than one service on a multi-family building.
- understand the provisions limiting the number of service disconnecting means to only six.
- be familiar with the provision requiring that each multi-family dwelling occupant have access to the unit's service disconnecting means.
- understand that each service supplying a building must be connected to the same grounding electrode.
- be familiar with service conductor vertical clearance requirements.
- be able to determine when Table 310.15(B)(6) can be used instead of Table 310.16 to size service and feeder conductors.
- know that outdoor receptacles are permitted, but not required in multi-family dwellings.
- be able to choose the appropriate voltage drop formula, insert the proper data, and compute the correct answer.

Introduction

A multi-family dwelling is a building containing three or more dwelling units. Apartments or condominiums are examples of multi-family dwellings. They can be any size, from one story structures containing three units to high-rise buildings containing hundreds of units. One-family dwelling units (covered in Section 2 of this book) contain fundamental provisions essential to all dwelling units, whether one-, two-, or multi-family. Most of the general installation requirements (such as receptacle and switch placement), as well as most of the specific provisions (such as kitchen and bathroom regulations) pertain to all dwelling occupancies. A few requirements, such as outdoor receptacle outlets, are specific to one- and two-family dwellings. Service provisions pertaining to multi-family dwellings, such as the maximum number of service disconnecting means, are explained. This unit also includes a brief description and explanation of voltage drop formulas. A thorough understanding of one-family dwelling requirements is essential to understanding those for multi-family dwellings.

PLANS (BLUEPRINTS)

Partial Site Plan

A Dwelling unit conductors, as listed in Table 310.15(B)(6), are permitted as 120/240, 3-wire, single-phase service-entrance conductors, service-lateral conductors, and feeder conductors serving as the main power feeder to a dwelling unit and installed in raceway or cable with or without an equipment grounding conductor »310.15(B)(6)«.

In a multiple-occupancy building, each occupant must have access to the unit's service disconnecting means, unless electric service and electrical maintenance are provided by building management »230.72(C)«

B A laundry receptacle is not required, if one of the applicable exceptions is met »210.52(F)«.

C With few exceptions, general lighting provisions are the same for both one- and multi-family dwellings »210.70«.

D Use 220.30 to perform optional method load calculations for individual feeders in each unit of multi-family dwellings.

E Clubhouse and swimming pool load calculations are determined in accordance with Article 220, Part II provisions.

The clubhouse is exempt from dwelling unit(s) provisions.

If any portion of the clubhouse is designed, or intended, for the assembly of more than 99 persons, that portion must comply with Article 518 provisions.

F General receptacle installation provisions for multi-family dwellings are almost identical to those for one-family dwellings »210.52«.

G Outdoor outlets are not required for multi-family dwellings »210.52(E)«.

H Use Part II of Article 220 to perform standard method load calculations for both feeders and services.

Use 220.32 to perform multi-family dwelling optional load calculations for feeders or services consisting of 3 or more units.

I Swimming pool provisions are contained in Article 680. Swimming pool motor loads are computed in accordance with 430.22 and 430.24 requirements.

J Air-conditioning and refrigeration equipment provisions are found in Article 440.

> **NOTE**
>
> *House loads must be computed in accordance with Article 220, Part II and must be in addition to the dwelling unit loads computed by Table 220.32 demand factors »220.32(B)«.*

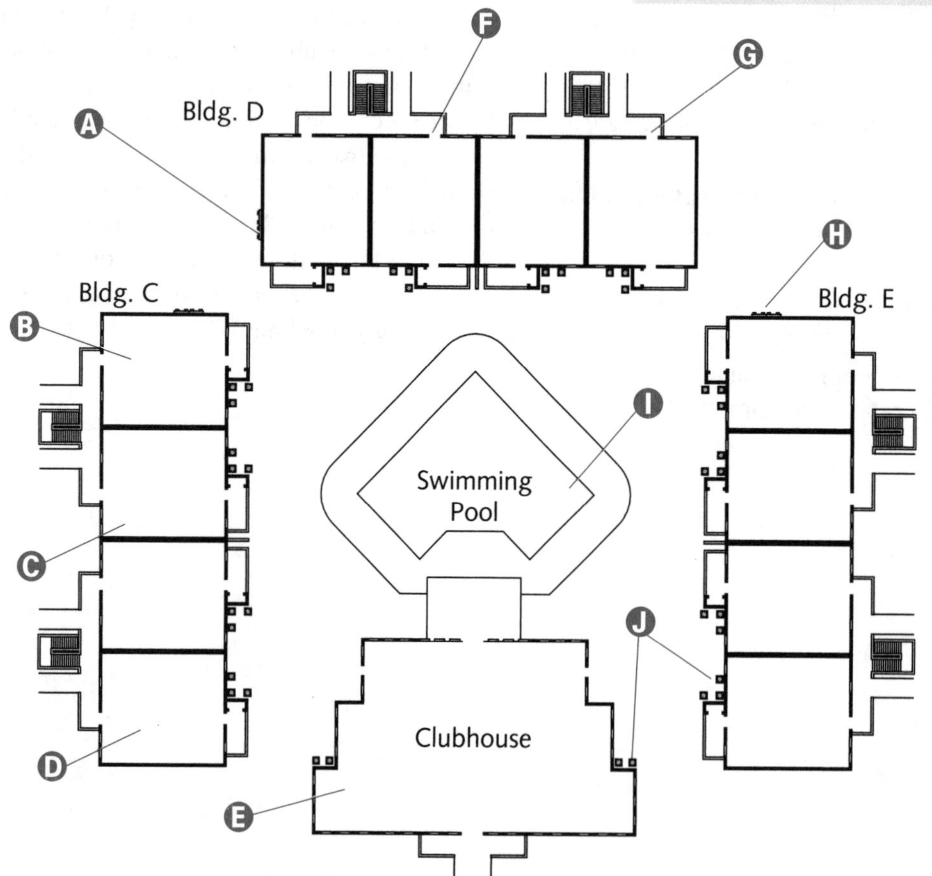

Electrical Plan

Ⓐ Equipment servicing receptacle fed from house panel.

Ⓑ Photo-cell controlled exterior hallway and stairway lighting connected to house panel.

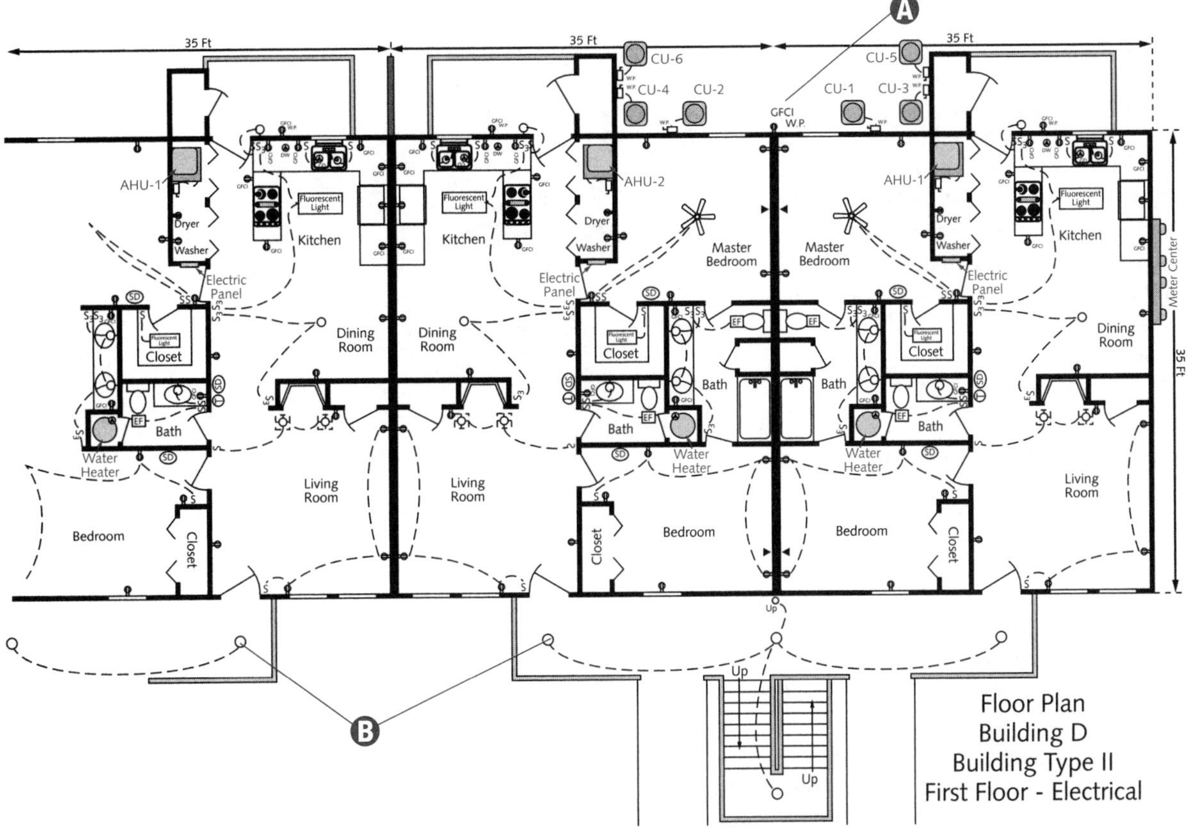

SERVICES

Maximum Number of Services

Ⓐ Vertical service-drop conductor clearances are covered in Unit 9 of this book.

Ⓑ A building (or other structure) shall be supplied by only one service unless permitted in 230.2(A) through (D) »230.2«.

Ⓒ Each service disconnecting means, permitted by 230.2, whether for a service or for a set of service-entrance conductors permitted by 230.40 (*Exception Nos. 1, 3, 4, or 5*) shall consist of not more than six switches (or sets of circuit breakers) mounted in a single enclosure, a group of separate enclosures, or in a switchboard »230.71(A)«.

Ⓓ Auxiliary gutter provisions are located in Article 366.

Ⓔ The conductors and equipment which deliver energy from the electricity supply system to the wiring system of the premises are called the service »Article 100«.

Ⓕ Wiring spaces can be supplemented by auxiliary gutters at meter centers, distribution centers, switchboards, and similar points and may enclose conductors or busbars. Auxiliary gutters, however, must not be used to enclose switches, overcurrent devices, appliances, or similar equipment »366.2«.

NOTE

Unit 9 explains equipment working space, 110.26.

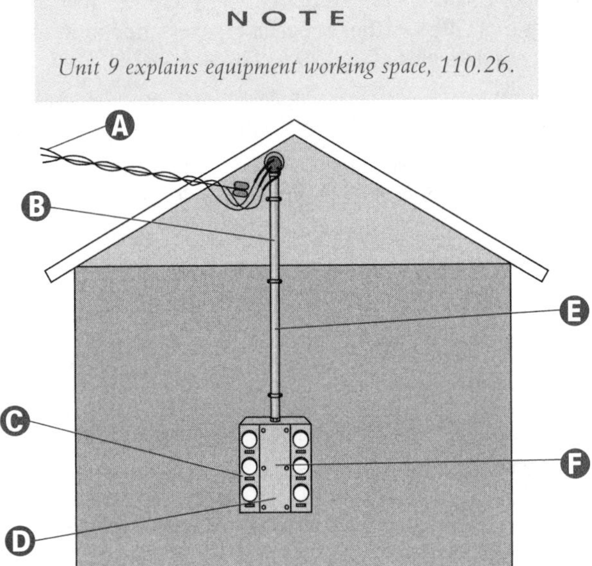

Maximum Number of Disconnects Per Service

A 230.40 basic requirements state that each service drop (or lateral) must supply only one set of service-entrance conductors. However, several exceptions exist.

B Two to six service disconnecting means in separate enclosures, grouped at one location and supplying separate loads from one service drop or lateral, can be supplied by one set of service-entrance conductors »230.40 *Exception No. 2*«.

C There shall be no more than six disconnects per service grouped in a single location »230.71(A)«.

D The two to six disconnects, as permitted in 230.71, must be grouped »230.72(A)«.

E Each disconnect must be marked to indicate the load served »230.72(A)«.

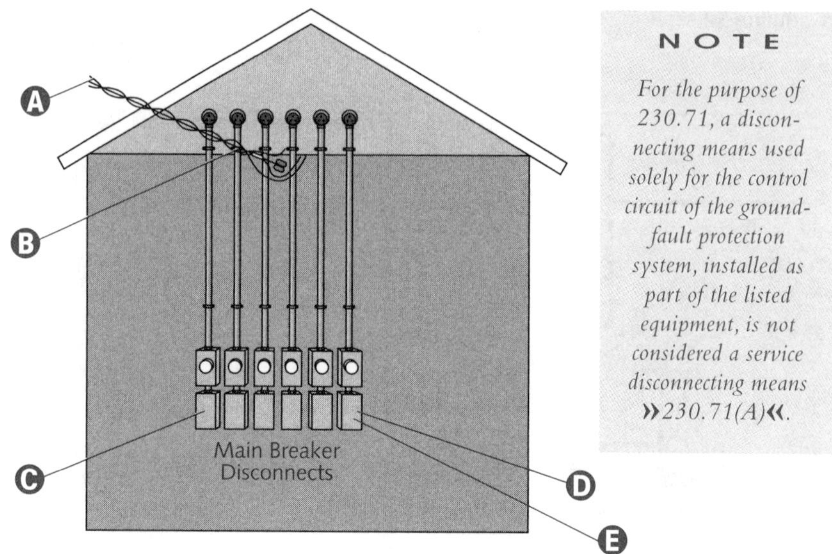

NOTE

For the purpose of 230.71, a disconnecting means used solely for the control circuit of the ground-fault protection system, installed as part of the listed equipment, is not considered a service disconnecting means »230.71(A)«.

Two Disconnects Supplying Twelve Units

A Conductors must have adequate mechanical strength and sufficient ampacity to carry the current for the load as computed in accordance with Article 220 »230.23(A)«.

B Many possible combinations exist for these twelve units: 1 main disconnect feeding all 12 units; 3 main disconnects feeding 12 units (4 units per disconnect); and 6 main disconnects feeding 12 units (2 units per disconnect), to name a few. The arrangement of service disconnects is not critical, as long as no more than six operations of the hand disconnect all of the service conductors.

C There must be no more than six disconnects per service grouped in a single location »230.71(A)«.

D If the service disconnecting means consists of more than one switch or circuit breaker, as permitted by 230.71, the combined ratings of all the switches or circuit breakers used must meet the rating requirements of 230.79 »230.80«.

E In a multiple-occupancy building, each occupant must have access to their unit's service disconnecting means »230.72(C)«.

F Each service disconnect must be permanently marked to identify it as a service disconnect »230.70(B)«.

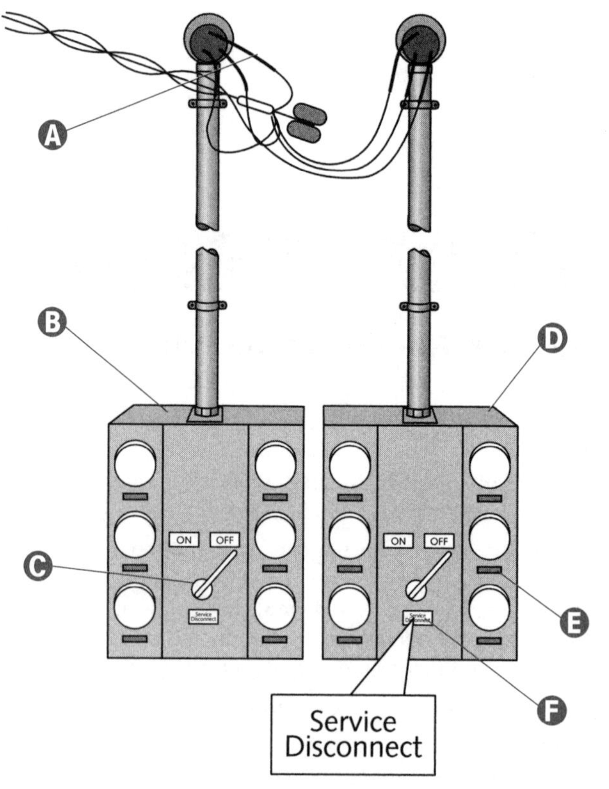

NOTE

In a multiple-occupancy building where electric service and electrical maintenance are provided by building management, under continuous supervision, the service disconnecting means supplying more than one occupancy can be accessible only to authorized management personnel »230.72(C) Exception«.

More Than One Service

A building or other structure can have another service for fire pumps where a separate service is required »230.2(A)(1)«.

A building or other structure can have another service for emergency, standby, or parallel power production systems where a separate service is required »230.2(A)(2) through (5)«.

By special permission, in multiple-occupancy buildings where there is no space available, service equipment does not have to be accessible to all occupants »230.2(B)(1)«. Special permission is the written consent of the AHJ »Article 100«.

Multiple services are permitted where the capacity requirements are in excess of 2000 amperes at a supply voltage of 600 volts or less »230.2(C)(1)«.

Ⓐ If the load requirements of a single-phase installation are greater than the serving agency normally supplies through one service, multiple services are permitted »230.2(C)(2)«.

Ⓑ Where a building or structure is supplied by multiple services, or by any combination of branch-circuits, feeders, and services, a permanent plaque (or directory) must be installed at each service disconnect location denoting all other services, feeders, and branch-circuits supplying that building as well as the area served by each »230.2(E)«.

Ⓒ Six is the maximum number of service disconnects for any one location »230.71(A)«.

WARNING

If a building supplied by separate services requires connection to a grounding electrode, the same grounding electrode must be used for all services. Two or more grounding electrodes that are effectively bonded together constitute a single grounding electrode system in this sense »250.58«.

NOTE

One set of service-entrance conductors can supply the circuits of a two-family or multi-family dwelling as covered in 210.25 » 230.40 Exception No. 5«.

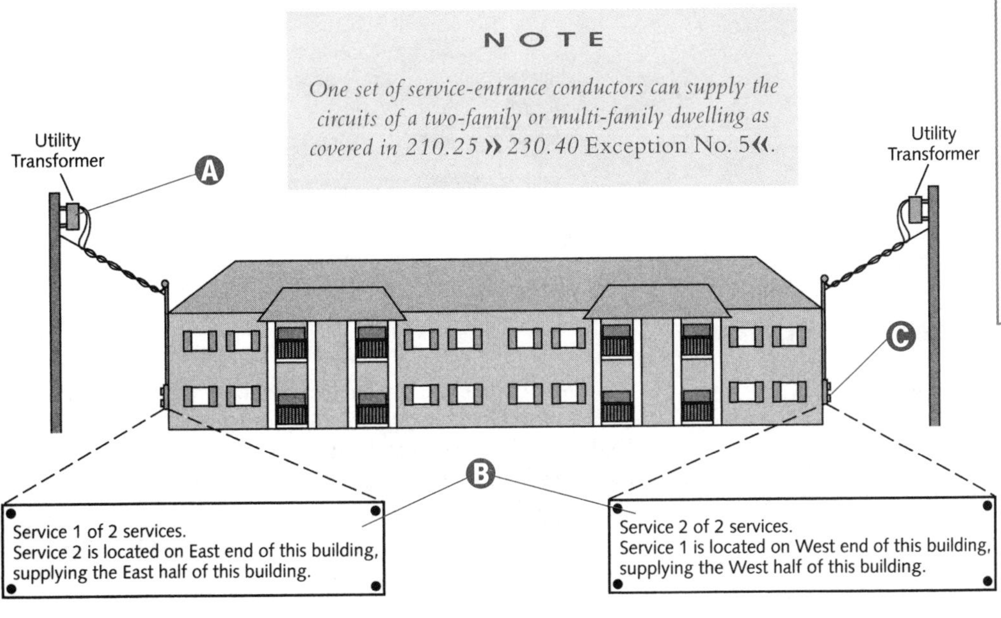

One Service Lateral—One Location

The *service lateral* is the underground service conductors between the main and the first point of connection to the service-entrance conductors in a terminal box, meter base, etc. (inside or outside the building wall). This includes any risers at a pole, from transformers, etc. »Article 100«.

Ⓐ One set of service-entrance conductors can supply two to six service disconnecting means where located in separate enclosures, grouped at one location, and supply separate loads from one service lateral. The one set of service-entrance conductors can supply each or several such service equipment enclosures »230.40 *Exception No. 2*«.

Ⓑ Only as applied to 230.40, *Exception No. 2*, underground sets of conductors, 1/0 AWG and larger, running to the same location and connected together at their supply end (but not at their load end) constitute one service lateral »230.2«.

Ⓒ Maximum of six disconnects

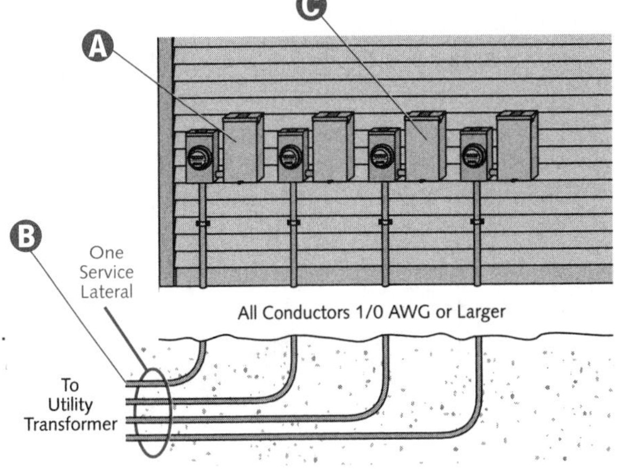

One Service Lateral—Six Disconnects Maximum

A Six is the maximum number of disconnects per service grouped in any one location »230.71(A)«.

B Even though the service conductors are not connected to the same meter socket enclosure, this installation is defined as one service (or service lateral) »230.2«.

C These are size 1/0 AWG and larger conductors, running to the same location. Underground conductors are connected together at the supply (in the utility transformer.)

D Underground service-lateral conductor provisions are found in 230.30 through 33.

CAUTION *Since the maximum number of disconnects is 6, it is a violation to install 2 six-gang meter socket enclosures (each without a main service disconnect) in one location. (Two six-gang meter socket enclosures times 6 service disconnects each equals 12 disconnects.)*

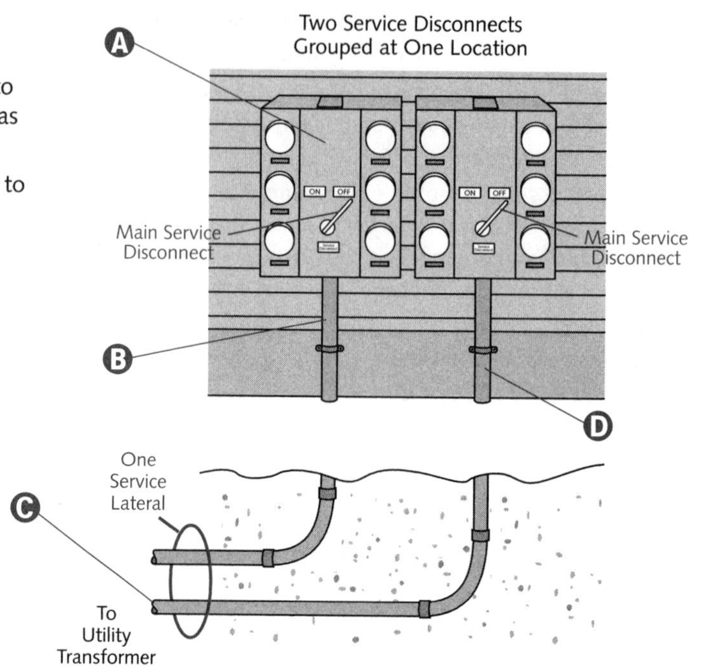

More Than Six Disconnects Per Building

A Here, two services supply one building as permitted by 230.2(C)(2).

B The installation of 2 six-gang meter socket enclosures (each without a main service disconnect) in two locations is not a violation.

C Where a building or structure is supplied by multiple services, or by any combination of branch-circuits, feeders, and services, a permanent plaque (or directory) must be installed at each service disconnect location denoting all other services, feeders, and branch-circuits supplying that building, as well as the area served by each »230.2(E)«.

D 230.71(A) permits a maximum of 6 disconnects per service, not per building. If more than one service feeds a building, each service can have 6 disconnects.

NOTE

Each service must be connected to the same grounding electrode »250.58«.

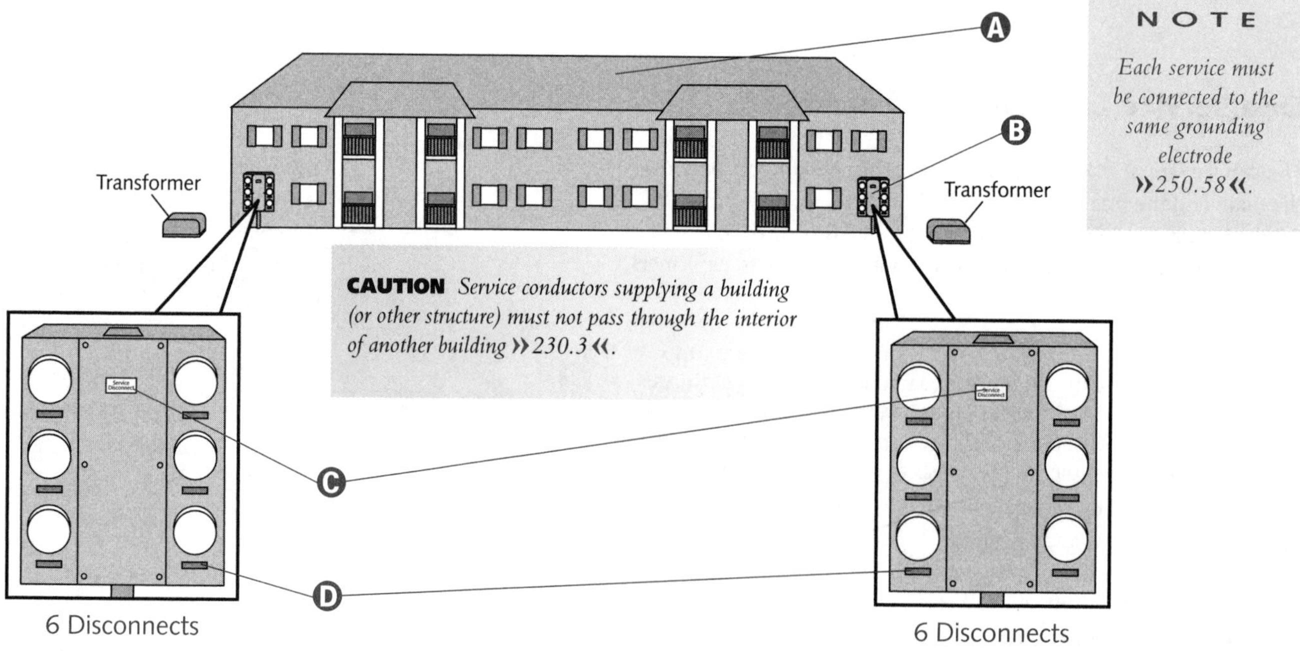

CAUTION *Service conductors supplying a building (or other structure) must not pass through the interior of another building »230.3«.*

Service Conductors

A Table 310.15(B)(6) can be used to size 120/240, 3-wire, single-phase service-entrance conductors and service-lateral conductors, up to a rating of 400 amperes »310.15(B)(6)«.

B This book explains standard and optional multi-family load calculations in Unit 11.

C According to Table 310.15(B)(6), the minimum size THHN copper service conductors for a 400-ampere rating is 400 kcmil. Table 310.16 requires a 500-kcmil conductor to attain the same rating.

D Two main service disconnecting means supplying power to twelve units must be computed on the basis of six units per service. Performing a service calculation on twelve units and then dividing by two is incorrect.

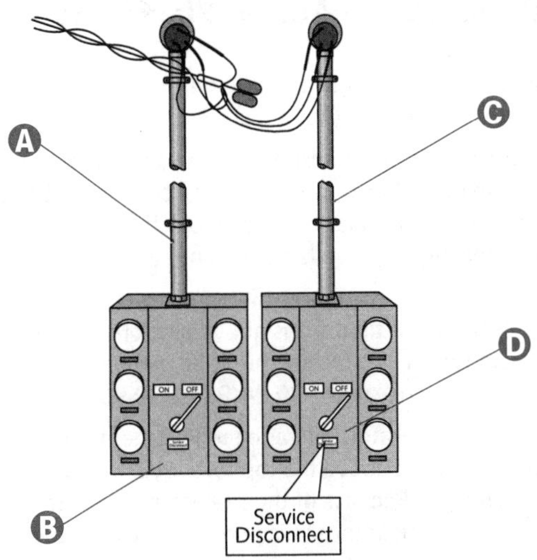

SERVICE WIRING CLEARANCES

Conductor Clearance Above Flat Roofs

A A service mast supporting service-drop conductors must be strong enough to safely withstand the strain imposed by the service drop. Supplemental braces or guys may be necessary »230.28«.

B Service-drop conductors must not be readily accessible »230.24«.

C Conductors must have a minimum vertical clearance of 8 ft (2.5 m) above the roof surface »230.24(A)«.

D The vertical clearance above roof level must be maintained for a distance of at least 3 ft (900 mm) from all edges of the roof »230.24(A)«.

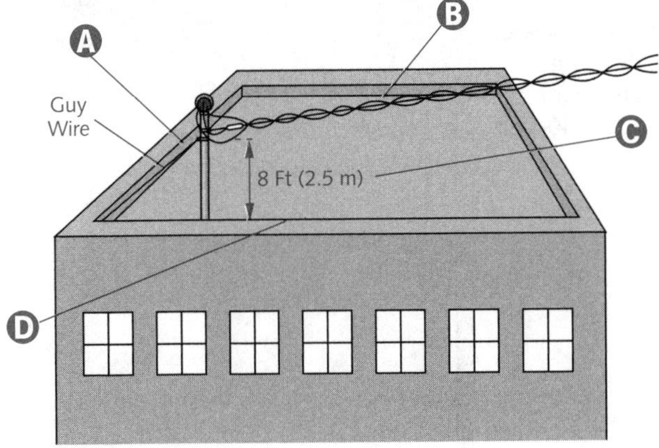

NOTE

Vertical clearances of service-drop conductors, covered in 230.24, pertain to every type of occupancy. Refer to Unit 9 (of this book) for detailed illustrations explaining vertical clearances.

Clearance from Openings

A Open service conductors or multiconductor cable (having no overall outer jacket), must have a clearance of at least 3 ft (900 mm) from windows (designed to open), doors, porches, balconies, ladders, stairs, fire escapes, and similar locations »230.9(A)«.

B Public streets, alleys, roads, parking areas subject to truck traffic, and non-residential driveways require a minimum vertical clearance of 18 ft (5.5 m) »230.24(B)(4)«.

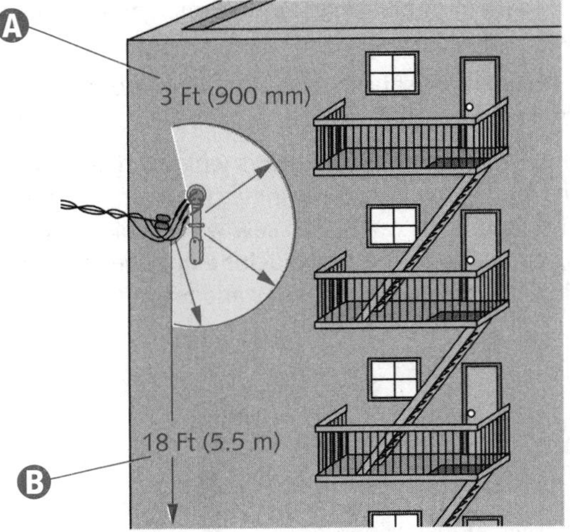

NOTE

Where buildings exceed three stories or 50 ft (15 m) in height, branch-circuit and feeder overhead lines must be arranged, where practicable, so that a clear area at least 6 ft (1.8 m) wide will be left adjacent to the buildings or beginning within 8 ft (2.5 m) to facilitate the raising of fire-fighting ladders »225.19(E)«.

Vertical Clearance Above Platforms

A A drip loop's lowest point has a minimum vertical clearance of 10 ft (3.0 m) »230.24(B)(1)«.

B Overhead conductors, accessible only to pedestrian traffic, must have a minimum 10-ft (3.0-m) vertical clearance above finished grade, sidewalks, platforms, or projections from which they might be reached. The voltage must not exceed 150 to ground »230.24(B)(1)«.

C Vertical clearance of final spans above platforms, projections, or surfaces from which conductors might be reached, must be maintained in accordance with 230.24(B) »230.9(B)«.

D Here, the 3-ft (900-mm) clearance requirement does not apply to conductors run above the top level of a window »230.9 *Exception*«. Because of the fire escape platform, the 10-ft (3.0-m) vertical clearance takes precedence over the window clearance.

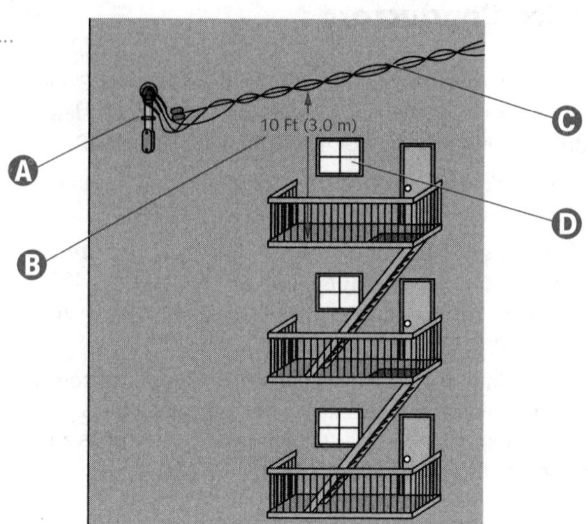

PANELBOARDS AND EQUIPMENT

Access to Overcurrent Devices

Where electric service and electrical maintenance are provided by building management, service and feeder overcurrent devices supplying more than one occupancy can be accessible only to authorized management personnel in the following: (1) in multiple occupancy buildings, and (2) for guest rooms of hotels and motels that are intended for transient occupancy »240.24(B) *Exception No. 1*«.

A In damp or wet locations, surface-type meter socket enclosures must be so situated and/or equipped as to prevent moisture from accumulating within the enclosure. There must be at least ¼-in. (6-mm) airspace between the enclosure and the wall »312.2(A)«.

B Each occupant must have ready access to all overcurrent devices protecting the conductors supplying their particular occupancy »240.24(B)«.

C Table 310.15(B)(6) lists conductors which can be used as service-entrance, service-lateral, and feeder conductors serving as the main power feeder for the dwelling unit(s). The conductors can be in raceways or cables with (or without) equipment grounding conductors. These conductor ampacities can be used only in 120/240-volt, 3-wire, single-phase dwelling services and feeders »Table 310.15(B)(6)«.

CAUTION *Overcurrent devices must not be located in close proximity to easily ignitible material (such as in clothes closets) »240.24(D)«.*

NOTE

Where the service overcurrent devices are locked, sealed, or otherwise not readily accessible to the occupant, branch-circuit overcurrent devices must be: (1) installed on the load side of the service overcurrent device; (2) mounted in a readily accessible location; and (3) of lower ampere rating than the service overcurrent device »230.92«.

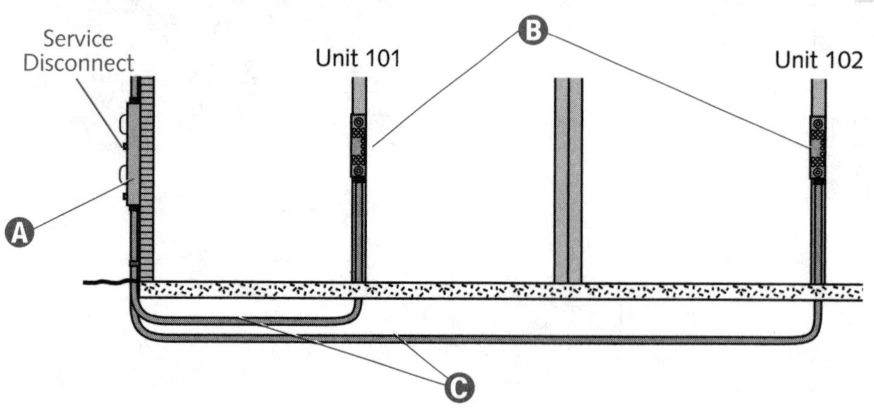

Panelboards Supplying Individual Units

A Noncurrent-carrying metal parts of equipment, raceways, and other enclosures, must be grounded (where required) by one of the following methods: (A) by any of the equipment grounding conductors permitted by 250.118; or (B) by an equipment grounding conductor contained within the same raceway or cable, or otherwise run with the circuit conductors »250.134«.

B Install the bonding jumper (screw, strap, etc.) only in the service disconnecting means, not in subpanels.

C At this point, the neutral bar is isolated from the equipment grounding system.

D The front edge of cabinets (panelboards) situated in walls constructed of noncombustible material (concrete, tile, etc.) must be within ¼ in. (6 mm) of, or extend beyond, the finished surface »312.3«.

E If all of the qualifications are met, Table 310.15(B)(6) can be used to size feeder conductors.

F A grounded circuit conductor cannot be used for grounding noncurrent-carrying metal parts of equipment on the load side of the service disconnecting means, unless an exception has been met »250.142(B)«.

G A grounding connection must *not* be made to any grounded circuit conductor on the load side of the service disconnecting means except as otherwise allowed in Article 250 »250.24(A)(5)«.

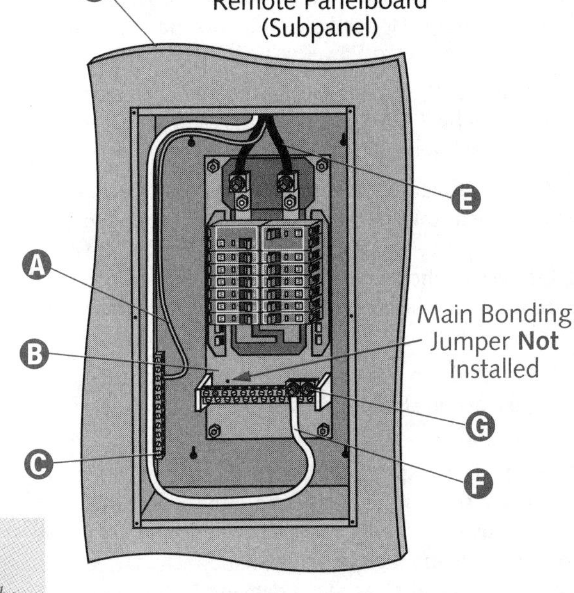

NOTE

A circuit directory must be affixed to the face (or inside) of the panel door(s) indicating the purpose of all circuits and any circuit modifications »408.4«.

CAUTION *Cabinets in walls constructed of combustible material must be flush with, or extend beyond, the finished surface »312.3«.*

Rooftop Heating and Air-Conditioning Equipment

A All 125-volt, single-phase, 15- and 20-ampere dwelling unit outdoor receptacles must have GFCI protection »210.8(A)(3)«.

B A 125-volt, single-phase, 15- or 20-ampere-rated receptacle outlet must be installed at an accessible location for the servicing of heating, air-conditioning, and refrigeration equipment on rooftops, in attics, and in crawl spaces »210.63«.

C The receptacle must be located on the same level, and within 25 ft (7.5 m), of the heating, air-conditioning, and refrigeration equipment »210.63«. The provision does not stipulate that a receptacle must be located at every unit. One centrally located receptacle could meet the 210.63 requirement for all pieces of equipment located within the 25-ft (7.5-m) radius.

D Article 440 contains provisions pertaining to air-conditioning and refrigeration equipment.

CAUTION *The receptacle outlet must not be connected to the load side of the equipment disconnecting means »210.63«.*

Feeder Conductors

A Table 310.15(B)(6) can be used to size 120/240, 3-wire, single-phase feeder conductors serving as the main power feeder to a dwelling unit »310.15(B)(6)«.

B If each unit contains a 100-ampere panelboard, the feeders can be 4 AWG copper (or 2 AWG aluminum) conductors »Table 310.15(B)(6)«.

C Load calculations are computed for each individual unit having its own panelboard. Although located in a multi-family dwelling, each unit is calculated by one of the one-family methods. It is incorrect to perform a load calculation on multiple units and then divide by the number of units.

D Standard and optional one-family load calculations are explained in Unit 8 of this book.

E The AHJ may require a diagram showing feeder details prior to installation. Such a diagram must show the area in sq ft of the building (or other structure) supplied by each feeder, the total connected load before applying demand factors, the demand factors used, the computed load after applying demand factors, and the size and type of conductors that will be used »215.5«.

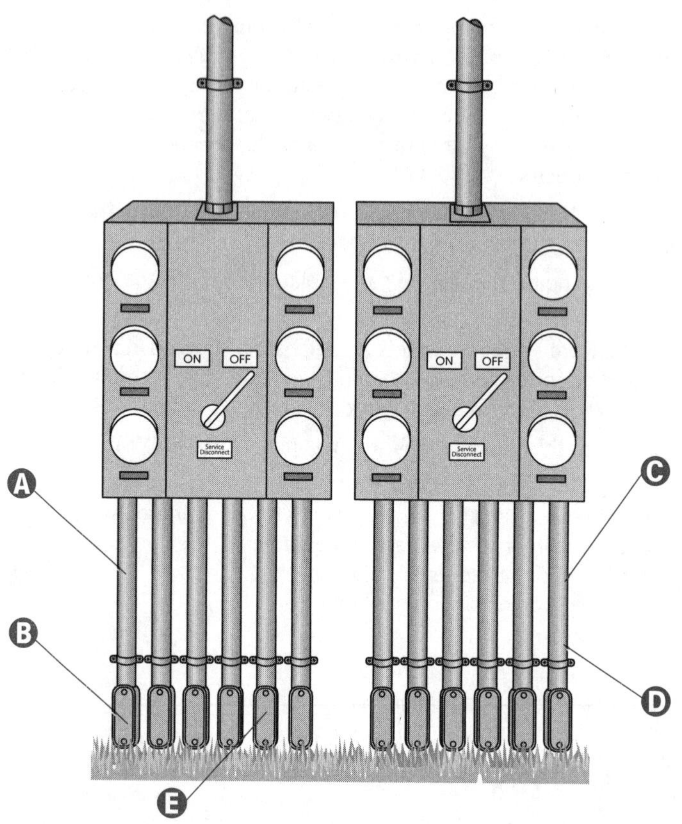

> **CAUTION** *On long runs it may be necessary to increase the feeder conductors' size, to compensate for voltage drop »215.2(A)(4) FPN No. 2«.*

BRANCH-CIRCUITS

Optional Receptacles

A Multi-family dwelling outdoor receptacle outlets, while not required, must be GFCI-protected, if installed »210.52(E) and 210.8(A)(3)«.

B Dwelling-unit branch-circuits shall supply only loads within, or associated only with, that dwelling unit »210.25«.

C In a multi-family dwelling where laundry facilities are provided on the premises, available to all building occupants, a laundry receptacle is not required in each individual unit »210.52(F) *Exception No. 1*«.

D In other than one-family dwellings without laundry facilities, a laundry receptacle is not required »210.52(F) *Exception No. 2*«.

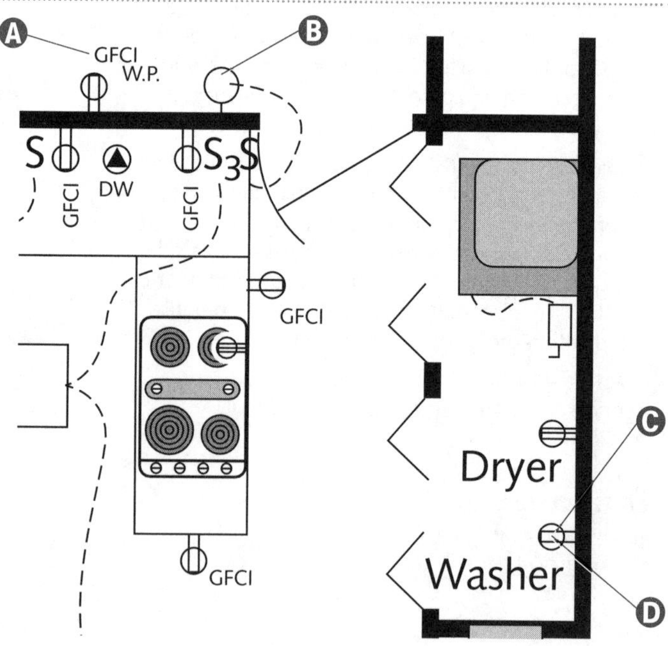

Receptacles

Ⓐ Twenty-four (24) in. (600 mm) is the maximum distance to any receptacle, measured horizontally, along the wall line »210.52(C)(1)«.

Ⓑ Dwelling-unit branch-circuits shall supply only loads within, or associated only with, that dwelling unit »210.25«.

Ⓒ Receptacles must be installed so that every 6 ft (1.8 m) of wall space, measured horizontally along the floor line, contains a receptacle outlet »210.52(A)(1)«.

Ⓓ *All* receptacles serving kitchen countertops must have GFCI protection »210.8(A)(6)«.

Ⓔ Kitchen receptacles not serving countertops (such as receptacles behind refrigerators) do not require GFCI protection.

Ⓕ All receptacles in the kitchen, pantry, breakfast room, dining room, or similar areas must be supplied from 20-ampere small appliance branch-circuits (except for refrigeration equipment) »210.52(B)(1)«.

Ⓖ Circuits feeding receptacles in kitchens, pantries, breakfast rooms, dining rooms, or similar areas must not feed receptacles outside these areas »210.52(B)(2)«.

Ⓗ GFCI protection is required for every 125-volt receptacle located in a bathroom area »210.8(A)(1)«.

Ⓘ A receptacle is required for any wall space, 2 ft (600 mm) or more in width (including space measured around corners), unbroken along the floor line by doorways, fireplaces, and similar openings »210.52(A)(2)(1)«.

Ⓙ Locate at least one wall receptacle within 36 in. (900 mm) of the outside edge of each basin (lavatory or sink). The receptacle outlet must be located on a wall adjacent to the basin location »210.52(D)«.

Ⓚ A receptacle is required for each peninsular counter space [at least 12 by 24 in. (300 x 600 mm) in size], which is separated from other counter space by a range top, refrigerator, or sink »210.52(C)(3)«.

Ⓛ A circuit providing power to bathroom receptacles cannot provide power to any receptacles (or lighting) outside the bathroom, unless meeting an exception »210.11(C)(3)«.

NOTE

Only a few differences exist between one-family and multi-family dwelling receptacle requirements. Refer to Units 6 and 7 for more information pertaining to general as well as specific receptacle provisions.

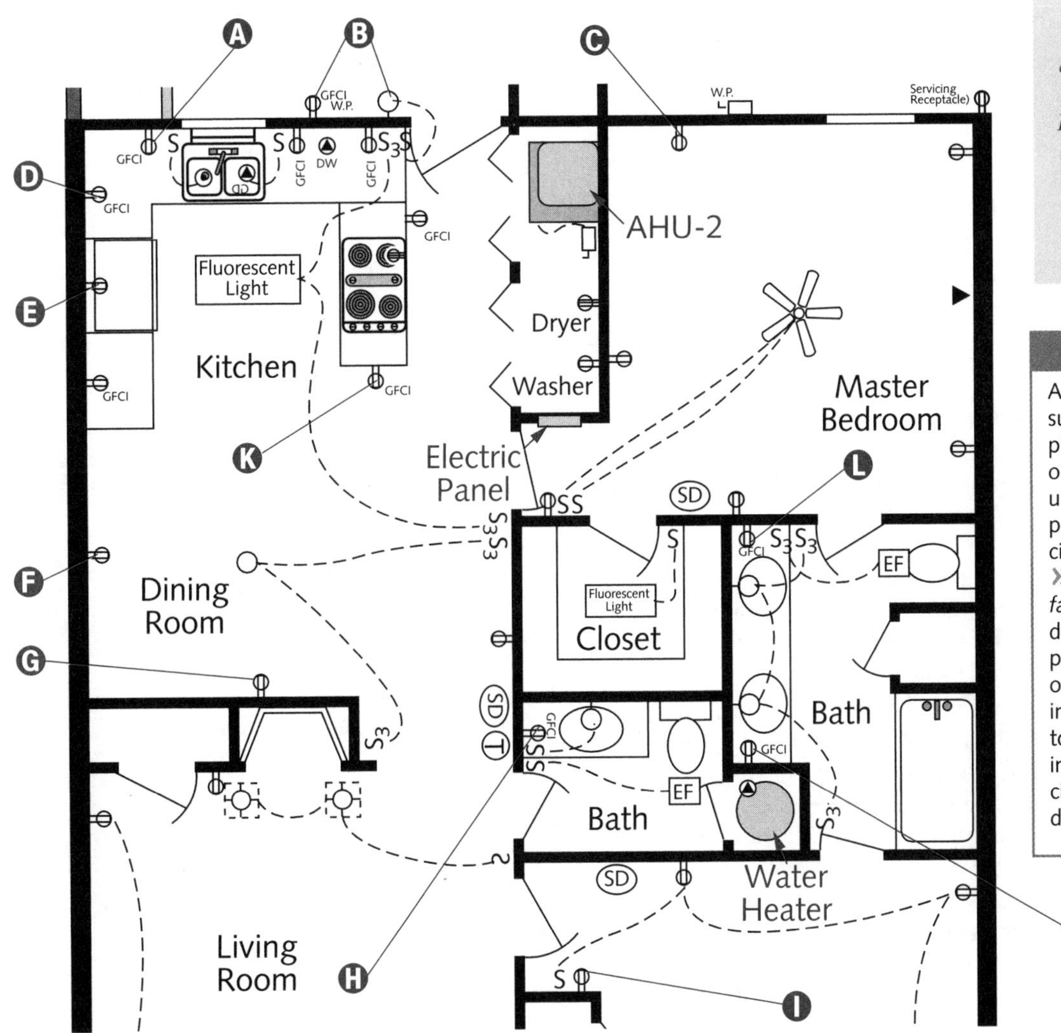

WARNING

All branch-circuits that supply 125-volt, single-phase, 15- and 20-ampere outlets located in dwelling unit bedrooms must be protected by an arc-fault circuit interrupter(s) »210.12(B)«. An *arc-fault circuit interrupter* is a device intended to provide protection from the effects of arc faults by recognizing characteristics unique to arcing and by functioning to de-energize the circuit when an arc fault is detected »210.12(A)«.

General Lighting

A minimum of one wall-switch-controlled lighting outlet is required for dwelling hallways » 210.70(A)(1) «.

Ⓐ At least one wall-switch-controlled lighting outlet must be installed in every habitable room of a dwelling » 210.70(A)(1) «.

Ⓑ Clothes closet lighting outlet provisions are found in 410.8.

Ⓒ At least one wall-switch-controlled lighting outlet must be installed in every dwelling bathroom » 210.70(A)(1) «.

Ⓓ Except for kitchens and bathrooms, every habitable room is allowed one or more wall-switch-controlled receptacle(s) in lieu of a lighting outlet » 210.70(A)(1) *Exception No. 1* «.

Ⓔ In hallways, stairways, and outdoor entrances, control of lighting can be remote, central, or automatic » 210.70(A)(2) *Exception* «.

Ⓕ Branch-circuits required for the purpose of lighting, central alarm, signal, communication, or other needs for public or common areas of a two-family or multi-family dwelling cannot be provided from equipment that supplies an individual dwelling unit » 210.25 «.

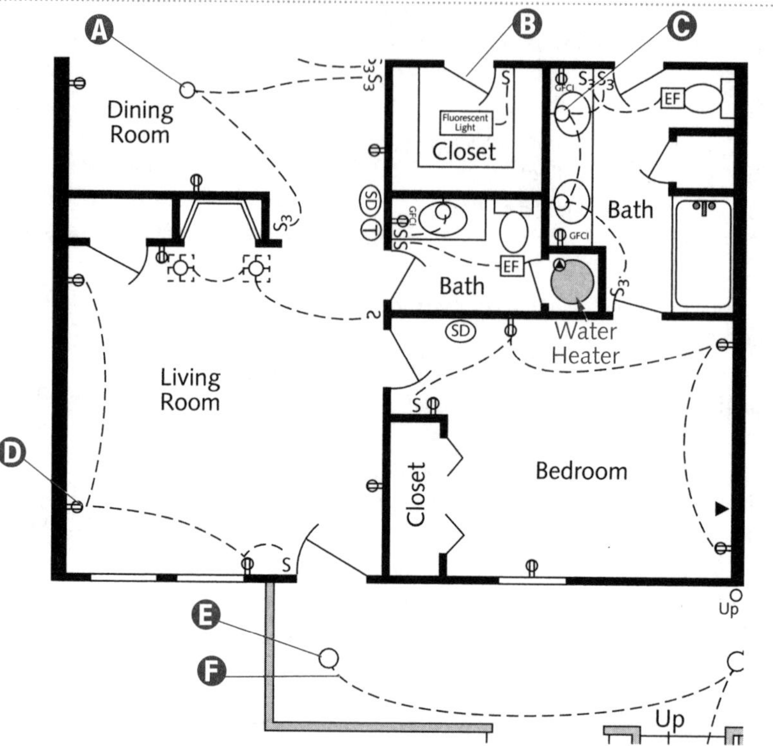

NOTE

A listed box is required for a ceiling-suspended (paddle) fan where the box provides the only support » 314.27(D) «.

Only a few differences exist between one-family and multi-family dwelling lighting requirements. Refer to Units 6 and 7 for more information pertaining to general as well as specific lighting provisions.

Appliance Disconnecting Means

Ⓐ A permanently connected appliance, rated over 300 volt-amperes or ⅛ horsepower, must have a means to disconnect all ungrounded (hot) conductors from the appliance » 422.30 «.

Ⓑ An appliance switch marked with an "off" position that effectively disconnects all ungrounded conductors can serve as the disconnecting means required by Article 422 where other means for disconnection are provided in 422.34(A) through (D) » 422.34 «.

Ⓒ A disconnecting means is not required if the branch-circuit switch or circuit breaker can be locked in the open (off) position » 422.31(B) «.

Ⓓ In multi-family dwellings, the other appliance disconnecting means (as described in 422.34) must be within the dwelling unit, or on the same floor as the dwelling unit in which the appliance is installed, and can also control lamps and other appliances » 422.34(A) «.

Ⓔ A branch-circuit supplying a fixed storage-type water heater with a capacity of 120 gallons (450 L) or less must have a rating of at least 125% of the water heater's nameplate rating » 422.13 «.

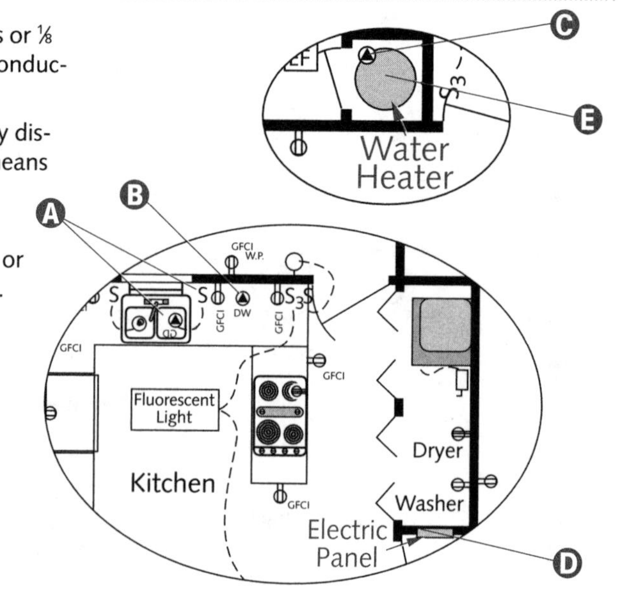

Heating and Air-Conditioning Equipment

A For a hermetic refrigerant motor-compressor, the rated-load current marked on the equipment nameplate in which the motor-compressor is employed must be used in determining the rating or ampacity of the disconnecting means, the branch-circuit conductors, the controller, the branch-circuit short-circuit and ground-fault protection, and the separate motor overload protection »440.6(A)«.

B The disconnecting means for air-conditioning or refrigerating equipment must be readily accessible and within sight of the equipment »440.14«.

C All electric equipment must be surrounded by sufficient, unobstructed space to allow ready and safe equipment operation and maintenance »110.26«. Working space is mandatory for equipment likely to require examination, adjustment, servicing, or maintenance while energized »110.26(A)«.

D The ampacity of branch-circuit conductors and the rating or setting of overcurrent protective devices that supply fixed electric space-heating equipment (consisting of resistance elements with or without a motor) must be at least 125% of the total load of the motors and heaters »424.3(B)«.

E A hermetic refrigerant motor-compressor consists of a compressor and motor, both enclosed in the same housing, with no external shaft or shaft seals, and the motor operating in the refrigerant »440.2«.

F On long runs it may be necessary to increase the branch-circuit conductor's size, to compensate for voltage drop »210.19(A)(1) FPN No. 4«.

G Means must be provided to disconnect all ungrounded conductors from the heater, motor controller(s), and supplementary overcurrent protective device(s) of all fixed electric space-heating equipment »424.19«.

H In multi-family dwellings, the other disconnecting means, as described in 424.19(C), must be within the dwelling unit, or on the same floor as the dwelling unit in which the fixed heater is installed, and can also control lamps and other appliances »424.19(C)(1)«.

NOTE

Fixed heater switch(es) marked with an "off" position which effectively disconnects all ungrounded conductors can serve as the disconnecting means required by Article 424 where other means for disconnection are provided in 424.19(C)(1) through (4) »424.19(C)«.

NOTE

The size of the branch-circuit conductors and the overcurrent protective devices supplying fixed electric space-heating equipment, including a hermetic refrigerant motor-compressor with or without resistance units, must be computed in accordance with 440.34 and 440.35 »424.3(B)«.

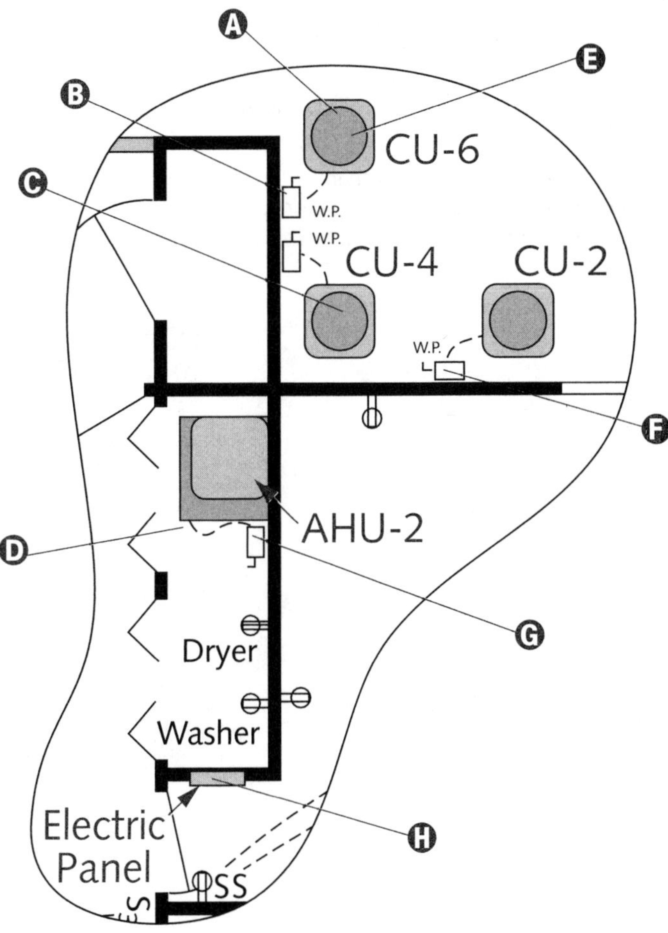

VOLTAGE DROP

Conductor Properties (Chapter 9, Tables 8 and 9)

While skin effect and induction change the resistance in ac circuits, they do not affect dc circuits. The resistance of a conductor is slightly higher in an ac circuit. Table 9 contains the resistance of conductors in PVC, aluminum, and steel raceways. The factors that change resistance in an ac circuit, such as skin effect, are negligible in smaller conductors.

A The larger the circular mil area, the smaller the resistance.

B 6 AWG has an area of 26,240 circular mils, listed in Table 8.

C 3/0 AWG has an area of 167,800 circular mils and contains 19 strands.

D The abbreviated term "kcmil" stands for: 1000 (k) circular (c) mills (mil).

E Each conductor size larger than 4/0 AWG is actually the circular mil size. A 250-kcmil conductor has a 250,000 circular mil area. A 500-kcmil conductor has a circular mil area of 500,000.

F Conductor overall diameter is listed in Table 8. This diameter is the actual wire and excludes conductor insulation.

G Conductor strand quantities are listed in Table 8; 6 AWG contains 7 strands.

H Table 8 lists dc resistance values (in ohms per 1000 ft) at an ambient temperature of 167°F (75°C).

I 500-kcmil conductors have a quantity of 37 strands. The overall diameter is 0.813 in., with each strand having a diameter of 0.116 in.

J The dc resistance for 1000 ft of 6 AWG copper is 0.491 ohms.

K The longer the conductor, the larger the resistance.

L The dc resistance for 100 ft of 6 AWG copper is 0.0491 ohms (0.491) 1000 = 0.000491 × 100 = 0.0491).

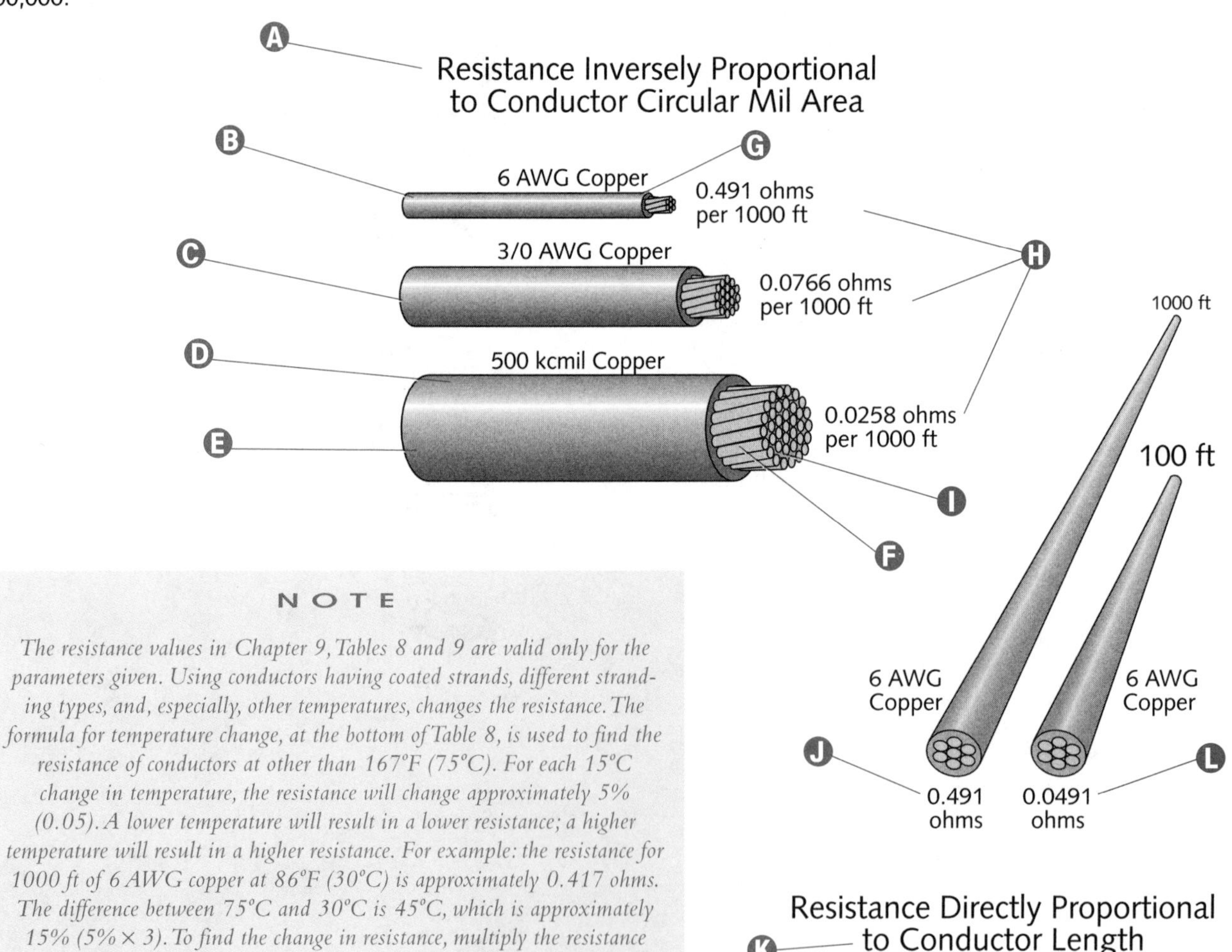

NOTE

The resistance values in Chapter 9, Tables 8 and 9 are valid only for the parameters given. Using conductors having coated strands, different stranding types, and, especially, other temperatures, changes the resistance. The formula for temperature change, at the bottom of Table 8, is used to find the resistance of conductors at other than 167°F (75°C). For each 15°C change in temperature, the resistance will change approximately 5% (0.05). A lower temperature will result in a lower resistance; a higher temperature will result in a higher resistance. For example: the resistance for 1000 ft of 6 AWG copper at 86°F (30°C) is approximately 0.417 ohms. The difference between 75°C and 30°C is 45°C, which is approximately 15% (5% × 3). To find the change in resistance, multiply the resistance (0.491) by 15% (0.491 × 15% = .07365 = .074). Since the temperature is lower, the resistance will be lower (0.491 – 0.074 = 0.417).

Voltage Drop Recommendation

A Voltage drop considerations are only recommendations; compliance is optional.

B The recommended maximum voltage drop for the combined feeder and branch-circuit is 5%. However, the voltage drop for feeders is not necessarily 2%. If the branch-circuit is 3%, the feeder is 2%, but if the branch-circuit is 1%, the feeder can be 4%. Any combination is possible, as long as the branch-circuit does not exceed 3%, and the combined branch-circuit and feeder does not exceed 5%.

C There is no recommended maximum voltage drop for service conductors.

D The conductor voltage that is dropped is a percentage of the source voltage.

E The recommended voltage drop for a branch-circuit is 3% or less. For example: the maximum allowable voltage drop for a 120-volt branch-circuit is 3.6 volts (120 × 3%), and the maximum allowable for a 240-volt branch-circuit is 7.2 volts (240 × 3%).

F The higher the resistance through a conductor, the higher the voltage drop. Such a conductor voltage drop could result in a reduction of voltage supplying the load.

G A single-phase feeder (or branch-circuit) conductor will have a total resistance twice that of one conductor. For example: the length of one conductor must be doubled in order to find the total resistance of the circuit. The total length for a three-phase circuit is found by multiplying the length of one conductor by the square root of 3 (1.732).

H Branch-circuit conductors, as defined in Article 100, sized to prevent a voltage drop exceeding 3% at the farthest outlet of power, heating, and lighting loads (or combinations of such loads) and where the maximum total voltage drop on both feeders and branch-circuits to the farthest outlet does not exceed 5%, will provide reasonable efficient operation »210.19(A)(1) FPN No. 4 and 215.2(A)(4) FPN No. 2«.

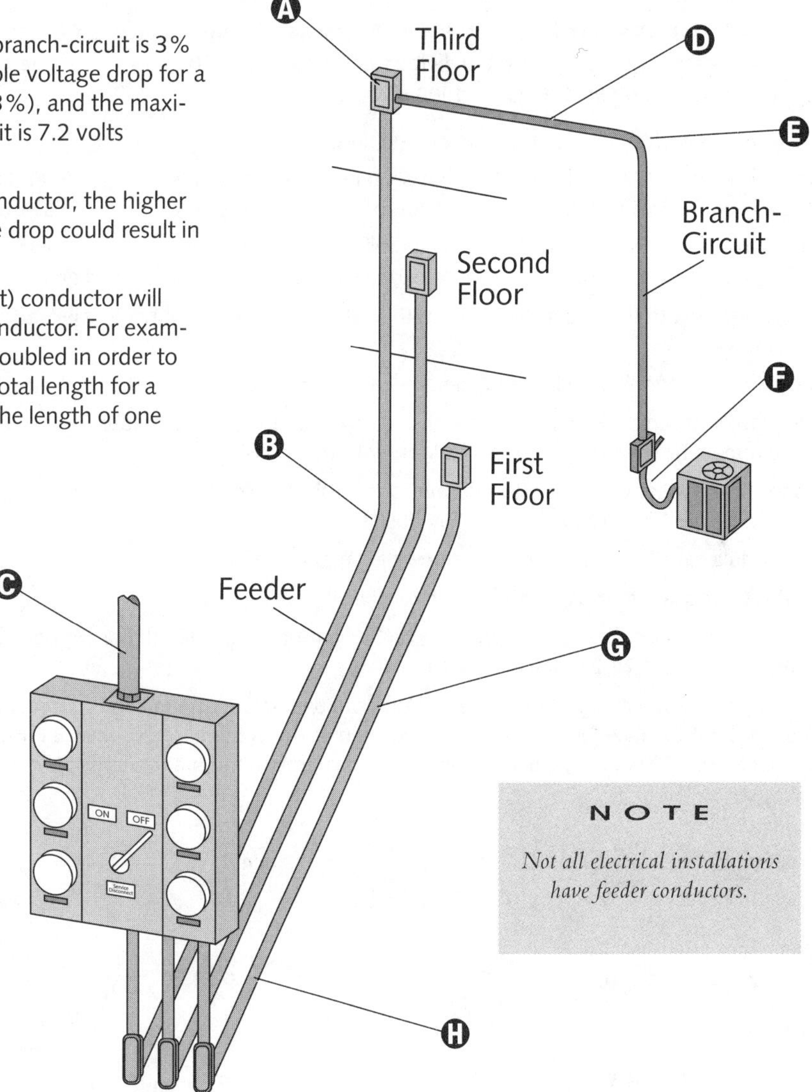

NOTE

Not all electrical installations have feeder conductors.

Single-Phase Voltage Drop Formulas

Formula Definitions:

- **V_D = Volts dropped from a circuit.**
- **2 = Multiplying factor for single-phase (the 2 represents the conductor length in a single-phase circuit.)**
- **K = Resistivity of the conductor material. The resistance value (measured in ohms) is based on a mil ft at a given temperature. A mil ft is a piece of wire 1 ft long and one mil in diameter. Each conductor resistivity (K) factor can be found by multiplying the circular mil area (cmil) by the resistance (R) for one ft. For example: the K factor for a 6 AWG copper conductor at 167°F (75°C) is 12.88384, because 26,240 (area circular mills) × 0.000491 (ohms per ft) = 12.88384. At 75°C, the approximate K for copper is 12.9; and for aluminum is 21.2.**
- **L = Length, *ONE WAY* only. The distance from the voltage source to the load.**
- **I = Actual current used by the load. Do not use 125% for motors and continuous loads.**
- **cmil = Circular mil area of the conductor (Chapter 9, Table 8).**
- **R = Conductor resistance for one ft. *Note:* The resistance values, listed in Chapter 9, Tables 8 and 9, are for 1000 ft. To find the resistance for one ft, simply divide the resistance by 1000. Example: the resistance for one foot of 6 AWG copper (at 75°C) is 0.000491 (0.491 ÷ 1000 = 0.000491).**

(A) One formula used to find voltage drop in a single-phase circuit.

(B) The conductor voltage that is dropped is a percentage of the source voltage. To find the percentage, divide "V_D" by the source voltage. Example: 2.4 volts dropped from a 120-volt circuit is 2% of the source voltage (2.4 ÷ 120 = 0.02 = 2%)

(C) Here is another formula for finding voltage drop in a single-phase circuit. This method utilizes the Ohm's law formula. Resistance (for one ft of the conductor) is used instead of the K factor and circular mil area.

(D) Use this formula to find the maximum length (distance) from the source to the load. Because of the "2" in the formula, the length is **one way**.

(E) This Ohm's law formula finds the maximum length from the source to the load.

(F) Conductor size can be found using this formula. The computation provides the minimum circular mil wire size that must be installed to remain below the "V_D" that was input into the formula.

(G) This formula applies Ohm's law to find the conductor size. The result yields the minimum conductor resistance (for one ft) in order to keep the voltage drop below the "V_D" used in the formula.

(H) This formula calculates the maximum current in amperes.

(I) Use this Ohm's law formula to find the current in amperes.

(J) Since the conductor size is unknown, the exact K cannot be calculated. For copper, at 75°C, use **12.9** (approximate K). For aluminum, at 75°C, use **21.2** (approximate K).

(K) "V_D" represents the actual volts that can drop from the circuit. The maximum (recommended) percentage for a branch-circuit is 3%. Be careful not to place "3%" in the "V_D" position. If the voltage is 120, then 3.6 is the actual volts that can drop in the branch-circuit. For 240 voltage, the actual volts that can drop in the branch-circuit is 7.2.

(A), (B): $$V_D = \frac{2 \times K \times L \times I}{cmil}$$

(F), (J), (K): $$cmil = \frac{2 \times K \times L \times I}{V_D}$$

(C): $$V_D = 2 \times R \times L \times I$$

(G): $$R = \frac{V_D}{2 \times L \times I}$$

(D): $$L = \frac{cmil \times V_D}{2 \times K \times I}$$

(H): $$I = \frac{cmil \times V_D}{2 \times K \times L}$$

(E): $$L = \frac{V_D}{2 \times R \times I}$$

(I): $$I = \frac{V_D}{2 \times R \times L}$$

Three-Phase Voltage Drop Formulas

A "1.732" is the multiplying factor for three-phase (the square root of 3 represents the conductor length in a three-phase circuit). The only difference between the single-phase and three-phase formulas, is that "1.732" has replaced "2".

B To find voltage drop in a three-phase circuit:

C To find the maximum length (distance) from the source to the load in a three-phase circuit: (Because of the "1.732" in the formula, the length is **one way**.)

D Variations of three-phase formulas

E To compute the conductor size in a three-phase circuit:

F To determine the resistance of the conductor in a three-phase circuit:

G To calculate the maximum current in amperes in a three-phase circuit:

H The single-phase formula can be change to a three-phase formula by simply placing ".866" on the line containing the "2". The purpose of putting ".866" on the same line as "2" is simple: 2 × .866 = 1.732.

> **NOTE**
>
> *The formula definitions for three-phase correspond with the single-phase definitions.*

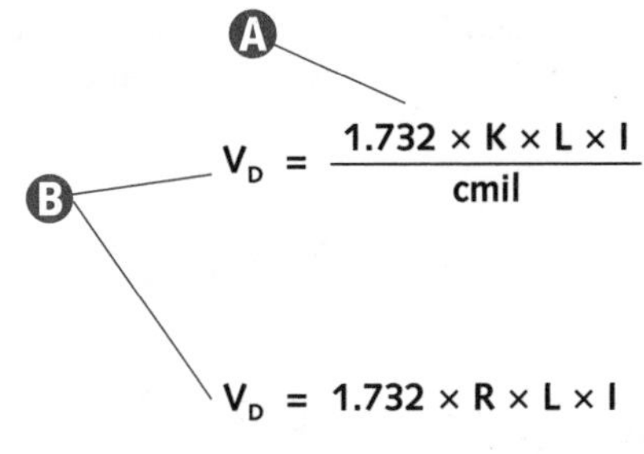

E

$$V_D = \frac{1.732 \times K \times L \times I}{V_D}$$

F

$$R = \frac{V_D}{1.7322 \times L \times I}$$

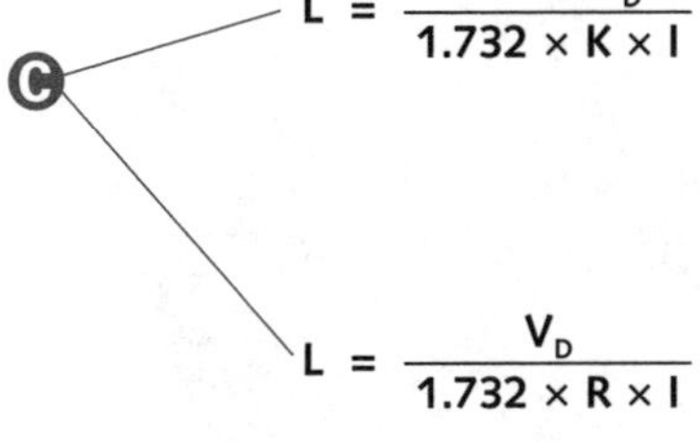

G

$$I = \frac{cmil \times V_D}{1.732 \times K \times L}$$

$$I = \frac{V_D}{1.732 \times R \times L}$$

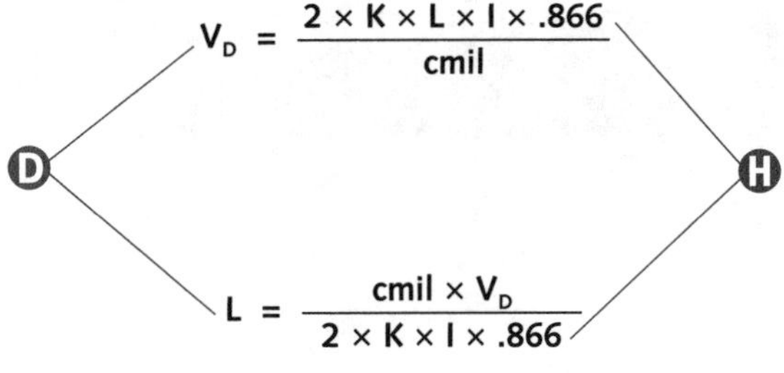

Single-Phase Voltage Drop Example

At 75°C, what size conductors are required to feed a 2-HP, 230-volt, single-phase motor that is 150 ft from the source? (Do not exceed *NEC®* recommendations.)

A Both the supply and load side of this equipment disconnecting means are branch-circuit conductors; therefore, the maximum recommended voltage drop is 3% of the source voltage.

B Table 430.148 lists full-load current (in amperes) for single-phase ac motors.

C Voltage drop formula used to find the minimum size conductor

D Approximate K for copper at 75°C

E Current in amperes for a 2-HP, 230-volt, single-phase motor »Table 430.148«.

F This is the minimum circular mil area required. Notice that a 12 AWG conductor (6530 circular mils) is too small for this installation.

G The minimum size conductor recommended for this installation.

H Maximum amount of voltage drop on a 230-volt branch-circuit that remains within the 3% recommendation (230 × 3% = 6.9)

I An alternative formula equally effective in determining the minimum size conductor

J In multiplying the 1 ft value by 1000, the resistance values will correspond with Tables 8 and 9, which contain ohms per 1000 ft values.

K The maximum conductor resistance for 1 ft

L The conductor chosen cannot exceed this resistance number. Both solid and stranded 12 AWG conductors have a higher resistance; therefore, 10 AWG conductors (solid or stranded) must be used.

NOTE

If the total distance of the branch-circuit was less than 145 ft, a 12 AWG copper conductor could be installed.

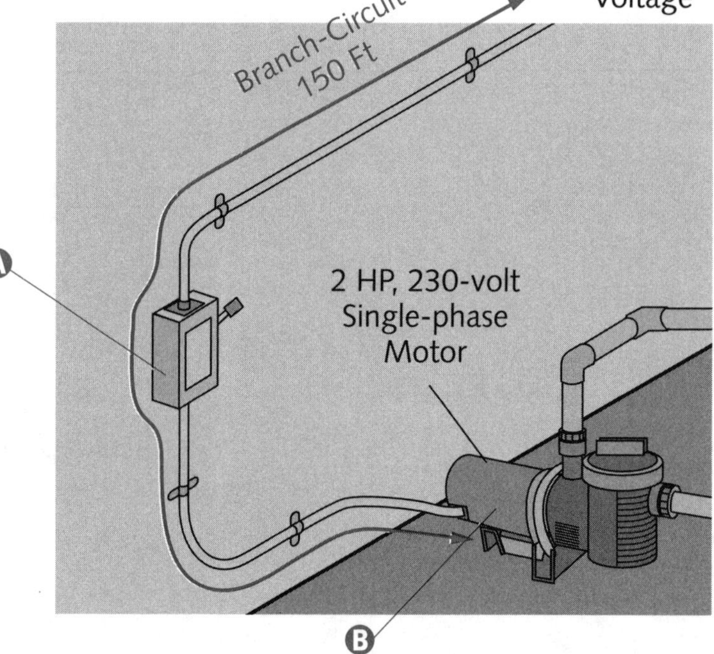

C D E F G H

$$\text{cmil} = \frac{2 \times K \times L \times I}{V_D} = \frac{2 \times 12.9 \times 150 \times 12}{6.9} = 6730 = \text{10 AWG copper}$$

I K J L

$$R = \frac{V_D}{2 \times L \times I} = \frac{6.9}{2 \times 150 \times 12} = 0.001917 \times 1000 = 1.917 = \text{10 AWG copper}$$

Summary

- Overall receptacle installation requirements, with few exceptions, are the same for both one-family and multi-family dwellings.
- While outdoor receptacles are not required for multi-family dwellings, GFCI protection must be provided, if installed.
- Under certain conditions, a laundry receptacle is not required.
- A multi-family clubhouse (or portion thereof), designed for the assembly of more than 99 persons, must comply with Article 518 provisions.
- Generally, a building is supplied by only one service.
- There can be no more than six disconnects per service grouped in any single location.
- Where separate services supply a building, and a grounding electrode is required, the same electrode must be used.
- Separate service laterals, grouped in one location, constitute one service.
- Use Table 310.15(B)(6) to size 120/240, 3-wire, single-phase service-entrance conductors, service-lateral conductors, and feeder conductors serving as the dwelling's main power feeder.
- Service and open conductor clearances pertain to all dwellings.
- Conductor properties are found in Chapter 9, Tables 8 and 9.
- Voltage drop compliance is discretionary.
- The recommended voltage drop for a branch-circuit is 3% or less.
- The recommended voltage drop for combined feeder and branch-circuits is 5% or less.
- Voltage drop formulas, complete with explanations, are found near the end of this unit.

Unit 10 Competency Test

***NEC®* Reference** **Answer**

1. _____ is a general term including material, fittings, devices, appliances, luminaires (fixtures), apparatus, and the like used as a part of, or in connection with, an electrical installation.
2. Where a building is supplied by more than one service, or combination of branch-circuits, feeders, and services, a permanent _____ or _____ shall be installed at each service disconnect location denoting all other services, feeders, and branch-circuits supplying that building and the area served by each.
3. Where the overcurrent device is rated over _____ amperes, the ampacity of the conductors it protects shall be equal to or greater than the rating of the overcurrent device.
4. A(n) _____ is a device, or group of devices, or other means by which the conductors of a circuit can be disconnected from their source of supply.
5. A(n) _____ is a device intended to provide protection from the effects of arc faults by recognizing characteristics unique to arcing and by functioning to de-energize the circuit when an arc fault is detected.
6. Service-drop conductors shall have a vertical clearance of not less than _____ ft above the roof surface.
7. What size copper equipment grounding conductor is required for a circuit having a 40-ampere overcurrent device?
8. Where a change occurs in the size of the _____, a similar change shall be permitted to be made in the size of the grounded conductor.

NEC® Reference	Answer	
________	________	9. _____ in dwelling units shall supply only loads within that dwelling unit or loads associated only with that dwelling unit.
________	________	10. A metal elbow that is installed in an underground installation of RNC and is isolated from possible contact by a minimum cover of _____ in. to any part of the elbow shall not be required to be grounded.
________	________	11. Conductors shall be considered outside of a building or other structure where installed under not less than _____ in. of concrete beneath a building or other structure.
________	________	12. What is the service demand load (in kW) for an apartment complex containing ten each of the following: 15-kW ranges, 13-kW ranges, 11-kW ranges, and 9-kW ranges? (The service is 120/240 single-phase.)
________	________	13. In _____, remote, central, or automatic control of lighting shall be permitted.
________	________	14. Service conductors installed as open conductors shall have a clearance of not less than _____ ft from fire escapes.
________	________	15. A grounded circuit conductor shall be permitted to ground noncurrent-carrying metal parts of equipment on the _____ side of the ac service disconnecting means.
________	________	16. Each unit of a thirty-unit apartment building has a 4.5-kW clothes dryer. What is the service demand load contribution (in kW) for these clothes dryers? (The service is 120/240 single-phase.)
________	________	17. The maximum distance between receptacle outlets in a multi-family dwelling is _____ ft.
________	________	18. Overcurrent devices shall not be located in the vicinity of _____, such as in clothes closets.
________	________	19. Cabinets and cutout boxes shall have _____ to accommodate all conductors installed in them without crowding.
________	________	20. Where conductors carrying alternating current are installed in metal enclosures or metal raceways, they shall be arranged so as to avoid heating the surrounding metal by _____.
________	________	21. _____ is the highest current at rated voltage that a device is intended to interrupt under standard test conditions.
________	________	22. Where conductors are adjusted in size to compensate for voltage drop, equipment grounding conductors, where installed, shall be adjusted proportionately according to:
________	________	23. Enclosures for overcurrent devices shall be mounted in a _____ position.
________	________	24. A metal pole supporting a luminaire (lighting fixture), 18 ft above grade, must have a handhole of at least _____ in.
________	________	25. Fittings and connectors shall be used only with the specific wiring methods for which they are _____.
________	________	26. _____ is a combination consisting of a compressor and motor, both of which are enclosed in the same housing, with no external shaft or shaft seals, the motor operating in the refrigerant.
________	________	27. For permanently connected appliances rated over _____ volt-amperes or _____ horsepower the branch-circuit switch or circuit breaker shall be permitted to serve as the disconnecting means where the switch or circuit breaker is within sight from the appliance or is capable of being locked in the open position.
________	________	28. The ampacity of the branch-circuit conductors and the rating or setting of overcurrent protective devices supplying fixed electric space-heating equipment consisting of resistance elements with or without a motor shall not be less than _____ % of the total load of the motors and the heaters.

NEC® Reference	Answer	
________	________	29. _____ is the necessary equipment, usually consisting of a circuit breaker(s) or switch(es) and fuse(s) and their accessories, connected to the load end of service conductors to a building or other structure, or an otherwise designated area, which is intended to constitute the main control and cutoff of the supply.
________	________	30. No hanging luminaire (fixture) parts can be located within a zone measured _____ ft vertically from the top of the bathtub rim and _____ ft vertically from the top of the shower stall threshold.
________	________	31. What is the ac resistance of 500 feet of 1000 kcmil aluminum wire in a steel conduit?
________	________	32. At 75°C, what size copper conductors are required (without exceeding NEC® recommendations) to feed a 1⁄4-horsepower, 115-volt, single-phase motor located 230 ft from the source?
________	________	33. Interior metal water piping located more than _____ ft from the point of entrance to the building shall not be used as a part of the grounding electrode system.
________	________	34. What is the minimum neutral demand load (in kW) for twelve apartments, each containing an 8-kW range?
________	________	35. What is the voltage drop percentage on two 10 AWG THW copper, stranded, branch-circuit conductors, 120 ft long, supplying a 21-ampere, 240-volt load?

SECTION THREE: MULTI-FAMILY DWELLINGS

Load Calculations

Objectives

After studying this unit, the student should:

- know that the number of dwelling units is essential in performing a multi-family calculation.
- be able to calculate the general lighting load in a multi-family dwelling.
- correctly identify minimum volt-ampere requirements for small appliance branch-circuits.
- thoroughly understand the exceptions pertaining to optional laundry branch-circuits.
- know how to apply demand factors to the general lighting load, as well as to fastened-in-place appliances.
- be able to calculate demand loads for household clothes dryers, cooking equipment, and heating and air-conditioning systems.
- understand the standard method for calculating a multi-family dwelling service.
- successfully calculate a multi-family dwelling service (using the optional method.)
- be able to calculate a feeder for one unit in a multi-family dwelling, using either method.
- know how to size service and feeder conductors.
- be familiar with paralleled conductor provisions.
- correctly calculate and select the appropriate size neutral conductor.
- understand how grounding electrode conductors, as well as an equipment grounding conductors, are selected.

Introduction

Strong similarities exist between multi-family dwelling load calculations and one-family load calculations. Both can be performed by a standard method as well as an optional method. In fact, the similarities are so strong that this is reflected in the very design of the load calculation forms. It is, therefore, obviously beneficial to have a clear understanding of one-family dwelling load calculations before studying this unit (see Unit 8). It is absolutely essential to these calculations, to know the total number of dwelling units being supplied by the service. To illustrate this point: two main service disconnecting means supplying power to twelve units must be computed on the basis of six units per service. Performing a service calculation on twelve units and then dividing the final figure by two, would render inaccurate results.

The multi-family dwelling load calculations performed in this unit are based on the floor plan found in Unit 10. The statistical information for each individual dwelling unit remains unchanged for the purpose of load calculation. This data can be found in the floor-plan diagram immediately following this introduction. Examples include a multi-family dwelling load calculation for both a 12-unit and a 6-unit service. Each will be calculated by the standard method as well as the optional. Feeder and panelboard sizing for individual units is demonstrated by a standard one-family dwelling calculation. For comparison, the optional one-family dwelling is shown after the optional multi-family dwelling computation.

COMPILING LOAD CALCULATION INFORMATION

Load Calculation Information for Each Unit

A The smaller of two (or more) noncoincident loads can be omitted, as long as they are never energized simultaneously » 220.21 «.

B Although a clothes dryer is not a requirement for a load calculation, be sure to include a dryer load in units containing clothes dryer outlets.

C A laundry branch-circuit is optional in multi-family dwellings » 210.52(F) *Exception No. 1 and 2* «.

D Gather all load calculation information needed to compute each service. For a service consisting of 12 units (apartment or condominium), compile the essential information for all 12 units. If the service consists of only 6 units, gather the data for those 6 units. Since most multi-family dwellings have a limited number of floor plans, compile the information pertaining to each "Unit Type" (style, floor plan, etc.). Be sure that each unit grouped in a unit type contains the same information, i.e. total square feet, small appliance branch-circuits, laundry branch-circuits, appliances, etc. For ease of demonstration in this section, each unit is identical in size and content.

E Calculate the floor area using the outside dimensions » 220.3(A) «.

F Include household cooking appliances rated over 1¾ kW.

G Include all fastened-in-place appliances.

H Include at least 2 small appliance branch-circuits » 210.11(C)(1) «.

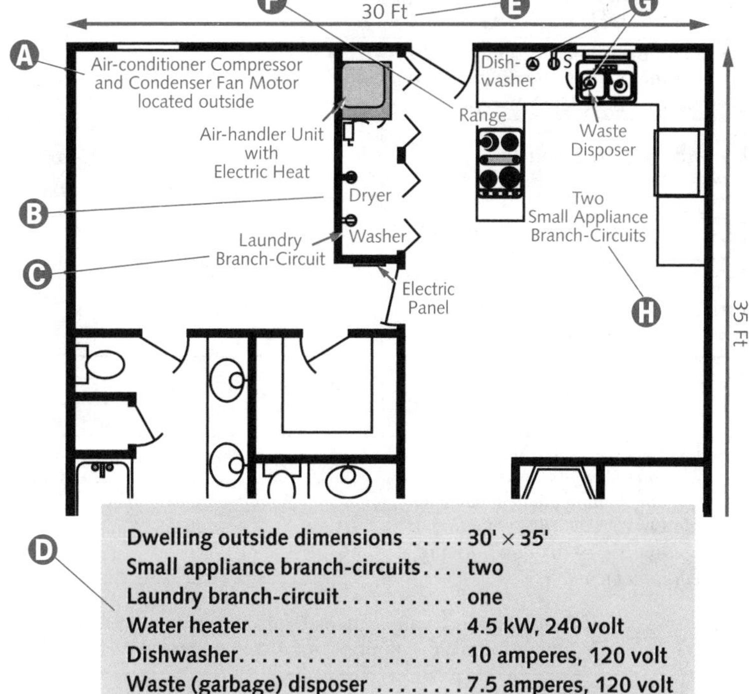

Dwelling outside dimensions . . . . . 30' × 35'
Small appliance branch-circuits . . . . two
Laundry branch-circuit . . . . one
Water heater . . . . 4.5 kW, 240 volt
Dishwasher . . . . 10 amperes, 120 volt
Waste (garbage) disposer . . . . 7.5 amperes, 120 volt
Clothes dryer . . . . 5.5 kW, 240 volt
Range . . . . 12 kW, 240 volt
Electric heat . . . . 5 kW, 240 volt
Air handler (blower motor) . . . . 3 amperes, 115 volt
Air-conditioner compressor . . . . 14 amperes, 230 volt
Condenser fan motor . . . . 2 amperes, 115 volt

Assume water heater, clothes dryer, range, and electric heat kW ratings equivalent to kVA.

STANDARD METHOD: MULTI-FAMILY DWELLINGS

Load Calculation for Twelve Units

A The first multi-family load calculation form is based on 12 units. Each unit is identical, as shown above.

B In a multiple-occupancy building, each occupant must have access to the unit's service disconnecting means » 230.72(C) «.

C There must not be more than six disconnects (switches or circuit breakers) per service grouped in a single location » 230.71(A) «.

D One service disconnecting switch controls all 12 units.

E Each unit's feeder (and overcurrent protection) must be calculated individually, using the one-family dwelling form (standard or optional method).

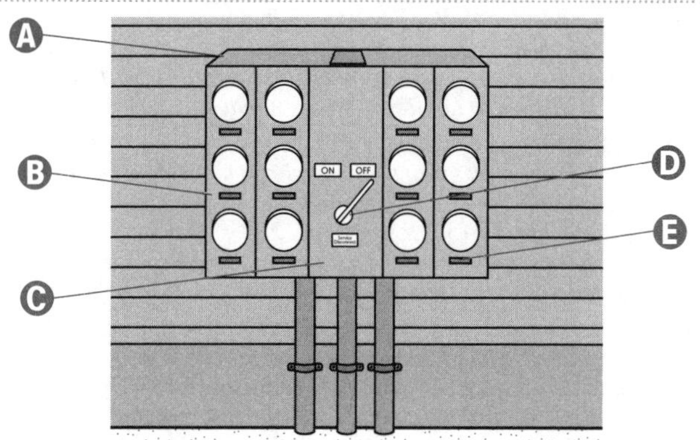

Standard Method Load Calculation for Multi-Family Dwellings

1 General Lighting and Receptacle Loads 220.3(A)
Do not include open porches, garages, and unused or unfinished spaces not adaptable for future use.
3 × ________ (sq ft outside dimensions) × ________ (number of units) = **1**

2 Small Appliance Branch-Circuits 220.16(A)
At least ***two*** *small appliance branch-circuits must be included. 210.11(C)(1)*
1500 × ________ (minimum of two) × ________ (number of units) = **2**

3 Laundry Branch-Circuit(s) 220.16(B)
Include at least ***one*** *laundry branch-circuit* ****unless*** *meeting 210.52(F) Exception No. 1 or 2. 210.11(C)(2)*
1500 × ________ (minimum of two) × ________ (number of units) = **3**

4 Add lines 1, 2, and 3 **4**

Lines 5 through 8 utilize the demand factors found in Table 220.11.

5 ________ (line 4) – 3000 = **5** (if 117,000 or less, skip to line 8)

6 ________ (line 5, if more than 117,000) – 117,000 = **6**

7 ________ (line 6) × 25% = **7**

8 ________ (smaller of line 5 or 117,000) × 35% = **8**

9 Total General Lighting and Receptacle Load 3000 + ________ (line 7) + ________ (line 8) = **9**

10 Fastened-In-Place Appliances 220.17
Use the nameplate rating.
Do not include electric ranges, clothes dryers, space-heating equipment, or air-conditioning equipment.

water heaters /	dishwashers /	disposers /
/	/	/
/	/	/

If fewer than four units, put total volt-amperes on line 10.
If four or more units, multiply total volt-amperes by 75%. ________ (volt-amps of four or more) × 75% = **10**

11 Clothes Dryers 220.18
(If present, otherwise skip to line 12.) Use 5000 watts or the nameplate rating, whichever is larger.
(The neutral demand load is 70% for feeders. 220.22) **11**

12 Ranges, Ovens, Cooktops, and Other Household Cooking Appliances Over 1750 Watts 220.19
(If present, otherwise skip to line 13.) Use Table 220.19 and all of the applicable Notes.
(The neutral demand load is 70% for feeders. 220.22) **12**

13 Heating or Air-Conditioning System (Compare the heat and A/C, and omit the smaller.) 220.21
Include the air handler when using either one. For heat pumps, include the compressor and the maximum amount of electric heat that can be energized while the compressor is running. **13**

14 Largest Motor (one motor only) 220.14 and 430.24
Multiply the volt-amperes of the largest motor by 25%. ________ (volt-amps of largest motor) × 25% = **14**

15 Total Volt-Ampere Demand Load: *Add lines 9 through 14 to find the minimum required volt-amperes.* **15**

16 Minimum Amperes
Divide the total volt-amperes by the voltage ________ (line 15) ÷ ________ (voltage) = **16** (minimum amperes)

17 **Minimum Size Service and/or Feeder 240.6(A)** **17**

18 Size the Service and/or Feeder Conductors.
Use 310.15(B)(6) to find the service conductors up to 400 amperes.
Ratings in excess of 400 amperes shall comply with Table 310.16.
310.15(B)(6) also applies to feeder conductors serving as the main power feeder.
Minimum Size Conductors **18**

19 Size the Neutral Conductor 220.22
310.15(B)(6) states that the neutral service and/or feeder conductor can be smaller than the ungrounded (hot) conductors, provided the requirements of 215.2, 220.22, and 230.42 are met.
250.24(B)(1) states that the neutral cannot be smaller than the required grounding electrode conductor specified in Table 250.66.
Minimum Size Neutral Conductor **19**

20 Size the Grounding Electrode Conductor (for Service) 250.66
Using line 18 to find the grounding electrode conductor in Table 250.66.
Size the Equipment Grounding Conductor (for Feeder) 250.122
Use line 17 to find the equipment grounding conductor in Table 250.122.
Equipment grounding conductor types are listed in 250.118.
Minimum Size Grounding Electrode Conductor . . . *or* . . . Equipment Grounding Conductor **20**

Line 1—General Lighting and Receptacle Loads

A The residential dwelling unit's lighting load is 3 volt-amperes (VA) per sq ft »Table 220.3(A)«.

B Each multi-family dwelling unit has an outside dimension of 30 ft × 35 ft (1050 sq ft).

C Insert the number of units (apartments or condominiums) that constitute one service. Additional space will be needed for computations if the units have different sq ft dimensions.

D Each unit's floor area shall be computed using the outside dimensions. The total floor area does not include open porches, garages, or unused (including unfinished) spaces which are not adaptable for future use »220.3(A)«.

> **NOTE**
>
> *All 15- or 20-ampere general-use receptacle outlets (except for small appliance and laundry), are considered part of the general lighting load. The "a" footnote at the bottom of Table 220.3(A) refers to 220.3(B)(10), which states that no additional load calculations are required for these outlets.*

1 General Lighting and Receptacle Loads 220.3(A)
Do not include open porches, garages, and unused or unfinished spaces not adaptable for future use.
3 × 1050 (sq ft outside dimensions) × 12 (number of units) = **1** 37,800

Line 2—Small Appliance Branch-Circuits

A Include small appliance branch-circuits when calculating a dwelling unit service »220.16(A)«.

B Each small appliance branch-circuit is calculated at no less than 1500 volt-amperes »220.16(A)«.

C If any unit(s) has a different number of small appliance branch-circuits, calculate each separately.

D Each dwelling unit must have at least two small appliance branch-circuits »210.11(C)(1)«.

E These loads can be included with the general lighting load and are subject to Table 220.11 demand factors »220.16(A)«.

> **NOTE**
>
> *A refrigeration equipment individual branch-circuit [permitted by 210.52(B)(1)* Exception No. 2*] can be excluded from this calculation »220.16(A)* Exception«.

2 Small Appliance Branch-Circuits 220.16(A)
*At least **two** small appliance branch-circuits must be included. 210.11(C)(1)*
1500 × 2 (minimum of two) × 12 (number of units) = **2** 36,000

Line 3—Laundry Branch-Circuit(s)

A Include laundry branch-circuit(s) when calculating a dwelling unit service »220.16(B)«.

B Each laundry branch-circuit is calculated at no less than 1500 volt-amperes »220.16(B)«.

C Include only the number of units having a laundry branch-circuit. If only 6 of 12 units contains a laundry branch-circuit, put 6 as the number of units. In this calculation, each unit has a laundry branch-circuit.

D In a multi-family building where laundry facilities are provided on the premises, available to all occupants, a laundry receptacle is not required in each individual unit »210.52(F) *Exception No. 1*«.

E A laundry receptacle is still not required, even in multi-family dwellings without common laundry facilities »210.52(F) *Exception No. 2*«.

F Unless one of the exceptions is met, include at least one laundry branch-circuit »210.11(C)(2)«.

G These loads can be included with the general lighting load and are subject to the Table 220.11 demand factors »220.16(B)«.

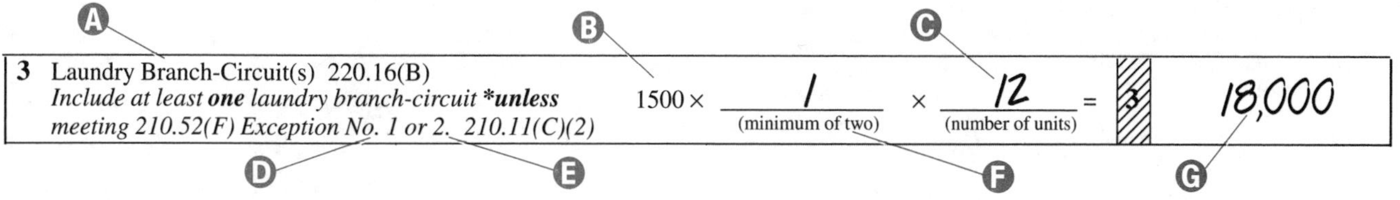

Lines 4 through 8—Applying Demand Factors Found in Table 220.11

A Line 4 represents the total of lines 1 through 3.

B Lines 5 through 8 outline the derating procedure, using Table 220.11 demand factors.

C Line 5 is the result of subtracting 3000 from the number in Line 4.

D The first 3000 (or less) volt-amperes must remain at 100% »Table 220.11«.

E If Line 5 is 117,000 or less, skip Lines 6 and 7.

F If Line 4 is 120,000 or less, then Lines 6 and 7 remain empty.

G Insert the smaller of Line 5 or 117,000.

H Multiply the next 117,000 (from 3001 to 120,000) by 35% »Table 220.11«.

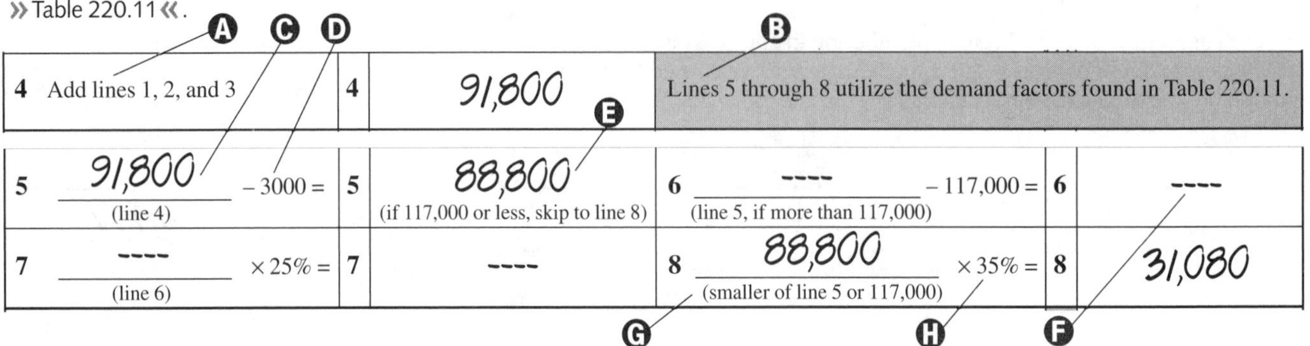

4 Add lines 1, 2, and 3	4	91,800	Lines 5 through 8 utilize the demand factors found in Table 220.11.		
5 91,800 (line 4) – 3000 =	5	88,800 (if 117,000 or less, skip to line 8)	6 ---- (line 5, if more than 117,000) – 117,000 =	6	----
7 ---- (line 6) × 25% =	7	----	8 88,800 (smaller of line 5 or 117,000) × 35% =	8	31,080

Line 9—Total General Lighting (and Receptacle) Load

A Line 9 is the total of general lighting, general-use receptacles, small appliance branch-circuits, and laundry branch-circuit(s) after derating.

B If Line 7 is blank, leave this blank also.

C Insert the result of line 8, not the number from Line 5.

D This shaded block draws attention to the first of six lines (boxes) that will be added together resulting in the total volt-ampere demand load for the one-family dwelling.

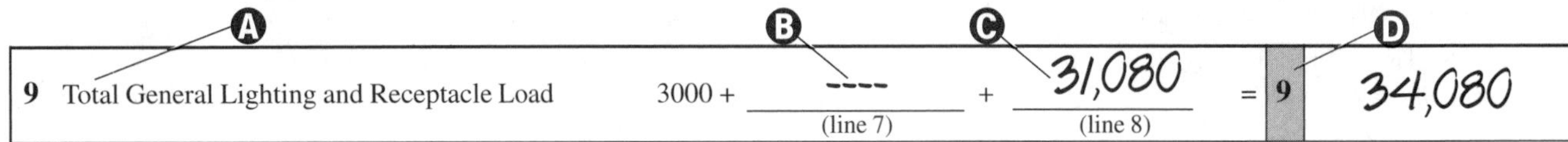

9 Total General Lighting and Receptacle Load	3000 + ---- (line 7)	+ 31,080 (line 8)	=	9	34,080

Line 10—Fastened-in-Place Appliances

A The term **appliance** designates utilization equipment (commonly built in standardized types and sizes) installed as a unit to perform specific function(s) »Article 100«.

B For the purpose of this load calculation, a kW rating is the same as a kVA rating.

C Additional space may be needed to calculate total appliance volt-amperes. (1200 × 12 = 14,400) and (900 × 12 = 10,800)

D Derating is allowed for a total of 4 (or more) fastened-in-place appliances. It is not necessary that the 4 fastened-in-place appliances be of different types. This derating factor can be used even if all units do not contain 4 appliances.

E If the calculation contains at least 4 fastened-in-place appliances, the combined volt-ampere rating of those appliances is multiplied by 75% and placed on Line 10 »220.17«.

NOTE

Do not include electric ranges, dryers, space-heating-equipment, and air-conditioning equipment as fastened-in-place appliances »220.17«.

Household cooking appliances, individually rated in excess of 1750 watts, are derated under 220.19 provisions.

10 Fastened-In-Place Appliances 220.17 *Use the nameplate rating. Do not include electric ranges, clothes dryers, space-heating equipment, or air-conditioning equipment.*	water heaters / 54,000	dishwashers / 14,400	disposers / 10,800
	microwave ovens / 14,400	/	/
	/	/	/
If fewer than four units, put total volt-amperes on line 10. If four or more units, multiply total volt-amperes by 75%.	93,600 (volt-amps of four or more) × 75% =	10	70,200

Line 11—Clothes Dryers

A A clothes dryer is not a requirement for a load calculation. Skip this line if there is no clothes dryer.

B The dryer rating must be at least 5000 watts (volt-amperes) » 220.18 «.

C Table 220.18 can be used to derate five (or more) clothes dryer loads. First, find the demand factor percent across from the number of dryers, in this case the percent is 45. Next, multiply the total dryer load by that percentage and put the result in Line 11. (5000 × 12 = 60,000 × 45% = 27,000)

D A clothes dryer neutral load on a feeder is calculated at 70% » 220.22 «.

E If the nameplate rating exceeds 5000 watts (volt-amperes), use the larger number » 220.18 «.

NOTE

For 4 (or fewer) clothes dryers the demand is 100%.

11 Clothes Dryers 220.18
(If present, otherwise skip to line 12.) Use 5000 watts or the nameplate rating, whichever is larger.
(The neutral demand load is 70% for feeders. 220.22) | **11** |

Line 12—Household Cooking Appliances

A Household cooking appliances (ranges, wall-mounted ovens, counter-mounted cooking units, etc.) are not required in a load calculation. If no cooking appliances are rated over 1¾ kW, skip this line.

B Individual household cooking appliances rated more than 1750 watts (volt-amperes) can be derated using Table 220.19 demand factors (and all applicable notes).

C The demand for twelve 12 kW ranges comes from Table 220.19, Column C. First, look at number of appliances column and find the number twelve. Next, follow that row across to the last column (Column C) and find the kW demand load, which is 27. Finally, multiply this number by 1000 and place in Line 12. Since each range is 12 kW, Columns A and B cannot be used.

D For household electric ranges, wall-mounted ovens, and counter-mounted cooking units the neutral load on a feeder is calculated at 70% » 220.22 «.

12 Ranges, Ovens, Cooktops, and Other Household Cooking Appliances Over 1750 Watts 220.19
(If present, otherwise skip to line 13.) Use Table 220.19 and all of the applicable Notes.
(The neutral demand load is 70% for feeders. 220.22) | **12** | 27,000

Line 13—Heating or Air-Conditioning System

A The smaller of two (or more) **noncoincident loads** can be omitted, as long as they are never energized simultaneously » 220.21 «. By definition, noncoincident means not occurring at the same time. A dwelling's electric heating system and air-conditioning system may be noncoincident loads.

B The air handler (blower motor) has a rating of 345 volt-amperes (3 × 115 = 345). Since both loads are energized simultaneously, add the air handler to the 5 kW of electric heat.

C Fixed electric space heating loads are computed at 100% of the total connected load » 220.15 «.

D Because the air handler (blower) works with either the heating or the air conditioning, add it to both calculations.

E Each unit contains an air-conditioner compressor (14 × 230 = 3,220 VA), condenser fan motor (2 × 115 = 230 VA), and blower motor (3 × 115 = 345 VA). The combined load is 3795 volt-amperes. Because this is less than the heating load omit the A/C compressor and the condenser fan motor. Multiply each unit's heat load by the number of units. (5345 × 12 = 64,140)

NOTE

A heat pump with supplementary heat is not considered a noncoincident load. Add the compressor full-load current to the maximum amount of heat that can be energized while the compressor is running. Remember to include all associated motor(s), i.e., blower, condenser, etc.

13 Heating or Air-Conditioning System (Compare the heat and A/C, and omit the smaller.) 220.21
Include the air handler when using either one. For heat pumps, include the compressor and the maximum amount of electric heat that can be energized while the compressor is running. | **13** |

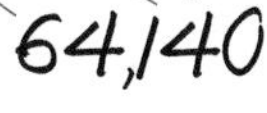

Line 14—Largest Motor

A Multiply the largest motor in the dwelling calculation by 25% »220.14 and 430.24«.

B This means a total of one motor, not one motor per unit.

C The largest motor is usually the air-conditioner compressor. If the air conditioner load has been omitted, use the second largest motor. In this case, use the 7.5-ampere garbage disposer motor. (7.5 × 120 = 900)

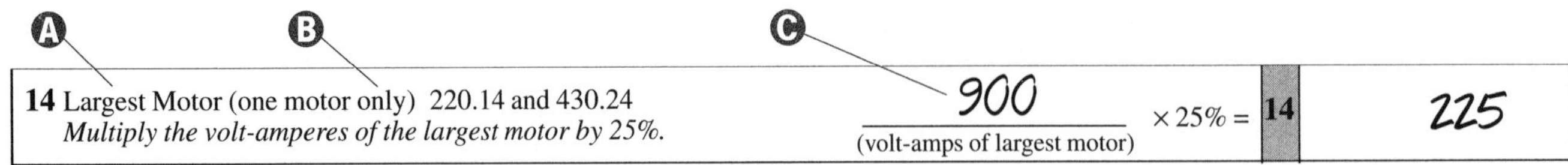

14 Largest Motor (one motor only) 220.14 and 430.24 *Multiply the volt-amperes of the largest motor by 25%.*	900 (volt-amps of largest motor) × 25% =	14	225

Line 15—Total Demand Load

A Find the minimum volt-ampere load by adding Lines 9 through 14.

15 Total Volt-Ampere Demand Load: *Add lines 9 through 14 to find the minimum required volt-amperes.*	15	222,645

Lines 16 and 17—Minimum Service and/or Feeder

A Place the total volt-ampere figure found in Line 15 here.

B Write down the source voltage supplying the service equipment.

C Fractions of an ampere 0.5 and higher are rounded up, while fractions less than 0.5 are dropped »220.2(B)«.

D The service (or feeder) overcurrent protection must be higher than the number in Line 16. 240.6(A) provides a list of standard ampere ratings for fuses and circuit breakers.

E Divide the volt-amperes by the voltage to determine the amperage.

F This is the minimum amperage rating required for the service and/or feeder being calculated.

G The next standard size fuse or breaker above 928 amperes is 1000 »240.6(A)«.

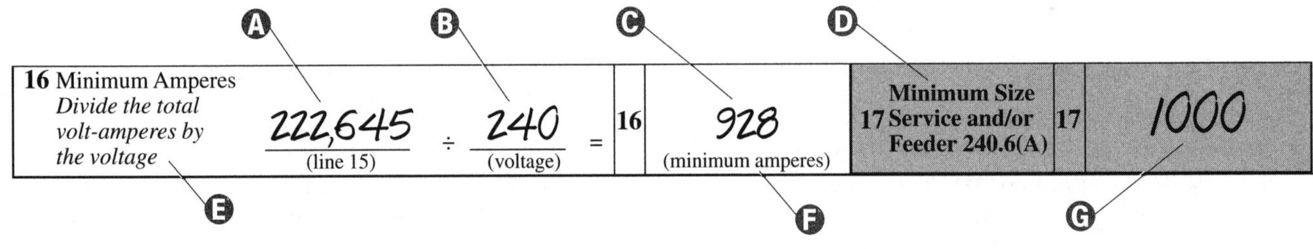

16 Minimum Amperes *Divide the total volt-amperes by the voltage*	222,645 (line 15) ÷ 240 (voltage) =	16	928 (minimum amperes)	Minimum Size 17 Service and/or Feeder 240.6(A)	17	1000

Line 18—Minimum Size Conductors

A Use 310.15(B)(6) to find service-entrance and service-lateral conductors.

B 310.15(B)(6) also applies to feeder conductors serving as the dwelling unit's main power feeder »310.15(B)(6)«.

C Use Table 310.15(B)(6) to size conductors up to a 400-ampere rating.

D When running **three** sets of paralleled copper conductors (for a 1000 ampere overcurrent device), the minimum size is 400 kcmil, unless the equipment is listed and identified for use with 194°F (90°C) conductors. Even if a 90°C conductor (such as THHN) is installed, the ampacity rating cannot exceed the 167°F (75°C) ampacity which is 335 »110.14(C)(2)«. Since the overcurrent protection is more than 800 amperes, the conductor ampacity must match (or exceed) the 1000 ampere rating (1000 ÷ 3 = 333.3 minimum amperes, each conductor) »240.4(C)«.

E Other parallel conductor combinations are possible, such as four sets of 250 kcmil Cu, four sets of 350 kcmil Al, five sets of 3/0 AWG Cu, five sets of 250 kcmil Al, etc. The smallest conductor that can be run in parallel is 1/0 AWG »310.4«.

F Since the rating is more than 400 amperes, use Table 310.16 to find the minimum size conductors.

G The number and type of paralleled conductor sets is a design consideration, not necessarily a *Code* issue. Exercise extreme care when designing a paralleled conductor installation, without violating provisions such as 110.14(A), 110.14(C), 240.3(C), 250.66, 250.122(F) 300.20(A), 310.4, etc.

H 1/0 AWG and larger conductors can be connected in parallel (electrically joined at both ends to form a single conductor) »310.4«.

CAUTION *Conductors carrying alternating current, installed in metal enclosures or* ***metal raceways****, must be arranged to avoid heating of the surrounding metal by induction »300.20(A)«.*

NOTE

The paralleled conductors in each phase, neutral, or grounded circuit conductor must share the same characteristics »310.4«. For example, all paralleled, Phase A conductors must:

(1) be of the same length,
(2) have the same conductor material,
(3) be the same size in circular mil area,
(4) have the same insulation type,
(5) be terminated in the same manner.

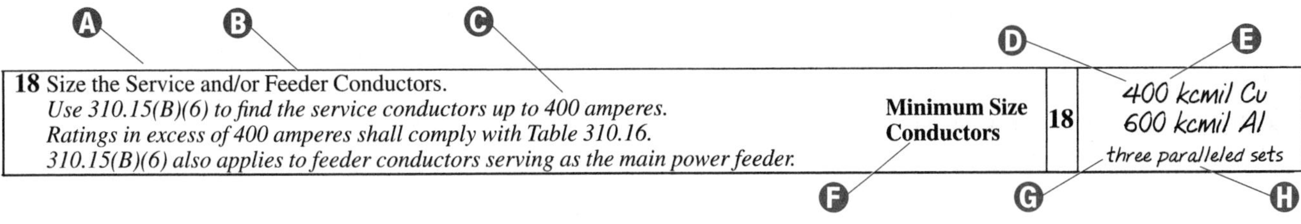

Line 19—Neutral Conductor

A For the purpose of this load calculation, the terms **Neutral** and **Grounded** are synonymous.

B The feeder (or service) neutral load is the maximum unbalance of the load determined by Article 220 »220.22«.

C The neutral (grounded) conductor can be smaller than the ungrounded (hot) conductors provided certain requirements in 215.2, 220.22, and 230.42 are met »310.15(B)(6)«.

D Find the maximum unbalanced load in order to size the neutral.

E Although the neutral conductor requires only 151 amperes each (for three paralleled sets), it cannot be smaller than the appropriate grounding electrode conductor (see Line 20).

NOTE

The smallest paralleled grounded (neutral) conductor is 1/0 AWG »250.24(B)(2)«.

CAUTION *The neutral must not be smaller than the required grounding electrode conductor specified in Table 250.66 »250.24(B)(1)«.*

A B C D E

19 Size the Neutral Conductor 220.22 *310.15(B)(6) states that the neutral service and/or feeder conductor can be smaller than the ungrounded (hot) conductors, provided the requirements of 215.2, 220.22, and 230.42 are met. 250.24(B)(1) states that the neutral cannot be smaller than the required grounding electrode conductor specified in Table 250.66.*	**Minimum Size Neutral Conductor**	19	3/0 Cu 250 kcmil Al three paralleled sets

Line 19—Neutral Conductor (*continued*)

(A) No neutral conductor is connected to the water heaters in this calculation.

(B) All fastened-in-place appliances utilizing a grounded conductor must be included.

(C) The electric heat and A/C compressor are 240-volt loads (no neutral).

(D) Because all of these motors can be energized simultaneously, thereby contributing to the neutral load, include them in the neutral calculation.

(E) To convert to amperes, simply divide volt-amperes by voltage.

(F) The minimum conductor ampacity for three paralleled neutral conductors is 151 amperes. Although the minimum size copper conductor at 75°C is 2/0 AWG, the neutral conductor cannot be smaller than the required grounding electrode conductor.

(G) All numbers are shown as volt-amperes.

(H) Since all of the general lighting and receptacle loads have a rating of 120 volts, the total volt-ampere rating (from Line 9) must be included in the neutral calculation.

(I) A demand factor of 75% can be applied to the nameplate rating load of 4 (or more) fastened-in-place appliances »220.17«.

(J) Multiply the clothes dryer load (Line 11) by 70% to determine the neutral load »220.21«.

(K) Multiplying Line 12 by 70% provides the neutral cooking appliances load »220.21«.

(L) Multiply the largest 120- (115-) volt motor (one motor only) by 25% and include in the calculation.

(M) The minimum neutral load is 108,705 volt-ampere.

General lighting and receptacles (Line 9) . . . . 34,080 (G) (H)
Fastened-in-place appliances (A)
Water heaters 0
Dishwashers (1200 × 12). . . . . . . 14,400 (B)
Waste disposers (7.5 × 120 × 12) 10,800
Microwave ovens (1200 × 12). . . 14,400
Total 39,600
Total fastened-in-place appliances (39,600 × 75%) 29,700 (I)
Clothes dryers (Line 11 × 70%) . . . . 18,900 (J)
Ranges (Line 12 × 70%) . . . . 18,900 (K)
Electric heat. . . . . 0 (C)
Air-conditioner compressors. 0
Air handlers (3 × 115 × 12). 4140 (D)
Condenser fan motors (2 × 115 × 12) 2760
Largest motor (900 × 25%) . . . . 225 (L)
TOTAL . . . . 108,705 (M)

(E) **108,705 ÷ 240 = 452.9 = 453 minimum neutral ampacity**
453 ÷ 3 = 151 minimum amperes per neutral conductor (F)

Line 20—Grounding Electrode Conductor

(A) Table 250.66 has two headings—service conductor size and grounding electrode conductor size. Each heading contains two columns: one for copper and one for aluminum (or copper-clad aluminum). Because of the four columns, take extreme care in selecting the correct grounding electrode conductor.

(B) The grounding electrode conductor is determined by the largest service-entrance conductor (or the equivalent area) if the service-entrance conductors are paralleled. The equivalent area for three parallel sets in this calculation is 1200 kcmil copper (3 × 400) *or* 1800 kcmil aluminum (3 × 600).

(C) The minimum size grounding electrode conductor is 3/0 AWG copper (or 250 kcmil aluminum) »Table 250.66«.

20 Size the Grounding Electrode Conductor (for Service) 250.66 *Using line 18 to find the grounding electrode conductor in Table 250.66.* Size the Equipment Grounding Conductor (for Feeder) 250.122 *Use line 17 to find the equipment grounding conductor in Table 250.122.* *Equipment grounding conductor types are listed in 250.118.*	**Minimum Size Grounding Electrode Conductor . . . *or* . . . Equipment Grounding Conductor**	**20**	3/0 Cu 250 kcmil Al

The following page shows the completed standard method load calculation for a 12-unit, multi-family dwelling, having all units combined in a single service.

Standard Method Load Calculation for Multi-Family Dwellings

Line	Description	Calculation		Result
1	General Lighting and Receptacle Loads 220.3(A) *Do not include open porches, garages, and unused or unfinished spaces not adaptable for future use.*	3 × 1050 (sq ft outside dimensions) × 12 (number of units) =	1	37,800
2	Small Appliance Branch-Circuits 220.16(A) *At least **two** small appliance branch-circuits must be included. 210.11(C)(1)*	1500 × 2 (minimum of two) × 12 (number of units) =	2	36,000
3	Laundry Branch-Circuit(s) 220.16(B) *Include at least **one** laundry branch-circuit *unless meeting 210.52(F) Exception No. 1 or 2. 210.11(C)(2)*	1500 × 1 (minimum of two) × 12 (number of units) =	3	18,000

4	Add lines 1, 2, and 3	4	91,800	Lines 5 through 8 utilize the demand factors found in Table 220.11.		
5	91,800 (line 4) − 3000 =	5	88,800 (if 117,000 or less, skip to line 8)	6 ---- (line 5, if more than 117,000) − 117,000 =	6	----
7	---- (line 6) × 25% =	7	----	8 88,800 (smaller of line 5 or 117,000) × 35% =	8	31,080

Line	Description	Calculation		Result
9	Total General Lighting and Receptacle Load	3000 + ---- (line 7) + 31,080 (line 8) =	9	34,080
10	Fastened-In-Place Appliances 220.17 *Use the nameplate rating. Do not include electric ranges, clothes dryers, space-heating equipment, or air-conditioning equipment.* water heaters / 54,000; dishwashers / 14,400; disposers / 10,800; microwave ovens / 14,400; / ; / ; / ; / ; / If fewer than four units, put total volt-amperes on line 10. If four or more units, multiply total volt-amperes by 75%.	93,600 (volt-amps of four or more) × 75% =	10	70,200
11	Clothes Dryers 220.18 *(If present, otherwise skip to line 12.) Use 5000 watts or the nameplate rating, whichever is larger.* (The neutral demand load is 70% for feeders. 220.22)		11	27,000
12	Ranges, Ovens, Cooktops, and Other Household Cooking Appliances Over 1750 Watts 220.19 *(If present, otherwise skip to line 13.) Use Table 220.19 and all of the applicable Notes.* (The neutral demand load is 70% for feeders. 220.22)		12	27,000
13	Heating or Air-Conditioning System (Compare the heat and A/C, and omit the smaller.) 220.21 *Include the air handler when using either one. For heat pumps, include the compressor and the maximum amount of electric heat that can be energized while the compressor is running.*		13	64,140
14	Largest Motor (one motor only) 220.14 and 430.24 *Multiply the volt-amperes of the largest motor by 25%.*	900 (volt-amps of largest motor) × 25% =	14	225
15	Total Volt-Ampere Demand Load: *Add lines 9 through 14 to find the minimum required volt-amperes.*		15	222,645
16	Minimum Amperes *Divide the total volt-amperes by the voltage*	222,645 (line 15) ÷ 240 (voltage) = 16: 928 (minimum amperes)		
17		**Minimum Size Service and/or Feeder 240.6(A)**	17	1000
18	Size the Service and/or Feeder Conductors. *Use 310.15(B)(6) to find the service conductors up to 400 amperes. Ratings in excess of 400 amperes shall comply with Table 310.16. 310.15(B)(6) also applies to feeder conductors serving as the main power feeder.*	**Minimum Size Conductors**	18	400 kcmil Cu 600 kcmil Al three paralleled sets
19	Size the Neutral Conductor 220.22 *310.15(B)(6) states that the neutral service and/or feeder conductor can be smaller than the ungrounded (hot) conductors, provided the requirements of 215.2, 220.22, and 230.42 are met. 250.24(B)(1) states that the neutral cannot be smaller than the required grounding electrode conductor specified in Table 250.66.*	**Minimum Size Neutral Conductor**	19	3/0 Cu 250 kcmil Al three paralleled sets
20	Size the Grounding Electrode Conductor (for Service) 250.66 *Using line 18 to find the grounding electrode conductor in Table 250.66.* Size the Equipment Grounding Conductor (for Feeder) 250.122 *Use line 17 to find the equipment grounding conductor in Table 250.122. Equipment grounding conductor types are listed in 250.118.*	**Minimum Size Grounding Electrode Conductor . . . or . . . Equipment Grounding Conductor**	20	3/0 Cu 250 kcmil Al

SIX UNIT MULTI-FAMILY DWELLING CALCULATION

Two Meter Centers, Each Supplying Six Units

A Each service drop (or lateral) supplies only one set of service-entrance conductors, unless two to six service-disconnecting means (in separate enclosures) are grouped at one location and supply separate loads from one service drop »230.40 *Exception No. 2*«.

B Because service equipment installations can vary greatly, the arrangement must be known before a load calculation is performed. In this configuration, each service (conductors and equipment) must be able to carry the calculated load for six dwelling units.

C The following multi-family dwelling, standard method, load calculation is based on 6 units with identical configurations. Turn to page 209 for dwelling unit details.

D Service-entrance conductors must be able to carry the ampacity determined in accordance with Article 220 »230.42(A)«.

E No more than six disconnects per service can be grouped in one location »230.71(A)«.

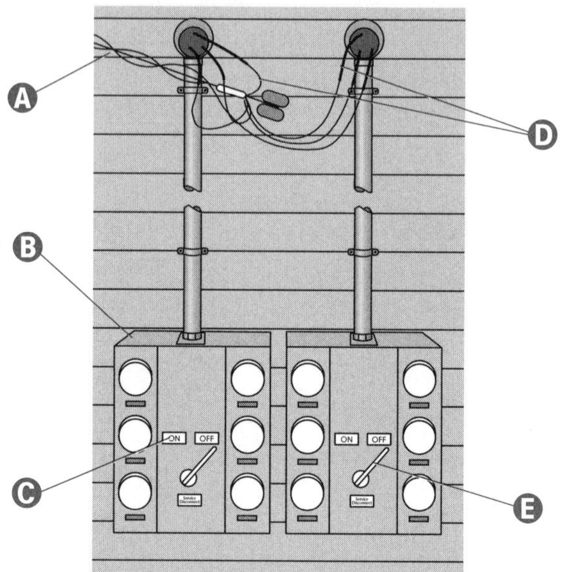

Lines 1 through 9—General Lighting and Receptacle Load

A Fill in the first three lines using six dwelling units instead of 12.

B Insert the result of Line 8, not the number from Line 5.

C The general lighting and receptacle load for these six dwelling units is 18,015 volt-amperes. Compare this Line 9 with the Line 9 for twelve units (34,080). Obviously, this is not half of Line 9 for twelve units; therefore, a calculation must be performed for each different application.

D Multiply the result of Line 5 by 35% and place on Line 8.

Line	Description	Calculation	Line	Result
1	General Lighting and Receptacle Loads 220.3(A) *Do not include open porches, garages, and unused or unfinished spaces not adaptable for future use.*	3 × 1050 (sq ft outside dimensions) × 6 (number of units) =	1	18,900
2	Small Appliance Branch-Circuits 220.16(A) *At least **two** small appliance branch-circuits must be included. 210.11(C)(1)*	1500 × 2 (minimum of two) × 6 (number of units) =	2	18,000
3	Laundry Branch-Circuit(s) 220.16(B) *At least **one** laundry branch-circuit must be included. 210.11(c)(2)*	1500 × 1 (minimum of two) × 6 (number of units) =	3	9000

Line	Calculation	Line	Result	Line	Calculation	Line	Result
4	Add lines 1, 2, and 3	4	45,900		Lines 5 through 8 utilize the demand factors found in Table 220.11.		
5	45,900 (line 4) − 3000 =	5	42,900 (if 117,000 or less, skip to line 8)	6	---- (line 5, if more than 117,000) − 117,000 =	6	----
7	---- (line 6) × 25% =	7	----	8	42,900 (smaller of line 5 or 117,000) × 35% =	8	15,015

Line	Description	Calculation	Line	Result
9	Total General Lighting and Receptacle Load	3000 + ---- (line 7) + 15,015 (line 8) =	9	18,015

Lines 11 and 12—Clothes Dryers and Ranges

A Use Table 220.18 to derate five (or more) clothes dryer loads. First, find the demand factor percent across from the number of dryers, in this case 70%. Next, multiply the total dryer load by that percent and record the result on Line 11. (5000 × 6 = 30,000 × 70% = 21,000)

B The demand for six 12 kW ranges comes from Table 220.19, Column C. First, look at the number of appliances column and find the number six. Next, follow that row across to Column C and find the kW demand load, which is 21. Finally, multiply this number by 1000 and place in Line 12. Since each range is 12 kW, Columns A and B cannot be used.

11 Clothes Dryers 220.18 *(If present, otherwise skip to line 12.) Use 5000 watts or the nameplate rating, whichever is larger.* (The neutral demand load is 70% for feeders. 220.22)	**11**	21,000 (A)
12 Ranges, Ovens, Cooktops, and Other Household Cooking Appliances Over 1750 Watts 220.19 *(If present, otherwise skip to line 13.) Use Table 220.19 and all of the applicable Notes.* (The neutral demand load is 70% for feeders. 220.22)	**12**	21,000 (B)

Lines 15 Through 17—Minimum Amperes

A The minimum volt-ampere load for these six dwelling units is 127,410.

B The service-entrance conductors and equipment must have a rating of at least 531 amperes.

C The next standard size fuse or circuit breaker above 531 is 600 amperes »240.6(A)«.

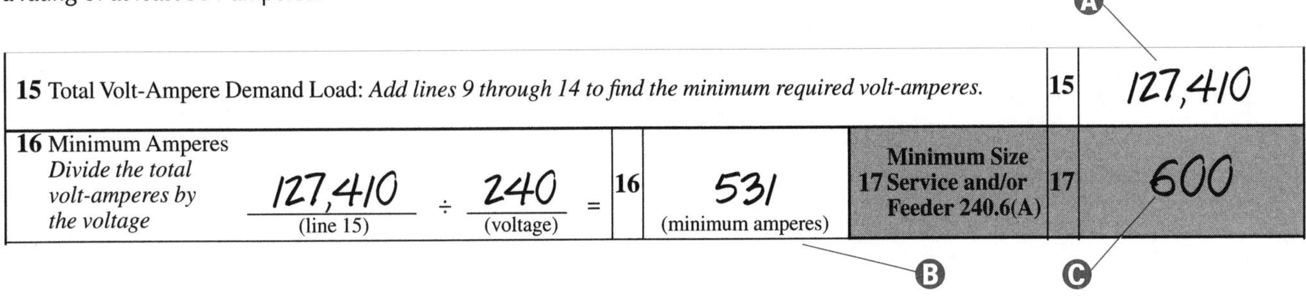

15 Total Volt-Ampere Demand Load: *Add lines 9 through 14 to find the minimum required volt-amperes.*				**15**	127,410 (A)
16 Minimum Amperes *Divide the total volt-amperes by the voltage*	127,410 (line 15) ÷ 240 (voltage) =	**16**	531 (minimum amperes) (B)	**Minimum Size 17 Service and/or Feeder 240.6(A)** **17**	600 (C)

Line 18—Minimum Size Conductors

A At 75°C, the minimum size copper conductors, for two parallel sets, matching or exceeding 531 amperes, are 300 kcmil. Since the overcurrent protection is less than 800 amperes, the conductor ampacity can be less than the 600-ampere rating. Where the conductor ampacity does not correspond to a standard size fuse or breaker, and 240.4(B)1 through 3 conditions are met, use the next higher standard overcurrent device »240.4(B)«. The combined rating of two 300-kcmil copper conductors is 570 amperes. 250-kcmil copper conductors cannot be used, because their combined rating of only 510 amperes is below the minimum required conductor ampacity of 531.

B Other parallel conductor combinations are possible, such as three sets of 3/0 AWG Cu and three sets of 4/0 AWG Al.

18 Size the Service and/or Feeder Conductors. *Use 310.15(B)(6) to find the service conductors up to 400 amperes.* *Ratings in excess of 400 amperes shall comply with Table 310.16.* *310.15(B)(6) also applies to feeder conductors serving as the main power feeder.*	**Minimum Size Conductors**	**18**	300 kcmil Cu (A) 400 kcmil Al two paralleled sets (B)

Line 19—Neutral Conductor

A The minimum neutral load is 65,940 volt-ampere.

B The minimum conductor ampacity for two paralleled neutral conductors is 138 amperes. At 75°C, the minimum size conductor is 1/0 AWG copper *or* 3/0 AWG aluminum

> **NOTE**
>
> *Even with a lower neutral ampacity, the conductors must remain this size, because the neutral conductor cannot be smaller than the required grounding electrode conductor* »*250.24(B)(1)*«.

General lighting and receptacles (Line 9) 18,015
Fastened-in-place appliances
- **Water heaters 0**
- **Dishwashers (1200 × 6) 7200**
- **Waste disposers (7.5 × 120 × 6). . . 5400**
- **Microwave ovens (1200 × 6) 7200**
- **Total 19,800**

Total fastened-in-place appliances (19,800 × 75%) 14,850
Clothes dryers (Line 11 × 70%) 14,700
Ranges (Line 12 × 70%) 14,700
Electric heat 0
Air-conditioner compressors. 0
Air handlers (3 × 115 × 6). 2070
Condenser fan motors (2 × 115 × 6) 1380
Largest motor (900 × 25%) 225
TOTAL 65,940

65,940 ÷ 240 = 274.75 = 275 minimum neutral ampacity
275 ÷ 2 = 137.5 = 138 minimum amperes per neutral conductor

19 Size the Neutral Conductor 220.22 *310.15(B)(6) states that the neutral service and/or feeder conductor can be smaller than the ungrounded (hot) conductors, provided the requirements of 215.2, 220.22, and 230.42 are met.* *250.24(B)(1) states that the neutral cannot be smaller than the required grounding electrode conductor specified in Table 250.66.*	**Minimum Size Neutral Conductor**	**19**	1/0 Cu 3/0 Al two paralleled sets (B)

Line 20—Grounding Electrode Conductor

A The equivalent area for two parallel sets of service entrance conductors, in this calculation, is
600 kcmil copper (2 x 300) . . . or . . .
800 kcmil aluminum (2 x 400).

B The minimum size grounding electrode conductor is 1/0 AWG copper *or* 3/0 AWG aluminum »Table 250.66«.

20 Size the Grounding Electrode Conductor (for Service) 250.66 *Using line 18 to find the grounding electrode conductor in Table 250.66.* Size the Equipment Grounding Conductor (for Feeder) 250.122 *Use line 17 to find the equipment grounding conductor in Table 250.122.* *Equipment grounding conductor types are listed in 250.118.*	**Minimum Size Grounding Electrode Conductor . . . *or* . . . Equipment Grounding Conductor**	**20**	1/0 Cu 3/0 Al

The following page shows the complete standard method load calculation for a six-unit, multi-family dwelling.

Standard Method Load Calculation for Multi-Family Dwellings

Line	Description	Calculation	Line	Result
1	General Lighting and Receptacle Loads 220.3(A) *Do not include open porches, garages, and unused or unfinished spaces not adaptable for future use.*	3 × 1050 (sq ft outside dimensions) × 6 (number of units) =	1	18,900
2	Small Appliance Branch-Circuits 220.16(A) *At least **two** small appliance branch-circuits must be included. 210.11(C)(1)*	1500 × 2 (minimum of two) × 6 (number of units) =	2	18,000
3	Laundry Branch-Circuit(s) 220.16(B) *At least **one** laundry branch-circuit must be included. 210.11(c)(2)*	1500 × 1 (minimum of two) × 6 (number of units) =	3	9000
4	Add lines 1, 2, and 3		4	45,900
	Lines 5 through 8 utilize the demand factors found in Table 220.11.			
5	45,900 (line 4) – 3000 =		5	42,900 (if 117,000 or less, skip to line 8)
6	---- (line 5, if more than 117,000) – 117,000 =		6	----
7	---- (line 6) × 25% =		7	----
8	42,900 (smaller of line 5 or 117,000) × 35% =		8	15,015
9	Total General Lighting and Receptacle Load	3000 + ---- (line 7) + 15,015 (line 8) =	9	18,015
10	Fastened-In-Place Appliances 220.17 *Use the nameplate rating. Do not include electric ranges, clothes dryers, space-heating equipment, or air-conditioning equipment.*	water heaters / 27,000; dishwashers / 7200; disposers / 5400; microwave ovens / 7200. If fewer than four units, put total volt-amperes on line 10. If four or more units, multiply total volt-amperes by 75%. 46,800 (volt-amps of four or more) × 75% =	10	35,100
11	Clothes Dryers 220.18 *(If present, otherwise skip to line 12.) Use 5000 watts or the nameplate rating, whichever is larger.* (The neutral demand load is 70% for feeders. 220.22)		11	21,000
12	Ranges, Ovens, Cooktops, and Other Household Cooking Appliances Over 1750 Watts 220.19 *(If present, otherwise skip to line 13.) Use Table 220.19 and all of the applicable Notes.* (The neutral demand load is 70% for feeders. 220.22)		12	21,000
13	Heating or Air-Conditioning System (Compare the heat and A/C, and omit the smaller.) 220.21 *Include the air handler when using either one. For heat pumps, include the compressor and the maximum amount of electric heat that can be energized while the compressor is running.*		13	32,070
14	Largest Motor (one motor only) 220.14 and 430.24 *Multiply the volt-amperes of the largest motor by 25%.*	900 (volt-amps of largest motor) × 25% =	14	225
15	Total Volt-Ampere Demand Load: *Add lines 9 through 14 to find the minimum required volt-amperes.*		15	127,410
16	Minimum Amperes *Divide the total volt-amperes by the voltage*	127,410 (line 15) ÷ 240 (voltage) = 531 (minimum amperes)	16	531
17	Minimum Size Service and/or Feeder 240.6(A)		17	600
18	Size the Service and/or Feeder Conductors. *Use 310.15(B)(6) to find the service conductors up to 400 amperes. Ratings in excess of 400 amperes shall comply with Table 310.16. 310.15(B)(6) also applies to feeder conductors serving as the main power feeder.*	**Minimum Size Conductors**	18	300 kcmil Cu 400 kcmil Al two paralleled sets
19	Size the Neutral Conductor 220.22 *310.15(B)(6) states that the neutral service and/or feeder conductor can be smaller than the ungrounded (hot) conductors, provided the requirements of 215.2, 220.22, and 230.42 are met. 250.24(B)(1) states that the neutral cannot be smaller than the required grounding electrode conductor specified in Table 250.66.*	**Minimum Size Neutral Conductor**	19	1/0 Cu 3/0 Al two paralleled sets
20	Size the Grounding Electrode Conductor (for Service) 250.66 *Using line 18 to find the grounding electrode conductor in Table 250.66.* Size the Equipment Grounding Conductor (for Feeder) 250.122 *Use line 17 to find the equipment grounding conductor in Table 250.122. Equipment grounding conductor types are listed in 250.118.*	**Minimum Size Grounding Electrode Conductor . . . *or* . . . Equipment Grounding Conductor**	20	1/0 Cu 3/0 Al

STANDARD LOAD CALCULATION FOR EACH UNIT OF A MULTI-FAMILY DWELLING

Calculating Feeders for Individual Dwelling Units

A A load calculation, on an individual unit in a multi-family dwelling, must be computed in accordance with one of the one-family-dwelling load-calculation methods (standard or optional).

B The service disconnecting means must have a rating of at least the calculated load, determined in accordance with Article 220 »230.79«.

C If the service disconnecting means consists of more than one switch or circuit breaker (permitted by 230.71), the combined ratings of all switches or circuit breakers used must not be less than that required by 230.79 »230.80«.

D A load calculation must be performed on each unique unit, i.e., different from any unit already calculated.

E A grounded circuit conductor (neutral) cannot be used for grounding noncurrent-carrying metal equipment on the load side of the service disconnecting means »250.142(B)«. In other words, the neutral must be isolated from the equipment ground in a remote panelboard (subpanel).

F Use Table 310.15(B)(6) to size the dwelling unit's main power feeder(s) »310.15(B)(6)«.

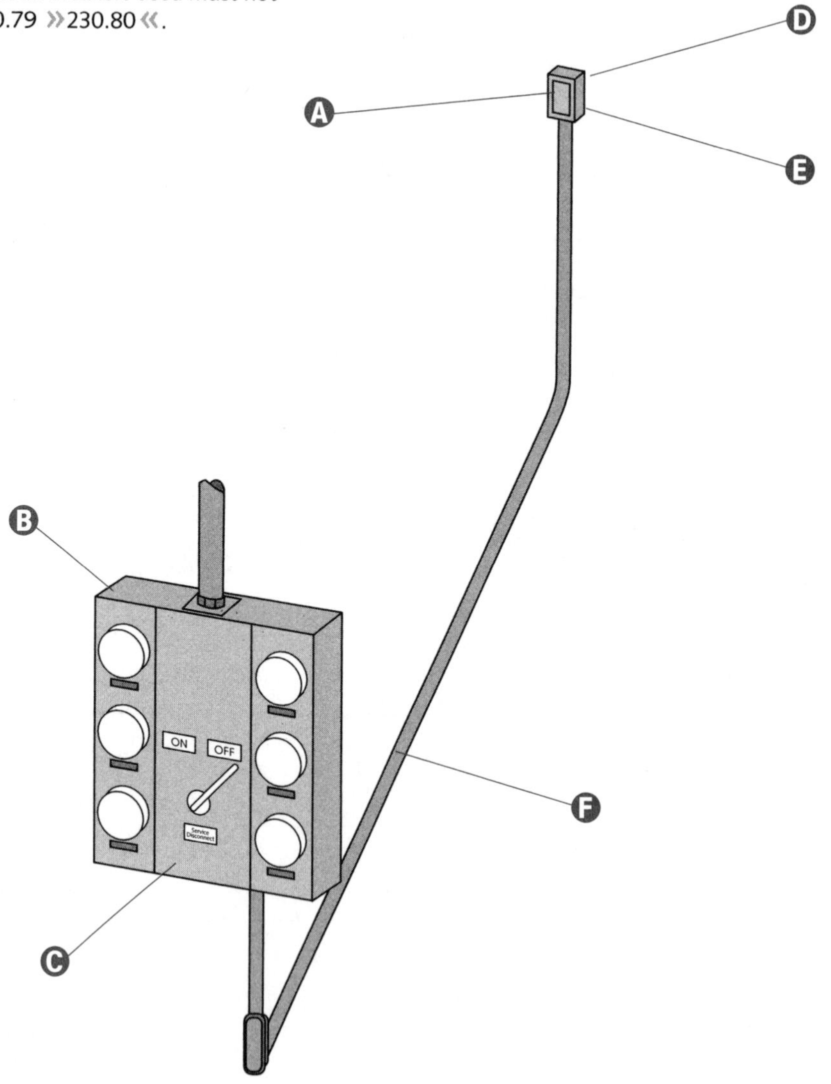

Standard Method—One Unit

A Each identically configured unit, i.e., sq ft dimensions, small appliance and laundry branch-circuits, fastened-in-place appliances, etc., results in the same size feeder.

B Although located in a multi-family dwelling, this dwelling unit contains one feeder and one panelboard. It is therefore computed in accordance with one of the one-family dwelling methods.

C A load calculation must be performed on each dissimilar unit.

D Using the standard method, the minimum required volt-amperes for this single unit is 29,048.

> **NOTE**
>
> *Load calculations are computed for each individual unit containing a panelboard (subpanel). Although located in a multi-family dwelling, calculate each unit by one of the methods for one-family dwellings ››Article 220‹‹. Performing a load calculation on multiple units, and then dividing by the number of units, will give an incorrect result.*

Standard Method Load Calculation for One-Family Dwellings

Line	Description	Calculation	Line	Result
1	General Lighting and Receptacle Loads 220.3(A) *Do not include open porches, garages, and unused or unfinished spaces not adaptable for future use.*	3 × 1050 (sq ft using outside dimensions) =	1	3150
2	Small Appliance Branch-Circuits 220.16(A) *At least* ***two*** *small appliance branch-circuits must be included. 210.11(C)(1)*	1500 × 2 (minimum of two) =	2	3000
3	Laundry Branch-Circuit(s) 220.16(B) *At least* ***one*** *laundry branch-circuit must be included. 210.11(C)(2)*	1500 × 1 (minimum of one) =	3	1500
4	Add lines 1, 2, and 3	7650		Lines 5 through 8 utilize the demand factors found in Table 220.11.
5	7650 (line 4) − 3000 =	4650 (if 117,000 or less, skip to line 8)	6	---- (line 5, if more than 117,000) − 117,000 = ----
7	---- (line 6) × 25% =	----	8	4650 (smaller of line 5 or 117,000) × 35% = 1628
9	Total General Lighting and Receptacle Load	3000 + ---- (line 7) + 1628 (line 8) =	9	4628
10	Fastened-In-Place Appliances 220.17 *Use the nameplate rating. Do not include electric ranges, clothes dryers, space-heating equipment, or air-conditioning equipment.* water heater / 4500; dishwasher / 1200; disposer / 900; compactor / ; microwave oven / 1200; / ; / ; / ; / If fewer than four units, put total volt-amperes on line 10. If four or more units, multiply total volt-amperes by 75%.	7800 (volt-amps of four or more) × 75% =	10	5850
11	Clothes Dryers 220.18 *(If present, otherwise skip to line 12.) Use 5000 watts or the nameplate rating, whichever is larger.* (The neutral demand load is 70% for feeders. 220.22)		11	5000
12	Ranges, Ovens, Cooktops, and Other Household Cooking Appliances Over 1750 Watts 220.19 *(If present, otherwise skip to line 13.) Use Table 220.19 and all of the applicable Notes.* (The neutral demand load is 70% for feeders. 220.22)		12	8000
13	Heating or Air-Conditioning System (Compare the heat and A/C, and omit the smaller.) 220.21 *Include the air handler when using either one. For heat pumps, include the compressor and the maximum amount of electric heat that can be energized while the compressor is running.*		13	5345
14	Largest Motor (one motor only) 220.14 and 430.24 *Multiply the volt-amperes of the largest motor by 25%.*	900 (volt-amps of largest motor) × 25% =	14	225
15	Total Volt-Ampere Demand Load: *Add lines 9 through 14 to find the minimum required volt-amperes.*		15	29,048

Minimum Size Feeder and Overcurrent Protection

A Feeder conductors must have an ampacity not less than required to supply the load as computed in Article 220, Parts II, III, and IV. The minimum feeder conductor size, before the application of any adjustment or correction factors, must have an allowable ampacity equal to, or greater than, the noncontinuous load plus 125% of the continuous load »215.2(A)«.

B Feeders must be protected against overcurrent in accordance with the Article 240, Part I provisions »215.3«.

C The minimum ampere rating for this one unit is 121.

D Use Table 310.15(B)(6) to determine the minimum size feeder conductors serving as the dwelling unit's main power feeders »310.15(B)(6)«.

16 Minimum Amperes *Divide the total volt-amperes by the voltage* 29,048 (line 15) ÷ 240 (voltage) =	16	121 (minimum amperes)	**Minimum Size 17 Service and/or Feeder 240.6(A)** (A)	17	125 (B)
18 Size the Service and/or Feeder Conductors. *Use 310.15(B)(6) to find the service conductors up to 400 amperes. Ratings in excess of 400 amperes shall comply with Table 310.16. 310.15(B)(6) also applies to feeder conductors serving as the main power feeder.* (C)			**Minimum Size Conductors** (D)	18	2 copper 1/0 aluminum

Minimum Size Neutral and Equipment Grounding Conductors

A The minimum neutral-conductor ampacity is 74.

B At 75°C, the minimum size conductor is 4 AWG copper *or* 3 AWG aluminum.

C If a feeder supplies branch-circuits requiring equipment grounding conductors, it must include (or provide) a grounding means to which the equipment grounding conductors of the branch-circuits shall be connected, in accordance with 250.134 provisions »215.6«.

D Copper, aluminum, or copper-clad aluminum equipment grounding conductors of the wire type must not be smaller than shown in Table 250.122, but do not have to be larger than the circuit conductors supplying the equipment »250.122(A)«.

> **NOTE**
>
> *A raceway or cable sheath acting as the equipment grounding conductor, as provided in 250.118 and 250.134(A), must comply with 250.2(D) »250.122(A)«.*

General lighting and receptacles (Line 9) 4628
Fastened-in-place appliances
- Water heater 0
- Dishwasher 1200
- Waste (garbage) disposer......... 900
- Microwave oven 1200
- Total......................... 3300

Total fastened-in-place appliances (3 units = 100%) . 3300
Clothes dryer (Line 11 × 70%)................. 3500
Range (Line 12 × 70%)....................... 5600
Electric heat.................................. 0
Air-conditioner compressor 0
Air handler (3 × 115)......................... 345
Condenser fan motor (2 × 115) 230
Largest motor (900 × 25%) 225
TOTAL 17,828

(A) 17,828 ÷ 240 = 74.28 minimum neutral-conductor ampacity

19 Size the Neutral Conductor 220.22 *310.15(B)(6) states that the neutral service and/or feeder conductor can be smaller than the ungrounded (hot) conductors, provided the requirements of 215.2, 220.22, and 230.42 are met. 250.24(B)(1) states that the neutral cannot be smaller than the required grounding electrode conductor specified in Table 250.66.*	**Minimum Size Neutral Conductor**	19	4 copper 3 aluminum (B)
20 Size the Grounding Electrode Conductor Table 250.66 *Using line 18 to find the grounding electrode conductor in Table 250.66.* Size the Equipment Grounding Conductor (for Feeder) 250.122 *Use line 17 to find the equipment grounding conductor in Table 250.122. Equipment grounding conductor types are listed in 250.118.*	**Minimum Size Grounding Electrode Conductor . . . *or* . . . Equipment Grounding Conductor** (C)	20	6 copper 4 aluminum (D)

The following page shows the complete one family dwelling load calculation for one multi-family dwelling-unit feeder and remote panelboard (subpanel), according to the standard method.

Standard Method Load Calculation for One-Family Dwellings

1 General Lighting and Receptacle Loads 220.3(A)
Do not include open porches, garages, and unused or unfinished spaces not adaptable for future use.
3 × 1050 (sq ft using outside dimensions) = **1** 3150

2 Small Appliance Branch-Circuits 220.16(A)
*At least **two** small appliance branch-circuits must be included. 210.11(C)(1)*
1500 × 2 (minimum of two) = **2** 3000

3 Laundry Branch-Circuit(s) 220.16(B)
*At least **one** laundry branch-circuit must be included. 210.11(C)(2)*
1500 × 1 (minimum of one) = **3** 1500

4 Add lines 1, 2, and 3 — **4** 7650

Lines 5 through 8 utilize the demand factors found in Table 220.11.

5 7650 (line 4) – 3000 = **5** 4650 (if 117,000 or less, skip to line 8)

6 ---- (line 5, if more than 117,000) – 117,000 = **6** ----

7 ---- (line 6) × 25% = **7** ----

8 4650 (smaller of line 5 or 117,000) × 35% = **8** 1628

9 Total General Lighting and Receptacle Load 3000 + ---- (line 7) + 1628 (line 8) = **9** 4628

10 Fastened-In-Place Appliances 220.17
Use the nameplate rating. Do not include electric ranges, clothes dryers, space-heating equipment, or air-conditioning equipment.

water heater / 4500	dishwasher / 1200	disposer / 900
compactor /	microwave oven / 1200	/
/	/	/

If fewer than four units, put total volt-amperes on line 10.
If four or more units, multiply total volt-amperes by 75%.
7800 (volt-amps of four or more) × 75% = **10** 5850

11 Clothes Dryers 220.18
(If present, otherwise skip to line 12.) Use 5000 watts or the nameplate rating, whichever is larger.
(The neutral demand load is 70% for feeders. 220.22) — **11** 5000

12 Ranges, Ovens, Cooktops, and Other Household Cooking Appliances Over 1750 Watts 220.19
(If present, otherwise skip to line 13.) Use Table 220.19 and all of the applicable Notes.
(The neutral demand load is 70% for feeders. 220.22) — **12** 8000

13 Heating or Air-Conditioning System (Compare the heat and A/C, and omit the smaller.) 220.21
Include the air handler when using either one. For heat pumps, include the compressor and the maximum amount of electric heat that can be energized while the compressor is running. — **13** 5345

14 Largest Motor (one motor only) 220.14 and 430.24
Multiply the volt-amperes of the largest motor by 25%.
900 (volt-amps of largest motor) × 25% = **14** 225

15 Total Volt-Ampere Demand Load: *Add lines 9 through 14 to find the minimum required volt-amperes.* — **15** 29,048

16 Minimum Amperes
Divide the total volt-amperes by the voltage
29,048 (line 15) ÷ 240 (voltage) = **16** 121 (minimum amperes)

17 Minimum Size Service and/or Feeder 240.6(A) — **17** 125

18 Size the Service and/or Feeder Conductors.
Use 310.15(B)(6) to find the service conductors up to 400 amperes. Ratings in excess of 400 amperes shall comply with Table 310.16. 310.15(B)(6) also applies to feeder conductors serving as the main power feeder.
Minimum Size Conductors — **18** 2 copper / 1/0 aluminum

19 Size the Neutral Conductor 220.22
310.15(B)(6) states that the neutral service and/or feeder conductor can be smaller than the ungrounded (hot) conductors, provided the requirements of 215.2, 220.22, and 230.42 are met. 250.24(B)(1) states that the neutral cannot be smaller than the required grounding electrode conductor specified in Table 250.66.
Minimum Size Neutral Conductor — **19** 4 copper / 3 aluminum

20 Size the Grounding Electrode Conductor Table 250.66
Using line 18 to find the grounding electrode conductor in Table 250.66.
Size the Equipment Grounding Conductor (for Feeder) 250.122
Use line 17 to find the equipment grounding conductor in Table 250.122. Equipment grounding conductor types are listed in 250.118.
Minimum Size Grounding Electrode Conductor . . . *or* . . . Equipment Grounding Conductor — **20** 6 copper / 4 aluminum

OPTIONAL METHOD: MULTI-FAMILY DWELLINGS

Optional Method Load Calculation for Multi-Family Dwellings

1 General Lighting and Receptacle Loads 220.32(C)(2) *Do not include open porches, garages, and unused or unfinished spaces not adaptable for future use.*	3 × ______ (sq ft outside dimensions) × ______ (number of units) =		1	
2 Small Appliance Branch-Circuits 220.2(C)(1) *At least **two** small appliance branch-circuits must be included. 210.11(C)(1)*	1500 × ______ (minimum of two) × ______ (number of units) =		2	
3 Laundry Branch-Circuit(s) 220.32(C)(1) *At least **one** laundry branch-circuit must be included. 210.11(C)(2)*	1500 × ______ (*minimum of two) × ______ (number of units) =		3	
4 through 11 Appliances and Motors 220.32(C)(3) and (4) *Use the nameplate rating of **ALL** appliances (fastened-in-place, permanently connected, or connected to a specific circuit), ranges, wall-mounted ovens, counter-mounted cooking units, motors, water heaters, and clothes dryers.* *Convert any nameplate rating given in amperes to volt-amperes by multiplying the amperes by the rated voltage.* ***Do not include any heating or air-conditioning equipment in this section.***	water heater / ______ (volt-amperes each) × ______ (number) =		4	
	dishwashers / ______ (volt-amperes each) × ______ (number) =		5	
	disposers / ______ (volt-amperes each) × ______ (number) =		6	
	clothes dryers / ______ (volt-amperes each) × ______ (number) =		7	
	ranges / ______ (volt-amperes each) × ______ (number) =		8	
	microwave ovens / ______ (volt-amperes each) × ______ (number) =		9	
	______ / ______ (volt-amperes each) × ______ (number) =		10	
	______ / ______ (volt-amperes each) × ______ (number) =		11	
12 Heating or Air-Conditioning System (Compare the heat and A/C, and omit the smaller.) 220.32(C)(5) *Include the air handler when using either one.* *For heat pumps, include the compressor and the maximum amount of electric heat that can be energized while the compressor is running.*	______ (volt-amperes each) × ______ (number) =		12	
13 Total Volt-Ampere Demand Load: *Multiply total VA by Table 220.32 demand factor percent.*	______ (total volt-amperes from lines 1 through 12) + ______ (Table 220.32 demand factor) =		13	
14 House Load *(If present, otherwise skip to line 15)* *Compute in accordance with Article 220, Part II.* *Do not include in Table 220.32 demand factors.*			14	
15 Minimum Amperes *Divide the total volt-amperes by the voltage*	______ (lines 13 and 14) ÷ ______ (voltage) = **15** ______ (minimum amperes)	**16** Minimum Size Service and/or Feeder 240.6(A)	16	
17 Size the Service and/or Feeder Conductors. *Use 310.15(B)(6) to find the service conductors up to 400 amperes.* *Ratings in excess of 400 amperes shall comply with Table 310.16.* *310.15(B)(6) also applies to feeder conductors serving as the main power feeder.*		**Minimum Size Conductors**	17	
18 Size the Neutral Conductor 220.22 **Note: There is no optional method for calculating the neutral conductor.** *310.15(B)(6) states that the neutral service and/or feeder conductor can be smaller than the ungrounded (hot) conductors, provided the requirements of 215.2, 220.22, and 230.42 are met.* *250.24(B)(1) states that the neutral cannot be smaller than the required grounding electrode conductor specified in Table 250.66.*		**Minimum Size Neutral Conductor**	18	
19 Size the Grounding Electrode Conductor Table 250.66 *Using line 17 to find the grounding electrode conductor in Table 250.66.* Size the Equipment Grounding Conductor (for Feeder) 250.122 *Use line 16 to find the equipment grounding conductor in Table 250.122.* *Equipment grounding conductor types are listed in 250.118.*		**Minimum Size Grounding Electrode Conductor . . . *or* . . . Equipment Grounding Conductor**	19	

Lines 1 Through 3—General Lighting and Receptacle Load

A In the **Optional Method**, Lines 1, 2, and 3 are calculated exactly as described in the **Standard Method**. (See explanations on page 211.)

B The feeder or service multi-family dwelling load can be computed in accordance with Table 220.32 (instead of Article 220, Part II) where all of the following conditions are met: (1) no dwelling unit is supplied by more than one feeder; (2) each dwelling unit is equipped with electric cooking equipment; and (3) each dwelling unit is equipped with either electric space-heating or air-conditioning, or both »220.32(A)«.

C This optional load calculation is computed on the same 12-unit multi-family dwelling used for the standard method (located in the first part of this unit).

NOTE

When the computed load for multi-family dwellings, without electric cooking, in Article 220, Part II exceeds that computed under Part III for the identical load, plus electric cooking (based on 8 kW per unit), use the lesser of the two loads »220.32(A)(2) Exception«.

Optional Method Load Calculation for Multi-Family Dwellings

1	General Lighting and Receptacle Loads 220.32(C)(2) *Do not include open porches, garages, and unused or unfinished spaces not adaptable for future use.*	3 × 1050 (sq ft outside dimensions)	× 12 (number of units)	=	1	37,800
2	Small Appliance Branch-Circuits 220.32(C)(1) *At least **two** small appliance branch circuits must be included. 210.11(C)(1)*	1500 × 2 (minimum of two)	× 12 (number of units)	=	2	36,000
3	Laundry Branch-Circuit(s) 220.32(C)(1) *At least **one** laundry branch-circuit must be included. 210.11(C)(2)*	1500 × 1 (*minimum of two)	× 12 (number of units)	=	3	18,000

Lines 4 through 11—All Appliances

A This Section includes appliances (fastened-in-place, permanently connected, or connected to a specific circuit), ranges, wall-mounted ovens, counter-mounted cooking units, clothes dryers, and water heaters »220.32(C)(3)«.

B Low-power-factor loads and motors (except air-conditioning motors), are also included »220.32(C)(4)«.

C Use the exact clothes dryer nameplate rating, if any, even if less than 5000 volt-amperes. Do not apply the demand factors in Table 220.19 to household cooking equipment (ranges, cooktops, ovens, etc.); Instead, list the nameplate ratings as they appear.

D Insert the number of appliances (such as water heaters) having the same rating, which may not necessarily be the number of dwelling units. Here, all dwelling units contain identical equipment, so the number of each type of appliance is the same as the number of dwelling units. In another application, if only half of the units are equipped with laundry hook-ups, the number of units is not the same as the number of a given type of appliance.

E Heating and air-conditioning systems are *not* included in this computation.

F Additional lines have been provided for listing appliances.

G Inserting numbers in all of the lines is not necessary. However, it is advisable to insert dashes, or some other indicator, to show that the line is intentionally blank.

NOTE

If water-heater elements are interlocked so that all elements cannot be activated simultaneously, the nameplate load is considered the maximum possible load »220.32(C)(3)«.

A B C D

4 through 11 Appliances and Motors 220.32(C)(3) and (4)	Appliance	(volt-amperes each)	× (number)	Line	
*Use the nameplate rating of **ALL** appliances (fastened-in-place, permanently connected, or connected to a specific circuit), ranges, wall-mounted ovens, counter-mounted cooking units, motors, water heaters, and clothes dryers.*	water heaters	4500	12	4	54,000
	dishwashers	1200	12	5	14,400
Convert any nameplate rating given in amperes to volt-amperes by multiplying the amperes by the rated voltage.	disposers	900	12	6	10,800
	clothes dryers	5000	12	7	60,000
Do not include any heating or air-conditioning equipment in this section.	ranges	12,000	12	8	144,000
	microwave ovens	1200	12	9	14,400
				10	----
				11	----

E F G

Line 12—Heating or Air-Conditioning System

A Include only the larger of the heating or air-conditioning loads (except for heat pumps).

B Electric heating loads (including associated motors) are computed at 100% of the total connected load. Because the air handler (blower) works with either the heating or the air conditioning, add it to both calculations.

C Air-conditioning loads (including associated motors) are calculated at 100% of the total connected load. No derating is allowed. Each unit contains an air-conditioner compressor (14 × 230 = 3220 VA), condenser fan motor (2 × 115 = 230 VA), and blower motor (3 × 115 = 345 VA). The combined load is 3795 volt-amperes. Because this is less than the heating load, omit the A/C compressor and the condenser fan motor.

D The air handler (blower motor) has a rating of 345 volt-amperes (3 × 115 = 345). Since both loads are energized simultaneously, add the air handler to the 5 kW of electric heat (345 + 5000 = 5345 volt-amperes).

> **NOTE**
>
> *A heat pump with supplementary heat is not considered a noncoincident load. Add the compressor full-load current to the maximum amount of heat that can be energized while the compressor is running. Remember to include all associated motor(s), i.e., blower, condenser, etc.*

A B C

12 Heating or Air-Conditioning System (Compare the heat and A/C, and omit the smaller.) 220.32(C)(5)
Include the air handler when using either one.
For heat pumps, include the compressor and the maximum amount of electric heat that can be energized while the compressor is running.

5345 (volt-amperes each) × 12 (number) = **12** 64,140

D

Line 13—Total Volt-Ampere Demand Load

A Adding Lines 1 through 12 results in the volt-ampere load before derating.

B Table 220.32 contains demand factors (percentages) for derating three (or more) multi-family dwelling units. Find the number of dwelling units in the first column and follow that row across to the demand factor. The demand factors listed in the second column are percentages. The demand factor for a 12-unit multi-family dwelling is 41 (41%).

C Total computed load excluding house loads (if any).

A B C

13 Total Volt-Ampere Demand Load:
Multiply total VA by Table 220.32 demand factor percent.

453,540 (total volt-amperes from lines 1 through 12) + 41% (Table 220.32 demand factor) = **13** 185,951

Line 14—House Load

A House loads must be computed as stipulated in Article 220, Part II, and must be in addition to the dwelling-unit loads computed in accordance with Table 220.32 »220.32(B)«.

B House loads include areas such as: clubhouse, office, hallway and stairway lighting, parking lot lighting, etc.

C Insert a line (or dashes) in spaces left empty intentionally.

A B

14 House Load *(If present, otherwise skip to line 15)*
Compute in accordance with Article 220, Part II.
Do not include in Table 220.32 demand factors.

14 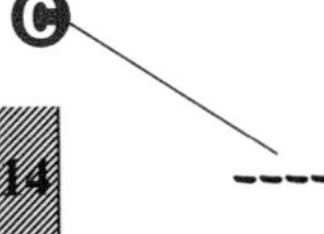

Lines 15 and 16—Minimum Service and/or Feeder

A Place the sum of Lines 13 and 14 (total volt-amperes) here. If no house load, insert only Line 13.

B Fill in the source voltage supplying the service equipment.

C Fractions of an ampere 0.5 and greater are rounded up, while fractions less than 0.5 are dropped (185,951 ÷ 240 = 774.8 = 775) »220.2(B)«. The minimum ampere rating required for conductors and overcurrent protection, whether for the service or feeder, is 775.

D The service (or feeder) overcurrent protection must be higher than the number found in Line 15. 240.6(A) lists standard ampere ratings for fuses and circuit breakers.

E The next standard size fuse or breaker above 775 amperes is 800 »240.6(A)«.

15 Minimum Amperes *Divide the total volt-amperes by the voltage*	185,951 (lines 13 and 14)	÷	240 (voltage)	=	15	775 (minimum amperes)	16 Minimum Size Service and/or Feeder 240.6(A)	16	800

Line 17—Minimum Size Conductors

A The total combined rating of paralleled conductors must meet, or exceed, the minimum calculated load.

B At 75°C, the minimum size copper conductors, for three parallel sets, matching or exceeding 775 amperes, are 300 kcmil. Although three paralleled sets of 250-kcmil copper conductors (at 75°C) are sufficient for an 800-ampere overcurrent protective device, they are not sufficient for the 775-ampere calculated load. (Three times 255 is only 765 amperes, short of the necessary 775 rating.)

17 Size the Service and/or Feeder Conductors. *Use 310.15(B)(6) to find the service conductors up to 400 amperes.* *Ratings in excess of 400 amperes shall comply with Table 310.16.* *310.15(B)(6) also applies to feeder conductors serving as the main power feeder.*	Minimum Size Conductors	17	300 kcmil Cu 400 kcmil Al three paralleled sets

Minimum Size Neutral Conductors

Ⓐ There is no optional method for neutral conductor calculation. This neutral calculation is performed in accordance with Article 220, Part II.

Ⓑ Lines 11 and 12 from the Standard Method Load Calculation form for Multi-family Dwellings.

Ⓒ The minimum computed volt-ampere load is 108,705.

Ⓓ The minimum neutral ampacity is 453.

Ⓔ The minimum rating for three paralleled sets of neutral conductors is 151 amperes each.

Ⓕ Line 9 on the Multi-family Dwelling Standard Method Load Calculation form.

Ⓖ At 75°C, the minimum size copper conductor is 2/0 AWG, with an individual rating of 175 amperes »Table 310.16«.

Ⓗ At 75°C, the minimum size aluminum conductor, if determined by ampacity alone, would be 3/0 AWG, with a rating of 155 amperes each. However, 3/0 AWG would not be acceptable because the neutral cannot be smaller than the required grounding electrode conductor, which in this example is 4/0 AWG »250.24(B)(1)«.

General lighting and receptacles (Line 9) . . . . 34,080 Ⓕ
Fastened-in-place appliances Ⓐ
- **Water heaters 0**
- **Dishwashers (1200 × 12). 14,400**
- **Waste disposers (7.5 × 120 × 12) 10,800**
- **Microwave ovens (1200 × 12). . . 14,400**
- **Total 39,600**

Total fastened-in-place appliances (39,600 × 75%) 29,700
Clothes dryers (Line 11 × 70%) . . . . 18,900 Ⓑ
Ranges (Line 12 × 70%) . . . . 18,900 Ⓑ
Electric heat 0
Air-conditioner compressors. 0
Air handlers (3 × 115 × 12). 4140
Condenser fan motors (2 × 115 × 12) 2760
Largest motor (900 × 25%) 225
TOTAL . . . . 108,705 Ⓒ

108,705 ÷ 240 = 452.9 = 453 minimum neutral ampacity Ⓓ
453 ÷ 3 = 151 minimum amperes per neutral conductor Ⓔ

18 Size the Neutral Conductor 220.22 **Note: There is no optional method for calculating the neutral conductor.** *310.15(B)(6) states that the neutral service and/or feeder conductor can be smaller than the ungrounded (hot) conductors, provided the requirements of 215.2, 220.22, and 230.42 are met.* *250.24(B)(1) states that the neutral cannot be smaller than the required grounding electrode conductor specified in Table 250.66.*	**Minimum Size Neutral Conductor**	**18**	Ⓖ 2/0 Cu Ⓗ 4/0 Al three paralleled sets

Line 19—Grounding Electrode Conductor

Ⓐ The grounding electrode conductor size is determined by the largest service-entrance conductor (or the equivalent area) if the service-entrance conductors are paralleled. The equivalent area for three parallel sets in this calculation is 900-kcmil copper (3 × 300) *or* 1200-kcmil aluminum (3 × 400).

Ⓑ The minimum size grounding electrode conductor is 2/0 AWG copper *or* 4/0 AWG aluminum »Table 250.66«.

19 Size the Grounding Electrode Conductor Table 250.66 *Using line 17 to find the grounding electrode conductor in Table 250.66.* Size the Equipment Grounding Conductor (for Feeder) 250.122 *Use line 16 to find the equipment grounding conductor in Table 250.122.* *Equipment grounding conductor types are listed in 250.118.*	Ⓐ **Minimum Size Grounding Electrode Conductor . . . *or* . . . Equipment Grounding Conductor**	**19**	Ⓑ 2/0 Cu 4/0 Al

The following page shows the complete optional method for a twelve-unit, multi-family dwelling, having all units combined into a single service.

Optional Method Load Calculation for Multi-Family Dwellings

1 General Lighting and Receptacle Loads 220.32(C)(2) *Do not include open porches, garages, and unused or unfinished spaces not adaptable for future use.*	3 × 1050 (sq ft outside dimensions) × 12 (number of units) =	1	37,800
2 Small Appliance Branch-Circuits 220.32(C)(1) *At least **two** small appliance branch circuits must be included. 210.11(C)(1)*	1500 × 2 (minimum of two) × 12 (number of units) =	2	36,000
3 Laundry Branch-Circuit(s) 220.32(C)(1) *At least **one** laundry branch-circuit must be included. 210.11(C)(2)*	1500 × 1 (*minimum of two) × 12 (number of units) =	3	18,000
4 through 11 Appliances and Motors 220.32(C)(3) and (4) *Use the nameplate rating of **ALL** appliances (fastened-in-place, permanently connected, or connected to a specific circuit), ranges, wall-mounted ovens, counter-mounted cooking units, motors, water heaters, and clothes dryers.* *Convert any nameplate rating given in amperes to volt-amperes by multiplying the amperes by the rated voltage.* ***Do not include any heating or air-conditioning equipment in this section.***	water heaters / 4500 (volt-amperes each) × 12 (number) =	4	54,000
	dishwashers / 1200 (volt-amperes each) × 12 (number) =	5	14,400
	disposers / 900 (volt-amperes each) × 12 (number) =	6	10,800
	clothes dryers / 5000 (volt-amperes each) × 12 (number) =	7	60,000
	ranges / 12,000 (volt-amperes each) × 12 (number) =	8	144,000
	microwave ovens / 1200 (volt-amperes each) × 12 (number) =	9	14,400
	/ (volt-amperes each) × (number) =	10	----
	/ (volt-amperes each) × (number) =	11	----
12 Heating or Air-Conditioning System (Compare the heat and A/C, and omit the smaller.) 220.32(C)(5) *Include the air handler when using either one.* *For heat pumps, include the compressor and the maximum amount of electric heat that can be energized while the compressor is running.*	5345 (volt-amperes each) × 12 (number) =	12	64,140
13 Total Volt-Ampere Demand Load: *Multiply total VA by Table 220.32 demand factor percent.*	453,540 (total volt-amperes from lines 1 through 12) + 41% (Table 220.32 demand factor) =	13	185,951
14 House Load *(If present, otherwise skip to line 15)* *Compute in accordance with Article 220, Part II.* *Do not include in Table 220.32 demand factors.*		14	----
15 Minimum Amperes *Divide the total volt-amperes by the voltage*	185,951 (lines 13 and 14) ÷ 240 (voltage) = **15** 775 (minimum amperes) **16 Minimum Size Service and/or Feeder 240.6(A)**	16	800
17 Size the Service and/or Feeder Conductors. *Use 310.15(B)(6) to find the service conductors up to 400 amperes.* *Ratings in excess of 400 amperes shall comply with Table 310.16.* *310.15(B)(6) also applies to feeder conductors serving as the main power feeder.*	**Minimum Size Conductors**	17	300 kcmil Cu 400 kcmil Al three paralleled sets
18 Size the Neutral Conductor 220.22 **Note: There is no optional method for calculating the neutral conductor.** *310.15(B)(6) states that the neutral service and/or feeder conductor can be smaller than the ungrounded (hot) conductors, provided the requirements of 215.2, 220.22, and 230.42 are met.* *250.24(B)(1) states that the neutral cannot be smaller than the required grounding electrode conductor specified in Table 250.66.*	**Minimum Size Neutral Conductor**	18	2/0 Cu 4/0 Al three paralleled sets
19 Size the Grounding Electrode Conductor Table 250.66 *Using line 17 to find the grounding electrode conductor in Table 250.66.* Size the Equipment Grounding Conductor (for Feeder) 250.122 *Use line 16 to find the equipment grounding conductor in Table 250.122.* *Equipment grounding conductor types are listed in 250.118.*	**Minimum Size Grounding Electrode Conductor . . . *or* . . . Equipment Grounding Conductor**	19	2/0 Cu 4/0 Al

SIX-UNIT MULTI-FAMILY DWELLING CALCULATION (OPTIONAL METHOD)

Lines 1 Through 12

A This optional method calculation is based on six units instead of twelve.

B Since everything is computed at 100%, Lines 1 through 12 are exactly half of that calculated for twelve units.

Optional Method Load Calculation for Multi-Family Dwellings

Item	Description	Calculation	Line	Result
1	General Lighting and Receptacle Loads 220.32(C)(2) *Do not include open porches, garages, and unused or unfinished spaces not adaptable for future use.*	3 × 1050 (sq ft outside dimensions) × 6 (number of units) =	1	18,900
2	Small Appliance Branch-Circuits 220.32(C)(1) *At least **two** small appliance branch-circuits must be included. 210.11(C)(1)*	1500 × 2 (minimum of two) × 6 (number of units) =	2	18,000
3	Laundry Branch-Circuit(s) 220.32(C)(1) *At least **one** laundry branch-circuit must be included. 210.11(C)(2)*	1500 × 1 (*minimum of two) × 6 (number of units) =	3	9000
4 through 11	Appliances and Motors 220.32(C)(3) and (4)	water heaters / 4500 (volt-amperes each) × 6 (number) =	4	27,000
	*Use the nameplate rating of **ALL** appliances (fastened-in-place, permanently connected, or connected to a specific circuit), ranges, wall-mounted ovens, counter-mounted cooking units, motors, water heaters, and clothes dryers.*	dishwashers / 1200 (volt-amperes each) × 6 (number) =	5	7200
		disposers / 900 (volt-amperes each) × 6 (number) =	6	5400
		clothes dryers / 5000 (volt-amperes each) × 6 (number) =	7	30,000
	Convert any nameplate rating given in amperes to volt-amperes by multiplying the amperes by the rated voltage.	ranges / 12,000 (volt-amperes each) × 6 (number) =	8	72,000
	Do not include any heating or air-conditioning equipment in this section.	microwave ovens / 1200 (volt-amperes each) × 6 (number) =	9	7200
		/ (volt-amperes each) × (number) =	10	----
		/ (volt-amperes each) × (number) =	11	----
12	Heating or Air-Conditioning System (Compare the heat and A/C, and omit the smaller.) 220.32(C)(5) *Include the air handler when using either one. For heat pumps, include the compressor and the maximum amount of electric heat that can be energized while the compressor is running.*	5345 (volt-amperes each) × 6 (number) =	12	32,070

Line 13—Total Volt-Ampere Demand Load

A Add Lines 1 through 12 to find the volt-ampere load before derating.

B Find the demand factor, located in Table 220.32, and insert here. The demand factor for a six-unit multi-family dwelling is 44 (44%).

Item	Description	Calculation	Line	Result
13	Total Volt-Ampere Demand Load: *Multiply total VA by Table 220.32 demand factor percent.*	226,770 (total volt-amperes from lines 1 through 12) + 44% (Table 220.32 demand factor) =	13	99,779

Lines 15 and 16—Minimum Service and/or Feeder

A Place the sum of Lines 13 and 14 (total volt-amperes) here. If no house load, insert only Line 13.

B Insert the source voltage supplying the service equipment.

C The minimum amperage rating required for conductors and overcurrent protection is 416 (99,779 ÷ 240 = 415.75 = 416).

D The minimum service (or feeder), determined by the optional method, for these six units is 450 amperes.

E The next standard size fuse or breaker above 416 amperes is 450 »240.6(A)«.

> **NOTE**
>
> *The same six units, calculated by the standard method, requires a rating of 600 amperes.*

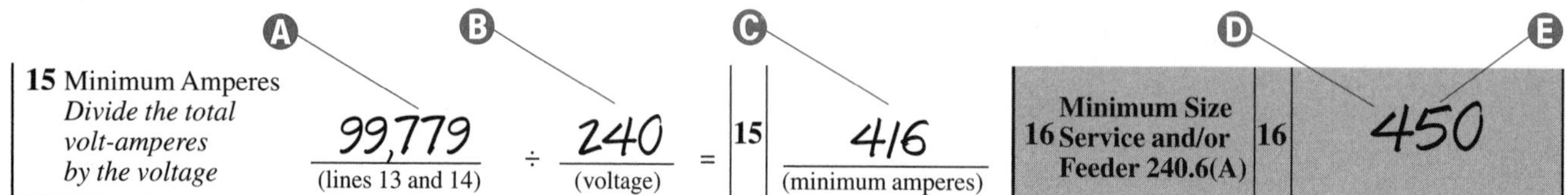

15 Minimum Amperes *Divide the total volt-amperes by the voltage*	99,779 (lines 13 and 14) ÷ 240 (voltage) =	15	416 (minimum amperes)	**Minimum Size 16 Service and/or Feeder 240.6(A)**	16	450

Line 17—Minimum Size Conductors

A Ensure that the total combined rating of paralleled conductors meets, or exceeds, the minimum calculated load.

B At 75°C, the minimum size conductors, for two parallel sets, matching or exceeding 416 amperes, are 4/0 AWG copper or 300-kcmil aluminum. The minimum rating for each conductor is 208 amperes.

17 Size the Service and/or Feeder Conductors. *Use 310.15(B)(6) to find the service conductors up to 400 amperes. Ratings in excess of 400 amperes shall comply with Table 310.16. 310.15(B)(6) also applies to feeder conductors serving as the main power feeder.*	**Minimum Size Conductors**	17	4/0 Cu 300 kcmil Al two paralleled sets

Minimum Size Neutral and Grounding Conductors

A Since there is no optional method for the neutral conductor calculation, refer to page 220, which gives the neutral for six units. The result requires a minimum rating of 275 amperes, or 138 amperes each for two paralleled conductors.

B At 75°C, the minimum size paralleled conductors are 1/0 AWG copper (with a rating of 175 amperes) *or* 3/0 AWG aluminum (with a rating of 155 amperes) »Table 310.16«.

C The equivalent area for two paralleled conductors in this calculation is 423.2-kcmil copper (2 × 211,600 ÷ 1000) *or* 600-kcmil aluminum (2 × 300). The minimum size grounding electrode conductor is 1/0 AWG copper *or* 3/0 AWG aluminum »Table 250.66«.

18 Size the Neutral Conductor 220.22 **Note: There is no optional method for calculating the neutral conductor.** *310.15(B)(6) states that the neutral service and/or feeder conductor can be smaller than the ungrounded (hot) conductors, provided the requirements of 215.2, 220.22, and 230.42 are met. 250.24(B)(1) states that the neutral cannot be smaller than the required grounding electrode conductor specified in Table 250.66.*	**Minimum Size Neutral Conductor**	18	1/0 Cu 3/0 Al two paralleled sets
19 Size the Grounding Electrode Conductor Table 250.66 *Using line 17 to find the grounding electrode conductor in Table 250.66.* Size the Equipment Grounding Conductor (for Feeder) 250.122 *Use line 16 to find the equipment grounding conductor in Table 250.122. Equipment grounding conductor types are listed in 250.118.*	**Minimum Size Grounding Electrode Conductor . . . or . . . Equipment Grounding Conductor**	19	1/0 Cu 3/0 Al

The following page shows the complete optional method for a six-unit, multi-family dwelling load calculation.

Optional Method Load Calculation for Multi-Family Dwellings

Line	Item	Calculation		Result
1	General Lighting and Receptacle Loads 220.32(C)(2) *Do not include open porches, garages, and unused or unfinished spaces not adaptable for future use.*	3 × 1050 (sq ft outside dimensions) × 6 (number of units) =	1	18,900
2	Small Appliance Branch-Circuits 220.32(C)(1) *At least **two** small appliance branch-circuits must be included. 210.11(C)(1)*	1500 × 2 (minimum of two) × 6 (number of units) =	2	18,000
3	Laundry Branch-Circuit(s) 220.32(C)(1) *At least **one** laundry branch-circuit must be included. 210.11(C)(2)*	1500 × 1 (*minimum of two) × 6 (number of units) =	3	9000
4 through 11	Appliances and Motors 220.32(C)(3) and (4)	water heaters / 4500 (volt-amperes each) × 6 (number) =	4	27,000
	*Use the nameplate rating of **ALL** appliances (fastened-in-place, permanently connected, or connected to a specific circuit), ranges, wall-mounted ovens, counter-mounted cooking units, motors, water heaters, and clothes dryers.*	dishwashers / 1200 (volt-amperes each) × 6 (number) =	5	7200
		disposers / 900 (volt-amperes each) × 6 (number) =	6	5400
		clothes dryers / 5000 (volt-amperes each) × 6 (number) =	7	30,000
	Convert any nameplate rating given in amperes to volt-amperes by multiplying the amperes by the rated voltage.	ranges / 12,000 (volt-amperes each) × 6 (number) =	8	72,000
	Do not include any heating or air-conditioning equipment in this section.	microwave ovens / 1200 (volt-amperes each) × 6 (number) =	9	7200
		/ (volt-amperes each) × (number) =	10	----
		/ (volt-amperes each) × (number) =	11	----
12	Heating or Air-Conditioning System (Compare the heat and A/C, and omit the smaller.) 220.32(C)(5) *Include the air handler when using either one. For heat pumps, include the compressor and the maximum amount of electric heat that can be energized while the compressor is running.*	5345 (volt-amperes each) × 6 (number) =	12	32,070
13	Total Volt-Ampere Demand Load: *Multiply total VA by Table 220.32 demand factor percent.*	226,770 (total volt-amperes from lines 1 through 12) + 44% (Table 220.32 demand factor) =	13	99,779
14	House Load *(If present, otherwise skip to line 15) Compute in accordance with Article 220, Part II. Do not include in Table 220.32 demand factors.*		14	----
15	Minimum Amperes *Divide the total volt-amperes by the voltage*	99,779 (lines 13 and 14) ÷ 240 (voltage) = 15: 416 (minimum amperes)		
16	**Minimum Size Service and/or Feeder 240.6(A)**		16	450
17	Size the Service and/or Feeder Conductors. *Use 310.15(B)(6) to find the service conductors up to 400 amperes. Ratings in excess of 400 amperes shall comply with Table 310.16. 310.15(B)(6) also applies to feeder conductors serving as the main power feeder.*	**Minimum Size Conductors**	17	4/0 Cu 300 kcmil Al two paralleled sets
18	Size the Neutral Conductor 220.22 **Note: There is no optional method for calculating the neutral conductor.** *310.15(B)(6) states that the neutral service and/or feeder conductor can be smaller than the ungrounded (hot) conductors, provided the requirements of 215.2, 220.22, and 230.42 are met. 250.24(B)(1) states that the neutral cannot be smaller than the required grounding electrode conductor specified in Table 250.66.*	**Minimum Size Neutral Conductor**	18	1/0 Cu 3/0 Al two paralleled sets
19	Size the Grounding Electrode Conductor Table 250.66 *Using line 17 to find the grounding electrode conductor in Table 250.66.* Size the Equipment Grounding Conductor (for Feeder) 250.122 *Use line 16 to find the equipment grounding conductor in Table 250.122. Equipment grounding conductor types are listed in 250.118.*	**Minimum Size Grounding Electrode Conductor . . . or . . . Equipment Grounding Conductor**	19	1/0 Cu 3/0 Al

OPTIONAL LOAD CALCULATION FOR EACH MULTI-FAMILY DWELLING UNIT

Optional Method—One Unit

A The first three lines of both methods (optional and standard) of one-family dwelling calculations are identical.

B Although located in a multi-family dwelling, this unit contains one feeder and one panelboard. Therefore, it is computed in accordance with a one-family dwelling method.

C Under certain conditions, a laundry branch-circuit is optional in multi-family dwellings »210.52(F)«.

B

Optional Method Load Calculation for One-Family Dwellings

1	General Lighting and Receptacle Loads 220.30(B)(2) *Do not include open porches, garages, and unused or unfinished spaces not adaptable for future use.*	3 × 1050 (sq ft using outside dimensions)	= 1	3150
2	Small Appliance Branch-Circuits 220.30(B)(1) *Include at least **two** small appliance branch-circuits. 210.11(C)(1)*	1500 × 2 (minimum of two)	= 2	3000
3	Laundry Branch-Circuit(s) 220.30(B)(1) *Include at least **one** laundry branch circuit. 210.11(C)(2)*	1500 × 1 (minimum of one)	= 3	1500

C

Line 4—Appliances

A This Section includes appliances (fastened-in-place, permanently connected, or connected to a specific circuit), ranges, wall-mounted ovens, counter-mounted cooking units, clothes dryers, and water heaters »220.30(B)(3)«.

B Low-power-factor loads and motors (except air-conditioning motors) are also found in this Section »220.30(B)(4)«.

C Heating and air-conditioning systems are **not** part of this computation.

D Use the exact nameplate rating of the clothes dryer, even if less than 5000 volt-amperes.

> **NOTE**
>
> *Do not apply the demand factors in Table 220.19 to household cooking equipment (ranges, cook tops, and ovens). Instead, list the nameplate ratings as they appear.*
>
> *Do not increase motor loads by 25% »220.30(B)(4)«.*

A B C D

4	Appliances 220.30(B)(3) and (4) *Use the nameplate rating of **ALL** appliances (fastened-in-place, permanently connected, or connected to a specific circuit), ranges, ovens, cook tops, motors, and clothes dryers. Convert any nameplate rating given in amperes to volt-amperes by multiplying the amperes by the rated voltage.*	*Do not include any heating or air-conditioning equipment in this section.*	Total volt-amperes of all appliances LISTED BELOW	4 24,800
		water heater / 4500	clothes dryer / 5500	range / 12,000
		dishwasher / 1200	disposer / 900	microwave oven / 1200
		/	/	/
		/	/	/
		/	/	/

Line 5—Applying Demand Factors to Lines 1 through 4

A This demand factor is found in the last two lines of Table 220.30.

B The newly calculated load includes everything except the heating and air-conditioning systems. (3152 + 3000 + 1500 + 24,800)

C Because the first 10,000 volt-amperes are calculated at 100%, 10,000 is subtracted here.

D The remainder of the load is calculated at 40%.

E The result on Line 5 will be part of Line 7's computation.

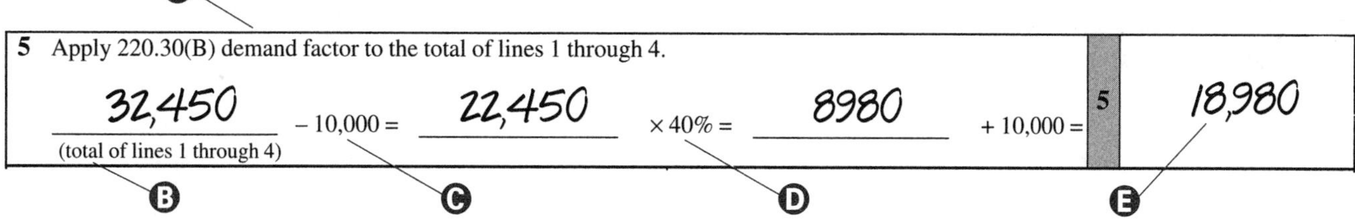

5 Apply 220.30(B) demand factor to the total of lines 1 through 4.

32,450 (total of lines 1 through 4) – 10,000 = 22,450 × 40% = 8980 + 10,000 = | 5 | 18,980

Line 6—Heating and/or Air-Conditioning Systems

A The air-conditioner compressor (14 × 230 = 3220 VA), condenser fan motor (2 × 115 = 230 VA), and blower motor (3 × 115 = 345 VA) have a combined load of 3795 volt-amperes. Calculate the air-conditioning system(s) at 100%.

B The electric heat (5000) added to the blower motor (3 × 115 = 345) is 5345 volt-amperes. The heating system is multiplied by 65% and the result placed on Line 6c.

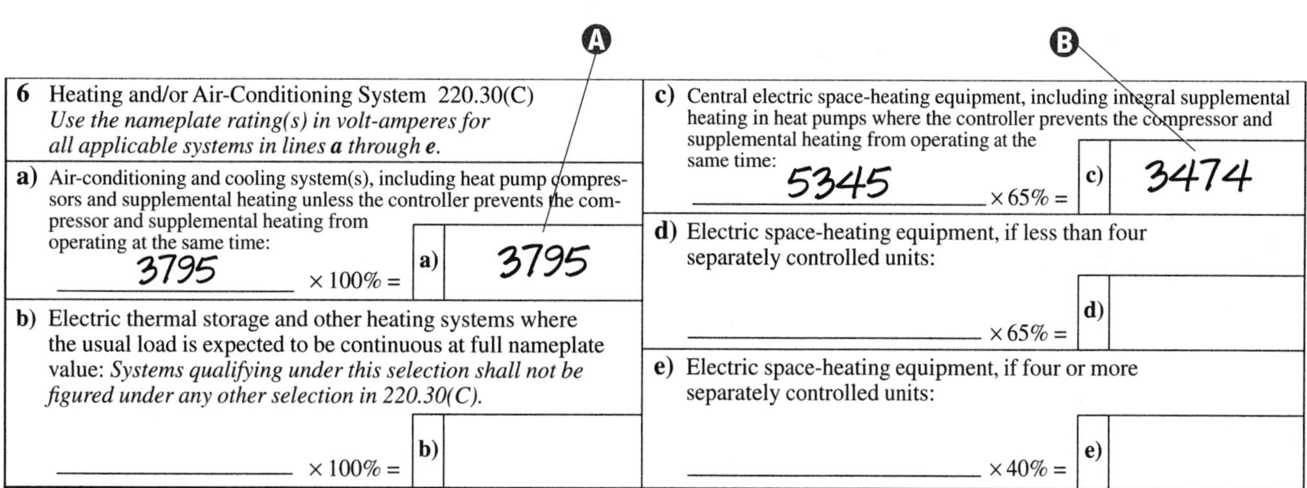

6 Heating and/or Air-Conditioning System 220.30(C)
*Use the nameplate rating(s) in volt-amperes for all applicable systems in lines **a** through **e**.*

a) Air-conditioning and cooling system(s), including heat pump compressors and supplemental heating unless the controller prevents the compressor and supplemental heating from operating at the same time: 3795 × 100% = | a) | 3795

b) Electric thermal storage and other heating systems where the usual load is expected to be continuous at full nameplate value: *Systems qualifying under this selection shall not be figured under any other selection in 220.30(C).* ______ × 100% = | b) |

c) Central electric space-heating equipment, including integral supplemental heating in heat pumps where the controller prevents the compressor and supplemental heating from operating at the same time: 5345 × 65% = | c) | 3474

d) Electric space-heating equipment, if less than four separately controlled units: ______ × 65% = | d) |

e) Electric space-heating equipment, if four or more separately controlled units: ______ × 40% = | e) |

Line 7—Total Volt-Ampere Demand Load

A The largest volt-ampere rating of the heating and/or air-conditioning system(s), after application of demand factors.

B The volt-ampere demand load from Line 5.

C The total volt-ampere load calculated by the optional method. (For comparison, this one-family dwelling, calculated by the standard method, is 29,048 volt-amperes.)

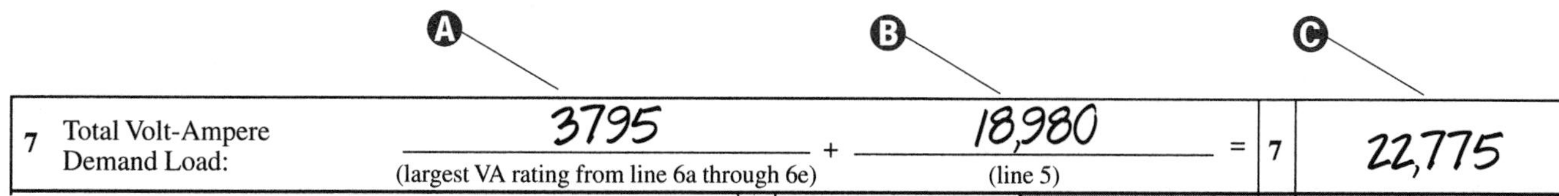

7 Total Volt-Ampere Demand Load: 3795 (largest VA rating from line 6a through 6e) + 18,980 (line 5) = | 7 | 22,775

Lines 8 And 9—Minimum Feeder

A The volt-ampere load from Line 7.

B The source voltage.

C Since the fraction is more than .5, the number 94.9 is rounded up to 95.

D The overcurrent protection chosen for this feeder must be higher than the number found on Line 8. 240.6(A) lists standard ampere ratings for fuses and circuit breakers.

E This dwelling unit, calculated in accordance with the optional method, requires a 100-ampere feeder and panelboard.

F A 100-ampere minimum restriction has been placed on this optional method for both services and feeders »220.30(A)«. Feeders computed in accordance with Article 220, Part II, are not restricted to a minimum rating of 100 amperes.

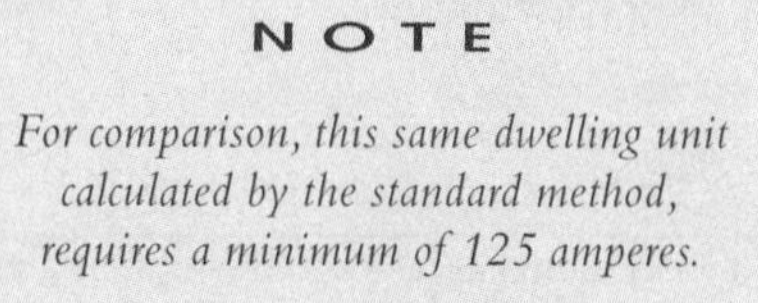

NOTE

For comparison, this same dwelling unit calculated by the standard method, requires a minimum of 125 amperes.

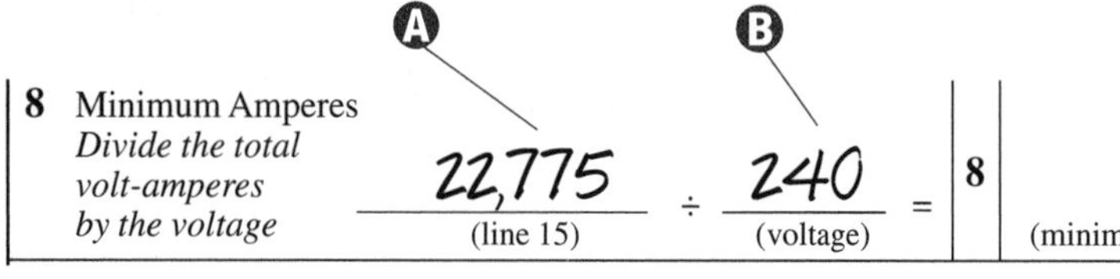

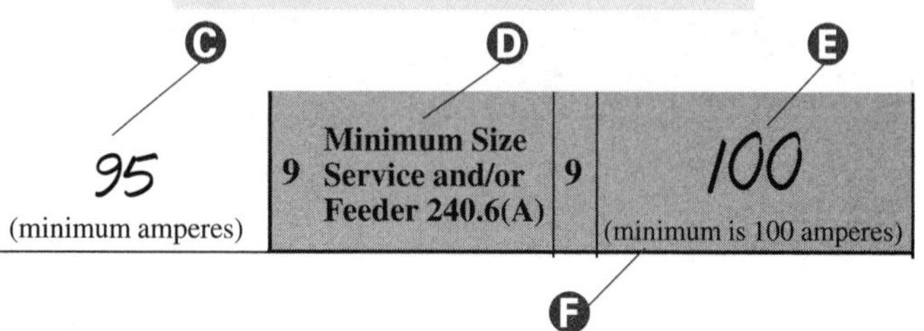

Line 10—Minimum Size Conductors

A Use 310.15(B)(6) to select feeder conductors serving as the main power feeder.

B The minimum size conductors, for this dwelling unit, are 4 AWG copper *or* 2 AWG aluminum »Table 310.15(B)(6)«.

10 Size the Service and/or Feeder Conductors. *Use 310.15(B)(6) to find the service conductors up to 400 amperes.* *Ratings in excess of 400 amperes shall comply with Table 310.16.* *310.15(B)(6) also applies to feeder conductors serving as the main power feeder.*	**Minimum Size Conductors**	10	4 copper 2 aluminum

Minimum Size Neutral and Equipment Grounding Conductors

A Since there is no optional method for neutral calculation, refer to page 224 for a detailed analysis.

B At 75°C, the minimum size conductor is 4 AWG copper *or* 3 AWG aluminum.

C Because these are feeder, not service, conductors, use Table 250.122 to size the equipment grounding conductor.

D The minimum size equipment grounding conductor for a 100-ampere overcurrent protective device is 8 AWG copper *or* 6 AWG aluminum »Table 250.122«.

11 Size the Neutral Conductor 220.22 **Note: There is no optional method for calculating the neutral conductor.** *310.15(B)(6) states that the neutral service and/or feeder conductor can be smaller than the ungrounded (hot) conductors, provided the requirements of 215.2, 220.22, and 230.42 are met.* *250.24(B)(1) states that the neutral cannot be smaller than the required grounding electrode conductor specified in Table 250.66.*	**Minimum Size Neutral Conductor**	11	4 copper 3 aluminum
12 Size the Grounding Electrode Conductor Table 250.66 *Using line 10 to find the grounding electrode conductor in Table 250.66.* Size the Equipment Grounding Conductor (for Feeder) 250.122 *Use line 9 to find the equipment grounding conductor in Table 250.122.* *Equipment grounding conductor types are listed in 250.118.*	**Minimum Size Grounding Electrode Conductor . . . *or* . . . Equipment Grounding Conductor**	12	8 copper 6 aluminum

The following page shows the complete optional method one-family dwelling load calculation for one multi-family dwelling-unit feeder and remote panelboard (subpanel).

Optional Method Load Calculation for One-Family Dwellings

1	General Lighting and Receptacle Loads 220.30(B)(2) *Do not include open porches, garages, and unused or unfinished spaces not adaptable for future use.*	3 × 1050 (sq ft using outside dimensions) =	1	3150
2	Small Appliance Branch-Circuits 220.30(B)(1) *Include at least **two** small appliance branch-circuits. 210.11(C)(1)*	1500 × 2 (minimum of two) =	2	3000
3	Laundry Branch-Circuit(s) 220.30(B)(1) *Include at least **one** laundry branch circuit. 210.11(C)(2)*	1500 × 1 (minimum of one) =	3	1500
4	Appliances 220.30(B)(3) and (4) *Use the nameplate rating of **ALL** appliances (fastened-in-place, permanently connected, or connected to a specific circuit), ranges, ovens, cook tops, motors, and clothes dryers.* *Convert any nameplate rating given in amperes to volt-amperes by multiplying the amperes by the rated voltage.*	*Do not include any heating or air-conditioning equipment in this section.* Total volt-amperes of all appliances LISTED BELOW water heater / 4500; clothes dryer / 5500; range / 12,000 dishwasher / 1200; disposer / 900; microwave oven / 1200	4	24,800
5	Apply 220.30(B) demand factor to the total of lines 1 through 4.	32,450 (total of lines 1 through 4) − 10,000 = 22,450 × 40% = 8980 + 10,000 =	5	18,980

6 Heating and/or Air-Conditioning System 220.30(C)
*Use the nameplate rating(s) in volt-amperes for all applicable systems in lines **a** through **e**.*

a)	Air-conditioning and cooling system(s), including heat pump compressors and supplemental heating unless the controller prevents the compressor and supplemental heating from operating at the same time: 3795 × 100% =	a)	3795
b)	Electric thermal storage and other heating systems where the usual load is expected to be continuous at full nameplate value: *Systems qualifying under this selection shall not be figured under any other selection in 220.30(C).* ______ × 100% =	b)	
c)	Central electric space-heating equipment, including integral supplemental heating in heat pumps where the controller prevents the compressor and supplemental heating from operating at the same time: 5345 × 65% =	c)	3474
d)	Electric space-heating equipment, if less than four separately controlled units: ______ × 65% =	d)	
e)	Electric space-heating equipment, if four or more separately controlled units: ______ × 40% =	e)	

7	Total Volt-Ampere Demand Load:	3795 (largest VA rating from line 6a through 6e) + 18,980 (line 5) =	7	22,775
8	Minimum Amperes *Divide the total volt-amperes by the voltage*	22,775 (line 15) ÷ 240 (voltage) = **8** 95 (minimum amperes)		
9	**Minimum Size Service and/or Feeder 240.6(A)**		9	100 (minimum is 100 amperes)
10	Size the Service and/or Feeder Conductors. *Use 310.15(B)(6) to find the service conductors up to 400 amperes. Ratings in excess of 400 amperes shall comply with Table 310.16. 310.15(B)(6) also applies to feeder conductors serving as the main power feeder.*	**Minimum Size Conductors**	10	4 copper 2 aluminum
11	Size the Neutral Conductor 220.22 **Note: There is no optional method for calculating the neutral conductor.** *310.15(B)(6) states that the neutral service and/or feeder conductor can be smaller than the ungrounded (hot) conductors, provided the requirements of 215.2, 220.22, and 230.42 are met.* *250.24(B)(1) states that the neutral cannot be smaller than the required grounding electrode conductor specified in Table 250.66.*	**Minimum Size Neutral Conductor**	11	4 copper 3 aluminum
12	Size the Grounding Electrode Conductor Table 250.66 *Using line 10 to find the grounding electrode conductor in Table 250.66.* Size the Equipment Grounding Conductor (for Feeder) 250.122 *Use line 9 to find the equipment grounding conductor in Table 250.122. Equipment grounding conductor types are listed in 250.118.*	**Minimum Size Grounding Electrode Conductor . . . *or* . . . Equipment Grounding Conductor**	12	8 copper 6 aluminum

Summary

- Multi-family dwelling calculations are a progression of the one-family dwelling calculations (standard and optional).
- Some of the information needed for each individual dwelling unit includes: total sq-ft area, the number of small-appliance branch-circuits and laundry branch-circuits (if any), appliance ratings, dryer ratings (if any), cooking equipment ratings, and heating and air-conditioning loads.
- Identifying the number of dwelling units supplied by each service and/or disconnecting means is essential to the performance of multi-family load calculations.
- In certain instances, the laundry branch-circuit may be omitted from both the dwelling unit and the calculation.
- Laundry (if installed) and small appliance branch-circuits are calculated using a rating of at least 1500 volt-amperes each.
- Table 220.11 demand factors are applied to general lighting, small appliance, and laundry loads.
- The total volt-ampere rating of four (or more) fastened-in-place appliances is multiplied by 75% in the standard method calculation.
- Demand factors for household electric clothes dryers are found in Table 220.18.
- Table 220.19 lists the demand loads for household cooking appliances (over 1¾-kW rating).
- The heating and air-conditioning loads are compared, and the larger of the two is used.
- For heat pumps, include the compressor and the maximum amount of electric heat that can be energized while the compressor is running.
- Two multi-family load calculation methods are provided in Article 220: **Standard** and **Optional**.
- The neutral conductor can be smaller than the ungrounded (hot) conductors, provided the requirements of 215.2, 220.22, and 230.42 are met.
- Table 250.66 is used to size grounding electrode conductors.
- Remote panelboards (subpanels) require equipment grounding conductors, which can be one, more, or a combination of the types listed in 250.118.
- Equipment grounding conductors are sized in accordance with Table 250.122.

Unit 11 Competency Test

***NEC*® Reference**	**Answer**	
________	________	1. One unit in a multi-family dwelling has a 125-ampere, 240-volt panelboard fed with 75°C conductors. What is the minimum size aluminum equipment grounding conductor required?
________	________	2. What is the dryer demand load (in kW) for a 25-unit multi-family dwelling with a 4.5-kW clothes dryer in each unit? (The service is 120/240-volt, single-phase.)
________	________	3. A four-unit apartment has the following ranges: a 15 kW, a 14 kW, a 10 kW, and a 9 kW. What is the kW demand load added to the service by these ranges?
________	________	4. An 18-unit multi-family dwelling contains the following appliances in each unit: a 1-kVA, 115-volt dishwasher; a ⅓-HP, 115-volt waste (garbage) disposer; and a 4500-watt, 230-volt water heater. Using the standard method, what is the service neutral load contribution, in volt-amperes, for these appliances? (Assume water heater watt rating equivalent to volt-ampere.)
________	________	5. A multi-family dwelling's service-entrance conductors consist of four paralleled sets of 3/0 AWG THHN copper conductors. What size copper grounding electrode conductor is required?
________	________	6. A 5.4-kVA, 240-volt clothes dryer will contribute _____ amperes to the neutral load, when calculating the service by the standard method.
________	________	7. What is the minimum kW service demand load for twenty 6.5-kW ranges in a multi-family dwelling?

NEC® Reference	Answer	
________	________	8. An apartment has 3000 watts allocated for general lighting and receptacles. How many 15-ampere circuits are required? (The apartment contains laundry facilities.)
________	________	9. Two paralleled sets of 500-kcmil copper service-entrance conductors supply a multi-family dwelling disconnecting means. What is the minimum size aluminum grounding electrode conductor required?
________	________	10. A 50-unit apartment building has a 4.5-kW clothes dryer in each apartment. By using the optional method, what is the service demand load?

Questions 11 through 18 are based on a 120/240-volt, 3-wire, single-phase, eight-unit multi-family dwelling. The eight identical units contain the following:

Floor area . . . . 1200 sq ft
Range. 8¾ kW, 240 volt
Water heater . . . . 4 kW, 240 volt
Dishwasher . . . . 1 kW, 120 volt
Clothes dryer . . . . 5 kW, 240 volt
Electric heat. . . . . 4.2 kW, 240 volt
Air handler (blower motor) . . . . ¼ HP, 115 volt
Air-conditioner compressor . . . . 3 HP, 230 volt
Condenser fan motor . . . . ⅙ HP, 115 volt

(Assume water heater, clothes dryer, range, and electric heat kW ratings equivalent to kVA.)

Using Article 220, Part II (the standard method):

________ ________ 11. What is the service demand load for the clothes dryers?

________ ________ 12. What is the service demand load for the cooking appliances?

________ ________ 13. Of the heating or air-conditioning (and associated motors), which is omitted?

________ ________ 14. What is the minimum rating (in amperes) for the service overcurrent device?

________ ________ 15. A parallel run (2 sets) of service-entrance conductors will be installed in two raceways. What is the minimum size THWN copper ungrounded conductors that can be used?

________ ________ 16. What is the minimum rating (in amperes) for the neutral conductors?

________ ________ 17. What is the minimum size THWN copper neutral (grounded) conductors that can be installed?

________ ________ 18. What is the minimum size copper grounding electrode conductor?

Questions 19 through 22 are based on a 120/240-volt, 3-wire, single-phase, six-unit multi-family dwelling. The six identical units contain the following:

Floor area . . . . 1350 sq ft
Range. 10.6 kW, 240 volt
Water heater . . . . 4.2 kW, 240 volt
Dishwasher . . . . 1.2 kW, 120 volt
Garbage disposer. . . . . ⅓ HP, 115 volt
Clothes dryer . . . . 4 kW, 240 volt
Electric heat (2 banks @ 3 kW each) . . . . . . 6 kW, 240 volt
Air handler (blower motor) . . . . ⅓ HP, 115 volt
Air-conditioner compressor . . . . 3 HP, 230 volt
Condenser fan motor . . . . ⅙ HP, 115 volt

(Assume water heater, clothes dryer, range, and electric heat kW ratings equivalent to kVA.)

Using the Optional Method:

________ ________ 19. What is minimum rating (in amperes) for the service overcurrent device?

________ ________ 20. What is the minimum size THWN copper ungrounded conductors that can be installed? (Do not parallel the conductors.)

________ ________ 21. What is the minimum size THWN copper neutral (grounded) conductor that can be installed?

________ ________ 22. What is the minimum acceptable size copper grounding electrode conductor?

section 4

COMMERCIAL LOCATIONS

UNIT 12

General Provisions

Objectives

After studying this unit, the student should:

- know that commercial occupancy receptacle placement differs from that of one-family and multi-family dwellings.
- understand that commercial bathrooms do not require receptacle outlets.
- be aware that each commercial building (or occupancy) accessible to pedestrians must have at least one sign outlet.
- be familiar with Article 430 motor provisions.
- know that air-conditioning and refrigeration provisions are found in Article 440.
- be able to determine receptacle volt-ampere ratings for single, duplex, quad, etc.
- know the maximum number of receptacles permitted on 15- and 20-ampere branch-circuits.
- be aware that at least two receptacle outlets must be readily accessible in guest rooms of hotels, motels, and similar occupancies.
- thoroughly understand showcase and show window provisions.
- be able to determine an occupancy's general lighting load based on sq-ft area.
- be familiar with fluorescent, HID, recessed, and track-lighting provisions.
- understand that while certain luminaires (lighting fixtures) can be used as raceways, the number of branch-circuits permitted therein is limited.
- know that a metal pole supporting a luminaire(s) (lighting fixture[s]) requires a handhole, unless an exception is met.

Introduction

The broad topic of commercial wiring encompasses many types of structures—office buildings, restaurants, stores, schools, and warehouses, to name a few. In fact, most of the occupancies found in Table 220.3(A) are indeed commercial. Many of the electrical requirements that apply to one-family and multi-family dwellings also apply to commercial and industrial locations. Because this overlap of requirements is common, it is especially important to begin with a solid understanding of not only some, but all of, the preceding material. Of particular importance are Units 3 through 5 of Section 1, which contain fundamental principles applicable to all types of occupancies. In addition, Unit 9 (Services, Feeders, and Electric Equipment) explains one-family dwelling provisions that extend into the realm of commercial wiring systems. Unit 12 (General Provisions) likewise covers overlapping topics, such as: branch-circuits, receptacles, and lighting. Because it would be very difficult, if not impossible, to cover such a large field as commercial wiring in the framework of this text, only a small, but representative, portion is presented. The next unit explains non-dwelling (inclusive of commercial) load calculations, which are helpful in determining service and feeder loads. Complete subsequent units, and portions of others, apply to commercial wiring as well.

The inception of most commercial electrical projects includes multi-faceted plans (blueprints) and detailed specifications. As a reference, illustrations for a retail store's electrical and lighting floor plans are provided. Various general provisions (branch-circuits, receptacles, lighting, etc.) and numerous specific provisions (required receptacle outlets, show windows, showcases, etc.) are discussed. Proper receptacle placement is demonstrated through the use of a hotel/motel guest-room floor plan, including furniture placement. Receptacle and lighting branch-circuit load calculations are also part of this unit.

While this unit is by no means a complete reference for commercial electrical wiring, taken together with the extensive information presented throughout this text, it does provide a substantial foundation toward understanding commercial electrical wiring systems.

BRANCH-CIRCUITS

Required Sign Outlet

The disconnecting means must be within sight of the sign or outline lighting system that it controls. If out of the line of sight from any section that may be energized, the disconnecting means must be capable of locking in the open position »600.6(A)(1)«.

Signs or outline lighting systems operated by external electronic or electromechanical controllers can have a disconnecting means located within sight of the controller or within the controller enclosure. The disconnecting means must disconnect both the sign (or outline lighting system) and the controller from all ungrounded supply conductors. It must be designed so that no pole operates independently and it must be capable of locking in the open position »600.6(A)(2)«.

NOTE

Service hallways or corridors are not considered accessible to pedestrians »600.5(A)«.

Switches, flashers, and similar devices controlling transformers and electronic power supplies must either be rated for controlling inductive loads, or have a current rating not less than twice the transformer's current rating »600.6(B)«.

Ⓐ Each commercial building and occupancy open to pedestrians must have at least one accessible outlet at every entrance to each tenant space for sign or outline lighting systems use »600.5(A)«.

Ⓑ The outlet(s) must be supplied by a branch-circuit rated 20 amperes or more and that supplies no other load »600.5(A)«.

Ⓒ Each sign and outline lighting system, or feeder or branch-circuit supplying a sign or outline lighting system, must be controlled by an externally operable switch (or circuit breaker) that opens all ungrounded conductors, unless: (1) the sign is an exit directional sign located within a building; or (2) the sign is cord-connected with an attachment plug »600.6«.

Ⓓ Sign and outline lighting outlets must be computed at a minimum of 1200 volt-amperes per required branch-circuit specified in 600.5(A) »220.3(B)(6)«.

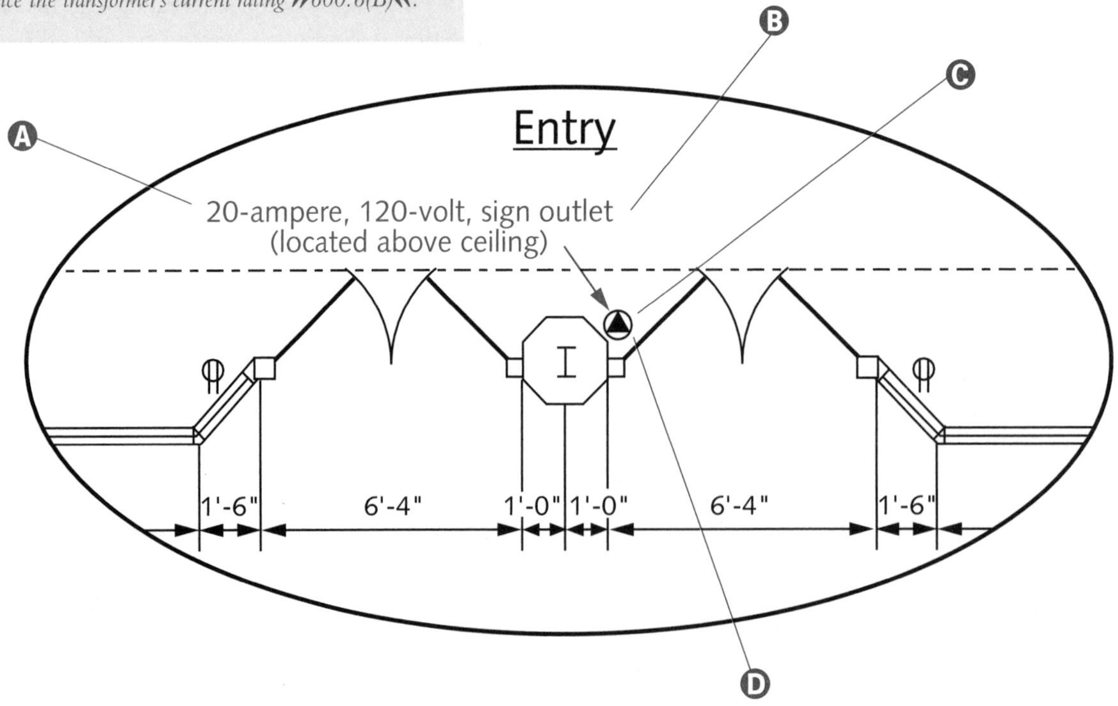

Commercial (Retail) Occupancy

Branch-circuits recognized by Article 220 must be rated in accordance with the maximum permitted ampere rating (or setting) of the overcurrent device. Circuits, other than individual branch-circuits, must be rated 15, 20, 30, 40, and 50 amperes. If higher ampacity conductors are used for any reason, the ampere rating (or setting) of the specified overcurrent device determines the circuit rating » 210.3 «.

Branch-circuit conductors must have an ampacity not less than the maximum load served. In a branch-circuit supplying continuous loads, or any combination of continuous and noncontinuous loads, the minimum branch-circuit conductor size (before the application of any adjustment or correction factors) must have an allowable ampacity equal to, or greater, than the noncontinuous load plus 125% of the continuous load » 210.19(A)(1) «.

Equipment and branch-circuit conductors must be protected by overcurrent protective devices rated in compliance with 210.20(A) through (D) » 210.20 «.

Outlet devices must have an ampere rating not less than the load served, and must comply with 210.21(A) and (B) » 210.21 «.

The requirements for multiple outlet or receptacle are summarized in Table 210.24 » 210.24 «.

The minimum number of branch-circuits is determined from the total computed load and the size (or rating) of the circuits used. In all installations, the number of circuits must be sufficient to supply the load served. In no case can the load on any circuit exceed the maximum specified by 220.4 » 210.11(A) «.

Article 210 covers branch-circuits with the exception of branch-circuits supplying only motor loads. These are covered in Article 430. Articles 210 and 430 provisions apply to branch-circuits that have combination loads » 210.1 «.

A A unit load at least equal to that shown in Table 220.3(A) for a specified occupancy constitutes the minimum lighting load for each square foot of floor area. Floor area (for each floor) is computed using the outside dimensions of the building, dwelling unit, or other area involved. The minimum general lighting load for a store is 3 volt-amperes per sq ft » 220.3(A) «. The minimum general lighting load for this retail space is 3726 volt-amperes. ($27 \times 46 = 1242 \times 3 = 3726$)

B No receptacle wall placement provisions exist for commercial occupancies. The "6-ft rule," which is the maximum distance to a receptacle (measured horizontally along the wall), applies only to dwelling units » 210.52(A) «.

C Receptacles are not required in non-dwelling bathrooms. If installed however, they must be GFCI protected » 210.8(B)(1) «.

D In a building having more than one nominal voltage system, each accessible ungrounded (hot) conductor of a multiwire branch-circuit, must be identified by phase and system. The identification method can be separate color coding, marking, tape, tagging, or other approved means, and must be permanently posted at each branch-circuit panelboard » 210.4(D) «.

NOTE

Refer to 210.2 for a list of other Articles (and Sections) pertaining to branch-circuit requirements. The provisions for branch-circuits supplying equipment in that list amend or supplement Article 210 provisions and apply to the referenced branch-circuits »210.2«.

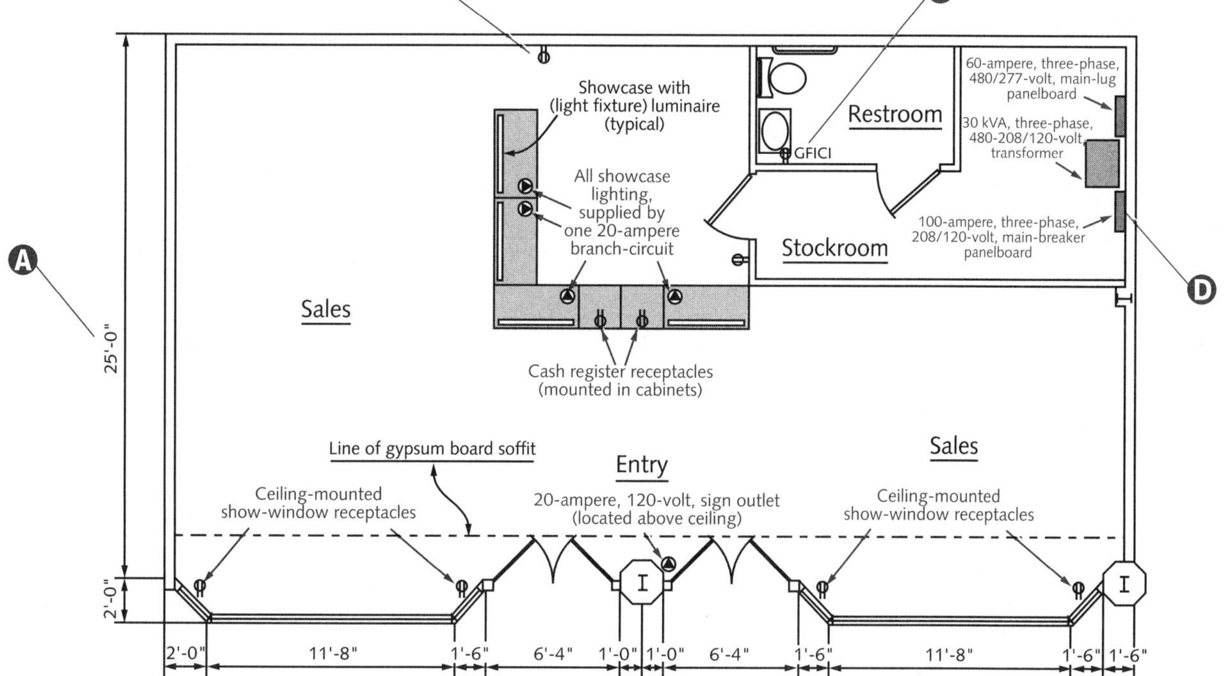

Showcases

A Individual moveable showcases can be connected by flexible cord to permanently installed receptacles. Groups of not more than six such showcases can be coupled together (by flexible cord and separable locking-type connectors) provided one of the group is connected (by flexible cord) to a permanently installed receptacle. The installation must comply with 410.29(A) through (E) »410.29 «.

B Flexible cords must be secured to showcase undersides so that: (1) wiring will not be exposed to mechanical damage; (2) cases cannot separate more than 2 in. (50 mm), nor can the first case extend more than 12 in. (300 mm) from the supply receptacle; and (3) the free lead at the end of the showcase group has a female fitting that does not extend beyond the case »410.29(C) «.

C In cord-connected showcases, the secondary circuit(s) of each electric-discharge lighting ballast is (are) limited to one showcase »410.29(E) «.

D Flexible cord must be hard-service type, have conductors not smaller than the branch-circuit overcurrent device, and have an equipment grounding conductor »410.29(A) «.

E Receptacles, connectors, and attachment plugs must be of a listed grounding type, rated 15 or 20 amperes »410.29(B) «.

F Unless used in wiring of chain-supported luminaires (lighting fixtures) or as supply cords for portable lamps and other merchandise being displayed or exhibited, flexible cords used in show cases and show windows must be: Type S, SE, SEO, SEOO, SJ, SJE, SJEO, SJEOO, SJO, SJOO, SJT, SJTO, SJTOO, SO, SOO, ST, STO, or STOO »400.11 «.

NOTE

Other equipment must not be electrically connected to showcases »410.29(D)«.

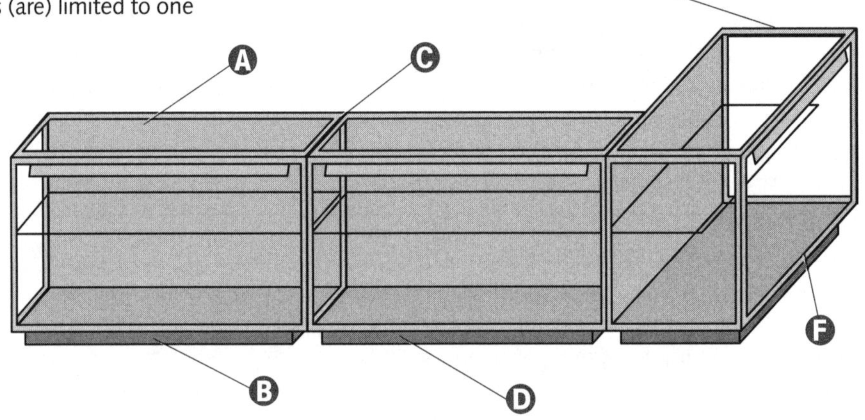

Motor Loads

A Motor load outlets must be computed in accordance with 430.22, 430.24 and 440.6 »220.3(B)(3) «.

B If a circuit supplies only motor-operated loads, apply Article 430 »220.4(A) «.

C Branch-circuit conductors supplying a single motor, used in a continuous duty application, must have an ampacity of not less than 125% of the motor's full-load current rating as determined by 430.6(A)(1) »430.22(A) «.

NOTE

Conductors used for a motor in a short-time, intermittent, periodic, or varying duty application must have an ampacity of not less than the percentage of the motor nameplate current rating shown in Table 430.22(E), unless the AHJ grants special permission for lower ampacity conductors »430.22(E)«.

Any motor application is considered continuous duty unless the nature of the apparatus driven is such that the motor will not operate continuously with load under any condition of use »Note under Table 430.22(E)«.

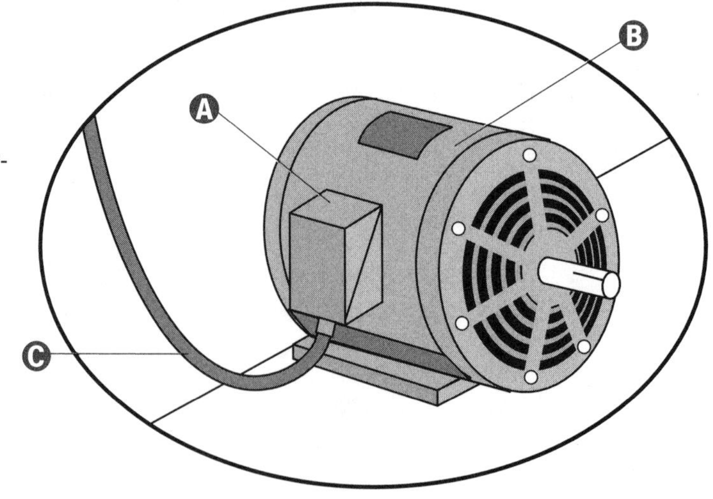

Motor-Operated and Combination Loads

A For circuits supplying loads consisting of motor-operated utilization equipment (fastened in place with a motor larger than ⅛ HP) in combination with other loads, the total computed load must be based on 125% of the largest motor load plus the sum of the other loads »220.4(A)«.

B Because equipment such as room air-conditioners, household refrigerators and freezers, drinking water coolers, and beverage dispensers are considered appliances, Article 422 provisions also apply »440.3(C)«.

C For cord- and plug-connected appliances, an accessible, separable connector or an accessible plug and receptacle can serve as the disconnecting means. Where the separable connector or plug and receptacle are not accessible, cord- and plug-connected appliances must be provided with a disconnecting means in accordance with 422.31 »422.33(A)«.

D The rating of a receptacle or a separable connector must not be less than the rating of any connected appliance »422.33(C)«.

E Combination branch-circuit, consisting of two duplex receptacles.

F If supplied by a 20-ampere circuit, this drinking fountain's maximum rating is 10 amperes (20 × 50%). If supplied by a 15-ampere circuit, the maximum rating is 7.5 (15 × 50%).

G Receptacle outlet for cord- and plug-connected, non fastened-in-place utilization equipment.

H Dedicated branch-circuit.

I If supplied by a 20-ampere circuit, the maximum rating for this drinking fountain cooler is 16 amperes (20 × 80%). If supplied by a 15-ampere circuit, the maximum rating is 12 amperes (15 × 80%).

J For cord-connected equipment such as room air-conditioners, household refrigerators (and freezers), drinking water coolers, and beverage dispensers, a separable connector or an attachment plug and receptacle can serve as the disconnecting means »440.13«.

NOTE

A 15- or 20-ampere branch-circuit can supply lighting units, other utilization equipment, or a combination of both. The rating of any one cord- and plug-connected utilization equipment must not exceed 80% of the branch-circuit ampere rating. The total rating of utilization equipment fastened in place (other than luminaires [lighting fixtures]) cannot exceed 50% of the branch-circuit ampere rating in the event that lighting units, cord- and plug-connected utilization equipment (not fastened in place), or both, are also supplied »210.23(A)«.

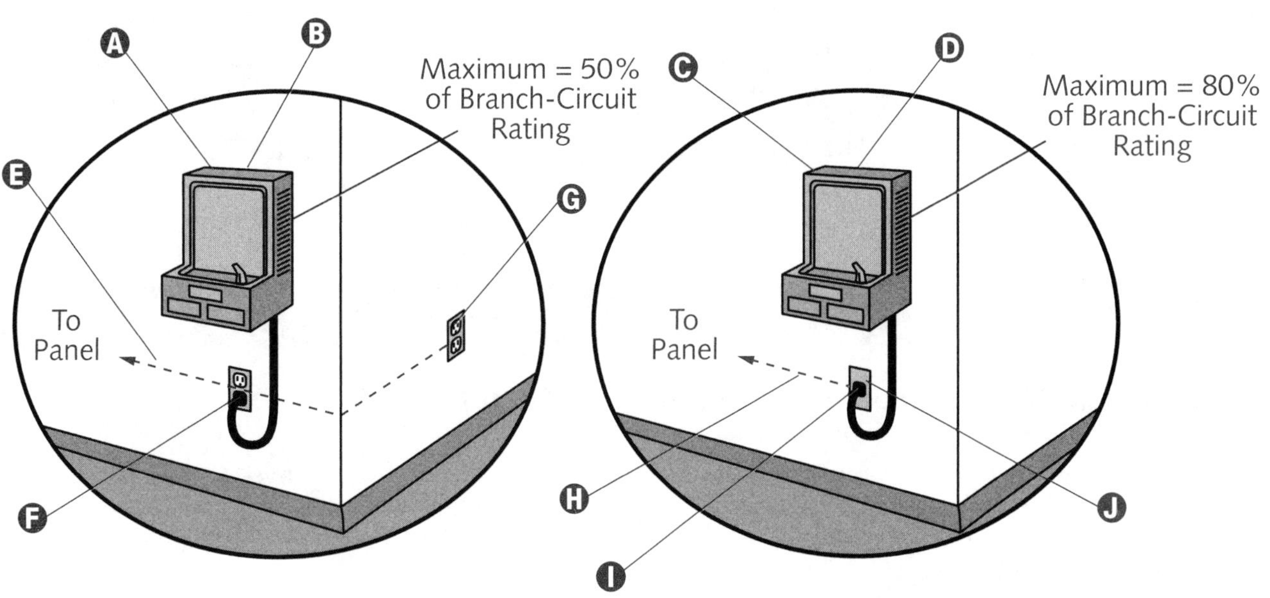

Air-Conditioning and/or Refrigeration Equipment

Article 440, Part II provisions require a means capable of disconnecting air-conditioning and refrigerating equipment (including motor-compressors and controllers) from the circuit conductors »440.11«.

The provisions of Article 440, Part III specify devices intended to protect the branch-circuit conductors, control apparatus, and motors in circuits supplying hermetic refrigerant motor-compressors against overcurrent due to short circuits and grounds. These provisions amend and/or supplement those in Article 240 »440.21«.

Article 310 and Article 440, Part IV, provisions specify conductor ampacities required to carry the motor current (without overheating) under the conditions specified, except as modified in 440.6(A), *Exception No. 1* »440.31«.

Article 440, Part VI specifies devices intended to protect the motor-compressor against excessive heating due to motor overload and failure to start. See 240.4(G) and 430.31 »440.51«.

Ⓐ Motor load outlets must be computed in accordance with Article 440 requirements »220.3(B)(3)«.

Ⓑ Where a circuit supplies only air-conditioning equipment, refrigeration equipment, or both, apply Article 440 »220.4(A)«.

Ⓒ Disconnecting means must be located within sight from, and readily accessible from, the air-conditioning or refrigerating equipment. The disconnecting means can be installed on or within the equipment »440.14«.

Ⓓ For a hermetic refrigerant motor-compressor, the rated-load current marked on the nameplate of the equipment in which the motor-compressor is employed must be used in determining the rating or ampacity of the disconnecting means, the branch-circuit conductors, the controller, the branch-circuit short-circuit and ground-fault protection, and the separate motor overload protection. Where no rated-load current is shown on the equipment nameplate, use the rated-load current shown on the compressor nameplate »440.6(A)«.

Ⓔ Article 440 provisions apply to electric motor-driven air-conditioning and refrigerating equipment, and to their branch-circuits and controllers. It provides the special considerations necessary for circuits supplying hermetic refrigerant motor-compressors and for any air-conditioning or refrigerating equipment supplied from a branch-circuit that also supplies a hermetic refrigeration motor-compressor »440.1«.

CAUTION *The disconnecting means must not be located on panels that are designed to allow access to the air-conditioning or refrigeration equipment »440.14«.*

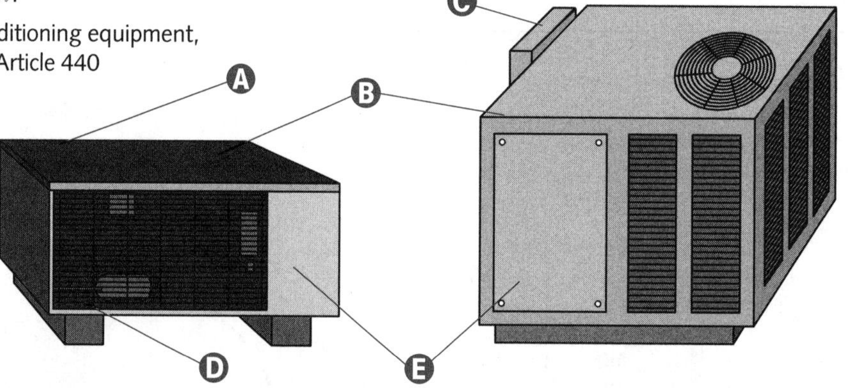

RECEPTACLES

Maximum Number of Receptacles on a Branch-Circuit

Ⓐ Ten receptacles are permitted on a 15-ampere overcurrent protective device (15 ÷ 1.5 = 10).

Ⓑ Receptacles are calculated at 180 volt-amperes per strap »220.3(B)(9)«. Therefore, when the source is 120 volts, the rating of each receptacle is 1.5 amperes (180 ÷ 120 = 1.5).

Ⓒ Thirteen receptacles are permitted on a 20-ampere overcurrent protective device (20 ÷ 1.5 = 13.3 or 13).

Ⓓ A load having the maximum level of current sustained for three hours or more is referred to as a continuous load. A general purpose receptacle is not a continuous load »Article 100«.

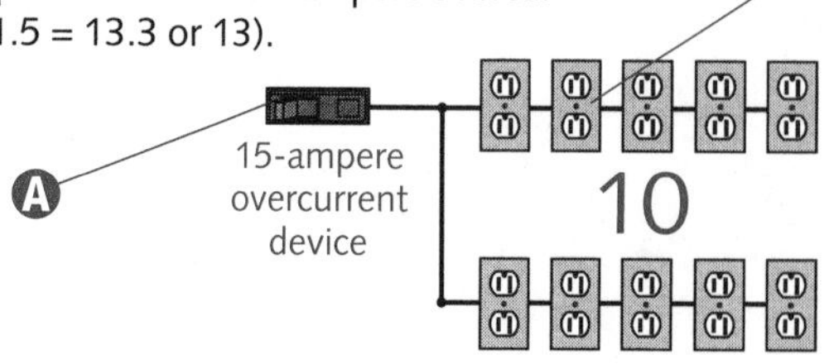

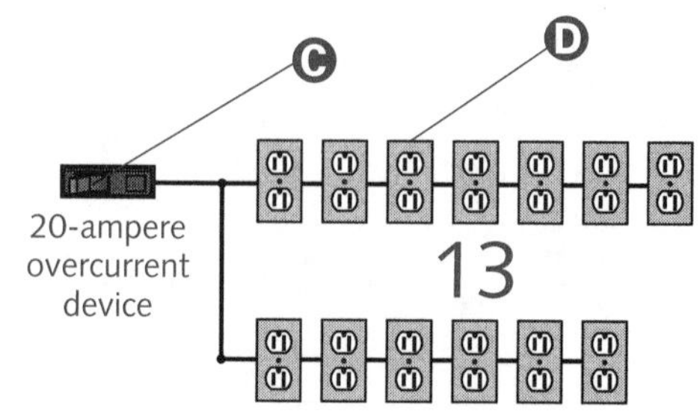

Non-Dwelling Receptacle Placement

A Dwelling unit receptacles must be installed so that no point along the floor line in any wall space is more than 6 ft (1.8 m), measured horizontally, from a receptacle outlet in that space »210.52(A)(1)«.

B Receptacle outlets in hotel, motel, and similar occupancy guest-rooms must be installed in accordance with 210.52 (A) and 210.52(D) »210.60(A)«.

C Although no provisions pertain to non-dwelling general receptacle placement (except hotel and motel guest rooms), certain requirements exist for specific occupancies and/or areas. Examples include, but are not limited to: show windows »210.62«; rooftop, attic, and crawl space receptacle outlets for servicing heating, air-conditioning, and refrigeration equipment »210.63«; and health care patient bed location receptacles »517.18(B)«.

D Generally speaking, there are no receptacle placement provisions for non-dwelling occupancies.

Dwelling Occupancy

6 Ft (1.8 m) 6 Ft (1.8 m) 6 Ft (1.8 m) 6 Ft (1.8 m)

A B

C — Non-dwelling Occupancy

24 Ft (7.3 m)

No Receptacle Outlet Required D

Bathroom Receptacles

A Non-dwelling bathrooms do not require receptacles. A bathroom receptacle is required only in dwelling units and hotel or motel guest rooms »210.52(D) and 210.60(A)«.

B Where installed, 125-volt, single-phase, 15- and 20-ampere bathroom receptacles must have GFCI protection for personnel »210.8(B)«.

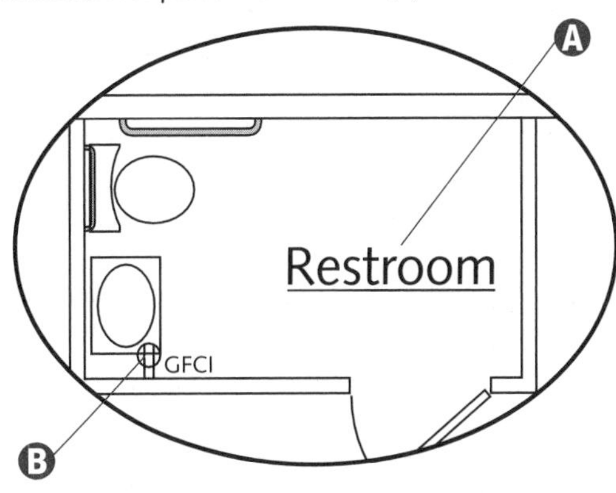

NOTE

GFCIs are designed to trip when a 4-through 6-milliampere (0.004-through 0.006-ampere) difference between the ungrounded (hot) conductor and grounded conductor occurs.

Fixed Multioutlet Assemblies

A Where appliances are unlikely to be used simultaneously, each 5 ft (1.5 m) or fraction thereof (of separate and continuous length) shall be considered one outlet of not less than 180 volt-amperes »220.3(B)(8)(1)«.

B Because fixed multioutlet assemblies used in dwelling units or hotel and motel guest rooms are included in the general lighting load calculation, no additional load calculation is required »220.3(B)(8)«.

C Where appliances are likely to be used simultaneously, each 1 ft (300 mm) or fraction thereof shall be considered one outlet of not less than 180 volt-amperes »220.3(B)(8)(2)«.

D It is not stipulated that each receptacle be rated 180 volt-amperes. This illustration shows one receptacle per foot; therefore, each receptacle is rated 180 volt-amperes. Multioutlet assemblies having two receptacles per ft [6 in. (150 mm) on center], have a rating of 180 volt-amperes for two receptacles.

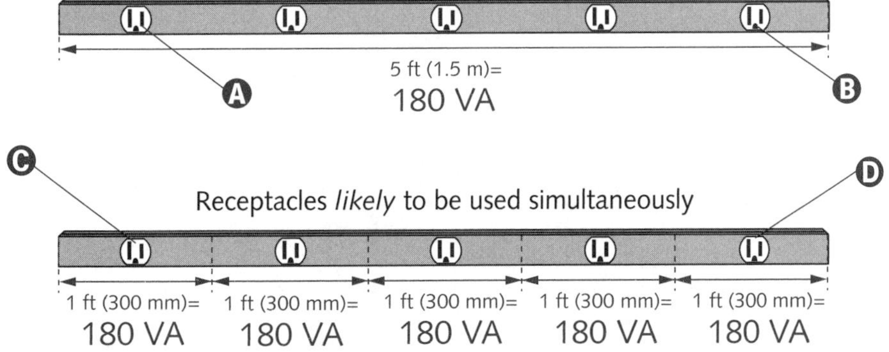

Receptacle Volt-Ampere Rating

Ⓐ Except as covered in one-, two-, and multi-family dwellings, and in hotel and motel guest rooms, receptacle outlets must be computed at not less than 180 volt-amperes for each single, or multiple, receptacle on one yoke »220.3(B)(9)«.

Ⓑ A single piece of equipment (consisting of four or more receptacles) must be computed at not less than 90 volt-amperes per receptacle »220.3(B)(9)«.

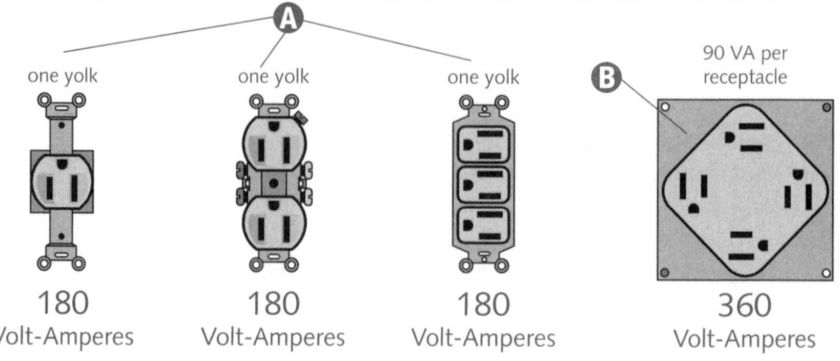

Provisions for Hotel and Motel Guest Rooms

When performing service (or feeder) calculations, use Table 220.11 to derate hotel and motel (include apartment houses without cooking equipment) guest room lighting and receptacle loads. Do not apply Table 220.11 demand factors to areas where all lighting will probably be used at one time. Examples include, but are not limited to: office, lobby, restaurant, hallway, parking area, etc.

Ⓐ GFCI protection is required for every 125-volt bathroom receptacle »210.8(A)(1)«.

Ⓑ Hallways of less than 10 ft (3.0 m) in length do not require a receptacle »210.52(H)«.

Ⓒ At least one wall-switch-controlled lighting outlet or receptacle must be installed in hotel, motel, or similar occupancy guest rooms »210.70(B)«.

Ⓓ Receptacles must be installed so that no point along the floor line in any wall space is more than 6 ft (1.8 m), measured horizontally, from an outlet in that space »210.52(A)(1)«. (See 210.60(B) provisions allowing a greater distance due to permanent luminaire [fixture] layout.)

Ⓔ Guest rooms in hotels, motels, and similar occupancies must have receptacle outlets installed in accordance with 210.52 (dwelling unit receptacle provisions) »210.60(A)«. (See Units 6 and 7, in this book, for dwelling unit receptacle requirements.)

Ⓕ Locate at least one wall receptacle within 36 in. (900 mm) of the outside edge of each basin (lavatory or sink). The receptacle outlet must be located on a wall adjacent to the basin »210.52(D)«.

Ⓖ 410.4(D) contains bathtub and shower area lighting provisions. (See illustrated explanation in Unit 7 of this book.)

Ⓗ Receptacles installed behind the bed must either be positioned so that the bed does not contact any installed attachment plug, or the receptacle must include a suitable guard »210.60(B)«.

Ⓘ The total number of receptacle outlets must comply with the minimum number of receptacles provision of 210.52(A). These receptacle outlets can be located conveniently for permanent furniture layout »210.60(B)«. This receptacle could be located between the beds, even though the distance to the receptacle behind the desk exceeds 12 ft (3.7 m).

Ⓙ At least two receptacle outlets must be readily accessible »210.60(B)«.

Ⓚ Branch-circuit conductors supplying more than one receptacle for cord- and plug-connected portable loads must have an ampacity at least equal to the rating of the branch-circuit »210.19(A)(2)«.

NOTE

Hotels, motels, and similar occupancies having guest room kitchen facilities must meet 210.52(B) and (C) provisions.

It is not required that occupants in hotel/motel guest rooms (intended for transient use) have ready access to all overcurrent devices protecting conductors supplying their room »240.24(B)«.

CAUTION *In dwelling units and guest rooms of hotels and motels, overcurrent devices (other than supplementary overcurrent protection) must not be located in bathrooms as defined in Article 100 »240.24(E)«.*

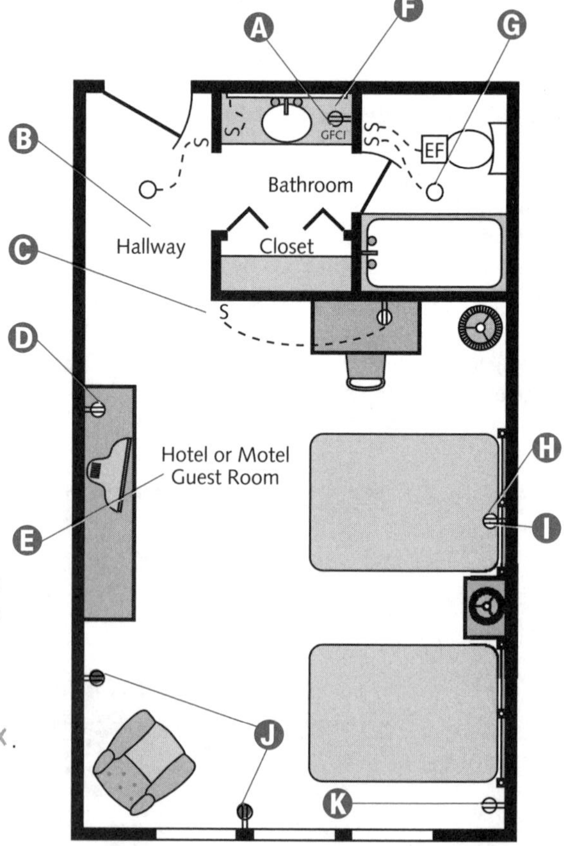

Show Windows

A *show window* is any window designed or used for the display of goods or advertising material, whether it is fully or partly enclosed, entirely open at the rear, and whether or not it has a platform raised higher than the street floor level »Article 100«.

Unless used in wiring of chain-supported luminaires (lighting fixtures), or as supply cords for portable lamps and other merchandise being displayed or exhibited, flexible cords used in show cases and show windows must be Type S, SE, SEO, SEOO, SJ, SJE, SJEO, SJEOO, SJO, SJOO, SJT, SJTO, SJTOO, SO, SOO, ST, STO, or STOO »400.11«.

A Show-window branch-circuit loads must be computed either by: (a) the unit load per outlet as required by other provisions of 220.3(B); or (B) at 200 volt-amperes per linear ft »220.3(B)(7)«.

B At least one receptacle outlet must be installed directly above a show window for each 12 linear ft (3.7 m) —or major fraction thereof—of window area measured horizontally at its widest point »210.62«.

C The second method for computing show-window branch-circuit loads is to multiply each show-window linear ft (or major fraction thereof) by 200 volt-amperes. (2 + 11.5 + 2 = 15.5 = 16 linear ft . . . 16 × 200 = 3200 volt-amperes) »220.3(B)(7)«.

D One of two methods for calculating show-window branch-circuit loads is to multiply each receptacle by 180 volt-amperes (180 × 2 = 360 volt-amperes) »220.3(B)(7) and 220.3(B)(9)«.

E Complete show-window feeder and/or service loads using a minimum rating of 200 volt-amperes for each linear ft of window, measured horizontally along its base (16 × 200 = 3200 volt-amperes) »220.12(A)«.

NOTE

*Where the AHJ judges them free from likely exposure to physical damage, moisture, and dirt, boxes located in **elevated floors of show windows** and similar locations can be other than those listed for floor applications. Receptacles and covers must be listed as an assembly for this type of location* »*314.27(C)* Exception«.

Electric signs (including neon tubing) and associated wiring within show windows must comply with Article 600.

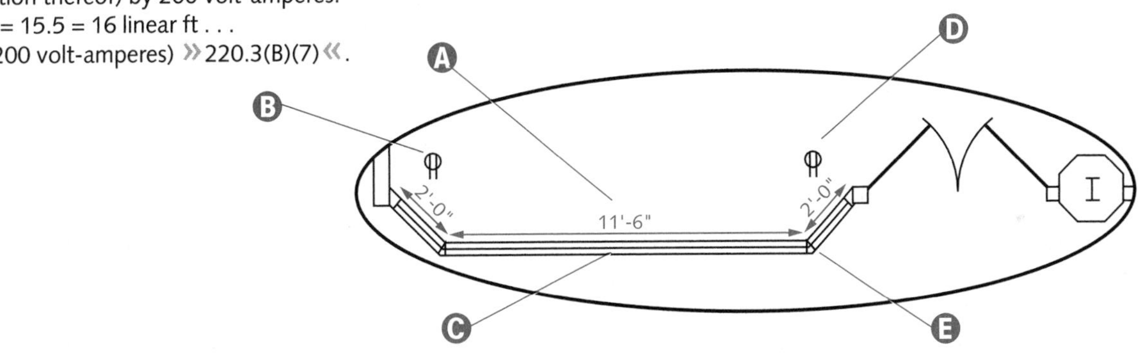

LIGHTING

Lampholder Installations

A Screw-shell type lampholders must be installed only for use as lampholders »410.47«.

B If the supply circuit has a grounded conductor, it must be connected to the screw shell »410.47«.

NOTE

Only weatherproof-type lampholders can be installed in wet or damp locations »*410.49*«.

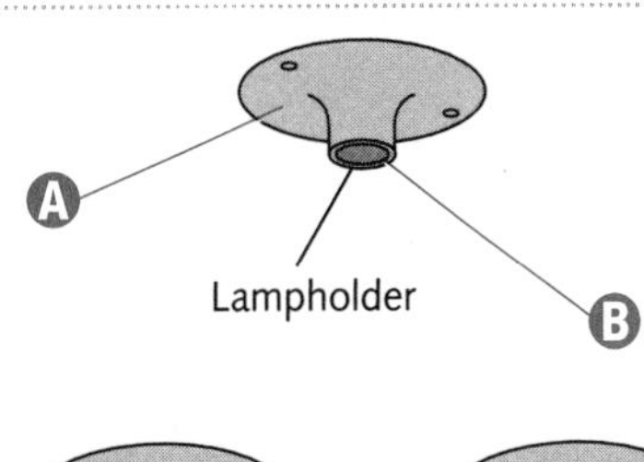

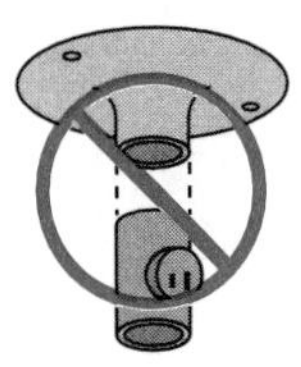

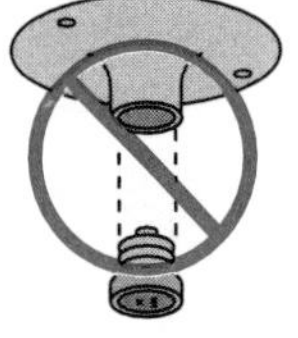

Reflective Ceiling/Lighting Plan

Branch-circuits recognized by Article 210 must be rated in accordance with the maximum permitted overcurrent device ampere rating (or setting) »210.3«.

Branch-circuits for lighting and appliances (including motor-operated appliances) must supply the loads in accordance with 220.3 »210.11«.

Branch-circuit conductor ampacity must meet or exceed the maximum load being served. Where a branch-circuit supplies continuous loads, or any combination of continuous and noncontinuous loads, the minimum *branch-circuit conductor* size, before the application of any adjustment or correction factors, must have an allowable ampacity equal to or greater than the noncontinuous load plus 125% of the continuous load »210.19(A)«.

Where a branch-circuit supplies continuous loads, or any combination of continuous and noncontinuous loads, the rating of the *overcurrent device* must not be less than the noncontinuous load plus 125% of the continuous load »210.20(A)«.

A 15- or 20-ampere branch-circuit can supply lighting units or other utilization equipment, or a combination of both »210.23(A)«.

Wall-switch-controlled lighting outlets are not required in commercial occupancies, except for attic and underfloor spaces containing equipment which requires servicing »210.70(C)«.

A unit load meeting Table 220.3(A) specifications for listed occupancies constitutes the minimum lighting load for each sq ft of floor area. The area for each floor must be computed using the outside dimensions of the building, dwelling unit, or other area involved »220.3(A)«.

Supplementary overcurrent protection used for luminaires (lighting fixtures), appliances, and other equipment (including internal circuits and components) must not be used as a substitute for the branch-circuit protection specified in Article 210. Ready accessibility is not required for supplementary overcurrent devices »240.10«.

A 30-, 40- or 50-ampere branch-circuit can supply fixed lighting units with heavy-duty lampholders in other than dwelling unit(s) »210.23(B) and (C)«.

Lighting outlet loads must not be supplied by branch-circuits larger than 50 amperes »210.23(D)«.

A For circuits of over 250 volts to ground, the electrical continuity of metal raceways and cables with metal sheaths, containing any conductor other than service conductors, must be ensured by one or more of the methods specified for services in 250.94(2) through (4) »250.97«.

B Circuit breakers used as switches in 120-volt and 277-volt fluorescent lighting circuits must be listed and must be marked SWD or HID »240.83(D)«. All switches and circuit breakers used as switches must be located so that they are operable from a readily accessible place. The center of the operating-handle grip of the switch or circuit breaker, when in its highest position, must not be more than 6 ft, 7 in. (2.0 m), above the floor, or working platform »404.8(A)«.

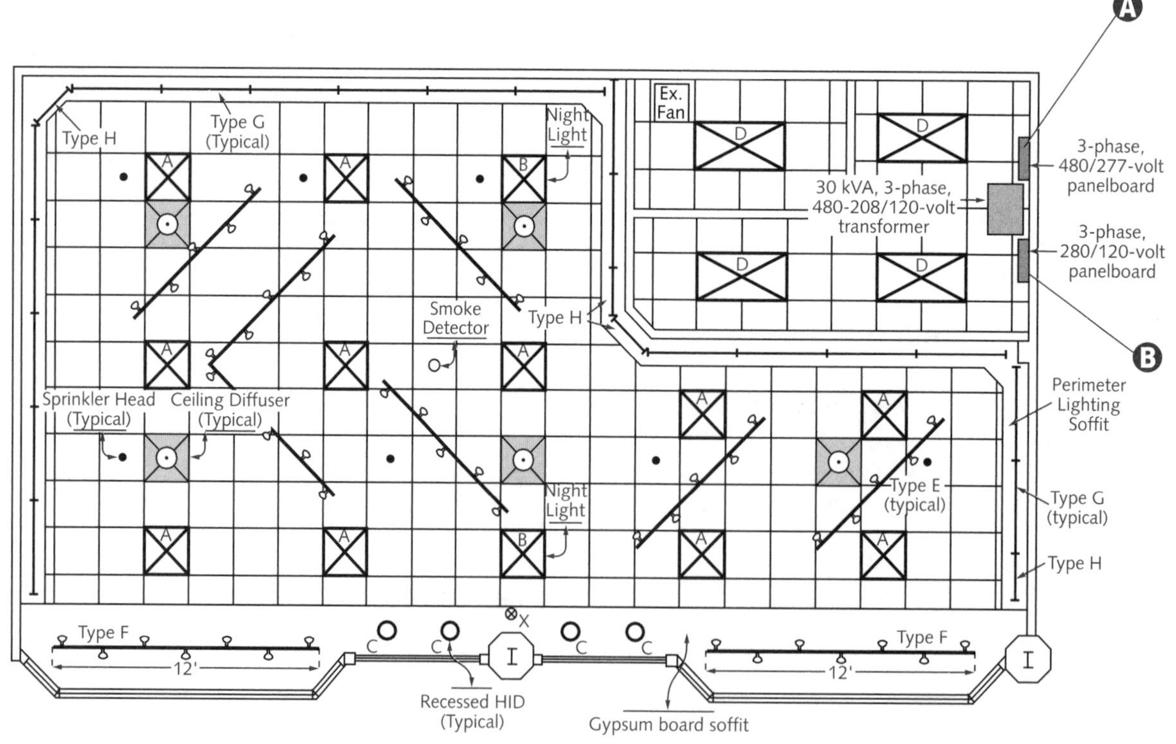

Luminaire (Lighting Fixture) Schedule

The minimum number of branch-circuits is determined from the total computed load, and the size (or rating) of the circuits used. In all installations, the number of circuits must be sufficient to supply the load served. In no case shall the load on any circuit exceed the maximum specified by 220.4 »210.11(A)«.

The load calculated on a volt-ampere per sq ft basis, must be adequately served by the wiring system up to and including the branch-circuit panelboard(s). This load must be evenly proportioned among multioutlet branch-circuits within the panelboard(s). Branch-circuits and overcurrent devices must only be installed to serve the connected load »210.11(B)«.

Ⓐ Conductors of circuits rated no more than 600 volts, nominal, ac circuits, and dc circuits can occupy the same equipment wiring enclosure, cable, or raceway. The conductors' insulation rating must equal or exceed the maximum circuit voltage applied to any conductor within the enclosure, cable, or raceway »300.3(C)(1)«.

Ⓑ For lighting units that have ballasts, transformers, or autotransformers, the supply circuit's computed load is based on the total ampere ratings of such units, not on the lamps' total wattage »220.4(B)«. Obtain the ampere rating from the manufacturer or from the information listed on the ballast. If the luminaire (fixture) contains more than one ballast, add the ampere ratings. Multiply the total ampere rating by the voltage to find the luminaire's (fixture's)volt-ampere rating.

Ⓐ Ⓑ

LUMINAIRE (LIGHTING FIXTURE) SCHEDULE					LAMP SCHEDULE		
Type	Manufacturer	Volts	Catalog Number	Comments	Lamps per Luminaire (Fixture)	Watts per Lamp	Lamp Size
A	Lighthouse	277	P622-232U277	2′ × 2′ lay-in fluorescent deep cell parabolic	2	32	F32-T8-U
B	Lighthouse	277	P622-232U277-EM	emergency light	2	32	F32-T8-U
C	Lighthouse	120	10R-CM70MH120	10″ recessed, 70W metal halide	1	70	MH70/C/U/M
D	Lighthouse	277	RT24–432EB277	2′ × 4′ lay-in fluorescent	4	32	F32-T8
E	Lighthouse	120	8001BK/HT308	8′ lighting track with 5 heads	5	75	75W-PAR30FL
F	Lighthouse	120	8001BK/HT308	12′ lighting track with 7 heads	7	75	75W-PAR30FL
G	Lighthouse	120	SL48-32EB120	4′ Single lamp fluorescent strip	1	32	F32-T8
H	Lighthouse	120	SL24-20EB120	2′ Single lamp fluorescent strip	1	20	F20-T8
X	Lighthouse	120	EX7B1LEDR	Exit light with battery	Lamps are included in the luminaire (fixture).		

NOTE

The total load on any branch-circuit conductor must not exceed 80% of its rating, after the application of any adjustment or correction factors, where the load will continue for three hours or more, i.e. a continuous load »210.19(A)(1)«. Most commercial lighting is considered continuous.

Fluorescent Luminaires (Lighting Fixtures)

Luminaires (fixtures) must be equipped with shades (or guards), constructed, or installed so that combustible material will not be subjected to temperatures exceeding 90°C »410.5«.

Securely support all luminaires (fixtures), lampholders, and receptacles »410.15(A)«.

Exposed metal parts must be: (1) grounded or insulated from ground and other conducting surfaces; or (2) inaccessible to unqualified personnel. Grounding is not required for lamp tie wires, mounting screws, clips, and decorative bands on glass spaced at least 1½ in. (38 mm) from lamp terminals »410.18(A)«.

Raceway fittings supporting a luminaire(s) (lighting fixture[s]) must be capable of supporting the combined weight of the luminaire (fixture) assembly and lamp(s) »410.16(F)«.

Ⓐ Suspended-ceiling system framing members used to support luminaires (fixtures) must be securely fastened to each other and to the building structure at appropriate intervals. Luminaires (fixtures) must be fastened securely to the ceiling framing member by mechanical means such as bolts, screws, or rivets. Listed clips identified for use with the type of ceiling framing member(s) and luminaire(s) (fixture[s]) are also permitted »410.16(C)«.

Ⓑ Branch-circuits recognized by Article 210 can be used in multiwire circuits. A multiwire branch-circuit can be thought of as multiple circuits. All conductors must originate from the same panelboard »210.4(A)«. Unless an exception is met, multiwire branch-circuits can only supply line-to-neutral loads »210.4(C)«.

Ⓒ A 3-phase, 4-wire, wye-connected power system supplying nonlinear loads may necessitate a design allowing for the possibility of high harmonic neutral currents »210.4(A) FPN«.

Ⓓ Luminaires (fixtures)and equipment are considered grounded if mechanically connected to an equipment grounding conductor per 250.118 and sized in accordance with 250.122 »410.21«.

Ⓔ Wiring on, or within, luminaires (fixtures) must be neatly arranged and must not be exposed to physical damage. Avoid excess wiring. Arrange conductors so they are not subjected to temperatures greater than those for which they are rated »410.22«.

NOTE

Install luminaires (fixtures) so that the connections between luminaire (fixture) conductors and circuit conductors can be inspected without having to disconnect any of the wiring, unless the luminaires (fixtures) are connected by attachment plugs and receptacles »410.16(B)«.

***Luminaire**, the international term for "lighting fixture", is defined as a complete lighting unit consisting of lamp(s), parts designed to distribute the light, parts positioning and protecting the lamp(s), and parts connecting the lamp(s) to the power supply »Article 100«.*

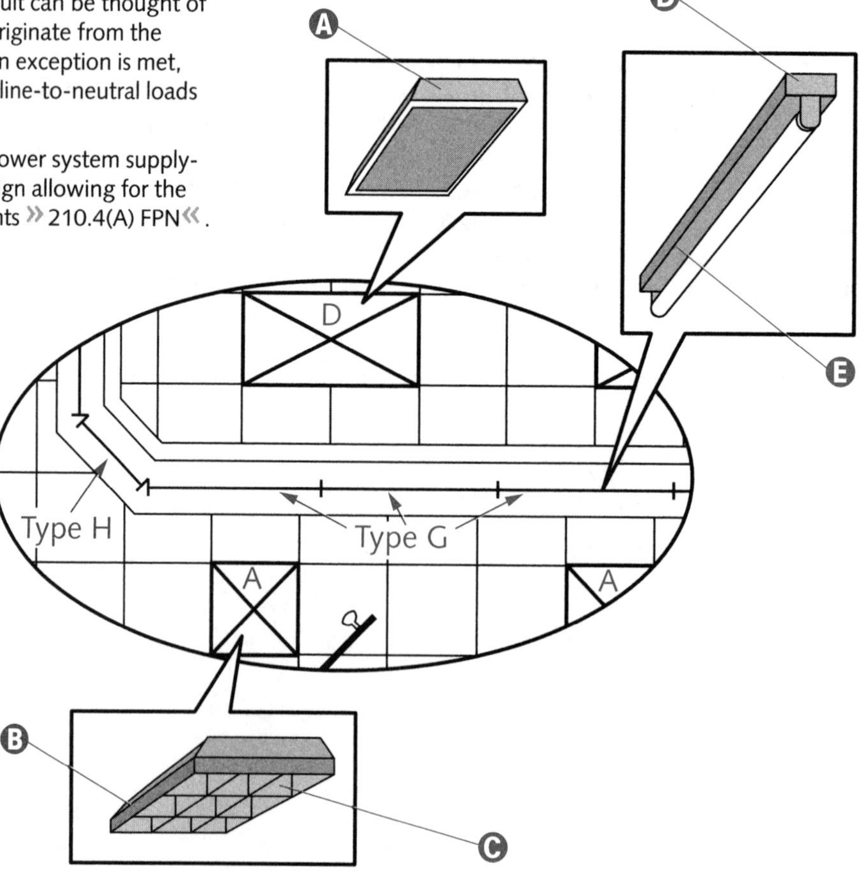

Show-Window and Track Lighting

Heavy-duty lighting track is identified for use exceeding 20 amperes. Each fitting attached to a heavy-duty lighting track must have individual overcurrent protection »410.103«.

A Lighting track is a manufactured assembly designed to support and energize luminaires (lighting fixtures) that can be readily repositioned along the track. Its length may be altered by the addition or subtraction of track sections »410.100«.

B Lighting track must be permanently installed as well as permanently connected to a branch-circuit. Install only lighting track fittings on the lighting track. Fittings equipped with general-purpose receptacles cannot be used on lighting track »410.101(A)«.

C Lighting track must be securely mounted so that each fastening suitably supports the maximum weight of luminaires (fixtures) that can be installed. Unless identified for greater support intervals, a single section 4 ft (1.2 m) or less in length must have two supports. If installed in a continuous row, each individual section of not more than 4 ft (1.2 m) in length must have one additional support »410.104«.

D Lighting track must be grounded in accordance with Article 250. Track sections must be securely coupled to maintain continuity of the circuitry, polarization, and grounding throughout »410.105(B)«.

E Track system ends must be insulated and capped »410.105(A)«.

F The load connected to a lighting track must not exceed the track's rating. The rating of the branch-circuit which supplies lighting track cannot exceed the track rating »410.101(B)«.

G Fittings identified for use on lighting track must be designed specifically for the track on which they are installed. They must be securely fastened to the track, maintain polarization and grounding, and be designed to be suspended directly from the track »410.101(D)«.

H When computing feeder and/or service track lighting in other than dwelling units or hotel/motel guest rooms, include an additional load of 150 volt-amperes for every 2 ft (610 mm) of lighting track or fraction thereof »220.12(B)«.

I No externally wired luminaire (fixture), except chain-supported, can be used in show windows »410.7«.

J When computing feeder and/or service show-window lighting, include a unit load of not less than 200 volt-amperes per linear foot of show window, measured horizontally along the base »220.12(A)«.
(2 + 11.5 + 2 = 15.5 = 16 × 200 = 3200 volt-amperes)

NOTE

Do not install lighting track in the following locations:

1. *Where physical damage is likely*
2. *In wet or damp locations*
3. *Where subject to corrosive vapors*
4. *In storage battery rooms*
5. *In hazardous (classified) locations*
6. *Where concealed*
7. *Where extended through walls or partitions*
8. *Less than 5 ft (1.5 m) above the finished floor except where protected from physical damage or where the track operates at less than 30 volts rms open-circuit voltage*
9. *Within the zone measured 3 ft (900 mm) horizontally and 8 ft (2.5 m) vertically from the top of a bathtub rim »410.101(C)«.*

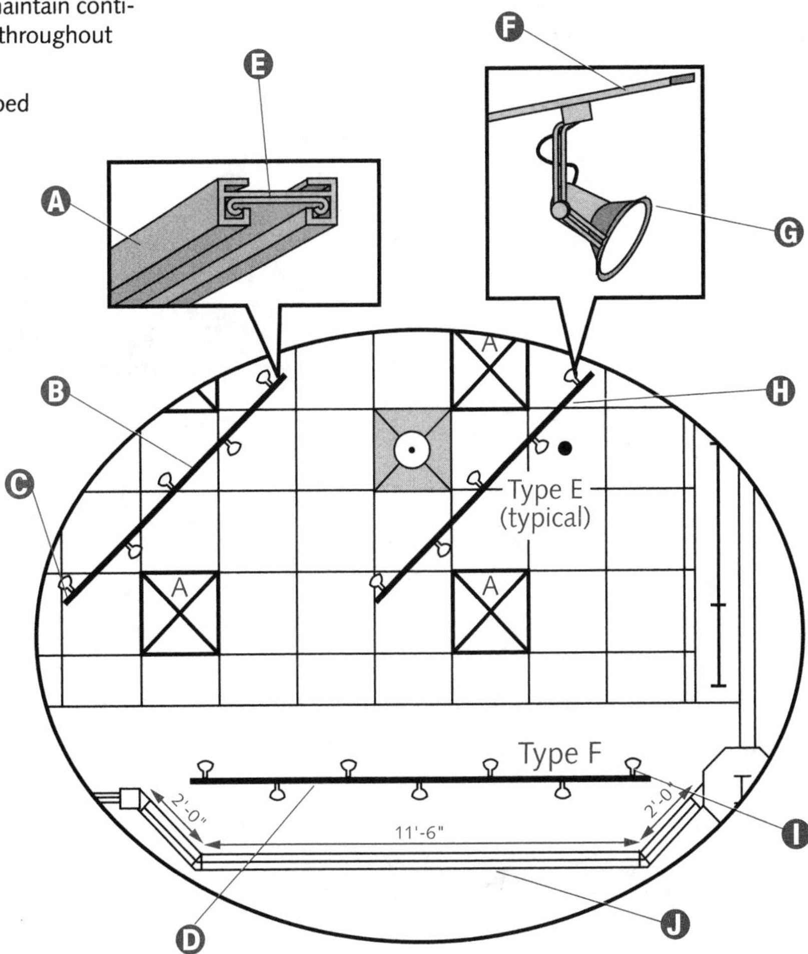

Recessed Luminaires (Fixtures)

A recessed luminaire (fixture), not identified for contact with insulation, must have all recessed parts spaced at least ½ in. (13 mm) from combustible materials. Support points and the trim finishing off the ceiling or wall opening can contact combustible materials »410.66(A)(1)«.

A recessed luminaire (fixture) identified for contact with insulation, Type IC, can contact combustible materials at recessed parts, points of support, and portions passing through or finishing off the structural opening »410.66(A)(2)«.

A recessed HID luminaire (fixture) identified for use and installed in poured concrete does not require thermal protection »410.73(F)(3)«.

Ⓐ A luminaire (fixture) can be recessed in fire-resistant material in a building of fire-resistant construction, subject to temperatures between 194°F (90°C) and 302°F (150°C), provided the luminaire (fixture) is plainly marked as listed for that service »410.65(B)«.

Ⓑ Install luminaires (fixtures) so that adjacent combustible material will not be subjected to temperatures greater than 194°F (90°C) »410.65(A)«.

Ⓒ Luminaires (fixtures) installed in recessed cavities (walls or ceilings) must comply with 410.65 through 410.72 »410.64«.

Ⓓ Do not install thermal insulation above a recessed luminaire (fixture) or within 3 in. (75 mm) of the luminaire's (fixture's) enclosure (wiring compartment, or ballast) unless it is identified for contact with insulation, Type IC »410.66(B)«.

Ⓔ High-intensity discharge (HID) luminaires (fixtures) supply power to HID lamps. HID lamps produce light from gaseous discharge arc tubes. HID lamps include: mercury lamps, metal halide lamps, and high-pressure sodium lamps.

Ⓕ A recessed remote ballast for a HID luminaire (fixture) must have integral thermal protection and be identified as such »410.73(F)(4)«.

Ⓖ Recessed high-intensity luminaires (fixtures) designed for installations in wall or ceiling cavities must have thermal protection and be identified as such »410.73(F)(1)«. Thermal protection is not required in a recessed high-intensity luminaire (fixture) whose design, construction, and thermal performance characteristics are equivalent to a thermally protected luminaire (fixture), and are identified as inherently protected »410.73(F)(2)«.

> **NOTE**
>
> *Install luminaires (fixtures) with exposed ballasts or transformers so that such ballasts or transformers will not contact combustible material »410.76(A)«.*

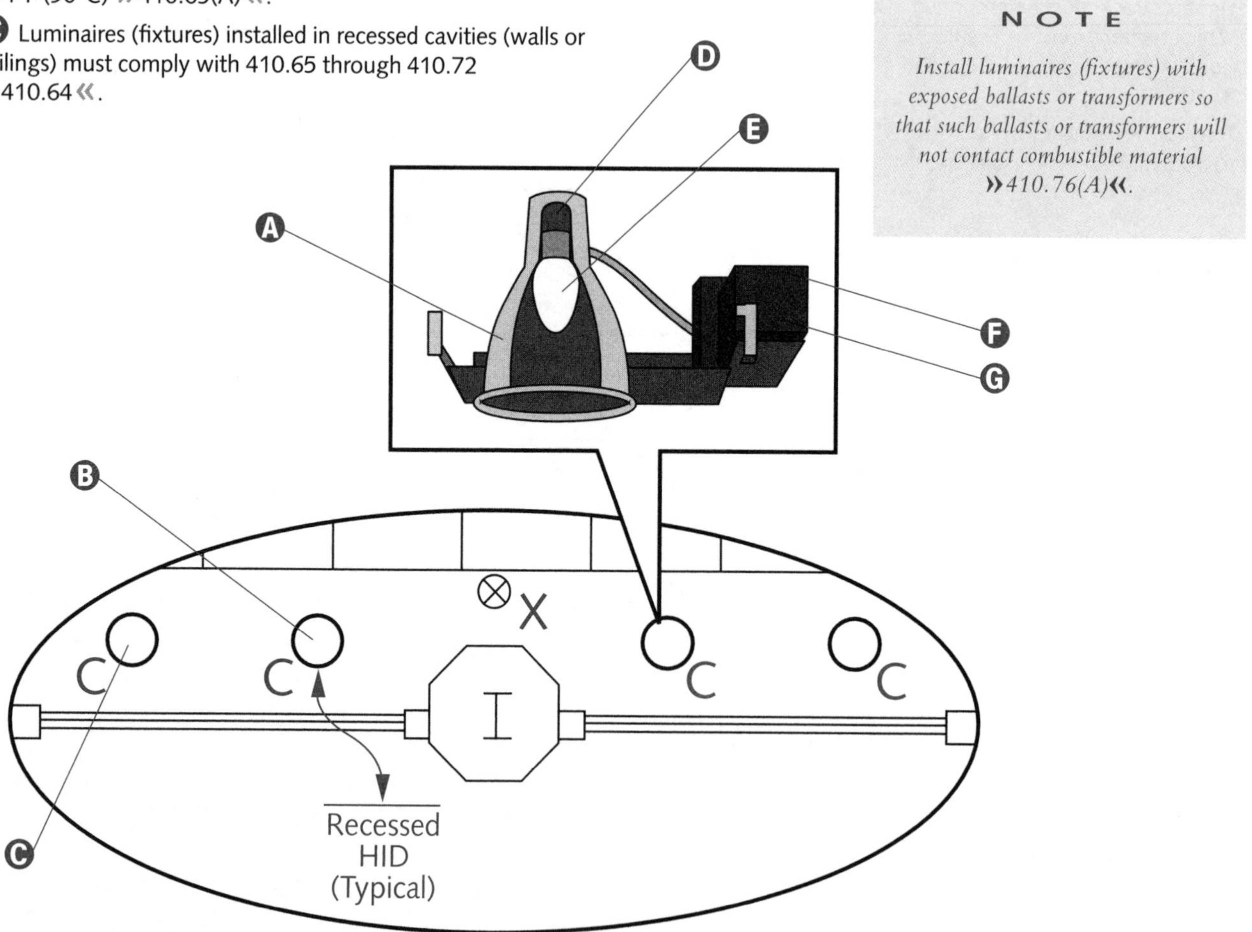

Exit Signs and Emergency Lighting

A ballast in a fluorescent exit luminaire (fixture) shall not have thermal protection »410.73(E)(3)«.

Ⓐ Suspended ceiling systems framing members supporting luminaires (fixtures) must be securely fastened to each other and to the building structure at appropriate intervals. Luminaires (fixtures) must be securely fastened to the ceiling framing member by mechanical means such as bolts, screws, or rivets. Listed clips identified for use with the type of ceiling framing member(s) and luminaires (fixture[s]) are also permitted »410.16(C)«.

Ⓑ The ballast of a fluorescent luminaire (fixture) installed indoors must have integral thermal protection. Replacement ballasts must also have similar thermal protection »410.73(E)(1)«.

Ⓒ A fluorescent luminaire (fixture) ballast used for egress lighting, energized only during an emergency, shall not have thermal protection »410.73(E)(4)«.

Ⓓ An exit directional sign located within a building does not require a disconnecting means »600.6 *Exception No. 1*«.

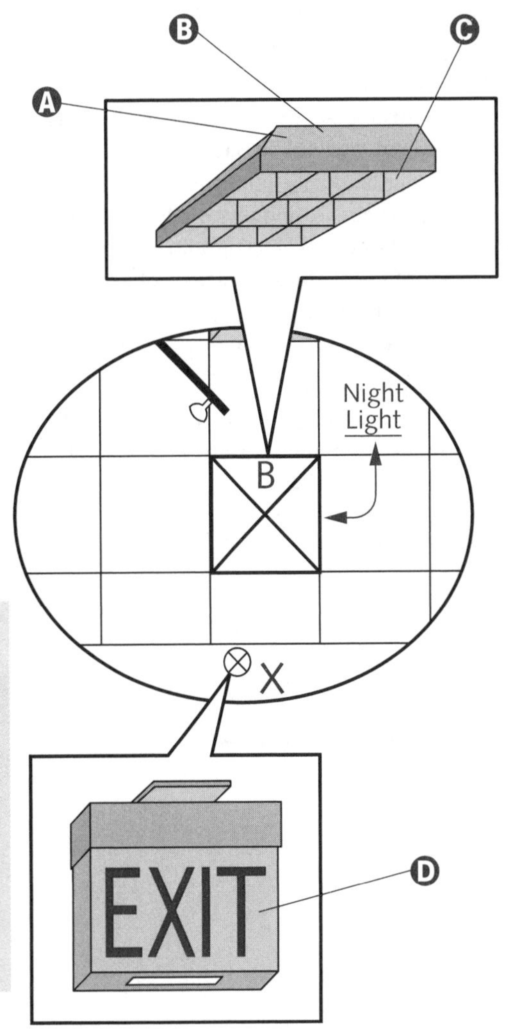

NOTE

Article 700 provisions apply to the electrical safety of emergency systems installation, operation, and maintenance. These systems consist of circuits and equipment intended to supply, distribute, and control electricity for facility illumination, power, or both, whenever the normal electrical service is interrupted. Emergency systems are legally mandated by municipal, state, federal, or other codes, or by any governmental agency having jurisdiction. The purpose of these systems is to automatically supply illumination, power, or both, to designated areas and equipment in the event of failure of the normal supply, or in the event of accidental damage to systems elements intended to supply, distribute, and control power and illumination essential for safety to human life »700.1«.

Adjustable Luminaires (Fixtures)

Ⓐ Two conduits can support an enclosure containing a device under the conditions found in 314.23(F): (1) The box must not exceed 100 cubic in. (1650 cm^3); (2) The box has either threaded entries or hubs identified for the purpose; (3) The box must be supported by two or more conduits threaded wrenchtight into the enclosure or hubs; (4) Each conduit must be secured within 18 in. (450 mm) of the enclosure provided all entries are on the same side.

Ⓑ Equipment grounding conductors that are part of flexible cords, or used with luminaire (fixture) wires in accordance with 240.5, cannot be smaller than 18 AWG copper and not smaller than the circuit conductors.

Ⓒ A wet location receptacle, where the product plugged in is unattended while in use, requires an enclosure that is weatherproof at all times (plug inserted or not) »406.8(B)(2)(a)«.

Ⓓ Luminaires (fixtures) that require adjusting or aiming after installation do not require an attachment plug or cord connector, provided the exposed cord is of the hard or extra-hard usage type and is not longer than that required for maximum adjustment. The cord must not be subject to strain or physical damage »410.30(B)«.

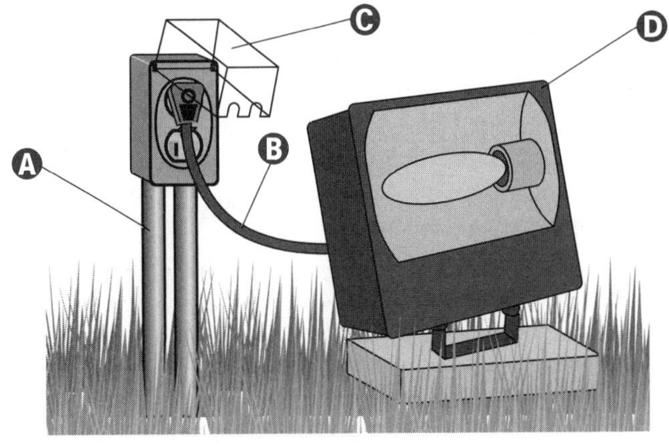

Luminaires (Fixtures) Used as Raceways

In a completed installation, a cover must be provided for each outlet box, unless covered by means of a luminaire (fixture) canopy, lampholder, receptacle, or similar device »410.12«.

Branch-circuit conductors must have an ampacity at least equal to the maximum load being served. In a branch-circuit supplying continuous loads, the minimum branch-circuit conductor size (before the application of any adjustment or correction factors) must have an allowable ampacity equal to or greater than 125% of the continuous load »210.19(A)(1)«.

The overcurrent device, on a branch-circuit supplying continuous loads, must be rated at least 125% of the load »210.20(A)«.

Auxiliary equipment for electric-discharge lamps must be enclosed in noncombustible cases and be treated as heat sources »410.54(A)«.

Ⓐ Luminaires (fixtures) designed for end-to-end connection, thereby forming a continuous assembly, or luminaires (fixtures) connected by recognized wiring methods, are permitted to contain the supply conductors of two 2-wire or one multiwire branch-circuit supplying the luminaires (fixtures) »410.32«.

Ⓑ One additional two-wire branch-circuit separately supplying one or more of the connected luminaires (fixtures) is also permitted »410.32«.

Ⓒ Branch-circuit conductors, within 3 in. (75 mm) of a ballast must have an insulation temperature rating no lower than 194°F (90°C) »410.33«.

Ⓓ Luminaires (fixtures) must not be used as a raceway for circuit conductors unless listed for such use »410.31«.

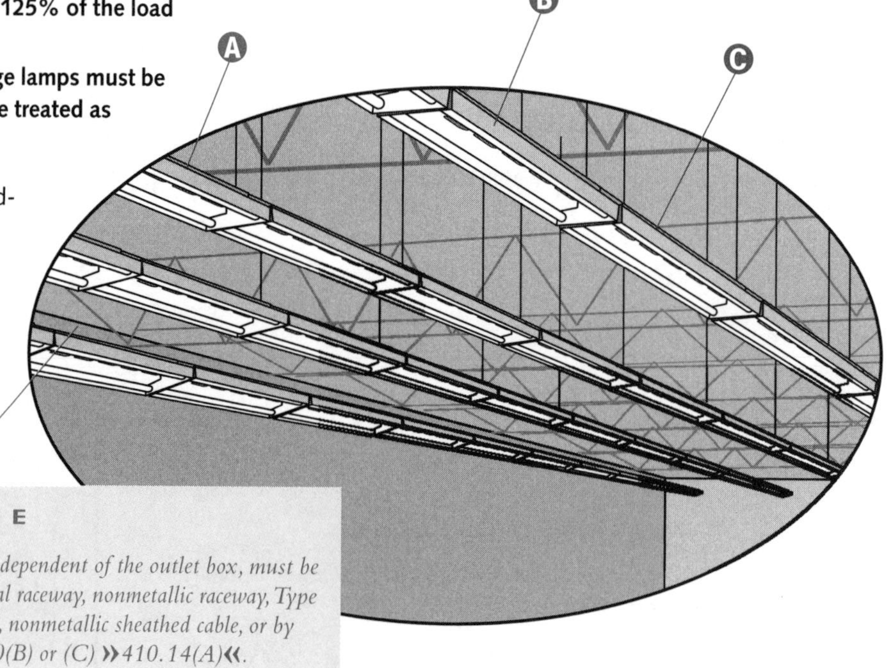

NOTE

Luminaires (lighting fixtures), supported independent of the outlet box, must be connected to the branch-circuit through metal raceway, nonmetallic raceway, Type MC cable, Type AC cable, Type MI cable, nonmetallic sheathed cable, or by flexible cord as permitted in 410.30(B) or (C) »410.14(A)«.

Hinged Metal Poles, 20 Ft (6.0 m) or Less, Supporting Luminaires (Lighting Fixtures)

Ⓐ A hinged-base pole longer than 20 ft (6.0 m), supporting a luminaire(s) (lighting fixture[s]), requires a handhole.

Ⓑ A metal pole must have an accessible grounding terminal within the base »410.15(B)(3)(b)«.

Ⓒ A metal pole 20 ft (6.0 m) or less in height above grade that has a hinged base does not require a handhole »410.15(B)(1) *Exception No. 2*«.

Ⓓ If equipped with a hinged base, the pole and base must be bonded together »410.15(B)(4)«.

Ⓔ Metal raceways or other equipment grounding conductors must be bonded to the pole by means of an equipment grounding conductor recognized by 250.118 and sized in accordance with 250.122 »410.15(B)(5)«.

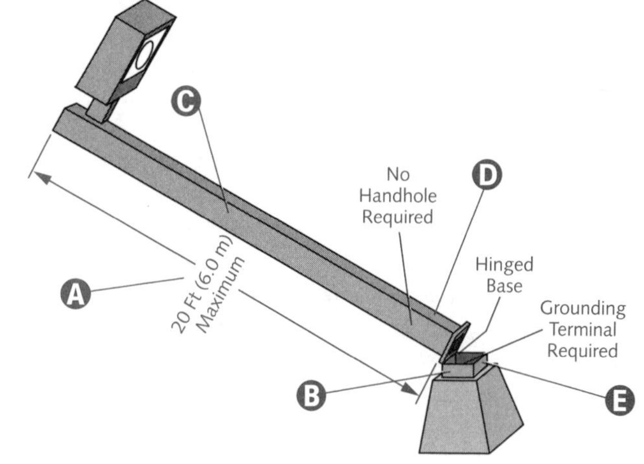

Cord Connected Lampholders and Luminaires (Fixtures)

The inlet of a metal lampholder attached to a flexible cord must be equipped with an insulating bushing that, if threaded, is not smaller than nominal ⅜-in. pipe size. The cord hole must be of an appropriate size, and all burrs/fins must be removed, providing a smooth bearing surface for the cord »410.30(A)«.

Ⓐ Electric-discharge luminaires (lighting fixtures) provided with mogul-base, screw-shell lampholders can be connected to branch-circuits of 50 amperes or less by cords that comply with 240.4. While receptacles and attachment plugs can be of a lower ampere rating than the branch-circuit, they cannot be less than 125% of the luminaire (fixture) load current »410.30(C)(2)«.

Ⓑ A listed luminaire (fixture) or a listed assembly can be cord connected if:

1. the luminaire (fixture) is located directly below the outlet box or busway, and
2. the flexible cord must be:
   a. Visible over its entire length outside the luminaire(fixture)
   b. Not subject to strain or physical damage
   c. Terminated in a grounding-type attachment plug cap or busway plug or have a luminaire (fixture) assembly with a strain relief and canopy »410.30(C)(1)«.

Ⓒ For circuits which supply lighting units with ballasts, transformers, or autotransformers, the computed load must be based on the total ampere rating of such units and not on the total lamp(s) wattage »220.4(B)«.

Ⓓ A load expected to continue at maximum current for 3 hours or more is a continuous load »Article 100«.

Ⓔ Electric-discharge luminaires (lighting fixtures) having a flanged surface inlet can be supplied by cord pendants equipped with cord connectors. Inlets and connectors can be of a lower ampere rating than the branch-circuit but not less than 125% of the luminaire (fixture) load current »410.30(C)(3)«.

Ⓕ Cord-connected luminaires (fixtures) must be located directly below the outlet box or busway.

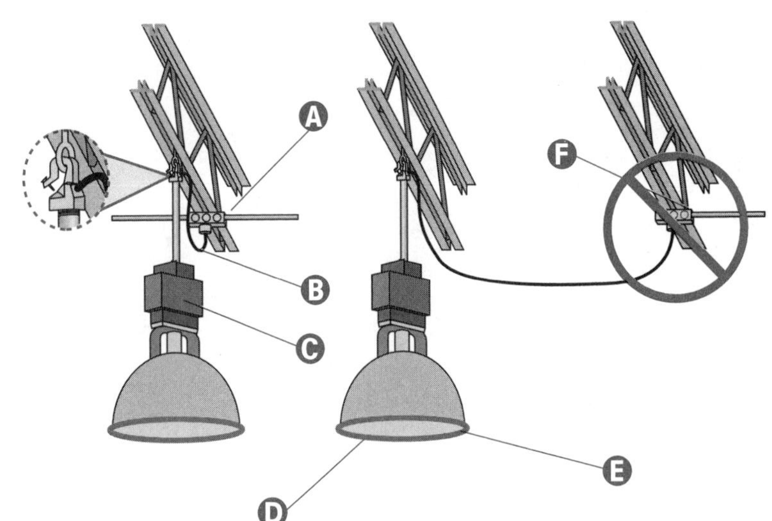

Portable Lamps

Ⓐ Portable lamps must be wired with flexible cord, recognized by 400.4, and a polarized or grounding type attachment plug »410.42(A)«.

Ⓑ If used with Edison-base lampholders, the grounded conductor must be identified and attached to the screw shell as well as the attachment plug's identified blade »410.42(A)«.

Ⓒ In addition to the provisions of 410.42(A), portable handlamps must comply with 410.42(B)(1) through (5):

1. Metal shell, paper-lined lampholders must not be used.
2. Handlamps must be equipped with a handle of molded composition or other insulating material.
3. Handlamps must be equipped with a substantial guard attached to the lampholder or handle.
4. Metallic guards must be grounded by means of an equipment grounding conductor run with circuit conductors within the power-supply cord.
5. Portable handlamps supplied through an isolating transformer, with an ungrounded secondary of not over 50 volts, do not require grounding.

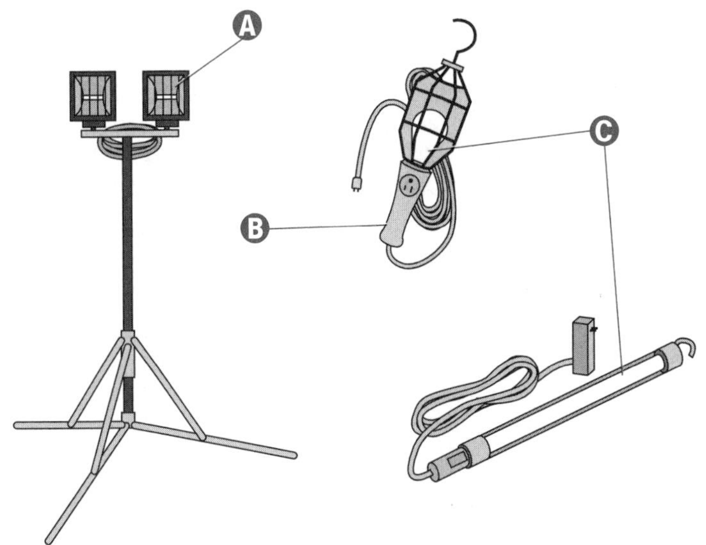

Metal Poles, 8 Ft (2.5 m) or Less, Supporting Luminaires (Lighting Fixtures)

A No handhole is required in a pole no more than 8 ft (2.5 m) in height (above grade) where the supply wiring method continues without splice or pull point and where the pole interior and splices are accessible by removing the luminaire (fixture) »410.15(B)(1) *Exception No. 1*«.

B Splices must remain accessible

C The wiring method must continue without splice to the luminaire (lighting fixture) termination point.

D A pole longer than 8 ft (2.5 m), supporting a luminaire(s) (lighting fixture[s]), requires a handhole.

E No grounding terminal is required in a pole 8 ft (2.5 m), or less, in height above grade where the supply wiring method continues without splice or pull point, and where the pole interior and splices are accessible by luminaire (fixture) removal »410.15(B)(3) *Exception*«.

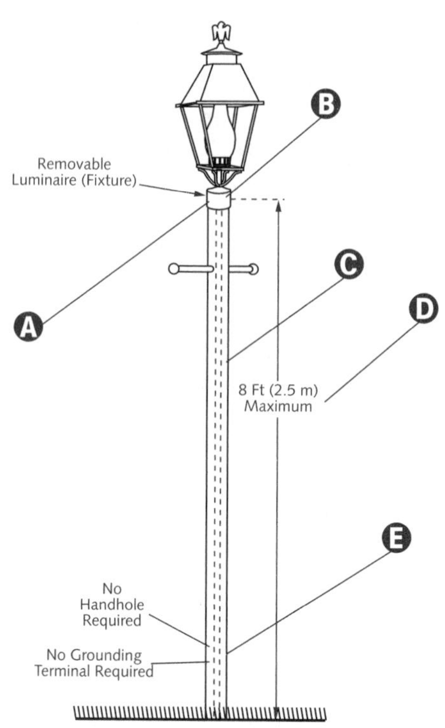

NOTE

A receptacle installed in a pole, supporting a luminaire (lighting fixture), may require GFCI protection (See 210.8). A wet-location receptacle where the product will be attended while in use (e.g. portable tools, etc.) must have a weatherproof enclosure when the attachment plug is removed »406.8(B)(2)(b)«.

Required Handhole in a Metal Pole

A Metal pole more than 8 ft (2.5 m) tall above grade

B Metal raceways or other equipment grounding conductors must be bonded to the pole by means of an equipment grounding conductor recognized by 250.118 and sized in accordance with 250.122 »410.15(B)(5)«.

C Support conductors in vertical metal poles used as raceway as provided in 300.19 »410.15(B)(6)«.

D A metal pole must have a handhole not less than 2 in. by 4 in. (50 mm x 100 mm) with a raintight cover to provide access to the supply terminations within the pole or pole base »410.15(B)(1)«.

E A metal pole must be provided with a handhole accessible grounding terminal »410.15(B)(3)(a)«.

F Direct-buried cable, conduit, or other raceways must meet Table 300.5 minimum cover requirements »300.5(A)«.

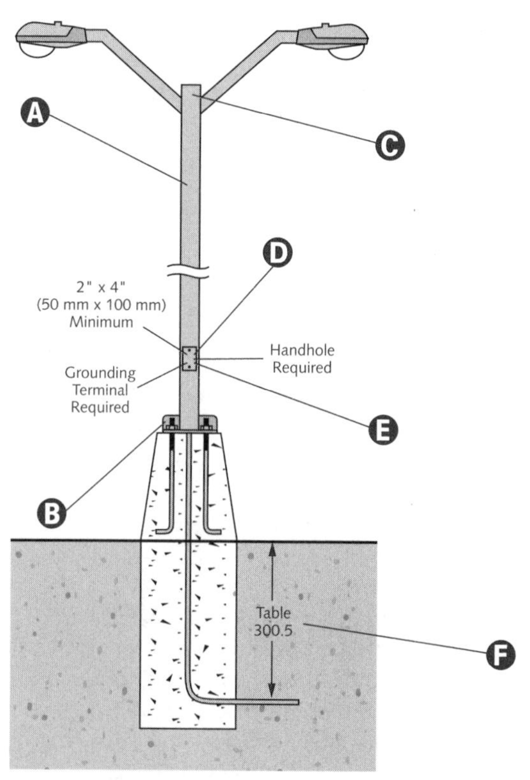

NOTE

Where conductor size is adjusted to compensate for voltage drop, equipment grounding conductors must be adjusted proportionately according to circular mil area »250.122(B)«.

Supply Conductors Not Installed Within the Pole

Ⓐ Where cable or raceway risers are not installed within the pole, a threaded fitting (or nipple) must be brazed (or welded) to the pole, opposite the handhole for the supply connection »410.15(B)(2)«.

Ⓑ Installed raceways subject to physical damage may require additional protection.

Ⓒ Raceways, fittings, boxes, etc. must be securely fastened in place »300.11(A)«. Follow specific raceway support provisions for the type of raceway used.

Ⓓ See Unit 5 for raceway provisions.

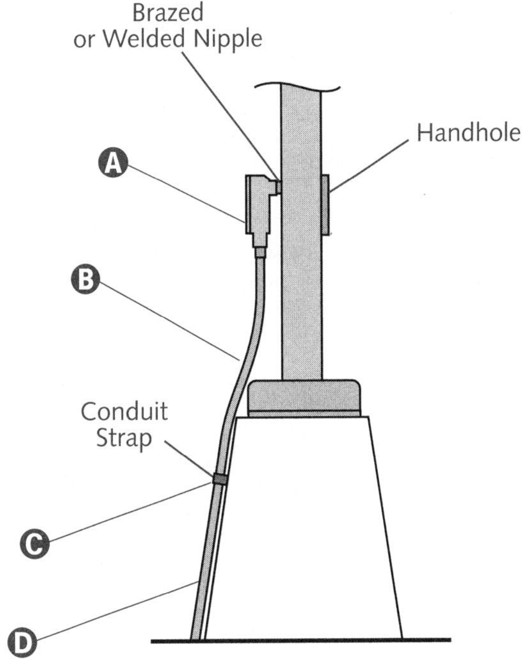

Summary

- Receptacle placement provisions for one-family and multifamily dwellings do not apply to commercial occupancies.
- Although a bathroom receptacle is not required in a commercial occupancy, it must be GFCI protected if installed.
- Each commercial building/occupancy open to pedestrians must be provided with at least one 20-ampere outlet in an accessible location at each entrance to every tenant space for sign or outline lighting systems use.
- One similarity in calculating branch-circuit conductors for continuous loads and motor loads is that each is multiplied by 125%.
- Article 440 contains air-conditioning and refrigeration equipment requirements.
- Commercial receptacle(s) are computed at not less than 180 volt-amperes per strap.
- No more than 10 receptacles are permitted on a 15-ampere branch-circuit, and 13 on a 20-ampere branch-circuit.
- At least 2 receptacle outlets must be readily accessible in guest rooms of hotels, motels, and similar occupancies.
- Table 220.3(A) gives the volt-ampere (per sq ft) lighting load for listed occupancies.
- Compute feeder and/or service track lighting at 150 volt-amperes for every 2 ft (610 mm) of lighting track or fraction thereof.
- Calculate feeder and/or service show-window lighting at not less than 200 volt-amperes per linear ft of show window.
- Certain luminaires (lighting fixtures), such as end-to-end connected fluorescent lights, have a limited use as raceways.
- A handhole is required in a metal pole taller than 8 ft (2.5 m)above grade, unless the pole is no more than 20 ft (6.0 m) with a hinged base.

Unit 12 Competency Test

***NEC*® Reference** **Answer**

______ ______ 1. In computing the load of lighting units that employ an autotransformer, the load shall be based on _____.

______ ______ 2. _____ luminaires (fixtures) used in a show window can be externally wired.

______ ______ 3. Luminaires (fixtures) shall be wired with conductors having insulation suitable for the _____ to which the conductors will be subjected.

______ ______ 4. Systems of illumination utilizing fluorescent lamps, HID lamps, or neon tubing are defined as _____.

______ ______ 5. The international term for a lighting fixture is _____ and is defined as a complete lighting unit consisting of lamp(s) together with the parts designed to distribute the light, to position and protect the lamps, and to connect the lamps to the power supply.

______ ______ 6. The total rating of utilization equipment fastened in place, other than luminaires (lighting fixtures), shall not exceed _____ % of the branch-circuit ampere rating where lighting units, cord- and plug-connected utilization equipment not fastened in place, or both, are also supplied.

______ ______ 7. A metal pole supporting a luminaire (lighting fixture) must have a handhole not less than _____ in. with a raintight cover to provide access to the supply terminations within the pole or pole base.

______ ______ 8. Branch-circuits that supply neon tubing installations shall not be rated in excess of _____ amperes.

______ ______ 9. Multioutlet assembly shall not be permitted:

I. to be run through hoistways

II. to be run within dry partitions

III. to be run through dry partitions

a) II only b) III only c) I and II only d) I, II, and III

______ ______ 10. Where portable lamps are used with Edison-base lampholders, the grounded conductor shall be identified and attached to the _____ of the attachment plug.

______ ______ 11. When computing track lighting loads for a commercial service, an additional load of _____ volt-amperes shall be included for every _____ ft of lighting track or fraction thereof.

______ ______ 12. The terminal of an electric-discharge lamp shall be considered as a live part where any lamp terminal is connected to a circuit of over _____.

______ ______ 13. _____ luminaires (fixtures) surface mounted over concealed outlet, pull, or junction boxes shall be installed with suitable openings in back of the luminaire (fixture) to provide access to the boxes.

______ ______ 14. When computing show-window lighting loads for a commercial service, a load of not less than _____ shall be included for each linear foot of show window, measured horizontally along its base.

______ ______ 15. Twenty-four ft (3 eight-ft sections) of track lighting must be installed in a continuous row in a retail store. What is the minimum number of supports required?

______ ______ 16. No handhole shall be required in a pole, supporting a luminaire (lighting fixture), that is _____ ft or less in height above grade where the supply wiring method continues without splice or pull point, and where the interior of the pole and any splices are accessible by removing the luminaire (fixture).

***NEC®* Reference**	**Answer**	
______	______	17. A hermetic refrigerant motor-compressor's branch-circuit short-circuit and ground-fault protective device shall be capable of carrying _____ of the motor's starting current. a) 100% b) 125% c) 150% d) 175%
______	______	18. Heavy-duty lighting track is lighting track identified for use exceeding _____.
______	______	19. At least _____ receptacle outlet(s) shall be readily accessible in hotel and motel guest rooms.
______	______	20. Flexible cords shall be secured to the undersides of showcases so that a separation between cases not in excess of _____in(es)., nor more than _____ in(es). between the first case and the supply receptacle, will be ensured.
______	______	21. Branch-circuits that supply sign and outline lighting systems containing incandescent and fluorescent forms of illumination shall be rated not to exceed _____.
______	______	22. No handhole shall be required in a metal pole, supporting a luminaire (lighting fixture), that is _____ ft or less in height above grade that is provided with a hinged base.
______	______	23. A single piece of equipment consisting of a multiple receptacle comprised of four or more receptacles shall be computed at not less than _____ per receptacle.
______	______	24. Luminaires (lighting fixtures) shall not be used as a raceway for circuit conductors unless _____ for use as a raceway.
______	______	25. At least one receptacle outlet shall be installed above a show window for each _____ linear ft or major fraction thereof of show-window area measured horizontally at its maximum width.
______	______	26. On hermetic refrigerant motor-compressors, the value of branch-circuit selection current will always be _____ the marked rated-load current. I. less than II. equal to III. greater than a) I only b) I or II c) II or III d) III only
______	______	27. Electric-discharge luminaires (lighting fixtures) provided with mogul-base, screw-shell lampholders shall be permitted to be connected to branch-circuits of _____ amperes or less by cords complying with 240.4.
______	______	28. The load for the required sign branch-circuit shall be computed at a minimum of _____.
______	______	29. A(n) _____ branch-circuit consists of two or more ungrounded conductors having a potential difference between them, and a grounded conductor having equal potential difference between it and each ungrounded conductor of the circuit and that is connected to the neutral or grounded conductor of the system.
______	______	30. Lighting track shall be securely mounted so that each fastening will be suitable for supporting the maximum weight of luminaires (fixtures) that can be installed. Unless identified for supports at greater intervals, a single section _____ ft or shorter in length shall have _____ supports.
______	______	31. Switches, flashers, and similar devices controlling sign transformers and electronic power supplies shall be rated for controlling inductive loads or have a current rating not less than _____ of the transformer's current rating. a) 80% b) 100% c) 125% d) 200%

NEC® Reference	Answer	
________	________	32. Conductors of circuits rated 600 volts, nominal, or less, ac circuits, and dc circuits shall be permitted to occupy the same enclosure if: a) all conductors within the enclosure have an insulation rating equal to at least 600 volts. b) all conductors operating at different voltage levels within the enclosure are separated by a permanent barrier. c) all conductors have an insulation rating equal to at least the maximum circuit voltage applied to any conductor within the enclosure. d) each conductor has an insulation rating equal to at least the maximum circuit voltage applied to each individual conductor within the enclosure.
________	________	33. Branch-circuit conductors within _____ in(es). of a ballast shall have an insulation temperature rating not lower than 90°C, unless supplying a luminaire (fixture) listed and identified as suitable for a different insulation temperature.
________	________	34. The screw shell of all sign lampholders supplied by a grounded circuit shall be connected to the _____ conductor.
________	________	35. Where connected to a branch-circuit having a rating in excess of _____, lampholders shall be of the heavy-duty type.
________	________	36. A recessed luminaire (lighting fixture) that is not identified for contact with insulation shall have all recessed parts spaced at least _____ in(es). from combustible materials.

UNIT 13

SECTION FOUR: COMMERCIAL LOCATIONS

Non-Dwelling Load Calculations

Objectives

After studying this unit, the student should:

- have a good understanding of the elements required to perform a non-dwelling load calculation.
- know that receptacle outlets are counted, unlike in dwelling-unit load calculations.
- understand how to apply Table 220.13 demand factor to receptacle loads in excess of 10 kVA.
- be able to compute the receptacle load for banks and office buildings where the actual number of receptacle outlets is unknown.
- be familiar with volt-ampere unit loads for different types of occupancies, and even for different areas within certain occupancies.
- know when and how to apply Table 220.11 demand factors.
- be aware that track lighting is computed in addition to the general lighting load.
- know when to include a sign and/or outline lighting outlet in a load calculation.
- understand the method for calculating show-window lighting loads.
- be able to determine whether a load is continuous or noncontinuous.
- understand that continuous loads require the inclusion of an additional 25% volt-ampere rating in the load calculation.
- know that the load calculation rating represents only a minimum rating.

Introduction

Unit 13 contains a non-dwelling load calculation form, along with a detailed explanation of each line. Dwelling calculations are covered in Unit 8 (one-family) and Unit 11 (multi-family) of this text. Unlike dwelling unit load calculations, receptacle outlets are counted. Receptacle outlets (if known) and fixed multioutlet assemblies (if any) are entered into the calculation, and if the load is great enough, a demand factor is applied. As with dwelling-unit load calculations, general lighting is computed using outside dimensions. Other items, such as a sign outlet (where required) and show-window(s) (if present) are part of the non-dwelling load calculation. Continuous Loads (Line 10) is a very important computation. All continuous load ratings must be increased by 25%. While Kitchen Equipment (Line 11) is not included in every type of calculation, be aware that kitchen equipment is not limited to restaurants. For instance, kitchen equipment could be a portion of a load calculation for a school. Line 14 (All Other Loads) is a catchall for any load not included otherwise in the calculation.

The load calculation form has little room for listing individual items, such as motors, equipment, etc. Depending upon the size of the occupancy, the calculation could contain hundreds, if not thousands, of individual items. Space is also limited for certain calculations. When the need arises, simply attach additional sheets of paper containing extra items and/or calculations. Some procedure(s) will not apply to certain load calculations. It is recommended that a line not be left completely empty. Some predetermined marking should fill the space; i.e., a dashed line, the letters NA, etc. One of the procedures (Line 4) is applicable only in banks and office buildings where the actual number of receptacle outlets is unknown.

The load calculation form results (overcurrent protection and conductors) represent only a *minimum* requirement. No consideration is given for the addition of future electrical loads. The size service and/or feeder is not restricted to the form's calculated size. For example, an electrician might install a 200-ampere service in an occupancy where the load calculation only required a 125-ampere rating.

Because certain cities, states, etc. require that some (if not all) electrical installations be designed by a licensed Electrical Engineer, caution is advised. Check with local authorities to determine these, as well as other, requirements.

NON-DWELLING LOAD CALCULATIONS

Line 1—Receptacle Load

A load of 15,300 volt-amperes would be placed in Line 1 for a commercial occupancy having 75 duplex and 10 single receptacles. (75 + 10 = 85 × 180 = 15,300)

Ⓐ General purpose receptacles are not continuous loads.

Ⓑ Receptacle outlets are computed at a minimum of 180 volt-amperes for each single (or multiple) receptacle on one yoke (or strap) »220.3(B)(9)« (See Unit 12).

Ⓒ A single piece of equipment (consisting of four or more receptacles) must be computed at no less than 90 volt-amperes per receptacle »220.3(B)(9)« (See Unit 12).

> NOTE
>
> *Because receptacles located in hotel/motel guest rooms are included in the general lighting load calculation, no additional load calculation is required.*

1 Receptacle Load (noncontinuous) 220.3(B)(9)
Multiply each single or multiple receptacle on one strap by 180 volt-amperes.
Multiply each single piece of equipment comprised of 4 or more receptacles by 90 VA per receptacle.

Line 2—Fixed Multioutlet Assembly Load

A commercial occupancy has 75 linear ft of fixed multioutlet assembly, with 15 ft of the assembly subject to simultaneous use. A load of 4860 volt-amperes is placed in Line 2, because 75 – 15 = 60 ft (non-simultaneous); 60 ÷ 5 = 12 × 180 = 2160 volt-amperes for non-simultaneous use multioutlet assembly; 15 × 180 = 2700 volt-amperes for simultaneous use multioutlet assembly; and 2160 + 2700 = 4860 volt-amperes total for multioutlet assembly.

Ⓐ Where simultaneous use of appliances is unlikely, each 5 ft (1.5 m) or fraction thereof (of separate and continuous lengths) are considered one outlet of no less than 180 volt-amperes »220.3(B)(8)(a)«. The number of receptacles within the 5-ft (1.5-m) measurement is irrelevant.

Ⓑ Where simultaneous use of appliances is likely, each 1 ft (300 mm) or fraction thereof shall be considered one outlet of no less than 180 volt-amperes »220.3(B)(8)(b)«.

> NOTE
>
> *Because fixed multioutlet assemblies in hotel/motel guest rooms are included in the general lighting load calculation, no additional load calculation is needed.*

2 Fixed Multioutlet Assemblies (noncontinuous) 220.3(B)(8)
Where not likely to be used simultaneously, multiply each 5-ft section by 180 volt-amperes.
Where likely to be used simultaneously, multiply each 1-ft section by 180 volt-amperes.

Non-Dwelling Feeder/Service Load Calculation

1	Receptacle Load (noncontinuous) 220.3(B)(9) *Multiply each single or multiple receptacle on one strap by 180 volt-amperes.* *Multiply each single piece of equipment comprised of 4 or more receptacles by 90 VA per receptacle.*	1	
2	Fixed Multioutlet Assemblies (noncontinuous) 220.3(B)(8) *Where not likely to be used simultaneously, multiply each 5-ft section by 180 volt-amperes.* *Where likely to be used simultaneously, multiply each 1-ft section by 180 volt-amperes.*	2	
3	Receptacle Load Demand Factor (for nondwelling receptacles) 220.13 *If the receptacle load is more than 10,000 volt-amperes, apply the demand factor from Table 220.13.* *Add lines 1 and 2. Multiply the first 10 kVA or less by 100%. Then, multiply the remainder by 50%.*	3	
4	Unknown Receptacle Load (Banks and Office buildings only) *Where the actual number of general purpose receptacle outlets is unknown, include 1 volt-ampere per sq ft Table 220.3(A) footnote [b]* 1 × ____________ (sq ft outside dimensions) =	4	
5	General Lighting Load Table 220.3(A) *Multiply the volt-ampere unit load (for the type of occupancy) by the sq ft outside dimensions.* ________ (VA unit load) × ____________ (sq ft outside dimensions) =	5	
6	Lighting Load Demand Factors 220.11 . . . *Apply Table 220.11 demand factors to certain portions of hospitals, hotels, motels, apartment houses (without provisions for cooking), and storage warehouses. Do not include areas in hospitals, hotels, and motels where the entire lighting will be used at one time.*	6	
7	Track Lighting (in addition to general lighting) 220.12(B) *Include 150 volt-amperes for every 2 ft, or fraction thereof, of lighting track.* ________ (total linear ft) ÷ 2 ________ × 150 =	7	
8	Sign and/or Outline Lighting Outlet (where required) 220.3(B)(6) *Each commercial building (or occupancy) accessible to pedestrians must have at least one outlet per tenant space entrance. 600.5(A) Each outlet must be at least 1200 volt-amperes.*	8	
9	Show-Window Lighting 220.12(A) *Include at least 200 volt-amperes for each linear ft, measured horizontally along the show window's base.* ____________ (total linear ft of show window) × 200 =	9	
10	Continuous Loads 215.2(A), 215.3, and 230.42(A) . . . *Multiply the continuous load volt-amperes (listed above) by 25%. (General purpose receptacles are not considered continuous.)* ____________ (total continuous load volt-amperes) × 25% =	10	
11	Kitchen Equipment 220.20 *Multiply three or more pieces of equipment by the Table 220.20 demand factor (%).* Use Table 220.19 for household cooking equipment used in instructional programs. Table 220.19 Note 5	11	
12	Noncoincident Loads 220.21 . . . *The smaller of two (or more) noncoincident loads can be omitted, as long as they will never be energized simultaneously (such as certain portions of heating and A/C systems). Calculate fixed electric space-heating loads at 100% of the total connected load. 220.15*	12	
13	Motor Loads 220.4(A), 430.24, 430.25, 430.26, and Article 440 *Motor-driven air-conditioning and refrigeration equipment is found in Article 440.* *Multiply the largest motor (one motor only) by 25% and add it to the load.*	13	
14	All Other Loads . . . *Add all other noncontinuous loads into the calculation at 100%.* *Multiply all other continuous loads (operating for 3 hours or more) by 125%.*	14	
15	Total Volt-Ampere Demand Load: *Add Lines 3 through 14 to find the minimum required volt-amperes.*	15	
16	Minimum Amperes *Divide the total volt-amperes by the voltage* ________ (line 15) ÷ ________ (voltage) = **16** ________ (minimum amperes) **17 Minimum Size Service and/or Feeder 240.6(A)**	17	
18	Size the Service and/or Feeder Conductors. Tables 310.16 through 310.22 *Use the tables along with 310.15(B)(1) through (7) to determine conductor size. If the overcurrent device is rated more than 800 amperes, the conductor ampacity must be equal to, or greater than, the rating of the overcurrent device. 240.4(C)* **Minimum Size Conductors**	18	
19	Size the Neutral Conductor 220.22 *The neutral service and/or feeder conductor can be smaller than the ungrounded (hot) conductors, but not smaller than the maximum unbalanced load determined by Article 220. 250.24(B)(1) states that the neutral cannot be smaller than the required grounding electrode conductor specified in Table 250.66. A further demand factor is permitted for any neutral load over 200 amperes.* **Minimum Size Neutral Conductor**	19	
20	Size the Grounding Electrode Conductor (for Service) 250.66 *Use line 18 to find the grounding electrode conductor in Table 250.66.* Size the Equipment Grounding Conductor (for Feeder) 250.122 *Use line 17 to find the equipment grounding conductor in Table 250.122.* *Equipment grounding conductor types are listed in 250.118.* **Minimum Size Grounding Electrode Conductor . . . *or* . . . Equipment Grounding Conductor**	20	

Line 3—Receptacle Load Demand Factor

A commercial occupancy has a receptacle load of 15,300 volt-amperes and a fixed multioutlet assembly load of 4860. The total receptacle load is 20,160 volt-amperes. First, subtract 10,000 from the total receptacle load (20,160 – 10,000 = 10,160). Next, multiply the remainder by 50% (10,160 × 50% = 5080). Finally, add the result back to the original 10,000 volt-amperes (5080 + 10,000 = 15,080). The receptacle load, after demand, is then entered into Line 3.

Ⓐ Receptacle loads in other than dwelling units are computed at no more than 180 volt-amperes per outlet in accordance with 220.3(B)(9) and fixed multioutlet assemblies are computed per 220.3(B)(8). Both can be added to the lighting load and made subject to the demand factors given in Table 220.11, or they can be made subject to Table 220.13 demand factors »220.13«.

Ⓑ If the total receptacle load is no more than 10,000 volt-amperes, insert the number directly in Line 3.

Ⓒ If the total receptacle load is more than 10,000 volt-amperes, subtract 10,000 from the total, and multiply the remainder by 50%. Finally, add that number to the original 10,000 and place the total in Line 3.

NOTE

General purpose receptacles are not considered continuous loads

Ⓐ Ⓑ

3 Receptacle Load Demand Factor (for nondwelling receptacles) 220.13 *If the receptacle load is more than 10,000 volt-amperes, apply the demand factor from Table 220.13. Add lines 1 and 2. Multiply the first 10 kVA or less by 100%. Then, multiply the remainder by 50%.*	3	

Ⓒ

Line 4—Unknown Receptacle Load

It is not known how many receptacles are in an 8500-sq-ft bank. A load of 8500 volt-amperes should be placed in Line 4 (8500 × 1 = 8500). Since the actual number of receptacle outlets is unknown, Line 3 would not contain a number.

Ⓐ In addition, a unit load of 1 volt-ampere per sq ft must be included for general purpose receptacle outlets where the actual number is unknown »Table 220.3(A) footnote [b]«.

Ⓑ If the actual number of general purpose receptacle outlets in a bank or office building is known, Line 4 would not contain a number.

Ⓒ Line 4 would contain a figure only if the occupancy is a bank or office building.

CAUTION *Line 4 applies only to bank and office building occupancies.*

4 Unknown Receptacle Load (Banks and Office buildings only) *Where the actual number of general purpose receptacle outlets is unknown, include 1 volt-ampere per sq ft Table 220.3(A) footnote [b]*	1 × ________ = (sq ft outside dimensions)	4	

Line 5—General Lighting Load

An 8500-sq-ft bank has a general lighting load of 29,750 volt-amperes (8500 × 3.5 = 29,750). The receptacle load, whether known or unknown, is not used here.

A 30,000-sq-ft store has a general lighting load of 90,000 volt-amperes (30,000 × 3 = 90,000).

Ⓐ A unit load which meets or exceeds that specified in Table 220.3(A) for occupancies listed therein constitutes the minimum lighting load for each sq ft of floor area »220.3(A)«.

Ⓑ Find the correct volt-ampere unit load located across from the occupancy type, and insert it into the calculation »Table 220.3(A)«.

Ⓒ Each floor's area must be computed using the building's (or area's) outside dimensions »220.3(A)«.

> **NOTE**
>
> *In all Table 220.3(A) occupancies (except one-family dwellings and individual dwelling units of two-family and multifamily dwellings), specific areas can be separately multiplied by different volt-ampere unit loads. For example: assembly halls and auditoriums have a unit load of 1; halls, corridors, closets, and stairways have a unit load of 0.5; and storage spaces have a unit load of 0.25 volt-amperes per sq ft.*

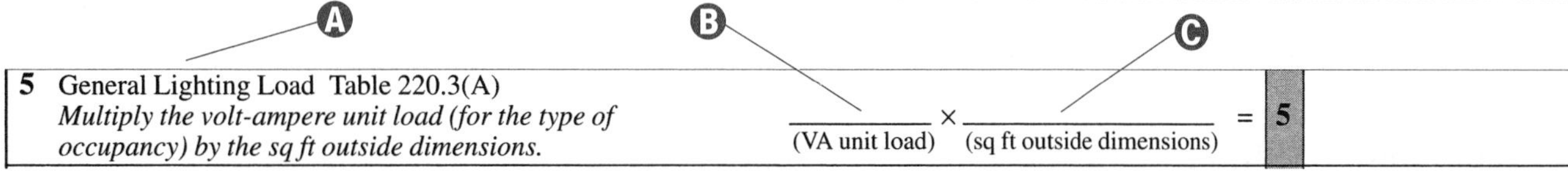

Line 6—Lighting Load Demand Factors

After demand, the general lighting load for hospital patients' rooms where the room dimensions total 100,000 sq ft is 30,000 volt-amperes. First, multiply 100,000 by the hospital volt-ampere unit load (2), found in Table 220.3(A) (100,000 × 2 = 200,000). Next, multiply the first 50,000 by 40% (50,000 × 40% = 20,000). Then, multiply the remainder by 20% (100,000 – 50,000 = 50,000 × 20% = 10,000). Finally, add the two figures (20,000 + 10,000 = 30,000).

Ⓐ Demand factors apply to certain areas in hospitals, hotels, motels, apartment houses (without cooking provisions), and warehouses.

Ⓑ Do not include areas in hospitals, hotels, and motels where all lighting is subject to simultaneous use. Primarily, such lighting is considered a continuous load. Continuous loads must not be derated by Table 220.11, but instead, are increased by 25% (Line 10).

> **NOTE**
>
> *Do not use Table 220.11 to determine the total number of branch-circuits »220.11«.*

Ⓐ

6	Lighting Load Demand Factors 220.11 . . . *Apply Table 220.11 demand factors to certain portions of hospitals, hotels, motels, apartment houses (without provisions for cooking), and storage warehouses. Do not include areas in hospitals, hotels, and motels where the entire lighting will be used at one time.*	6	

Line 7—Track Lighting

A store having 80 ft of lighting track has an additional lighting load of 6000 volt-amperes. First, divide the total length of lighting track by two (80 ÷ 2 = 40). Then, multiply the result by 150 volt-amperes (40 × 150 = 6000).

Ⓐ Do not consider track lighting in dwelling unit (or hotel/motel guest room) service and/or feeder load calculations »220.12(B)«.

Ⓑ Track lighting is calculated in addition to the occupancy general lighting load, found in Table 220.3(A).

Ⓒ Multiply each 2-ft section, or fraction thereof, by 150 volt-amperes »220.12(B)«.

7	Track Lighting (in addition to general lighting) 220.12(B) *Include 150 volt-amperes for every 2 ft, or fraction thereof, of lighting track.*	_______ ÷ 2 _______ × 150 = (total linear ft)	7	

Line 8—Sign and Outline Lighting

A Each commercial building (or occupancy) open to pedestrians must be provided with at least one outlet in a location accessible to each tenant space entrance for sign or outline lighting system use »600.5(A)«.

B Each sign and outline lighting outlet is computed at a minimum of 1200 volt-amperes »220.3(B)(6)«.

CAUTION *If the rating is more than 1200 volt-amperes, use the actual rating of the sign and/or outline lighting.*

NOTE

Because service hallways or corridors are not considered accessible to pedestrians, no circuit is required »600.5(A)«.

8 Sign and/or Outline Lighting Outlet (where required) 220.3(B)(6) *Each commercial building (or occupancy) accessible to pedestrians must have at least one outlet per tenant space entrance. 600.5(A) Each outlet must be at least 1200 volt-amperes.*	**8**	

Line 9—Show-Window Lighting

A store having 75 linear ft of show-window space has a feeder/service load of 15,000 volt-amperes (75 × 200 = 15,000).

A If an occupancy has show window(s), include at least 200 volt-amperes for each linear ft, measured horizontally along the show window's base »220.12(A)«.

NOTE

Calculate show-window branch-circuits in accordance with 220.3(B)(7).

9 Show-Window Lighting 220.12(A) *Include at least 200 volt-amperes for each linear ft, measured horizontally along the show window's base.*	________ (total linear ft of show window) × 200 =	**9**	

Line 10—Continuous Loads

A A load where the maximum current is expected to continue for 3 hours or more is a continuous load »Article 100«.

B Feeder conductors and overcurrent protection must be sized at 100% of noncontinuous loads *plus* 125% of continuous loads »215.2(A) and 215.3«. In other words, an additional 25% of the volt-ampere rating must be added to the continuous loads. This form simplifies that computation. Add all of the continuous loads (listed above Line 10) and multiply by 25%. This rule also applies to service entrance conductors »230.42(A)«.

C General purpose receptacles and multioutlet assemblies are not included in this calculation.

D Total all of the *continuous* loads listed in Lines 5 through 9. These lines are not automatically considered continuous. Only if the load is expected to continue for 3 hours or more is it a continuous load. Examples may include, but are not limited to: general lighting, track lighting, show-window lighting, signs, and outline lighting.

10 Continuous Loads 215.2(A), 215.3, and 230.42(A) . . . *Multiply the continuous load volt-amperes (listed above) by 25%. (General purpose receptacles are not considered continuous.)*	________ (total continuous load volt-amperes) × 25% =	**10**	

Line 11—Kitchen Equipment

Ⓐ Compute the load for commercial electric cooking equipment, dishwasher booster heaters, water heaters, and other kitchen equipment in accordance with Table 220.20. Apply the demand factors to all equipment that: (1) has thermostatic control, or; (2) is intermittently used as kitchen equipment » 220.20«.

Ⓑ Table 220.20 is relatively simple. No derating is allowed for only 1 or 2 pieces of equipment. The demand factor for 3 pieces of equipment is 90%; for 4 pieces, 80%; and for 5 pieces, 70%. The demand factor for more than 5 pieces of equipment is 65%.

Ⓒ For household cooking appliances rated over 1¾ kW and used in instructional programs, refer to Table 220.19 » Table 220.19 Note 5«. For example, a feeder supplying twenty 12-kW household electric ranges, in a school home-economics classroom, requires a minimum rating of only 35 kW « Table 220.19, Column C«.

NOTE

Table 220.20 does not apply to space-heating, ventilating, or air-conditioning equipment.

CAUTION *The feeder or service demand load can never be smaller than the combined rating of the two largest kitchen equipment loads » 220.20«.*

11 Kitchen Equipment 220.20 *Multiply three or more pieces of equipment by the Table 220.20 demand factor (%).* Use Table 220.19 for household cooking equipment used in instructional programs. Table 220.19 Note 5	11

Line 12—Noncoincident Loads

Ⓐ Where it is unlikely that multiple noncoincident loads will be used simultaneously, use only the largest load(s) that will be used at one time, in computing the total feeder/service load » 220.21«.

Ⓑ While fixed electric space-heating loads are computed at 100% of the total connected load, in no case shall a feeder or service load current rating be less than the rating of the largest branch-circuit supplied » 220.15«.

NOTE

Where reduced loading of the conductors results from units operating on duty-cycle, intermittently, or from all units not operating at one time, the AHJ may grant permission for feeder and service conductors to have an ampacity less than 100%, provided the conductors have an ampacity for the load so determined »220.15 Exception«.

12 Noncoincident Loads 220.21 . . . *The smaller of two (or more) noncoincident loads can be omitted, as long as they will never be energized simultaneously (such as certain portions of heating and A/C systems). Calculate fixed electric space-heating loads at 100% of the total connected load. 220.15*	12

Line 13—Motor Loads

Where one or more of the motors of the group are used for short-time, intermittent, periodic, or varying duty, the ampere rating of such motors used in the summation shall be determined in accordance with 430.22(B). For the highest-rated motor, the greater of either the ampere rating from 430.22(B) or the largest continuous-duty motor full-load current multiplied by 1.25 must be used in the summation »430.24 *Exception No.1*«.

Where interlocked circuitry prevents operation of selected motors or other loads at the same time, the conductor ampacity can be based on the summation of the currents of the motors and other loads being operated at the same time resulting in the highest total current »430.24 *Exception No.3*«.

Ⓐ The ampacity of the conductors supplying multimotor and combination-load equipment must not be less than the minimum circuit ampacity marked on the equipment per 430.7(D). If the individual equipment nameplates are visible [per 430.7(D)(2)], but the equipment is not factory-wired, use 430.24 to determine conductor ampacity »430.25«.

Ⓑ Where conductor heating is reduced as a result of motors operating on duty-cycle, intermittently, or from all motors not operating at one time, the AHJ may grant permission for feeder conductors to have an ampacity less than specified in 430.24, provided the conductors have sufficient ampacity for the number of motors supplied and the nature of their loads and duties »430.26«.

Ⓒ Conductors supplying several motors, or a motor(s) and other load(s), must have an ampacity at least equal to the sum of the full-load current rating of all the motors (as determined by 430.6(A), *plus* 25% of the highest-rated motor in the group »430.24«.

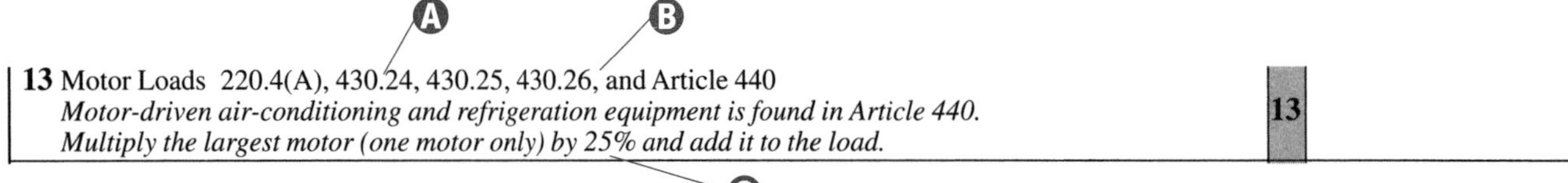

Line 14—All Other Loads

Ⓐ This line is a catchall for loads not falling into one of the previous categories.

Ⓑ Calculate all noncontinuous loads (not yet input) at 100% of their volt-ampere rating.

Ⓒ Total all continuous loads (not yet calculated) and multiply by 125%. Add the result of the continuous loads to that for noncontinuous loads and put the total on Line14.

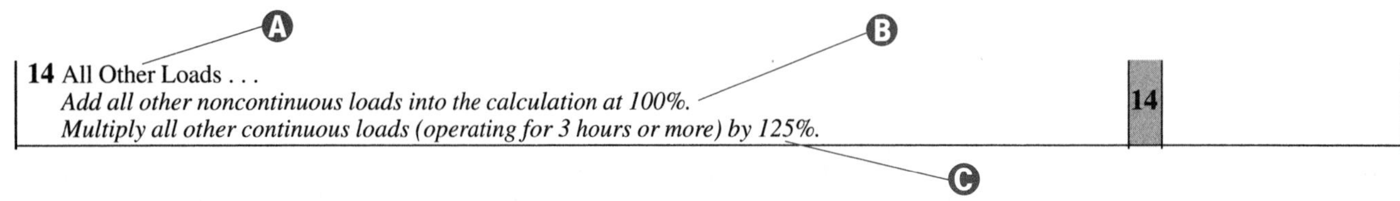

Line 15—Total Volt-Ampere Demand Load

Ⓐ Add all volt-ampere loads listed in Lines 3 through 14 and place the result in Line 15.

> **NOTE**
>
> *Not all lines will contain volt-ampere loads. Be sure to add only the lines containing loads. For example, if the actual number of receptacles in a bank is known, Line 4 will not contain a figure.*

15 Total Volt-Ampere Demand Load: *Add Lines 3 through 14 to find the minimum required volt-amperes.*	**15**	

Lines 16 and 17—Minimum Size Service and/or Feeder

Ⓐ Place the total volt-ampere amount found in Line 15 here.

Ⓑ Write down the source voltage that supplies the feeder/service.

Ⓒ Fractions of an ampere 0.5 and higher are rounded up, while fractions less than 0.5 are dropped »220.2(B)«.

Ⓓ The service (or feeder) overcurrent protection must be higher than the number found in Line 16. Refer to 240.6(A) for a list of fuse and circuit breaker standard ampere ratings.

Ⓔ Divide the volt-amperes by the voltage to find the amperage.

Ⓕ Three-phase voltage is found by multiplying the voltage (single-phase) by 1.732. For example, the source voltage for 208-volt, three-phase is 360 (208 × 1.732).

Ⓖ The result on Line 16 is the minimum amperage rating required for the service and/or feeder being calculated.

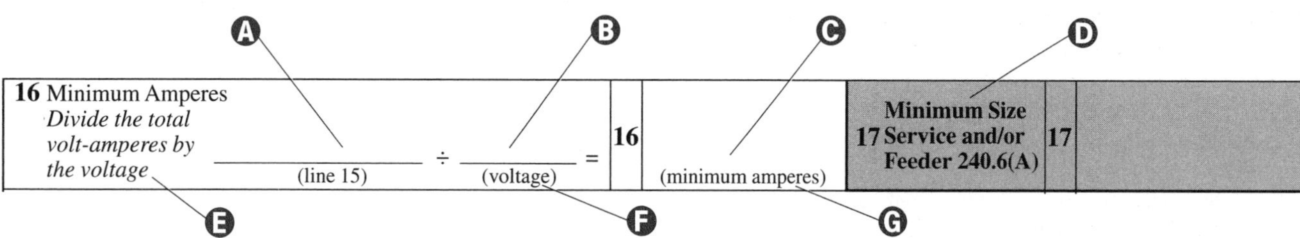

Line 18—Minimum Size Conductors

1/0 AWG and larger conductors can be connected in parallel (electrically joined at both ends to form a single conductor) »310.4«.

The number and type of paralleled conductor sets is a design consideration, not necessarily a *Code* issue. Exercise extreme care when designing a paralleled conductor installation without violating provisions such as 110.14(A), 110.14(C), 240.3(C), 250.66, 250.122(F), 300.20(A), 310.4, etc.

Ⓐ Using Tables 310.16 through 310.22, choose a conductor size which equals (or exceeds) Line 16's minimum ampacity rating. The conductor's ampacity rating does not have to equal or exceed the overcurrent device rating unless its rating exceeds 800 amperes »240.4(B)«.

Ⓑ Ampacity adjustment (correction) factors for more than three current-carrying conductors are located in Table 310.15(B)(2)(a).

NOTE

The paralleled conductors in each phase, neutral, or grounded circuit conductor must share the same characteristics »310.4«. For example, all paralleled, Phase A conductors must:

(1) be of the same length,
(2) have the same conductor material,
(3) be the same size in circular mil area,
(4) have the same insulation type, and
(5) be terminated in the same manner.

WARNING

For overcurrent devices rated over 800 amperes, the ampacity of the conductors protected must equal or exceed the rating of the overcurrent device per 240.6 »240.4(C)«. Do not exceed the temperature limitations outlined in 110.14(C).

Ⓐ

18 Size the Service and/or Feeder Conductors. Tables 310.16 through 310.22 *Use the tables along with 310.15(B)(1) through (7) to determine conductor size.* *If the overcurrent device is rated more than 800 amperes, the conductor ampacity must be equal to, or greater than, the rating of the overcurrent device. 240.4(C)*	**Minimum Size Conductors**	**18**	

Line 19—Neutral Conductor

Include all lighting and receptacle loads having a 120-volt rating in the neutral calculation.

All appliances, equipment, motors, etc. utilizing a grounded conductor are included in the neutral calculation.

A For the purpose of this load calculation, the terms **Neutral** and **Grounded** are synonymous.

B The feeder (or service) neutral load is the maximum unbalance of the load determined by Article 220 »220.22«.

C In addition to 220.22 demand factors, a 70% demand factor is permitted for that portion of the unbalanced load above 200 amperes »220.22«.

D The neutral is sized by finding the maximum unbalanced load.

NOTE

The neutral must not be smaller than the required grounding electrode conductor specified in Table 250.66 »250.24(B)(1)«.

CAUTION *Reduction of neutral capacity is not permitted for that portion of the load that consists of nonlinear loads supplied from a four-wire, wye-connected, three-phase system nor the grounded conductor of a three-wire circuit consisting of two phase wires and the neutral of a four-wire, three-phase, wye-connected system »220.22«.*

19 Size the Neutral Conductor 220.22 *The neutral service and/or feeder conductor can be smaller than the ungrounded (hot) conductors, but not smaller than the maximum unbalanced load determined by Article 220. 250.24(B)(1) states that the neutral cannot be smaller than the required grounding electrode conductor specified in Table 250.66. A further demand factor is permitted for any neutral load over 200 amperes.*	**Minimum Size Neutral Conductor**	**19**	

Line 20—Grounding Conductor

Where the grounding electrode conductor connects to made electrode, and that conductor portion is the sole grounding electrode connection, the maximum size required is 6 AWG copper or 4 AWG aluminum »250.66(A)«.

For a grounding electrode conductor connected to a concrete-encased electrode, serving as the only grounding electrode connection, the maximum size required is 4 AWG copper »250.66(B)«.

If the grounding electrode conductor is connected to a ground ring, and that conductor portion is the sole connection to the grounding electrode, the maximum size required is the size used for the ground ring »250.66(C)«.

A Grounding electrode conductor size is based on the largest service-entrance conductor or equivalent area for parallel conductors »Table 250.66«. (The minimum size service-entrance conductors were determined on Line 18.)

B While Table 250.66 is used for service installations, it is also used in other wiring applications, such as separately derived systems »250.30(A)(2)«, and connections at separate buildings or structures »250.32«.

C While the equipment bonding jumper on the load side of the service overcurrent devices must be sized, as a minimum, in accordance with Table 250.122 sizes, they do not have to be larger than the circuit conductors supplying the equipment »250.102(D)«.

D The equipment grounding conductor size is based on the rating (or setting) of the circuit's automatic overcurrent device ahead of equipment, conduit, etc. »Table 250.122«.

20 Size the Grounding Electrode Conductor (for Service) 250.66 *Use line 18 to find the grounding electrode conductor in Table 250.66.* Size the Equipment Grounding Conductor (for Feeder) 250.122 *Use line 17 to find the equipment grounding conductor in Table 250.122. Equipment grounding conductor types are listed in 250.118.*	**Minimum Size Grounding Electrode Conductor . . . *or* . . . Equipment Grounding Conductor**	**20**	

SAMPLE LOAD CALCULATION—STORE

Load Calculation Information

(A) Load calculation data for a small retail store.

(B) Calculate the floor area from the outside dimensions. In this example, the sq-ft area is already provided.

(C) A three-phase transformer steps down the voltage from 480 to 208Y/120 volts.

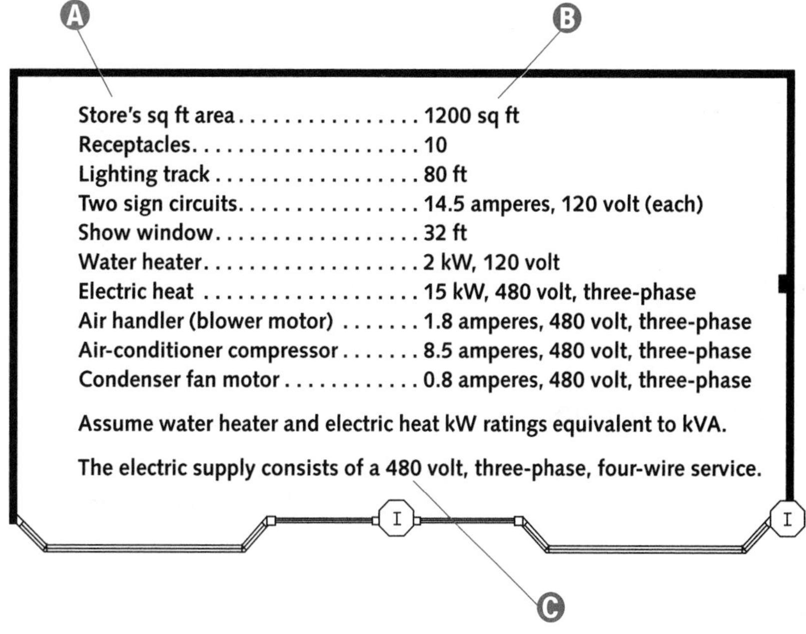

Lines 1 and 2—Store

(A) The different style of box on Line 1 (unlike Lines 3 through 14) indicates that the load will become part (if not all) of Line 3. Line 1 is not included in Line 15's total volt-ampere demand load.

(B) Since there are 10 receptacles, the calculated load is 1800 volt-amperes (10 × 180).

(C) As with Line 1, Line 2 has a different style box from Lines 3 through 14 because the load (if any) becomes part of Line 3. Line 2 is not included in Line 15's total volt-ampere demand load.

(D) Because this store has no fixed multioutlet assembly, a dashed line occupies Line 2.

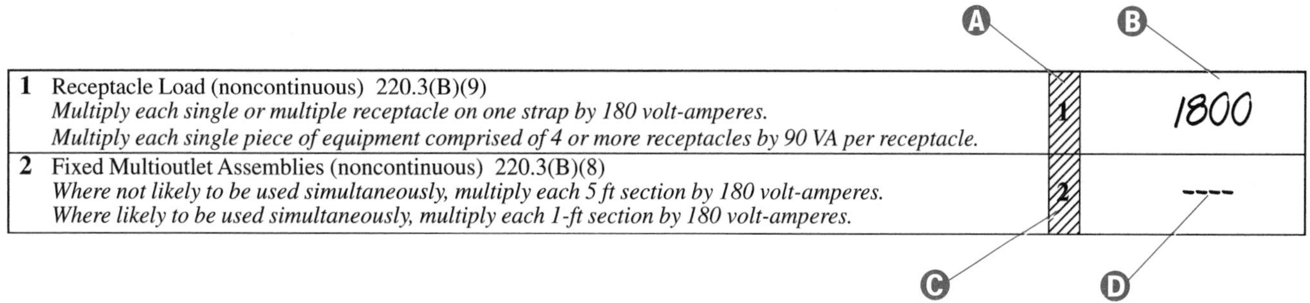

1	Receptacle Load (noncontinuous) 220.3(B)(9) *Multiply each single or multiple receptacle on one strap by 180 volt-amperes.* *Multiply each single piece of equipment comprised of 4 or more receptacles by 90 VA per receptacle.*	1	1800
2	Fixed Multioutlet Assemblies (noncontinuous) 220.3(B)(8) *Where not likely to be used simultaneously, multiply each 5 ft section by 180 volt-amperes.* *Where likely to be used simultaneously, multiply each 1-ft section by 180 volt-amperes.*	2	----

Lines 3 and 4—Store

A Line 3 is the first of twelve lines whose values are combined to determine the total volt-ampere demand load.

B Because there is no multioutlet assembly, and the receptacle load is less than 10 kVA, the 1800 volt-ampere load from Line 1 is inserted in Line 3.

C Use Line 4 only when the actual number of receptacles, in banks and office buildings, is unknown.

D A dashed line has been drawn in Line 4 showing that it has intentionally been left blank.

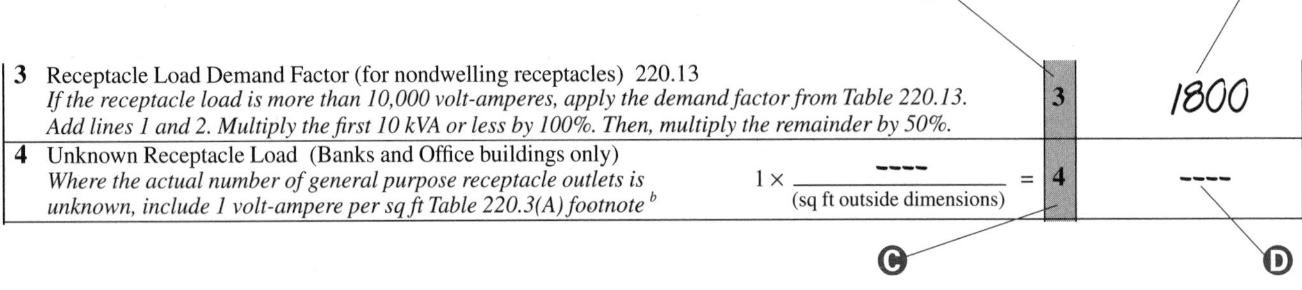

3	Receptacle Load Demand Factor (for nondwelling receptacles) 220.13 *If the receptacle load is more than 10,000 volt-amperes, apply the demand factor from Table 220.13. Add lines 1 and 2. Multiply the first 10 kVA or less by 100%. Then, multiply the remainder by 50%.*	3	1800
4	Unknown Receptacle Load (Banks and Office buildings only) *Where the actual number of general purpose receptacle outlets is unknown, include 1 volt-ampere per sq ft Table 220.3(A) footnote [b]* 1 × ---- (sq ft outside dimensions) =	4	----

Lines 5 and 6—Store

A The volt-ampere unit load for a store is 3 »Table 220.3(A)«.

B Since the store's sq ft area is provided, no calculation is required.

C Store lighting loads are continuous. The result of Line 5 becomes part of the continuous volt-ampere load calculation (Line 10).

D Use Line 6 only for certain portions of hospitals, hotels, motels, storage warehouses, and apartment houses without provisions for cooking.

E The marking in Line 6 indicates that it has intentionally been left blank.

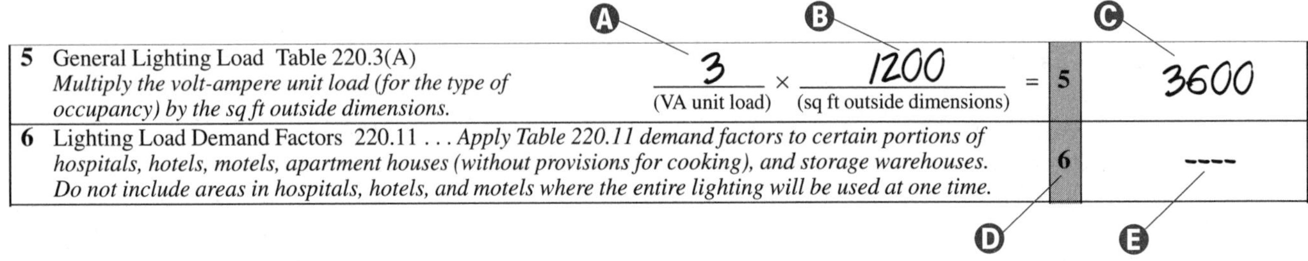

5	General Lighting Load Table 220.3(A) *Multiply the volt-ampere unit load (for the type of occupancy) by the sq ft outside dimensions.* 3 (VA unit load) × 1200 (sq ft outside dimensions) =	5	3600
6	Lighting Load Demand Factors 220.11 . . . *Apply Table 220.11 demand factors to certain portions of hospitals, hotels, motels, apartment houses (without provisions for cooking), and storage warehouses. Do not include areas in hospitals, hotels, and motels where the entire lighting will be used at one time.*	6	----

Lines 7 and 8—Store

A Except for dwelling units or hotel/motel guest rooms, track lighting loads are additional to Table 220.3(A)'s general lighting loads »220.12(B)«.

B This store contains 80 ft of lighting track.

C Base the calculation on every 2 ft, or fraction thereof, of track.

D A store's track lighting load is continuous. Line 7's result becomes part of the continuous volt-ampere load calculation (Line 10).

E At least one outlet, having a minimum rating of 1200 volt-amperes, is required.

F This store's actual sign load is 3420 volt-amperes (2 × 14.5 × 120). This store has two sign circuits.

G Since a store's sign load is continuous, the result of Line 8 becomes part of the continuous volt-ampere load calculation (Line 10).

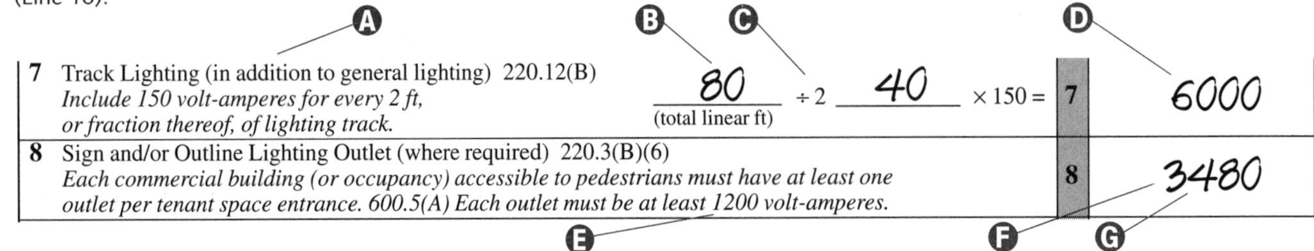

7	Track Lighting (in addition to general lighting) 220.12(B) *Include 150 volt-amperes for every 2 ft, or fraction thereof, of lighting track.* 80 (total linear ft) ÷ 2 40 × 150 =	7	6000
8	Sign and/or Outline Lighting Outlet (where required) 220.3(B)(6) *Each commercial building (or occupancy) accessible to pedestrians must have at least one outlet per tenant space entrance. 600.5(A) Each outlet must be at least 1200 volt-amperes.*	8	3480

Line 9—Store

Ⓐ Not all commercial occupancies have show windows.

Ⓑ The store in this example has 32 linear ft of show window.

Ⓒ Because show-window load is continuous, the result of Line 9 becomes part of the continuous volt-ampere load calculation (Line 10).

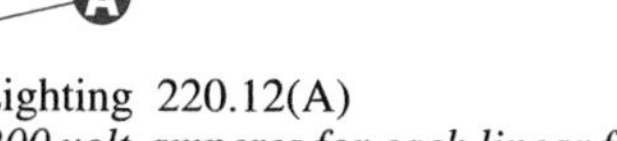
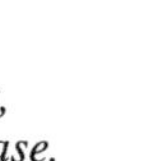
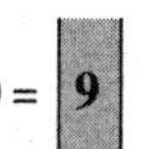

9 Show-Window Lighting 220.12(A) *Include at least 200 volt-amperes for each linear ft, measured horizontally along the show window's base.*	32 (total linear ft of show window) × 200 =	**9**	6400

Lines 10 and 11—Store

Ⓐ Lines 5, 7, 8, and 9 represent the store's continuous loads.

Ⓑ 3600 + 6000 + 3480 + 6400 = 19,480

Ⓒ An additional 25% volt-ampere load has been added because of the continuous loads.

Ⓓ The marking in Line 11 indicates a lack of kitchen equipment.

10 Continuous Loads 215.2(A), 215.3, and 230.42(A) . . . *Multiply the continuous load volt-amperes (listed above) by 25%. (General purpose receptacles are not considered continuous.)*	19,480 (total continuous load volt-amperes) × 25% =	**10**	4870
11 Kitchen Equipment 220.20 *Multiply three or more pieces of equipment by Table 220.20 demand factor (%).* Use Table 220.19 for household cooking equipment used in instructional programs. Table 220.19 Note 5		**11**	----

Lines 12 and 13—Store

Ⓐ Because the combined electric heat and air handler load is larger than the combined compressor, condenser fan motor, and air handler, omit the compressor and condenser fan motor load.

Ⓑ Find the largest motor. In this example that would be the air-handler motor (1.8 × 480 × 1.732 = 1,496.45). Increase the largest motor by 25% (1,496.45 × 25% = 374).

12 Noncoincident Loads 220.21 . . . *The smaller of two (or more) noncoincident loads can be omitted, as long as they will never be energized simultaneously (such as certain portions of heating and A/C systems). Calculate fixed electric space-heating loads at 100% of the total connected load. 220.15*	**12**	16,456
13 Motor Loads 220.4(A), 430.24, 430.25, 430.26, and Article 440 *Motor-driven air-conditioning and refrigeration equipment is found in Article 440. Multiply the largest motor (one motor only) by 25% and add to load.*	**13**	374

Lines 14 and 15—Store

Ⓐ The only load not previously calculated is the water heater. Under normal conditions, a water heater is a noncontinuous load. Add noncontinuous loads to the calculation at 100%.

Ⓑ Add Lines 3 through 14 to find the minimum volt-ampere load (Lines 4, 6, and 11 are blank)

14 All Other Loads . . . *Add all other noncontinuous loads into the calculation at 100%. Multiply all other continuous loads (operating for 3 hours or more) by 125%.*	**14**	2000
15 Total Volt-Ampere Demand Load: *Add Lines 3 through 14 to find the minimum required volt-amperes.*	**15**	44,980

Lines 16 and 17—Store

A Line 15's minimum volt-ampere load

B The total voltage for a 480-volt, three-phase system is 831 (480 × 1.732).

C The minimum ampacity for a 480-volt, three-phase service is 54 amperes.

D The next-higher standard size fuse (or breaker) above 54 amperes is 60 »240.6(A)«.

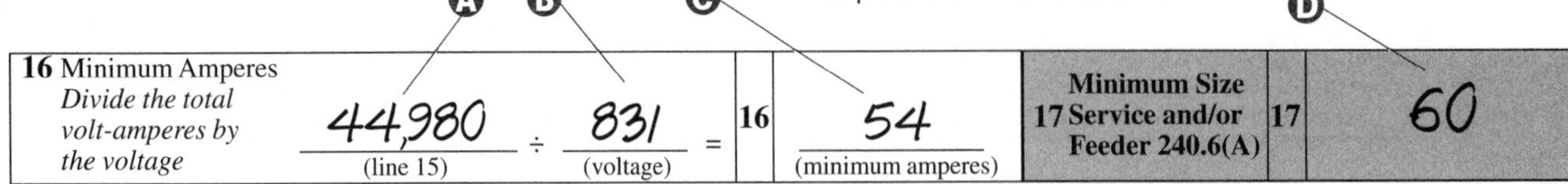

16 Minimum Amperes
Divide the total volt-amperes by the voltage

44,980 (line 15) ÷ 831 (voltage) = **16** 54 (minimum amperes)

Minimum Size 17 Service and/or Feeder 240.6(A) | **17** | 60

Line 18—Store

A The minimum 60/75°C copper service conductor is 6 AWG »Table 310.16«.

B The minimum 60/75°C aluminum service conductor is 4 AWG »Table 310.16«.

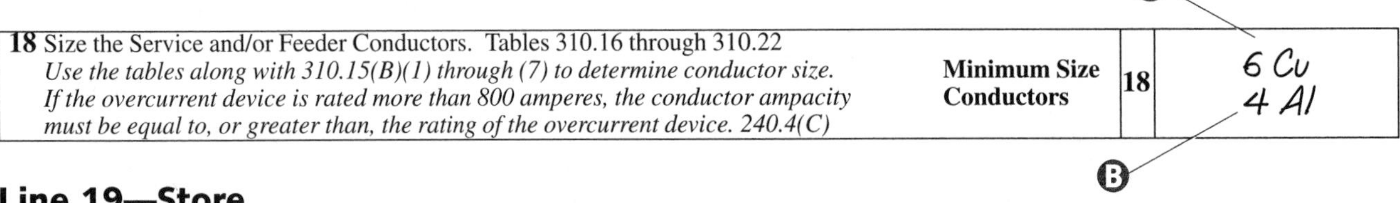

18 Size the Service and/or Feeder Conductors. Tables 310.16 through 310.22
Use the tables along with 310.15(B)(1) through (7) to determine conductor size. If the overcurrent device is rated more than 800 amperes, the conductor ampacity must be equal to, or greater than, the rating of the overcurrent device. 240.4(C)

Minimum Size Conductors | **18** | 6 Cu, 4 Al

Line 19—Store

A Although there will be little (if any) load on the neutral, a neutral conductor is required. Where an ac system (operating at less than 1000 volts) is grounded at any point, the grounded conductor(s) must run to each service disconnecting means, and must bond to each disconnecting means enclosure »250.24(B)«.

B While the grounded conductor must not be smaller than the required grounding electrode conductor specified in Table 250.66, it does not have to be larger than the largest ungrounded service-entrance phase conductor »250.24(B)(1)«.

C Since the minimum size copper grounding electrode conductor is 8 AWG, the neutral must also be 8 AWG.

D Because the minimum size aluminum grounding electrode conductor is 6 AWG, the neutral must also be 6 AWG.

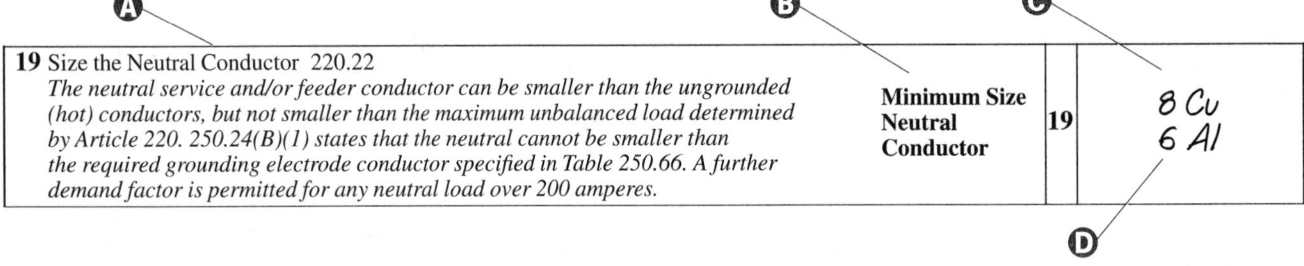

19 Size the Neutral Conductor 220.22
The neutral service and/or feeder conductor can be smaller than the ungrounded (hot) conductors, but not smaller than the maximum unbalanced load determined by Article 220. 250.24(B)(1) states that the neutral cannot be smaller than the required grounding electrode conductor specified in Table 250.66. A further demand factor is permitted for any neutral load over 200 amperes.

Minimum Size Neutral Conductor | **19** | 8 Cu, 6 Al

Line 20—Store

A The grounding electrode conductor's size is based on the service-entrance conductor »Table 250.66«.

B The minimum size copper grounding conductor is 8 AWG.

C The minimum size aluminum grounding conductor is 6 AWG.

20 Size the Grounding Electrode Conductor (for Service) 250.66
Using line 18 to find the grounding electrode conductor in Table 250.66.
Size the Equipment Grounding Conductor (for Feeder) 250.122
Use line 17 to find the equipment grounding conductor in Table 250.122. Equipment grounding conductor types are listed in 250.118.

Minimum Size Grounding Electrode Conductor . . . *or* . . . Equipment Grounding Conductor | **20** | 8 Cu, 6 Al

The following page shows the store's completed load calculation.

Non-Dwelling Feeder/Service Load Calculation

Line	Description		No.	Value
1	Receptacle Load (noncontinuous) 220.3(B)(9) *Multiply each single or multiple receptacle on one strap by 180 volt-amperes.* *Multiply each single piece of equipment comprised of 4 or more receptacles by 90 VA per receptacle.*		1	1800
2	Fixed Multioutlet Assemblies (noncontinuous) 220.3(B)(8) *Where not likely to be used simultaneously, multiply each 5 ft section by 180 volt-amperes.* *Where likely to be used simultaneously, multiply each 1-ft section by 180 volt-amperes.*		2	----
3	Receptacle Load Demand Factor (for nondwelling receptacles) 220.13 *If the receptacle load is more than 10,000 volt-amperes, apply the demand factor from Table 220.13.* *Add lines 1 and 2. Multiply the first 10 kVA or less by 100%. Then, multiply the remainder by 50%.*		3	1800
4	Unknown Receptacle Load (Banks and Office buildings only) *Where the actual number of general purpose receptacle outlets is unknown, include 1 volt-ampere per sq ft Table 220.3(A) footnote* [b]	1 × ---- (sq ft outside dimensions) =	4	----
5	General Lighting Load Table 220.3(A) *Multiply the volt-ampere unit load (for the type of occupancy) by the sq ft outside dimensions.*	3 (VA unit load) × 1200 (sq ft outside dimensions) =	5	3600
6	Lighting Load Demand Factors 220.11 . . . *Apply Table 220.11 demand factors to certain portions of hospitals, hotels, motels, apartment houses (without provisions for cooking), and storage warehouses. Do not include areas in hospitals, hotels, and motels where the entire lighting will be used at one time.*		6	----
7	Track Lighting (in addition to general lighting) 220.12(B) *Include 150 volt-amperes for every 2 ft, or fraction thereof, of lighting track.*	80 (total linear ft) ÷ 2 40 × 150 =	7	6000
8	Sign and/or Outline Lighting Outlet (where required) 220.3(B)(6) *Each commercial building (or occupancy) accessible to pedestrians must have at least one outlet per tenant space entrance. 600.5(A) Each outlet must be at least 1200 volt-amperes.*		8	3480
9	Show-Window Lighting 220.12(A) *Include at least 200 volt-amperes for each linear ft, measured horizontally along the show window's base.*	32 (total linear ft of show window) × 200 =	9	6400
10	Continuous Loads 215.2(A), 215.3, and 230.42(A) . . . *Multiply the continuous load volt-amperes (listed above) by 25%. (General purpose receptacles are not considered continuous.)*	19,480 (total continuous load volt-amperes) × 25% =	10	4870
11	Kitchen Equipment 220.20 *Multiply three or more pieces of equipment by Table 220.20 demand factor (%).* Use Table 220.19 for household cooking equipment used in instructional programs. Table 220.19 Note 5		11	----
12	Noncoincident Loads 220.21 . . . *The smaller of two (or more) noncoincident loads can be omitted, as long as they will never be energized simultaneously (such as certain portions of heating and A/C systems). Calculate fixed electric space-heating loads at 100% of the total connected load. 220.15*		12	16,456
13	Motor Loads 220.4(A), 430.24, 430.25, 430.26, and Article 440 *Motor-driven air-conditioning and refrigeration equipment is found in Article 440.* *Multiply the largest motor (one motor only) by 25% and add to load.*		13	374
14	All Other Loads . . . *Add all other noncontinuous loads into the calculation at 100%.* *Multiply all other continuous loads (operating for 3 hours or more) by 125%.*		14	2000
15	Total Volt-Ampere Demand Load: *Add Lines 3 through 14 to find the minimum required volt-amperes.*		15	44,980
16	Minimum Amperes *Divide the total volt-amperes by the voltage*	44,980 (line 15) ÷ 831 (voltage) = **16** 54 (minimum amperes)	**17** Minimum Size Service and/or Feeder 240.6(A)	60
18	Size the Service and/or Feeder Conductors. Tables 310.16 through 310.22 *Use the tables along with 310.15(B)(1) through (7) to determine conductor size. If the overcurrent device is rated more than 800 amperes, the conductor ampacity must be equal to, or greater than, the rating of the overcurrent device. 240.4(C)*	**Minimum Size Conductors**	18	6 Cu 4 Al
19	Size the Neutral Conductor 220.22 *The neutral service and/or feeder conductor can be smaller than the ungrounded (hot) conductors, but not smaller than the maximum unbalanced load determined by Article 220. 250.24(B)(1) states that the neutral cannot be smaller than the required grounding electrode conductor specified in Table 250.66. A further demand factor is permitted for any neutral load over 200 amperes.*	**Minimum Size Neutral Conductor**	19	8 Cu 6 Al
20	Size the Grounding Electrode Conductor (for Service) 250.66 *Using line 18 to find the grounding electrode conductor in Table 250.66.* Size the Equipment Grounding Conductor (for Feeder) 250.122 *Use line 17 to find the equipment grounding conductor in Table 250.122. Equipment grounding conductor types are listed in 250.118.*	**Minimum Size Grounding Electrode Conductor . . . *or* . . . Equipment Grounding Conductor**	20	8 Cu 6 Al

SAMPLE LOAD CALCULATION—BANK

Load Calculation Information

A Load calculation data for a bank

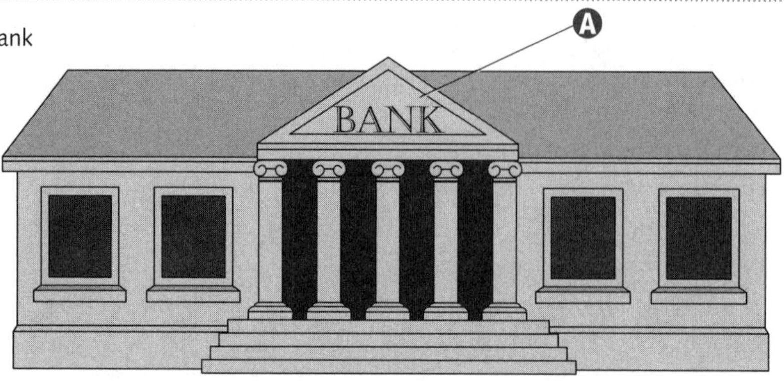

Bank's outside dimensions .... **50' × 100'**
Actual inside connected lighting load .. **16,200 volt-amperes**
Receptacles .... **unknown**
Sign .... **9 amperes, 120 volt**
Show window .... **none**
Parking lot lighting .... **57 amperes, 120 volt**
Water heater .... **4 kW, 208 volt**
Electric heating unit #1 .... **6 kW, 208 volt, three-phase**
Electric heating unit #2 .... **10 kW, 208 volt, three-phase**
Air handler (blower motor) #1 .... **4 amperes, 208 volt, three-phase**
Air handler (blower motor) #2 .... **4 amperes, 208 volt, three-phase**
Air-conditioner compressor #1 .... **22 amperes, 208 volt, three-phase**
Air-conditioner compressor #2 .... **22 amperes, 208 volt, three-phase**
Condenser fan motor #1 .... **2.5 amperes, 208 volt, three-phase**
Condenser fan motor #2 .... **2.5 amperes, 208 volt, three-phase**

Assume water heater and electric heat kW ratings equivalent to kVA.

The electric supply consists of a 208 volt, three-phase, four-wire service.

NOTE

Because the actual number of receptacles is unknown, Line 1 does not contain a number.

Line 2 will not contain a number, since the bank has no fixed multioutlet assemblies.

Lines 3 and 4—Bank

A Since Lines 1 and 2 are blank, Line 3 also does not contain a number. A dashed line has been drawn in Line 3 showing that it has been left blank intentionally.

B A bank (or office building), with an unknown number of general purpose receptacle outlets, requires a rating of 1 volt-ampere per sq ft.

C Since the bank is 5000 sq ft (50 × 100), the receptacle load is 5000 volt-amperes.

NOTE

Use 220.3(B)(8) and (9) for banks (or office buildings) having a known number of receptacles.

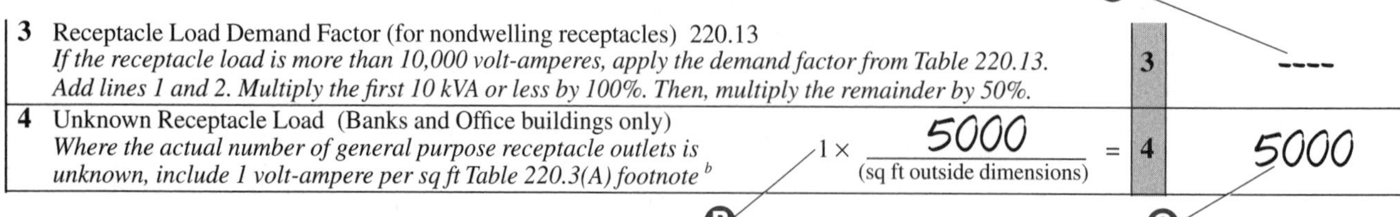

3	Receptacle Load Demand Factor (for nondwelling receptacles) 220.13 *If the receptacle load is more than 10,000 volt-amperes, apply the demand factor from Table 220.13. Add lines 1 and 2. Multiply the first 10 kVA or less by 100%. Then, multiply the remainder by 50%.*		3	----
4	Unknown Receptacle Load (Banks and Office buildings only) *Where the actual number of general purpose receptacle outlets is unknown, include 1 volt-ampere per sq ft Table 220.3(A) footnote [b]*	1 × 5000 (sq ft outside dimensions) =	4	5000

Lines 5 and 6—Bank

A The volt-ampere unit load for a bank is 3.5 »Table 220.3(A)«.

B This bank, with outside dimensions of 50' × 100', has a total calculated area of 5000 sq ft.

C A bank's lighting load is continuous. The result of Line 5 becomes part of the continuous volt-ampere load calculation (Line 10).

D Because Line 6 does not apply to this occupancy, a dashed line has been drawn showing that it has been left blank intentionally.

> **NOTE**
>
> *This bank has no lighting track. The dashed line drawn shows that it has intentionally been left blank.*
>
> *Although the actual connected lighting load, excluding exterior lighting, is 16,200 volt-amperes, Table 220.3(A) requires a minimum lighting load of 17,500 volt-amperes (3.5 × 50 × 100). The greater value of 17,500 must be used in this calculation. Use the actual connected lighting load (if any) only when it is greater than the value computed from Table 220.3(A). Actual versus computed lighting loads are both continuous and must be increased by 25%. This function is performed in Line 10 as a separate computation. Another example can be found in Annex D, Example No. D3.*

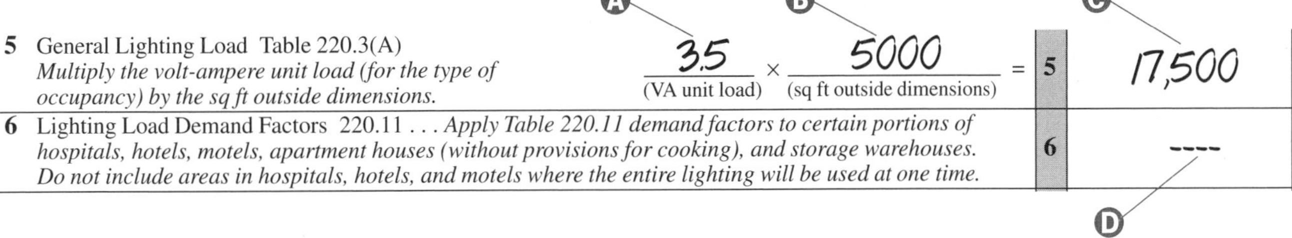

5	General Lighting Load Table 220.3(A) *Multiply the volt-ampere unit load (for the type of occupancy) by the sq ft outside dimensions.*	3.5 (VA unit load) × 5000 (sq ft outside dimensions) =	**5**	17,500
6	Lighting Load Demand Factors 220.11 . . . *Apply Table 220.11 demand factors to certain portions of hospitals, hotels, motels, apartment houses (without provisions for cooking), and storage warehouses. Do not include areas in hospitals, hotels, and motels where the entire lighting will be used at one time.*		**6**	----

Line 8—Bank

A Although the bank's actual sign load is 1080 volt-amperes (9 × 120), each sign outlet must have a minimum rating of 1200 volt-amperes. Use the actual sign and/or outline load(s) only if the rating is greater than the required minimum.

B This bank's sign load is continuous. Therefore, the rating is part of the continuous volt-ampere load calculation (Line 10).

> **NOTE**
>
> *Because this bank has no show window, a show-window lighting load is not required.*

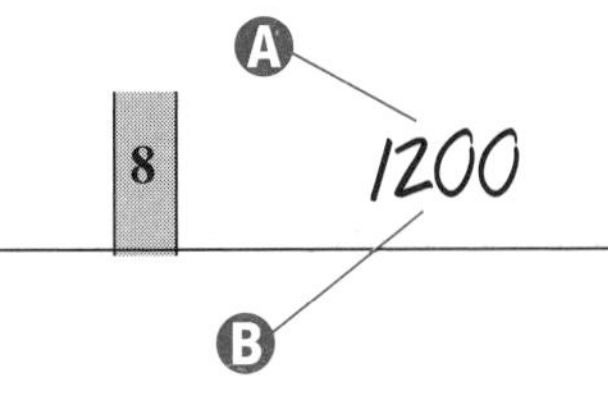

8	Sign and/or Outline Lighting Outlet (where required) 220.3(B)(6) *Each commercial building (or occupancy) accessible to pedestrians must have at least one outlet per tenant space entrance. 600.5(A) Each outlet must be at least 1200 volt-amperes.*	**8**	1200

Line 10—Bank

A This store's continuous loads are found on Lines 5 and 8.

B 17,500 + 1200 = 18,700

C An additional 25% volt-ampere load is added to the computation because of the continuous loads.

> **NOTE**
>
> *Line 11 contains a dashed line because there is no kitchen equipment.*

10	Continuous Loads 215.2(A), 215.3, and 230.42(A) . . . *Multiply the continuous load volt-amperes (listed above) by 25%. (General purpose receptacles are not considered continuous.)*	18,700 (total continuous load volt-amperes) × 25% =	**10**	4675

Lines 12 and 13—Bank

Each air handler load is 1441 volt-amperes (4 × 208 × 1.732). Each condenser-fan motor load is 901 volt-amperes (2.5 × 208 × 1.732). Each compressor motor load is 7926 volt-amperes (22 × 208 × 1.732).

The total combined compressor, condenser-fan motor, and air handler load is 20,536 volt-amperes (7926 + 7926 + 901 + 901 + 1441 + 1441).

The total combined electric heat and air handler load is 18,882 (6000 + 10,000 + 1441 + 1441).

A Because the combined compressor, condenser-fan motor, and air handler load is larger than the combined electric heat and air handler, omit the electric heat and air handler load.

B Find the largest motor. In this example, that is the compressor motor (22 × 208 × 1.732 = 7,925.63). Increase the largest motor by 25% (7,925.63 × 25% = 1981).

12 Noncoincident Loads 220.21 . . . *The smaller of two (or more) noncoincident loads can be omitted, as long as they will never be energized simultaneously (such as certain portions of heating and A/C systems). Calculate fixed electric space heating loads at 100% of the total connected load. 220.15*	12	20,536 (A)
13 Motor Loads 220.4(A), 430.24, 430.25, 430.26, and Article 440 *Motor-driven air-conditioning and refrigeration equipment is found in Article 440. Multiply the largest motor (one motor only) by 25% and add to load.*	13	1981 (B)

Lines 14 and 15—Bank

A At this point, two loads have not been calculated: one continuous and one noncontinuous load. Because parking-lot lighting is continuous, multiply by 125% (57 × 120 × 125% = 8550 volt-amperes). The water heater is a noncontinuous load. These two loads have a combined total rating of 12,550 volt-amperes (8550 + 4000).

B Add Lines 3 through 14 to determine the minimum volt-ampere load (Lines 3, 6, 7, 9, and 11 are blank).

14 All Other Loads . . . *Add all other noncontinuous loads into the calculation at 100%. Multiply all other continuous loads (operating for 3 hours or more) by 125%.*	14	12,550 (A)
15 Total Volt-Ampere Demand Load: *Add Lines 3 through 14 to find the minimum required volt-amperes.*	15	63,442 (B)

Lines 16 and 17—Bank

A Line 15's minimum volt-ampere load

B The total voltage for a 208 volt, three-phase system is 360 (208 × 1.732).

C The minimum ampacity for a 208 volt, three-phase service is 176 amperes.

D The next higher standard size fuse (or breaker) above 176 amperes is 200 »240.6(A)«.

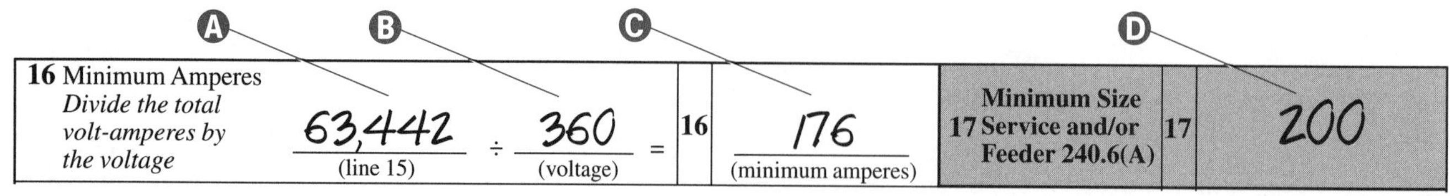

16 Minimum Amperes *Divide the total volt-amperes by the voltage*	63,442 (A) (line 15)	÷	360 (B) (voltage)	=	16	176 (C) (minimum amperes)	**Minimum Size 17 Service and/or Feeder 240.6(A)**	17	200 (D)

Line 18—Bank

A The minimum 75°C copper service conductor is 3/0 AWG »Table 310.16«.

B The minimum 75°C aluminum service conductor is 4/0 AWG »Table 310.16«.

18 Size the Service and/or Feeder Conductors. Tables 310.16 through 310.22
Use the tables along with 310.15(B)(1) through (7) to determine conductor size. If the overcurrent device is rated more than 800 amperes, the conductor ampacity must be equal to, or greater than, the rating of the overcurrent device. 240.4(C)

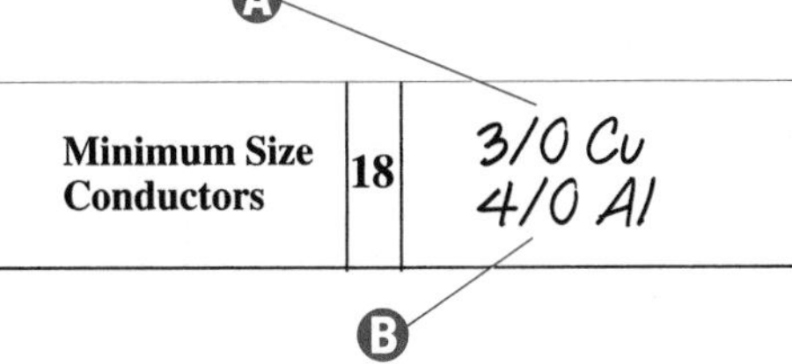

Line 19—Bank

A The neutral conductor computation requires only loads having a neutral or grounded conductor termination.In this example, include Lines 4, 5, 8, 10, and part of Line 14. Line 14's neutral load calculation is 8550 volt-amperes (57 × 120 × 125%). The total neutral load is 36,925 volt-amperes (5000 + 17,500 + 1200 + 4675 + 8550). The minimum required neutral rating is 103 amperes (36,925 ÷ 360).

B The minimum 75°C copper neutral conductor is 2 AWG »Table 310.16«

C The minimum 75°C aluminum neutral conductor is 1/0 AWG »Table 310.16«.

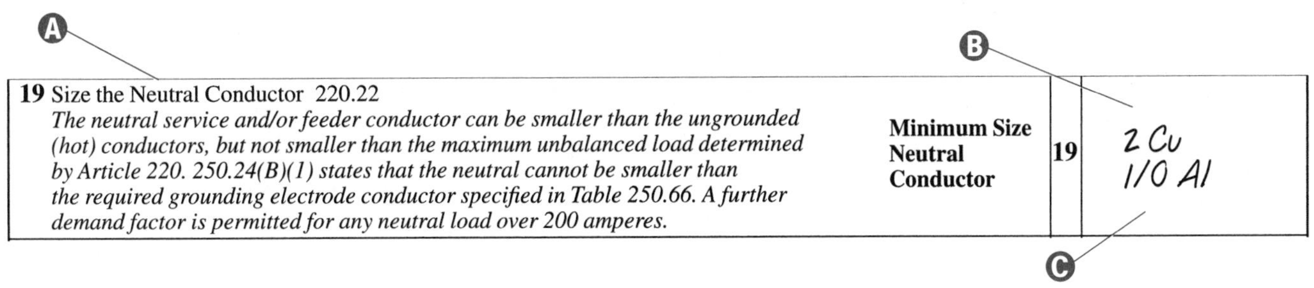

19 Size the Neutral Conductor 220.22
The neutral service and/or feeder conductor can be smaller than the ungrounded (hot) conductors, but not smaller than the maximum unbalanced load determined by Article 220. 250.24(B)(1) states that the neutral cannot be smaller than the required grounding electrode conductor specified in Table 250.66. A further demand factor is permitted for any neutral load over 200 amperes.

Minimum Size Neutral Conductor	**19**	2 Cu 1/0 Al

Line 20—Bank

A The grounding electrode conductor's size is based on the service-entrance conductor »Table 250.66«.

B The minimum size copper grounding conductor is 4 AWG.

C The minimum size aluminum grounding conductor is 2 AWG.

20 Size the Grounding Electrode Conductor (for Service) 250.66
Using line 18 to find the grounding electrode conductor in Table 250.66.
Size the Equipment Grounding Conductor (for Feeder) 250.122
Use line 17 to find the equipment grounding conductor in Table 250.122. Equipment grounding conductor types are listed in 250.118.

Minimum Size Grounding Electrode Conductor . . . *or* . . . Equipment Grounding Conductor	**20**	4 Cu 2 Al

The following page shows the bank's completed load calculation.

Non-Dwelling Feeder/Service Load Calculation

Line	Description	Line	Value
1	Receptacle Load (noncontinuous) 220.3(B)(9) *Multiply each single or multiple receptacle on one strap by 180 volt-amperes* *Multiply each single piece of equipment comprised of 4 or more receptacles by 90 VA per receptacle.*	1	----
2	Fixed Multioutlet Assemblies (noncontinuous) 220.3(B)(8) *Where not likely to be used simultaneously, multiply each 5-ft section by 180 volt-amperes.* *Where likely to be used simultaneously, multiply each 1-ft section by 180 volt-amperes.*	2	----
3	Receptacle Load Demand Factor (for nondwelling receptacles) 220.13 *If the receptacle load is more than 10,000 volt-amperes, apply the demand factor from Table 220.13.* *Add lines 1 and 2. Multiply the first 10 kVA or less by 100%. Then, multiply the remainder by 50%.*	3	----
4	Unknown Receptacle Load (Banks and Office buildings only) *Where the actual number of general purpose receptacle outlets is unknown, include 1 volt-ampere per sq ft Table 220.3(A) footnote* [b] 1 × 5000 (sq ft outside dimensions) =	4	5000
5	General Lighting Load Table 220.3(A) *Multiply the volt-ampere unit load (for the type of occupancy) by the sq ft outside dimensions.* 3.5 (VA unit load) × 5000 (sq ft outside dimensions) =	5	17,500
6	Lighting Load Demand Factors 220.11 . . . *Apply Table 220.11 demand factors to certain portions of hospitals, hotels, motels, apartment houses (without provisions for cooking), and storage warehouses. Do not include areas in hospitals, hotels, and motels where the entire lighting will be used at one time.*	6	----
7	Track Lighting (in addition to general lighting) 220.12(B) *Include 150 volt-amperes for every 2 ft, or fraction thereof, of lighting track.* ---- (total linear ft) ÷ 2 ---- × 150 =	7	----
8	Sign and/or Outline Lighting Outlet (where required) 220.3(B)(6) *Each commercial building (or occupancy) accessible to pedestrians must have at least one outlet per tenant space entrance. 600.5(A) Each outlet must be at least 1200 volt-amperes.*	8	1200
9	Show-Window Lighting 220.12(A) *Include at least 200 volt-amperes for each linear ft, measured horizontally along the show window's base.* ---- (total linear ft of show window) × 200 =	9	----
10	Continuous Loads 215.2(A), 215.3, and 230.42(A) . . . *Multiply the continuous load volt-amperes (listed above) by 25%. (General purpose receptacles are not considered continuous.)* 18,700 (total continuous load volt-amperes) × 25% =	10	4675
11	Kitchen Equipment 220.20 *Multiply three or more pieces of equipment by Table 220.20 demand factor (%).* Use Table 220.19 for household cooking equipment used in instructional programs. Table 220.19 Note 5	11	----
12	Noncoincident Loads 220.21 . . . *The smaller of two (or more) noncoincident loads can be omitted, as long as they will never be energized simultaneously (such as certain portions of heating and A/C systems). Calculate fixed electric space heating loads at 100% of the total connected load. 220.15*	12	20,536
13	Motor Loads 220.4(A), 430.24, 430.25, 430.26, and Article 440 *Motor-driven air-conditioning and refrigeration equipment is found in Article 440.* *Multiply the largest motor (one motor only) by 25% and add to load.*	13	1981
14	All Other Loads . . . *Add all other noncontinuous loads into the calculation at 100%.* *Multiply all other continuous loads (operating for 3 hours or more) by 125%.*	14	12,550
15	Total Volt-Ampere Demand Load: *Add Lines 3 through 14 to find the minimum required volt-amperes.*	15	63,442
16	Minimum Amperes *Divide the total volt-amperes by the voltage* 63,442 (line 15) ÷ 360 (voltage) = 16: 176 (minimum amperes) **Minimum Size 17 Service and/or Feeder 240.6(A)**	17	200
18	Size the Service and/or Feeder Conductors. Tables 310.16 through 310.22 *Use the tables along with 310.15(B)(1) through (7) to determine conductor size. If the overcurrent device is rated more than 800 amperes, the conductor ampacity must be equal to, or greater than, the rating of the overcurrent device. 240.4(C)* **Minimum Size Conductors**	18	3/0 Cu 4/0 Al
19	Size the Neutral Conductor 220.22 *The neutral service and/or feeder conductor can be smaller than the ungrounded (hot) conductors, but not smaller than the maximum unbalanced load determined by Article 220. 250.24(B)(1) states that the neutral cannot be smaller than the required grounding electrode conductor specified in Table 250.66. A further demand factor is permitted for any neutral load over 200 amperes.* **Minimum Size Neutral Conductor**	19	2 Cu 1/0 Al
20	Size the Grounding Electrode Conductor (for Service) 250.66 *Using line 18 to find the grounding electrode conductor in Table 250.66.* Size the Equipment Grounding Conductor (for Feeder) 250.122 *Use line 17 to find the equipment grounding conductor in Table 250.122. Equipment grounding conductor types are listed in 250.118.* **Minimum Size Grounding Electrode Conductor . . . *or* . . . Equipment Grounding Conductor**	20	4 Cu 2 Al

Summary

- Article 220 contains load calculation procedures.
- Unlike dwelling unit load calculations, receptacle outlets must be included.
- Receptacle loads, including fixed multioutlet assemblies, can be derated if the load is greater than 10,000 volt-amperes.
- If the actual number of receptacle outlets in a bank or office building is unknown, a load of one volt-ampere per sq ft must be included.
- Various occupancy types have different volt-ampere unit loads per Table 220.3(A).
- One similarity between dwelling unit and non-dwelling unit load calculations is that a volt-ampere unit load must be multiplied by the square footage using outside dimensions.
- Table 220.11 contains lighting load demand factors for certain portions of hospitals, hotels, motels, storage warehouses, and apartment houses without provisions for cooking.
- Track lighting loads must be included in addition to the general lighting load.
- Some occupancies require a sign or outline lighting outlet.
- Show-window minimum volt-ampere lighting loads are based on a linear ft measurement.
- Continuous load calculations must be increased by an additional 25% volt-ampere rating.
- The volt-ampere rating, calculated in accordance with Article 220, represents only the minimum rating.

Unit 13 Competency Test

NEC® Reference	Answer	
________	________	1. What is the demand load contribution for a commercial laundry with ten 5-kVA clothes dryers and ten 1.5-kVA washing machines?
________	________	2. What is the volt-ampere unit load per sq ft for a beauty salon?
________	________	3. Outlets for heavy-duty lampholders shall be computed at a minimum of _____ volt-amperes.
________	________	4. What is the minimum general lighting load for an 11,750-sq-ft nightclub with an actual connected lighting load of 22,800 volt-amperes? (For this question, do not apply the continuous load demand factor.)
________	________	5. A restaurant contains the following commercial kitchen equipment: two 12-kW ovens, one 10-kW grill, two 8-kW deep fryers, one 1.2-kW disposer, one 1.5-kW dishwasher, one 10-kW booster heater, and one 4.5-kW water heater. What is the feeder demand load (in kW) for this equipment?
________	________	6. For track lighting in a church, a load of not less than _____ must be included for every _____ ft of lighting track or fraction thereof.
________	________	7. What is the receptacle demand load for a 9600-sq-ft bank with fifty-two receptacles?
________	________	8. Where it is unlikely that two or more _____ loads will be in use simultaneously, it shall be permissible to use only the largest load(s) that will be used at one time in computing the total load of a feeder or service.
________	________	9. What is the receptacle load (after demand factors) for an office building with 250 receptacle outlets?
________	________	10. What is the minimum general lighting load for an 8000-sq-ft assembly hall? (Do not apply the continuous load demand factor.)
________	________	11. A manufacturing plant has 320 ft of multioutlet assemblies, of which 60 ft will contain equipment subject to simultaneous use. What is the volt-ampere demand load contribution for these multioutlet assemblies?

NEC® Reference	Answer	
________	________	12. A school home-economics class has twelve 8-kW household ranges. The kitchen equipment is supplied by a 230-volt, single-phase panelboard. What is the minimum size THW copper conductors required for this panelboard?
________	________	13. A motel has fifty guest rooms, each room measuring 18' x 30'. Each room contains 8 duplex receptacles, of which one is GFCI-protected. Exterior hallways have a total area of 3000 sq ft. What is the lighting and receptacle load (after demand factors) for this motel?
________	________	14. Fixed electric space-heating loads shall be computed at _____ % of the total connected load; however, in no case shall a feeder or service load current rating be less than the rating of the largest branch-circuit supplied.
________	________	15. An outlet supplying recessed luminaire(s) (lighting fixture[s]) shall be computed based on _____.
________	________	16. For circuits supplying loads consisting of motor-operated utilization equipment that is fastened in place and that has a motor larger than ⅛ hp in combination with other loads, the total computed load shall be based on _____.
________	________	17. In a commercial storage garage, duplex receptacle outlets shall be computed at not less than _____.
________	________	18. A restaurant contains the following commercial kitchen equipment: one 12-kW range, one 2.8-kW mixer, one 4.2-kW deep fryer, two 0.8-kW soup wells, and one 6-kW water heater. What is the feeder demand load (in kW) for this equipment?
________	________	19. A hospital has 150 patient's rooms, each room measuring 15' × 25'. What is the lighting load (after demand factors) for these rooms?
________	________	20. For circuits supplying lighting units that have ballasts, transformers, or autotransformers, the computed load shall be based on _____.

Questions 21 through 30 are based on an office containing the following:

Office's total sq ft area	19,000 sq ft
Receptacle outlets	200
Multioutlet assemblies	80 ft
Multioutlet assemblies (simultaneously used)	25 ft
Sign	15 amperes, 120 volt
Track lighting	48 ft
Water heater	4 kW, 208 volt
Four electric heating units	10 kW, 208 volt (each)
Four air handlers (blower motors)	4 amperes, 208 volt, three-phase (each)
Four air-conditioner compressors	24.6 amperes, 208 volt, three-phase (each)
Eight condenser fan motors	2.2 amperes, 208 volt, three-phase (each)

Assume water heater and electric heat kW ratings equivalent to kVA.

The electric supply is a 208-volt, three-phase, four-wire service.

________	________	21. What is the total receptacle (including multioutlet assembly) volt-ampere service load after demand factors are applied?
________	________	22. What is the general lighting load for this 19,000-sq-ft office? (Do not apply the continuous load demand factor.)

NEC® Reference	Answer	
___________	___________	23. Without applying the continuous load demand factor, what track lighting load (if any) must be contributed to this office calculation?
___________	___________	24. What volt-ampere load must be included for the sign? (For this question, do not apply the continuous load demand factor.)
___________	___________	25. Which load(s), if any, can be omitted?
___________	___________	26. What is the volt-ampere load contribution for the heating and/or air-conditioning system(s)? (Do not include 25% of the largest motor.)
___________	___________	27. What is the minimum size service overcurrent protective device required for this office?
___________	___________	28. A parallel run (2 sets) of service-entrance conductors will be installed in two raceways. What is the minimum size THWN copper ungrounded conductors that can be used?
___________	___________	29. What is the minimum size THWN copper neutral (grounded) conductors that can be installed?
___________	___________	30. What is the minimum size copper grounding electrode conductor?

SECTION FOUR: COMMERCIAL LOCATIONS

UNIT 14 Services, Feeders, and Equipment

Objectives

After studying this unit, the student should:

- know the required minimum vertical wiring clearances for each electrical installation.
- be aware that other associated equipment, above or below the electrical installation, cannot extend more than 6 in. (152 mm) beyond the front of that electrical equipment.
- be familiar with the dedicated space above panelboards, switchboards, and motor control centers.
- be aware that a space equal to the equipment's width and depth, above panelboards, switchboards, and motor control centers, must be clear of foreign systems unless protection is provided.
- know that working space clearances vary depending on the conditions and voltages.
- understand the conditions that require two equipment working space entrances.
- know where to connect grounded and grounding conductors on both the service supply side and load side.
- understand the conditions that require ground-fault protection of equipment.
- have a good understanding of transformer and generator provisions.
- know that, while grounding and bonding connection points must be made in the same place, this can be either the panel or the transformer.
- have an extensive understanding of busway provisions.

Introduction

Unit 14 contains electrical provisions relative to commercial type occupancies. Before beginning this unit, a review of Unit 9 (Services and Electrical Equipment) is recommended. Some of the same provisions that apply to one-family dwelling likewise apply to commercial installations. Vertical wiring clearances, found in Articles 225 and 230, apply to all electrical feeder and service installations. Also applicable to both dwelling and commercial areas is 110.26, which contains minimum working space provisions (600 volts or less) above, below, and in front of electrical equipment. Some requirements, such as for equipment rated 1200 amperes or more, over 6 ft (1.8 m) wide, and containing devices (overcurrent, switching, or control), are not included in one-family dwellings. This unit contains separately derived system provisions that pertain to certain battery, solar photovoltaic, generator, transformer, and converter winding-system installations. Due to space limitations, only transformer illustrations are used to explain these provisions. In addition to the separately derived system provisions, transformers have their own article (450), which covers all transformer installations. Following the transformer material is an illustration briefly touching on generator provisions. Although generators have their own article (445), other generator provisions are found throughout the *Code*. Generators and their associated wiring and equipment must also adhere to applicable provisions of Articles 695, 700, 701, 702, and 705. Busway provisions close this unit. Busways can also be found in multiple occupancy types, i.e., multi-family, commercial, and industrial.

Since state and local jurisdictions may modify *NEC*® provisions, it is expedient to obtain a copy of local rules and regulations.

CLEARANCES AND WORKING SPACE

Vertical Clearances

A 10-ft (3.0-m) minimum vertical clearance is required from any platform or projection, from which overhead conductors (of 150 volts, or less, to ground) might be reached »230.24(B) and 225.18«.

A minimum 12-ft (3.7-m) vertical clearance is required over commercial areas (not subject to truck traffic) provided conductor voltage does not exceed 300 volts to ground »230.24(B) and 225.18«.

The minimum vertical clearance increases from 12 ft (3.7 m) to 15 ft (4.5 m) for conductors whose voltage exceeds 300 volts to ground »230.24(B) and 225.18«.

A drip loop's lowest point has a minimum vertical clearance of 10 ft (3.0 m) »230.24(B)«.

Ⓐ The point of attachment of the service-drop conductors to a building (or other structure) must meet 230.24 minimum clearance specifications. This point of attachment must always be at least 10 ft (3.0 m) above finished grade »230.26«.

Ⓑ Public streets, alleys, roads, parking areas subject to truck traffic, driveways on non-residential property, and other land traversed by vehicles (such as cultivated, grazing, forest, and orchard) require a minimum 18-ft (5.5-m) vertical clearance »230.24(B) and 225.18«.

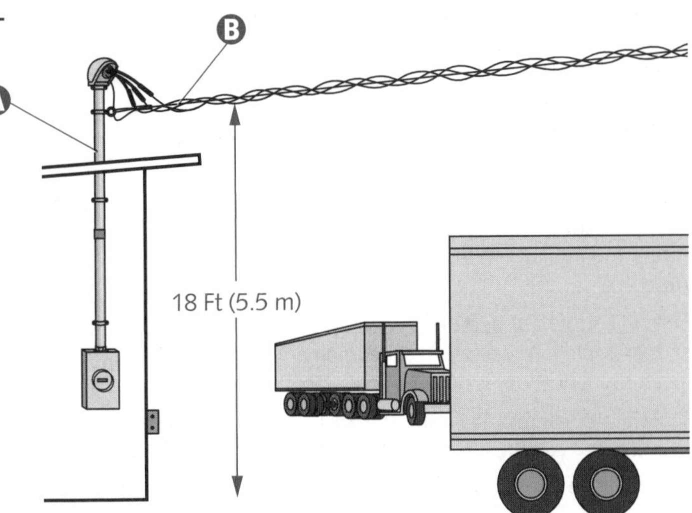

Maximum Depth of Associated Equipment

Ⓐ The working space depth, in the direction of access to live parts, must not be less than indicated by Table 110.26(A) »110.26(A)(1)«.

Ⓑ All electrical equipment must be constantly surrounded by sufficient access and work space permitting ready and safe equipment operation and maintenance. Electrical apparatus enclosures, controlled by lock and key, are considered accessible to qualified persons »110.26«.

Ⓒ Wireways are troughs (sheet-metal or nonmetallic) with hinged or removable covers used to house as well as protect electric wires and cables. A wireway must be installed as a complete system before conductors are laid in place »376.2 and 378.2«.

Ⓓ Other associated equipment, located above or below electrical equipment, cannot extend more than **6 in. (150 mm)** beyond the front of that electrical equipment »110.26(A)(3)«.

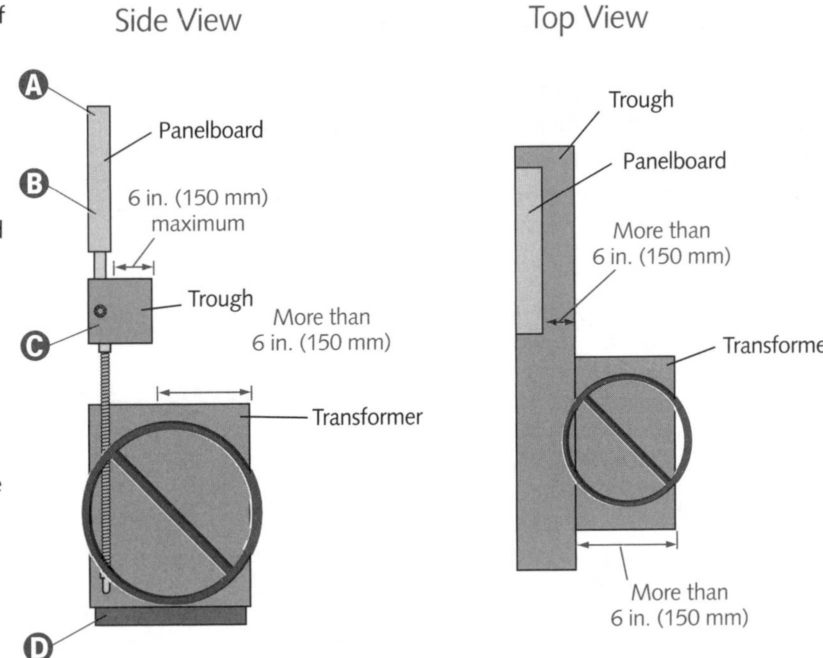

Dedicated Equipment Space

Sprinkler protection is permitted where piping complies with 110.26 »110.26(F)(1)(c)«.

A The area above the dedicated space required by 110.26(F)(1) can contain foreign systems, provided protection is installed to avoid damage to the electrical equipment from condensation, leaks, or breaks in such foreign systems »110.26(F)(1)(b)«.

B The space equal to the equipment's width and depth extending from the floor to a height of 6 ft (1.8 m) above the equipment or to the structural ceiling (whichever is lower) is dedicated to the electrical installation. No piping, ducts, or foreign equipment shall be located in this zone »110.26(F)(1)(a)«.

C 110.26(F) applies only to motor control centers and equipment within the scope of Article 408.

D 110.26(F)(1)(b) pertains to piping, ducts, etc. that could damage electrical equipment due to condensation, leaks, or breaks.

E Foreign systems may enter the area above the dedicated electrical space if the electrical equipment is protected from condensation, leaks, or breaks in such foreign systems »110.26(F)(1)(b)«.

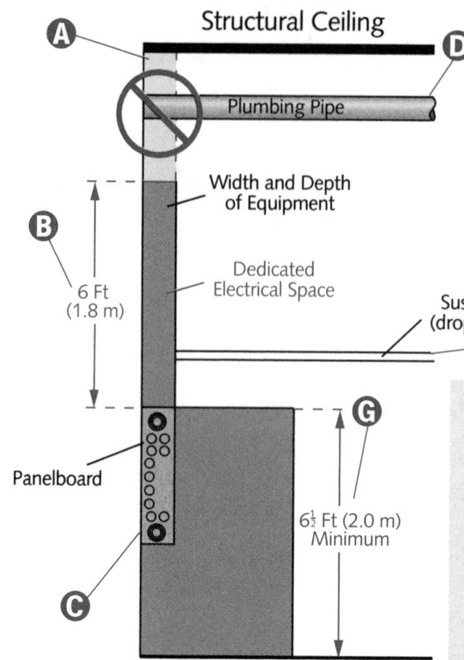
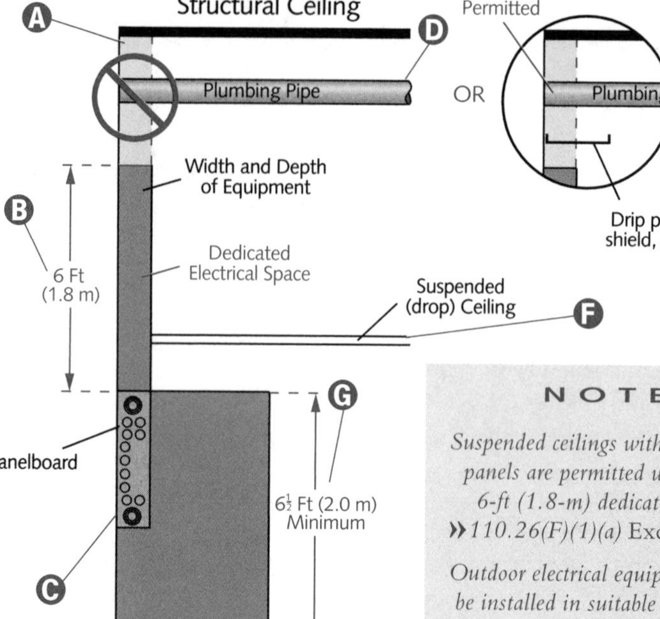

F A dropped, suspended, or similar ceiling not adding strength to the building structure is not a structural ceiling »110.26(F)(1)(d)«.

G The minimum working space headroom about service equipment, switchboards, panelboards, or motor control centers must be 6½ ft (2.0 m) in height, or the height of the equipment, whichever is greater »110.26(E)«.

NOTE

Suspended ceilings with removable panels are permitted within the 6-ft (1.8-m) dedicated space »110.26(F)(1)(a) Exception«.

Outdoor electrical equipment must be installed in suitable enclosures and must be protected from accidental contact by unauthorized personnel, by vehicular traffic, or by accidental piping system leakage. The working clearance space includes the zone described in 110.26(A). No architectural appurtenance or other equipment can occupy this zone »110.26(F)(2)«.

Working Space Access and Entrances

One access means is permitted where the location provides a continuous and unobstructed exit »110.26(C)(2)(a)«.

When normally enclosed live parts are exposed for inspection or servicing, any surrounding work space, in a passageway or general open area, must be suitably guarded »110.26(B)«.

A Equipment rated 1200 amperes (or more) and over 6 ft (1.8 m) wide containing overcurrent devices, switching devices, or control devices, requires an entrance at least 24 in. (610 mm) wide and 6½ ft (2.0 m) high at each end of the working space »110.26(C)«.

NOTE

Equipment operating over 600 volts, nominal, must comply with Article 110, part III.

B If the work space specified by 110.26(A) is doubled, only one entrance is required. The edge of the entrance nearest the equipment must meet the minimum clear distance given in Table 110.26(A) »110.26(C)(2)(b)«.

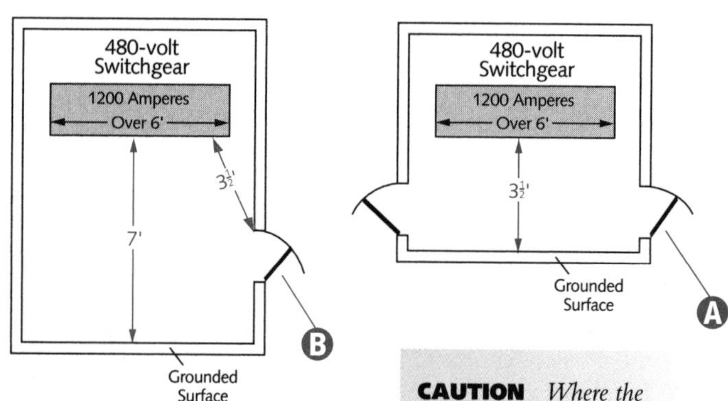

WARNING

Do not use working space required by 110.26 for storage »110.26(B)«.

CAUTION *Where the entrance has a personnel door(s), the door(s) must open in the direction of egress and be equipped with panic bars, pressure plates, or other devices that are normally latched but will open under simple pressure »110.26(B)(2)«.*

Working Space

Condition 1—Exposed live parts on one side with no live or grounded parts on the other side of the working space, or exposed live parts on both sides effectively guarded by suitable wood or other insulating materials. Insulated wire/busbars operating at 300 volts or less to ground are not considered live parts »Table 110.26(A)«.

Ⓐ The working space depth in the direction of access to live parts must meet Table 110.26(A) specifications. If exposed, measure the distance from the live part(s). Where enclosed, distance measurement begins from the enclosure front (or opening) »110.26(A)(1)«.

Ⓑ Sufficient access and working space must be continuously maintained about all electric equipment, permitting ready, safe equipment operation and maintenance. Enclosures housing electrical apparatus, controlled by lock and key, are considered accessible to qualified persons »110.26«.

Ⓒ Working space is not required behind or beside assemblies, such as dead-front switchboards or motor control centers, having no renewable or adjustable parts (such as fuses or switches) on the back or sides, provided all connections are accessible from other locations »110.26(A)(1(a)«.

Ⓓ Where rear access is required to work on nonelectrical parts at the back of enclosed equipment, a minimum working space of 30 in. (762 mm) horizontally is required »110.26(A)(1)(a)«.

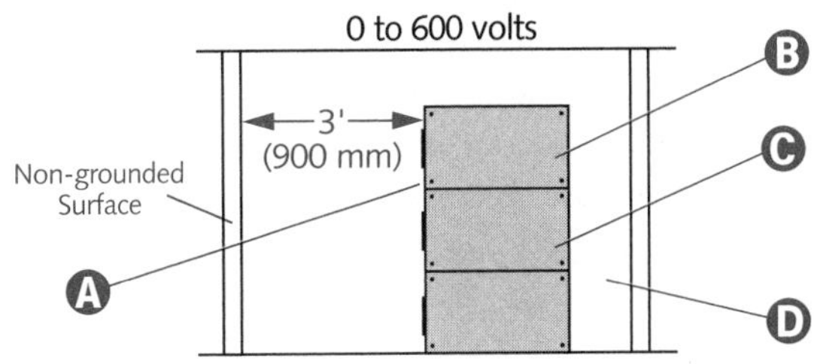

Condition 2—Exposed live parts on one side with grounded parts on the other side. Concrete, brick, or tile walls qualify as grounded »Table 110.26(A)«.

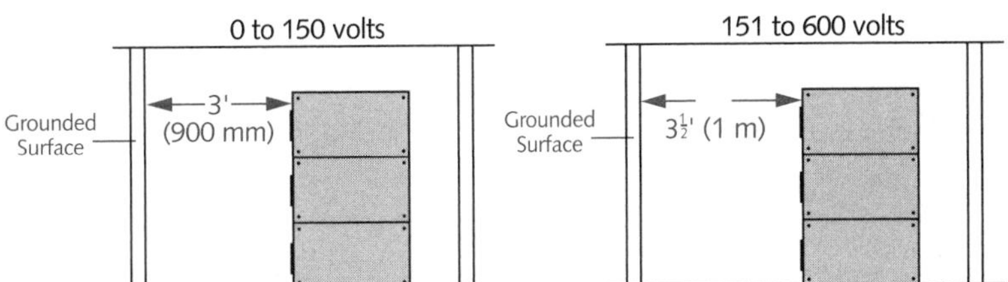

Condition 3—The work space having exposed live parts on both sides (not guarded as provided in Condition 1) with the operator between »Table 110.26(A)«.

Ⓔ In replacing equipment within existing buildings, Condition 2 working clearance is permitted between dead-front switchboards, panelboards, or motor control centers located across the aisle from each other where: (1) conditions of maintenance and supervision ensure that written procedures have been adopted to prohibit equipment on both sides of the aisle from being open at the same time; and (2) authorized qualified personnel will service the installation »110.26(A)(1)(c)«.

CAUTION *Working space does not apply only to panelboards and switchboards . . . Working space for equipment operating at 600 volts, nominal, or less to ground and likely to require examination, adjustment, servicing, or maintenance while energized must comply with 110.26(A)(1) through (3) dimensions, or as required/permitted elsewhere in the Code »110.26(A)«.*

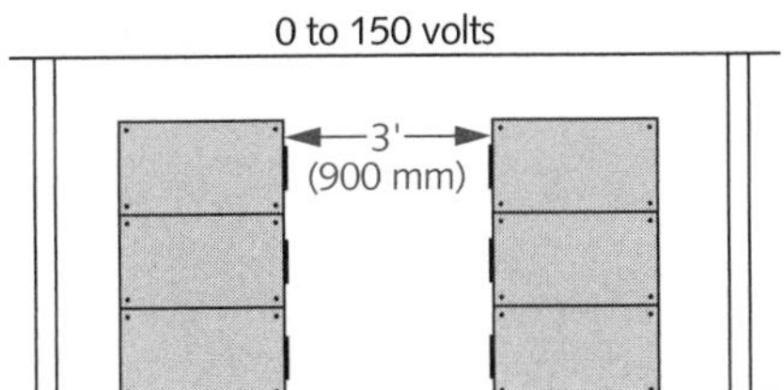

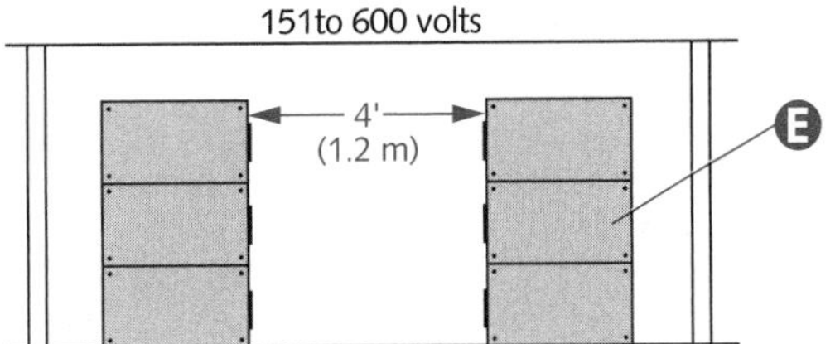

NOTE

By special permission, smaller spaces are allowed where all uninsulated parts are no greater than 30 volts rms, 42 peak, or 60 volts dc »110.26(A)(1)(b)«.

Equipment operating over 600 volts, nominal, must comply with Article 110, part III.

Guarding and Warning

Except as required (or permitted) elsewhere by the *NEC*®, live parts of electric equipment operating at 50 volts or more must be guarded against accidental contact by approved enclosures or by any of the following means listed in 110.27(A):

(1) By location in a room, vault, or similar enclosure accessible only to qualified persons.

(2) By suitable permanent, substantial partitions (or screens) arranged so that only qualified persons have access to the space within reach of live parts. Any openings in such partitions (barriers) must be sized and located so that accidental human contact with the live parts and/or conducting objects is unlikely.

(3) By locations on a suitable balcony, gallery, or elevated platform whose arrangement excludes unqualified persons.

(4) By elevations of 8 ft (2.4 m) or more above the floor or other working surface.

Where potential for physical damage to electric equipment exists, enclosures (or guards) must be of sufficient strength and placement to prevent such damage »110.27(B)«.

A At least one entrance of sufficient area must be provided, giving access to the electrical equipment's working space »110.26(C)«.

B Entrances to rooms and other guarded locations containing exposed live parts must be marked with conspicuous warning signs forbidding entrance of unqualified persons »110.27(C)«.

C Where working on nonelectric parts on the back of enclosed equipment requires rear access, a minimum working space of 30 in. (762 mm) horizontally must be provided »110.26(A)(1)(a)«.

D Condition 3 in Table 110.26(A)

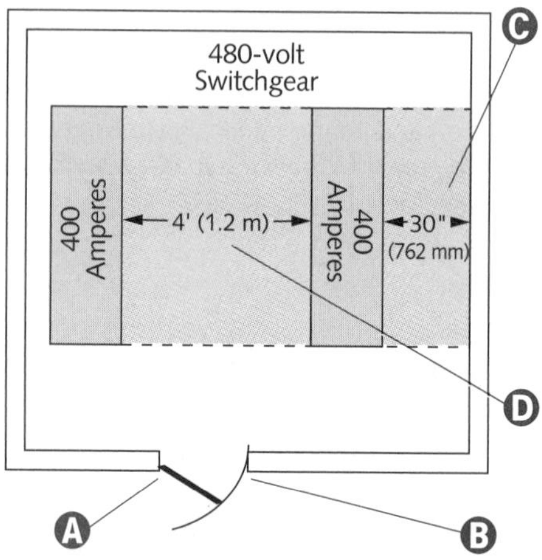

SWITCHBOARDS AND PANELBOARDS

Phase Arrangement

The phase arrangement on three-phase busses are A, B, C from front to back, top to bottom, or left to right, as viewed from the front of the switchboard or panelboard. The B phase must be that phase having the higher voltage to ground on three-phase, four-wire, delta-connected systems. Other clearly marked busbar arrangements are permitted for existing installation additions »408.3(E)«.

Equipment sharing the same single section or multisectional switchboard or panelboard as the meter (on three-phase, four-wire, delta-connected systems) is allowed to have the same phase configuration as the metering equipment »408.3(E) *Exception*«.

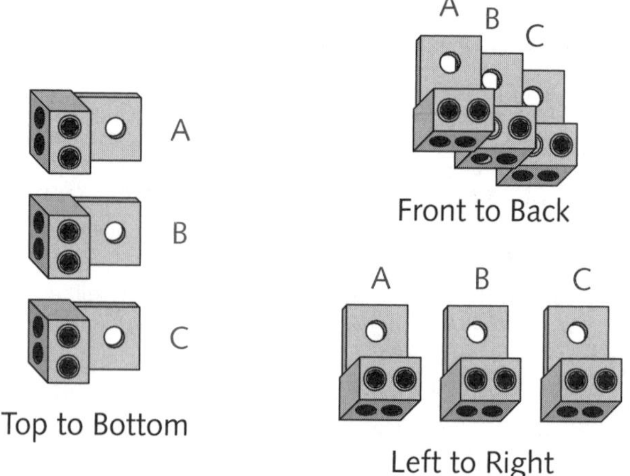

NOTE

A switchboard or panelboard supplied from a four-wire, delta-connected system where the midpoint of one phase winding is grounded, requires the phase busbar or conductor having the higher voltage to ground to be durably and permanently marked by an outer orange finish, or by other effective means »408.3(E)«.

CAUTION *The area electric utility company may require a different phase arrangement on three-phase, four-wire, delta-connected systems. Obtain a copy of the utility's phase arrangement requirements.*

Switchboards

A switchboard is a large single panel, frame, or assembly of panels with front and/or rear mounted, switches, protective devices (overcurrent, etc.), buses, and usually instruments. Switchboards are generally accessible from both the front and the rear and are not intended for cabinet installation »Article 100«.

Each switchboard or panelboard used as service equipment must be provided with a main bonding jumper sized according to 250.28(D) or the equivalent placed within the panelboard or one of the switchboard sections for the grounded service conductor connection (on its supply side) to the switchboard or panelboard frame. All switchboard sections must be bonded together using an equipment-grounding conductor sized per Table 250.122 »408.3(C)«.

Switchboards having exposed live parts must be: (1) located in permanently dry locations; (2) under competent supervision, and (3) accessible only to qualified persons. Switchboards must be located to minimize the probability of damage from equipment or processes »408.5«.

Anything other than a totally enclosed switchboard requires a space not less than 3 ft (900 mm) between the top of the switchboard and any combustible ceiling, unless a noncombustible shield separates them »408.8(A)«.

Conduit or other raceway entering a switchboard, floor-standing panelboard, or similar enclosure from the bottom, requires sufficient space to permit installation of conductors in the enclosure. The wiring space must meet Table 408.10 specifications where the conduit or raceway enters or leaves the enclosure below the busbars, their supports, or other obstructions. The conduit or raceway, including end fittings, must not rise more than 3 in. (75 mm) above the bottom of the enclosure »408.10«.

Position switchboards to reduce the probability of communicating fire to adjacent combustible materials to a minimum. Where installed over a combustible floor, suitable protection must be provided »408.7«.

NOTE

A bushing, or terminal fitting, with an integral bushed opening must be used at the end of a conduit or other raceway terminating underground where the conductors or cables emerge as a direct burial wiring method. A seal incorporating the same physical protection characteristics can be used in lieu of a bushing »300.5(H)«.

Power Panelboard

All panelboards must have a rating not less than the minimum feeder capacity required for the load computed according to Article 220 »408.13«.

Panelboards must be durably marked by the manufacturer with the voltage, current rating, and the number of phases for which they are designed. The manufacturer's name or trademark must also be visible after installation, without disturbing the interior parts or wiring »408.13«.

A Panelboard classifications are either: (1) lighting and appliance branch-circuit; or (2) power »408.14«.

B The purpose (or use) of a panelboard's circuits (and circuit modifications) must be legibly identified on a circuit directory affixed to the panel door(s) face or inside »408.4«. The marking must durably withstand surrounding environmental conditions »110.22«.

C In addition to 408.13 requirements, a power panelboard whose supply conductors include a neutral, having more than 10% of its overcurrent devices protecting branch-circuits rated 30 amperes (or less), must be protected on the supply side by an overcurrent protective device with a rating not greater than that of the panelboard »408.16(B)«.

D A three-pole circuit breaker counts as three overcurrent devices when determining the maximum number allowed »408.15«.

E No more than 10% of a power panelboard's overcurrent devices protect lighting and appliance branch-circuits »408.14(B)«.

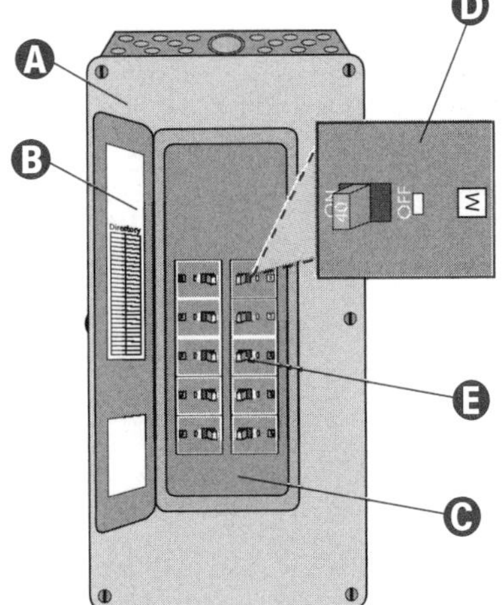

NOTE

A power panelboard, used as service equipment, with multiple disconnecting means per 230.71, does not require an individual overcurrent protective device » 408.16(B) Exception«.

Lighting and Appliance Branch-Circuit Panelboards

Lighting and appliance branch-circuit panelboards physically inhibit the installation of excess overcurrent devices. These panelboards accommodate only that number for which the panelboard was designed, rated and approved »408.15«.

A A two-pole circuit breaker counts as two overcurrent devices for 408.15 purposes.

B Individual protection for a lighting and appliance panelboard is not required if the feeder has overcurrent protection not exceeding the panelboard's rating »408.16(A) *Exception No. 1*«.

C A grounding connection must not be made to any grounded circuit conductor on the load side of the service disconnecting means, except as otherwise permitted in Article 250 »250.24(A)(5)«.

D More than 10% of a lighting and appliance branch-circuit panelboard's overcurrent devices protect lighting and appliance branch-circuits »408.14(A)«.

E A maximum of 42 overcurrent devices (excluding those provided for in the mains) can be installed in any one cabinet/cutout box of a lighting and appliance branch-circuit panelboard »408.15«.

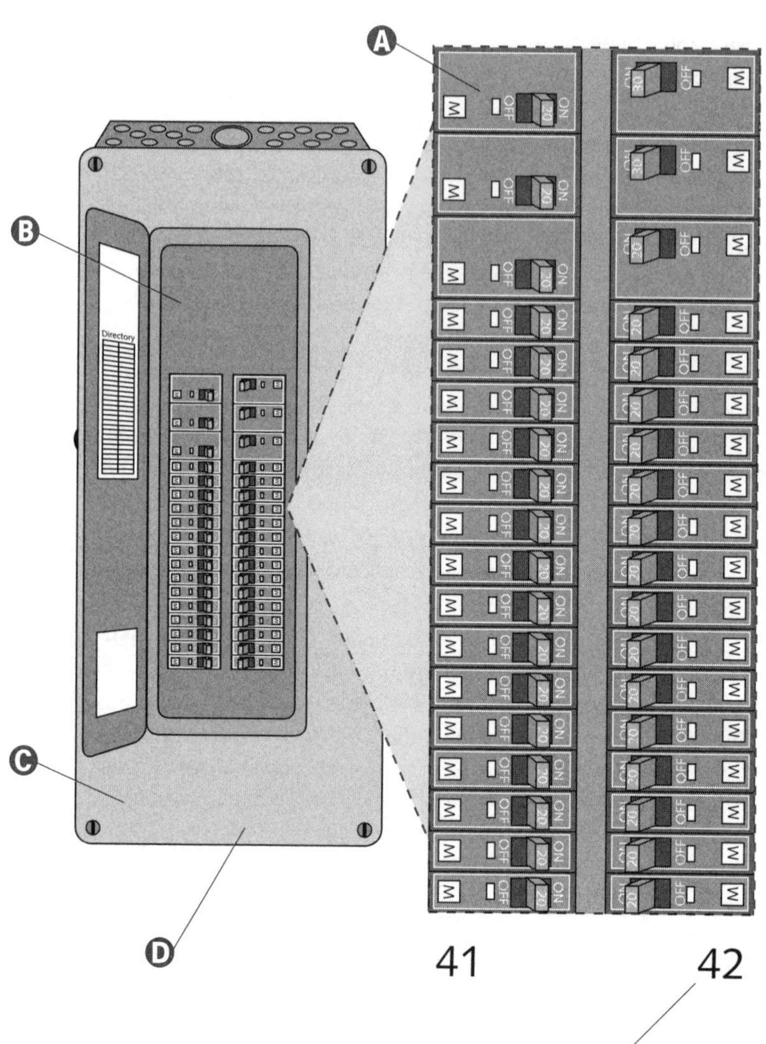

NOTE

Panelboards used as service equipment in pre-existing single residential occupancy installations do not require individual protection for lighting and appliance branch-circuit panelboards »408.16(A) Exception No. 2«.

Each lighting and appliance branch-circuit panelboard must be individually protected on the supply side by no more than two main circuit breakers (or two sets of fuses) having a combined rating not greater than that of the panelboard »408.16(A)«.

Grounding—Supply Side of Service

The bonding jumper must meet the size requirements shown in Table 250.66 for grounding electrode conductors. If the service-entrance phase conductors are larger than 1100-kcmil copper (or 1750-kcmil aluminum), the bonding jumper must have an area not less than 12½% of the largest phase conductor's area. If the phase conductors and the bonding jumper are of different materials (copper or aluminum), the minimum size of the bonding jumper is based on the assumed use of the same material and with an ampacity equal to that of the installed phase conductors. Where the service-entrance conductors are paralleled in multiple raceways or cables, the equipment bonding jumper, where routed with the raceway or cables, must also run in parallel. The bonding jumper size for each raceway or cable must be based on the size of the service-entrance conductors in each raceway or cable »250.102(C)«.

A Size the grounding electrode conductor according to 250.66 for the derived phase conductors »250.30(A)(2)«.

B The connection between the grounded circuit conductor (neutral) and the equipment grounding conductor at the service is the main bonding jumper »Article 100«. If a screw serves as the main bonding jumper, it must be identified with a green finish that remains visible after installation »250.28(B)«.

C 250.92(A) requires that the noncurrent-carrying metal equipment parts indicated below must be effectively bonded together:

(1) Service raceways, cable trays, cablebus framework, auxiliary gutters, or service cable armor/sheath except as 250.84 permits.

(2) All service enclosures containing service conductors, including meter fittings, boxes, or the like, interposed in the service raceway or armor.

(3) Any metallic raceway or armor enclosing a grounding electrode conductor as specified in 250.64(B). Bonding applies at each end and to all intervening raceways, boxes, and enclosures between the service equipment and the grounding electrode.

D Two to six circuit breakers (or sets of fuses) are permitted as the overcurrent device providing overload protection for service equipment. The sum of the circuit breakers (or fuses) ratings can exceed the service conductor's ampacity, provided the calculated load does not exceed the ampacity of the service conductors »230.90(A) *Exception No. 3*«.

E Each service-disconnecting means permitted by 230.2, or for each set of service-entrance conductors permitted by 230.40 Exception Nos. 1, 3, 4, or 5, must consist of no more than six switches and sets of circuit breakers mounted within a single enclosure, in a group of separate enclosures, or in (or on) a switchboard. The maximum number of disconnects per service, grouped in any one location, is six »230.71(A)«.

F A grounded circuit conductor can ground noncurrent-carrying metal equipment parts, raceways, and other enclosures at any of the following locations listed in 250.142(A):

(1) On the supply side, or within the enclosure, of the ac service-disconnecting means

(2) On the supply side, or within the enclosure, of the main-disconnecting means for separate buildings as provided in 250.32(B)

(3) On the supply side, or within the enclosure, of the main-disconnecting means or overcurrent devices of a separately derived system where 250.30(A)(1) permits.

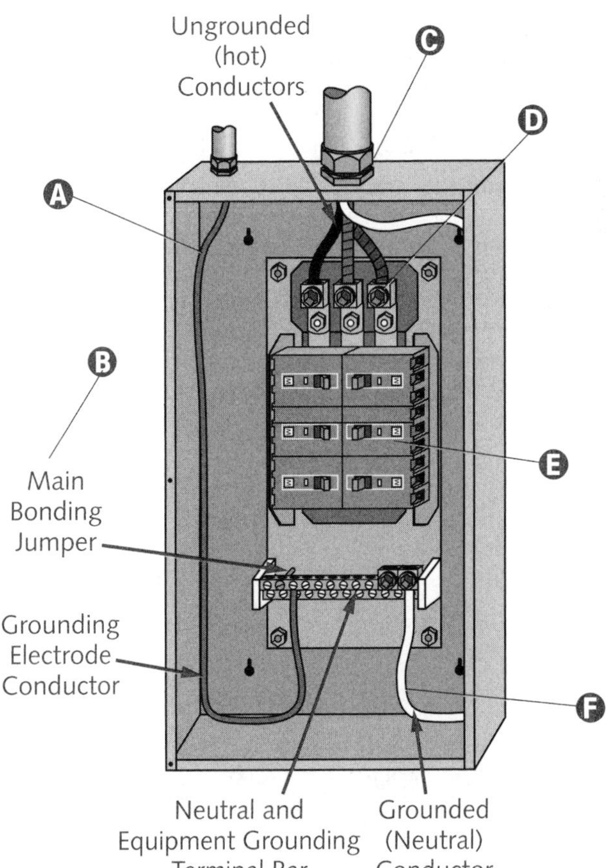

NOTE

Metal water piping system(s) installed in, or attached to a building or structure, must be bonded to either the service equipment enclosure, the grounded conductor at the service, the grounding electrode conductor (if of sufficient size), or the grounding electrode(s) used »250.104(A)(1)«.

Grounding—Load Side of Service

Where conductor sizes are increased in size to compensate for voltage drop, equipment grounding conductors (if installed) must be proportionate, according to circular mil area »250.122(B)«.

A Equipment bonding jumpers (on the load side of the service overcurrent devices) must be: (1) sized, as a minimum, in accordance with Table 250.122; but (2) are not required to be larger than the circuit conductors supplying the equipment; and (3) cannot be smaller than 14 AWG copper »250.102(D)«.

B Do *not* make a grounding connection to any grounded circuit conductor on the load side of the service disconnecting means except as otherwise allowed in Article 250 »250.24(A)(5)«.

C Equipment grounding conductors can be bare, covered, or insulated. Individually covered (or insulated) equipment grounding conductors must have a continuous outer finish, either green or green with yellow stripe(s), unless otherwise permitted by Article 250, Part IV »250.119«.

D Metal panelboard cabinets and frames must physically contact each other and be grounded. If the panelboard is used with nonmetallic raceway (or cable), or if separate grounding conductors are provided, secure a terminal bar for the grounding conductors within the cabinet. The terminal bar must be bonded to the metal cabinet and panelboard frame. Otherwise, connect it to the grounding conductor that runs with the conductors feeding the panelboard »408.20«.

E While equipment grounding conductors of wire type copper, aluminum, or copper-clad aluminum must not be less than shown in Table 250.122, they are not required to be larger than the circuit conductors supplying the equipment. A raceway, cable armor, or sheath used as the equipment grounding conductor [as provided in 250.118 and 250.134(A)], must comply with 250.4(A)(5) or 250.4(B)(4) »250.122(A)«.

F Unless grounded by connection to the grounded circuit conductor (permitted by 250.32, 250.140, and 250.142), noncurrent-carrying metal parts of equipment, raceways, and other enclosures, must be grounded (where required) by one of the following methods: (a) by any of the equipment grounding conductors permitted by 250.118; or (b) by an equipment grounding conductor contained within the same raceway, cable, or otherwise run with the circuit conductors »250.134«.

G Except as permitted in 250.30(A)(1) and 250.32(B), a grounded circuit conductor cannot be used for grounding noncurrent-carrying metal equipment parts on the load side of the service disconnecting means, or on the load side of a separately derived system disconnecting means, or the overcurrent device for a separately derived system not having a main disconnecting means, unless an exception has been met »250.142(B)«.

NOTE

Paralleled conductors within multiple raceways or cables (permitted in 310.4) require that the equipment grounding conductors (where used) also run in parallel within each raceway or cable. Each parallel equipment grounding conductor must be sized on the basis of the ampere rating of the overcurrent device protecting the circuit conductors in the raceway or cable per Table 250.122, unless ground-fault protection of equipment is implemented »250.122(F)«.

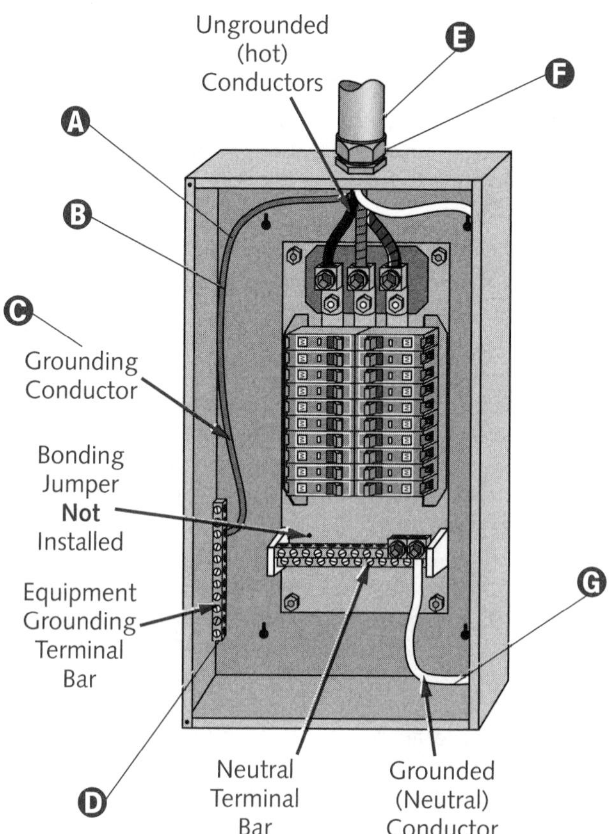

Ground-Fault Protection of Equipment

If a switch and fuse combination is used, the fuses employed must be capable of interrupting any current higher than the interrupting capacity of the switch when the ground-fault protective system will not cause the switch to open »230.95(B)«.

Ground-fault protection that opens the service disconnect will afford no protection from faults on the line side of the protective element. It serves only to limit damage to conductors and equipment only on the load side of the protective element in the event of an arcing ground fault »230.95(C) FPN No. 1«.

This added protective apparatus at the service equipment may necessitate a review of the overall wiring system for proper selective overcurrent protection coordination. Additional installations of ground-fault protective equipment may be needed on feeders and branch-circuits requiring maximum continuity of electrical service »230.95(C) FPN No. 2«.

Where ground-fault protection is provided for the service disconnect, and another supply system is interconnected by a transfer device, a method may be needed to ensure proper sensing by the ground-fault protection equipment »230.95(C) FPN No. 3«.

A Ground-fault protection of equipment must be provided for solidly grounded wye electrical systems of more than 150 volts to ground, but not exceeding 600 volts phase-to-phase for each service disconnect rated 1000 amperes or more. **Solidly Grounded** means that the grounded conductor is grounded without implementing any resistor or impedance device »230.95 and 240.13«.

B The service disconnect rating is considered to be the rating of the largest fuse that can be installed, or the highest continuous current trip setting for which the actual overcurrent device installed in a circuit breaker is rated (or can be adjusted) »230.95«.

C Ground-fault protection systems, when activated, must cause the service disconnect to open all ungrounded conductors of the faulted circuit. The maximum ground-fault setting is 1200 amperes »230.95(A)«.

D One second is the maximum time delay for ground-fault currents of 3000 amperes or more »230.95(A)«.

E **Ground-Fault Protection of Equipment** is a system intended to provide protection of equipment from damaging line-to-ground fault currents by causing the disconnecting means to open all ungrounded conductors of the faulted circuit. This protection is provided at current levels less than those required to protect conductors from damage through the operation of a supply circuit overcurrent device »Article 100«.

NOTE

230.95 ground-fault protection provisions do not apply to fire pumps »*230.95* Exception No. 2 *and 240.13(3)*«.

The ground-fault protection provisions of 230.95 do not apply to a service disconnect for a continuous industrial process where a nonorderly shutdown introduces additional or increased hazards »*230.95* Exception No. 1 *and 240.13(1)*«.

CAUTION *Bonding for circuits over 250 volts to ground must comply with 250.97.*

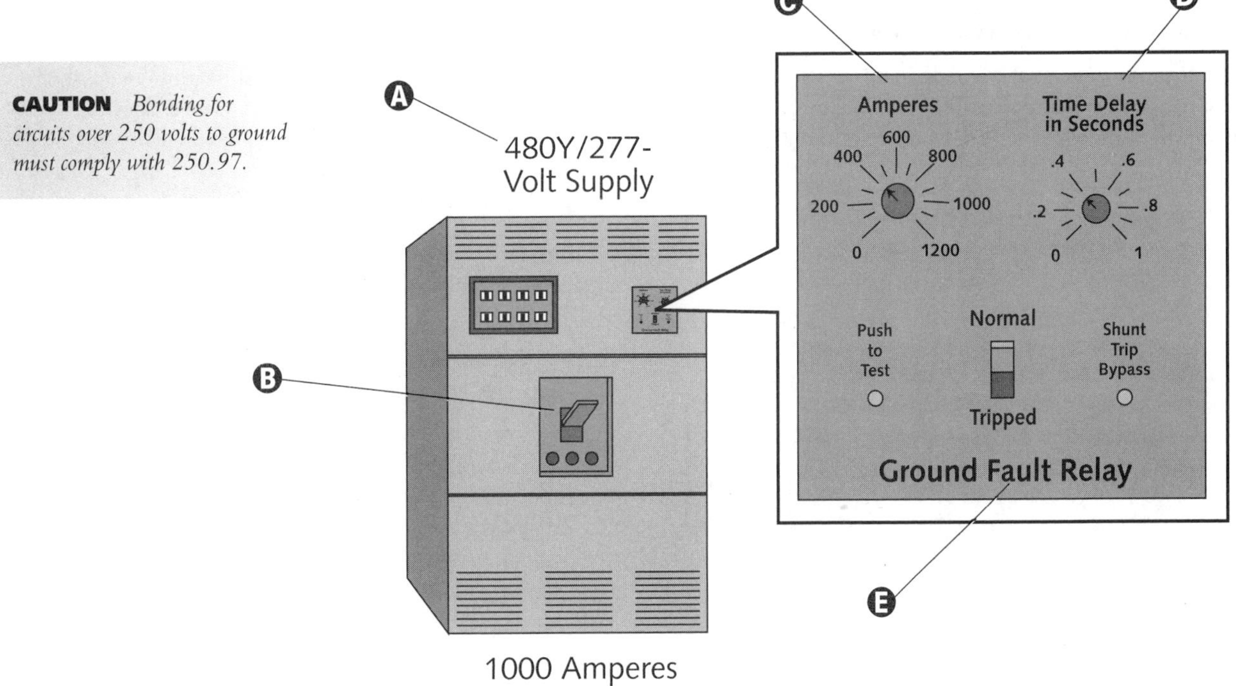

SEPARATELY DERIVED SYSTEMS

Transformers

Article 450 covers the installation of all transformers except:

(1) Current transformers.

(2) Dry-type transformers that constitute a component of other apparatus and comply with these requirements.

(3) Transformers that are an integral part of X-ray, high-frequency, or electrostatic-coating apparatus.

(4) Transformers used with Class 2 and Class 3 circuits in compliance with Article 725.

(5) Sign and outline lighting transformers that comply with Article 600.

(6) Transformers for electric-discharge lighting per Article 410.

(7) Power-limited fire alarm circuit transformers that comply with Article 760, Part III.

(8) Transformers used for research, development, or testing, where effective safeguards protect persons from contacting energized parts »450.1 *Exception Nos. 1 through 8*«.

Article 450 covers the installation of transformers dedicated to supplying power to a fire pump installation as modified by Article 695 »450.1«.

Article 450 covers the installation of transformers in hazardous (classified) locations as modified by Articles 501 through 504 »450.1«.

Autotransformers are covered in 450.4 and 5.

Appropriate provisions must be made to minimize the possibility of damage to transformers from external causes where the transformers are exposed to such dangers »450.8(A)«.

(A) The intended meaning of **transformer** is an individual transformer, single- or polyphase, identified by a single nameplate, unless otherwise indicated by Article 450 »450.2«.

(B) Overcurrent protection for transformers over 600 volts, nominal, must be provided in accordance with Table 450.3(A) »450.3(A)«.

(C) Each transformer must have a nameplate giving the manufacturer's name; rated kilovolt-amperes; frequency; primary and secondary voltage; impedance of transformers (25 kVA and larger); required clearances for transformers with ventilating openings; and the amount and kind of insulating liquid (where used). In addition, the nameplate of each dry-type transformer must include the insulation system's temperature class »450.11«.

(D) Transformers with ventilation openings must be installed so that such openings are not blocked by walls or other obstructions. The required clearances must be legibly marked on the transformer »450.9«.

(E) Dry-type transformers rated 600 volts, nominal, or less, and not exceeding 50 kVA, are permitted in hollow spaces of buildings that are not permanently closed in, provided they meet 450.9 ventilation requirements and 450.21(A) separation from combustible material requirements. Transformers so installed do not have to be readily accessible »450.13(B)«.

(F) It is not required that dry-type transformers rated 600 volts, nominal, or less, located in the open on walls, columns, or structures, be readily accessible »450.13(A)«.

(G) Exposed noncurrent-carrying metal parts of transformer installations (fences, guards, etc.), must be grounded, where required, under the conditions and in a manner specified for electric equipment and other exposed metal parts in Article 250 »450.10«.

NOTE

All transformers and their vaults must be readily accessible to qualified personnel for inspection and maintenance, or must otherwise meet 450.13(A) or (B) requirements »450.13«.

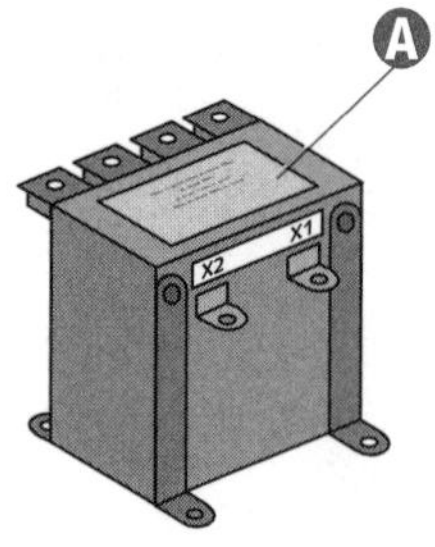

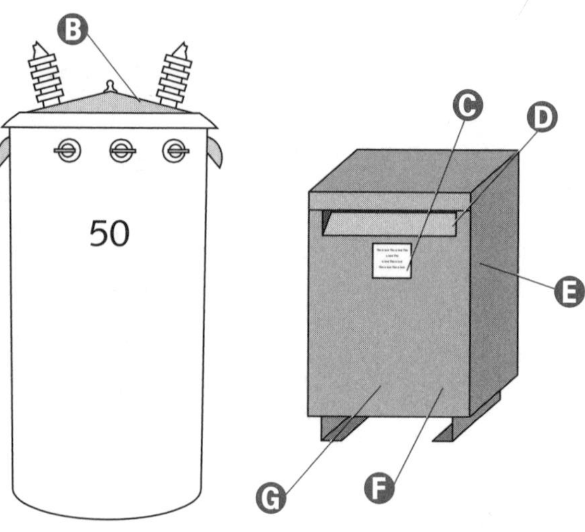

Transformer—Primary Only Protection 9 Amperes or More—600 Volts or Less

Overcurrent protection for transformers 600 volts or less having a rating less than 9 amperes must be provided in accordance with Table 450.3(B) »450.3(B)«.

Conductors supplied by the secondary side of a single-phase transformer having a two-wire (single-voltage) secondary, or a three-phase, delta-delta connected transformer having a three-wire (single-voltage) secondary, can be protected by overcurrent protection situated on the primary (supply) side of the transformer, provided this protection is in accordance with 450.3 and does not exceed the value determined when multiplying the secondary conductor ampacity by the secondary to primary transformer voltage ratio »240.21(C)(1)«.

Ⓐ Where 125% of this current does not correspond to a standard fuse or nonadjustable circuit breaker rating, the next higher rating described in 240.6 is permitted »Table 450.3(B) Note 1«.

Ⓑ This represents either a single-phase transformer having a two-wire (single-voltage) secondary, or a three-phase, delta-delta connected transformer having a three-wire (single-voltage) secondary.

Ⓒ No secondary overcurrent protection is required for certain transformers 600 volts, nominal, or less, with currents of at least 9 amperes and a maximum primary overcurrent protection of 125%. [The absence of secondary protection applies only to certain panelboards. See 408.16(D) *Exception*.]

Ⓓ Single-phase (other than two-wire) and multiphase (other than delta-delta, three-wire) transformer secondary conductors are not thought of as protected by the primary overcurrent protective device »240.21(C)(1)«.

Ⓔ Overcurrent protection for transformers rated 600 volts, nominal, or less must be provided in accordance with Table 450.3(B), unless the transformer is installed as a motor-control circuit transformer in accordance with 430.72(C)(1) through (5) »450.3(B)«.

WARNING

Although Table 450.3(B) does not require secondary overcurrent protection if the primary overcurrent protection is limited to 125%, 408.16(D) requires secondary overcurrent for most panelboards. Where a panelboard is supplied through a transformer, locate the overcurrent protection in 408.16(A), (B), and (C) on the transformer's secondary side »408.16(D)«. A panelboard supplied by the secondary side of a transformer is considered as protected by the overcurrent protection provided on the transformer's primary side, where that protection complies with 240.21(C)(1) »408.16(D) *Exception*«.

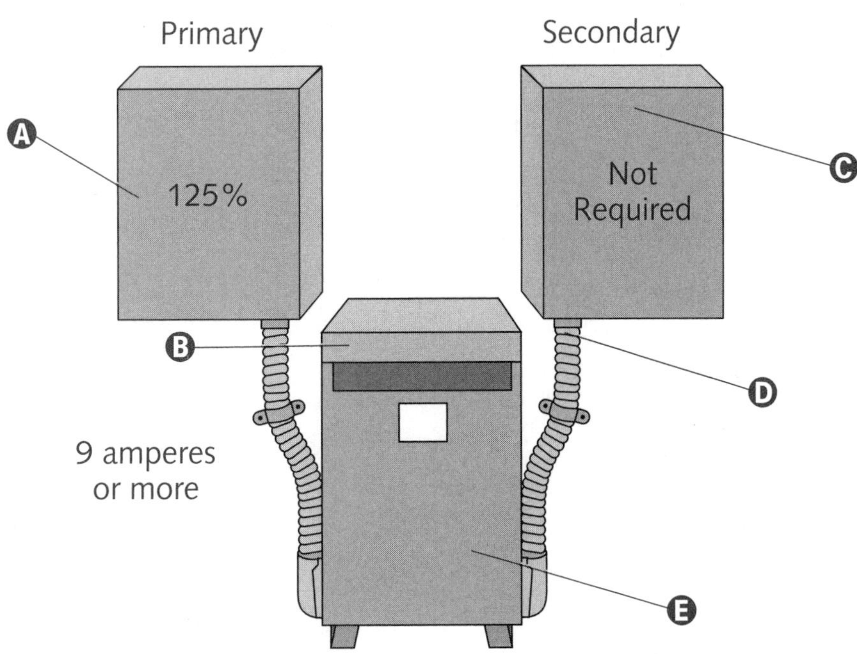

Transformer—Primary and Secondary Protection 9 Amperes or More—600 Volts or Less

A Unlike the note allowing the next standard rating above 125%, the **maximum** rating or setting is 250%.

B A transformer equipped with coordinated thermal overload protection (by the manufacturer) and arranged to interrupt the primary current can have primary overcurrent protection rated (or set) at a current value not more than six times the rated current of transformers having not more than 6% impedance and not more than four times the rated current of transformers having more than 6%, but not more than 10%, impedance »Table 450.3(B) Note 3 «.

C Overcurrent protection for transformers rated 600 volts, nominal, or less must be provided according to Table 450.3(B), unless the transformer is installed as a motor-control circuit transformer per 430.72(C)(1) through (5) »450.3(B) «.

D Where 125% of this current does not correspond to a standard fuse or nonadjustable circuit breaker rating, the next higher rating described in 240.6 is permitted »Table 450.3(B) Note 1 «.

E If secondary overcurrent protection is required, these devices are permitted to consist of no more than six circuit breakers (or six sets of fuses) grouped in one location. Where multiple overcurrent devices are implemented, the total of all the device ratings must not exceed the single overcurrent device allowed value. If both breakers and fuses constitute the overcurrent device, the total of the device ratings must not exceed that allowed for fuses »Table 450.3(B) Note 2 «.

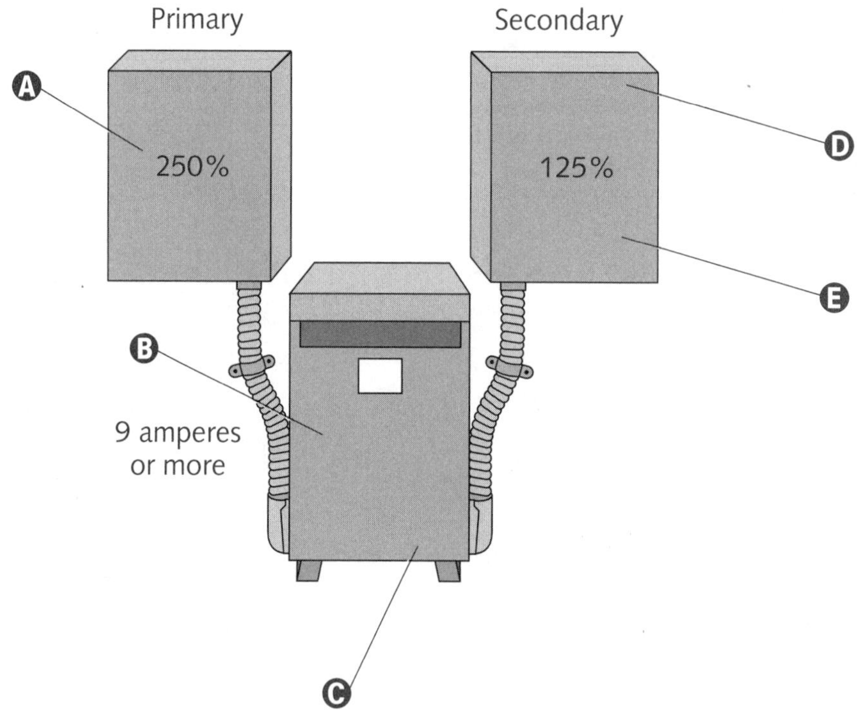

Grounding and Bonding—Terminating in the Panel

Premises wiring deriving power from systems having no direct electrical connection (such as generators, transformers, or converter windings) to supply conductors originating in another system, and any system having direct electrical connection (such as a battery, or solar photovoltaic system) is a *separately derived system*. Grounded circuit conductors connecting systems do not constitute a direct electrical connection »Article 100«.

A grounded circuit conductor can be used to ground noncurrent-carrying metal parts of equipment, raceways, and other enclosures on the supply side, or within the enclosure of, the main disconnecting means (or overcurrent devices) of a separately derived system as permitted by 250.30(A)(1) »250.142(A)(3)«.

A A bonding jumper complying with 250.28(A) through (D), and sized for the derived phase conductors, must connect the equipment grounding conductors of the separately derived system to the grounded conductor. Except as allowed by 250.24(A)(3), this connection is made at any point on the separately derived system from the source to the first system disconnecting means (or overcurrent devices), or made at the source of a separately derived system that has no disconnecting means (or overcurrent device) »250.30(A)(1)«.

B The bonding jumper point of connection must be the same as the grounding electrode conductor per 250.30(A)(2) requirements »250.30(A)(1)«.

C For vertically operated circuit-breaker handles, the "up" position of the handle must be the "on" position »240.81«.

D A grounded separately derived ac system must comply with 250.30(A)(1) through (6), unless it is a high-impedance grounded neutral system »250.30(A)«.

E Size the equipment grounding conductor in accordance with 250.122. The equipment grounding conductor must be of the types listed in 250.118.

F A bonding jumper at both the source and the first disconnecting means is permitted, if doing so does not establish a parallel path for the grounded circuit conductor. While grounded conductors used in this manner must not be smaller than the size specified for the bonding jumper, they do not have to be larger than the ungrounded conductor(s). In applying this exception, connection through the earth does not qualify as providing a parallel path »250.30(A)(1) *Exception No. 1*«.

G Grounding (and bonding) terminations made at the system disconnecting means or overcurrent device, requires that the transformer's grounded (neutral) bus be insulated from the transformer enclosure. If the neutral bus is bonded to the enclosure and the raceway to the panelboard is metallic, then part of the neutral current flows on the raceway. 250.6(A) prohibits objectionable current flow over grounding paths or grounding conductors.

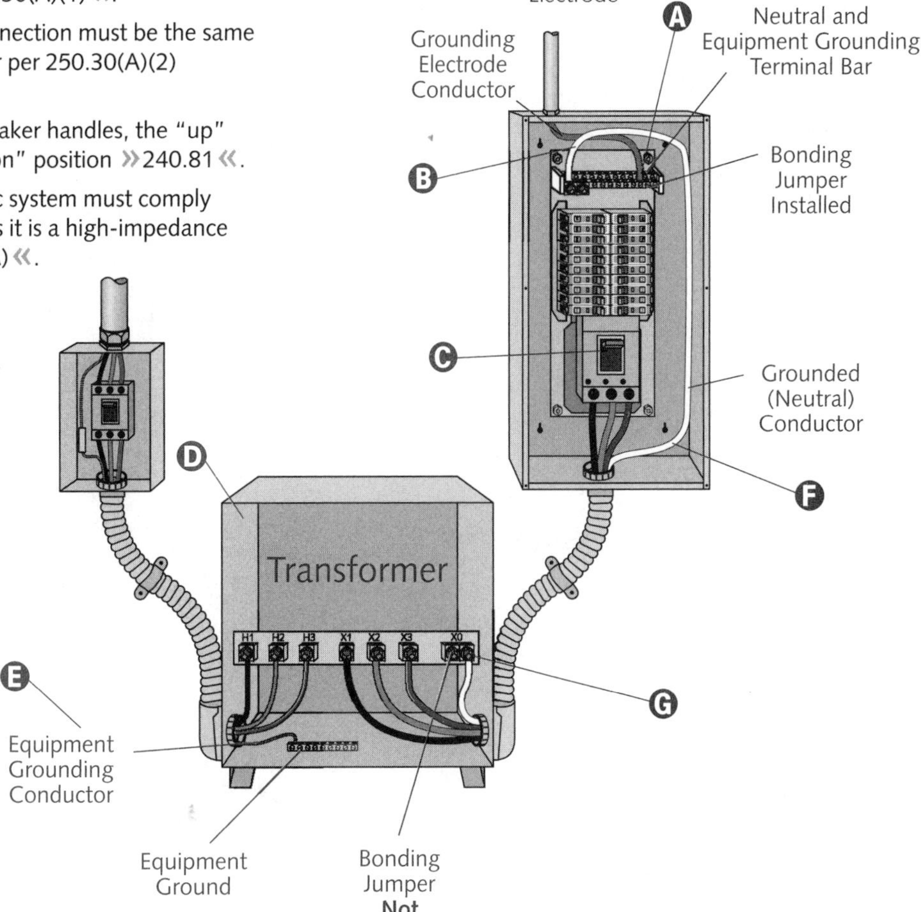

NOTE

Transformers, which are only one type of separately derived system, will be used to illustrate and explain 250.30.

Grounding and Bonding—Terminating in the Transformer

The equipment of an ungrounded separately derived system must be grounded as specified in 250.30(B)(1) and (2).

While a three-phase separately derived system is used for this illustration, the same rules apply to single-phase systems.

A system supplying power to a Class 1, Class 2, or Class 3 circuit, having the power derived from a transformer rated 1000 volt-amperes or less, must have a bonding jumper sized no smaller than the derived phase conductors. The smallest size allowed is 14 AWG copper or 12 AWG aluminum »250.30(A)(1) *Exception No. 2*«.

A system that supplies a Class 1, 2, or 3 circuit and is derived from a transformer rated no more than 1000 volt-amperes does not require a grounding electrode conductor, provided the system grounded conductor is bonded to the transformer frame (or enclosure) by means of a jumper sized in accordance with 250.30(A)(1), *Exception No. 2*, and the transformer frame (or enclosure) is grounded by one of the means specified in 250.134 »250.30(A)(2) *Exception*«.

A If the grounding (and bonding) terminations are made at the transformer, the system disconnecting means (or overcurrent device) grounded (neutral) terminal bar must be insulated from both the panelboard enclosure and the equipment grounding conductor. If the neutral terminal bar is bonded to the enclosure as well as the equipment grounding conductor, part of the neutral current will flow through the raceway (if metallic) and the equipment grounding conductor. 250.6(A) strictly prohibits such a condition.

B Except as permitted in 250.30(A)(1), a grounded circuit conductor cannot be used for grounding noncurrent-carrying metal equipment parts on the load side of a separately derived system disconnecting means (or the overcurrent devices) for a separately derived system having no main disconnecting means »250.142(B)«.

C The bonding jumper connection can be made at any point on the separately derived system, i.e., in either the panel or transformer »250.30(A)(1)«.

D The bonding jumper and the grounding electrode conductor must share the same point of connection as required in 250.30(A)(2) »250.30(A)(1)«.

E Use a bonding jumper that complies with 250.28(A) through (D) to connect the equipment grounding conductors of the separately derived system to the grounded conductor. The bonding jumper size, determined by the derived phase conductors, must not be smaller than Table 250.66 grounding electrode conductor sizes »250.30(A)(1)«.

F The grounding electrode conductor installation must comply with 250.64.

G If the main bonding jumper (specified in 250.28) is a wire or busbar, installed from the neutral bar (or bus) to the equipment grounding terminal bar (or bus), the grounding electrode conductor can be connected to the same equipment grounding terminal bar (or bus) as the main bonding jumper »250.24(A)(4)«.

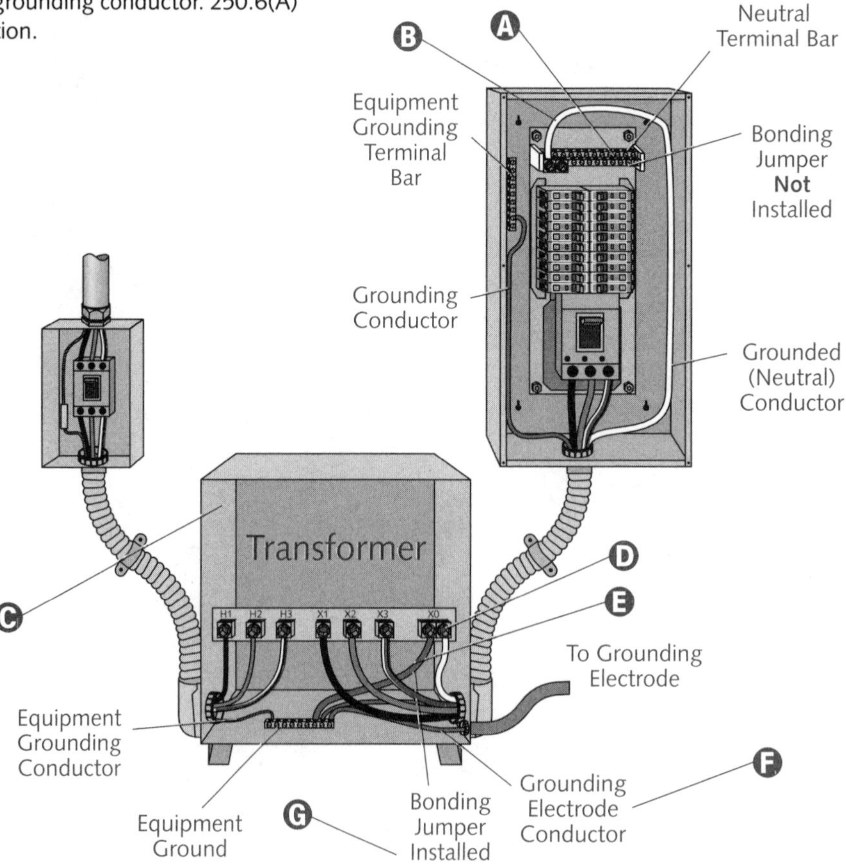

Grounding Electrode and Grounding Electrode Conductor

Where a separately derived system originates in listed equipment suitable for use as service equipment, the service equipment's grounding electrode can serve as the separately derived system's grounding electrode, provided it is sufficiently sized for the separately derived system. Where the equipment ground bus (internal to the service equipment) is not smaller than the required grounding electrode conductor, the grounding electrode connection for the separately derived system can be made to that bus »250.30(A)(4) *Exception*«.

Ⓐ The grounding electrode must be as near as practicable to, and preferably in the same area as, the grounding electrode conductor connection to the system. The grounding electrode must be the nearest one of the following:

(1) An effectively grounded metal member of the structure.

(2) An effectively grounded metal water pipe within 5 ft (1.5 m) of the entrance point into the building (in industrial and commercial buildings where maintenance and supervision conditions ensure that only qualified persons will service the installation and the entire length of the interior metal water pipe is exposed, the connection is permitted at any point along the water pipe system.)

(3) Other electrodes as specified in 250.52 where the electrodes specified by 250.30(A)(4)(1) or (A)(4)(2) are not available »250.30(A)(4)«.

Ⓑ Grounding electrode conductor metal enclosures must be electrically continuous from the point of attachment to cabinets (or equipment) to the grounding electrode, and must be securely fastened to the ground clamp (or fitting). Metal enclosures that are not physically continuous from cabinet (or equipment) to the grounding electrode must be made electrically continuous by bonding *each end* to the grounding conductor »250.64(E)«.

Ⓒ A grounding electrode conductor for a single separately derived system must be sized according to 250.66 for the derived phase conductors and must connect the grounding conductor of the derived system to the grounding electrode per 250.30(A)(4) specifications. Except as permitted by 250.24(A)(3) or (A)(4), this connection must be made at the same point on the separately derived system where the bonding jumper is installed »250.30(A)(2)(a)«.

Ⓓ The connection of a grounding electrode conductor to a grounding electrode must be accessible »250.68(A)«.

Ⓔ Each separately derived system's grounded conductor must be bonded to the nearest available point of interior metal water piping system(s) in the area served by each separately derived system. This connection must be made at the same point on the separately derived system as the grounding electrode conductor connection. Size the bonding jumper per Table 250.66 »250.104(A)(4)«.

Ⓕ Equipment requiring grounding and having a nonconductive coating (such as paint, lacquer, and enamel) must have the coating removed from threads and other contact surfaces to ensure good electrical continuity (unless connected by fittings that are designed so as to make the removal unnecessary) »250.12«.

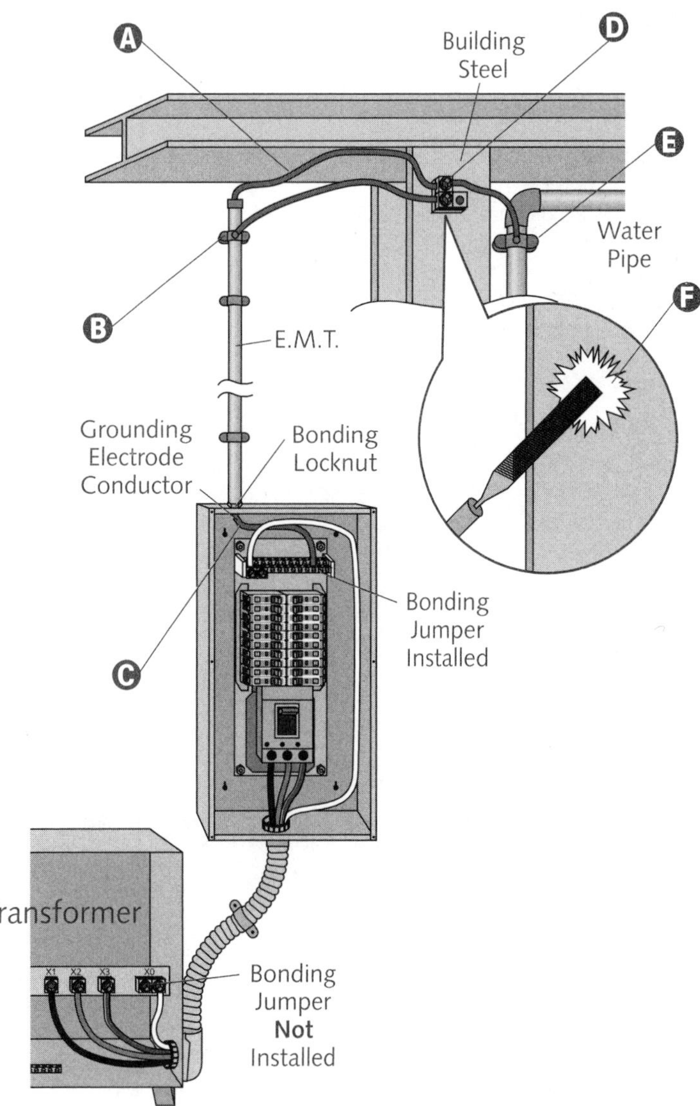

Generators

For systems not separately derived and not required to be grounded as specified in 250.30, see 445.13 to determine the minimum size of conductors that must carry fault current »250.20(D) FPN No. 2«.

Live generator parts operating at more than 50 volts to ground must not be exposed to accidental contact by unqualified persons »445.14«.

While legally required standby system requirements for generators and associated equipment are located in Article 701, optional standby system provisions are found in Article 702.

Ⓐ The conductor ampacity from the generator terminals to the first distribution device(s) containing overcurrent device must not be less than 115% of the generator's nameplate current rating »445.13«.

Ⓑ An alternate ac power source (such as an on-site generator) is not a separately derived system if the neutral is solidly interconnected to a service-supplied system neutral »250.20(D) FPN No. 1«.

Ⓒ Generators, together with associated wiring and equipment, must also comply with the applicable provisions of Articles 695, 700, 701, 702, and 705 »445.3«.

Ⓓ The type of generator used must be suitable for the installed location. The motor requirements in 430.14 must also be met. Generators installed in hazardous (classified) locations as described in Articles 500 through 503, 505, and in other locations as described in Articles 510 through 517, 518, 520, 525, 530, 665, and 695 must also comply with the applicable provisions of those articles »445.10«.

Ⓔ Depending on the generator type, overcurrent protection must meet 445.12 specifications.

Ⓕ Generators must have a disconnect means by which the generator, all protective devices, and control apparatus can be entirely disconnected from the circuits supplied by the generator, except where: (1) The generator's driving means can be readily shut down; and (2) The generator is not arranged to operate in parallel with another generator or other source of voltage »445.18«.

Ⓖ Where wires pass through an enclosure, conduit box, or barrier opening, a bushing must protect the conductors from any sharp edges. The bushing surfaces must be smooth and well-rounded where they may contact the conductors. If oils, grease, or other contaminants are present, the bushing must be made of a material not harmfully affected »445.16«.

Ⓗ Each generator must have a nameplate identifying the manufacturer; the rated frequency; power factor; number of phases (if of alternating current); the rating in kilowatts or kilovolt amperes; the normal volts and amperes corresponding to the rating; rated revolutions per minute; insulation system class and rated ambient temperature (or rated temperature rise); and time rating »445.11«.

> **NOTE**
>
> *Portable and vehicle-mounted generators must be grounded in accordance with 250.34.*
>
> *A generator meeting the qualifications of a separately derived system must be grounded per 250.30 specifications.*

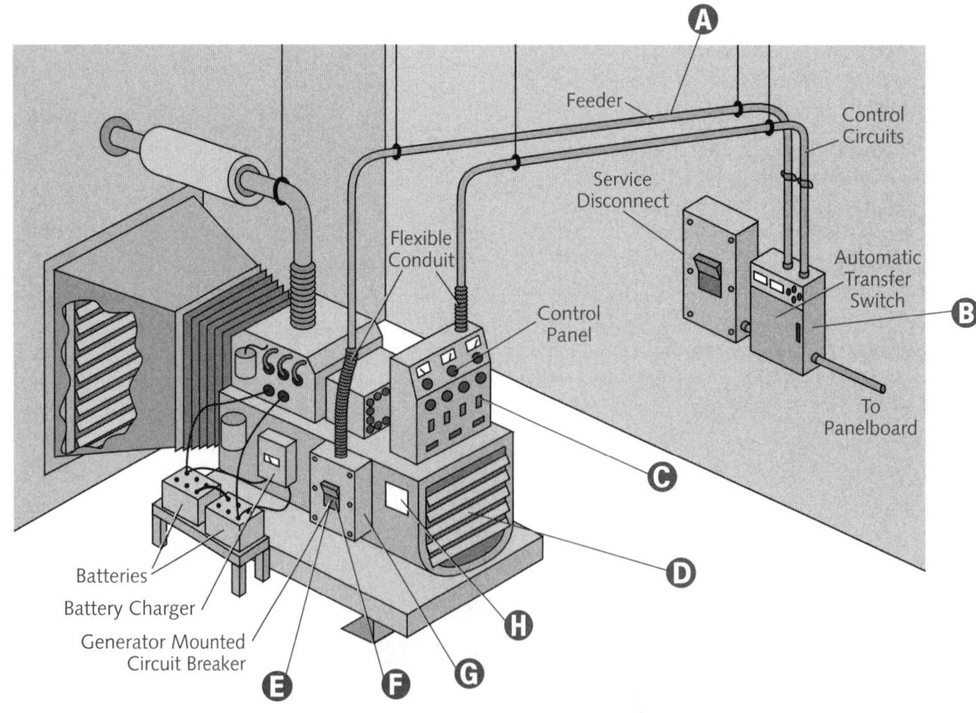

BUSWAYS

General Installation Provisions

Busways must not be installed as follows: (1) Where subject to severe physical damage or corrosive vapors; (2) In hoistways; (3) In any hazardous (classified) location, unless specifically approved for such use; and (4) Outdoors or in wet or damp locations unless identified for such use »368.4(B)«.

Cord and cable installations must comply with 368.8(B) and (C).

Overcurrent protection is required where busways are reduced in ampacity, unless the industrial establishment exception is met »368.11«.

A busway used as a branch-circuit must be protected against overcurrent in accordance with 210.20. Where so used, the circuit must comply with the applicable requirements of Articles 210, 430, and 440 »368.13«.

Ⓐ Article 368 covers the service-entrance, feeder, branch-circuit busways, and associated fittings »368.1«. For Article 368 purposes, a busway is a grounded metal enclosure containing factory-mounted, bare, or insulated conductors of copper or aluminum bars, rods, or tubes »368.2«.

Ⓑ Busways must be securely supported at no more than 5-ft (1.5-m) intervals unless otherwise designed and so marked »368.5«.

Ⓒ Where a busway serves as a feeder, devices or plug-in connections for tapping off feeder or branch-circuits from the busway shall consist of an externally operable circuit breaker or fusible switch. If such devices contain a disconnecting means and are mounted out of reach, a suitable means (ropes, chains, sticks, etc.) must be provided to facilitate operation of the disconnecting means from the floor »368.12«. (This section also contains three exceptions.)

Ⓓ A busway must be protected against overcurrent in accordance with the busway's allowable current rating, unless applying the applicable provisions of 240.4 »368.10«.

Ⓔ Unbroken busway lengths can extend through dry walls »368.6(A)«.

Ⓕ A busway's dead end must be closed »368.7«.

Ⓖ Branches from busways must be made in accordance with Articles 320, 330, 332, 342, 344, 348, 350, 352, 356, 358, 362, 368, 384, 386, and 388 »368.8(A)«.

NOTE

Lighting and trolley busways must not be installed less than 8 ft (2.5 m) above the floor (or working platform), unless provided with a cover and identified for the purpose »368.4(B)«.

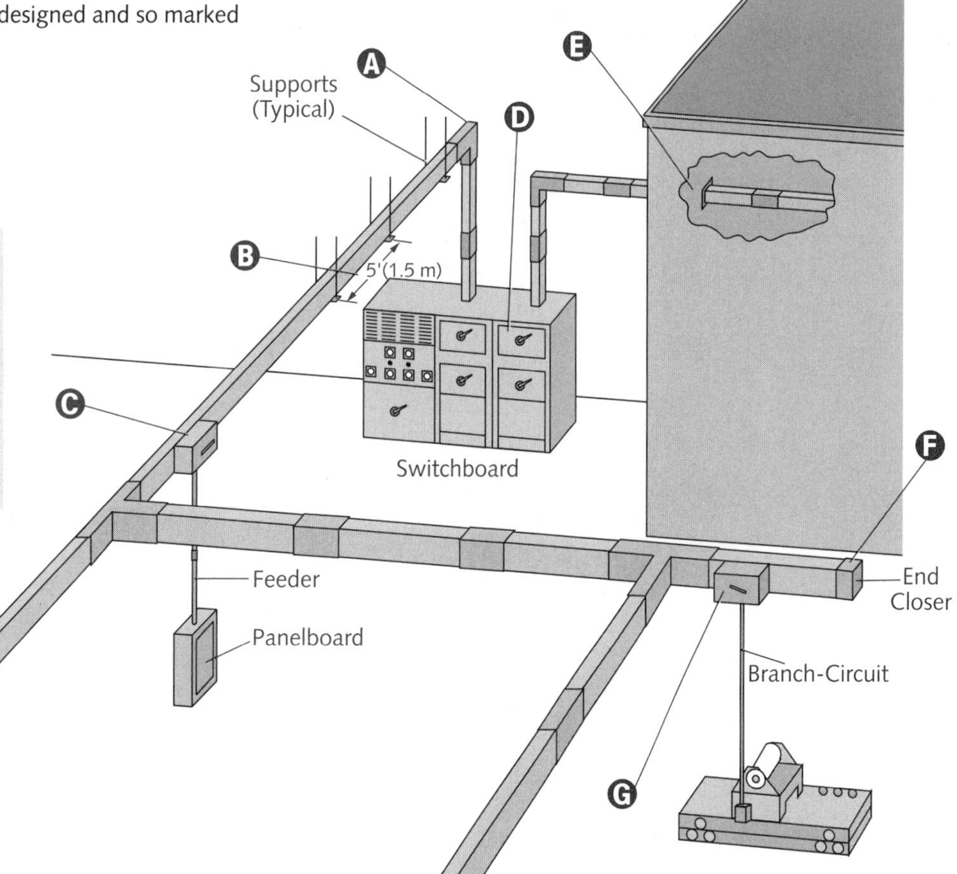

Vertical Installation Provisions

A Busways and busway taps must be protected against overcurrent in accordance with 368.10 through 368.13 »240.21(E)«.

B 368.6(B) requires that floor penetrations comply with (1) and (2):

(1) Vertical busways can extend through dry floors if totally enclosed (unventilated) where passing through, and for a minimum distance of 6 ft (1.8 m) above the floor, providing adequate physical damage protection.

(2) In other than industrial establishments, where a vertical riser penetrates multiple dry floors, a minimum 4-in. (100-mm) high curb must be installed around all busway riser floor openings to prevent liquid penetration. The curb must be installed within 12 in. (300 mm) of the floor opening. Locate electrical equipment so that it is not likely to be damaged by curb retained liquids.

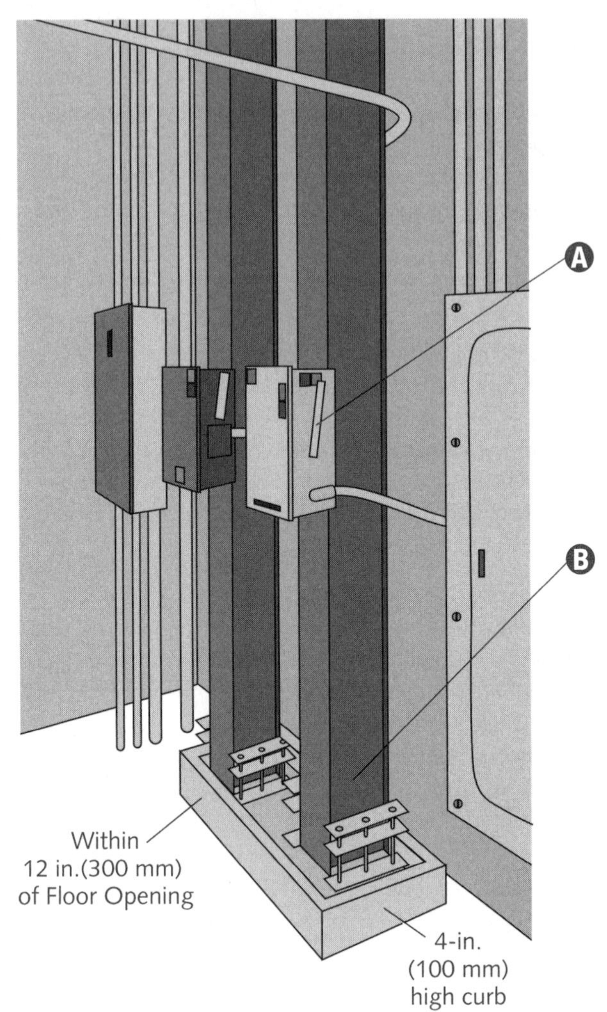

NOTE

Implement electrical installations in hollow spaces or vertical shafts so that the possible spread of fire, or products of combustion, will not be substantially increased. Openings around electrical penetrations through fire-resistant-rated walls, partitions, floors, or ceilings must be firestopped by approved methods to maintain the fire resistance rating »300.21«.

Summary

- 225.18 and 230.24(B) contain minimum vertical wiring clearance provisions.
- Associated equipment, located above or below electrical equipment, cannot extend more than 6 in. (150 mm) beyond the electrical equipment's front.
- A space equal to the width and depth of the equipment (panelboards, switchboards, and motor control centers), extending from the floor to a height of 6 ft (1.8 m) above the equipment (or to the structural ceiling, whichever is lower) must be dedicated to the electrical installation.
- The space equal to the width and depth of the equipment above panelboards, switchboards, and motor control centers must be clear of foreign systems unless protection is provided to avoid damage from condensation, leaks, or breaks.
- Working space clearances vary, depending on conditions and voltage.
- For equipment rated 1200 amperes or more, over 6 ft (1.8 m) wide, and containing overcurrent devices, switching devices, or control devices, there must be an entrance at each end of the working space.
- More than 10% of a lighting and appliance branch-circuit panelboard's overcurrent devices protect lighting and appliance branch-circuits.
- On the supply side of the service, the grounded circuit conductor and grounding conductors are connected together.
- A grounding connection must not be made to any grounded circuit conductor on the load side of the service disconnecting means except as otherwise allowed by Article 250.
- Transformer provisions are located in Article 450.
- 250.30 contains separately derived system grounding provisions.
- An alternate ac power source (such as an on-site generator) is not a separately derived system if the neutral is solidly interconnected to a service-supplied neutral.
- Article 368 covers service-entrance, feeder, and branch-circuit busways and associated fittings.

Unit 14 Competency Test

***NEC*® Reference** **Answer**

________ ________ 1. A separately derived system's grounding electrode conductor connection can be made at any point from the source of the separately derived system to the _____ system disconnecting means or overcurrent device.

________ ________ 2. Not more than _____ overcurrent devices (other than those provided for in the mains) of a lighting and appliance branch-circuit panelboard shall be installed in any one cabinet or cutout box.

________ ________ 3. Internal parts of electrical equipment, including busbars, wiring terminals, insulators, and other surfaces, shall not be _____ by foreign materials such as paint, plaster, cleaners, abrasives, or corrosive residues.

________ ________ 4. An alternate ac power source such as an on-site generator is not a separately derived system if the _____ is solidly interconnected to a service-supplied system _____.

________ ________ 5. For equipment rated 1200 amperes or more and over 6 ft (1.8 m) wide that contains overcurrent devices, switching devices, or control devices, there shall be _____ entrance(s).

________ ________ 6. A transformer, less than 600 volts, equipped with coordinated thermal overload protection by the manufacturer and arranged to interrupt the primary current, shall be permitted to have primary overcurrent protection rated or set at a current value that is not more than _____ the rated current of the transformer for a transformer having 8% impedance.

________ ________ 7. The width of the working space in front of the electric equipment shall be the width of the equipment or_____ in., whichever is greater.

NEC® **Reference** **Answer**

_______ _______ 8. Where wires pass through an opening in an enclosure, conduit box, or barrier, a _____ shall be used to protect the conductors from the edges of an opening having sharp edges within generators.

_______ _______ 9. On a switchboard or panelboard supplied from a four-wire, delta-connected system where the midpoint of one phase winding is grounded, that phase busbar or conductor having the higher voltage to ground shall be durably and permanently marked by an outer finish that is _____ in color or by other effective means.

_______ _______ 10. Ground-fault protection of equipment shall be provided for solidly grounded wye electrical systems of more than 150 volts to ground, but not exceeding 600 volts phase-to-phase for each individual device used as a building or structure main disconnecting means rated _____ or more.

_______ _______ 11. The _____ conductor of each separately derived system shall be bonded to the nearest available point of the interior metal water piping system(s) in the area served by each separately derived system.

_______ _______ 12. Busways shall be securely supported at intervals not exceeding _____ ft unless otherwise designed and marked.

_______ _______ 13. What is the minimum vertical clearance (in ft) for 208Y/120-volt, three-phase service-drop conductors in a commercial area subject to minimal truck traffic?

_______ _______ 14. Covered, shielded, fenced, enclosed, or otherwise protected by means of suitable covers, casings, barriers, rails, screens, mats, or platforms to remove the likelihood of approach or contact by persons or objects to a point of danger is the definition of _____.

_______ _______ 15. Where the overcurrent device is rated over _____ amperes, the ampacity of the conductors it protects shall be equal to or greater than the rating of the overcurrent device as defined in 240.6.

_______ _______ 16. The enclosure for a panelboard shall have the top and bottom wire-bending space sized in accordance with _____ for the largest conductor entering or leaving the enclosure.

_______ _______ 17. A(n) _____ is a premises wiring system whose power is derived from a battery, a solar photovoltaic system, a generator, a transformer, or converter windings, and that has no direct electrical connection, including a solidly connected grounded circuit conductor, to supply conductors originating in another system.

_______ _______ 18. Dry-type transformers 600 volts, nominal, or less and not exceeding _____ shall be permitted in hollow spaces of buildings not permanently closed in by structure.

_______ _______ 19. Where rear access is required to work on nonelectrical parts on the back of enclosed equipment, a minimum horizontal working space of _____ in. shall be provided.

_______ _______ 20. The ampacity of the conductors from the generator terminals to the first distribution device(s) containing overcurrent protection shall not be less than _____ % of the nameplate current rating of the generator.

_______ _______ 21. Where exposed to the weather, raceways enclosing service-entrance conductors shall be _____.

_______ _______ 22. For industrial establishments only, omission of overcurrent protection shall be permitted at points where busways are reduced in ampacity, provided that the length of the busway having the smaller ampacity does not exceed _____ ft and has an ampacity at least equal to one-third the rating or setting of the overcurrent device next back on the line, and provided that such busway is free from contact with combustible material.

***NEC®* Reference**	**Answer**	
______	______	23. Circuit breakers shall be marked with their ampere rating in a manner that will be _____ after installation.
______	______	24. The minimum headroom of working spaces about service equipment, switchboards, panelboards, or motor control centers shall be _____ft.
______	______	25. A 480Y/277-volt switchboard, having exposed live parts on one side and a concrete wall on the other side, requires a minimum working space depth of _____ in.
______	______	26. For systems that are not separately derived and are not required to be grounded as specified in 250.30, see _____ for the minimum size conductors that must carry fault current.
______	______	27. In other than industrial establishments, where a vertical riser penetrates two or more dry floors, a minimum _____ in.-high curb shall be installed around all floor openings for riser busways to prevent liquids from entering the opening.
______	______	28. _____ are generally accessible from the rear as well as from the front and are not intended to be installed in cabinets.
______	______	29. Generators shall be equipped with a _____ by means of which the generator and all protective devices and control apparatus are able to be disconnected entirely from the circuits supplied by the generator.
______	______	30. The space equal to the width and depth of switchboards, panelboards, and motor control centers and extending from the floor to a height of _____ ft above the equipment or to the structural ceiling, whichever is lower, shall be dedicated to the electrical installation.
______	______	31. For other than a totally enclosed switchboard, a space not less than _____ ft shall be provided between the top of the switchboard and any combustible ceiling, unless a noncombustible shield is provided between the switchboard and the ceiling.
______	______	32. Where the service-entrance phase conductors are larger than 1100-kcmil copper or 1750-kcmil aluminum, the main bonding jumper shall have an area that is not less than _____ of the area of the largest phase conductor.
______	______	33. _____ is defined as being surrounded by a case, housing, fence, or walls that prevent persons from accidentally contacting energized parts.
______	______	34. Busways shall be permitted to be extended vertically through dry floors if totally enclosed (unventilated) where passing through and for a minimum distance of _____ ft above the floor to provide adequate protection from physical damage.

SPECIAL OCCUPANCIES, AREAS, AND EQUIPMENT

section 5

UNIT 15

Hazardous (Classified) Locations

Objectives

After studying this unit, the student should:

- understand what constitutes Class I, II, and III locations, as well as the parameters for Divisions 1 and 2 within each class.
- have a good understanding of provisions pertaining to commercial garages, repair and storage facilities.
- understand that Article 513 applies to buildings/structures, or parts thereof, inside which aircraft are housed, stored, serviced, repaired, or altered.
- understand the distinction between Division 1's normal operating conditions, and Division 2's abnormal operations.
- know that Articles 500 through 504 cover requirements for electrical and electronic equipment/wiring, of all voltages, within hazardous (classified) locations with potential fire or explosion hazards due to the presence of flammable gases (or vapors), flammable liquids, combustible dust, or ignitible fibers (or flyings).
- know that a gasoline dispensing and service station involves the transfer of gasoline, or other volatile flammable liquids or liquefied flammable gases, to the fuel tanks of self-propelled vehicles or to approved containers.

Introduction

Hazardous (classified) locations are divided into certain categories and sub-categories, otherwise known as classes and divisions. Articles 500 through 504 cover the requirements for Class I, Divisions 1 and 2; Class II, Divisions 1 and 2; and Class III, Divisions 1- and 2-located electrical and electronic equipment/wiring, of all voltages, where potential fire or explosion hazards exist due to the presence of flammable gases (or vapors), flammable liquids, combustible dust, or ignitible fibers (or flyings). Locations containing quantities of airborne gases (or vapors), potentially sufficient to cause explosion or ignition are categorized as Class I. Locations hazardous due to the presence of combustible dust are designated Class II. Class III locations are hazardous because of the presence of easily ignitible fibers (or flyings), which, however, are not likely to be suspended in the air in quantities sufficient to produce ignitible mixtures. Under typical, i.e. normal, operating conditions, Division 1 locations contain ignitible or combustible elements (gases, dust, fibers, etc.) in quantities sufficient to produce explosive (or ignitible) mixtures. Most locations where airborne combustible or ignitible elements (gases, dust, fibers, etc.) are not ordinarily present in quantities sufficient to produce explosive (or ignitible) mixtures are classified Division 2. Within these Articles, many FPNs reference other standards for more information on specific applications. Similarly, this test refers the reader, in this instance, to the *Code* itself for Article 504 (*Intrinsically Safe Systems*) and Article 505 (*Class I, Zone 0, 1, and 2 Locations*). The fact that, for various reasons, this text could not address these topics is in no way intended to diminish their significance. Articles 511 through 516 cover occupancies, or parts thereof, that are or may be hazardous because of atmospheric concentrations of flammable liquids, gases, or vapors; or, because of deposits or accumulations of readily ignitible materials.

OVERVIEW

Class Categorizations

A Locations containing potentially sufficient quantities of air-borne gases (or vapors) to cause explosion or ignition are categorized as Class I locations. Class I locations include those specified in 500.5(B)(1) and (2) »500.5(B)«.

B Generally, under normal operating conditions, Division 1 locations contain ignitible or combustible elements (gases, dust, fibers, etc.) in quantities sufficient to produce explosive (or ignitible) mixtures.

C Equipment approved for Class I locations are permitted in Class I, Class II, and Class III locations. Equipment approved for Class II locations are permitted in Class II and Class III locations, but not in Class I locations. Equipment approved for Class III locations are not permitted in either Class I or Class II locations.

D Class II group classifications are covered in 500.6(B).

E The Class I temperature marking, specified in 500.8(B), must not exceed the existing gas or vapor ignition temperature »500.8(C)(1)«. (See FPN for the location of additional literature regarding gas/vapor ignition temperatures.)

F In most cases, locations where combustible or ignitible elements (gases, dust, fibers, etc.) are not ordinarily present in the air in quantities sufficient to produce explosive (or ignitible) mixtures, are Division 2 locations.

G Class I group classifications are covered in 500.6(A).

H The temperature marking for Class II locations, per 500.8(B), must be less than the ignition temperature of the existing specified dust. The temperature rating of any organic dusts, which may dehydrate or carbonize, must be the lower of the ignition temperature or 329°F (165°C) »500.8(C)(2)«.

Locations which are hazardous due to the presence of combustible dust are designated Class II locations. Class II locations include those specified in 500.5(C)(1) and (2) »500.5(C)«.

I Locations that are hazardous because of the presence of easily ignitible fibers (or flyings), but in which such fibers (or flyings) are not likely to be suspended in the air in quantities sufficient to produce ignitible mixtures, are designated Class III. Class III locations include those specified in 500.5(D)(1) and (2) »500.5(D)«.

J There are no groups within Class III locations.

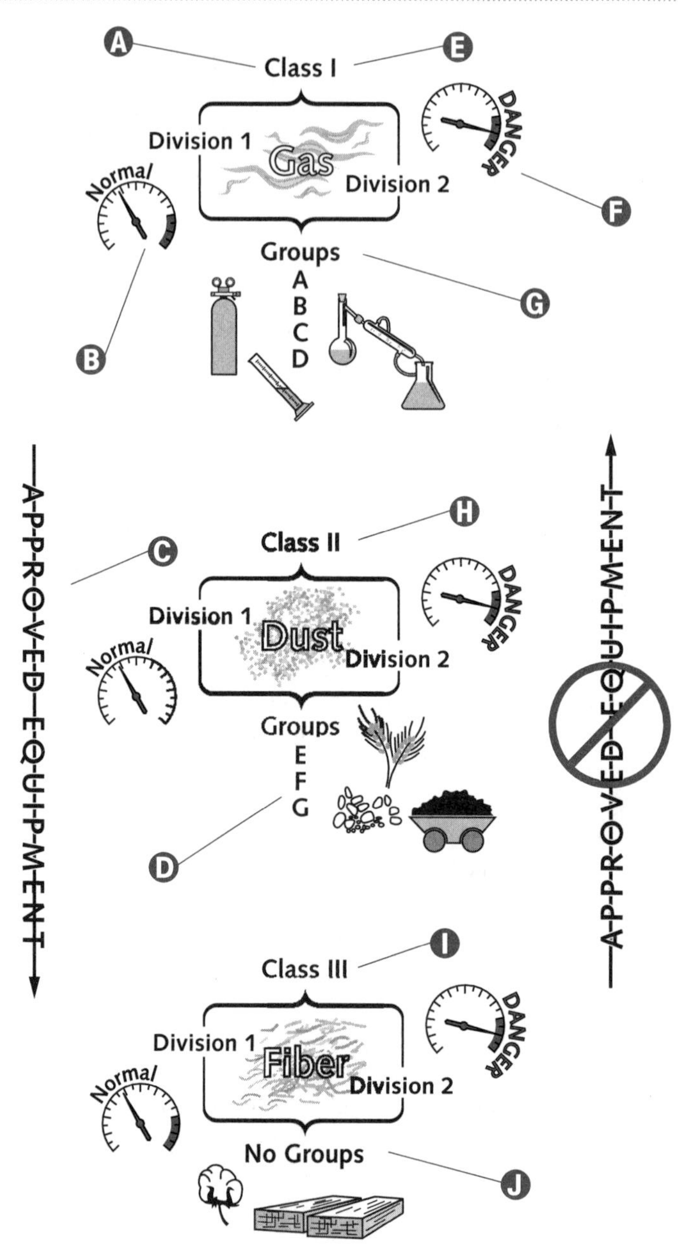

NOTE

Within a single class and group, equipment identified for Division 1 use is permitted in a Division 2 location »500.8(A)(2)«.

Threaded Conduit

Equipment having metric threaded entries must be so identified, or must provide listed adapters appropriate for connection to conduit or NPT-threaded fittings. Listed metric-threaded cable fittings are also acceptable »500.8(D(2)«.

Ⓐ Threaded conduit within hazardous (classified) locations must be threaded using an NPT standard conduit cutting die which renders a taper of 1 in 16 (¾-in. taper per ft) »500.8(D)«.

Ⓑ Threaded conduit must be wrenchtight to prevent sparking should fault current flow through the conduit system; and, to ensure explosion proof or flameproof integrity where applicable »500.8(D)«.

NOTE

Threaded conduit installations must also comply with Chapter 3's requirements, in particular, Article 344 for RMC and Article 342 for IMC.

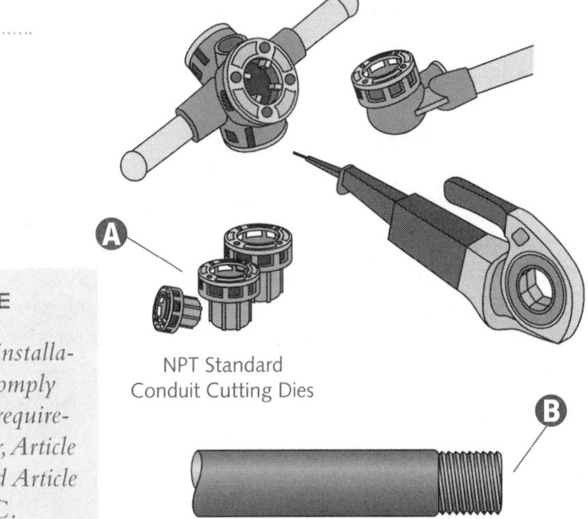

General

Article 505 details the electrical/electronic requirements for all voltages in Class, Zone 0, Zone 1, and Zone 2 hazardous (classified) locations where fire or explosion hazards may exist due to flammable gases (or vapors), or flammable liquids »500.1 FPN«.

Except as modified in Articles 500 through 504, all other applicable *Code* rules apply to electrical equipment/wiring installed in hazardous (classified) locations »500.3«.

Creative design and implementation of hazardous (classified) electrical systems may facilitate the placement of equipment in less hazardous, or even nonhazardous, areas, greatly reducing the risk as well as the need for special equipment »500.5(A) FPN«.

When making determinations regarding classification, ventilation, protection against static electricity, lightning, etc., the AHJ must be familiar with the recorded industrial experience, together with the standards of the National Fire Protection Association (NFPA), the American Petroleum Institute (API), and the Instrument Society of America (ISA) »500.4(B) FPN No. 1«.

Fiberoptic cable assemblies, containing potentially current-carrying conductors, must be installed in accordance with Articles 500 through 503, as applicable »500.8(E)«.

Articles 510 through 517 cover: garages, aircraft hangars, gasoline dispensing/service stations, bulk storage plants, spray applications, dipping/coating processes, and health care facilities »500.9«. (Health care facilities are covered in Unit 16 of this book.)

Ⓐ **Scope**—Articles 500 through 504 cover the requirements for electrical and electronic equipment/wiring for all voltages in Class I, Divisions 1 and 2; Class II, Divisions 1 and 2; and Class III, Divisions 1 and 2 locations where fire or explosion hazards may exist due to presence of flammable gases (or vapors), flammable liquids, combustible dust, or ignitible fibers (or flyings) »500.1«.

Ⓑ Each room, section or area within a hazardous (classified) location, is considered individually to determine the appropriate classification »500.5(A)«. (Here no specific classifications are assigned for each area; they are shown for illustration purposes only.)

Ⓒ What constitutes a hazardous (classified) location, and the requirements that apply, are determined contingent upon: (1) the properties of flammable vapors, liquids, gases, or combustible dusts/fibers that are present; and (2) the likelihood that a flammable/combustible concentration of such substances can occur »500.5(A)«.

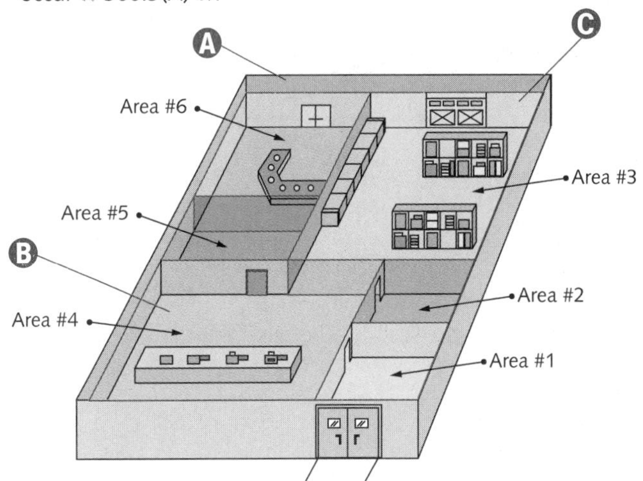

NOTE

Proper documentation must be maintained on each hazardous (classified) location. Such documentation must be available to those authorized to design, install, inspect, maintain, or operate electrical equipment at that particular location »500.4(A)«.

Class I Group Classifications

Hazardous (classified) locations divide atmosphere identifications for Class I, Divisions 1 and 2, locations into Groups A, B, C, and D »500.6(A)(1) through (4)«.

A Group A is comprised only of acetylene »500.6(A)(1)«.

B If a flammable/combustible gas (or liquid-produced vapor), when mixed with air, has a maximum experimental safe gap (MESG) of no more than 0.45 mm, or a minimum igniting current ratio (MIC ratio) of no more than 0.40, the material falls into Group B »500.6(A)(2)«.

An example of a Group B material is hydrogen »500.6(A)(2) FPN«.

C If a flammable/combustible gas (or liquid-produced vapor), when mixed with air, has either an MESG value greater than 0.45 mm (but not more than 0.75 mm), or a minimum MIC ratio greater than 0.40 (but no more than 0.80), the material is classified as Group C »500.6(A)(3)«.

Ethylene is a typical Group C material »500.6(A)(3) FPN«.

D Atmospheric Group D expands the limits of both the MESG (greater than 0.75 mm) and the MIC ratio (greater than 0.80) »500.6(A)(4)«.

An example of a Group D material is propane »500.6(A)(4) FPN«.

Class II Group Classifications

Groups E, F, and G comprise Class II hazardous atmospheres »500.6(B)(1) through (3)«.

E Group E atmospheres contain combustible metal dust whose particle size, abrasiveness, and conductivity present similar hazards in electrical equipment operation. Aluminum, magnesium, and their commercial alloys are Group E materials »500.6(B)(1)«.

As with Class I Groups, certain metal dust characteristics may require safeguards beyond those given here. Uranium dust, for example, may ignite at temperatures as low as 68°F (20°C) »500.6(B)(1) FPN«.

F Potentially explosive dusts, containing or yielding carbon, having more than 8% total entrapped volatiles, or having been sensitized through the action of other substances, are Group F materials. Coal, carbon black (used in the production of ink and rubber products), and charcoal are three examples of Group F items »500.6(B)(2)«.

G Group G is the catchall group whose atmospheres contain combustible dusts not included in Group E or F, including flour, grain, wood, plastic, and chemicals »500.6(B)(3)«.

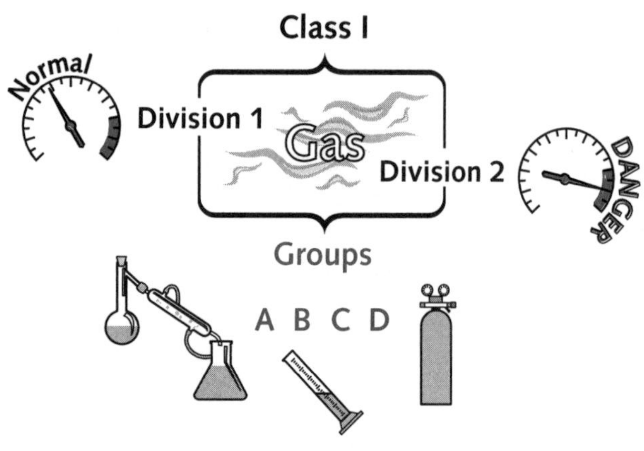

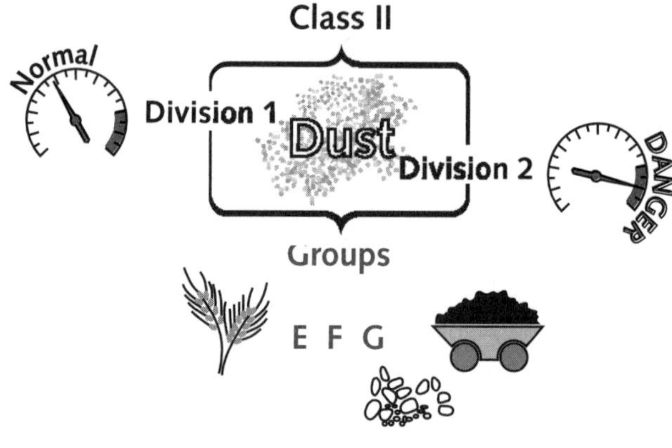

CAUTION *Equipment used in any location containing such gases/vapors must be approved for the class, as well as for the specific group, involved »500.6(A) FPN No. 2«.*

CAUTION *Equipment used in Class II locations must be approved not only for the class, but also for the specific group, of dust that will be present »500.6(B)(3) FPN No. 2«.*

Equipment

Inspection authorities and users are advised to exercise extraordinary care with regard to installation and maintenance »500.8 FPN No. 1«.

Any equipment preventing the entrance of flammable/ combustible fluids by means of a single compression seal, diaphragm, or tube must be identified for a Class I, Division 2 location. Equipment installed in a Class I, Division 1 location must be identified for the Class I, Division 1 location »500.8(A)(4)«.

Where Articles 501 through 503 specifically permit, general-purpose equipment or equipment in general-purpose enclosures can be installed in Division 2 locations if, under normal operating conditions, the equipment does not constitute a source of ignition »500.8(A)(3)«.

Attention must be given to the possibility of simultaneously occurring flammable gases and combustible dust when determining the safe operating temperature of installed electrical equipment »500.8(A)(6)«.

Ⓐ Equipment must be identified not only for the class of location but also for the explosive, combustible, or ignitible properties of the specific gas, vapor, dust, fiber, or flyings that are present. In addition, Class I equipment must not have any exposed surface operating at temperatures in excess of the ignition temperature of the specific gas/vapor. Class II equipment must not have an external temperature higher than that specified in 500.8. Class III equipment must not exceed the maximum surface temperature specified in 503.1 »500.8(A)(1)«.

Ⓑ Articles 500 through 504 require equipment construction and installation that will ensure safe performance under conditions of proper use and maintenance »500.8«.

Ⓒ At a minimum, equipment installed must be approved for all applicable levels of Class, Division, and Group, giving careful consideration to normal operating and ignition temperatures. Equipment exceeding the specified requirements is permitted. However, the reverse is not true; i.e., Division 1 approved equipment is permitted in a Division 2 location of the same class and group; but Division 2 approved equipment cannot be used in a Division 1 location »500.8(A)(2)«.

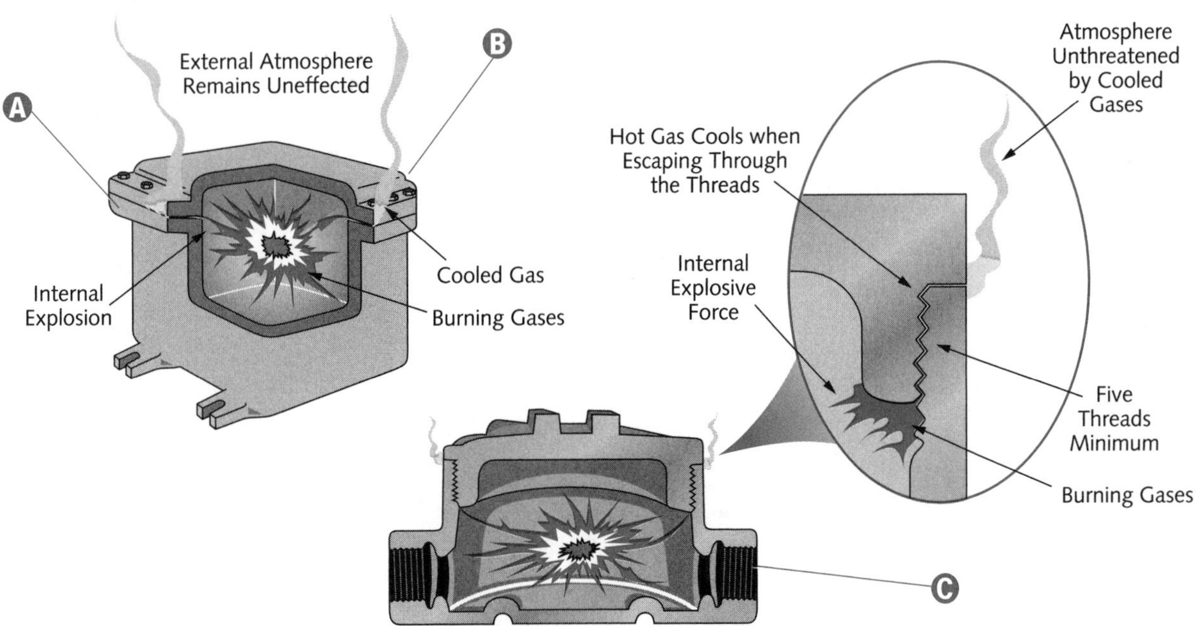

Marking

A Equipment must display markings of the class, group, and operating temperature or temperature class referenced to a 40°C ambient, unless provided otherwise in 500.8(B) *Exception No. 1* through *Exception No. 5* »500.8(B)«.

B The maximum safe operating temperature must be marked on equipment for Class I and Class II locations, having been determined by simultaneous exposure to the combined conditions for both classes »500.8(B)«.

C The temperature class, if provided, shall be indicated using the temperature class (*T Codes*) shown in Table 500.8(B). The *T Codes* marked on equipment nameplates must be in accordance with Table 500.8(B) »500.8(B)«.

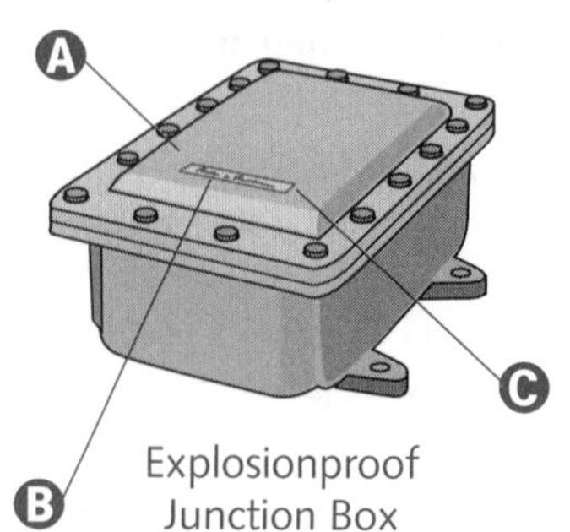

Explosionproof Junction Box

NOTE

Because of inconsistent relationships between explosion properties and ignition temperatures, the two are independent requirements »500.8(B) FPN«.

Protection Techniques

The Purged and Pressurized protection technique is permitted for equipment in any hazardous (classified) location for which it is identified »500.7(D)«.

Intrinsically safe apparatus and wiring is permitted in the specified hazardous (classified) location for which it is approved, provided the installation meets Article 504 requirements. Articles 501 through 503, and 510 through 516, are not applicable to such installations, unless required by Article 504 »500.7(E)«.

***Dusttight:* Constructed to prevent dust from entering the enclosed case, under stipulated test conditions »Article 100«.**

A Acceptable protection techniques for electrical and electronic equipment in hazardous (Classified) locations include: Explosionproof Apparatus; Dust ignitionproof; Dusttight; Purged and Pressurized; Intrinsically Safe Systems; Nonincendive (circuit and equipment); Oil Immersion; Hermetically Sealed; and other such techniques employing equipment specifically listed for a given location »500.7«.

B Dusttight designated equipment is permitted for Class II Division 2 and Class III locations for which it is identified »500.7(C)«.

C **Explosionproof Apparatus:** is encased by constructive design to withstand a potential internal explosion of a specified gas/vapor, effectively preventing the ignition of surrounding gas/vapors by internal sparks, flashes, or explosions, and operating at such an external temperature that any surrounding flammable atmosphere will not be ignited as a result »Article 100«.

D Explosionproof apparatus are required for Class I, Divisions 1 and 2 locations »500.7(A)«.

E Equipment of the Dust ignitionproof variety is acceptable for Class II, Divisions 1 and 2 locations »500.7(B)«.

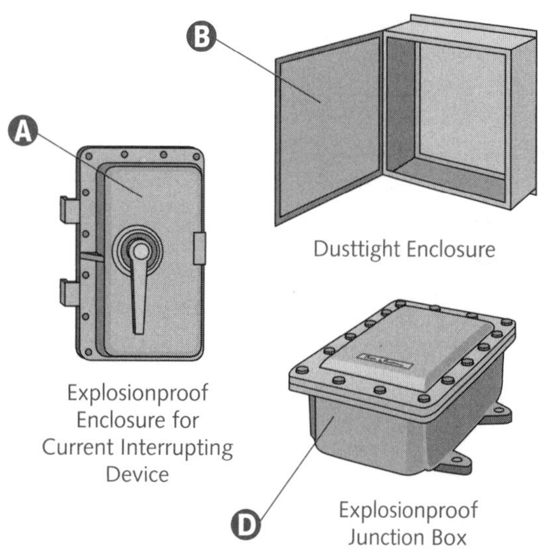

Explosionproof Enclosure for Current Interrupting Device

Dusttight Enclosure

Explosionproof Junction Box

Explosionproof Panelboard

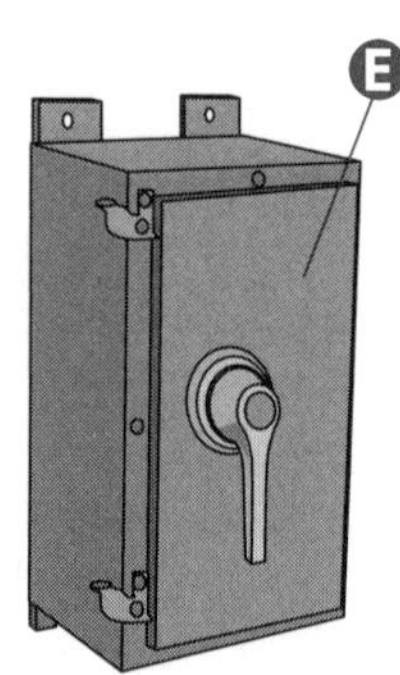

Dust Ignitionproof Enclosure for Current Interrupting Devices

Protection Techniques (*continued*)

Nonincendive Circuit: A circuit, other than field wiring, in which any arc or thermal effect produced under the equipment's intended operating conditions, is incapable, under specified test conditions, of igniting the flammable gas-, vapor-, or dust-air mixture »Article 100 and 500.2«.

Nonincendive Field Wiring: Wiring that enters or leaves an equipment enclosure and, under the equipment's normal operating conditions, is not capable, due to arcing or thermal effects, of igniting the flammable gas-, vapor-, or dust-air mixture. Normal operation includes opening, shorting, or grounding the field wiring »Article 100«.

Under normal operating conditions, nonincendive equipment's electrical or electronic circuitry is incapable of igniting a given flammable substance due to arcing or thermal means »500.2«.

A nonincendive component has a contacting mechanism capable of making and breaking an incendive circuit, which mechanism's construction renders the component incapable of igniting the specified flammable gas- or vapor-air mixture. A nonincendive component's housing does not exclude the flammable atmosphere nor does it contain an explosion. This protection technique is allowed for current-interrupting contacts in those Class I, Division 2 locations for which the component is approved »500.2 and 500.7(H)«.

The oil-immersion protective technique is valid for current-interrupting contacts in Class I, Division 2 locations as described in 501.6(B)(1)(b) »500.7(I)«.

Equipment can be hermetically sealed against the intrusion of an external atmosphere via a fusion seal, i.e., soldering, brazing, welding, or the fusion of glass to metal »500.2«.

Other protection techniques inherent to equipment listed for use in hazardous (classified) locations are allowed »500.7(L)«.

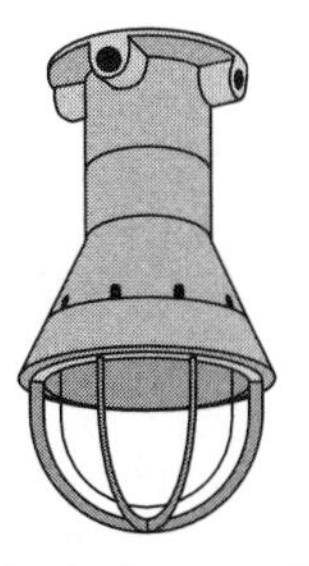

Explosionproof Incandescent Light Fixture

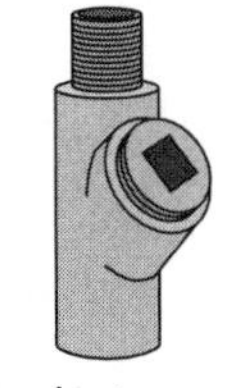

Explosionproof Vertical Seal

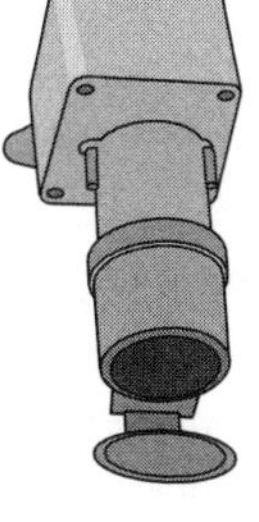

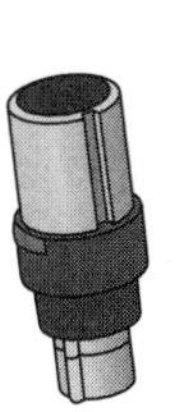

Explosionproof Attachment Plug and Receptacle

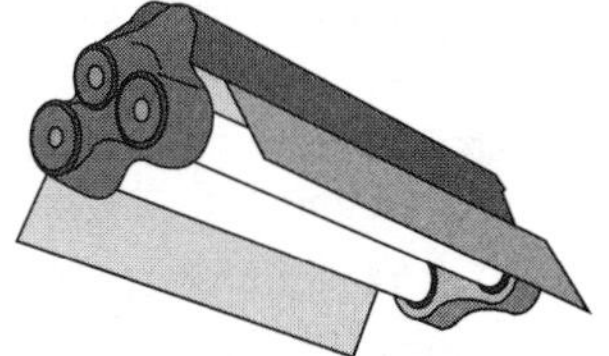

Explosionproof Fluorescent Light Fixture

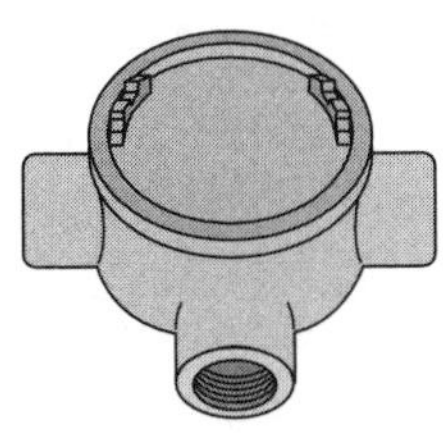

Explosionproof Outlet Box

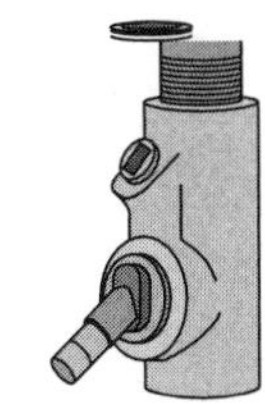

Explosionproof Vertical Seal with Drain Fitting

Explosionproof Flexible Coupling

CLASS I LOCATIONS

Class I, Division 1 Locations

In addition to the specific requirements found within Article 500, general *Code* rules apply to the electric wiring and equipment in Class I locations »501.1«.

In Class I, Division 1 locations, transformers and capacitors containing burnable liquid must be installed only in approved vaults compliant with 450.41 through 450.48, in addition to all the following stipulations: (1) the vault cannot have any form of communicating opening into the Division 1 location; (2) ample ventilation must continually remove flammable gases/vapors; (3) all vents/ducts lead to safe locations, outside of buildings; and (4) vent ducts and openings must be of sufficient area to relieve the vault's internal explosive pressures, having all portions of vent ducts (within the building) constructed of reinforced concrete »501.2(A)(1)«. Subsequently, transformers and capacitors not containing burnable liquids must either be installed in 501.2(A)(1) compliant vaults, or be approved for Class I locations »501.2(A)(2)«.

Approved enclosures must be provided for Class I, Division 1 located meters, instruments, and relays. This includes kilowatt-hour meters, instrument transformers, resistors, rectifiers, and thermionic tubes. Class I, Division 1 enclosures include explosionproof as well as purged and pressurized enclosures »501.3(A)«.

A Class I, Division 1 classifications usually include the following locations:

(1) Where volatile flammable liquids or liquefied flammable gases are transferred from one container to another
(2) Spray booth interiors and areas in the vicinity of spraying/painting operations where flammable solvents are used
(3) Locations containing open tanks/vats of volatile flammable liquids
(4) Drying rooms or components for the evaporation of flammable solvents
(5) Locations containing fat and oil extraction equipment using volatile flammable solvents
(6) Portions of cleaning and dyeing plants where flammable liquids are used
(7) Gas generator rooms and other portions of gas manufacturing plants where flammable gas may escape
(8) Inadequately ventilated pump rooms for flammable gas or for volatile flammable liquids
(9) Refrigerator/freezer interiors in which volatile flammable materials are stored in open, lightly stoppered, or easily ruptured containers
(10) All other locations where ignitible concentrations of flammable vapors/gases are likely to occur in the course of normal operation »500.5(B)(1) FPN No. 1«.

B In some Division 1 locations, ignitible concentrations of flammable gases/vapors may be present continuously or for extended periods of time. Examples include the following:

(1) The inside of inadequately vented enclosures containing instruments normally venting flammable gases/vapors within the enclosure
(2) The inside of vented tanks containing volatile flammable liquids
(3) The area between the inner and outer roof sections of a floating roof tank containing volatile flammable fluids
(4) Inadequately ventilated areas within spraying/coating operations using volatile flammable fluids
(5) The interior of an exhaust duct, used to vent ignitible concentrations of gases/vapors.

Avoiding the installation of electrical components in these areas altogether is strongly recommended. If relocation is unfeasible, use electric equipment or instrumentation approved specifically for the application, or consisting of what Article 504 describes as intrinsically safe systems »500.5(B)(1) FPN No. 2«.

C Class I, Division 1 locations contain ignitible concentrations of flammable gases/vapors which: (1) can exist during normal operation; (2) may often exist due to repair, maintenance, or leakage; and/or (3) might be released because operational failures cause simultaneous breakdowns of electrical equipment, capable of direct ignition »500.5(B)(1)«.

D Locations containing potentially sufficient quantities of airborne gases (or vapors) to cause explosion or ignition, are categorized as Class I locations »500.5(B)«.

> **NOTE**
>
> *Bonding and grounding must comply with 501.16(A) and (B) provisions.*

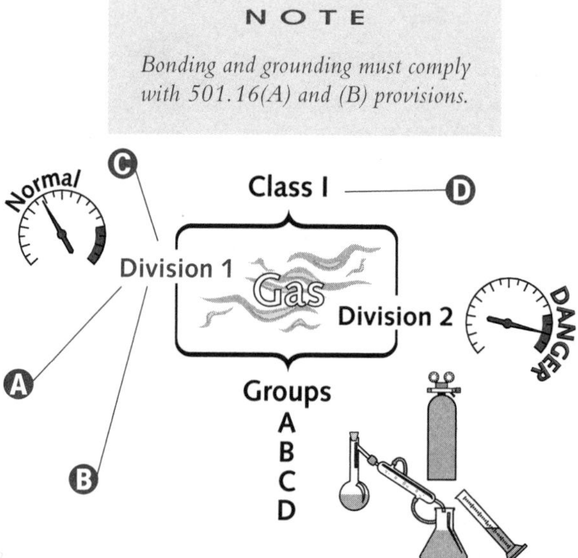

Class I, Division 2 Locations

Unless modified by this Article, 505.9 approved equipment, specifically listed for use in Class I, Zone 0, 1, or 2 locations, is acceptable in Class I, Division 2 applications for the same gas when appropriately rated for the temperature »500.1 and *Exception*«.

In Class I, Division 2 locations, transformers and capacitors must comply with 450.21 through 450.27 »501.2(B)«.

Switches, circuit breakers, make-and-break pushbutton contacts, relays, alarm bells and horns, must have Class I, Division 1 rated enclosures per 501.3(A) »501.3(B)(1)«. General-purpose enclosures are permitted if the current-interrupting contacts are: (1) immersed in oil; (2) enclosed within a hermetically sealed chamber; (3) in nonincendive circuits; or (4) part of a listed nonincendive component. »501.3(B)(1) *Exception*«.

Resistors, resistance devices, thermionic tubes, rectifiers, and similar equipment used in association with meters, instruments and relays must comply with 501.3(A) »501.3(B)(2)«.

Transformer windings, impedance coils, solenoids, etc., without sliding/make-or-break contacts must have enclosures, which may be of the general-purpose-type »501.3(B)(3)«.

A single general-purpose enclosure is allowed for any assembly whose individual components qualify for general-purpose enclosures according to 501.3(B)(1) through (3). The exterior of an assembly enclosure containing resistors and similar equipment must be clearly and permanently marked with the maximum obtainable surface temperature of any internal component. Approved equipment can be marked with Table 500.8(B) identification numbers specifying its suitable temperature class »501.3(B)(4)«.

Where general-purpose enclosures are allowed in 501.3(B)(1)–(4), they may contain overcurrent-protective fuses for instrument circuits (normally exempt from overloading), provided each fuse is preceded by a 501.3(B)(1) compliant switch »501.3(B)(5)«.

Flexible cord, attachment plug, and receptacle connections are acceptable for process control instruments, provided all of the following conditions are met:

(1) A 501.3(B)(1) compliant switch, rather than the attachment plug, provides for current interruption;

(2) Maximum current is 3 amperes at 120 volts, nominal;

(3) The power-supply cord is no more than 3 ft (914 mm) in length, is approved for extra-hard usage (where exposed) or for hard usage (if protected by location), and is supplied through a locking and grounding type attachment plug and receptacle;

(4) Only necessary receptacles are provided; and

(5) A receptacle label warns against unplugging under load »501.3(B)(6)«.

Ⓐ Class I, Division 2 locations: (1) are areas where volatile flammable liquids/gases are handled, processed, or used, but the potentially volatile substance (liquid, gas, vapor) is normally enclosed (containers or systems) and is subject to escape only during accidental rupture (or breakdown) of such enclosures or in case of abnormal conditions; (2) are equipped with positive mechanical ventilation that normally prevents the buildup of gas/vapor concentrations to an ignition level; and (3) adjoin a Class I, Division 1 location that might communicate ignitible gaseous concentrations unless prevented by adequate positive-pressure ventilation from a clean air source, with adequate safeguards against ventilation failure »500.5(B)(2)«.

Ⓑ Class I, Division 2 classification usually includes locations where volatile flammable liquids/gases, or vapors are used; but that, in the judgement of the AHJ, would become hazardous only in case of an accident or unusual operating condition. The quantity of flammable material accidentally escaping, adequacy of ventilating equipment, total area involved, and the industry/business record with respect to explosions (or fires) are all factors that merit consideration in determining each location's classification and extent »500.5(B)(2) FPN No. 1«.

Ⓒ Piping without valves, checks, meters, and similar devices would not ordinarily introduce a hazard even though used for flammable liquids or gases. Depending on factors such as quantity and size of containers and ventilation, locations used for the sealed-container storage of flammable liquids or liquefied or compressed gases may be considered either hazardous (classified) or unclassified locations »500.(B)(2) FPN No. 2«.

NOTE

Bonding and grounding must comply with 501.16(A) and (B) provisions.

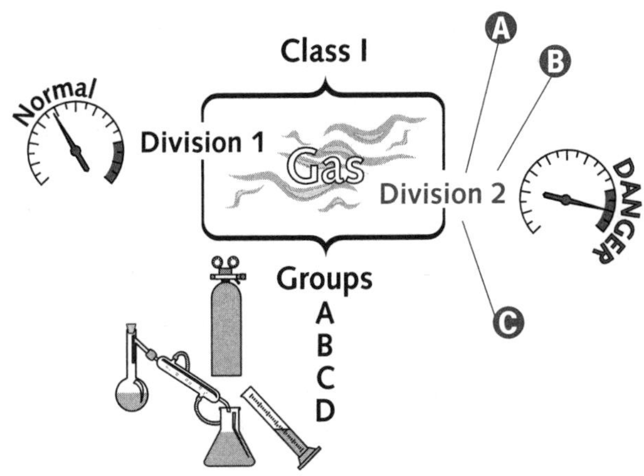

Wiring Methods—Class I, Division 1

Avoid tensile stress at Type MI cable termination fittings through proper installation and support methods »501.4(A)(1)(b)«.

RNC (complying with Article 352) can be used when first surrounded by at least 2 in. (50 mm) of concrete, followed by a minimum cover of 24 in. (600 mm) (top of conduit to grade). In some cases, 511.4, *Exception;* 514.8, *Exception No. 2*; and 515.8(A) permit the omission of the concrete encasement » 501.4(A)(1)(a) *Exception*«.

Threaded metal conduit (either rigid or steel intermediated) must comprise the last 24 in. (600 mm) of a rigid nonmetallic underground conduit. This measurement ends at either the point of emergence, or the point of aboveground raceway connection. An equipment grounding conductor is required for the dual purpose of providing raceway system electrical continuity, as well as grounding of noncurrent-carrying metal parts. »501.4(A)(1)(a) *Exception*«.

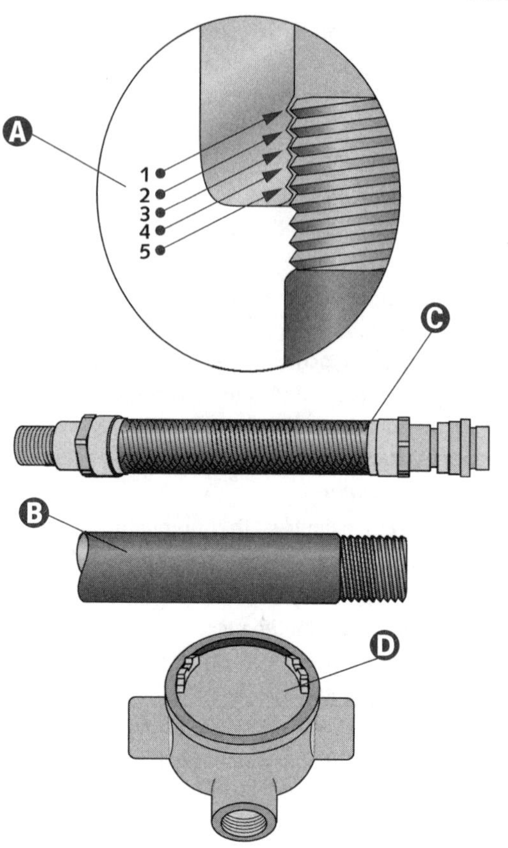

A Threaded joints must fully engage at least five threads »501.4(A)(1)(a)«.

B Fixed wiring in Class I, Division 1 locations must consist of threaded metal conduit, either rigid metal or steel intermediate metal, or Type MI cable with location-approved termination fittings »501.4(A)(1)«.

C Connections requiring flexibility, such as those at motor terminals, can employ Class I listed flexible fittings. Flexible cord, installed per 501.11, is also acceptable »501.4(A)(2)«.

D All boxes, fittings, and joints must be approved for Class I, Division I »501.4(A)(3)«.

Wiring Methods—Class I, Division 2

Explosionproof boxes, fittings, and joints are not mandatory unless required by 501.3(B)(1), 501.6(B)(1), and 501.14(B)(1) »501.4(B)(4)«.

Nonincendive field wiring is permitted using any of the methods suitable for wiring in ordinary locations »501.4(B)(3)«.

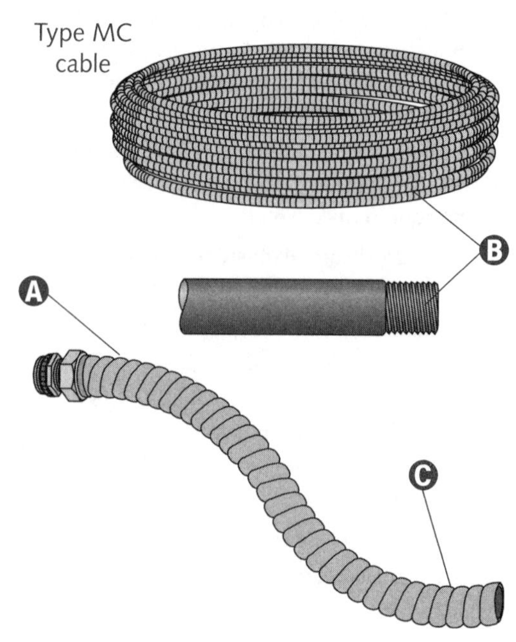

A Where limited flexibility is necessary (at motor terminals), one of the following must be used: flexible metal fittings, FMC with listed fittings, LFMC with listed fittings, LFNC with listed fittings, or flexible cord listed for extra-hard usage and provided with listed bushed fittings. Flexible cord must include an additional conductor for grounding »501.4(B)(2)«.

B In Class I, Division 2 locations, use threaded RMC, threaded steel IMC, enclosed gasketed busways, enclosed gasketed wireways, Type PLTC cable (per Article 725), or Type ITC cable in cable trays/raceways, supported by messenger wire, or directly buried where the cable is listed for this use; Type MI, MC, MV, or TC cable with termination fittings must be the wiring method employed. Type ITC, PLTC, MI, MC, MV, or TC cable that can be placed in cable tray systems must be installed in a manner to avoid tensile stress at the termination fitting »501.4(B(1)«.

C See 501.16(B) for grounding requirements where flexible conduit is used »501.4(B)(2) FPN«.

Motor Fuel Dispensing Facilities

If the AHJ can satisfactorily determine that flammable liquids having a flash point below 100°F (38°C), such as gasoline, will not be handled, such location does not require a hazardous classification »514.3(A)«.

A Class I location does not extend beyond any unpierced wall, roof, or other solid partition »514.3(B)(1)«.

All metal raceways, the metal armor or metallic sheath on cables, and all noncurrent-carrying metal parts of fixed/portable electrical equipment, regardless of voltage, must be grounded according to Article 250 »514.16«.

(A) A *motor fuel dispensing facility* is a location where gasoline or other volatile flammable liquids or liquefied flammable gases are transferred to the fuel tanks (including auxiliary fuel tanks) of self-propelled vehicles or approved containers »514.2«.

(B) Wiring and equipment above Class I locations classified in 514.3 must comply with 511.7 »514.7«.

(C) All electrical equipment/wiring within Class I locations classified in 514.3 must comply with the applicable provisions of Article 501 »514.4«.

NOTE

Table 514.3(B)(1) must be used to delineate and classify motor fuel dispensing facilities as well as any other location storing, handling, or dispensing Class I liquids »514.3(B)(1)«.

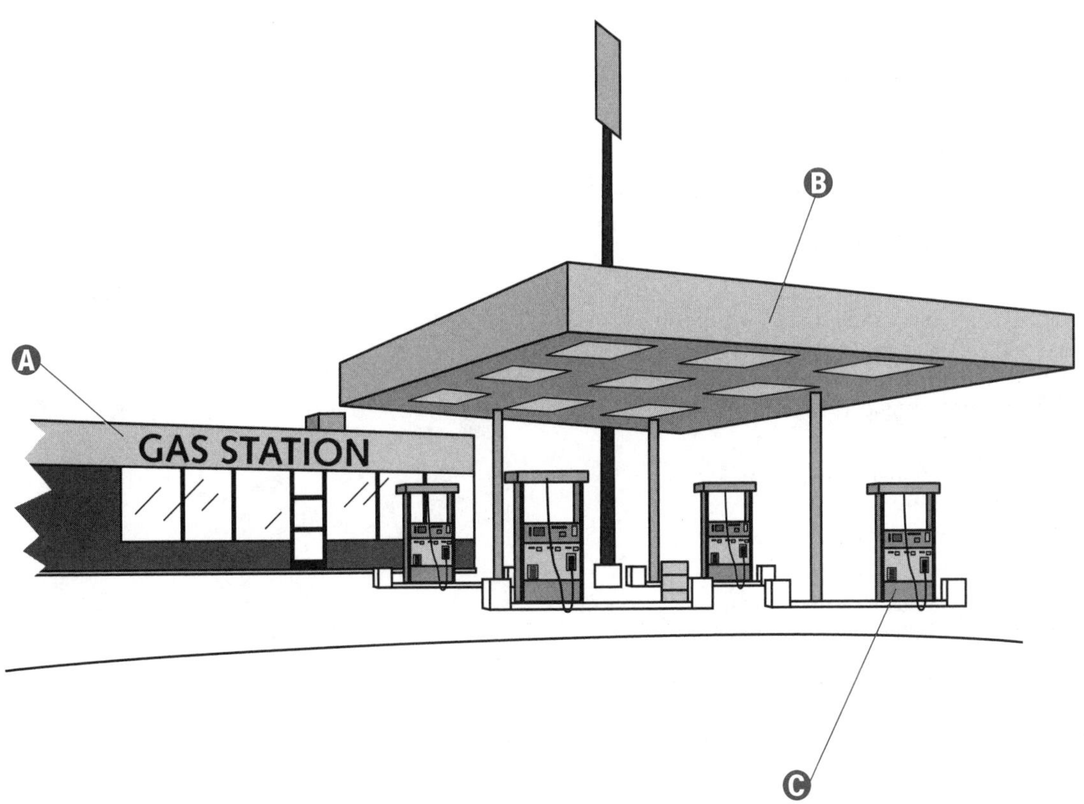

Conduit and Cable Seals—Class I, Divisions 1 and 2

Type MI cable termination fittings must contain sealing compound that excludes moisture/fluids from the cable insulation »501.5«.

In Class I, Division 1 locations, conduit seals must be placed in each conduit entry into an explosionproof enclosure where either: (1) the enclosure contains apparatus, such as switches, circuit breakers, fuses, relays, or resistors, that may produce arcs, sparks, or high temperatures that are an ignition source in normal operation; or (2) the entry is 2-in. size or larger and the enclosure contains terminals, splices, or taps »501.5(A)(1)«.

Class I, Division 2 connections to required explosionproof enclosures must have a conduit seal in accordance with 501.5(A)(1)(1) and (A)(3). The entire conduit run or nipple between the seal and such enclosure must comply with 501.4(A) »501.5(B)(1)«.

Splices/taps must not be made inside fittings intended only for sealing with compound, nor can other fittings that contain splices/taps be filled with compound »501.5(C)(4)«.

Ⓐ Seals in conduit and cable systems, within Class I, Divisions 1 and 2, must comply with 501.5(A) through (F) »501.5«.

Ⓑ In a seal, the conductors' cross-sectional area must not exceed 25% of the cross-sectional area of an RMC of the same trade size, unless it is specifically identified for a higher percentage of fill »501.5(C)(6)«.

Ⓒ If a probability exists that liquid or other condensed vapor may be trapped within control equipment enclosures, or at any point in the raceway system, install approved means to prevent accumulation or to permit periodic draining »501.5(F)(1)«.

Ⓓ Conduit/cable system seals minimize the passage of gasses/vapors and prevent the passage of flames from one portion of the electrical installation to another »501.5 FPN«.

Ⓔ In a completed seal, the sealing compound's minimum thickness must at least equal the trade size of the sealing fitting and must be in no case less than ⅝ in. (16 mm) »501.5(C)(3)«.

Ⓕ Sealing compound must: (1) provide a seal against passage of gas/vapors through the seal fitting; (2) not be affected by surrounding atmosphere or liquids; and (3) have a melting point of 200°F (93°C), or higher »501.5(C)(2)«.

Ⓖ Install a conduit seal in each conduit run leaving a Class I, Division 1 (or Division 2) location. The conduit seal is permitted on either side, and within 10 ft (3.05 m) of such location's boundary, and must be designated and installed to minimize the amount of gas/vapor within the Division 1 (or Division 2) portion of the conduit from being communicated to the conduit beyond the seal »501.5(A)(4) and 501.5(B)(2)«.

Ⓗ Except for listed explosionproof reducers at the conduit seal, there must be no union, coupling, box, or fitting between the conduit seal and the point at which the conduit leaves the Division 1 or 2 location »501.5(A)(4) and 501.5(B)(2)«.

CAUTION *Sealing fittings must be accessible »501.5(C)(1)«.*

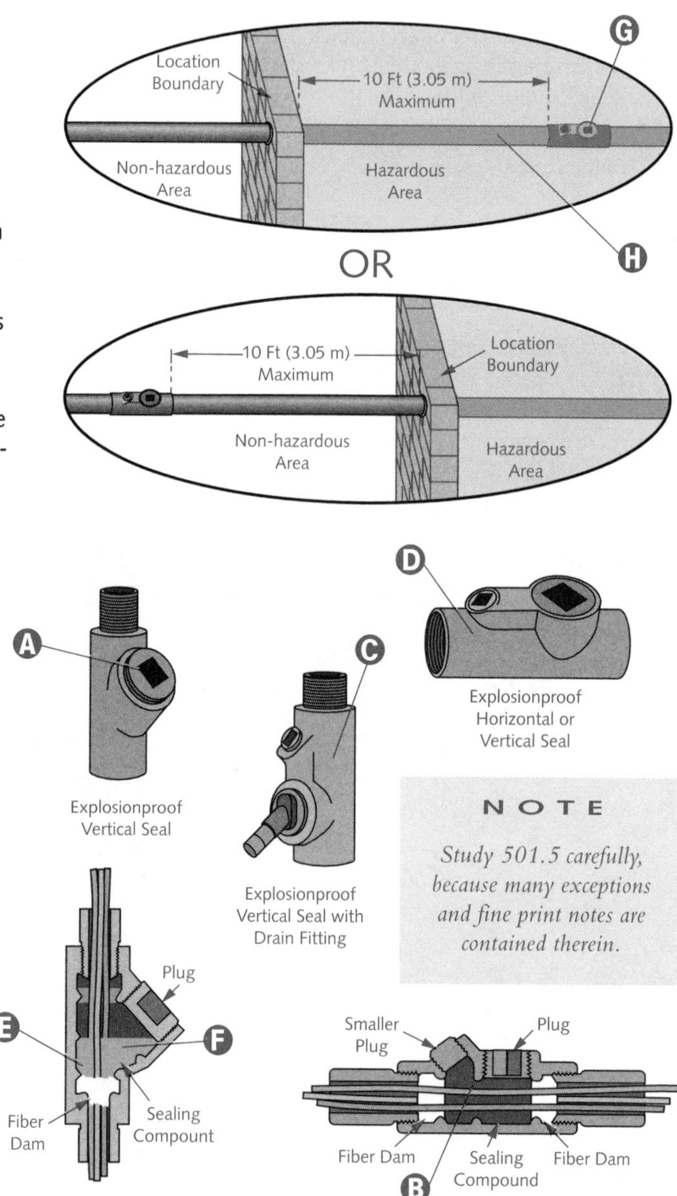

NOTE

Study 501.5 carefully, because many exceptions and fine print notes are contained therein.

Commercial Garages, Repair, and Storage

Any area where flammable fuel is transferred to vehicle fuel tanks must conform to Article 514 »511.3(B)«.

Article 514 requirements apply for fuel dispensing units (other than liquid petroleum gas, which is prohibited) located within buildings »511.4(B)(1)«.

Within Class I locations as defined in 511.3, wiring must conform to applicable Article 501 provisions. Raceways embedded in a masonry wall, or buried beneath a floor, are considered to be within the Class I location above the floor, if any connections or extensions lead into (or through) such areas »511.4(A) and (A)(1)«.

Seals conforming to 501.5 and 501.5(B)(2) provisions must be provided, and shall apply not only to horizontal but also to vertical boundaries of the defined Class I locations »511.9«.

Equipment less than 12 ft (3.7 m) above the floor level that may produce arcs, sparks, or hot metal particles (such as cutouts, switches, charging panels, generators, motors, or other equipment and excluding receptacles, lamps, and lampholders) having make-and-break or sliding contacts, must be of the totally enclosed type or constructed to prevent the escape of sparks or hot metal particles »511.7(B)(1)(a)«.

Ⓐ Areas adjoining defined locations, in which flammable vapor release is unlikely (such as stock rooms, switchboard rooms, and similar locations), are not classified if: (1) mechanically ventilated at a rate of four (or more) air changes per hour; or (2) effectively separated by walls or partitions »511.3(B)(4)«. Adjacent areas having ventilation, air pressure differentials, or physical spacing such that, in the opinion of the authority enforcing the *NEC*®, no ignition hazard exists, are unclassified »511.3(B)(5)«.

Ⓑ Article 511 occupancies include those used for service and repair of self-propelled vehicles (including, but not limited to, passenger automobiles, buses, trucks, and tractors) that use volatile flammable liquids for fuel or power »511.1«.

Ⓒ For pendants, flexible cord suitable for the type of service and listed for hard usage must be used »511.7(A)(2)«.

Ⓓ The entire area up to a level of 18 in. (450 mm) above each floor is a Class 1, Division 2 location »511.3(B)(1)«. The Division 2 classification may not be required if the enforcing agency determines that mechanical ventilation provides at least four air changes per hour »511.3(B)(1) *Exception*«.

Ⓔ A pit or depression, in which six air changes per hour are exhausted at the pit's floor level, can be judged by the enforcing agency to be a Class I, Division 2 location »511.3(B)(3) *Exception No. 1*«.

Ⓕ Fixed lighting lamps and lampholders that are located over lanes through which vehicles are commonly driven or that may otherwise be exposed to physical damage must be located at least 12 ft (3.7 m) above floor level, unless totally enclosed or constructed to prevent the escape of sparks or hot metal particles »511.7(B)(1)(b)«.

Ⓖ Any pit or depression below floor level, lacking adequate ventilation, is considered a Class I, Division 1 location that extends up to said floor level »511.3(B)(3)«.

B
C
A
OFFICE
D
18 In.
(450 mm)
Class I,
Division 2
Ventilated Pit
E

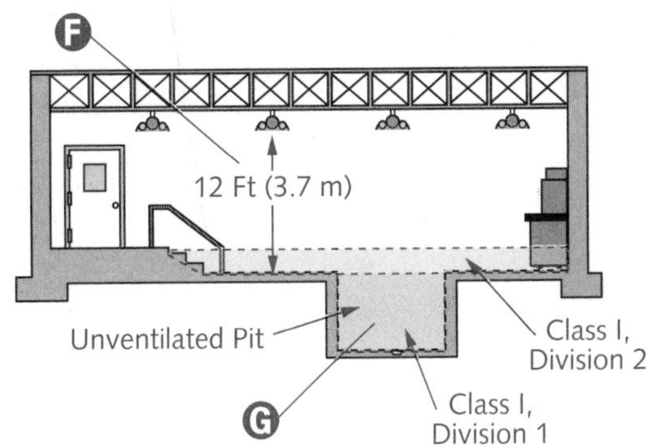

CAUTION *All 125-volt, single-phase, 15- and 20-ampere receptacles in areas where electrical diagnostic equipment, electrical hand tools, or portable lighting equipment are used must have GFCI protection for personnel »511.12«.*

Commercial Garages, Repair, and Storage *(continued)*

Any circuit, in a Class I location, supplying portables (or pendants) that include a grounded conductor (per Article 200), must have receptacles, attachment plugs, connectors, and similar devices of the grounding type, and the flexible cord's grounded conductor must either be connected to the screw shell of any lampholder or to the grounded terminal of any utilization equipment supplied. Grounding conductor continuity (between the fixed wiring system and the noncurrent-carrying metal portions of pendant luminaires (fixtures), portable lamps, and portable utilization equipment) must be maintained via approved means »511.16(B)(1) and (2)«.

Battery chargers, their control equipment, and batteries being charged, must not be located within areas classified in 511.3 »511.10(A)«.

All metal raceways, the metal armor or metallic sheath on cables, and all noncurrent-carrying metal parts of fixed/portable electrical equipment, regardless of voltage, must be grounded according to Article 250 »511.16(A)«.

A All fixed wiring above Class I locations must be placed within metal raceways, RNC, ENT, FMC, LFMC, or LFNC; or must be Type MC, MI, manufactured wiring systems; or PLTC cable in accordance with Article 725; or Type TC cable or Type ITC cable in accordance with Article 727»511.7(A)(1)«.

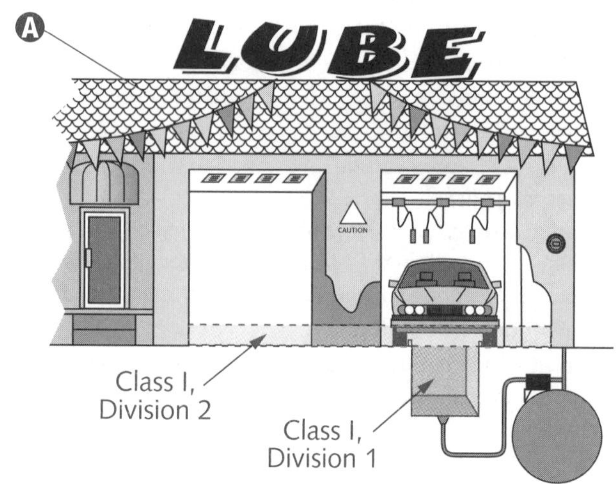

NOTE

Fixed position electrical equipment must be located above the level of any Class I defined location, or be identified for the location »511.7(B)(1)«.

Parking Garages

For further information, see *Standard for Parking Structures*, **NFPA 88A-1998, and** *Standard for Repair Garages*, **NFPA 88B-1997 »511.3(A) FPN «.**

A Garages used for parking or storage, where no repair work is done except exchange of parts and routine maintenance not requiring use of electrical equipment, open flame, welding, or volatile flammable liquids, are **not classified** »511.3(A)«.

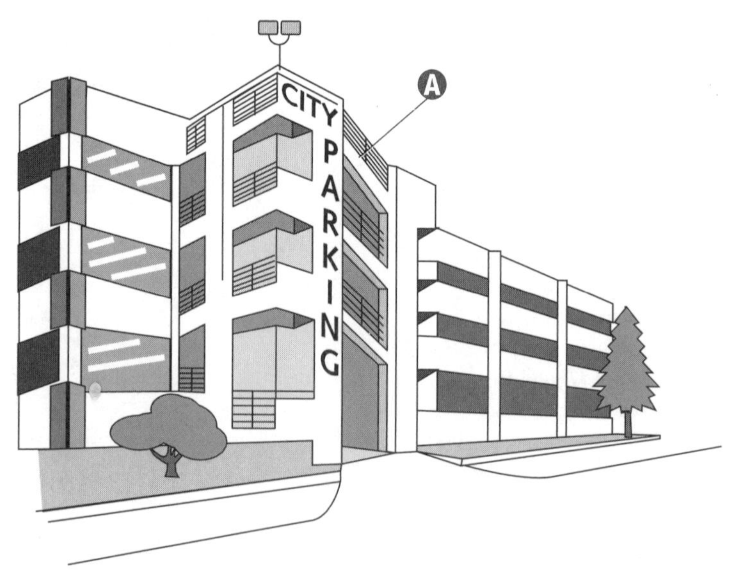

Aircraft Hangars

Mobile Equipment: **Equipment having electric components designed for movement via mechanical aids, or having wheels to facilitate movement by person(s) or powered devices »513.2«.**

Portable Equipment: **Equipment with electric components that can be moved by a single person without using mechanical aids »513.2«.**

All wiring and equipment intended for installation, installed or operated within any of the 513.3 defined Class I locations, must comply with the applicable provisions of Article 501 or Article 505 for the division or zone in which they are used »513.4(A)«.

Attachment plugs and receptacles in Class I locations must either be identified for Class I locations, or designed so that they cannot be energized during connection/disconnection »513.4(A)«.

All fixed wiring in a hangar, not within a Class I location as defined in 513.3, must be: (1) installed in metal raceways; or (2) Type MI, TC, or MC cable »513.7(A)«.

Portable utilization equipment and lamps require flexible cord suitable for the type of service and identified for extra-hard usage. Such cords must include a separate equipment grounding conductor »513.10(E)(1) and (2)«.

Adjacent areas in which flammable liquids or vapors are not likely to be released (stock rooms, electrical control rooms, etc.) are not classified, provided such areas are adequately ventilated and effectively separated from the hangar itself by walls or partitions »513.3(D)«.

Portable utilization equipment that can be, or is, used within a hangar must be of a type suitable for use in Class I, Division 2 or Zone 2 locations »513.10(E)(2)«.

Approved seals must be provided per 501.5 and 505.16. Sealing requirements specified apply both to horizontal and vertical Class I location boundaries. Raceways embedded within a concrete floor, or buried beneath the floor, are considered to be within the Class I location above the floor »513.9«.

All metal raceways, the metal armor or metallic sheath on cables, and all noncurrent-carrying metal parts of fixed/portable electrical equipment, regardless of voltage, must be grounded per Article 250. Grounding in Class I locations must comply with 501.16 or 505.25 »513.16(A)«.

(A) Article 513 applies to buildings/structures, or parts thereof inside which aircraft (containing Class I liquids or Class II liquids whose temperatures exceed their flash points) are housed, stored, serviced, repaired, or altered. It does not apply to locations used exclusively for aircraft that do not now, nor have ever, contained fuel »513.1«.

(B) The area within 5 ft (1.5 m) horizontally from aircraft power plants or fuel tanks is a Class I, Division 2 or Zone 2 location extending upward from the floor to a level 5 ft (1.5 m) above the upper surface of wings and engine enclosures »513.3(C)«.

(C) Any pit or depression below the hangar floor level is a Class I, Division 1 or Zone 1 location which extends up to floor level »513.3(A)«.

(D) The hangar's entire area, including any adjacent and communicating areas not suitably separated from the hangar, is designated as a Class I, Division 2 or Zone 2 location up to a level 18 in. (450 mm) above the floor »513.3(B)«.

> **NOTE**
>
> *In locations above those described in 513.3, equipment less than 10 ft (3.0 m) above aircraft wings and enclosures, which may produce arcs, sparks, or particles of hot metal (such as lamps and lampholders for fixed lighting, cutouts, switches, receptacles, charging panels, generators, motors, or other equipment having make-and-break or sliding contacts), must be of the totally enclosed type or by construction must prevent the escape of sparks or hot metal particles »513.7(C)«.*

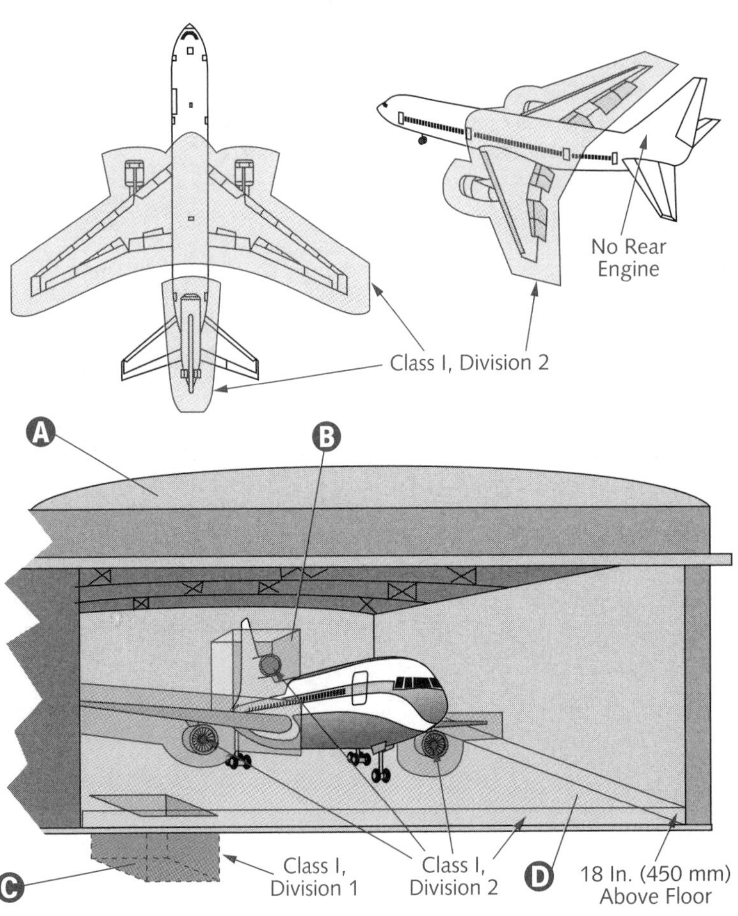

Dispensing Devices

Unattended **self-service motor fuel dispensing facility emergency controls as specified in 514.11(A) must be installed at a location acceptable to the AHJ, such location being more than 20 ft (6 m), but less than 100 ft (30 m) from dispensers. Additional emergency controls must either be installed on each group of dispensers, or on the outdoor equipment controlling the dispensers. Emergency controls must completely shut off power to all of the station's dispensing equipment. Controls must be manually reset only in a manner approved by the AHJ »514.11(C)«.**

Attended **self-service motor fuel dispensing facility emergency controls, as specified in 514.11(A), must be installed at a location acceptable to the AHJ and not more than 100 ft (30 m) from dispensers »514.11(B)«.**

Additional seals apply to horizontal as well as vertical boundaries of the Class I locations »514.9(B)«.

A Within 18 in. (450 mm) horizontally, extending in all directions from grade to the dispenser enclosure, or that portion of the dispenser enclosure containing liquid handling components, is a Class I, Division 2 location »Table 514.3(B)(1)«.

B The space inside the dispenser enclosure is classified as covered in *Power Operated Dispensing Devices for Petroleum Products*, ANSI/UL 87-1995 »Table 514.3(B)(1)«.

C Each circuit leading to (or through) dispensing equipment, including remote pumping system equipment, must have a clearly identified, readily accessible switch (or other acceptable means), located away from the dispensing devices, to simultaneously disconnect all circuit conductors (including the grounded conductor, if any) from the supply source »514.11(A)«.

D From grade level up to a height of 18 in. (450 mm) within 20 ft (6.0 m) horizontally of any enclosure edge, is a Class I, Division 2 location »Table 514.3(B)(1)«.

E All or part of any pit, box, or space below grade level within a Division 1 or 2 (or Zone 1 or 2) classified location, is designated Class I, Division 1 »Table 514.3(B)(1)«.

F A listed seal must be provided in each conduit run entering/leaving a dispenser or any cavities (or enclosures) in direct communication therewith. The sealing fitting must be the first fitting after the conduit emerges from the earth or concrete »514.9(A)«.

G Underground wiring must be installed either in threaded RMC or threaded steel IMC. Any portion of electrical wiring or equipment, below the surface of a Class I, Division 1 or 2 location qualifies as a Class I, Division 1 location, and extends at least to the point of emergence above grade »514.8«.

NOTE

Each dispensing device must have a means to remove all external voltage sources, including feedback, during periods of maintenance/service »514.13«.

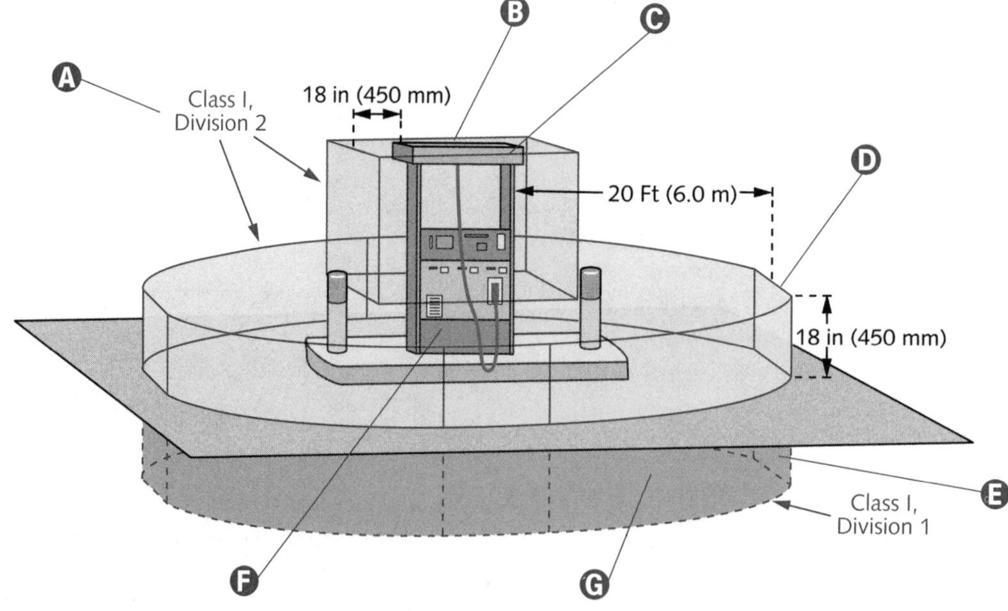

Bulk Storage Plants

Sealing requirements apply to horizontal as well as vertical boundaries of the defined Class I locations. Raceways buried under a defined Class I area are considered to be within a Class I, Division 1 or Zone 1 location »515.9«.

Underground wiring must be enclosed in either threaded RMC or threaded steel IMC; if buried under at least 2 ft (600 mm) of cover, RNC (or a listed cable) can be used. If RNC is used, threaded metal conduit (either rigid or steel intermediate) must be used for the last 2 ft (600 mm) of the conduit run to emergence or to the point of connection to the above ground raceway. Cable, if used, must also be enclosed in threaded metal conduit (either rigid or steel intermediate) from the point of lowest buried cable to the point of aboveground raceway connection »515.8(A)«.

If RNC or cable with a nonmetallic sheath is used, include an equipment grounding conductor to provide electrical continuity of the raceway system and grounding of noncurrent-carrying metal parts »515.8(C)«.

Where gasoline, or other volatile flammable liquids or liquefied flammable gases, are dispensed at bulk stations, the appropriate provisions of Article 514 apply »515.10«.

All metal raceways, the metal armor or metallic sheath on cables, and all noncurrent-carrying metal parts of fixed/portable electrical equipment, regardless of voltage, must be grounded according to Article 250 »515.16«.

(A) A **bulk storage plant** is that portion of a property where flammable liquids are received (by tank vessel, pipelines, tank car, or tank vehicle), stored or blended in bulk for the purpose of distribution by vessel, pipeline, tank car, tank vehicle, portable tank, or container »515.2«.

(B) Apply Table 515.3 where Class I liquids are stored, handled, or dispensed. Table 515.3 is also used to delineate and classify bulk storage plants. The Class I location does not extend beyond any floor, wall, roof, or other solid partition that has no communicating openings »515.3«.

(C) All electric wiring and equipment within the Class I locations, defined in 515.2, must comply with the applicable provisions of Articles 501 or 505 »515.4«.

(D) All fixed wiring above Class I locations must be in metal raceways or PVC Schedule 80 RNC (or equivalent) or be Type MI, TC, or MC cable. Fixed equipment that may produce arcs, sparks, or particles of hot metal (such as lamps and lampholders for fixed lighting, cutouts, switches, receptacles, motors, or other equipment having make-and-break or sliding contacts), must either be of the totally enclosed type or otherwise be constructed to prevent the escape of sparks or hot metal particles. Portable lamps or other utilization equipment, and associated flexible cords, must comply with Article 501's or Article 505's requirements for the class of location above which they are connected or used »515.7(A), (B), and (C)«.

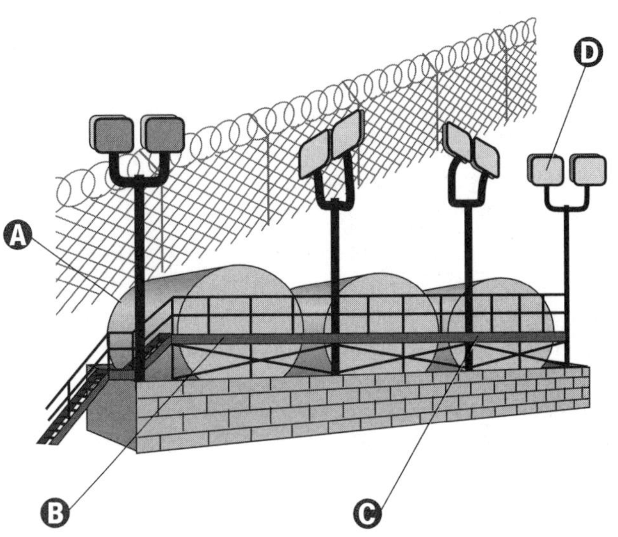

Spray Application, Dipping, and Coating Process

Classification is based on dangerous quantities of flammable vapors, combustible mists, residues, dusts, or deposits »516.3«.

The following spaces are considered Class I or Class II, Division 1 locations, as applicable:

(1) Spray booth/room interiors except as specifically provided in 516.3(D).

(2) Exhaust duct interiors.

(3) Any area in the direct path of spray operations.

(4) For dipping and coating operations, all space within a 5-ft (1.5 m) radius of the vapor source extending from the surface to the floor. The vapor source is liquid exposed during the process, which includes draining, and any dipped or coated object from which measurable vapor concentrations exceed 25% of the lower flammable limit within 1 ft (300 mm), in any direction, from the object.

(5) Pits within 25 ft (7.5 m) horizontally of the vapor source. Pits extending beyond 25 ft (7.5 m) must have a vapor stop, without which the entire pit is designated Class I, Division 1.

(6) The interior of any enclosed coating/dipping process »516.3(A)«.

Spray Application, Dipping, and Coating Process *(continued)*

The following spaces are considered Class I or Class II, Division 2 locations, as applicable:

(1) All spaces outside of open spacing, but within 20 ft (6 m) horizontally and 10 ft (3 m) vertically of the Class I, Division 1 location as defined in 516.3(A), and not separated by partitions.

(2) If spray application operations are conducted within a closed-top, open-face, or open-front booth/room, any electrical wiring or utilization equipment located outside of the booth/room, but within the Division 2 boundaries must be suitable for Class I, Division 2 or Class II, Division 2 locations, whichever is applicable.

(3) In open top spray booth operations, the space 3 ft (900 mm) vertically above the booth, and within 3 ft (900 mm) of all other booth openings, qualifies as Class I or Class II, Division 2.

(4) For enclosed spray booth/room operations, the space within 3 ft (900 mm) in all directions from any opening is considered Class I or Class II, Division 2 »516.3(B)«.

The space adjacent to an enclosed dipping (or coating) process/apparatus is unclassified »516.3(C)«.

Adjacent locations, cut off from the defined Class I or Class II locations by tight partitions without communicating openings, where release of flammable vapors or combustible powders is unlikely, are unclassified »516.3(D)«.

Locations using drying, curing, or fusion apparatus having (1) positive mechanical ventilation adequate to prevent accumulation of flammable vapor concentrations, and (2) effective interlocks to de-energize all electrical equipment (other than equipment approved for Class I locations) in case of inoperative ventilating equipment, are unclassified where the AHJ deems appropriate »516.3(E)«.

NOTE

Wiring and equipment above Class I and Class II locations must comply with 516.7 specifications.

While fixed electrostatic spraying and detearing equipment requirements are located in 516.10(A), electrostatic hand-spraying equipment provisions are found in 516.10(B).

All metal raceways, the metal armor or metallic sheath on cables, and all noncurrent-carrying metal parts of fixed/portable electrical equipment, regardless of voltage, must be grounded according to Article 250 »516.16«.

A Article 516 covers the regular (or frequent) spray application of flammable liquids, combustible liquids, and combustible powders, as well as the application of flammable/combustible liquids at temperatures above their flashpoint, by dipping, coating, or other means »516.1«.

B All space within a 5-ft (1.5 m) radius of the vapor source, extending from the surface to the floor, is designated Class I, Division 1 »516.3(A)(4)«.

C A space 3 ft (900 mm) above the floor, extending 20 ft (6 m) horizontally in all directions from the Class I, Division 1 location, is designated Class I, Division 2 »516.3(B)(5) and (6)«.

D A 3-ft (900-mm) space surrounding the dip-tank and drain-board's Class I, Division 1 location, is a Class I, Division 2 location »516.2(B)«.

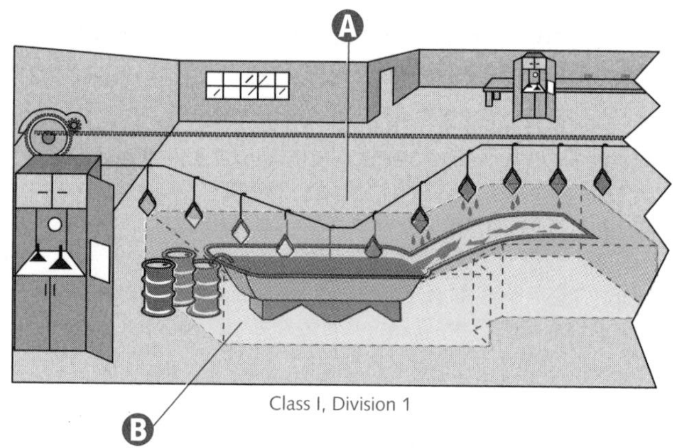

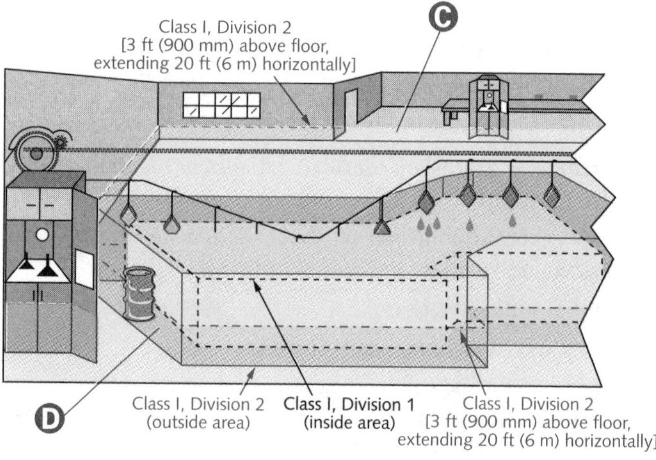

CLASS II LOCATIONS

Class II, Division 1 Locations

In addition to the requirements found in Article 500, wiring and equipment in Class II locations must meet Article 502's provisions ›› 502.1 ‹‹.

Dust ignitionproof means enclosed in a manner that excludes dust; and, where installed and protected in accordance with this *Code*, which does not permit arcs, sparks, or heat otherwise generated or liberated inside the enclosure to ignite exterior accumulations or atmospheric suspensions of a specified dust on or near the enclosure ›› 500.2 ‹‹.

Equipment in Class II locations must be able to function at full rating without developing surface temperatures high enough to excessively dehydrate or gradually carbonize any organic dust deposits that may occur ›› 502.1 ‹‹. Carbonized/excessively dry dust is highly susceptible to spontaneous ignition ›› 502.1 FPN ‹‹.

The type of equipment and wiring defined as explosionproof (in Article 100) is not required and therefore not acceptable for Class II locations unless identified for such use ›› 502.1 ‹‹.

All locations containing Class II, Group E dusts in hazardous quantities are Division 1 ›› 502.1 ‹‹.

In Class II, Division 1 locations, transformers and capacitors, containing burnable liquid, must be installed only in approved vaults compliant with 450.41 through 450.48, in addition to all the following stipulations: (A) Openings (doors, etc.) communicating with the Division 1 locations must have self-closing fire doors on both sides of the wall, carefully fitted and having suitable seals (such as weather stripping) to minimize the entrance of dust into the vault; (B) Vent openings and ducts must connect directly with outside air; and (C) Suitable pressure-relief openings to the outside air must be provided ›› 502.2(A)(1) ‹‹. Subsequently, transformers and capacitors not containing burnable liquids must either be installed in vaults complying with 450.41 through 450.48, or be approved as a complete assembly, including terminal connections for Class II locations ›› 502.2(A)(2) ‹‹.

No transformer or capacitor can be installed in a location where dust from magnesium, aluminum bronze powders, or other similarly hazardous metals may be present ›› 502.2(A)(3) ‹‹.

A A Class II, Division 1 location is an area: (1) in which combustible dust is airborne under normal operating conditions in quantities sufficient to produce explosive or ignitible mixtures; (2) where mechanical failure or abnormal operation of machinery/equipment might not only produce such explosive or ignitible mixtures, but also provide an ignition source through simultaneous failure of electric equipment, operation of protection devices, or by other causes; or (3) in which combustible dusts of an electrically conductive nature may exist in hazardous quantities ›› 500.5(C)(1) ‹‹.

B Combustible dusts, that are electrically nonconductive, include dust produced in the handling and processing of grain and grain products, pulverized sugar and cocoa, dried egg and milk powders, pulverized spices, starches and pastes, potato and wood-flour, oil meal from beans and seeds, dried hay, and other organic materials. Only Group E dusts are considered electrically conductive for classification purposes. Because dusts containing magnesium or aluminum are particularly hazardous, extreme precautions are necessary to avoid ignition and explosion ›› 500.5(C)(1) FPN ‹‹.

C Class II locations are hazardous because of the presence of combustible dust ›› 500.5(C) ‹‹.

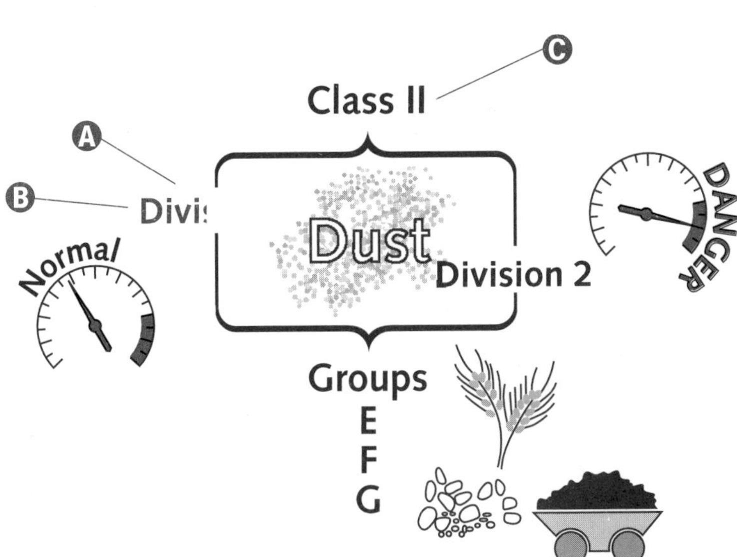

NOTE

Bonding and grounding must comply with 502.16(A) and (B) provisions.

Class II, Division 2 Locations

In Class II, Division 2 locations, transformers and capacitors containing burnable liquid must be installed within vaults compliant with 450.41 through 450.48 »502.2(B)(1)«.

Askarel-containing transformers, rated in excess of 25 kVA, must have: (A) pressure-relief vents; (B) a means for absorbing any gases generated by arcing inside the case, or chimney/flue connected pressure-relief vents must carry such gases outside the building; and (C) at least 6 in. (150 mm) of airspace between the transformer case and any adjacent combustible material »502.2(B)(2)«.

Dry-type transformers, operating at not over 600 volts nominal, must either be installed in vaults or must have their windings and terminal connections enclosed in tight metal housings without ventilation or other openings »502.2(B)(3)«.

Ⓐ A Class II, Division 2 location is an area: (1) where airborne combustible dust is not normally present in quantities sufficient to produce explosive or ignitible mixtures, and any dust accumulations typically do not interfere with normal operation of electrical equipment or other apparatus, but combustible dust may become airborne as a result of infrequent malfunctioning of processing/handling equipment; and (2) where combustible dust accumulations on, in, or near electrical equipment may be sufficient to interfere with safe heat dissipation from said equipment, or may be ignited by abnormal operation or failure of electrical equipment »500.5(C)(2)«.

Ⓑ The quantity of combustible dust that may be present, and the adequacy of dust removal systems, are factors that merit consideration in determining classification. Such factors may even result in a nonhazardous area »500.5(C)(2) FPN No. 1«.

Ⓒ Where the handling of substances such as seed produces low quantities of dust, the amount of dust deposited may not warrant classification »500.5(C)(2) FPN No. 2«.

NOTE

Bonding and grounding must comply with 502.16(A) and (B) provisions.

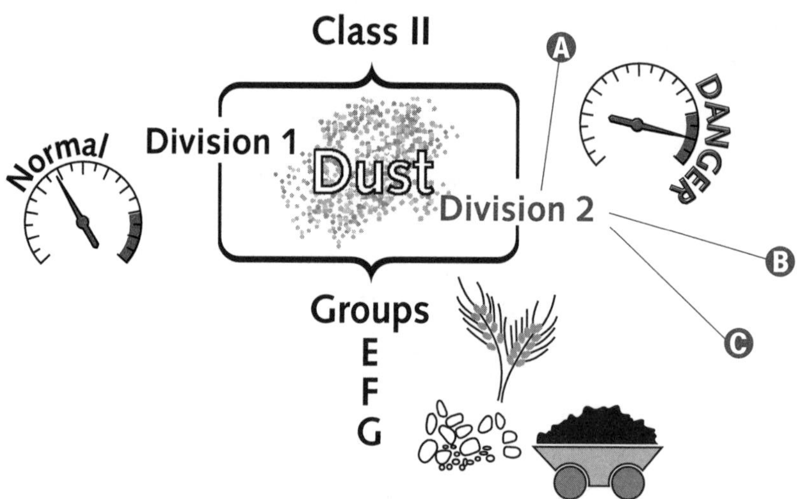

Wiring Methods—Class II, Divisions 1 and 2

Approved wiring methods in Class II, Division 1 locations include: threaded RMC, threaded steel IMC, or type MI cable with termination fittings listed for the location »502.4(A)«. In industrial establishments with limited public access, where maintenance and supervision conditions ensure that only qualified persons will service the installation, Class II, Division 1 listed Type MC cable with a gas/vaportight continuous corrugated aluminum sheath, an overall jacket of suitable polymeric material, separate grounding conductors (per 250.122), and termination fittings listed for the application are permitted »502.4(A)(1)(c)«.

Class II, Division 1 fittings and boxes must be provided with threaded bosses for conduit connections or cable terminations and must be dusttight. Fittings and boxes containing taps, joints, or terminal connections, or that are used in Group E locations must be identified for Class II locations »502.4(A)(1)(d)«.

Use only dusttight flexible connectors, LFMC with listed fittings, LFNC with listed fittings, or flexible cord listed for extra-hard usage and provided with bushed fittings in Class II, Divisions 1 and 2 locations necessitating flexible connections. Any flexible cords used must comply with 502.12. If flexible connections are subject to oil or other corrosive conditions, the conductor's insulation must either be of a type approved for the condition, or must be protected by means of a suitable sheath »502.4(A)(1)(e)«.

Sealing fittings must be accessible »502.5«.

Unless an exception is met, approved wiring methods in Class II, Division 2 locations include: RMC, IMC, EMT, dusttight wireways, Type MC or MI cable with listed termination fittings, Type PLTC in cable trays, Type ITC in cable trays, or Type MC, MI, or TC cable installed in cable trays (ladder, ventilated trough, or ventilated channel) in a single layer, with a space at least equal to the larger cable diameter between the two adjacent cables »502.4(B)(1)«.

All Class II, Division 2 boxes and fittings must be dusttight. »502.4(B)(4)«.

In Class II, Division 1 and 2 locations, where a raceway provides communication between an enclosure that is required to be dust ignitionproof and one that is not, suitable means must prevent the entrance of dust into the dust ignitionproof enclosure via the raceway. The following means are permitted: (1) a permanent and effective seal; (2) a horizontal raceway at least 10 ft (3.05 m) long; or (3) a vertical raceway not less than 5 ft (1.5 m) long that extends downward from the dust ignitionproof enclosure »502.5«.

Seals are not required where a raceway provides communication between an enclosure that is required to be dust ignitionproof and an enclosure in an unclassified location »502.5«.

In Class II, Division 1 and 2 locations, explosionproof seals are not required »502.5«.

Ⓐ Class II locations are hazardous because of the presence of combustible dust »500.5(C)«.

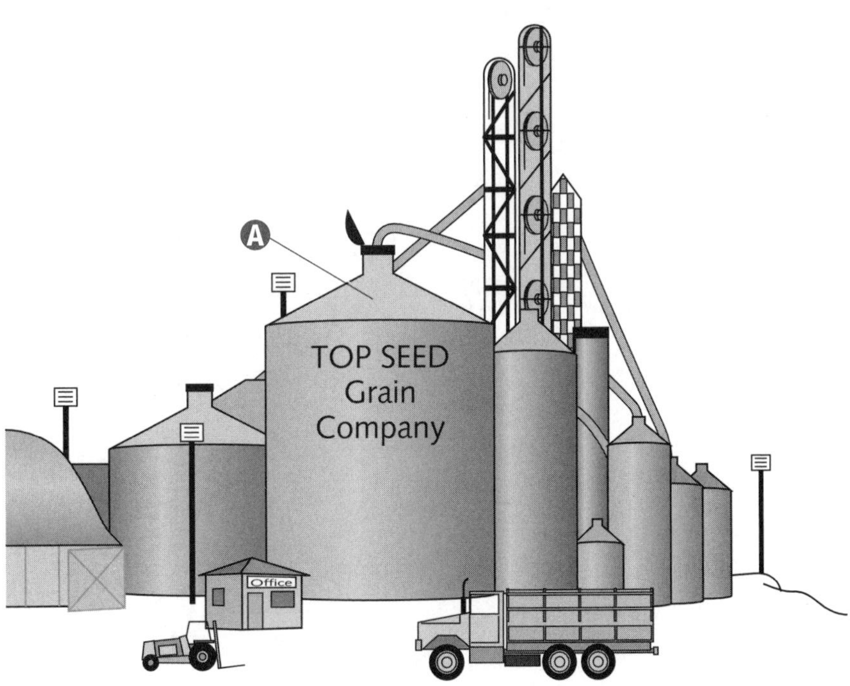

CLASS III LOCATIONS

Class III, Division 1 Locations

In addition to Article 500 requirements, wiring and equipment in locations designated Class III must meet Article 503's provisions »503.1«.

Class III located equipment must be able to function at full rating without developing surface temperatures high enough to cause excessive dehydration or gradual carbonization of accumulated fibers or flyings. Carbonized or excessively dried organic material is highly susceptible to spontaneous ignition. During operation, the maximum surface temperature must not exceed 329°F (165°C) for equipment not subject to overloading, and 248°F (120°C) for equipment that may be overloaded (such as motors or power transformers) »503.1«.

Transformers and capacitors in Class III, Division 1 locations, must comply with 502.2(B) »503.2«.

Ⓐ A Class III, Division 1 location is an area where easily ignitible fibers or materials producing combustible flyings are handled, manufactured, or used »500.5(D)(1)«.

Ⓑ Easily ignitible fibers and flyings include rayon, cotton (including cotton linters and cotton waste), sisal or henequen, istle, jute, hemp, tow, cocoa fiber, oakum, baled waste kapok, Spanish moss, excelsior, and other materials of similar nature »500.5(D)(1) FPN No. 2«.

Ⓒ Class III locations are hazardous because of the presence of easily ignitible fibers or flyings, which are not likely to be airborne in quantities sufficient to produce ignitible mixtures »500.5(D)«.

NOTE

Bonding and grounding must comply with 503.16(A) and (B) provisions.

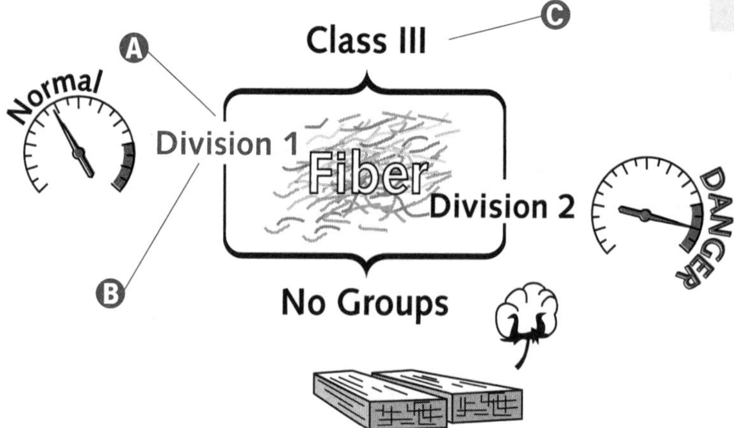

Class III, Division 2 Locations

In Class III, Division 2 locations, transformers and capacitors must comply with 502.2(B) »503.2«.

Ⓐ A Class III, Division 2 location is an area in which easily ignitible fibers are stored or handled other than in the manufacturing process »500.5(D)(2)«.

NOTE

Bonding and grounding must comply with 503.16(A) and (B) provisions.

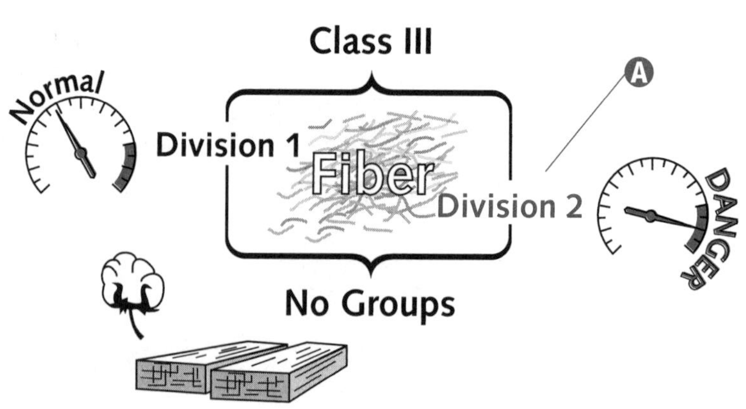

Wiring Methods—Class III, Division 1

Approved wiring methods for Class III, Division 1 locations include: RMC, RNC, IMC, electrical metallic tubing, dusttight wireways, and Type MC or MI cable with listed fittings »503.3(A)«.

All Class III, Division 1 boxes and fittings must be dusttight »503.3(A)(1)«.

Use only dusttight flexible connectors, LFMC with listed fittings, LFNC with listed fittings, or flexible cord per 503.10 where flexible connections are necessary »503.3(A)(2)«. For grounding requirements where flexible conduit is used, see 503.16(B).

Class III, Division 1 and 2 switches, circuit breakers, motor controllers, and fuses, (including pushbuttons, relays, and similar devices), must have dusttight enclosures »503.4«.

Class III, Division 1 and 2 flexible cords must comply with the following: (1) be of a type listed for extra-hard usage; (2) contain, in addition to the conductors of the circuit, a 400.23 compliant grounding conductor; (3) be connected to terminals or supply conductors in an approved manner; (4) be supported by clamps or other suitable means so that there will be no tension on the terminal connections; and (5) be provided with suitable means to prevent the entrance of fibers or flyings where the cord enters boxes or fittings »503.10«.

Class III, Division 1 and 2 receptacles and attachment plugs must: (1) be of the grounding type; (2) be designed to minimize the accumulation (or entry) of fibers or flyings; and (3) prevent the escape of sparks or molten particles »503.11«. If, in the judgement of the AHJ, only moderate accumulations of lint or flyings are likely to collect in the vicinity of a receptacle, and where such receptacle is readily accessible for routine cleaning, general-purpose grounding-type receptacles (mounted to minimize the entry of fibers or flyings) are permitted »503.11 *Exception*«.

Live parts must not be exposed except as provided in 503.13 »503.15«.

A Such locations usually include some parts of textile mills (rayon, cotton, etc.); cotton gins and cotton seed mills; flax-processing plants; clothing manufacturing facilities; woodworking plants; and establishments/industries involving similar hazardous processes or conditions »500.5(D)(1) FPN No. 1«.

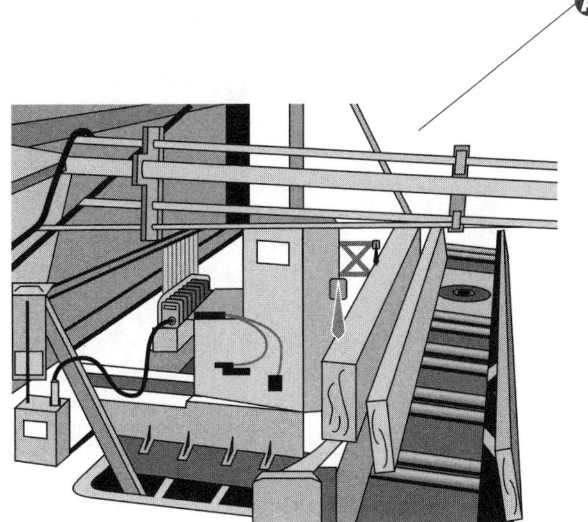

Woodworking Plant

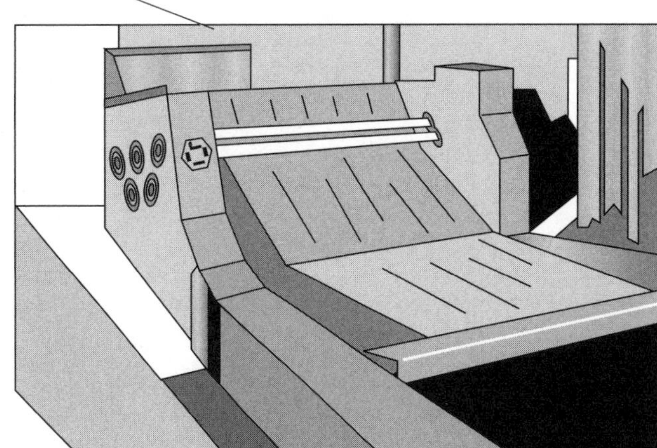

Textile Production Equipment

Wiring Methods—Class III, Division 2

Unless an exception is met, Class III, Division 2 location wiring methods must comply with 503.3(A) »503.3(B)«.

All Class III, Division 2 boxes and fittings must be dusttight »503.3(A)(1)«.

Class III, Division 1 and 2 pendant luminaires (fixtures) must be suspended by stems of threaded RMC, threaded IMC, threaded metal tubing of equivalent thickness, or by chains with approved fittings. Stems longer than 12 in. (300 mm), must be permanently and effectively braced against lateral displacement at a level no more than 12 in. (300 mm) above the lower end of the stem, or flexibility in the form of an approved fitting or a flexible connector must be provided no more than 12 in. (300 mm) from the point of attachment to the supporting box or fitting »503.9(C)«.

Class III, Division 1 and 2 portable lighting equipment must have both handles and substantial guards. Lampholders must be of the unswitched type, incapable of receiving attachment plugs. There must be no exposed current-carrying metal parts, and all exposed noncurrent-carrying metal parts must be grounded. In all other respects, portable lighting equipment must comply with 503.9(A) »503.9(D)«.

A Class III, Division 1 and 2 luminaires (lighting fixtures) for fixed lighting must provide enclosures for lamps and lampholders designed to minimize entrance of fibers and flyings and to prevent the escape of sparks, burning material, or hot metal. Each luminaire (fixture) must be clearly marked to show the maximum permitted lamp wattage without exceeding an exposed surface temperature of 329°F (165°C) under normal use conditions »503.9(A)«.

B A Class III, Division 2 location is an area in which easily ignitible fibers are stored or handled other than in the process of manufacture »500.5(D)(2)«.

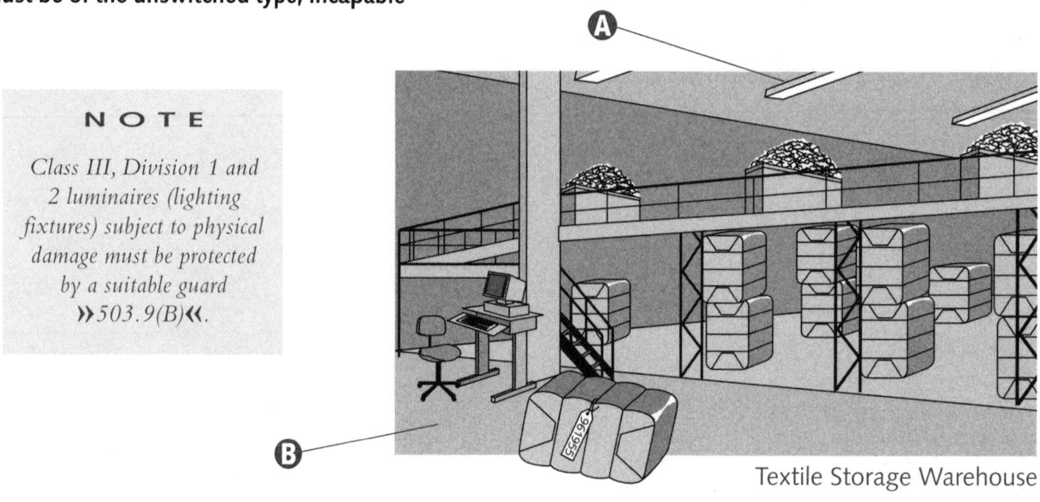

Textile Storage Warehouse

NOTE

Class III, Division 1 and 2 luminaires (lighting fixtures) subject to physical damage must be protected by a suitable guard »503.9(B)«.

Summary

- Hazardous locations are divided into three classifications (Class I, II, and III); each class is subdivided into two categories (Division 1 and 2).
- While Article 500 covers general requirements for all classes, Articles 501, 502, and 503 provide specific provisions for each class individually.
- Locations containing potentially sufficient quantities of air-borne gases (or vapors) to cause explosion or ignition are categorized as Class I locations.
- Class II locations are hazardous due to the presence of combustible dust.
- Class III locations are hazardous because of the presence of easily ignitible fibers or flyings, which are not likely to be airborne in quantities sufficient to produce ignitible mixtures.
- Under typical operating conditions, Division 1 locations contain ignitible or combustible elements (gases, dust, fibers, etc.) in quantities sufficient to produce explosive (or ignitible) mixtures.

- Locations where ordinarily present quantities of combustible or ignitible elements (gases, dust, fibers, etc.) in the air are insufficient to produce explosive (or ignitible) mixtures, are classified as Division 2.
- Article 511 occupancies include those used for service and repair of self-propelled vehicles (including, but not limited to, passenger automobiles, buses, trucks, and tractors) that use volatile flammable liquids for fuel or power.
- Article 513 applies to buildings/structures, or parts thereof, inside which aircraft (containing Class I or II liquids whose temperatures exceed their flashpoints) are housed, stored, serviced, repaired, or altered.
- Motor fuel dispensing facility requirements are in Article 514.
- Article 515 provisions address bulk storage plants.
- The regular (or frequent) spray application of flammable liquids, combustible liquids, and combustible powders, as well as the application of flammable/combustible liquids at temperatures above their flashpoint, by dipping, coating, or other means, are covered by Article 516.

Unit 15 Competency Test

NEC® Reference	Answer	
________	________	1. In aircraft hangars, any pit or depression below the level of the hangar floor shall be classified as a _____ location that shall extend up to said floor level.
________	________	2. Equipment, regardless of the classification of the location in which it is installed, that depends on a single compression seal, diaphragm, or tube to prevent flammable or combustible fluids from entering the equipment, shall be identified for a _____ location.
________	________	3. A _____ location is a location in which volatile flammable liquids or flammable gases are handled, processed, or used, but in which the liquids, vapors, or gases will normally be confined within closed containers or closed systems from which they can escape only in case of accidental rupture or breakdown of such containers or systems, or in case of abnormal operation of equipment.
________	________	4. All hazardous (classified) location threaded conduit shall be threaded with an NPT standard conduit cutting die that provides _____ in. taper per ft.
________	________	5. In bulk storage plants, aboveground tank vents shall be classified as _____ within 5 ft (1.5 m) of the open end, extending in all directions.
________	________	6. In Class I, Division 1 locations, threaded joints shall be made up with at least _____ threads fully engaged.
________	________	7. In Class II, Division 1 locations, fittings and boxes shall be provided with _____ for connection to conduit or cable terminations, and shall be dusttight.
________	________	8. A 1-in. threaded RMC is terminating into a 2-in. sealing fitting. The minimum thickness of the sealing compound shall be _____in.
________	________	9. In aircraft hangars, the area within _____ ft horizontally from aircraft power plants or aircraft fuel tanks shall be classified as a Class I, Division 2 or Zone 2 location that shall extend upward from the floor to a level _____ ft above the upper surface of wings and of the engine enclosures.
________	________	10. _____ is defined as equipment enclosed in a manner that will exclude dusts and, does not permit arcs, sparks, or heat otherwise generated or liberated inside of the enclosure to cause ignition of exterior accumulations or atmospheric suspensions of a specified dust on or in the vicinity of the enclosure.
________	________	11. Class I locations are those in which _____ are or may be present in the air in quantities sufficient to produce explosive or ignitible mixtures.

NEC® Reference	Answer	
________	________	12. In commercial garages, lamps and lampholders for fixed lighting that is located over lanes through which vehicles are commonly driven shall be located no less than _____ ft above the floor level.
________	________	13. A _____ location is a location in which easily ignitible fibers or materials producing combustible flyings are handled, manufactured, or used.
________	________	14. __________, __________, __________, __________, and __________ are five acceptable hazardous (classified) location protection techniques for electrical and electronic equipment.
________	________	15. A(n) _____ is defined as a circuit, other than field wiring, in which any arc or thermal effect produced under intended operating conditions of the equipment, is not capable, under specified test conditions, of igniting the flammable gas-, vapor-, or dust-air mixture.
________	________	16. In each conduit run passing from a Class I, Division 2 location into an unclassified area, the sealing fitting must be located within _____ ft of the boundary.
________	________	17. For dipping and coating operations, all space within a _____ ft radial distance from the vapor sources extending from these surfaces to the floor, shall be considered a Class I, Division 1 location.
________	________	18. In Class III, Division 2 locations, transformers and capacitors shall comply with *NEC®* _____.
________	________	19. An outdoor gasoline dispensing device is considered to be a Class I, Division 2 location up to _____ in. above grade level within _____ ft. horizontally of any edge of the enclosure.
________	________	20. In bulk storage plants, the space between _____ ft and _____ ft from the open end of an aboveground tank vent, extending in all directions, is a Class I, Division 2 or Zone 2 location.
________	________	21. A(n) _____ enclosure is constructed so that dust will not enter the enclosing case under specified test conditions.
________	________	22. For dip tanks and drain boards, the space _____ ft above the floor and extending _____ft horizontally in all directions from the Class I, Division 1 location, shall be considered a Class I, Division 2 location.
________	________	23. Aircraft energizers shall be designed and mounted so that all electric equipment and fixed wiring will be at least _____ in. above floor level.
________	________	24. A _____ location is a location in which combustible dust is in the air under normal operating conditions in quantities sufficient to produce explosive or ignitible mixtures.
________	________	25. For each commercial garage floor, the entire area up to a level of _____ in. above the floor shall be considered to be a Class I, Division 2 location.
________	________	26. A ½-in. threaded steel IMC is terminating into a ½-in. sealing fitting. The minimum thickness of the sealing compound shall be _____in.
________	________	27. The entire area of an aircraft hangar, including any adjacent and communicating areas not suitably cut off from the hangar, shall be classified as a Class I, Division 2 or Zone 2 location up to a level _____ in. above the floor.
________	________	28. For unattended self-service stations, emergency controls shall be more than _____ ft but less than _____ ft from the dispensers.

UNIT

16

SECTION FIVE: SPECIAL OCCUPANCIES, AREAS, AND EQUIPMENT

Health Care

Objectives

After studying this unit, the student should:

- have an adequate understanding of health care facility terminology.
- be familiar with general, as well as specific, hospital, nursing home, and limited care facility requirements.
- know the criteria for judging an ambulatory health care facility.
- know what area constitutes the patient vicinity.
- be familiar with grounding and bonding requirements.
- have a good understanding of patient care areas (general care, critical care, and wet locations).
- be familiar with general and critical care area branch-circuit and receptacle requirements.
- understand the receptacle/receptacle cover tamper resistant requirements for pediatric areas.
- understand the two types of systems (equipment and emergency), as well as the two types of branches (life safety and critical) required by hospitals.
- know the required nursing home and limited care facility branches (life safety and critical).
- have a thorough understanding of hazardous (classified) anesthetizing location requirements.
- be familiar with X-ray installation provisions, including both fixed (stationary) and mobile equipment.

Introduction

Health care facilities, by definition, are buildings, or portions thereof, which contain, at least in part, occupancies such as hospitals, nursing homes, limited care, supervisory care, clinics, medical/dental offices, and ambulatory care, whether permanent or movable. Electrical installation criteria and wiring methods for health care facilities, are far more specialized, than for a standard commercial or industrial project. Article 517 of the *NEC*® governs electrical installations within health care facilities. Applicable requirements of *NEC*® Chapters 1 through 4 must be met, unless specifically modified by Article 517. The provisions of Article 517, Parts II and III, apply not only to single-function buildings, but are also intended for individual application to their respective forms of occupancy within a multifunction structure (e.g., 517.10 requirements must be met by a doctor's examining room situated within a limited care facility). As with any specialized endeavor, a working knowledge of the pertinent terminology is highly recommended. Because most of these terms apply only to health care facilities, 517.2 provides a comprehensive list of the most commonly used words and phrases. Article 517, Part II *(Wiring and Protection)* covers patient care areas throughout the spectrum of health care facilities. *Essential Electrical Systems* (Part III) outlines electrical system requirements for specific occupancies, i.e., hospitals, nursing homes, etc. *Inhalation Anesthetizing Location* provisions are presented in Part IV of Article 517. Part V covers *X-Ray Installations* of all varieties: fixed, stationary, portable, and mobile, as well as transportable. In no way is Part V to be construed as specifying safeguards against the useful beam or stray X-ray radiation. As for *Communications, Signaling Systems, Data Systems, Fire Alarm Systems, and Systems Less than 120 Volts, Nominal,* see Part VI for particular provisions. And finally, Article 517, Part VII, applies to the installation of *Isolated Power Systems* that are comprised of an isolating transformer (or equivalent), a line isolation monitor, and its ungrounded circuit conductors. All in all, Article 517, together with the explanations found in this Unit, offers a reasonably complete picture of health care facility electrical system installations.

GENERAL

Nursing Home and Limited Care Facility

Part III, 517.40(C) through 517.44 requirements do not apply to freestanding buildings used as nursing homes and limited care facilities, provided that:

(a) Admitting and discharge policies preclude providing care to any patient (or resident) who may require electrical life-support.

(b) No surgical treatment requiring general anesthesia is offered.

(c) Automatic battery-operated system(s) or equipment is provided, effective for a minimum of 1½ hours, and otherwise in accordance with 700.12. The system must capably supply lighting for exit lights, exit corridors, stairways, nursing stations, medical preparation areas, boiler rooms, and communications areas, as well as power to operate all alarm systems »517.40(A) *Exception* «.

Nursing homes and limited care facilities that are contiguous with, or adjacent to, a hospital, can have essential electrical systems supplied by the hospital »517.40(C)«.

Ⓐ **Nursing Home:** A building (or part thereof) used for 24-hour lodging, boarding, and nursing care, of four (or more) persons who, because of mental or physical incapacity, may be unable to safely provide for their own needs without assistance. **Nursing home**, whenever used in the *NEC*®, includes nursing and convalescent homes, skilled nursing facilities, intermediate care facilities, and infirmaries of homes for the aged »517.2«.

Ⓑ Nursing homes and limited care facilities are included in the term **health care facilities**.

Ⓒ **Limited Care Facility:** A building (or part thereof) used for 24-hour housing of four or more persons incapable of self-preservation because of age; physical limitation, whether due to accident or illness; or mental limitations, such as retardation/developmental disability, mental illness, or chemical dependency »517.2«.

Ⓓ Part III, 517.40(C) through 517.44 requirements apply to nursing homes and limited care facilities »517.40(A)«.

NOTE

Nursing homes and limited care facilities providing inpatient hospital care must comply with the requirements of Part III, 517.30 through 517.35 »517.40(B)«.

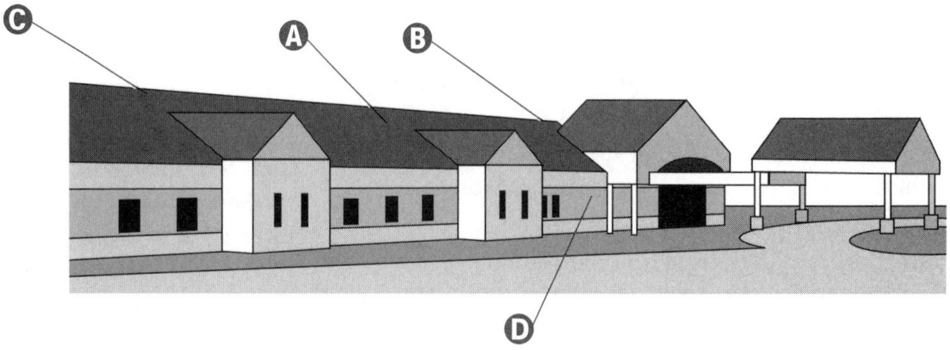

Hospital

Article 517 specifies the installation criteria and wiring methods that minimize electrical hazards by maintaining adequately low-potential differences between exposed conductive surfaces that are likely to become energized and are subject to patient contact »517.11«.

Psychiatric Hospital: A building used exclusively for 24-hour psychiatric care, with at least four inpatients »517.2«.

Refer to the appropriate health care facility documents for performance, maintenance, and testing criteria information »517.1 FPN«.

Parts II and III requirements apply not only to single-function buildings, but also to their respective forms of occupancy within a multifunction building (e.g., a doctor's examining room, located within a limited care facility, must meet 517.10 provisions) »517.1«.

Ⓐ **Hospital:** A building (or part thereof) used for 24-hour medical, psychiatric, obstetrical, or surgical care, of four or more inpatients. Hospital, wherever used in the Code, includes general hospitals, mental hospitals, tuberculosis hospitals, children's hospitals, and any such other facilities providing inpatient care »517.2«.

Ⓑ **Health Care Facilities:** Buildings (or portions of buildings) in which medical, dental, psychiatric, nursing, obstetrical, or surgical care are provided. Health care facilities include, but are not limited to, hospitals; nursing homes; limited care facilities; clinics; medical and dental offices; and ambulatory care centers, whether permanent or movable »517.2«.

Ⓒ **Patient Bed Location:** The location of an inpatient sleeping bed; or the bed (or procedure table) used in an area for critical patient care »517.2«.

Ⓓ Article 517, Part II, applies to patient care areas of all health care facilities »517.10(A)«.

Ⓔ **Patient Care Area:** Any portion of a health care facility wherein patients are examined or treated. Patient care areas of health care facilities are classified as *general care areas* or *critical care areas*, either of which may be a wet location »517.2«.

NOTE

Article 517, Part II does not apply to the following: (1) business offices, corridors, waiting rooms, and the like in clinics, medical and dental offices, and outpatient facilities; and (2) areas of nursing homes and limited care facilities, wired according to Chapters 1 through 4 of the NEC®, where these areas are used exclusively as patient sleeping rooms »517.10(B)«.

See 517.2 for definitions pertaining to health care facilities.

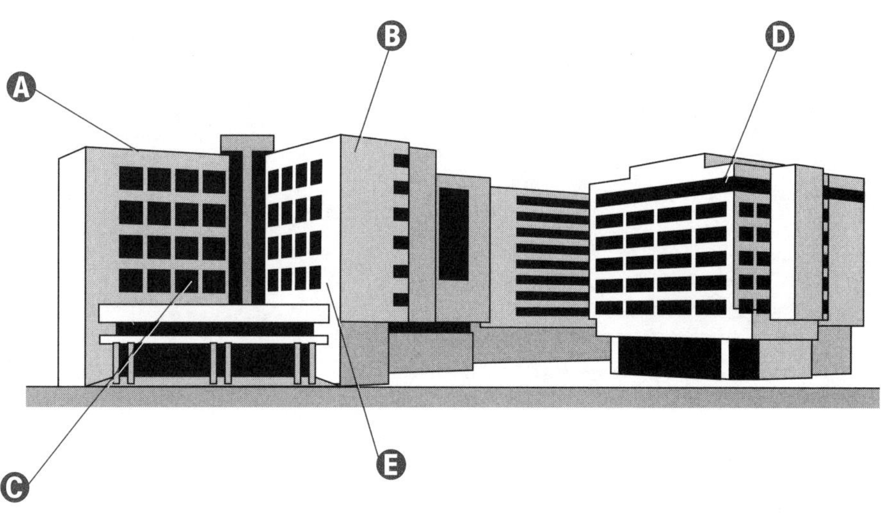

Patient Vicinity

A In an area where patient care normally occurs, the **patient vicinity** is the space having surfaces subject to contact by the patient or by an attendant who may touch the patient. In a patient room, this typically encloses a space within the room no less than 6 ft (1.8 m) beyond the bed's perimeter (in its normal location), and extending vertically no less than 7½ ft (2.3 m) from the floor »517.2«.

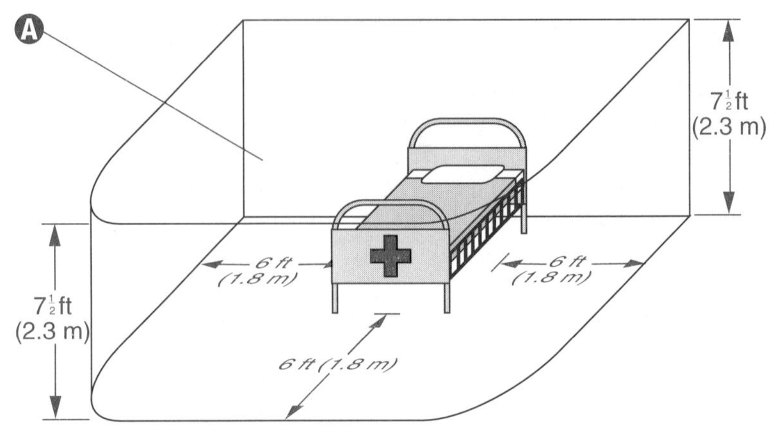

Ambulatory Health Care Facility

Where electrical life support equipment if required, the essential electrical distribution must comply with 517.30 through 517.35 »517.45(B)«.

Where critical care areas are present, the essential distribution must comply with 517.30 through 517.35 »517.45(C)«.

Battery systems must be installed per Article 700's requirements, and generator systems must comply with 517.30 through 517.35 »517.45(D)«.

A **Ambulatory Health Care Facility:** a building (or part thereof) used to provide simultaneous service (or treatment) to four or more patients, which meet either of the following criteria: (1) those facilities that provide treatment for outpatients that would render them incapable of self-preservation under emergency conditions without assistance (such as hemodialysis units or freestanding emergency medical units), or (2) those facilities that provide outpatient surgical treatment requiring general anesthesia »517.2«.

B The essential electrical distribution system must be a battery or generator system »517.45(A)«.

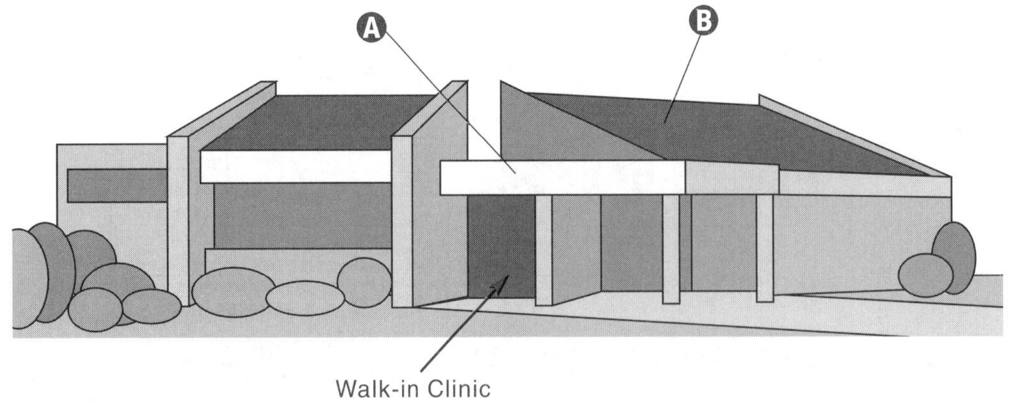

Electrical System Terminology

Electrical Life-Support Equipment: **Electrically powered equipment whose continuous operation is necessary to maintain a patient's life »517.2«.**

Isolated Power System: **A system comprising an isolating transformer (or equivalent), a line isolation monitor, and its ungrounded circuit conductors »517.2«. (See related definitions, which include: *Hazard Current, Fault Hazard Current, Monitor Hazard Current, Total Hazard Current,* and *Line Isolation Monitor.)***

Isolation Transformer: **A multiple-winding type transformer, with physically separated primary and secondary windings, which inductively couples its secondary winding to the grounded feeder systems energizing its primary winding »517.2«.**

Ⓐ Alternate Power Source: One or more generator sets (or battery systems where permitted) to provide power during normal electrical service interruption, or the public utility electrical service intended to provide power during interruption of service normally provided by on-site generating facilities »517.2«.

Ⓑ Life Safety Branch: An Article 700 compliant subsystem of the emergency system consisting of feeders and branch-circuits, providing adequate power to ensure patient and personnel safety, and which automatically connects to alternate power sources during interruption of normal power »517.2«.

Ⓒ Critical Branch: An emergency system subsystem consisting of feeders and branch-circuits supplying energy to task illumination, special power circuits, and selected receptacles serving patient care related areas and functions, and which connects to alternate power sources by transfer switch(es) during interruption of the normal power »517.2«.

Ⓓ Emergency System: A system of circuits and equipment intended to supply alternate power to a limited number of prescribed functions vital to the protection of life and safety »517.2«.

Ⓔ Equipment System: A system of circuits and equipment arranged for delayed, automatic, or manual connection to the alternate power source, primarily serving three-phase power equipment »517.2«.

Ⓕ Essential Electrical System: A system comprised of alternate power sources and all connected distribution systems (and ancillary equipment), designed both to ensure electrical power continuity to designated areas and functions of a health care facility during disruption of normal power sources, and to minimize disruption within the internal wiring system »517.2«.

Ⓖ The essential electrical system for these facilities shall comprise a system capable of supplying a limited amount of lighting and power service, considered essential for life safety and orderly cessation of procedures, during normal electrical service interruption. This includes clinics, medical/dental offices, hospitals, and other patient serving health care facilities »517.25«.

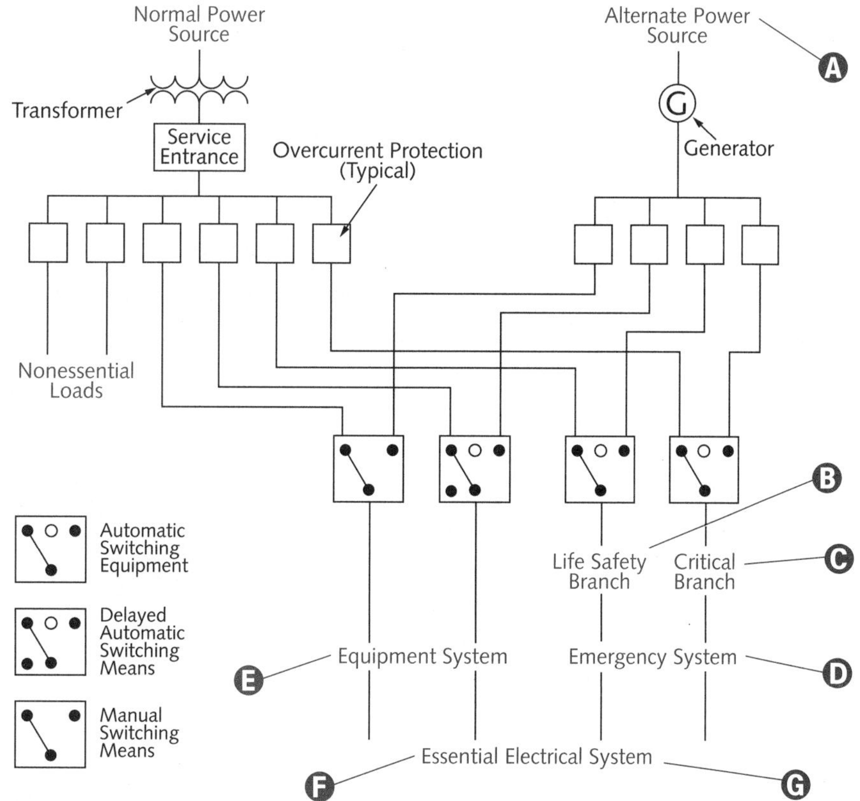

Grounding and Bonding

Metal faceplates can be grounded by means of metal mounting screw(s) securing the faceplate to a grounded outlet box (or grounded wiring device) »517.13(B) *Exception No. 1*«.

Luminaires (light fixtures) more than 7½ ft (2.3 m) above the floor, and switches located outside the patient vicinity, do not require grounding by an insulated grounding conductor »517.13(B) *Exception No. 2*«.

Patient Equipment Grounding Point: A jack or terminal bus that serves as the collection point for electric appliance redundant grounding in a patient vicinity, or for grounding other items to eliminate electromagnetic interference problems »517.2«.

Reference Grounding Point: The ground bus of the panelboard, or isolated power system panel, supplying a patient care area »517.2«.

Exposed Conductive Surfaces: Those unprotected, unenclosed, or unguarded,surfaces capable of carrying electric current and which permit personal contact. Paint, anodizing, and similar coatings do not provide suitable insulation, unless listed for such use »517.2«.

A All general care and critical care area receptacles must be listed and identified as "hospital grade" »517.18(B) and 517.19(B)(2)«.

B Receptacles with insulated grounding terminals, as permitted in 250.146(D), must be visibly identifiable after installation »517.16«.

C Normal and essential branch-circuit panelboard equipment grounding terminal buses, serving the same individual patient vicinity, must be bonded together with an insulated continuous copper conductor no smaller than 10 AWG. If more than two panels serve the same location, this conductor must be continuous from panel to panel, but can be broken in order to terminate on each panel's ground bus »517.14«.

D In a patient care area, all receptacle grounding terminals and all noncurrent-carrying conductive surfaces of fixed electric equipment likely to become energized that are subject to personal contact (operating at over 100 volts) must be grounded by an insulated copper conductor »517.13(B)«.

E The grounding conductor, sized per Table 250.122, must be enclosed in metal raceways together with the branch-circuit conductors supplying these receptacles (or fixed equipment), unless an exception is met »517.13(B)«.

F All branch-circuits serving patient care areas must have a ground path for fault current by installation in a metal raceway system (or cable assembly). The metal raceway system, or cable armor (or sheath) assembly, must qualify as an equipment grounding return path according to 250.118. Type AC, Type MC, and Type MI cables must have an outer metal armor/sheath identified as an acceptable grounding return path »517.13(A)«.

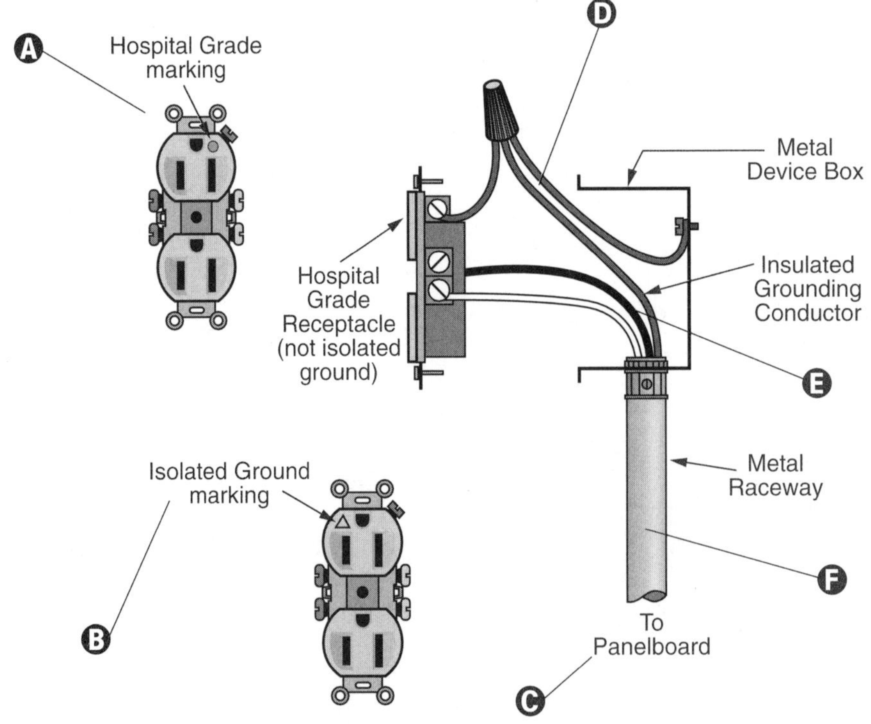

Ground-Fault Protection

Ground-fault protection for the service and feeder disconnecting means operation must be fully selective so that the feeder device, and not the service device, opens for ground faults on the feeder device's load side. A six-cycle minimum separation, between the service and feeder ground-fault tripping bands, must be provided. To achieve 100% selectivity, disconnecting device operating time must be considered when selecting the time spread between these two bands »517.17(B)«.

Upon installation of equipment ground-fault protection, each level must be performance tested to ensure compliance with 517.17(B) »517.17(C)«.

A If ground-fault protection is provided for service disconnecting means (or feeder disconnecting means) operation as specified by 230.95 or 215.10, an additional ground-fault protection step must be provided in the next level of feeder disconnecting means downstream toward the load. Such protection must consist of overcurrent devices and current transformers (or other equivalent protective equipment) that cause the feeder disconnecting means to open »517.17(A)«.

NOTE

Do not install additional levels of ground-fault protection: (1) on the load side of an essential electrical system transfer switch; or (2) between the on-site generating unit(s) described in 517.35(B) and the essential electrical system transfer switch(es); or (3) on electrical systems that are not solidly grounded wye systems with greater than 150 volts to ground, but no more than 600 volts phase-to-phase »517.17(A)«.

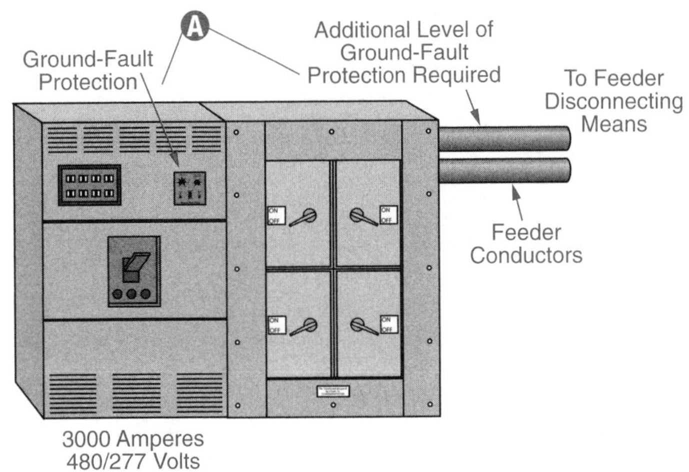

Wet Locations

Health care facility patient care areas are classified as *general care areas* or *critical care areas*, either of which may be a *wet location* »517.2«.

If an isolated power system is utilized, the equipment must be listed for the purpose and must be compliant with 517.160 requirements »517.20(B)«.

A **Wet Locations:** Are those patient care areas normally subject to wet conditions while patients are present. These include fluids either standing on the floor or drenching the work area. Routine housekeeping procedures and incidental liquid spillage do not constitute a wet location »517.2«.

B Branch-circuits (supplying only listed, fixed, therapeutic and diagnostic equipment) can be supplied from a normal grounded service, single- or three-phase system, provided that: (A) grounded and isolated circuit wiring does not occupy the same raceway; and (B) all conductive equipment surfaces are grounded »517.20(A) Exception«.

NOTE

All receptacles and fixed equipment within the wet location area must have GFCI protection for personnel if interruption of power under fault conditions can be tolerated. If such interruption cannot be tolerated, the receptacles must be served by an isolated power system »517.20(A)«.

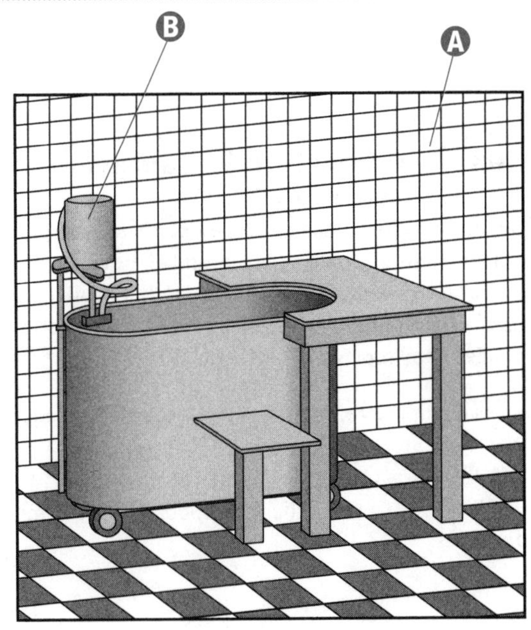

PATIENT CARE AREAS

Patient Bed Location Branch-Circuits (General Care Areas)

517.18(A) requirements do not apply to patient bed locations in clinics, medical/dental offices, and outpatient facilities; psychiatric, substance abuse, and rehabilitation hospitals; nursing home sleeping rooms; and limited care facilities where 517.10(B)(2) requirements are met »517.18(A) *Exception No. 2*«.

Ⓐ Branch-circuits serving only special-purpose outlets or receptacles (such as portable X-ray outlets) do not require service from the same distribution panel(s) »517.18(A) *Exception No. 1*«.

Ⓑ Each patient bed location must be supplied by at least two branch-circuits: one from the emergency system, one from the normal system. All normal system branch-circuits must originate within the same panelboard »517.18(A)«.

Ⓒ If served from two separate emergency system transfer switches, a general care location does not require circuits from the normal system »517.18(A) *Exception No. 3*«.

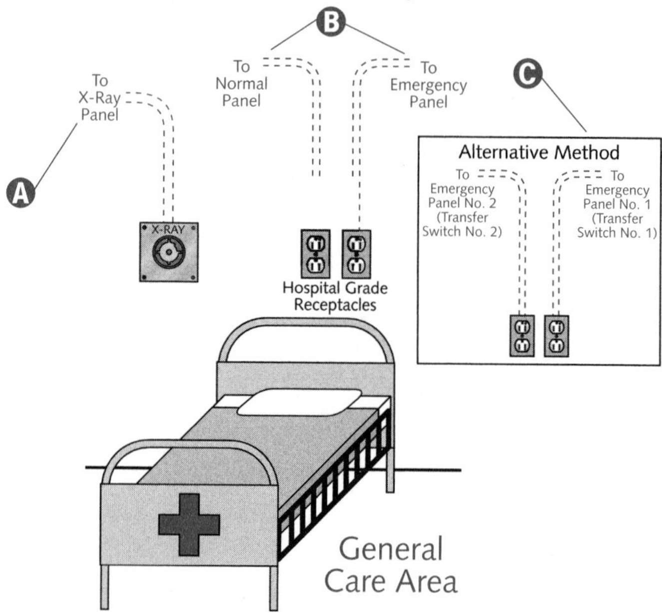

NOTE

General Care Areas are patient bedrooms, examining rooms, treatment rooms, clinics, and similar areas where the patient contacts ordinary appliances such as a nurse call system, electrical beds, examining lamps, a telephone, and entertainment devices. In such areas, patients may also be connected to electromedical devices (such as heating pads, electrocardiographs, drainage pumps, monitors, otoscopes, ophthalmoscopes, intravenous lines, etc.) »517.3«.

Patient Bed Location Receptacles (General Care Areas)

A total, immediate replacement of existing non-hospital grade receptacles is not necessary. It is intended, however, that non-hospital grade receptacles be replaced with hospital grade receptacles upon modification of use, during renovation, or as receptacles need replacing »517.18(B) FPN«.

Ⓐ Special purpose receptacles are not counted as required receptacles.

Ⓑ *Each* patient bed location must have a minimum of four receptacles. They can be either single, duplex, or a combination of both types »517.18(B)«.

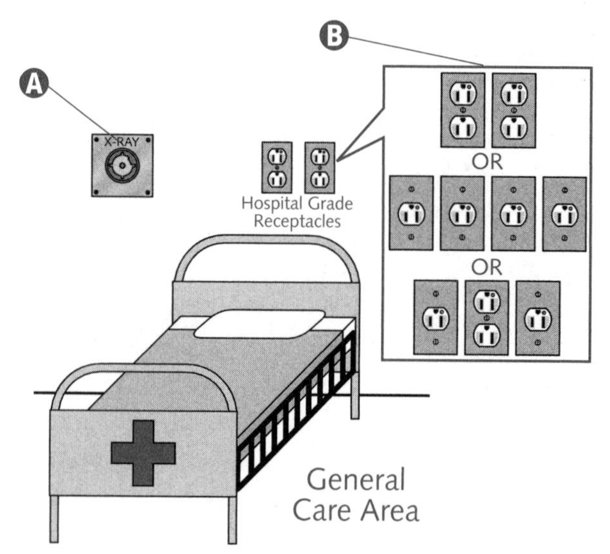

NOTE

All receptacles, must be listed and identified as "hospital grade." Each receptacle must be grounded by an insulated copper conductor sized per Table 250.122 »517.18(B)«.

Pediatric Location Receptacles (General Care Areas)

A Receptacles located within the patient care areas of pediatric wards, rooms, or areas must either be **listed** tamper resistant, or must employ a **listed** tamper resistant cover »517.18(C)«.

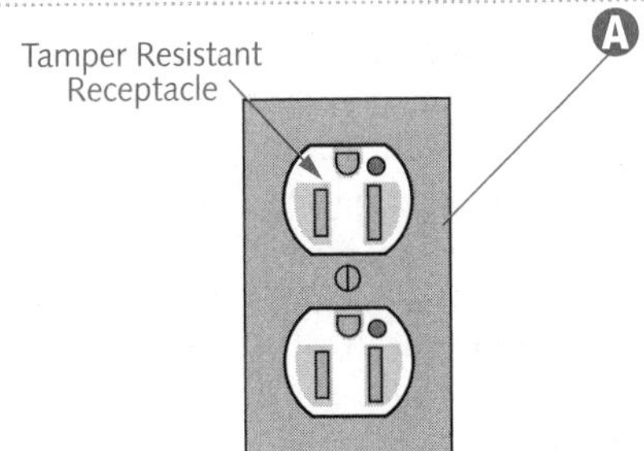

Patient Bed Location Receptacles (Critical Care Areas)

All receptacles, must be listed and identified as "hospital grade." Each receptacle must be grounded to the reference grounding point by means of an insulated copper equipment grounding conductor »517.19(B)(2)«.

Although not required, a patient vicinity can have a patient equipment grounding point. Where one is supplied, it can contain one (or more) jacks listed for the purpose. An equipment bonding jumper, no smaller than 10 AWG, must connect the grounding terminal of all grounding-type receptacles to the patient equipment grounding point. The bonding conductor can be arranged centrically or looped as convenient »517.19(C)«.

A The equipment grounding conductor for special-purpose receptacles (such as the operation of mobile X-ray equipment), must extend to the branch-circuit reference grounding points for all locations potentially served from such receptacles. Where an isolated ungrounded system serves such a circuit, the grounding conductor does not have to run with the power conductors; however, the special-purpose receptacle's equipment grounding terminal must connect to the reference grounding point »517.19(G)«.

B Each patient bed location must have a minimum of six receptacles. At least one of the receptacles must be connected to: (1) the normal system branch-circuit as required by 517.19(A); or (2) an emergency system branch-circuit supplied by a different transfer switch not associated with other receptacles at the same location »517.19(B)(1)«.

C The receptacles can be either single, duplex, or a combination of both types »517.19(B)(2)«.

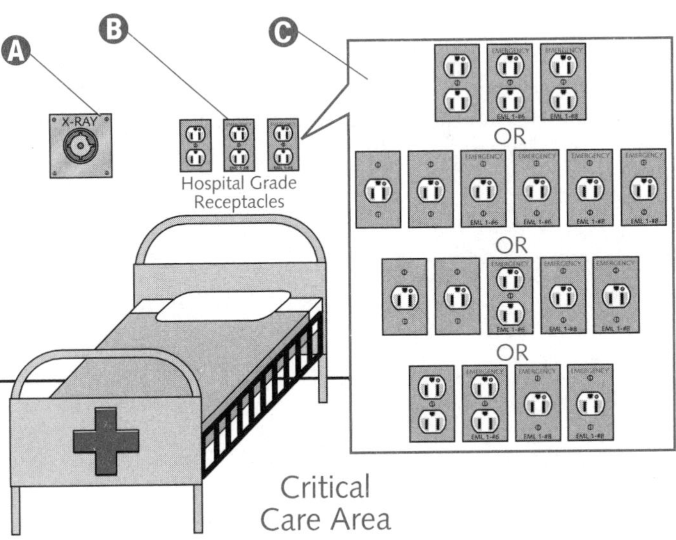

NOTE

A grounded electrical distribution system, having metal feeder raceway or Type MC or MI cable, requires that panelboard (or switchboard) grounding be ensured at ***each*** *termination (or junction point) of the raceway or Type MC or MI cable by one of the following means:*

(1) A grounding bushing and a continuous copper bonding jumper, sized according to 250.122, with the bonding jumper connected to the junction enclosure or the panel ground bus

(2) Connection of feeder raceways or Type MC or MI cables to threaded hubs or bosses on terminating enclosures

(3) Other approved devices such as bonding-type locknuts or bushings »517.19(D)«.

Patient Bed Location Branch-Circuits (Critical Care Areas)

A Branch-circuits serving only special-purpose receptacles, or equipment in critical care areas, can be served by other panelboards »517.19(A) *Exception No. 1*«.

B Each patient bed location must be supplied by at least two branch-circuits: one (or more) from the emergency system, and one (or more) from the normal system. At least one emergency system branch-circuit must supply outlet(s) at that bed location only »517.19(A)«.

C Emergency system receptacles must be identified, and must also indicate the supplying panelboard and circuit number »517.19(A)«.

D Critical care locations served from two separate emergency system transfer switches do not require circuits from the normal system »517.19(A) *Exception No. 2*«.

CAUTION *All normal system branch-circuits must originate from a single panelboard »517.19(A)«.*

NOTE

Critical Care Areas are those special care units, intensive care units, coronary care units, angiography laboratories, cardiac catheterization laboratories, delivery rooms, operating rooms, and similar areas where patients are subjected to invasive procedures while connected to line-operated, electromedical devices »517.2«.

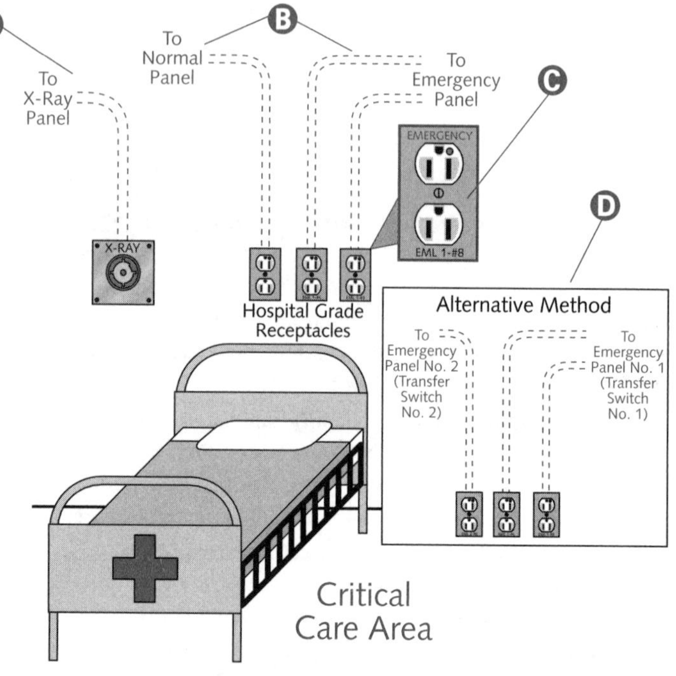

HOSPITALS

Life Safety Branch

A The emergency system life safety branch must supply power for lighting, receptacles, and equipment as listed in 517.32(A) through (G) »517.32«.

B Illumination of egress means is required, such as lighting for corridors, passageways, stairways, landings at exit doors, and all paths of approach to exits. Switching arrangement to transfer patient corridor lighting from general illumination circuits is permitted, provided only one of two circuits can be selected, and both circuits cannot be simultaneously extinguished »517.32(A)«.

C Task illumination battery charger for emergency battery-powered lighting unit(s) and selected receptacles at the generator set location »517.32(E)«.

D Exit and exit directional signs »517.32(B)«.

E Alarm and alerting systems, including: (1) fire alarms and (2) alarms required by systems used for piping nonflammable medical gases »517.32(C)«.

F Automatically operated doors used for building egress »517.32(G)«.

G Only functions listed in 517.32(A) through (G) can be connected to the life safety branch »517.32«.

H Hospital communication systems, used for issuing instructions during emergency conditions »517.32(D)«.

I Elevator cab lighting, control, communications, and signal systems »517.32(F)«.

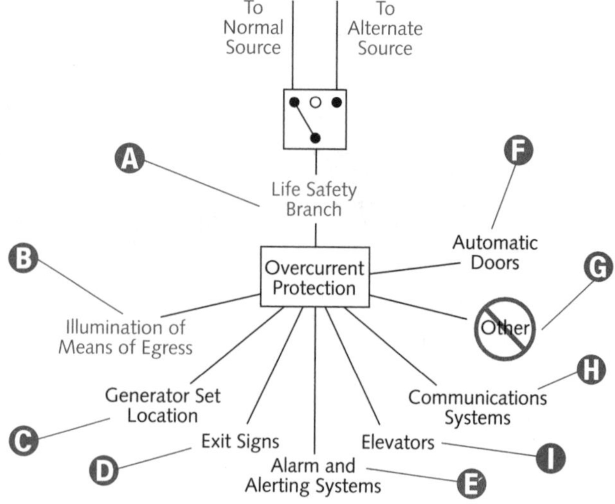

Essential Electrical Systems

A Hospital power sources and alternate power sources can serve the essential electrical systems of contiguous, or same site, facilities »517.30(B)(6)«.

B Loads, served by generating equipment not specifically named in Article 517, must be served by their own transfer switches so that these loads (1) are not transferred if the transfer will overload the generating equipment, and (2) are automatically shed upon generating equipment overloading »517.30(B)(5)«.

C A facility having a maximum essential electrical system demand of 150 kVA can have a single transfer switch serving one or more branches (critical and life safety) or systems (emergency and equipment) »517.30(B)(4)«.

D Limit the emergency system to only those circuits essential to life safety and critical patient care. These carry the *life safety branch* and *critical branch* designations »517.30(B)(2) and 517.31«.

E Emergency system branches must be installed and connected to the alternate power source so that all functions specified herein are automatically restored to operation within ten seconds after normal source interruption »517.31«.

F The equipment system supplies major electrical equipment necessary for patient care and for basic hospital operation »517.30(B)(3)«.

G The number of transfer switches used is based on reliability, design, and load considerations. Each essential electrical system (emergency and equipment) branch must be served by transfer switch(es) »517.30(B)(4)«.

H Essential electrical systems must be comprised of two separate systems, both capable of supplying a limited amount of the lighting and power service considered essential for life safety and effective hospital operation, during normal electrical service interruption, regardless of cause. These two systems are the *emergency system* and the *equipment system* »517.30(B)(1)«.

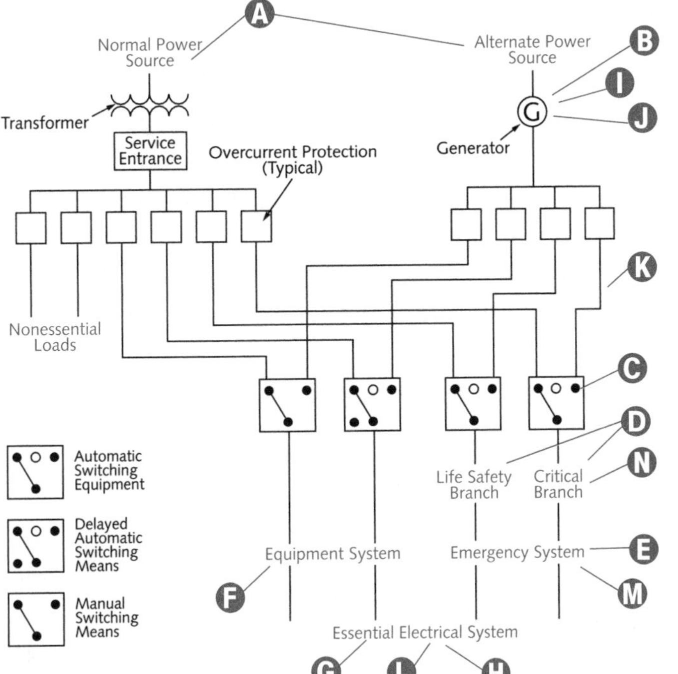

NOTE

Electrical system definitions can be found in the first part of this unit.

Wiring Requirements for Essential Electrical Systems

I Demand calculations for generator set(s) sizing must be based on the following: (1) prudent demand factors and historical data, or (2) connected load, or (3) Article 220 feeder calculation procedures or (4) any combination of the above »517.30(D)«.

J The generator set(s) must have sufficient capacity and be properly rated to meet the demand produced by the essential electrical system(s) load at any given time »517.30(D)«.

K Feeders must be sized per Articles 215 and 220 »517.30(D)«.

L The essential electrical system must capably meet the demand for the operation of all functions and equipment served by each system/branch »517.30(D)«.

M The hospital's emergency system wiring must be either mechanically protected by installation in nonflexible metal raceways, or must be wired with Type MI cable »517.30(C)(3)«.

N If isolated power systems are installed in any of the areas in 517.33(A)(1) and (A)(2), each system must be supplied by an individual circuit serving no other load »517.30(C)(2)«.

WARNING

The emergency system's life safety branch and critical branch must remain entirely independent of all other wiring (and equipment), and must not enter the same raceways, boxes, or cabinets with one another or any other wiring. Life safety branch and critical branch wiring can occupy the same raceways, boxes, or cabinets of other circuits (not part of the branch) where such wiring is:

(1) In transfer equipment enclosures, or
(2) In exit/emergency luminaires (lighting fixtures) supplied from two sources, or
(3) In a common junction box attached to exit/emergency luminaires (lighting fixtures) supplied from two sources, or
(4) For multiple emergency circuits supplied from the same branch.

The equipment system wiring can occupy the same raceways, boxes, or cabinets of other circuits that are not part of the emergency system »517.30(C)(1)«.

Receptacle Identification

Ⓐ Receptacles (or their cover plates) supplied from the emergency system must be distinctively colored or otherwise marked in a readily identifiable manner »517.30(E)«.

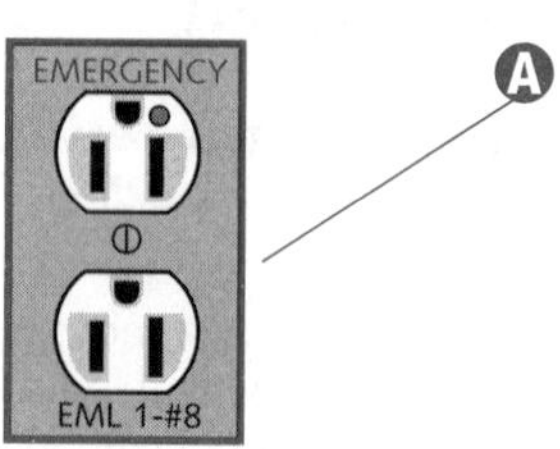

Critical Branch

Ⓐ Subdivisions of the critical branch (into two or more branches) is permitted »517.33(B)«.

Ⓑ Critical care areas that utilize anesthetizing gases—task illumination, selected receptacles, and fixed equipment »517.33(A)(1)«.

Ⓒ The isolating power systems in special environments »517.33(A)(2)«.

Ⓓ Additional specialized patient care task illumination and receptacles, where needed »517.33(A)(4)«.

Ⓔ Patient care areas — task illumination and selected receptacles in the following: (a) infant nurseries; (b) medication prep areas; (c) pharmacy dispensing areas; (d) selected acute nursing areas; (e) psychiatric bed areas (omit receptacles); (f) ward treatment rooms; (g) nurses' stations (unless adequately lighted by corridor luminaires [fixtures]) »517.33(A)(3)«.

Ⓕ The emergency system critical branch must supply power for task illumination, fixed equipment, selected receptacles, and special power circuits serving patient care areas and function as listed in 517.33(A)(1) through (9) »517.33(A)«.

Ⓖ Additional task illumination, receptacles, and selected power circuits needed for effective hospital operation (single-phase fractional horsepower motors can be connected to the critical branch) »517.33(A)(9)«.

Ⓗ Task illumination, selected receptacles, and selected power circuits for the following: (a) general care beds (at the least one duplex receptacle per patient bedroom); (b) angiographic labs; (c) cardiac catheterization labs; (d) coronary care units; (e) hemodialysis rooms or areas; (f) emergency room treatment areas (selected); (g) human physiology labs; (h) intensive care units; and (i) postoperative recovery rooms (selected) »517.33(A)(8)«.

Ⓘ Telephone equipment rooms and closets »517.33(A)(7)«.

Ⓙ Blood, bone, and tissue banks »517.33(A)(6)«.

Ⓚ Nurse call systems »517.33(A)(5)«.

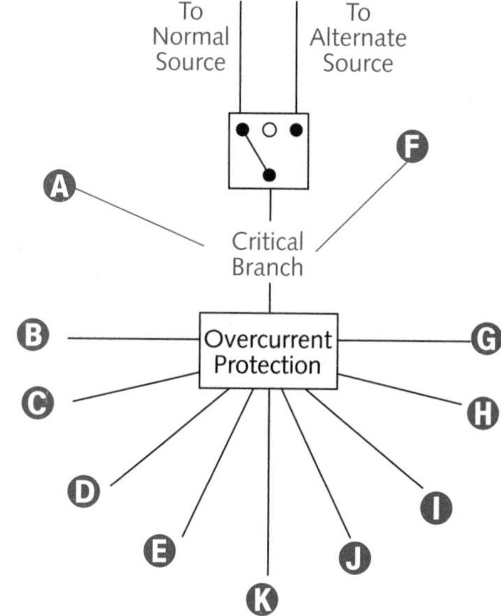

Power Sources

Facilities, whose normal power source is supplied by multiple separate central station-fed services, experience greater than normal electrical service reliability than those having only a single feed. Such a dual source of normal power consists of multiple electrical services, fed from separate generator sets, or a utility distribution network having multiple power input sources. These dual services are arranged to provide mechanical and electrical separation when a fault between the facility and the generating sources is unlikely to cause an interruption of more than one of the facility service feeders »517.35(C) FPN«.

NOTE

Carefully consider the location of the essential electrical system components (housing and spaces) to minimize interruptions caused by natural forces common to the area (e.g., storms, floods, earthquakes, or hazards created by adjoining structures or activities). Consideration must also be given to the potential normal electrical service interruptions resulting from similar causes, as well as possible service disruption due to internal wiring and equipment failures »517.35(C)«.

Power Sources *(continued)*

A Essential electrical systems must have a minimum of two independent power sources: a normal source generally supplying the entire system; and, one or more alternate sources for use during normal source interruption »517.35(A)«.

B The alternate power source can be on-site generator(s) driven by some form of prime mover(s) »517.35(B)«.

C The alternate power source can also be another generating unit(s) provided the normal source consists of on-site generating unit(s) »517.35(B)«.

D An external utility service can act as the alternate power source when the normal source consists of on-site generating unit(s) »517.35(B)«.

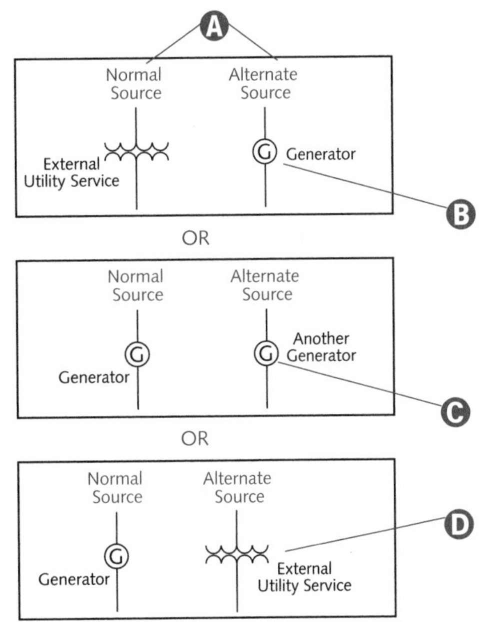

Equipment Systems

A The following equipment must have either delayed automatic or manual connection to the alternate power source:

(1) Heating equipment to providing heat for operating, delivery, labor, recovery, intensive care, coronary care, nurseries, infection/isolation rooms, emergency treatment spaces, general patient rooms, and pressure maintenance (jockey or make-up) pump(s) for water-based fire protection systems.

(2) Elevator(s) providing service to patient, surgical, obstetrical, and ground floors during normal power interruption. (Where interruption of normal power would result in other elevators stopping between floors, throw-over facilities must be provided to allow temporary operation of any elevator for patient/other person release.)

(3) Supply, returns, and exhaust ventilating systems for airborne infection/isolation rooms, protective environment rooms, exhaust fans for laboratory fume hoods, nuclear medicine areas where radioactive material is used, ethylene oxide evacuation, and anesthesia evacuation

(4) Hyperbaric facilities

(5) Hypobaric facilities

(6) Automatic doors

(7) Minimal electrically heated autoclaving equipment can be arranged for either automatic or manual connection to the alternate source.

(8) Controls for equipment listed in 517.34

(9) Other selected equipment can be served by the equipment system »517.34(B)«.

B Equipment system installation and connection to the alternate power source, for equipment described in 517.34(A), must automatically restore operation at appropriate time-lag intervals following emergency system activation. The arrangement must also provide for subsequent connection of equipment described in 517.34(B) »517.34«.

C Delayed automatic connection to the alternate power source must be provided for the following equipment:

(1) Central suction systems serving medical/surgical functions, including controls. (Such suction systems are permitted on the critical branch.)

(2) Sump pumps and other equipment required for the safe operation of major apparatus, including associated control systems and alarms

(3) Compressed air systems serving medical/surgical functions, including controls. (Such air systems are permitted on the critical branch.)

(4) Smoke control, stair pressurization systems, or both

(5) Kitchen hood supply, exhaust systems, or both, if operation is required during a fire in (or under) the hood »517.34(A)«.

NOTE

Sequential delayed automatic connection to the alternate power source, thereby preventing generator overload, is permitted where indicated by engineering studies »517.34(A) Exception«.

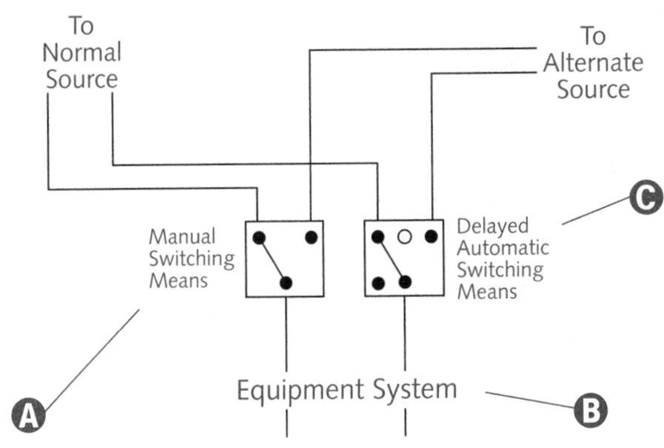

NURSING HOMES AND LIMITED CARE FACILITIES

Electrical Systems

Where the normal source consists of on-site generating units, the alternate source must be either another generator set, or an external utility service »517.44(B) *Exception No. 1*«.

Nursing homes or limited care facilities meeting the requirements of 517.40(A), *Exception*, can use a battery system or an equipment self-contained, integral battery »517.44(B) *Exception No. 2*«.

A Essential electrical systems, for nursing homes and limited care facilities, must be comprised of two separate branches, each capable of supplying a limited amount of lighting and power service, for those functions considered essential for life safety and effective institution operation during normal service interruption, regardless of cause. These are the *life safety branch* and the *critical branch* »517.41(A)«.

B The essential electrical system must be capable of meeting the demand for operation of all functions and equipment being served simultaneously by each branch »517.41(C)«.

C Essential electrical systems must have at least two independent power sources: a normal source generally supplying the entire system, and one (or more) alternate sources for use during normal source interruption »517.44(A)«.

D The alternate power source must be on-site generator(s) driven by some form of prime mover(s), unless an exception is met »517.44(B)«.

E The number of transfer switches used is based on reliability, design, and load considerations. Each essential electrical system branch (life safety and critical) must be served by one (or more) transfer switches. A facility whose maximum demand is 150 kVA can have a single transfer switch serving one or more branches/systems »517.41(B)«.

F Life safety branch installation and alternate power source connection must provide that all functions (specified herein) be automatically restored to operation within 10 seconds after interruption occurs 517.42«.

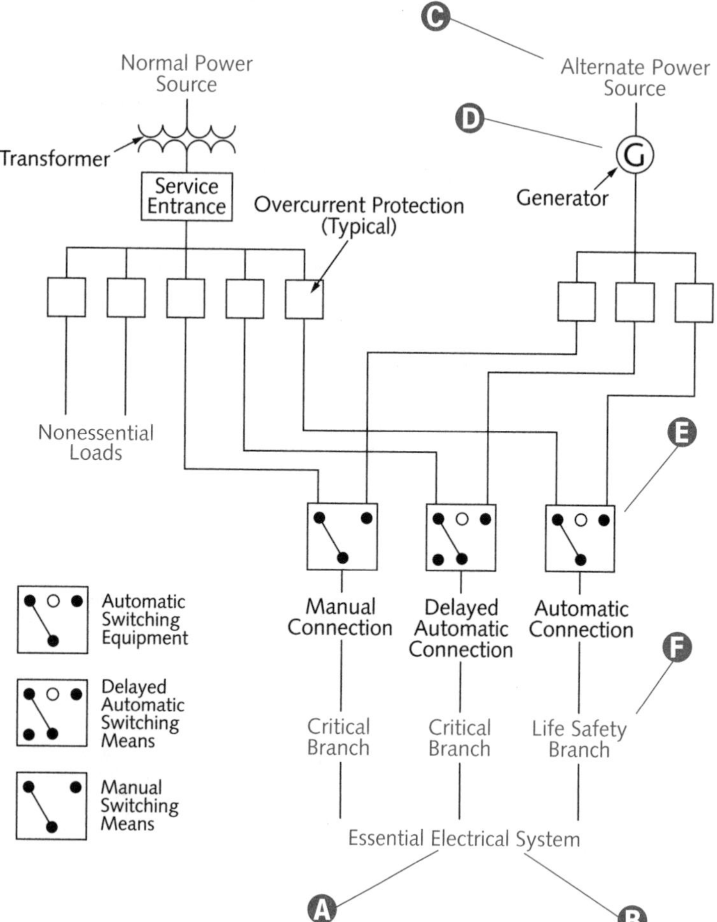

NOTE

Give careful consideration to the essential electrical system components location to minimize interruptions caused by natural forces common to the area (e.g., storms, floods, earthquakes, or hazards created by adjoining structures or activities). Consideration must also be given to possible normal services interruption resulting from similar causes, as well as, potential service disruption due to internal wiring and equipment failures »517.44(C)«.

WARNING

The life safety branch must be entirely independent of all other wiring (and equipment), and must not enter the same raceways, boxes, or cabinets with other wiring except as follows:

(1) In transfer switches
(2) In exit or emergency luminaires (lighting fixtures) supplied from two sources, or
(3) In a common junction box attached to exit or emergency luminaires (lighting fixtures) supplied from two sources

The critical branch wiring can occupy the same raceways, boxes, or cabinets of other circuits that are not part of the life safety branch »517.41(D)«.

Life Safety Branch

A Life safety branch installation and alternate power source connection must provide that all functions, specified in 517.42(A) through (G), are automatically restored to operation within 10 seconds after normal source interruption »517.42«.

B Illumination of egress pathways such as corridors, passageways, stairways, landings, exit doors, and all means of approach, is required. Patient corridor lighting can be transferred via switching arrangement from general illumination circuits provided only one circuit can be selected, and both circuits cannot be simultaneously extinguished »517.42(A)«.

C Task illumination and selected receptacles in the generator set location »517.42(F)«.

D Sufficient lighting in dining and recreation areas to provide exit way illumination »517.42(E)«.

E Exit and exit directional signs »517.42(B)«.

F Elevator cab lighting, control, communications, and signal systems »517.42(G)«.

G Connect only those functions listed in (A) through (G) to the life safety branch »517.42«.

H Communications systems, used for issuing emergency condition instructions »517.42(D)«.

I Alarm and alerting systems, including: (1) fire alarms, and (2) alarms required for systems piping nonflammable medical gases »517.42(C)«.

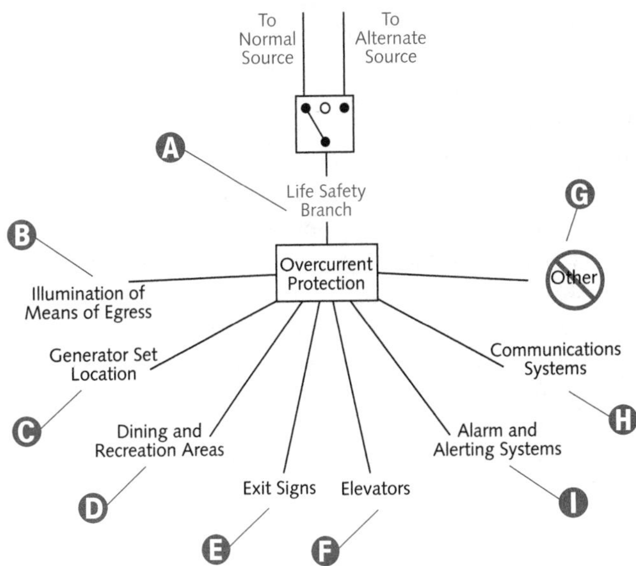

Critical Branch

A The following equipment connects to the critical branch, and must be arranged for either delayed automatic or manual connection to the alternate power source:

(1) Heating equipment providing patient room heating.

(2) Elevator service where power disruption would result in between-floor stops. Throw-over facilities must be provided to allow temporary operation of any elevator passenger release. See 517.42(G) for elevator cab lighting, control, and signal system requirements.

(3) Additional illumination, receptacles, and equipment can be connected only to the critical branch »517.43(B)«.

B Critical branch installation and alternate power source connection must provide that equipment listed in 517.43(A) is automatically restored to operation at appropriate time-lag intervals following the restoration of life safety branch operation. Arrangement must also provide for additional connection of equipment listed in 517.43(B), by either delayed automatic or manual operation »517.43«.

C The following equipment must be connected to the critical branch; and must be arranged for delayed automatic connection to the alternate power source:

(1) Patient care areas — task illumination and selected receptacles in: (a) medication prep areas; (b) pharmacy dispensing areas; and (c) nurses' stations (unless corridor luminaires provide adequate light)

(2) Sump pumps and other equipment required for major apparatus safe operation, as well as associated control systems/alarms

(3) Smoke control and stair pressurization systems

(4) Kitchen hood supply and/or exhaust systems, if operation is required during a fire in (or under) the hood

(5) Supply, return and exhaust ventilating systems for airborne infectious/isolation rooms »517.43(A)«.

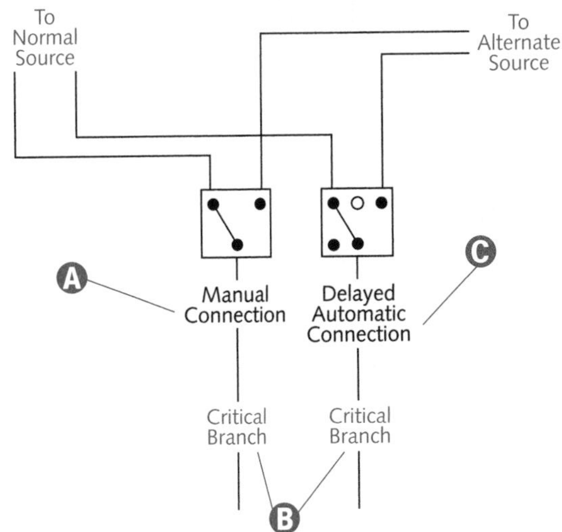

INHALATION ANESTHETIZING LOCATIONS

Hazardous (Classified) Anesthetizing Locations

Anesthetizing Location: Any facility area designated for use in administering any flammable inhalation anesthetic agent (during examination or treatment), including the use of such agents for relative analgesia »517.2«. (See also the definitions of *Flammable Anesthetics* and *Flammable Anesthetizing Location.)*

Unless otherwise allowed by 517.160, each power circuit within (or partially within) a flammable anesthetizing location (517.60) must be isolated from any distribution system by means of an isolated power system »517.61(A)(1)«.

Any inhalation anesthetizing location, designated for the exclusive use of nonflammable anesthetizing agents, is considered to be, other-than-hazardous (classified) location »517.60(B)«. Installations in other-than-hazardous (classified) locations must comply with 517.61(C).

Isolated power system equipment must be listed for the purpose; the system must be designed and installed so that it meets the provisions of Part VII »517.61(A)(2)«.

In hazardous (classified) locations referenced in 517.60, all fixed wiring/equipment, and all portable equipment (including lamps and other utilization equipment), operating at more than 10 volts between conductors, must comply with the requirements of 501.1 through 501.15 and 501.16(A) and (B) for Class I, Division 1 locations. All such equipment must be approved for the specific hazardous atmosphere involved »517.61(A)(3)«.

If a box, fitting, or enclosure is only partially within a hazardous (classified) location(s), the hazardous area extends to include the entire box, fitting, or enclosure »517.61(A)(4)«.

Receptacles and attachment plugs in a hazardous (classified) location(s) must be listed for use in Class I, Group C hazardous (classified) locations, and must provide a grounding conductor connection »517.61(A)(5)«.

Flexible cords used in hazardous (classified) locations for connection of portable utilization equipment (including lamps operating at more than 8 volts between conductors), must be of a type approved for extra-hard usage per Table 400.4 and must include an additional conductor for grounding »517.61(A)(6)«.

The flexible cord must include a storage device that does not subject the cord to less than a 3-in. (75 mm) bending radius »517.61(A)(7)«.

Ⓐ Any room (or location) storing flammable anesthetics or volatile flammable disinfecting agents is, from floor to ceiling, a Class I, Division 1 location »517.60(A)(2)«.

Ⓑ Wherever flammable anesthetics are employed, the entire area is a Class I, Division 1 location, extending from the floor upward to a level of 5 ft (1.52 m). The remaining space (up to the structural ceiling) is considered to be above a hazardous (classified) location »517.60(A)(1)«.

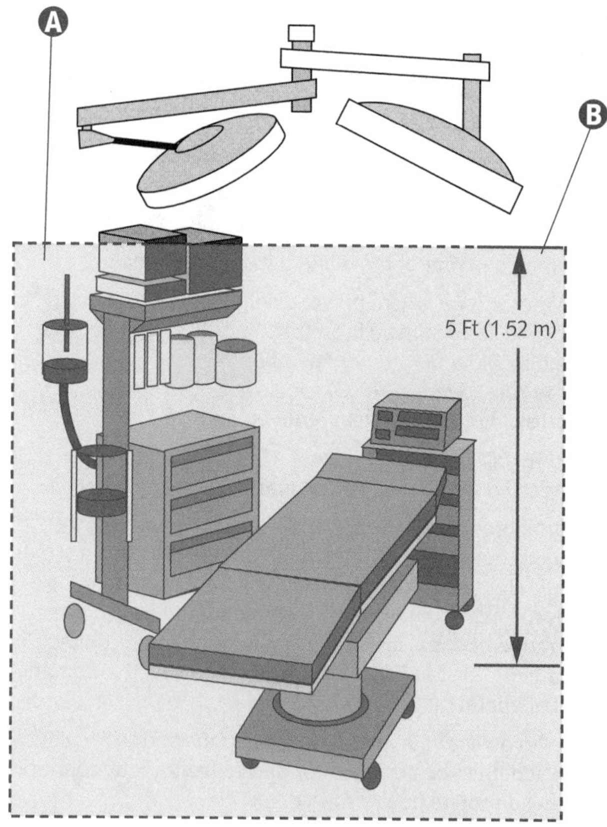

Above Hazardous (Classified) Anesthetizing Locations

Wiring above a hazardous (classified) location (referenced in 517.60) must be installed in: RMC, EMT, IMC, Type MI cable, or Type MC cable employing a continuous gas/vaportight metal sheath »517.61(B)(1)«.

Installed equipment that may produce arcs, sparks, or particles of hot metal (such as lamps and lampholders for fixed lighting, cutouts, switches, generators, motors, or other equipment having make-and-break or sliding contacts) must be of the totally enclosed type, or by design must prevent the escape of sparks or hot metal particles »517.61(B)(2)«. Wall-mounted receptacles, installed above the hazardous (classified) location in flammable anesthetizing locations, require neither total enclosure nor guarded or screened openings to prevent particle dispersion »517.61(B)(2) *Exception*«.

Approved seals must be provided in conformance with 501.5. 501.5(A)(4) applies to both horizontal and to vertical boundaries of the defined hazardous (classified) locations »517.61(B)(4)«.

Receptacles and attachment plugs, located above hazardous (classified) anesthetizing locations, must be listed for hospital use for services of prescribed voltage, frequency, rating, and number of conductors with grounding conductor connection provisions. This requirement applies to attachment plugs and receptacles of the two-pole, three-wire grounding type for single-phase, 120-volt, nominal, ac service »517.61(B)(5)«.

Plugs and receptacles rated 250 volts, for connection of 50- and 60-ampere ac medical equipment above hazardous (classified) locations, must be arranged so that the 60-ampere receptacle accepts either the 50- or 60-ampere plug. Fifty-ampere receptacles must *not* accept a 60-ampere attachment plug. The plugs must be on the two-pole, three-wire design, with a third contact connecting to the insulated (green or green with yellow stripe) electrical system equipment grounding conductor »517.61(B)(6)«.

Ⓐ Wiring above a hazardous (classified) location must meet 517.61(B) provisions.

Ⓑ Surgical and other luminaires (lighting fixtures) must conform to 501.9(B) »517.61(B)(3)«.

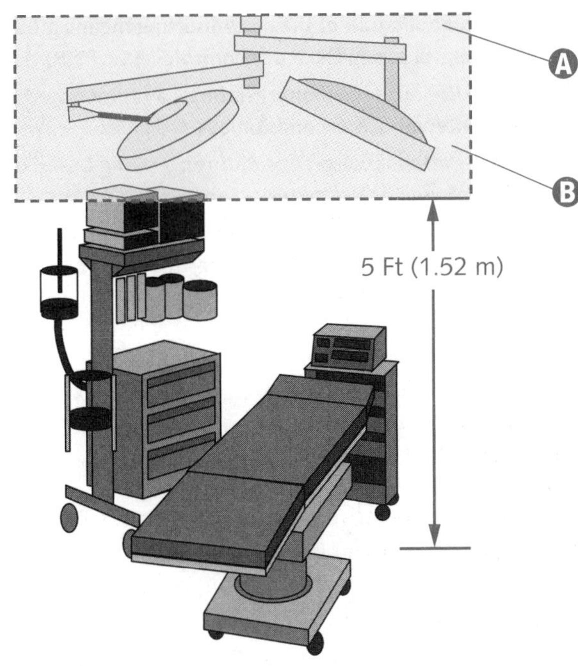

X-RAY INSTALLATIONS

Mobile, Portable, and Transportable X-Ray Equipment

Ⓐ **X-Ray Installations (Mobile):** X-ray equipment mounted on a permanent base with wheels, casters, or a combination of both, facilitating movement of equipment while fully assembled »517.2«.

X-Ray Installations (Portable): X-ray equipment designed to be hand carried »517.2«.

X-Ray Installations (Transportable): X-ray equipment whose design facilitates vehicular transport (via ready disassembly) and/or vehicular installation »517.2«.

Ⓑ Individual branch-circuits are not required for portable, mobile, and transportable medical X-ray equipment requiring a capacity of 60 amperes or less »517.71(B)«.

Ⓒ For portable X-ray equipment connected to a 120-volt branch-circuit of 30 amperes (or less), a properly rated grounding-type attachment plug and receptacle can serve as a disconnecting means »517.72(C)«.

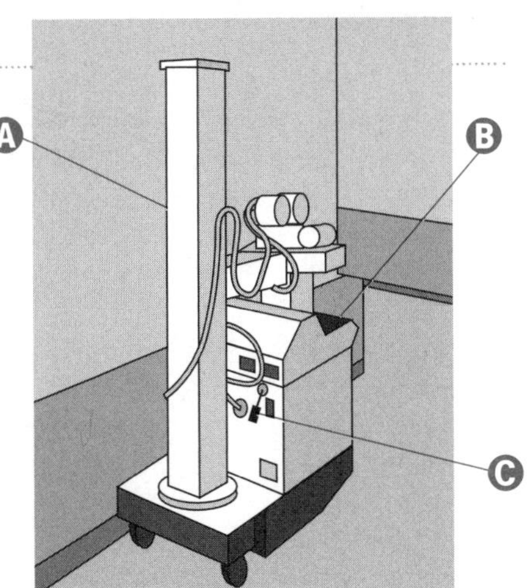

Fixed and Stationary X-Ray Equipment

Fixed and stationary X-ray equipment must be connected to the power supply by a wiring method that meets the *Code's* general requirements »517.71(A)«. Equipment, properly supplied by a branch-circuit rated at 30 amperes (or less), can be supplied through a suitable attachment plug and hard-service cable (or cord) »517.71 *Exception*«.

A supply circuit disconnecting means of adequate capacity for at least 50% of the input (for momentary rating) or 100% of the input (for long-time rating) of the X-ray equipment, whichever is greater, must be provided »517.72(A)«.

Location and operation of the disconnecting means must be readily accessible from the X-ray control »517.72(B)«.

X-Ray Installations (Momentary Rating): a rating based on an operating interval of 5 seconds or less »517.2«.

X-Ray Installations (Long-Time Rating): a rating based on an operating interval of 5 minutes or longer »517.2«.

Supply branch-circuit conductor ampacity and the current rating of overcurrent protective devices must not be less than 50% of the momentary rating or 100% of the long-time rating, whichever is greater »517.73(A)(1)«.

Supply feeder ampacity and the current rating of overcurrent protective devices supplying multiple branch-circuits feeding X-ray units must not be less than 50% of the largest unit's momentary demand rating, plus 25% of the momentary demand rating of the next largest unit, plus 10% of the momentary demand rating of each additional unit. If the X-ray unit performs simultaneous biplane examinations, the supply conductors and overcurrent protective devices must be 100% of each unit's momentary demand rating »517.73(A)(2)«. The manufacturer typically specifies minimum distribution transformer and conductor sizes, disconnecting means rating, and overcurrent protection »517.73(A)(2) FPN«.

Conductor ampacity and overcurrent protective device ratings must not be less than 100% of the current rating of medical X-ray *therapy* equipment »517.73(B)«.

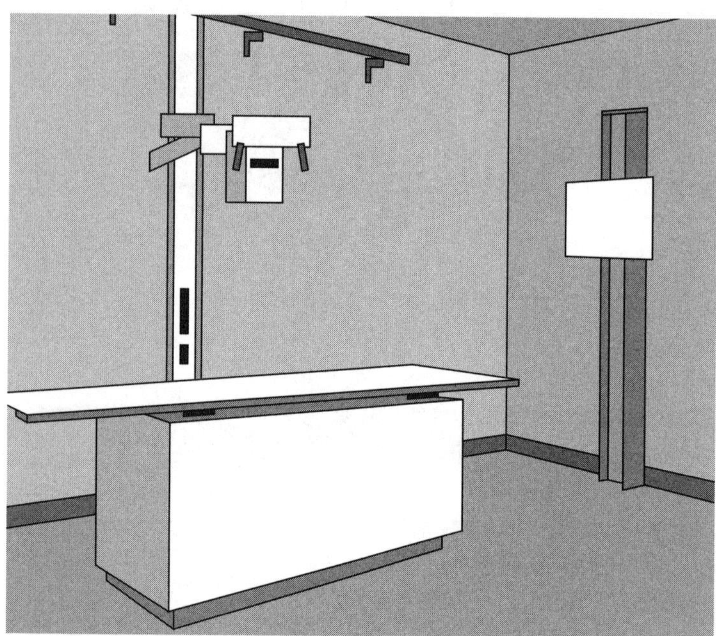

NOTE

Noncurrent-carrying metal parts of X-ray and associated equipment (controls, tables, X- ray tube supports, transformer tanks, shielded cables, X-ray tube heads, etc.) must be grounded per Article 250, as modified by 517.13(A) and (B) »517.78(C)«.

WARNING

Do not construe anything in Article 517, Part V, as specifying safeguards against the useful beam or stray X-ray radiation » Part V«. (See FPN Nos. 1 and 2 for information pertaining to safety and performance.)

Summary

- Article 517 provisions apply to health care facility electrical construction and installation.
- Many commonly used health care facility terms are defined in 517.2.
- Article 517's installation criteria and wiring methods serve to minimize electrical hazards.
- Nursing homes and limited care facilities share the same provisions, found in 517.40 through 517.44.
- Each patient bed location, in general care areas, must have at least four receptacles.
- Within critical care areas, each patient bed location requires a minimum of six receptacles.
- The required receptacles, in both general and critical care areas, can be of the single or duplex types or a combination of both.
- Receptacles located within pediatric ward patient care areas or rooms must be listed tamper resistant, or must employ a listed tamper resistant cover.
- A hospital essential electrical system is comprised of equipment system(s) and emergency system(s).
- The life safety branch and the critical branch make up a hospital's emergency system.
- An alternate power source for a hospital can be: (1) on-site generator(s) driven by some form of prime mover(s); (2) another generating unit(s) where generating unit(s) located on the premises is the normal source; or (3) an external utility service when the normal source consists of on-site generating unit(s).
- Essential electrical systems for nursing homes and limited care facilities are comprised of two separate branches: the life safety branch and the critical branch.
- The entire area surrounding any location employing flammable anesthetics is a Class I, Division 1 location, including the area extending upward from the floor to a level of 5 ft (1.52 m).
- Wiring and equipment within anesthetizing locations must comply with 517.61 requirements.
- X-ray installations must comply with 517, Part V requirements.

Unit 16 Competency Test

***NEC*® Reference** **Answer**

_________ _________ 1. X-ray low-voltage cables connecting to oil-filled units that are not completely sealed, such as transformers, condensers, oil coolers, and high-voltage switches, shall have insulation of the _____ type.

_________ _________ 2. The branches of the hospital emergency system shall be installed and connected to the alternate power source so that all functions specified herein for the emergency system shall be automatically restored to operation within _____ after interruption of the normal source.

_________ _________ 3. The equipment-grounding terminal buses of the normal and essential branch-circuit panelboards serving the same individual patient vicinity shall be bonded together with an insulated continuous copper conductor not smaller than _____.

_________ _________ 4. The _____ is a subsystem of the health care emergency system consisting of feeders and branch-circuits supplying energy to task illumination, special power circuits, and selected receptacles serving areas and functions related to patient care, and which are connected to alternate power sources by one or more transfer switches during interruption of the normal power source.

_________ _________ 5. The receptacles or the coverplates for receptacles supplied by the hospital's emergency system shall have a distinctive _____ so as to be readily identifiable.

_________ _________ 6. _____ is a jack or terminal bus that serves as the collection point for redundant grounding of electric appliances serving a patient vicinity or for grounding other items in order to eliminate electromagnetic interference problems.

NEC® Reference	Answer	
______	______	7. In any anesthetizing area, all metal raceways and metal-sheathed cables, and all noncurrent-carrying conductive portions of fixed electric equipment, shall be ______.
______	______	8. A disconnecting means of adequate capacity for at least ______ %t of the input required for the long-time rating or ______ % of the input required for the momentary rating of the X-ray equipment, whichever is greater, shall be provided in the supply circuit.
______	______	9. Each critical care area patient bed location shall be supplied by at least ______ branch-circuit(s).
______	______	10. Typically, in a patient room, the patient vicinity encloses a space within the room not less than ______ ft beyond the perimeter of the bed in its nominal location, and extending vertically not less than ______ ft above the floor.
______	______	11. The line isolation monitor shall not alarm for a fault hazard of less than ______ mA or for a total hazard current of less than ______mA.
______	______	12. Receptacles located within the patient care areas of pediatric wards, rooms, or areas shall be ______ resistant or shall employ a ______ resistant cover.
______	______	13. ______ are patient bedrooms, examining rooms, treatment rooms, clinics, and similar areas in which it is intended that the patient shall come in contact with ordinary appliances such as a nurse call system, electrical beds, examining lamps, telephone, and entertainment devices.
______	______	14. In locations where flammable anesthetics are employed, the entire area shall be considered to be a Class I, Division 1 location that shall extend upward to a level ______ ft above the floor.
______	______	15. A(n) ______ minimum separation between the service and feeder ground-fault tripping bands shall be provided in health care facilities.
______	______	16. For portable X-ray equipment connected to a 120-volt branch-circuit of ______ or less, a grounding-type attachment plug and receptacle of proper rating shall be permitted to serve as a disconnecting means.
______	______	17. The ______ is a subsystem of the health care emergency system consisting of feeders and branch-circuits, meeting the requirements of Article 700 and intended to provide adequate power needs to ensure safety to patients and personnel, and which are automatically connected to alternate power sources during interruption of the normal power source.
______	______	18. The ampacity of conductors and the rating of overcurrent protective devices shall not be less than ______ of the current rating of medical X-ray therapy equipment.
______	______	19. As a general rule, the wiring of a hospital's emergency system shall be mechanically protected by installation in ______ metal raceways, or shall be wired with Type ______ cable.
______	______	20. A(n) ______ is any area of the health care facility that has been designated to be used for the administration of any flammable inhalation anesthetic agents in the normal course of examination or treatment.
______	______	21. Each critical care area patient bed location shall be provided with a minimum of ______ receptacle(s).
______	______	22. In areas used for patient care, the grounding terminals of all receptacles and all noncurrent-carrying conductive surfaces of fixed electric equipment likely to become energized that are subject to personal contact, operating at over ______, shall be grounded by an insulated copper conductor.

NEC® Reference	Answer	
______	______	23. Wall-mounted remote-control stations for remote-control switches operating at ______ or less shall be permitted to be installed in any anesthetizing location.
______	______	24. A(n) ______ is a system comprised of alternate sources of power and all connected distribution systems and ancillary equipment, designed to ensure continuity of electrical power to designated areas and functions of a health care facility during disruption of normal power sources, and also designed to minimize disruption within the internal wiring system.
______	______	25. In critical care areas, a patient vicinity shall be permitted to have a patient equipment grounding point. An equipment-bonding jumper, not smaller than ______, shall be used to connect the grounding terminal of all grounding-type receptacles to the patient equipment grounding point.
______	______	26. Individual branch-circuits shall not be required for portable, mobile, and transportable medical X-ray equipment requiring a capacity of not over ______.
______	______	27. Each general care area patient bed location shall be provided with a minimum of ______ receptacle(s).
______	______	28. ______ areas are those special care units, intensive care units, coronary care units, angiography laboratories, cardiac catheterization laboratories, delivery rooms, operating rooms, and similar areas in which patients are intended to be subjected to invasive procedures and connected to line-operated, electromedical devices.
______	______	29. One transfer switch shall be permitted to serve one or more branches or systems in a limited care facility with a maximum demand on the essential electrical system of ______ kVA.
______	______	30. 18 or 16 AWG luminaire (fixture) wires as specified in 725.27 and flexible cords shall be permitted for the control and operating circuits of X-ray and auxiliary equipment where protected by not larger than ______ -ampere overcurrent devices.
______	______	31. A flexible cord storage device, within a hazardous (classified) anesthetizing location, shall not subject the cord to bending at a radius of less than ______ inch(es).
______	______	32. In an area used for patient care, Type MC cable and Type MI cable shall have an outer metal armor or sheath that is ______ as an acceptable grounding return path.
______	______	33. Essential electrical systems for nursing homes and limited care facilities shall be comprised of two separate branches capable of supplying a limited amount of lighting and power service, which is considered essential for the protection of life safety and effective operation of the institution during the time normal electrical service is interrupted for any reason. These two separate branches shall be the ______ branch and the ______ branch.
______	______	34. The ampacity of supply branch-circuit conductors and the current rating of overcurrent protective devices, for X-ray equipment, shall not be less than ______ % of the long-time rating or ______ % of the momentary rating, whichever is greater.
______	______	35. In critical care areas, the equipment grounding conductor for special-purpose receptacles, such as the operation of mobile X-ray equipment, shall be extended to the ______ of branch-circuits for all locations likely to be served from such receptacles.

UNIT 17

SECTION FIVE: SPECIAL OCCUPANCIES, AREAS, AND EQUIPMENT

Industrial Locations

Objectives

After studying this unit, the student should:

- be familiar with service entrance provisions with or without a single main disconnecting means.
- know that a raceway (or cable) cannot contain both service conductors and non-service conductors.
- have a good understanding of feeder tap rules.
- understand the provisions pertaining to conductors supplying a transformer.
- be familiar with transformer vault provisions.
- know how to correctly bond service raceways and equipment.
- be able to accurately connect taps to grounding electrode conductors.
- understand the different methods for sizing bonding conductors in paralleled service raceways.
- be familiar with the provision requiring a full size equipment grounding conductor (when used) in each paralleled raceway.
- have a thorough understanding of cable tray provisions.
- be familiar with both general and specific motor provisions, i.e., motor and branch-circuit overload protection, branch-circuit conductors, branch-circuit protection, feeder conductors, feeder protection, etc.
- understand specific equipment provisions for cranes and hoists, electric welders, electroplating equipment, industrial machinery, and capacitors.

Introduction

Because industrial wiring covers such a broad spectrum of electrical applications, one unit simply cannot contain all of the related provisions. As with other topics, this text's building block approach has laid the foundation for industrial locations, and has steadily built upon that foundation with layers of additional information. For instance, previous units address items pertinent to industrial projects, such as: raceways, services, clearances, busways, transformers, etc. Provisions for services and service equipment can be found in Units 9, 10, and 14, as well as here (as they apply more specifically to industrial applications).

In particular, this unit explains various feeder tap installation provisions, as found in 240.21. These include, among others: the 10- and 25-ft tap rules, plus transformer tap rules. While general transformer provisions were introduced in Unit 14, this unit expands that topic by discussing transformer vault requirements. Grounding and bonding provisions make up a part of this unit. Included are grounding electrode conductor taps, bonding service raceways, and equipment grounding conductors in paralleled raceways. Cable tray requirements are also addressed in Unit 17. Cable tray installations, of course, are not limited to industrial environments. They are often part of commercial projects, and are used in many other areas as well.

Article 430 (Motors, Motor Circuits, and Controllers) applies a great number of requirements to the installation of motors, motor branch-circuit and feeder conductors, their protection, motor overload protection, motor-control circuits, motor controllers, and motor control centers. The most commonly used of these provisions are illustrated and explained within this unit. Finally, specific equipment, typically found in industrial type locations, is presented. Such specific equipment includes: cranes and hoists, electric welders, electroplating equipment, industrial machinery, and capacitors.

The aim of this unit is to introduce, in a broad manner, the more common components of a very complex area. See Delmar's extensive list of offerings for a more in-depth study of industrial electrical wiring.

GENERAL

Service Entrance: One Main Disconnect

If an orderly shutdown is required to minimize the personnel/equipment hazard(s), a system of coordination based on the following two conditions is permitted: (1) coordinated short-circuit protection, and (2) overload indication based on monitoring systems or devices »240.12«.

Knife switches rated more than 1200 amperes at 250 volts or less, and at over 600 amperes at 251 to 600 volts, must be used only as isolating switches and must not be opened under load »404.13(A)«.

Use a circuit breaker or a switch of special design listed for such purpose to interrupt currents over 1200 amperes at 250 volts, nominal, or less; or, over 600 amperes at 251 to 600 volts, nominal »404.13(B)«.

Ⓐ The next higher standard overcurrent device rating (above the ampacity of the conductors being protected) is permitted, provided all the following conditions are met.

1. The protected conductors are not part of a multioutlet branch-circuit supplying receptacles for cord- and plug-connected portable loads.
2. The conductors' ampacity does not correspond with a fuse/circuit breaker's standard ampere rating without overload trip adjustments above its rating (but it can have other trip or rating adjustments).
3. The next higher standard rating does not exceed 800 amperes »240.4(B)«.

Ⓑ Service equipment rated no more than 600 volts must be identified by markings as suitable for service equipment »230.66«.

Ⓒ Each ungrounded circuit conductor must have overcurrent protection located at the point where the conductors receive their supply, except as specified in 240.21(A) through (G) »240.21«.

Ⓓ The number of disconnects on the load side of the service disconnecting means is not limited to six.

Ⓔ Cartridge fuse and fuseholder overcurrent protection provisions are located in 240.60 and 61.

Ⓕ All switches, and circuit breakers used as switches, must be located in a readily accessible and operable place. The center of the switch (or circuit breaker) operating handle's grip must not, in its highest position, be more than 6 ft 7 in. (2.0 m) above the floor/working platform »404.8«.

Ⓖ Comply with the appropriate dimensions from 376.23(A) where insulated conductors are deflected within a wireway, either at the ends or where conduits, fittings, or other raceways or cables enter or leave the wireway; or, where the wireway's direction deflects more than 30° »376.23(A)«.

> **NOTE**
>
> *See this book's Units 8 and 14 for electrical equipment working space clearances »110.26«.*

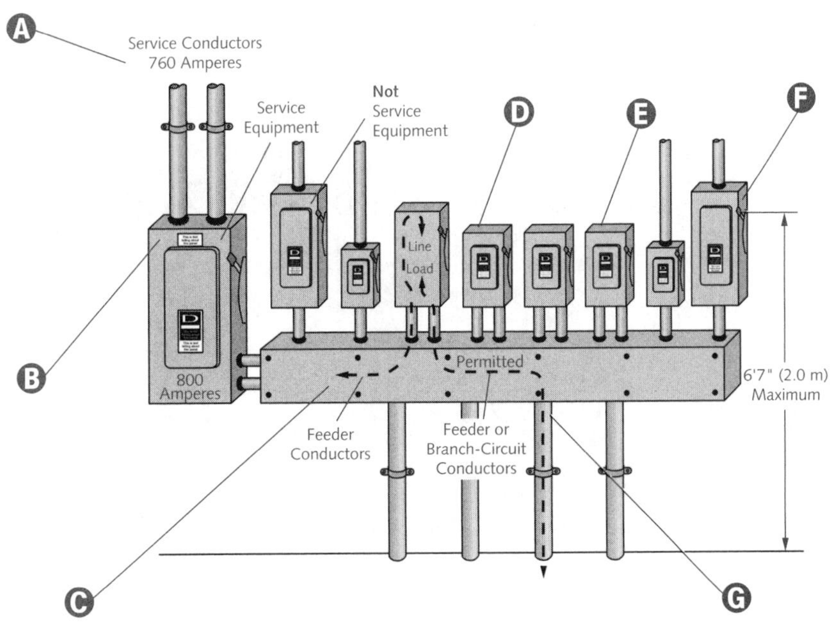

Service Entrance: Six Disconnects

Means must be provided to disconnect all conductors in a building (or other structure) from the service-entrance conductors »230.70«.

Install the service disconnecting means at a readily accessible location, either outside of a building/structure, or inside nearest the service conductors' entry point »230.70(A)(1)«.

A set of fuses is considered all the fuses required to protect all of the circuit's ungrounded conductors »230.90(A)«.

A The service disconnecting means for each service (permitted by 230.2), or for each set of service-entrance conductors (permitted by 230.40, *Exception Nos. 1, 3, 4, or 5*), must consist of no more than six switches/circuit breakers in a single enclosure, in a group of separate enclosures, or in (or on) a switchboard »230.71(A)«.

B Two to six disconnects, permitted by 230.71, must be grouped »230.72(A)«.

C Two to six circuit breakers (or sets of fuses) can serve as the overcurrent device providing overload protection. The sum of the circuit breakers'/fuses' ratings can exceed the service conductor's ampacity, provided the calculated load (according to Article 220) does not exceed the ampacity of the service conductors »230.90(A) *Exception No. 3*«.

D Each service disconnect must be permanently marked to identify it as a service disconnect »230.70(B)«.

E Where two to six separately enclosed service disconnecting means (supplying separate loads from one service drop or lateral) are grouped at one location, one set of service-entrance conductors can supply each, or several such, service equipment enclosures »230.40 *Exception No. 2*«.

F Article 376 contains metal wireway provisions.

G Service-entrance conductors can be spliced (or tapped) in accordance with 110.14, 300.5(E), 300.13, and 300.15 »230.46«.

H Only service conductors can be installed in the same service raceway or service cable »230.7«.

NOTE

Each service disconnect must simultaneously disconnect all ungrounded service conductors under its control from the premises wiring system »230.74«.

230.2 permits one (or more) additional service disconnecting means for fire pumps and legally required or optional standby services. These additional service disconnecting means must be installed at a sufficient distance from the normal (one to six) service disconnecting means in order to minimize the possibility of simultaneous interruption of supply »230.72(B)«.

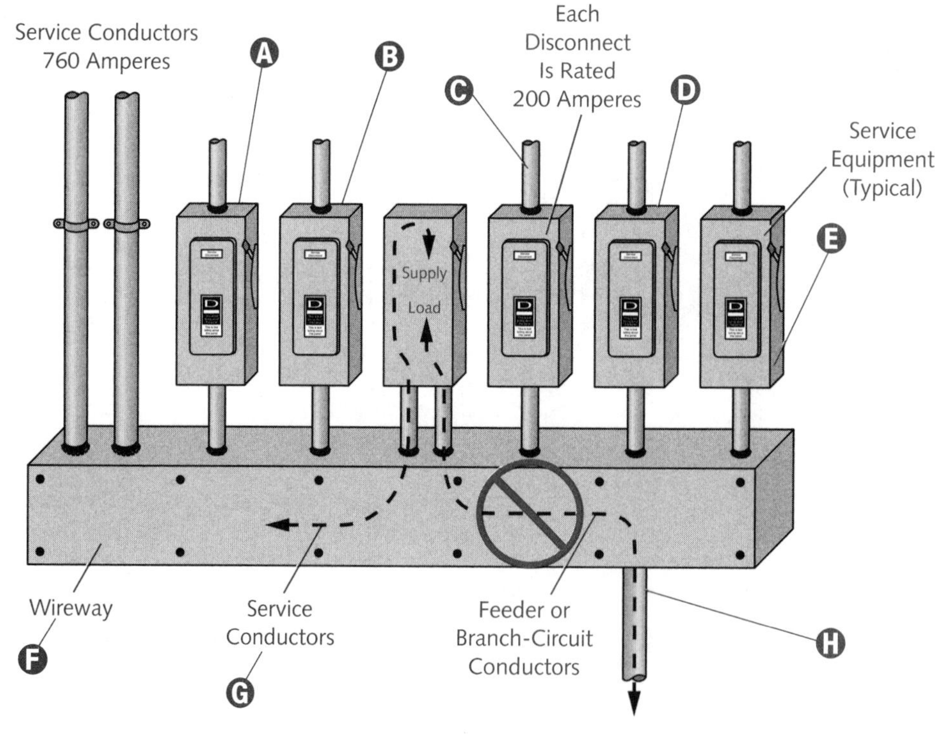

Feedor Taps

Because of dissimilar metals' characteristics, devices such as pressure terminal/splicing connectors and soldering lugs must be identified for the conductor's material, must be properly installed, and must be properly used. Conductors of dissimilar metals must not be intermixed in a terminal (or splicing) connector where physical contact occurs between dissimilar conductors (such as copper and aluminum, copper and copper-clad aluminum, etc.), unless the device is identified for the purpose and conditions of use »110.14«.

While single-phase system illustrations are used, be aware that the same rules also apply to 3-phase systems.

A Tap conductors can be protected against overcurrent in accordance with 210.19(C) and (D), 240.5(B)(2), 240.21, 368.11, 368.12, and 430.53(D) »240.4(E)«.

B Conductors can be tapped to a feeder, without having overcurrent protection at the tap, as specified in 240.21(B)(1) through(5) »240.21(B)«.

C Per Article 240, a **tap conductor** is a conductor other than a service conductor having overcurrent protection ahead of its point of supply, which exceeds the value permitted for similar conductors that are protected as otherwise described in Article 240 »240.2«.

D All conductor splices, joints, and free ends must either be covered with an insulation equivalent to that of the conductors, or with an insulating device identified for the purpose »110.14(B)«.

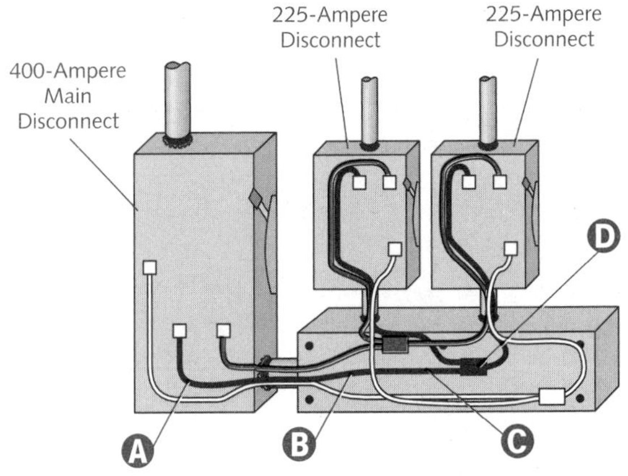

Ten-Ft (3.0-M) Feeder Tap Rule

A A 60-ampere rated conductor is the minimum size allowed if the overcurrent protection on the tap conductor's line side is 600 amperes (600 ÷ 10 = 60) »240.21(B)(1)(4)«.

B The tap conductor's ampacity cannot be less than the rating of the device supplied by the tap conductors or less than the overcurrent-protective device rating at the tap conductor's termination »240.21(B)(1)(1)(b)«.

C Lighting and appliance branch-circuit panelboards must be individually protected (on the supply side) by no more than two main circuit breakers (or two sets of fuses) having a combined rating no greater than the panelboard »408.16(A)«.

D The tap conductor's ampacity cannot be less than the combined computed loads on the circuits it supplies »240.21(B)(1)(1)(a)«.

E The tap conductors must not extend beyond the switchboard, panelboard, disconnecting means, or control devices they supply »240.21(B)(1)(2)«.

F Except where connected to the feeder, tap conductors must be enclosed in a raceway extending from the tap to the enclosure of an enclosed switchboard, panelboard, or control device, or to the back of an open switchboard »240.21(B)(1)(3)«.

G If the tap conductor's length does not exceed 10 ft (3 m), and all of 240.21(B)(1)(1) through (4) stipulations are satisfied, overcurrent protection at the tap to the feeder is not required »240.21(B)(1)«.

H Field installations with the tap conductors exiting the enclosure (or vault) where the tap is made require that the rating of the overcurrent device on the tap conductor's line side not exceed 10 times the tap conductor's ampacity »240.21(B)(1)(4)«.

> **NOTE**
>
> *210.19 specifies overcurrent protection locations for branch-circuit tap conductors meeting the Section's requirements »240.21(A)«.*

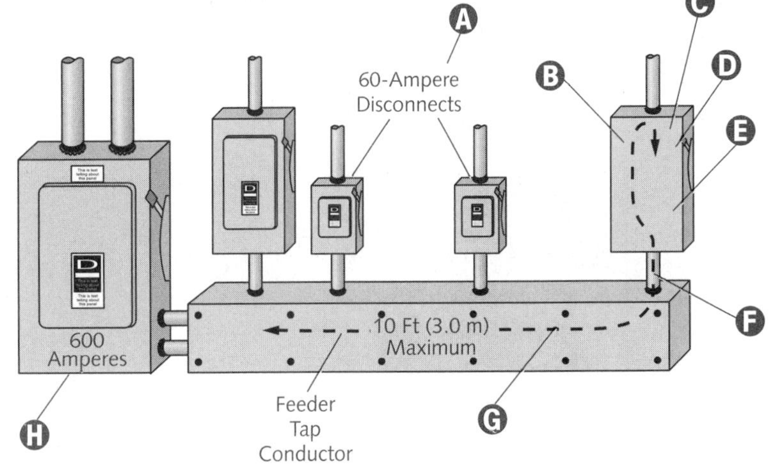

Twenty-Five Ft (7.5 M) Feeder Tap Rule

A A 200-ampere rated conductor is the minimum size allowed if the overcurrent protection on the tap conductor's line side is 600 amperes (600 ÷ 3 = 200) »240.21(B)(2)(1)«.

B The tap conductors must terminate in a single circuit breaker (or single set of fuses) that will limit the load of the tap conductor's ampacity. This device can supply unlimited additional load side overcurrent devices »240.21(B)(2)(2)«.

C The tap conductors must either be enclosed in a raceway, or otherwise suitably protected from physical damage »240.21(B)(2)(3)«.

NOTE

Tap conductors longer than 25 ft (7.5 m) must comply with 240.21(B)(4) specifications.

D If the tap conductor's length does not exceed 25 ft (7.5 m), and all of 240.21(B)(2)(1) through (3) stipulations are satisfied, overcurrent protection at the tap to the feeder is not required »240.21(B)(2)«.

E The tap conductors' ampacity must not be less than ⅓ of the feeder conductor's overcurrent device rating »240.21(B)(2)(1)«.

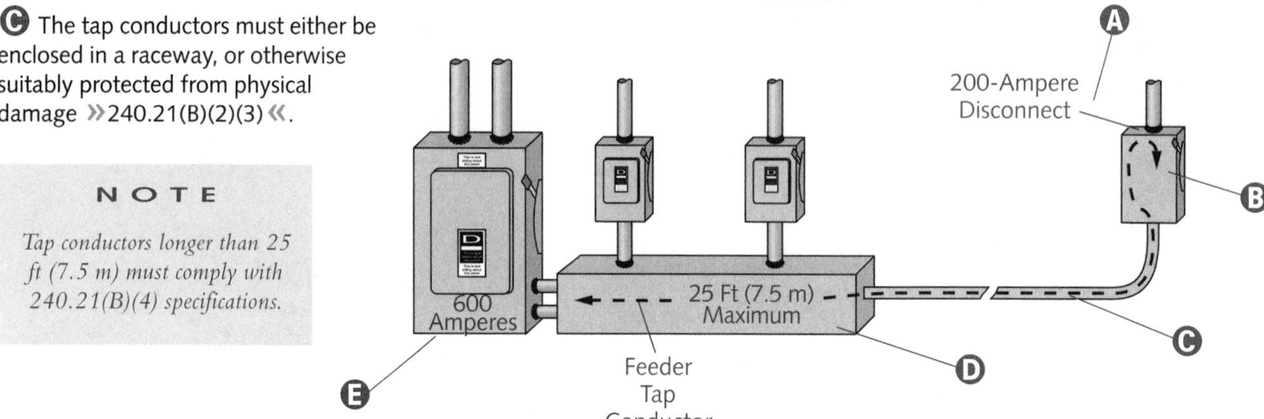

Bonding Service Raceways

A The bonding jumper must meet Table 250.66 size requirements for grounding electrode conductors. For service-entrance phase conductors larger than 1100-kcmil copper (or 1750-kcmil aluminum), the bonding jumper must have an area no less than 12½% of the largest phase conductor's area, unless the phase conductors and the bonding jumper are of different materials (copper or aluminum) »250.102(C)«.

B Here, one jumper is used to bond *all* service raceways. This bonding jumper's size is based on the total combined service-entrance phase (hot) conductors' area.

C Here, a jumper bonds *each individual* service raceway. Each bonding jumper is sized according to each individual raceway's service-entrance conductors.

D If the service-entrance conductors are paralleled in multiple raceways (or cables), and the equipment bonding jumper is routed with the raceways (or cables), it must also run in parallel. Each raceway's (or cable's) bonding jumper size must be based on the size of the service-entrance conductors it contains »250.102(C)«.

E The total phase conductors' area is 1400 kcmil (350 × 4). The bonding jumper must be at least 12½% of that area (1400 × 12.5% = 175 kcmil or 175,000 cmil). A 3/0 AWG copper conductor (with a circular mil area of 167,800) is insufficient. Therefore, a 4/0 AWG copper conductor (with a circular mil area of 211,600) meets the minimum size requirement.

F Service raceways must be effectively bonded together »250.92(A)(2)«.

G The minimum bonding jumper size for each raceway is 2 AWG copper »Table 250.66«.

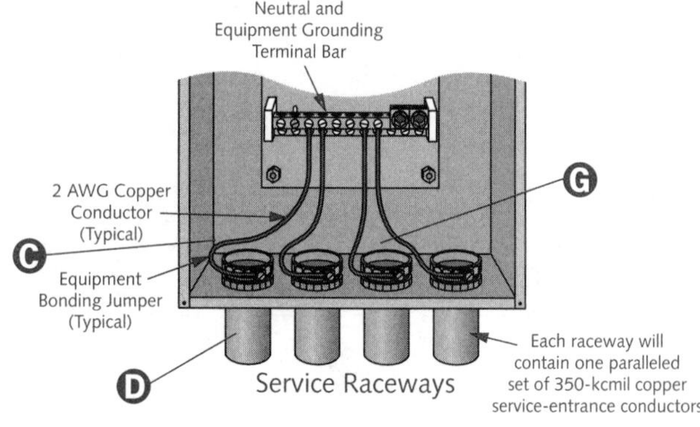

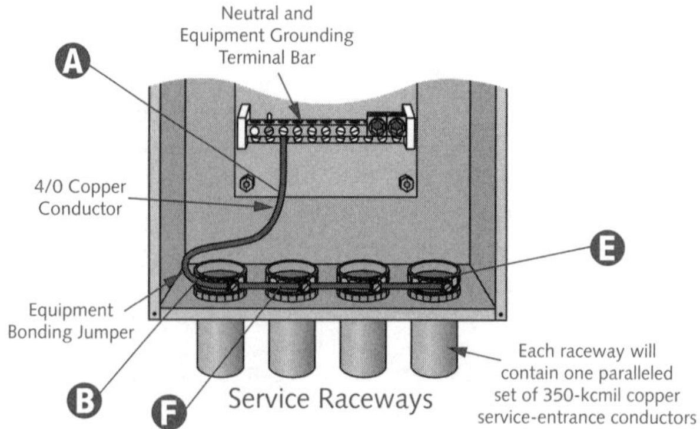

Tap Conductors Supplying a Transformer

A A transformer's primary conductors must have an ampacity at least one-third of the feeder conductors' overcurrent device rating »240.21(B)(3)(1)«.

B A 100-ampere rated conductor is the minimum size allowed if the overcurrent protection on the tap conductor's line side is 300 amperes (300 ÷ 3 = 100) »240.21(B)(3)(1)«.

C Primary and secondary conductors must be adequately protected from physical damage »240.21(B)(3)(4)«.

D Secondary conductors must terminate in a single circuit breaker or (set of fuses) that will limit the load current to no more than the conductor ampacity permitted by 310.15 »240.21(B)(3)(5)«.

E Conductors supplied by the transformer's secondary must have an ampacity that, when multiplied by the ratio of the secondary-to-primary voltage, is at least ⅓ of the feeder conductors' overcurrent device rating »240.21(B)(3)(2)«.

F The combined length of one primary plus one secondary conductor (excluding any portion of the primary conductor protected at its ampacity) must not exceed 25 ft (7.5 m) »240.21(B)(3)(3)«.

G Conductors, supplied from a set of fuses (or circuit breaker), feeding a transformer are feeder conductors, not feeder tap conductors.

H Refer to this book's Unit 14 for transformer provisions (overcurrent protection, grounding, etc.).

I Transformer secondary conductor provisions are found in 240.21(C).

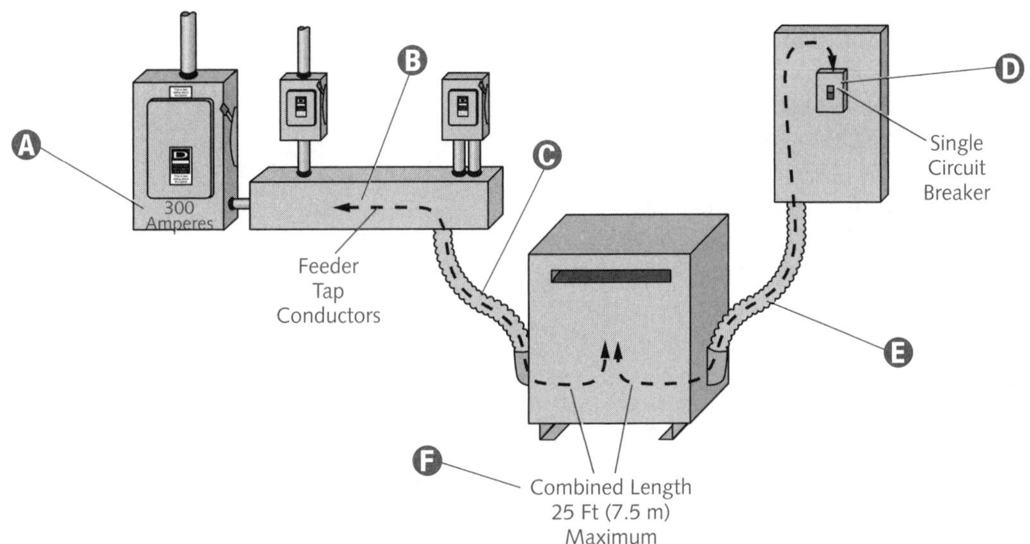

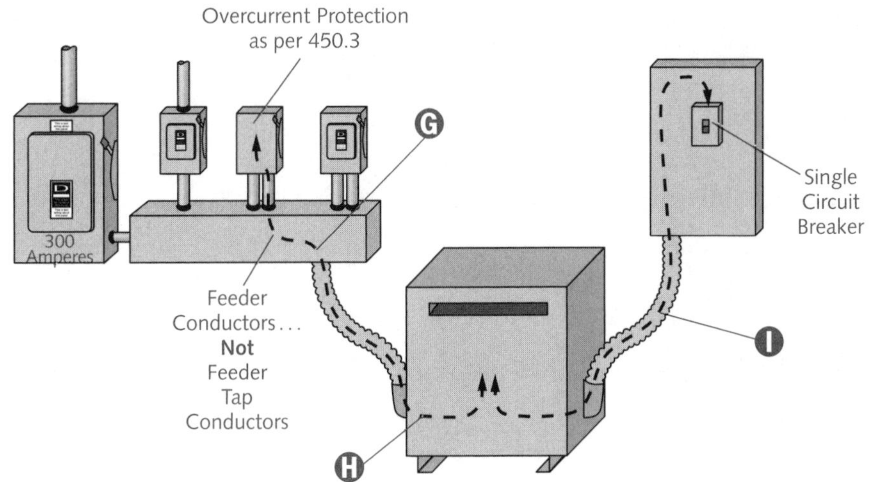

Transformer Vaults

Wherever practicable, transformers must be located so that ventilation to outside air does not rely on flues or ducts »450.41«.

Where required by 450.9, openings for ventilation must also comply with 450.45(A) through (F) »450.45«. That section includes provisions pertaining to location, arrangement, size, covering, dampers, and ducts.

Each doorway from the building interior into the vault must have a tight-fitting door with a minimum fire rating of 3 hours, unless the exception is met. The AHJ may require such a door for exterior wall openings, should conditions warrant »450.43(A)«.

Doors must be equipped with fully engaged locks, thereby limiting access to qualified persons only »450.43(C)«.

Personnel doors must swing out and be equipped with panic bars, pressure plates, or other devices that are normally latched but which open under simple pressure »450.43(C)«.

Vaults containing more than 100-kVA transformer capacity must have a drain or other means to carry off any oil or water accumulation, unless local conditions make this impracticable. The vault floor must pitch toward the drain, where provided »450.46«.

Non-electrical pipe or duct systems must neither enter nor pass through a transformer vault. Piping or other facilities providing vault fire protection, or transformer cooling, are allowed »450.47«.

Ⓐ Vault walls and roofs must be constructed of materials having adequate structural strength for the conditions and a minimum fire resistance of 3 hours. For this section's purpose, studs and wallboard construction are not acceptable »450.42«.

Ⓑ Transformers protected by automatic sprinkler, water spray, carbon dioxide, or halon systems, and 1-hour rated construction, are permitted »450.42 *Exception* and 450.43(A) *Exception*«.

Ⓒ A door sill (or curb) of a sufficient height, at least 4 in. (100 mm) to confine oil from the largest transformer within the vault must be provided »450.43(B)«.

Ⓓ Vault floors in contact with the earth must be of concrete, no less than 4 in. (100 mm) thick. Where the vault is constructed above a vacant space (or other stories), the floor must adequately support the load imposed thereon, and must have a minimum fire resistance of 3 hours »450.42«.

NOTE

Entrances to all buildings, rooms, or enclosures containing exposed live parts or exposed conductors operating at over 600 volts, nominal, must be kept locked, unless such entrances are constantly observed by qualified personnel. Where the voltage exceeds 600 volts, nominal, permanent and conspicuous warning signs must be posted, reading as follows: DANGER—HIGH VOLTAGE — KEEP OUT »*110.34(C)*«.

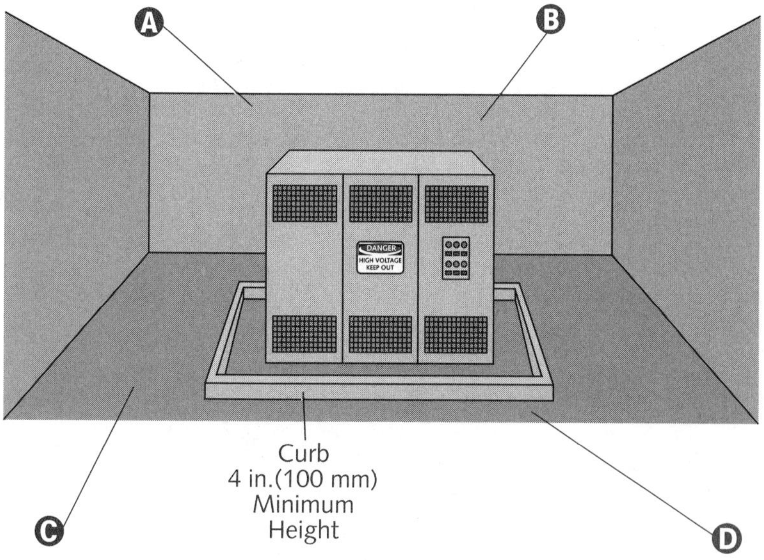

CAUTION *Transformer vaults must not be used for material storage »450.48«.*

Grounding Electrode Conductor Taps

The bonding jumper must meet Table 250.66 size requirements for grounding electrode conductors. If the service-entrance phase conductors are larger than 1100-kcmil copper (or 1750-kcmil aluminum), the bonding jumper must have an area not less than 12½% of the largest phase conductor's area, unless the phase conductors and the bonding jumper are of different materials (copper or aluminum) »250.102(C)«.

If the service-entrance conductors are paralleled in multiple raceways (or cables), and the equipment bonding jumper is routed with the raceways (or cables), it must also run in parallel. Base the bonding jumper size for each raceway (or cable) on the size of its corresponding service-entrance conductors »250.102(C)«.

Ⓐ Taps can connect the grounding electrode conductors in a service consisting of more than a single enclosure as permitted in 230.40, *Exception No. 2*. In such a configuration, the tap conductor must extend to the inside of the enclosure. The grounding electrode conductor must be sized according to 250.66, while the tap conductors can be sized in accordance with the grounding electrode conductors (also 250.66) for the largest conductor serving their respective enclosures »250.64(D)«.

Ⓑ Service enclosures containing service conductors (including meter fittings, boxes, or the like) that are interposed in the service raceway (or armor) must be effectively bonded together »250.92(A)(2)«.

Ⓒ Service raceways must be effectively bonded together »250.92(A)(1)«.

Ⓓ The grounding electrode conductor must be of one continuous length (without splice or joint), unless spliced by exclusive use of irreversible compression-type connections listed for the purpose, or by exothermic welding »250.64(C)«.

Ⓔ Any metallic raceway (or armor) that encloses a grounding electrode conductor, as specified in 250.64(B), must be effectively bonded together. Bonding applies at each end as well as to all intervening raceways, boxes, and enclosures between the service equipment and the grounding electrode »250.92(A)(3)«.

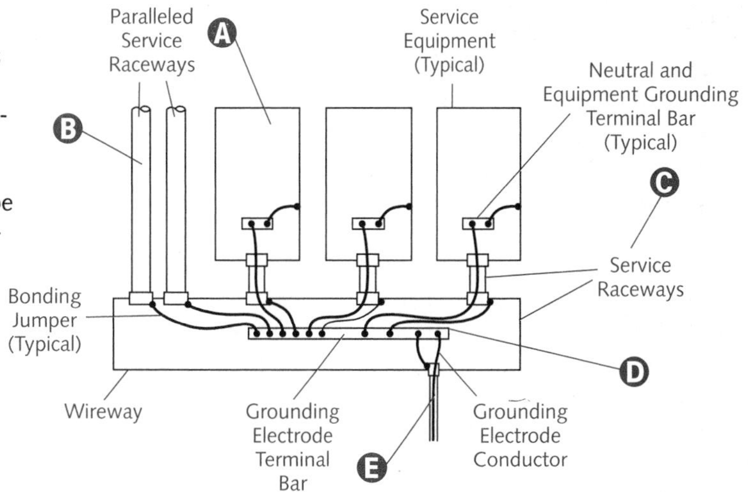

Equipment Grounding Conductors

If a single equipment grounding conductor is run with multiple circuits within the same raceway (or cable), it must be sized for the largest overcurrent device protecting the raceway's (or cable's) conductors »250.122(C)«.

While copper, aluminum, or copper-clad aluminum equipment grounding conductors of the wire type must not be smaller than shown in Table 250.122, they do not have to be larger than the circuit conductors supplying the equipment »250.122(A)«.

If a raceway or cable armor (or sheath) serves as the equipment grounding conductor, as provided in 250.118 and 250.134(A), it must comply with 250.4(A)(5) or 250.4(B)(4) »250.122(A)«.

Where ground-fault protection of equipment is installed, each parallel equipment grounding conductor in a multiconductor cable can be sized per Table 250.122 on the basis of the ground-fault protection's trip rating, provided 250.122(F)(2)(1) through (3) conditions are met »250.122(F)(2)«.

Ⓐ Since the overcurrent protection has a rating of 1200 amperes, the minimum size equipment grounding conductor is 3/0 AWG copper (or 250-kcmil aluminum) »Table 250.122«.

Ⓑ **Each** parallel equipment grounding conductor's size is based on the overcurrent device's ampere rating protecting the circuit conductors in the raceway (or cable) per Table 250.122 »250.122(F)(1)«.

Ⓒ Equipment grounding conductor provisions are not limited to feeders. For example, an alternative drawing could have shown one piece of equipment supplied by a paralleled set of branch-circuit conductors. Equipment grounding conductor provisions also apply to branch-circuits.

NOTE

If conductor size is adjusted to compensate for voltage drop, any equipment grounding conductors installed must be proportionately adjusted according to circular mil area »250.122(B)«.

If a single raceway supplies the equipment, only one 3/0 AWG copper (or 250-kcmil aluminum) equipment grounding conductor is required.

Equipment Grounding Conductors *(continued)*

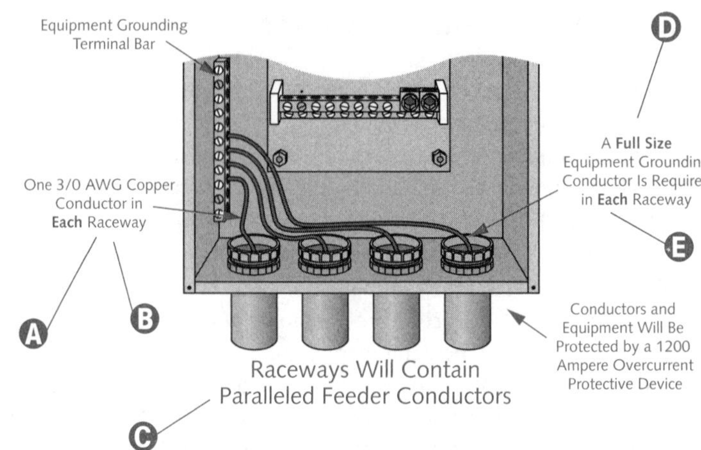

D Unlike service-equipment bonding jumpers, which can be sized according to each paralleled raceway's service-entrance conductor size, a **full size** equipment grounding conductor (when used) must be installed in **each** raceway.

E For paralleled conductors in multiple raceways (or cables) as permitted by 310.4, any equipment grounding conductors installed must also run in parallel within **each** raceway. Use one of the methods in 250.122(F) to ensure protection of the equipment grounding conductors »250.122(F)«.

Cable Trays

Cable tray installations are not limited to industrial establishments. 392.3(A) listed materials can be installed in cable tray systems, under the conditions described in their respective articles and sections »392.3«.

Only in industrial establishments, where maintenance and supervision conditions restrict cable tray system service to qualified personnel, can any of the cables in 392.3(B)(1) and (2) be installed in ladder, ventilated troughs, solid bottom, or ventilated channel cable trays »392.3(B)«.

Cable tray systems must be corrosion resistant. If made of ferrous material, the system must be protected from corrosion per 300.6 »392.5(C)«.

Steel or aluminum cable tray systems can be used as equipment grounding conductors provided all 392.7(B)(1) through (4) conditions are met.

While provisions pertaining to the number of multiconductor cables (rated 2000 volts or less) can be found in 392.9, 392.10 contains single conductor cable provisions.

A Cable trays can extend transversely through partitions and walls, or vertically through platforms and floors, in either wet or dry locations where the completed installation meets 300.21 requirements »392.6(G)«.

B Sufficient space must be provided and maintained about cable trays to permit adequate access for cable installation and maintenance »392.6(I)«.

C Cable trays must be of adequate strength and rigidity to provide support for all contained wiring »392.5(A)«.

D A cable tray system is a unit (or assembly of units or sections) and associated fittings that together form a structural system used to securely fasten/support cables and raceways »392.2«.

E Cable trays must be exposed and accessible except as permitted by 392.6(G) »392.6(H)«.

F Cable trays must include fittings, or other suitable means, to facilitate changes in direction and elevation »392.5(E)«.

G Cable trays must have side rails or similar structural members »392.5(D)«.

H Metallic cable trays can serve as equipment grounding conductors where: (1) continuous supervision ensures that only qualified persons will maintain the cable tray system, and (2) the cable tray complies with 392.7 provisions »392.3(C)«.

I Each cable tray run must be completed before cables are installed »392.6(B)«.

NOTE

Cable tray systems must not be used in hoistways, where subject to severe physical damage, or in environmental airspaces, except as permitted in 300.22, to support appropriate wiring methods »392.4«.

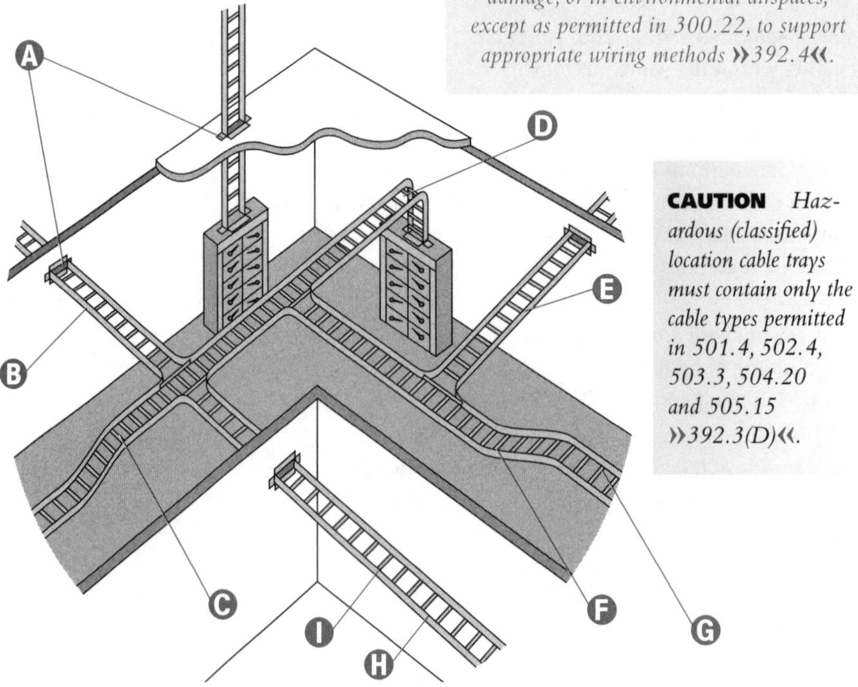

CAUTION *Hazardous (classified) location cable trays must contain only the cable types permitted in 501.4, 502.4, 503.3, 504.20 and 505.15 »392.3(D)«.*

MOTORS

General Motor Provisions

Article 430 covers motors and all of the following as they relate specifically to motors: branch-circuit and feeder conductors and their protection, overload protection, control circuits, controllers, and motor control centers »430.1«.

Wires passing through an opening in an enclosure, conduit box, or barrier require a bushing to protect the conductors from any sharp opening edges »430.13«.

Ⓐ Article 430, Part V specifies protective devices designed to protect feeder conductors supplying motors against overcurrents due to short circuits or grounds »430.61«.

Ⓑ 430.24 through 26 contain provisions for conductors supplying multiple motors, or a motor(s) and other load(s).These provisions are applicable to more than feeder conductors.

Ⓒ Article 430, Part IV lists devices that are used to protect motor branch-circuit conductors, motor-control apparatus, and motors against overcurrent due to short circuits or grounds. The rules add to, or amend, Article 240 provisions »430.51«.

Ⓓ Article 430, Part II specifies ampacities of conductors capable of carrying the motor current without overheating, under specified conditions »430.21«.

Ⓔ Article 430, Part III specifies overload devices intended to protect motors, motor-control apparatus, and motor branch-circuit conductors against excessive heating caused by motor overloads and failure to start. Electrical apparatus overload is an operating overcurrent, that when of sufficient duration, can result in damage or dangerous overheating of the apparatus. Short circuits or ground faults are not included »431.31«.

Ⓕ General motor provisions are found in 430.1 through 430.18 (Part I).

Ⓖ Part XII of Article 430 addresses the grounding of exposed noncurrent-carrying metal parts of motor and controller frames which are likely to become energized in order to prevent a voltage above ground should accidental contact between energized parts and frames occur. Insulating, isolating, or guarding are suitable alternatives to motor grounding under certain conditions »430.141«.

Ⓗ For general motor applications, base current ratings on 430.6(A)(1) and (2). (1) Use values given in Tables 430.147 through 430.150 (including notes) to determine conductor ampacity or ampere ratings of switches, branch-circuit short-circuit, and ground-fault protection. Do not use the actual current ratings marked on the motor nameplate, unless an exception applies. (2) Separate motor overload protection is based on the motor nameplate current rating »430.(6)(A)«.

Ⓘ Motor disconnecting means provisions are found in 430.101 through 430.113 (Part IX).

NOTE

Motor tables are located in Part XIII.

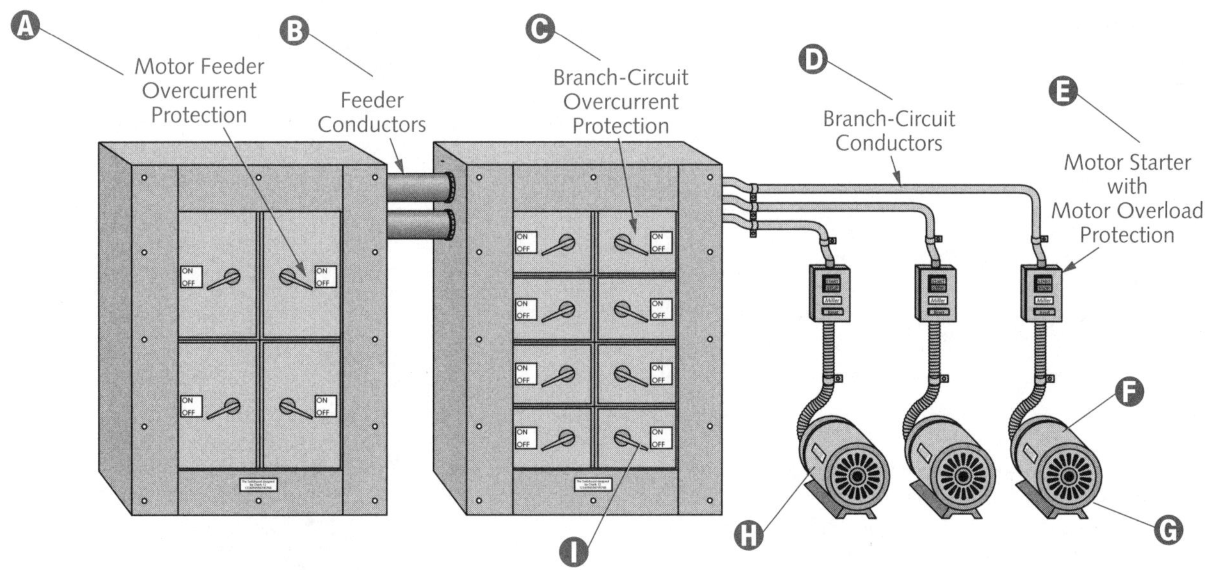

Motor and Branch-Circuit Overload Protection

Each continuous-duty motor, rated 1 hp or less, *not* permanently installed, nonautomatically started, and within sight from the controller, can be protected against overloads by the branch-circuit short-circuit and ground-fault device. Comply with branch-circuit overcurrent protection size specifications found in Article 430, Part IV »430.32(D)«.

Any automatically started motor, 1 hp or less, must be protected against overload by one of the methods found in 430.32(B)(1) through (4).

A motor controller can serve as an overload device if the number of overload units complies with Table 430.37, and, if these units are operative in both the starting and running position (dc motor), or in the running position (ac motor) »430.39«.

Overload relays, and other devices for motor overload protection incapable of opening short circuits or ground faults, must be protected by fuses (or circuit breakers) rated or set according to 430.52, or by a motor short-circuit protector per 430.52 »430.40«.

Overload protection for motors on general-purpose branch-circuits (permitted by Article 210) is required and must meet 430.42(A), (B), (C), or (D) specifications.

A As an approved method of protecting a motor against overload, 430.32(A)(1) lists a separate overload device, responsive to motor current. This device must be selected to trip, or must be rated at no more than the percentage of the motor nameplate full-load current rating shown below:

Motors with a marked service factor not less than 1.15 .... 125%

Motors with a marked temperature rise not over 40°C 125%

All other motors 115%

NOTE

Any motor application must be considered continuous duty unless the driven apparatus, by nature, means the motor cannot, under any conditions, operate continuously with the load »430.33«.

WARNING

430, Part III provisions *do not* require overload protection where it might introduce or increase hazards, as with fire pumps »430.31«. For fire pump supply conductor protection, see 695.6.

B Controllers must be marked according to 430.8 provisions.

C Article 430, Part III lists overload devices intended to protect motors, motor-control apparatus, and motor branch-circuit conductors against excessive heating due to motor overloads and failure to start »430.31«.

D Separate motor overload protection must be based on the actual motor nameplate current rating, not on the ratings listed in Tables 430.147 through 150 »430.6(A)(2)«.

E Motors (in usual applications) must be marked with the information listed in 430.7(A)(1) through (14).

F Each continuous-duty motor rated more than 1 hp must be protected against overload by a means listed in 430.32(A)(1) through (4).

G Overload in electrical apparatus is an operating overcurrent that, when of sufficient duration, causes damage or dangerous overheating of the involved apparatus. Short circuits or ground faults are not included »430.31«.

CAUTION *A motor overload device, which restarts a motor automatically after overload tripping, must not be installed if automatic restarting can result in injury to persons »430.43«.*

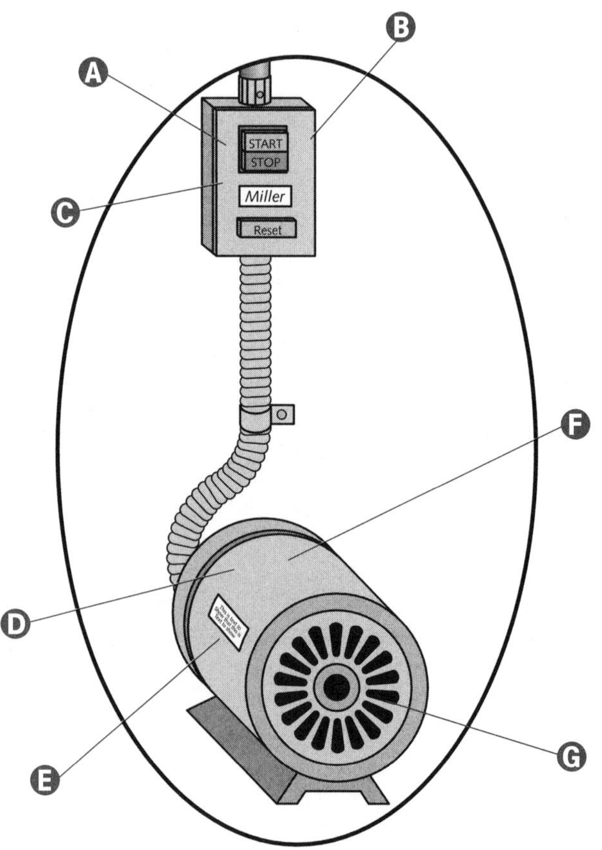

Motors on General-Purpose Branch-Circuits

Where a motor without individual overload protection [430.42(A)], is connected to a branch-circuit by means of an attachment plug and receptacle, the attachment plug and receptacle rating must not exceed 15 amperes at either 125 or 250 volts. Where individual overload protection is required per 430.42(B) for a motor or motor-operated appliance attached to a branch-circuit through an attachment plug and receptacle, the overload device must be an integral part of the motor/appliance. The rating of the attachment plug and receptacle determines the rating of the circuit to which the motor is connected, as provided in Article 210 » 430.42(C)«.

The branch-circuit short-circuit and ground-fault protective device of a circuit to which a motor (or motor-operated appliance) is connected must have a time delay sufficient to start the motor and allow it to accelerate » 430.42(D)«.

A One or more motors, without individual overload protection, can be connected to a general-purpose branch-circuit only if the installation complies with the limiting conditions of 430.32(B) and (D) and 430.53(A)(1) and (A)(2) » 430.42(A)«.

B Motors used on general-purpose branch-circuits (as permitted in Article 210) require overload protection per 430.42(A), (B), (C), or (D) specifications » 430.42«.

C Motors, of larger ratings than specified in 430.53(A), can be connected to general-purpose branch-circuits only if each motor has overload protection specifically listed for that motor as specified in 430.32. In the case of more than one motor, both the controller and the motor overload device must be approved for group installation, and the short-circuit and ground-fault protective device must be selected according to 430.53 » 430.42(B)«.

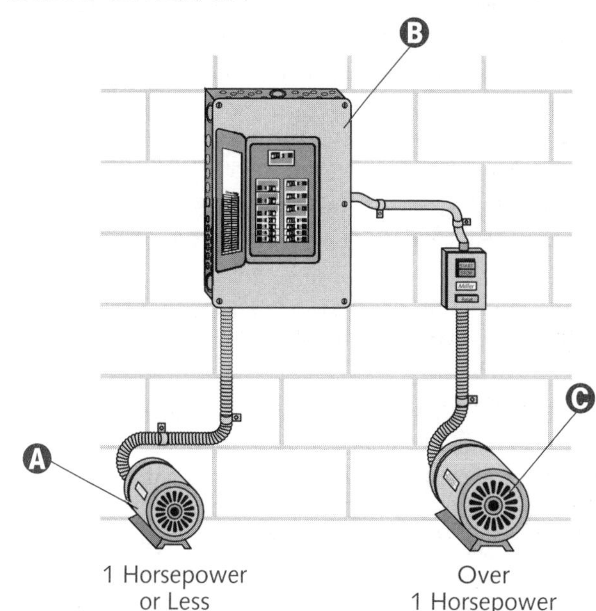

Motor Branch-Circuit Conductors

Conductors for motors used in short-time, intermittent, periodic, or varying duty applications must have an ampacity of no less than the percentage of the motor nameplate current rating shown in Table 430.22(E), unless the AHJ grants special permission for conductors of lower ampacity » 430.22(E)«.

Where motor circuits contain capacitors, conductors must comply with 460.8 and 460.9 » 430.27«.

For multispeed motors, base the selection of branch-circuit conductors (on the line side of the controller) on the current rating of the winding(s) energized by the conductors » 430.22(B)«.

A Branch-circuit conductors, supplying a single continuous-duty motor, must have an ampacity of at least 125% of the motor's full load current rating as determined by 430.6(A)(1) » 430.22(A)«.

B Annex D contains a motor application example (Example No. D8).

C Placement of motors must allow adequate ventilation and facilitate maintenance, such as bearing lubrication and brush replacement » 430.14(A)«.

NOTE

Conductors supplying several motors, or a motor(s) and other load(s), must comply with 430.24.

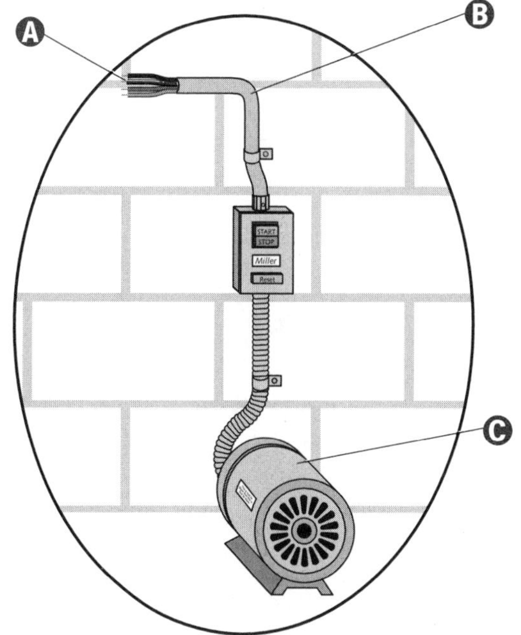

Motor Branch-Circuit Short-Circuit and Ground-Fault Protection

Part IV specifies protective devices for motor branch-circuit conductors, motor control apparatus, and motors against overcurrent due to short circuits or grounds »430.51«.

If the rating specified in Table 430.52, as modified by Exception No. 1, is not sufficient for the motor's starting current, apply 430.52(C)(1) *Exception No. 2* (a) through (d) provisions »430.52(C)(1) *Exception No. 2*«.

Suitable fuses are permitted in lieu of devices listed in Table 430.52 for power electronic devices in a solid-state motor controller system, provided replacement fuse markings are adjacent to the fuses »430.52(C)(5)«.

A listed self-protected combination controller may be substituted for the devices specified in Table 430.52 »430.52(C)(6)«.

Torque motor branch-circuits must be protected at the motor nameplate current rating according to 240.3(B) »430.52(D)«.

The branch-circuit short-circuit and ground-fault protective device rating, for multimotor and combination-load equipment, must not exceed the equipment's marked rating in accordance with 430.7(D) »430.54«.

Where fuses are used for motor branch-circuit short-circuit and ground-fault protection, the fuseholders must be of adequate size to accommodate the fuses specified by Table 430.52 »430.57«. (An exception is provided for fuses having a time delay appropriate for the motor's starting characteristics.)

A circuit breaker for motor branch-circuit short-circuit and ground-fault protection must have a current rating as determined by 430.52 and 430.110 »430.58«.

CAUTION

If the maximum branch-circuit short-circuit and ground-fault protective device ratings are shown in the manufacturer's overload relay table (for use with a motor controller), or are otherwise marked on the equipment, they must not be exceeded even if 430.52(C)(1) allows higher values »430.52(C)(2)«.

A Should the values for branch-circuit short-circuit and ground-fault protective devices (determined by Table 430.52) not correspond to a standard size or rating of fuses, nonadjustable circuit breakers, thermal protective devices, or possible settings of adjustable circuit breakers, then the next higher standard size, rating, or possible setting is permitted. This rating or setting must not exceed the next higher standard ampere rating »430.52(C)(1) *Exception No.1*«.

B Use a protective device whose rating (or setting) does not exceed the value calculated according to Table 430.52 »430.52(C)(1)«.

C The motor branch-circuit short-circuit and ground-fault protective device must comply with 430.52(B) and either (C) or (D), as applicable »430.52(A)«.

D Branch-circuit protective devices must comply with 240.20 provisions »430.56«.

NOTE

The motor branch-circuit short-circuit and ground-fault protective device must be able to carry the motor's starting current »430.52(B)«.

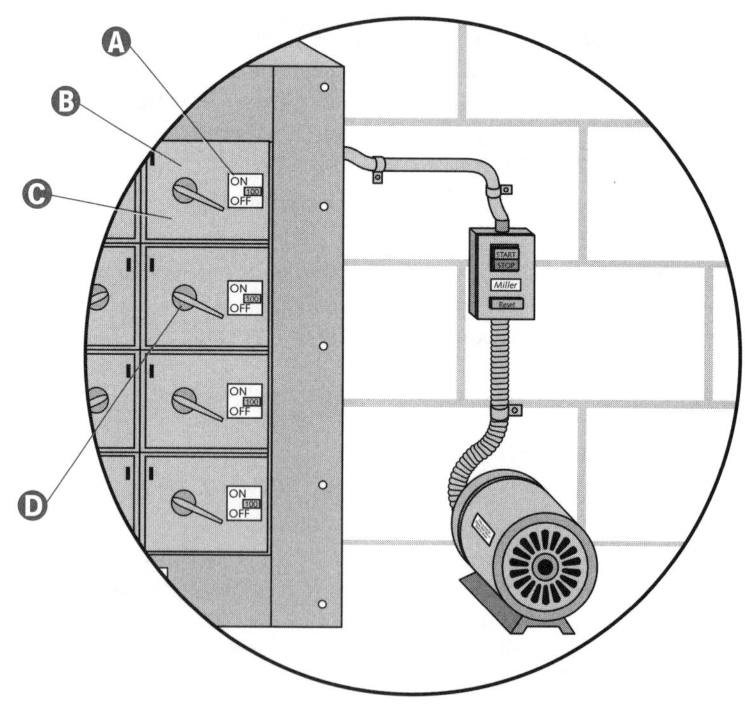

Motor Disconnecting Means

The disconnecting means must open all ungrounded supply conductors, and must be designed so that no pole operates independently. The disconnecting means can occupy the same enclosure as the controller »430.103«.

The disconnecting means must plainly indicate whether it is in the open (off) or closed (on) position »430.104«.

Motor-control circuits' arrangement must accommodate disconnection from all supply sources when the disconnecting means is in the open position »430.74(A)«.

Multiple motors used together, or one or more motors used in combination with other loads, must comply with 430.110(C) provisions.

A switch (or circuit breaker) can serve as both controller and disconnecting means if it complies with 430.111(A) and is one of a type listed in 430.111(B) »430.111«.

Every motor-circuit disconnecting means, between the feeder's point of attachment and the motor's connection point, must comply with the requirements of 430.109 and 430.110 »430.108«.

The disconnecting means must be a type specified in 430.109(A), unless otherwise permitted in (B) through (G), under the conditions given »430.109«.

Ⓐ Motor circuits rated 600 volts, nominal, or less, must have a disconnecting means with an ampere rating of at least 115% of the motor's full-load current rating »430.110(A)«.

Ⓑ According to Article 430, a controller is any switch (or device) normally used to start and stop a motor by making and breaking motor circuit current »430.81(A)«.

Ⓒ An individual, fully-functional disconnecting means must be provided for each controller. The disconnecting means must be located in sight from the controller, unless an exception is met »430.102(A)«.

Ⓓ A separate disconnecting means must be located in sight from the motor and driven machinery location »430.102(B)«.

Ⓔ Motor branch-circuit and ground-fault protection, and motor overload protection, can be combined into a single device, where the device's rating (or setting) provides the overload protection required by 430.32 »430.55«.

Ⓕ Article 430, Part IX requires a disconnecting means capable of breaking the connection between the motors/controllers and the circuit »430.101«.

Ⓖ The phrase "in sight from" indicates that specified items of equipment are visible and are no more than 50 ft (15 m) apart »Article 100«.

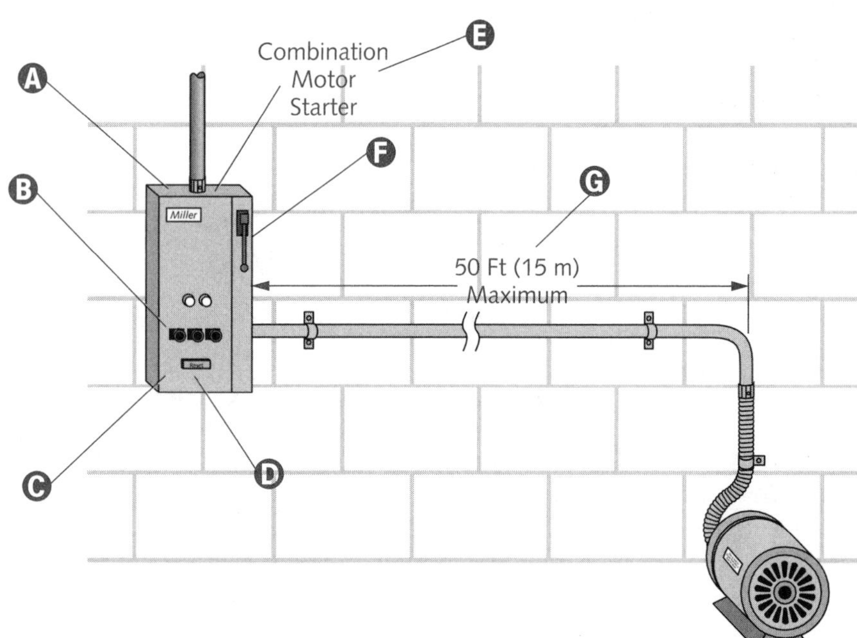

NOTE

At least one of the disconnecting means must be ***readily*** *accessible »430.107«.*

Motor Control Centers

At motor control center terminals, minimum wire bending space as well as minimum gutter space must comply with Article 312 »430.97(C)«.

Any motor control center used as service equipment must be provided with a single main disconnecting means to disconnect all ungrounded service conductors »430.95«.

Spacing between motor control center bus terminals and other bare metal parts must meet Table 430.97 specifications »430.97(D)«.

Barriers are required in all service-entrance motor control centers to isolate service busbars and terminals from the remainder of the motor control center »430.97(E)«.

Motor control centers must be marked according to 110.21, so that the markings remain plainly visible after installation. Common power bus current rating and motor control center short-circuit rating must be included in the marking »430.98(A)«.

(A) Article 430, Part VIII covers motor control centers installed for the control of motors, lighting, and power circuits »430.92«.

(B) Multisection motor control centers must be bonded together with an equipment grounding conductor, or an equivalent grounding bus, sized according to Table 250.122. Equipment grounding conductors must either terminate on the grounding bus, or to a grounding termination point within a single-section motor control center »430.96«.

(C) Motor control centers must have overcurrent protection based on the rating of the common power bus, in accordance with Parts I, II, and IX of Article 240. This protection must be provided by: (1) an overcurrent protective device located ahead of the motor control center, or (2) a main overcurrent protective device in the motor control center »430.94«.

(D) Horizontal travel through vertical sections is acceptable where a barrier isolates the conductors from the busbars »430.97(A) *Exception*«.

(E) Secure busbars firmly in place, and provide adequate protection from physical damage. Other than those required for interconnections and control wiring, only the conductors that terminate can be located within that vertical section »430.97(A)«.

NOTE

Motor control units within a motor control center must comply with 430.8 »430.98(B)«.

Motor control circuit installations must comply with Article 430, Part VI (430.71 through 74).

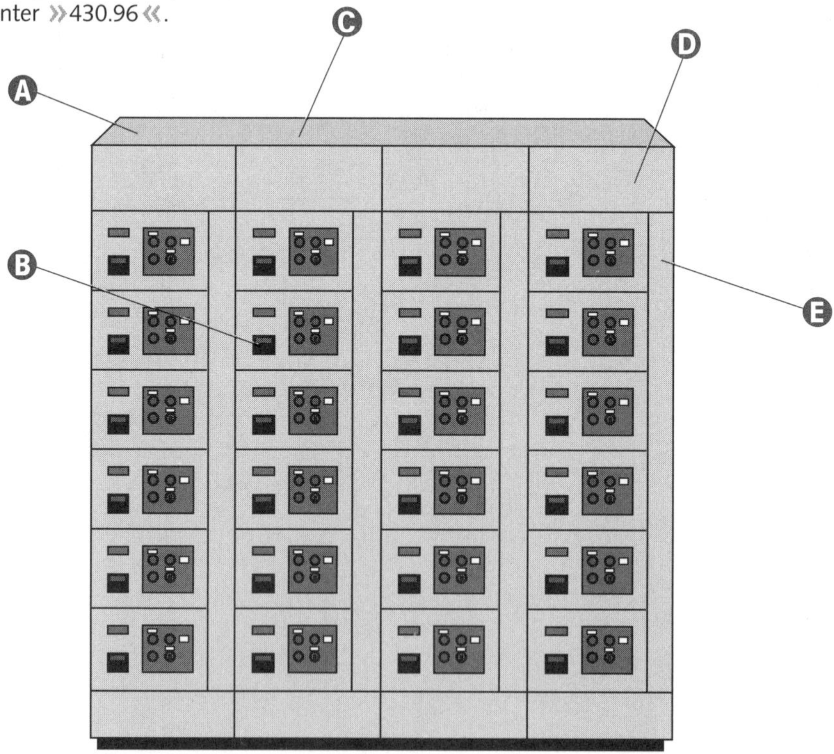

Feeder Conductors

Where one or more of the group's motors are used for short-time, intermittent, periodic, or varying duty, the ampere rating of such motors (used in the calculation) is determined by 430.22(E). For the highest rated motor, the greater of: (1) the ampere rating from 430.22(E), or (2) the largest continuous duty motor full-load current multiplied by 1.25, is used ›› 430.24 *Exception No. 1* ‹‹.

The ampacity of conductors supplying motor-operated fixed electric space-heating equipment must conform to 424.3(B) specifications ›› 430.24 *Exception No. 2* ‹‹.

Where interlocked circuitry prevents operation of selected motors (or other loads) at the same time, the conductor ampacity can be based on the highest possible total of motor and other load currents to be operated at the same time ›› 430.24 *Exception No. 3* ‹‹.

Feeder tap conductors must comply with 430.28 provisions.

Ⓐ Conductors supplying several motors, or a motor(s) and other load(s), must have an ampacity not less than 125% of the full-load current rating of the highest rated motor, plus the ampere ratings of all other motors in the group, per 430.6(A). Also include the ampacity required for other loads, if any. ›› 430.24 ‹‹. In other words, multiply the largest motor full-load current (FLC) rating by 125% and add the FLC ratings of all other motors (or loads) in the group.

Ⓑ Where reduced heating of the conductors results from duty-cycle operation, intermittent operation, or from all motors not operating at one time, the AHJ may allow a lower feeder-conductor ampacity than that specified in 430.24, provided the conductors' ampacity is sufficient for the maximum load determined according to the size and number of motors supplied and the character of their respective loads and duties ›› 430.26 ‹‹.

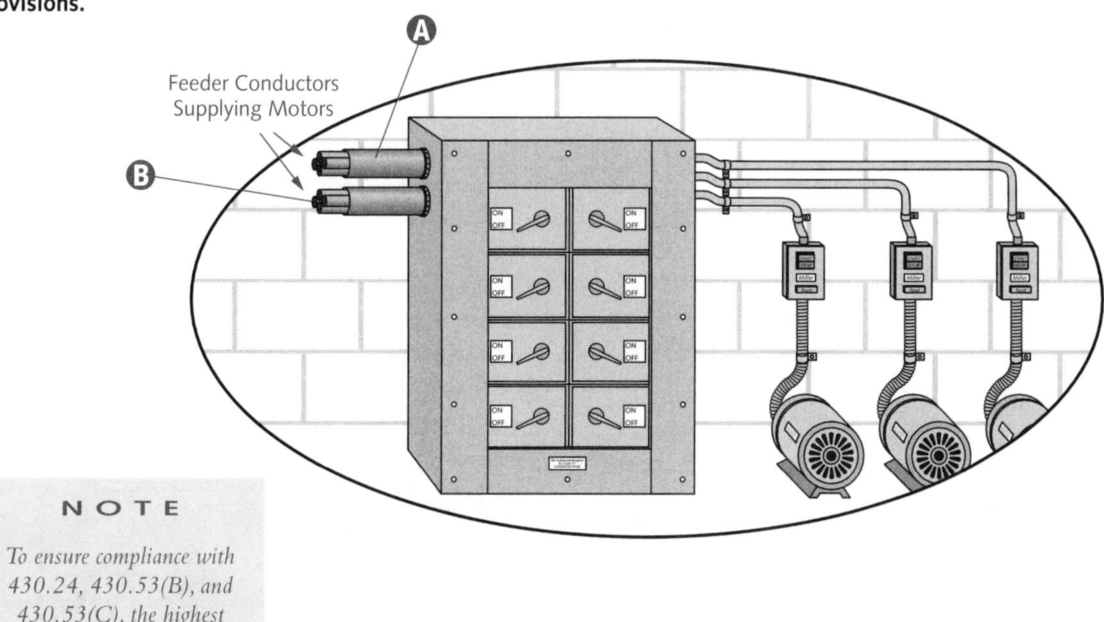

NOTE

To ensure compliance with 430.24, 430.53(B), and 430.53(C), the highest rated, or smallest rated, motor must be based on the rated FLC selected from Tables 430.147, 430.148, 430.149, and 430.150 ›› 430.17 ‹‹.

Motor Feeder Short-Circuit and Ground-Fault Protection

Where instantaneous trip circuit breaker(s), or motor short-circuit protector(s), are used for motor branch-circuit short-circuit and ground-fault protection, per 430.52(C), the above procedure for determining the maximum feeder protective device rating applies with the following provision. For calculation purposes, each instantaneous trip circuit breaker or motor short-circuit protector is assumed to have a rating not exceeding the maximum percentage of motor FLC permitted by Table 430.52 for the type of feeder protective device used »430.62(A) *Exception No. 1*«.

Where feeder conductor ampacity is greater than 430.24 requires, the rating (or setting) of the feeder overcurrent protective device can be based on the feeder conductor's ampacity »430.62(B)«.

If the same rating (or setting) of the branch-circuit short-circuit and ground-fault protective device is used on multiple branch-circuits supplied by the feeder, one of the protective devices is considered the largest for the above calculation »430.62(A)«.

A A feeder consisting of conductor sizes based on 430.24 that supplies a specific fixed motor load(s) must have a protective device with a rating (or setting) no greater than the largest rating (or setting) of the branch-circuit short-circuit and ground-fault protective device for any motor supplied by the feeder (based on the maximum permitted value for the specific protective device in accordance with 430.52 or 440.22(A) for hermetic refrigerant motor-compressors), plus the sum of the FLC's of the other motors in the group »430.62(A)«.

B Article 430, Part V, specifies devices for protection of feeder conductors supplying motors against overcurrents due to short circuits or grounds »430.61«.

NOTE

Where a feeder supplies a motor load as well as a lighting (or a lighting and appliance) load, the feeder and service protective device must have a rating sufficient to carry the lighting (or the lighting and appliance) load plus, for a single motor, the rating permitted by 430.52, or for a single motor comprised of a hermetic refrigerant motor-compressor, the rating permitted by 440.22, and, for multiple motors, the rating per 430.62 »430.63«.

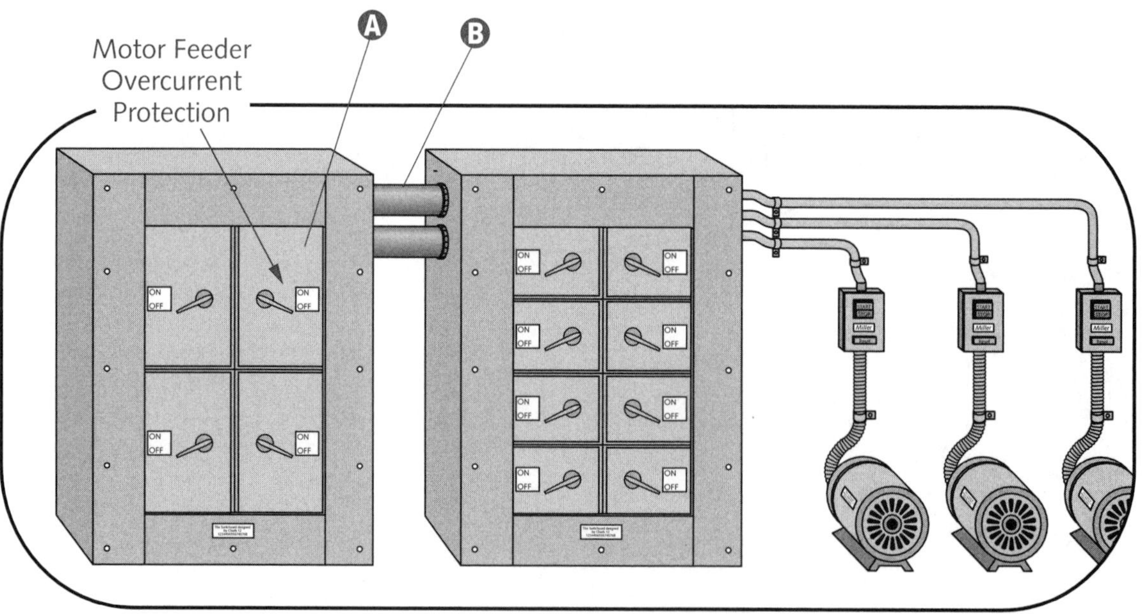

SPECIFIC EQUIPMENT

Cranes and Hoists

Conductors must either be enclosed in raceways or be Type AC cable with insulated grounding conductor, Type MC cable, or Type MI cable, unless 610.11(A) through (E) otherwise permits ›› 610.11 ‹‹.

If a crane, hoist, or monorail hoist operates above readily combustible material, the resistors must be located as outlined in 610.2(B)(1) and (2) ›› 610.2(B) ‹‹.

Conductors must comply with Table 310.13 unless otherwise permitted in 610.13(A) through (D) ›› 610.13 ‹‹.

Conductors exiting raceways/cables must comply with 610.12(A) or (B) ›› 610.12 ‹‹.

Table 610.14(A) dictates the allowable ampacities of conductors ›› 610.14(A) ‹‹.

Where the secondary resistor and the controller are separate, calculate the minimum size conductors (between controller and resistor) by multiplying the motor secondary current by the appropriate Table 610.14(B) factor, and select a wire from Table 610.14(A) ›› 610.14(B) ‹‹.

Motor and control external conductors must not be smaller than 16 AWG, unless otherwise permitted in 610.14(C)(1) and (2) ›› 610.14(C) ‹‹.

A disconnecting means with a continuous ampere rating not less than that computed in 610.14(E) and (F) must be provided between the runway contact conductors and the power supply. Such a disconnecting means must consist of a motor-circuit switch, circuit breaker, or molded case switch. This switch must meet 610.31(1) through (4) provisions ›› 610.31 ‹‹.

Leads from crane and monorail hoist runway contact conductors (or other power supply) must have a motor-circuit switch (or circuit breaker) that can be locked in the open position ›› 610.32 ‹‹.

610.32 requires that the continuous ampere rating of the switch or circuit breaker not be less than 50% of the combined short-time ampere rating of the motors, nor less than 75% of the sum of the short-time ampere rating of the motors required for any single motor ›› 610.33 ‹‹.

Each motor must have an individual controller unless 610.51(A) or (B) allows otherwise ›› 610.51 ‹‹.

A Article 610 covers electrical equipment and wiring installation for use with cranes, monorail hoist, hoists, and all runways ›› 610.1 ‹‹.

B All exposed noncurrent-carrying metal parts of cranes, monorail hoist, hoists, and accessories (including pendant controls) must be joined metallically into a continuous electrical conductor so that the entire crane or hoist will be grounded according to Article 250. Moving parts, other than removable accessories or attachments, having metal-to-metal bearing surfaces qualify as electrically connected (via bearing surfaces) for grounding purposes. The frames (trolley and bridge) are electrically grounded through the wheels and respective tracks, unless conditions, such as paint or other insulating material, prevent reliable metal-to-metal contact. In such cases, a separate bonding conductor must be provided ›› 610.61 ‹‹.

C Cranes, hoists, and monorail hoist overload protection must comply with 610.43.

D Crane, hoist, and monorail hoist motor branch-circuits must meet 610.42(A) requirements.

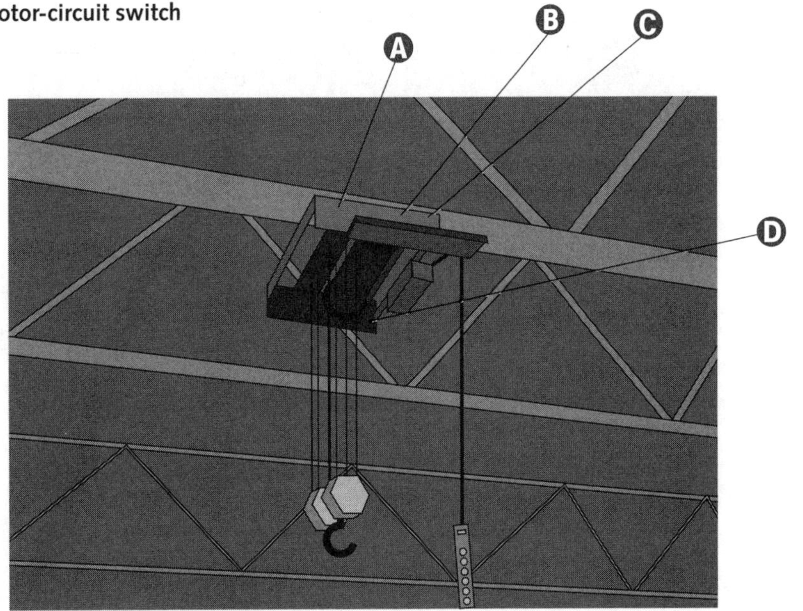

NOTE

All equipment operating in a hazardous (classified) location must comply with Article 500 ›› 610.2(A) ‹‹.

Electric Welders

Each arc welder must have overcurrent protection rated (or set) at no more than 200% of $I_{1\ max}$. If the $I_{1\ max}$ is not given, the overcurrent protection must be rated (or set) at no more than 200% of the welder's rated primary current »630.12(A)«.

An overcurrent device is not required for an arc welder having supply conductors protected by an overcurrent device rated (or set) at no more than 200% of $I_{1\ max}$, or the welder's rated primary current »630.12(A)«.

Supply conductors for an arc welder, protected by an overcurrent device rated (or set) at no more than 200% of $I_{1\ max}$ or rated primary welder current, do not require a separate overcurrent device »630.12(A)«.

A The ampacity of arc welder conductors supplying a group of welders shall be based on individual currents determined in 630.11(A) as the sum of 100% of the two largest welders, plus 85% of the third largest welder, plus 70% of the fourth largest, plus 60% of any other welder. »630.11(B)«.

B 630.12(A) and (B) requires arc welder overcurrent protection. Where the resulting values do not correspond with 240.6(A) standard ampere ratings; or the rating or setting specified results in unnecessary opening of the overcurrent device, the next higher standard rating or setting is acceptable »630.12«.

C Lower percentages are permitted in cases where the work is such that a high-operating duty cycle for individual welders is impossible »630.11(B) *Exception*«.

D Supply conductor ampacity, for a single arc welder, must not be less than the $I_{1\ eff}$ value on the rating plate. If the $I_{1\ eff}$ is not given, the supply conductor ampacity must meet 630.11 requirements »630.11(A)«.

E Conductors supplying one or more arc welders must have an overcurrent protective device rated (or set) not more than 200% of the conductor ampacity »630.12(B)«. (See 630.12(B) FPN for calculation explanation.)

F Article 630 covers electric arc welding, resistance welding apparatus, and similar welding equipment connected to an electric supply system »630.1«.

G Conductors used in the secondary circuit of electric welders must have flame-retardant insulation »630.41«.

H Arc welders must have a rating plate with the information listed in 630.14.

NOTE

Resistance welders must comply with Article 630, Part III (630.31 through 630.34) provisions.

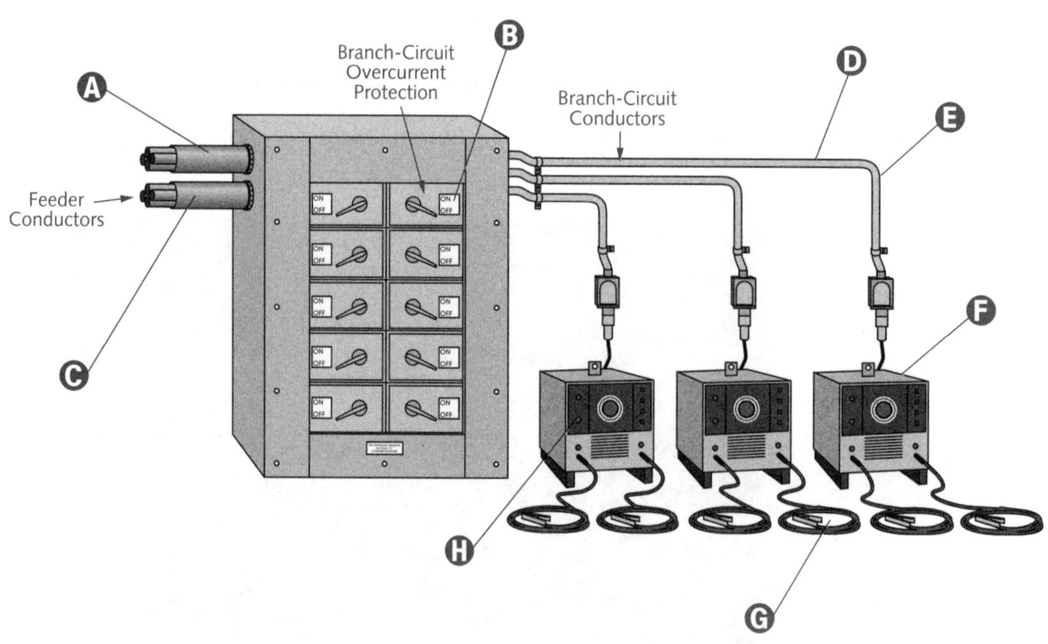

Electroplating

Unless modified by Article 669, electroplating wiring and equipment must comply with the applicable requirements of Chapters 1 through 4 »669.2«.

Equipment used in the electroplating process must be specifically identified for such service »669.3«.

Branch-circuit conductors supplying equipment unit(s) must have an ampacity of at least 125% of the total connected load. Busbar ampacities must meet 366.7 requirements »669.5«.

The following stipulations apply to conductors connecting the electrolyte tank equipment to the conversion equipment: (a) Insulated conductors in systems not exceeding 50 volts direct current can be run without insulated support, provided they are protected from physical damage. Bare copper (or aluminum) conductors are permitted where supported on insulators. (b) Insulated conductors in systems exceeding 50 volts direct current can be run on insulated supports, provided they are protected from physical damage. Bare copper (or aluminum) conductors are permitted where (1) supported on insulators and (2) guarded against accidental contact up to the termination point, per 110.27 »669.6«.

Where multiple power supplies serve the same dc system, the dc side of each power supply must have a disconnecting means »669.8(A)«.

Removable links/conductors can serve as the disconnecting means »669.8(B)«.

Direct-current conductors must be protected from overcurrent by at least one of the following: (1) fuses or circuit breakers; (2) a current-sensing device that operates a disconnecting means; or (3) other approved means »669.9«.

Ⓐ Article 669 provisions apply to the installation of the electrical components and accessory equipment that supply the power and controls for: electroplating, anodizing, electropolishing, and electrostripping, herein referred to simply as *electroplating* »669.1«.

CAUTION *Warning signs must be posted to indicate the presence of bare conductors »669.7«.*

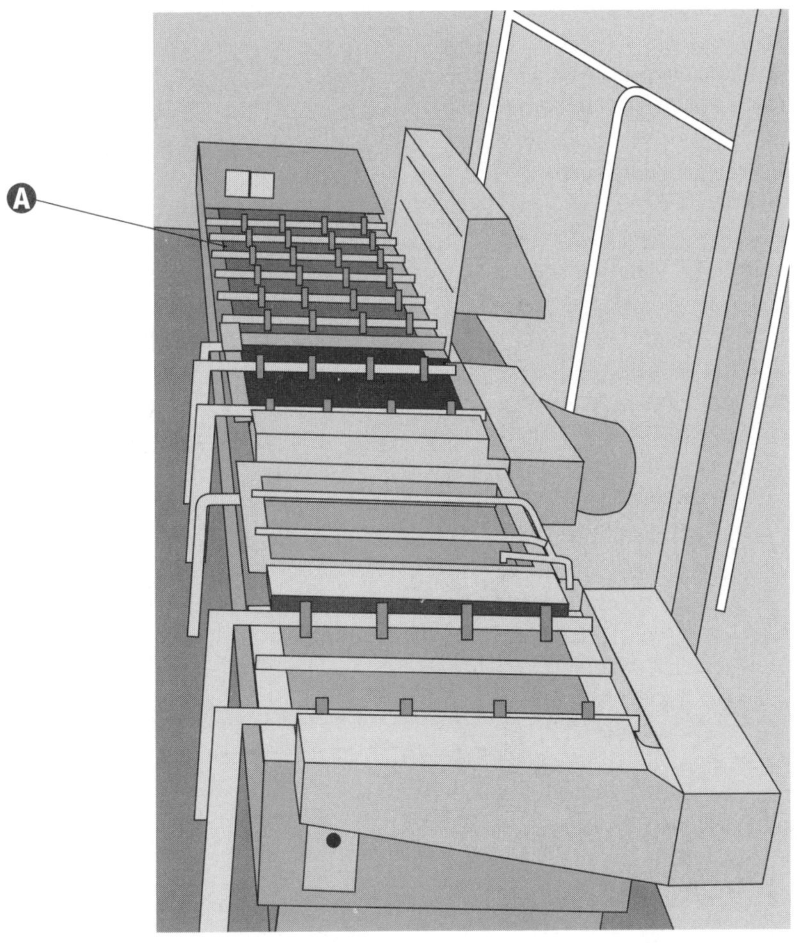

Industrial Machinery

A machine is considered to be an individual unit, and therefore must have a disconnecting means. Branch-circuits protected by either fuses (or circuit breakers) can be the disconnecting means. The disconnecting means is not required to incorporate overcurrent protection. Where overcurrent protection (single circuit breaker or set of fuses) is furnished as part of the machine, 670.3 required markings must be used and the supply conductors must be considered either as feeders (or taps), as covered by 240.21 »670.4(B)«.

An Industrial Manufacturing System is a systematic array of industrial machine(s) not portable by hand and including associated material handling, manipulating, gauging, measuring, or inspection equipment »670.2«.

If the machine has no branch-circuit short-circuit and ground-fault protective device, the overcurrent protective device's rating (or setting) must be based on 430.52 and 430.53, as applicable »670.4(B)«.

Where overcurrent protection is provided, per 670.4(B), the machine must be marked "overcurrent protection provided at machine supply terminals" »670.3(B)«.

If opening a machine's enclosure requires a tool, and only troubleshooting/diagnostic testing is involved on live parts, the clearances can be less than 2½ ft (750 mm) »670.5 *Exception*«.

A Article 670 covers the definition of, the nameplate data for, and the size and overcurrent protection of supply conductors to industrial machinery »670.1«.

B Where controls are enclosed, the cabinet door(s) must either open at least 90°, or be removable »670.5«.

C Article 110, Part II provisions apply to working space around electrical equipment operating at 600 volts, nominal, or less to ground.

D The overcurrent protective device's rating (or setting) for the circuit supplying the machine must comply with 670.4(B).

E The selected supply conductor size must have an ampacity not less than 125% of the FLC rating for all resistance heating loads, plus 125% of the highest rated motor's FLC rating, plus the sum of the FLC ratings of all other connected motors and apparatus that may be simultaneously operated »670.4(A)«.

F A permanent nameplate listing supply voltage, phase, frequency, FLC, maximum short-circuit and ground-fault protective device ampere rating, largest motor or load ampere rating, short-circuit interrupting rating of the machine overcurrent-protective device, if furnished, and diagram number must be attached to the control equipment enclosure/machine, remaining plainly visible after installation »670.3(A)«.

G **Industrial Machinery (Machine)**: is defined as a power-driven machine (or a group of machines working together in a coordinated manner), not portable by hand during operation, used to process material by cutting; forming; pressure; electrical, thermal, or optical techniques; lamination; or a combination of these processes. It can include associated equipment used to transfer material or tooling (including luminaires [fixtures]), assemble/disassemble, inspect, test, or package. (Associated electrical equipment, including logic controller[s] and associated software [or logic] together with the machine actuators and sensors, are all part of the industrial machine.) »670.2«

NOTE

Where maintenance and supervision conditions ensure that only qualified persons will service the installation, the working space dimensions in the direction of access to live parts (operating at not over 150 volts line-to-line or line-to-ground) likely to require examination, adjustment, servicing, or maintenance while energized, must be a minimum of 2½ ft (750 mm) »670.5«.

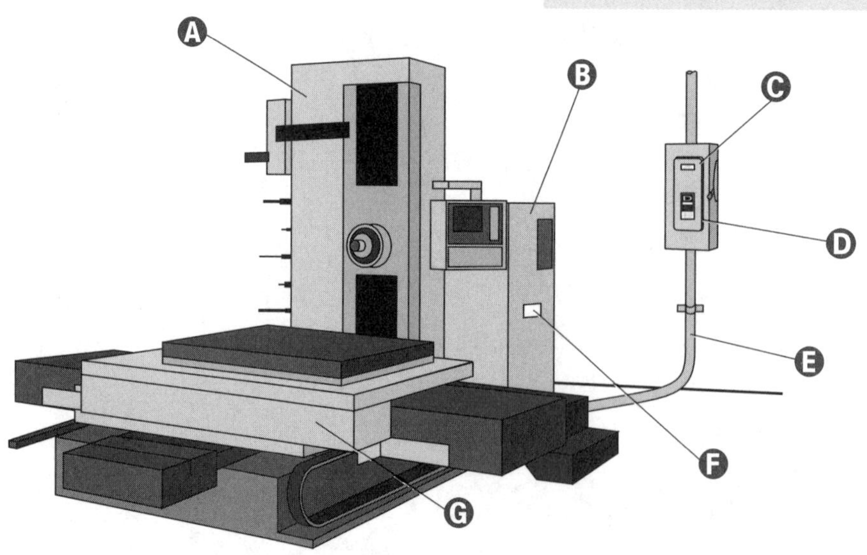

Capacitors

Surge capacitors, or capacitors included as a component part of other apparatus and conforming with such apparatus' requirements, are excluded from Article 460 requirements »460.1«.

Article 460 also covers capacitor installation in hazardous (classified) locations as modified by Articles 501 through 503 »460.1«.

Capacitors containing more than 3 gallons (11 L) of flammable liquid must be enclosed in vaults or outdoor fenced enclosures in order to comply with Article 110, Part III. This limit applies to any single unit in capacitor installations »460.2(A)«.

If a motor installation includes a capacitor connected on the load side of a motor overload device, the motor overload device's rating/setting must be based on the improved power factor of the motor circuit. Disregard the capacitor effect in determining the motor circuit conductor rating per 430.22 »460.9«.

A capacitor connected on the load side of a motor overload protective device does not require a separate overcurrent device »460.8(B) *Exception*«.

A separate disconnecting means is not required where a capacitor is connected on a motor controller's load side »460.8(C) *Exception*«.

Ⓐ Each capacitor must have a nameplate stating the manufacturer's name, rated voltage, frequency, kilovar or amperes, number of phases, and, if filled with a combustible liquid, the volume of liquid. Any nonflammable liquid filling must be indicated on the nameplate. The nameplate must also identify a capacitor having a discharge device inside the case »460.12«.

Ⓑ Capacitors must have a means of discharging stored energy »460.6«.

Ⓒ Article 460 covers the installation of capacitors on electric circuits »460.1«.

Ⓓ Capacitor cases must be grounded according to Article 250 stipulations »460.10«. Do not ground capacitor cases where capacitor units are supported on a structure designed to operate at other than ground potential »460.10 *Exception*«.

Ⓔ Capacitor residual voltage must be reduced to 50 volts, nominal, or less within 1 minute after the capacitor disconnects from the supply source »460.6(A)«. The discharge circuit shall either be permanently connected to the capacitor/capacitor-bank terminals, or have automatic means of connecting it to the capacitor-bank terminals upon line voltage removal. Manual means of switching or connecting the discharge circuit must not be used »460.6(B)«.

Ⓕ Each capacitor bank's ungrounded conductor must have a disconnecting means. The disconnecting means must simultaneously open all ungrounded conductors. It can disconnect the capacitor from the line as a regular operating procedure. The rating of the disconnecting means must be at least 135% of the capacitor's rated current »460.8(C)«.

Ⓖ The capacitor circuit conductor's ampacity must be at least 135% of the capacitor's rated current. The ampacity of the conductors that connect a capacitor to a motor's terminals, or to motor circuit conductors, must not be less than ⅓ of the motor circuit conductor's ampacity and in no case less than 135% of the capacitor's rated current »460.8(A)«.

Ⓗ Each capacitor bank's ungrounded conductor must have an overcurrent device. The overcurrent device's rating (or setting) must be as low as practicable »460.8(B)«.

NOTE

Capacitors, accessible to unauthorized/unqualified persons, must be enclosed, located, or guarded so that neither persons nor conducting materials can come into accidental contact with exposed energized parts, terminals, or associated buses. However, no additional guarding is required for enclosures accessible only to authorized, qualified persons »460.2(B)«.

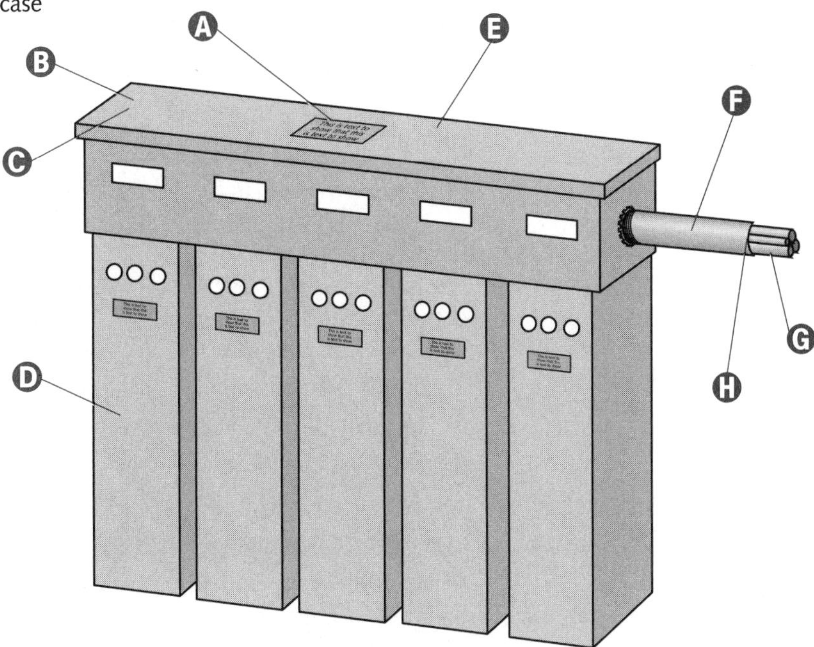

Summary

- A service may have up to six disconnecting means in a single enclosure; in a group of separate enclosures; or in (or on) a switchboard.
- A service-entrance conductor's raceway cannot contain non-service conductors.
- Conductors can be tapped, without overcurrent protection at the tap, to a feeder as specified in 240.21(B)(1) through (5).
- Conductors can connect to a transformer secondary, without overcurrent protection at the secondary, as specified in 240.21(C)(1) through (6).
- Article 450 not only covers transformers, but also includes transformer vault provisions.
- Taps can connect the grounding electrode conductors in a service consisting of more than a single enclosure.
- Cable tray provisions are located in Article 392.
- Article 430 covers motors; motor branch-circuit and feeder conductors (and their protection); motor overload protection; motor-control circuits; motor controllers; and motor control centers.
- Article 610 covers electrical equipment and wiring installations used in connection with cranes, monorail hoists, and all runways.
- Electric arc welding, resistance welding apparatus, and similar welding equipment provisions are located in Article 630.
- Article 669 requirements apply to the installation of the electrical components and accessory equipment supplying electroplating, anodizing, electropolishing, and electrostripping power and controls.
- Article 670 covers the definition of, the nameplate data for, and the size and overcurrent protection of supply conductors to industrial machinery.
- Provisions pertaining to the installation of capacitors on electric circuits is contained in Article 460.

Unit 17 Competency Test

***NEC*® Reference** **Answer**

__________ __________ 1. Terminals of motors and controllers shall be suitably marked or colored where necessary to indicate ______.

__________ __________ 2. Capacitors containing more than ______ gallons of flammable liquid shall be enclosed in vaults or outdoor fenced enclosures complying with Article 110, Part III.

__________ __________ 3. Cable tray systems shall not be used in ______ or where subject to severe physical damage.

I. hoistways

II. hazardous locations

III. corrosive areas

a) I only b) I and II only c) I and III only d) I, II, and III

__________ __________ 4. Wires that are used as crane (or hoist) contact conductors shall be secured at the ends by means of approved strain insulators and shall be mounted on approved insulators so that the extreme limit of displacement of the wire does not bring the latter within less than ______ in(es). from the surface wired over.

__________ __________ 5. Multispeed motors shall be marked with the code letter designating the locked-rotor kilovolt-ampere (kVA) per horsepower for the highest speed at which the motor ______.

__________ __________ 6. The walls and roofs of transformer vaults shall be constructed of materials that have adequate structural strength for the conditions with a minimum fire resistance of ______.

__________ __________ 7. Where cable trays support individual conductors and where the conductors pass from one cable tray to another, from a cable tray to raceways, or to equipment where the conductors are terminated, the distance between cable trays or between the cable tray and the equipment shall not exceed ____ft.

NEC® Reference	Answer	
______	______	8. The residual voltage of a capacitor shall be reduced to _____, nominal, or less, within _____ after the capacitor is disconnected from the source of supply.
______	______	9. The disconnecting means for motor circuits rated 600 volts, nominal, or less, shall have an ampacity rating of at least _____ of the FLC rating of the motor.
______	______	10. Steel cable trays shall not be used as equipment grounding conductors for circuits with ground-fault protection above _____.
______	______	11. Where the conditions of maintenance and supervision ensure that only qualified persons will service the industrial machinery installation, the dimensions of the working space in the direction of access to live parts operating at not over 150 volts line-to-line or line-to-ground that are likely to require examination, adjustment, servicing, or maintenance while energized must be a minimum of ____ ft.
______	______	12. Branch-circuit conductors supplying one or more units of electroplating equipment shall have an ampacity of not less than _____ of the total connected load.
______	______	13. Where practicable, vaults containing more than _____ transformer capacity shall be provided with a drain or other means that will carry off any accumulation of oil or water in the vault unless local conditions make this impracticable.
______	______	14. Conductors that supply one or more arc welders shall be protected by an overcurrent device rated or set at not more than _____ of the conductor ampacity.
______	______	15. The full-load current of a 25-horsepower, synchronous, 460-volt, 3-phase motor at unity power factor is _____.
______	______	16. The ampacity of capacitor circuit conductors shall not be less than _____ of the rated current of the capacitor.
______	______	17. Hoist and monorail hoist and their trolleys that are not used as part of an overhead traveling crane shall not require individual motor overload protection, provided the largest motor does not exceed _____ and all motors are under manual control of the operator.
______	______	18. A door sill or curb that is of sufficient height to confine the oil from the largest transformer within the vault shall be provided, and in no case shall the height be less than ____ in(es).
______	______	19. Branch-circuit conductors that supply a single motor used in a continuous duty application shall have an ampacity of not less than _____ of the motor's full-load current rating as determined by 430.6(A)(1).
______	______	20. A(n) _____ is a systematic array of one or more industrial machines not portable by hand and that includes any associated material handling, manipulating, gauging, measuring, or inspection equipment.
______	______	21. Cable trays containing welding cables must have a permanent sign attached to the cable tray at intervals not greater than ____ ft.
______	______	22. For raceway terminating at the tray, a listed ______________ or adapter shall be used to securely fasten the raceway to the cable tray system.
______	______	23. Where the motor-control circuit transformer rated primary current is less than 2 amperes, an overcurrent device rated or set at not more than _____ of the rated primary current shall be permitted in the primary circuit.
______	______	24. For cranes and hoists, the dimension of the working space in the direction of access to live parts that are likely to require examination, adjustment, servicing, or maintenance while energized shall be a minimum of ____ ft.

NEC® Reference	Answer	
________	________	25. All transformer vault ventilation openings to the indoors shall be provided with automatic closing fire dampers that operate in response to a vault fire. Such dampers shall possess a standard fire rating of not less than _____.
________	________	26. The ampacity of the supply conductors for a resistance welder that may be operated at different times values of primary current or duty cycle shall not be less than _____ of the rated primary current for manually operated nonautomatic welders.
________	________	27. Cable trays shall be permitted to extend _____ through partitions and walls, or _____ through platforms and floors in wet or dry locations where the installations, complete with installed cables, are made in accordance with the requirements of 300.21.
________	________	28. Conductors for a motor used in a short-time, intermittent, periodic, or varying duty application shall have an ampacity of not less than the percentages of the motor nameplate current ratings shown in _____.
________	________	29. Each doorway leading into a transformer vault from the building interior shall be provided with a tight-fitting door that has a minimum fire rating of _____.
________	________	30. _____ shall be placed in all service-entrance motor control centers to isolate service bus-bars and terminals from the remainder of the motor control center.
________	________	31. Crane, hoist, and monorail hoist motor branch circuits shall be protected by fuses or inverse-time circuit breakers that have a rating in accordance with _____.
________	________	32. For a nonautomatically started motor, the overload protection shall be permitted to be shunted or cut out of the circuit during the starting period of the motor if the device by which the overload protection is shunted or cut out cannot be left in the starting position and if fuses or inverse time circuit breakers rated or set at not over _____ of the FLC of the motor are located in the circuit so as to be operative during the starting period of the motor.
________	________	33. The discharge circuit of a capacitor or capacitor bank shall not be connected in which way(s)? I. permanently connected to the terminals II. automatic means of connecting it to the terminals III. manual means of connecting the discharge circuit a) I or II only b) I or III only c) III only d) I, II, or III
________	________	34. Each controller shall be capable of starting and stopping the motor it controls and shall be capable of interrupting the _____ of the motor.

UNIT

18

SECTION FIVE: SPECIAL OCCUPANCIES, AREAS, AND EQUIPMENT

Special Occupancies

Objectives

After studying this unit, the student should:

- know that portions of occupancies, intended for the assembly of more than 99 persons, are referred to as *Places of Assembly*.
- be familiar with definitions pertinent to motion picture (and television studio) audience areas, theaters, and similar locations.
- not only have a basic understanding of motion picture (and television studio) audience area, theater, and similar location general requirements, but also stage switchboards, stage equipment, and theater dressing room provisions.
- know the location of provisions pertaining to carnivals, circuses, exhibitions, fairs, traveling attractions, and similar functions.
- have a solid grasp of requirements for motion picture/television studios (and similar locations), including cellulose nitrate film storage vaults.
- be well acquainted with motion picture projector provisions.
- thoroughly understand manufactured building requirements.
- be familiar with agricultural building provisions including what constitutes a bonding and equipotential plane.
- be appraised of requirements for mobile homes, manufactured homes, and mobile home parks.
- know the location of provisions relating to recreational vehicles and recreational vehicle parks.
- have a good understanding of floating building requirements.
- be familiar with the provisions for marinas and boatyards.

Introduction

Unit 18, "Special Occupancies," covers a variety of mostly unrelated occupancies. This unit addresses most of the occupancy types found in Article 518 through Article 555. Unlike other Chapter 5 articles in the *NEC*®, *Places of Assembly* (Article 518) is not a specific occupancy type. Places of assembly are buildings, or portions of buildings, designed (or intended) for the assembly of 100 or more persons. *Theaters, Audience Areas of Motion Picture and Television Studios, Performance Areas, and Similar Locations* (Article 520) covers all buildings, or that part of a building or structure, indoor or outdoor, designed for presentation, dramatic, musical, motion picture projection, or similar purposes; and specific audience seating areas within motion picture or television studios. *Carnivals, Circuses, Fairs, and Similar Events* (Article 525) addresses portable wiring and equipment installations for carnivals, circuses, exhibitions, fairs, traveling attractions, and similar functions, including wiring in/on all structures. Article 530 (*Motion Picture and Television Studios and Similar Locations*) requirements apply to television studios and motion picture studios using either film or electronic cameras, except as provided in 520.1, and exchanges, factories, laboratories, stages, or a segment of the building in which film or tape [more than ⅞ inch (22 mm) in width] is exposed, developed, printed, cut, edited, rewound, repaired, or stored. Article 540 (*Motion Picture Projection Rooms*) provisions apply to motion picture projection rooms, motion picture projectors, and associated equipment (both professional and nonprofessional types) using incandescent, carbon arc, xenon, or other light source equipment that is known to develop hazardous gases, dust, or radiation. *Manufactured Buildings* (Article 545) details the requirements for manufactured buildings and building components. Article 547 (*Agricultural Buildings*) provisions apply to certain agricultural buildings, parts of buildings, and adjacent areas. Article 550 (*Mobile Homes, Manufactured Homes, and Mobile Home Parks*) requirements apply to the location types identified by the Article's title. Article 551 (*Recreational Vehicles and Recreational Vehicle Parks*) provisions apply to the electrical conductors and equipment installed within (or on) recreational vehicles, the conductors connecting recreational vehicles to an electrical supply, as well as the installation of equipment and devices related to electrical systems within a recreational vehicle park. *Floating Buildings* (Article 553) covers wiring, services, feeders, and grounding for, of course, floating buildings. Finally, Article 555 (*Marinas and Boatyards*) describes wiring and equipment installations in locations comprising fixed or floating piers, wharfs, docks, and other areas in marinas, boatyards, boat basins, boathouses, and similar occupancies that are, or can be, used for the purpose of repair, berthing, launching, storage, or fueling of small craft and the moorage of floating buildings.

MOTION PICTURE (AND TELEVISION STUDIO) AUDIENCE AREAS, PERFORMANCE AREAS, THEATERS, AND SIMILAR LOCATIONS

General Requirements

The fixed wiring methods employed must be metal raceway, nonmetallic raceway encased in 2 in. (50 mm) or more of concrete, Type MI cable, Type MC cable, or AC cable containing an insulated equipment grounding conductor sized per Table 250.122 »520.5(A)«.

Fixed wiring method stipulations in Article 640 apply to audio signal processing, amplification, and reproduction equipment; Article 800 to communication circuits; Article 725 to Class 2 and Class 3 remote control and signaling circuits; and Article 760 to fire alarm circuits »520.5(A) *Exception*«.

Portable switchboards, stage set lighting, stage effects, and other wiring without a fixed location is permitted with approved flexible cords and cables as provided elsewhere in Article 520. Such wiring cannot be fastened by uninsulated staples or nailing »520.5(B)«.

Nonmetallic-sheathed cable, Type AC cable, ENT, and RNC, can be installed where applicable building code does not require the building (or portions thereof) to be of fire-rated construction »520.5(C)«.

The number of conductors permitted in any metal conduit, RNC, or EMT for border or stage pocket circuits, or for remote-control conductors, must not exceed the percentage fill shown in Table 1 of Chapter 9. Where contained within an auxiliary gutter or wireway, the sum of the cross-sectional area of all internal conductors at any point must not exceed 20% of the interior cross-sectional area of the auxiliary gutter or wireway. The 30-conductor limitation of 366.6 and 376.22 does not apply »520.6«.

Live parts must be enclosed or guarded, preventing accidental contact by persons and objects. All switches must be of the externally operable type. Dimmers, including rheostats, must be installed in cases (or cabinets) that effectively surround all live parts »520.7«.

Portable stage/studio lighting equipment, and portable power distribution equipment, are permitted for temporary outdoor use, provided the equipment is (1) supervised by qualified personnel while energized and (2) barriered from the general public »520.10«.

Ⓐ Article 520 covers all buildings/structures, or parts thereof, indoor or outdoor, intended for presentation, dramatic, musical, motion picture projection, or similar purposes as well as specific audience seating areas within motion picture or television studios »520.1«.

CAUTION *Any size branch-circuit supplying receptacle(s) can also supply stage set lighting. The receptacle's voltage rating must not be less than the circuit voltage. Receptacle ampere rating and branch-circuit conductor ampacity must at least equal the branch-circuit overcurrent device ampere rating.* ***Table 210.21(B)(2) shall not apply*** *»520.9«.*

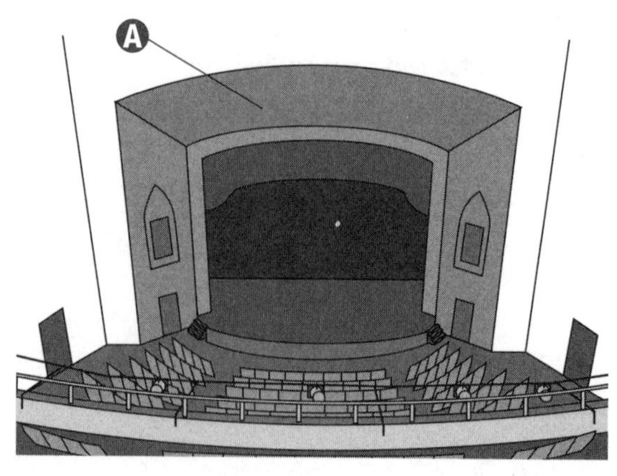

Definitions

Ⓐ Border Light: A permanently installed overhead strip light »520.2«.

Ⓑ Breakout assemblies can contain listed, hard usage (junior hard service) cords where all of the following conditions are met:

(1) The cords connect a single multipole connector (containing two or more branch circuits) and multiple two-pole, three-wire connectors.

(2) No cord in the breakout assembly exceeds 20 ft (6.0 m) in length.

(3) The breakout assembly's entire length is protected from physical damage by attachment to a pipe, truss, tower, scaffold, or other substantial support structure.

(4) All branch-circuits feeding the breakout assembly are protected by overcurrent devices rated at no more than 20 amperes »520.68(A)(4)«.

Ⓒ Breakout Assembly: An adapter used to join a multipole connector (containing two or more branch-circuits) to multiple individual branch-circuit connectors »520.2«.

Ⓓ Portable Equipment: Equipment intended for movement from place to place and fed with portable cords or cables »520.2«.

Ⓔ Flexible conductors, including cable extensions, supplying portable stage equipment must be listed extra-hard usage »520.68(A)(1)«.

Ⓕ Connector Strip: A metal wireway containing pendant or flush receptacles »520.2«.

Ⓖ Footlight: A border light installed on (or in) the stage »520.2«.

Ⓗ Portable Power Distribution Unit: A power distribution box containing receptacles and overcurrent devices »520.2«.

Ⓘ Portable power distribution units must comply with (a) through (e):

(a) The construction must prevent the exposure of current-carrying parts.

(b) Receptacles must comply with 520.45 and must have branch-circuit overcurrent protection within the box. Fuses and circuit breakers must be protected from physical damage. Cords/cables supplying pendant receptacles must be listed for extra-hard usage.

(c) Busbars must have an ampacity equal to the sum of the ampere ratings of all the connected circuits. Lugs must be provided for the master cable's connection.

(d) Power accepting flanged surface inlets (recessed plugs) must be rated in amperes.

(e) Cable arrangement must alleviate tension at the termination points and be adequately protected when passing through enclosures »520.62«.

Ⓙ Proscenium: The wall and arch separating the stage from the auditorium (house) »520.2«.

Ⓚ Stand Lamp (Work Light): A portable stand containing a general-purpose luminaire (lighting fixture), or guarded lampholder for stage or auditorium general illumination »520.2«.

Ⓛ Strip Light: A luminaire (lighting fixture) with a row of multiple lamps »520.2«.

Ⓜ Two-Fer: An adapter cable containing one male plug and two female cord connectors used to join two loads to one branch circuit »520.2«.

Ⓝ Adapters, two-fers, and other single/multiple circuit outlet devices must comply with all of the following:

(a) Each receptacle and corresponding cable must have the same current and voltage rating as its supply plug. Utilization in a stage circuit having a greater current rating is not allowed.

(b) All conductors must be wired in accordance with 520.67.

(c) Adapter and two-fer conductors must be listed either extra-hard usage or hard usage (junior hard service) cord. The overall length of hard usage (junior hard service) cord is restricted to 3.3 ft (1.0 m) »520.69«.

Ⓞ Stand lamps can be supplied by reinforced cord provided the cord is exempt from severe physical damage and is protected by an overcurrent device rated at not over 20 amperes »520.68(A)(2)«.

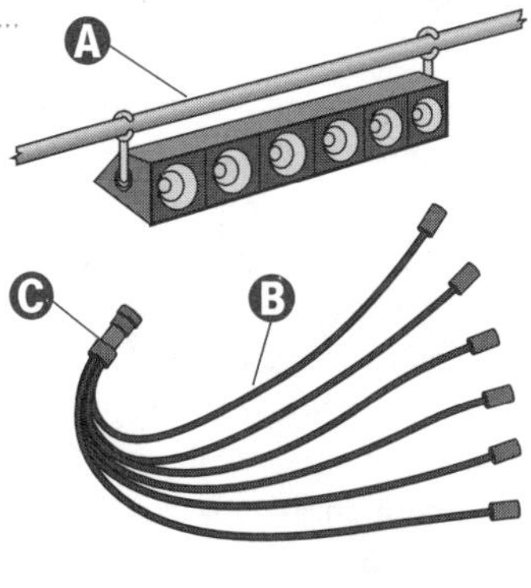

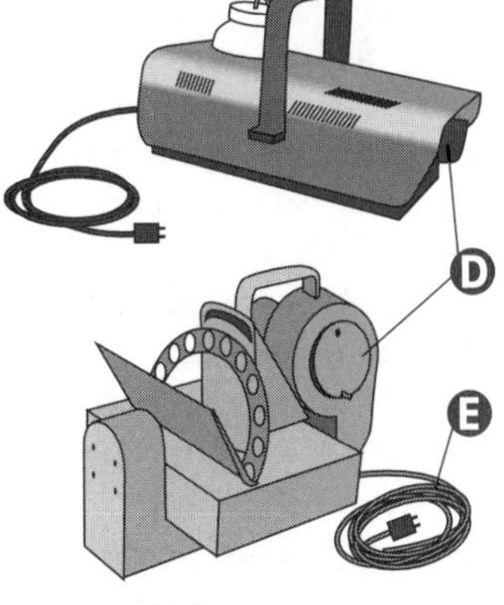

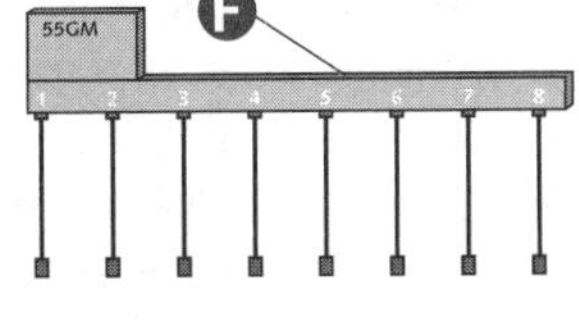

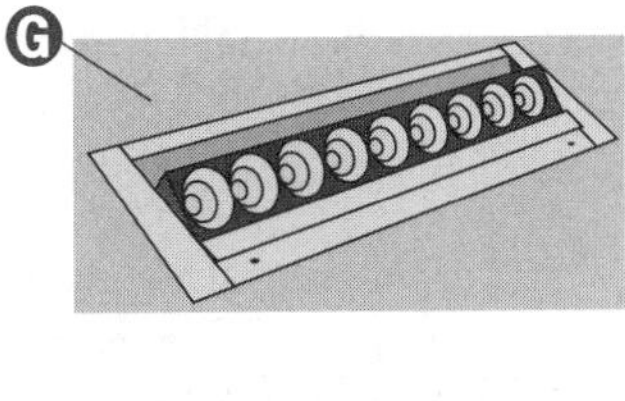

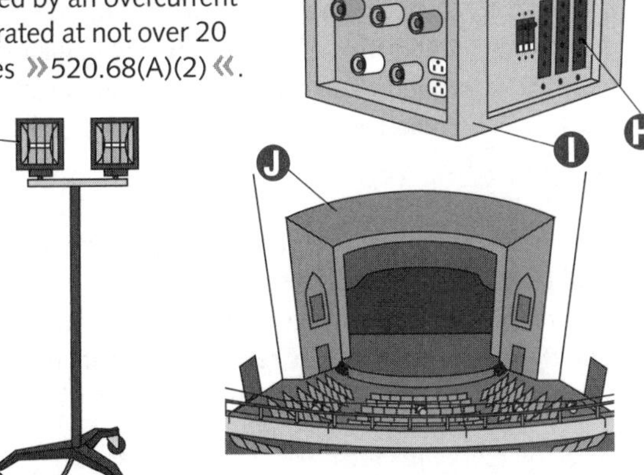

Fixed Stage Switchboards

A stage-lighting switchboard, to which load circuits are connected, must have a means of branch-circuit overcurrent protection. This includes branch-circuits supplying stage and auditorium receptacles used for cord- and plug-connected stage equipment. Where the stage switchboard contains nonstage lighting dimmers, the overcurrent protective devices for these branch-circuits can be located at the stage switchboard »520.23«.

A stage switchboard, not completely enclosed dead-front and dead-rear, or recessed into a wall, must have a full length metal hood to protect all equipment on the board from falling objects »520.24«.

Dimmers installed in ungrounded conductors must have overcurrent protection no greater than 125% of the dimmer rating, and must be disconnected from all ungrounded conductors when the dimmer's master or individual switch (or circuit breaker) is in the open position »520.25(A)«.

A stage switchboard must be one (or a combination) of the following types:

(A) Dimmers and switches operated by handles mechanically linked to control devices.

(B) Devices operated electrically from a pilot-type control console or panel. Pilot control panels can be part of the switchboard, or can be located elsewhere.

(C) A stage switchboard having interconnected circuits is a secondary switchboard (patch panel) or panelboard remote to the primary stage switchboard, and must contain overcurrent protection. If the dimmer panel provides the required branch-circuit overcurrent protection, it can be omitted from the intermediate switchboard »520.26«.

Stage switchboard supply feeders must be one of the following:

(1) A single feeder disconnected by a single disconnect device.

(2) Unlimited feeder quantities are permitted, provided all feeders are part of a single system. Where combined, neutral conductors in a given raceway must have sufficient ampacity to carry the maximum unbalanced current supplied by multiple feeder conductors within the same raceway. It is not necessary that the ampacity be greater than that of the neutral supplying the primary stage switchboard. Parallel neutral conductors must comply with 310.4.

(3) Separate feeders of a single primary stage switchboard must each have a disconnecting means. The primary stage switchboard must be permanently and conspicuously labeled with the number and location of disconnecting means. If the disconnecting means are located in multiple distribution switchboards, the primary stage switchboard must have barriers corresponding with these multiple locations »520.27(A)«.

When computing supply capacity to switchboards, it is permissible to consider the maximum intended switchboard load in a given installation, provided that: (1) all switchboard supply feeders are protected by an overcurrent device with a rating no greater than the feeder's ampacity, and (2) opening the overcurrent device does not affect proper operation of egress or emergency lighting systems »520.27(C)«.

CAUTION *Exposed live parts on the back of stage switchboards must be enclosed by building walls, wire mesh grills, or other approved methods. A self-closing door must cover the entrance »520.22«.*

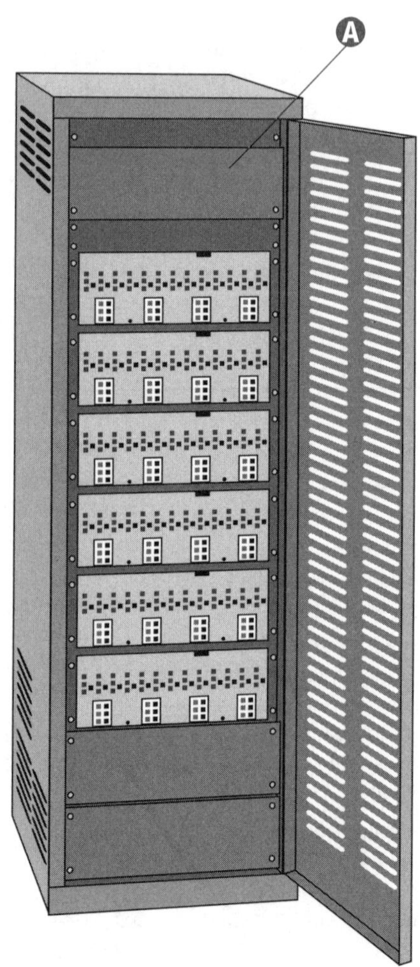

NOTE

The neutral of feeders supplying solid-state, three-phase, four-wire dimming systems qualifies as a current-carrying conductor »520.27(B)«.

(A) Stage switchboards must be of the dead-front type and must comply with Article 408, Part IV unless a qualified testing laboratory (via recognized test standards and principles) has approved it as suitable for such use »520.21«.

Portable Switchboards on Stage

Circuits from portable switchboards directly supplying equipment containing incandescent lamps of not over 300 watts must be protected by overcurrent protective devices having a rating or setting of not over 20 amperes. Circuits for lampholders over 300 watts are permitted if overcurrent protection complies with Article 210 »520.52«.

With the exception of busbars, all conductors within the switchboard must be stranded »520.53(F)(1)«.

Only listed extra-hard usage cords or cables can supply a portable switchboard. The supply cords (or cable) must terminate within the switchboard enclosure, in an externally operable fused master switch, circuit breaker, or a connector assembly identified for such purpose. The supply cords (or cable) and connection assembly must (1) have sufficient ampacity to carry the total switchboard-connected load, and (2) be protected by overcurrent devices »520.53(H)(1)«.

Single-pole portable cable connectors, where used, must be listed and of the locking type »520.53(K)«.

CAUTION *Unless an exception is met, only qualified personnel can route portable supply conductors, make/break supply connectors (and other supply connections), and energize/deenergize supply services. The portable switchboard must, by means of permanent and conspicuous markings, reflect this requirement »520.53(P)«.*

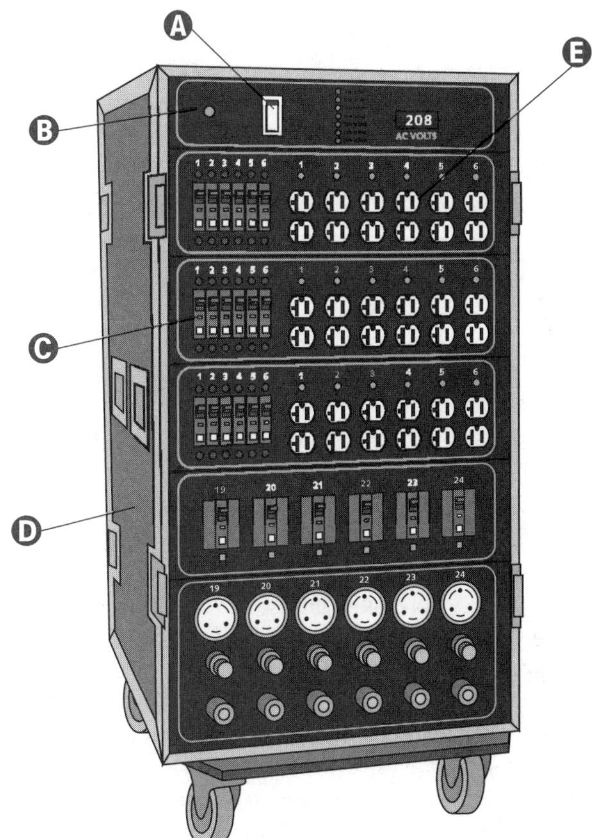

A The enclosure must contain a pilot light connected to the board's supply circuit so that opening the master switch does not cut off the lamp's supply. This lamp must be on an individual branch-circuit having overcurrent protection rated or set at no more than 15 amperes »520.53(G)«.

B Portable switchboards and feeders for on-stage use must comply with 520.53(A) through (P) »520.53«.

C Only externally operable, enclosed type switches and circuit breakers can be used »520.53(C)«.

D Portable switchboards must be housed within substantially constructed enclosures, which remain open during operation. Wooden enclosures must be fully lined with sheet metal of no less than 0.020 in. (0.51 mm). The sheet metal must either be properly galvanized, enameled, or otherwise coated to prevent corrosion, or be of a corrosion-resistant material »520.53(A)«.

E Road Show Connection Panel (A Type of Patch Panel): A panel designed to allow for road show connection of portable stage switchboards to fixed lighting outlets via permanently installed supplementary circuits. The panel, supplementary circuits, and outlets must comply with all of the following:

(a) Circuits must terminate in grounding-type polarized inlets whose current and voltage rating match the fixed-load receptacles.

(b) Circuits transferred between fixed and portable switchboards must transfer all circuit conductors simultaneously.

(c) Supplementary circuit supply devices must have branch-circuit overcurrent protection. The individual supplementary circuit, within the road show connection panel and theater, must be protected by branch-circuit overcurrent protective devices of suitable ampacity internal to the road show connection panel.

(d) Panel construction must be in accordance with Article 408 »520.50«.

NOTE

Supply conductors can have a maximum of three interconnections (mated connector pairs) where the total length from supply to switchboard does not exceed 100 ft (30 m). Should the total length from supply to switchboard exceed 100 ft (30 m), one additional supply conductor interconnection is permitted for each additional 100 ft (30 m) »520.53(J)«.

Stage Equipment Other Than Switchboards

Arrange footlights, border lights, and proscenium side lights so that no supplying branch circuit carries more than a 20-ampere load »520.41(A)«.

Article 210 provisions can be applied to heavy-duty lampholder circuits »520.41(B)«.

The conductors for foot, border, proscenium, or portable strip luminaires (light fixtures) and connector strips must have insulation suitable for the conductor's operating temperature, but in no case less than 257°F (125°C). The ampacity of the 257°F (125°C) conductors must be that of 140°F (60°C) conductors »520.42«.

All drops from connector strips must be 194°F (90°C) wire sized to the ampacity of 140°F (60°C) cords and cables, with no more than 6 in. (150 mm) of conductor extending into the connector strip. Section 310.15(B)(2)(a) does not apply »520.42«.

The ampacity of listed multiconductor extra-hard usage-type cords and cables not in direct contact with equipment containing heat-producing elements can be determined by Table 520.44. If so determined, the maximum load current in any conductor must not exceed the values in Table 520.44 »520.44(B)(2)«.

Receptacles for on-stage electrical equipment must be rated in amperes. Conductors supplying receptacles must comply with Articles 310 and 400 »520.45«.

Connector strip, drop box, floor pocket, and similar receptacles used to connect portable stage-lighting equipment must be pendant, or mounted in suitable pockets or enclosures, and must satisfy 520.45 requirements. Connector strip and drop box supply cables must meet 520.44(B) specifications »520.46«.

Bare-bulb lamps in backstage and ancillary areas, where contact with scenery is possible, must be located and guarded to remain free from physical damage. A minimum air space of at least 2 in. (50 mm) between such lamps and any combustible material must be provided »520.47«. Scenery-installed decorative lamps are not to be regarded as backstage lamps. »520.47 *Exception*«.

The operating circuit of an electrical device releasing stage smoke ventilators must be (1) normally closed and (2) controlled by at least two externally operable switches with one such switch being readily accessible on stage and the other positioned as dictated by the AHJ. The device must be designed for the connected circuit's full voltage, with no resistance inserted. The device must be located in the loft above the scenery and have been enclosed in a suitable metal box with a tight, self-closing door »520.49«.

A Borders and proscenium sidelights must be: (1) constructed per 520.43 specifications, (2) suitably secured and supported, and (3) designed so that the reflector flanges (or other adequate guards) protect the lamps from mechanical damage and from accidental contact with any combustible material »520.44(A)«.

B Cords and cables for the border lights' supply must be listed for extra-hard usage and must be suitably supported. Such cords and cables must be employed only where flexible conductors are necessary. Conductor ampacity must satisfy 400.5 provisions »520.44(B)(1)«.

C If the metal trough construction specified in 520.43(A) is not used, footlights must consist of individual outlets whose lampholders are wired with RMC, IMC, FMC, Type MC cable, or mineral-insulated, metal-sheathed cable. The circuit conductors must be soldered to the lampholder terminals »520.43(B)«.

D The current supply to disappearing footlights must automatically disconnect when the footlights are fully retracted »520.43(C)«.

E Metal trough footlights, containing circuit conductors, must be made of oxidation-resistant sheet metal no lighter than 0.032 in. (0.81 mm). Lampholder terminals must be spaced at least ½ in. (13 mm) from the metal trough. The circuit conductors must be soldered to the lampholder terminals »520.43(A)«.

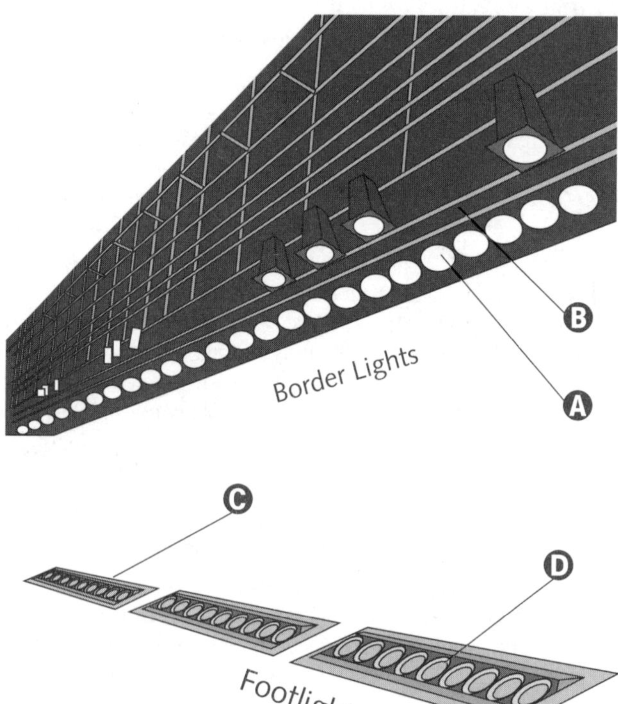

Theater Dressing Rooms

A All dressing room lights and receptacles adjacent to the mirror(s) and above the dressing table counter(s) must be controlled by wall switches within the dressing room(s) »520.73«.

B Each switch controlling receptacles adjacent to the mirror(s) and above the dressing table counter(s) must have a pilot light located outside the dressing room, adjacent to the door, indicating energized receptacles »520.73«. Note: although the pilot light must be installed outside the dressing room, the switch must be located inside the dressing room.

C In theater dressing rooms, all exposed incandescent lamps less than 8 ft (2.5 m) from the floor must be equipped with open-end guards riveted to the outlet box cover or otherwise appropriately secured »520.72«.

D Receptacles away from dressing table counter(s) and mirror(s) do not require switches »520.73«.

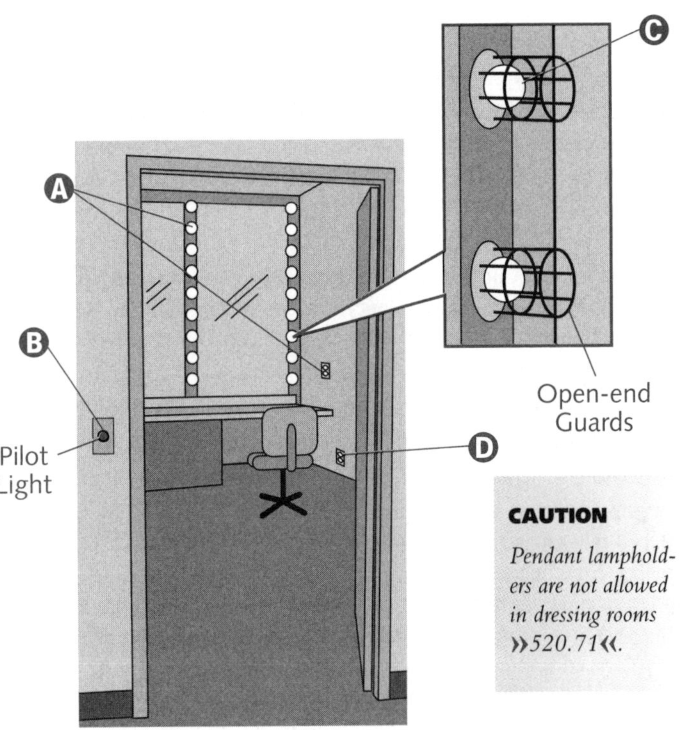

CAUTION

Pendant lampholders are not allowed in dressing rooms »520.71«.

NOTE

Fixed wiring in motion picture and television studio dressing rooms must be installed according to Chapter 3 methods »530.31«. (520.71 through 73 provisions apply only to theater dressing rooms, and not motion picture and television studio dressing rooms.)

CARNIVALS, CIRCUSES, FAIRS, AND SIMILAR EVENTS

Ground-Fault Circuit-Interrupter Protection

Egress lighting must not be connected to the GFCI receptacle's load side terminals. »525.23(A)«.

A All 125-volt, single-phase, 15- and 20-ampere receptacle outlets that are in use by personnel must have listed GFCI protection for personnel. The GFCI can be an integral part of the attachment plug, or located in the power supply cord, within 12 in. (300 mm) of the attachment plug. »525.23(A)«.

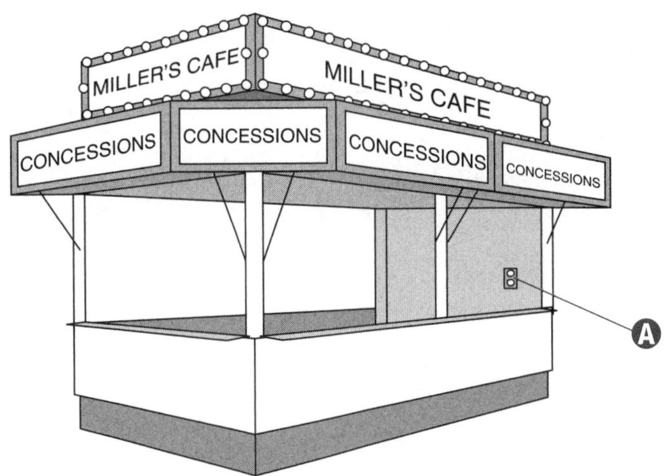

NOTE

Receptacles supplying items such as cooking and refrigeration equipment, that are not compatable with GFCI devices, do not require GFCI protection »525.23(B)«.

General Requirements

Permanent structures must comply with Articles 518 and 520 »525.3(B)«.

The following common-source connected equipment must be bonded: (1) metal raceways and metal sheathed cable; (2) metal enclosures of electric equipment; and (3) metal frames and metal parts of rides, concessions, trailers, trucks, or other equipment that contain or support electrical equipment »525.30«.

Approved nonconductive mats must cover publically accessible flexible cords (or cable) run on pavement or other hard and smooth surfaces. Arrangement of cables and mats must not present a tripping hazard »525.20(G)«.

No amusement ride, attraction, tent, or similar structure can support wiring for any other ride or structure, unless specifically designed for that purpose »525.20(F)«.

All equipment that requires grounding must be grounded by a 250.118 recognized equipment grounding conductor (type and size) installed per Article 250. The equipment grounding conductor must be bonded to the system grounded conductor at the service disconnecting means, or, in the case of a separately derived system such as a generator, at the generator or the generator's first disconnecting means. Do not connect the grounded circuit conductor to the equipment grounding conductor on the load side of the service disconnecting means, nor on the load side of a separately derived system disconnecting means »525.31«.

Ⓐ Conductors must have a vertical clearance to ground per 225.18. These clearances apply only to exterior wiring of tents and concessions (not within) »525.5(A)«.

Ⓑ Tent and concession interior electrical wiring for temporary lighting, must be securely installed, and mechanically protected where subject to physical damage. Protect all temporary general illumination lamps from accidental breakage by installing a suitable luminaire (fixture) or lampholder with a guard »525.21(B)«.

Ⓒ Article 525 covers the installation of portable wiring and equipment for carnivals, circuses, fairs,and similar functions, including all structure related wiring »525.1«.

Ⓓ Amusement rides/attractions must be at least 15 ft (4.5 m) in any direction away from overhead conductors operating at 600 volts or less, except for the ride/attraction supply conductors. Amusement rides/attractions must not be situated under, nor within 15 ft (4.5 m) horizontally, of conductors operating in excess of 600 volts »525.5(B)«.

Ⓔ Electrical equipment and wiring methods of rides, concessions, or other units must be mechanically protected where subject to physical damage »525.6«.

Ⓕ Each ride and concession must have a visible fused disconnect switch or circuit breaker within 6 ft (1.8 m) of the operator's station. The disconnecting means must be readily accessible to the operator, even during ride operation. If accessible to unqualified persons, the switch or circuit breaker enclosure must be lockable. A shunt trip device that opens the fused disconnect device or circuit breaker when a switch located in the ride operator's console is closed is an acceptable method of opening the circuit »525.21(A)«.

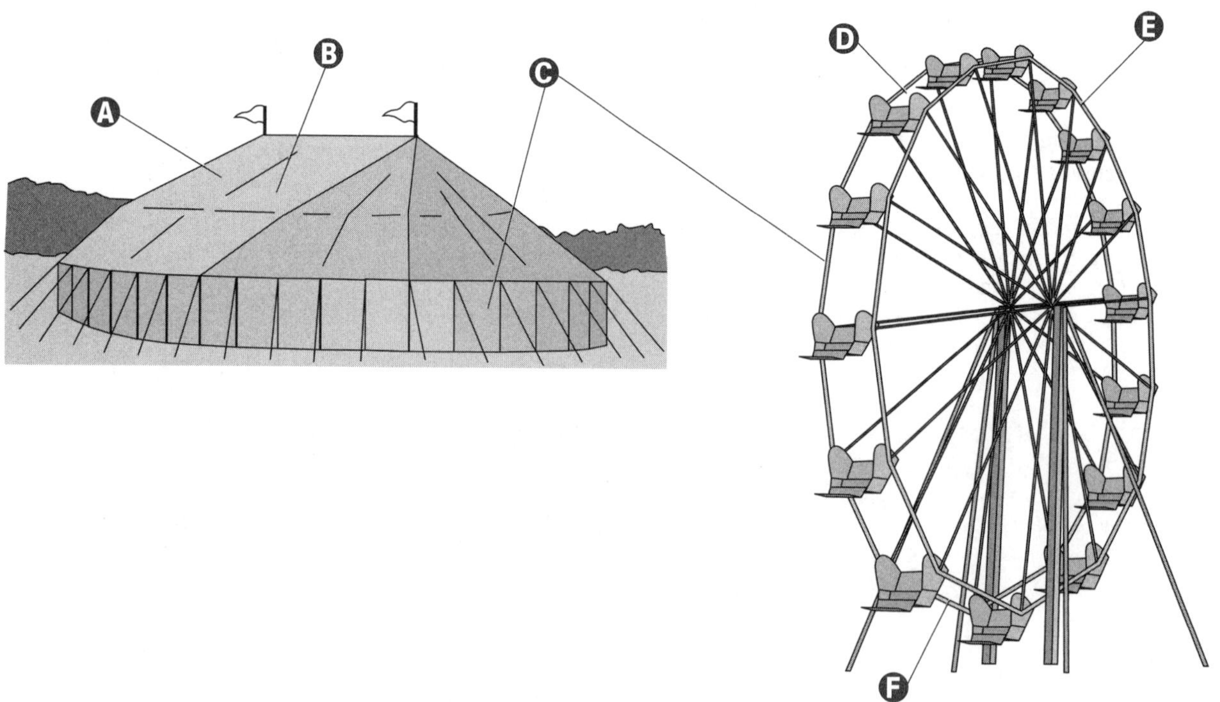

PLACES OF ASSEMBLY

Comprehensive Provisions

The wiring of any such building/area containing a projection booth, stage platform, or area for the presentation of theatrical/musical productions (fixed or portable) must comply with Article 520. These requirements also apply to associated audience seating, and all equipment used in the production, including portable equipment and associated wiring that is not connected to permanent wiring »518.2(C)«.

Only metal raceways, flexible metal raceways, nonmetallic raceways [encased in 2 in. (50 mm) or more of concrete], and Type MI, MC, or AC cable having an insulated equipment grounding conductor (sized per Table 250.122) can be used as a fixed wiring method »518.4(A)«.

ENT and RNC are not approved for environmental air spaces in accordance with 300.22(C) »518.4(C)«.

Portable switchboards and portable power distribution equipment must be supplied only from listed power outlets of sufficient voltage/ampere rating. These power outlets must be protected by overcurrent devices. Neither the overcurrent devices nor the power outlets can be publically accessible. A method for equipment grounding conductor connection must be provided. The neutral of feeders supplying solid-state, three-phase, four-wire dimmer systems qualifies as a current-carrying conductor »518.5«.

Ⓒ Nonmetallic-sheathed cable, Type AC cable, ENT, and RNC can be installed in those buildings (or portions thereof) for which the applicable building code does not require fire-rated construction »518.4(B)«.

Ⓓ ENT and RNC can be installed within club rooms, classrooms (college and university), conference and meeting rooms in hotels (or motels), courtrooms, drinking establishments, dining facilities, restaurants, mortuary chapels, museums, passenger stations and terminals of air, surface, underground, and marine public transportation facilities, libraries, and places of religious worship where:

(1) The ENT or RNC is concealed within walls, floors, and ceilings where the enclosing structure provides a thermal material barrier having at least a 15-minute finish rating as identified in fire-rated assembly listings.

(2) Such tubing or conduit is installed above suspended ceilings where the ceilings act as a thermal material barrier having at least a 15-minute finish rating according to fire-rated assembly listings »518.4(C)«.

Ⓔ Article 518 covers all buildings, portions of buildings, or structures serving as a place of assembly for more than 99 persons »518.1«.

Ⓐ Temporary wiring for display booths within exhibition halls (as in trade shows) must be installed in accordance with Article 527. Flexible cables and cords, approved for hard or extra-hard usage, laid on floors must be protected from contact by the general public. GFCI requirements of 527.6 do not apply »518.3(B)«.

Ⓑ Assemblage of fewer than 100 persons in any room (or space) within a building of other occupancy (where said assembly is incidental to the other occupancy), is part of the other occupancy by classification and is subject to the provisions applicable thereto »518.2(B)«.

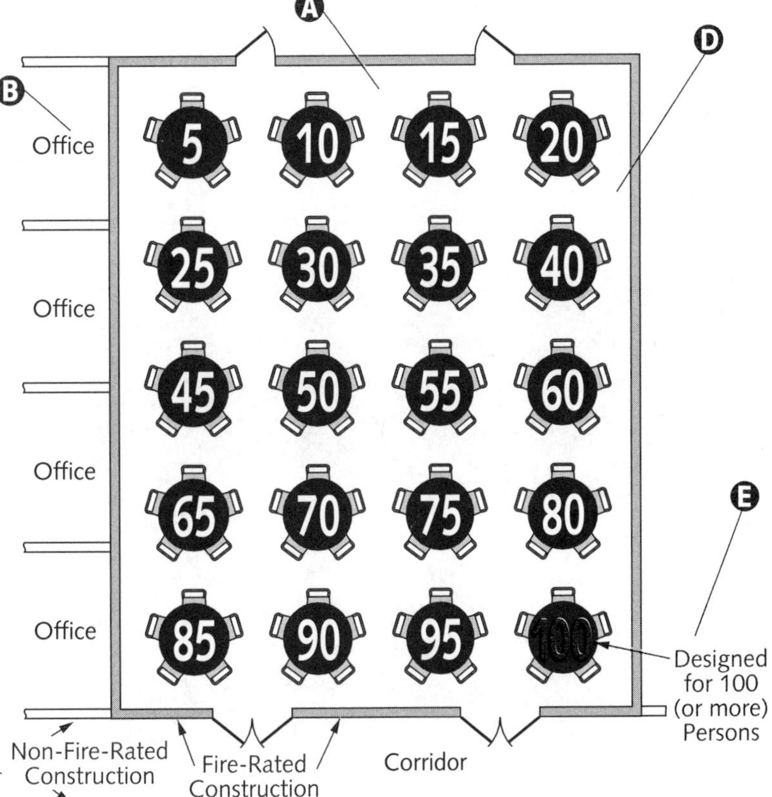

MOTION PICTURE (AND TELEVISION) STUDIOS AND SIMILAR LOCATIONS

Comprehensive Requirements

Portable stage/studio lighting equipment and portable power distribution equipment can be for temporary use outdoors, if (1) supervised by qualified personnel while energized and (2) barriered from the general public »530.6«.

Stage or set permanent wiring must be Type MC cable, Type AC cable containing an insulated equipment grounding conductor sized per Table 250.122, Type MI cable, or approved raceways »530.11«.

The wiring for stage set lighting, and other supply wiring not fixed as to location, must be achieved with listed hard usage flexible cords/cables. Where subject to physical damage, such flexible cords/cables must be of the listed extra-hard usage type. Cable splices or taps are allowed, provided the total connected load does not exceed the cable's maximum ampacity »530.12(A)«.

Stage effects and electrical equipment used as stage properties can be wired with single- or multi-conductor listed flexible cords or cables where protected from physical damage and secured to the scenery by approved cable ties or insulated staples. Splices (or taps) can only be made with listed devices in a circuit protected at no more than 20 amperes »530.12(B)«.

Use externally operable type switches for studio stage set lighting and effects whether on the stage, on the lot, or on location. Where contactors serve as fuse disconnecting means, and individual externally operable switches (such as tumbler switches) control each contactor, these switches must be located no more than 6 ft (1.8 m) from the contactor. Remote-control switches must also be provided. A single externally operable switch, capable of simultaneous disconnection of all contactors on any one location board, is permitted where located within 6 ft (1.8 m) of that location board »530.13«. (*Location Board* is defined in 530.2, definitions.)

Portable lamps and work lights must have flexible cords, composition or metal-sheathed porcelain sockets, and substantial guards »530.16«. Portable lamps used as properties in a motion picture or television stage set, on a studio stage or lot, or on location are, for this purpose, not considered portable lamps »530.16 *Exception*«.

Automatic overcurrent protective devices (circuit breakers or fuses) for motion picture studio stage set lighting and associated cables must comply with 530.18(A) through (G).

A Article 530 requirements apply to motion picture/television studios using either film or electronic cameras (except as provided in 520.1) as well as exchanges, factories, laboratories, stages, or a building segment in which film or tape wider than ⅞ in. (22 mm) is exposed, developed, printed, cut, edited, rewound, repaired, or stored »530.1«.

B **Stage Set:** A specific area set up with temporary scenery and properties planned for a particular motion picture or television production scene »530.2«.

C **Television Studio or Motion Picture Stage (Sound Stage):** All or part of a building, usually insulated from outside noise and natural light, used for motion picture, television, or commercial production purposes »530.2«.

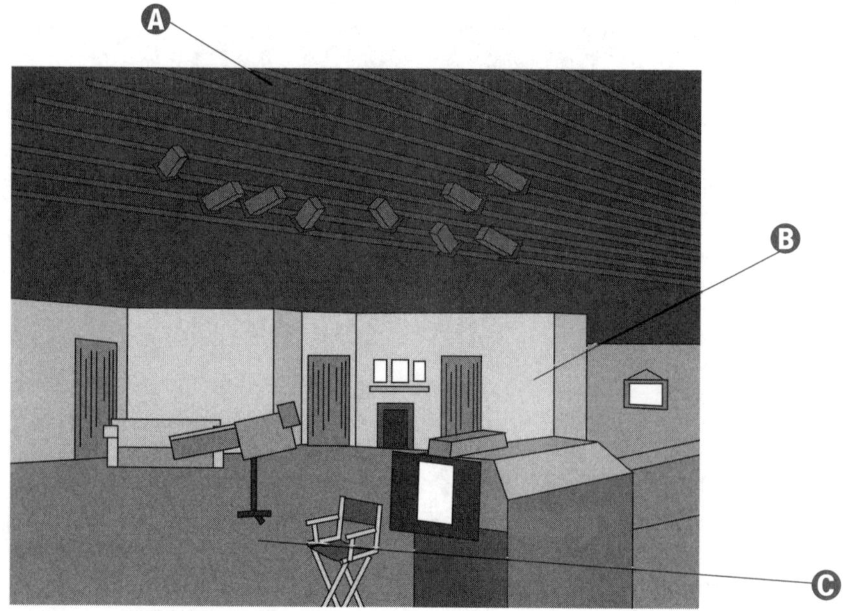

NOTE

Type MC cable, Type MI cable, metal raceways, and all noncurrent-carrying metal parts of appliances, devices, and equipment must be grounded as specified in Article 250. Pendant and portable lamps, stage lighting, and stage sound equipment operating at not over 150 volts dc to ground, are exempt »530.20«.

Comprehensive Requirements (*continued*)

The maximum ampacity allowed on a given conductor, cable, or cord size is dictated by the applicable tables of Articles 310 and 400 »530.18«.

It is acceptable to apply Table 530.19(A) demand factors to that portion of the maximum possible connected load for studio or stage set lighting for all permanently installed feeders between substations and stages and to all permanently installed feeders between the main stage switchboard and stage distribution centers (or location boards) »530.19(A)«.

Portable feeders can have a demand of 50% of maximum possible connected load »530.19(B)«.

Plugs and receptacles must be rated in amperes. The voltage rating of plugs and receptacles must be at least equal to the circuit voltage. All ac circuit plug and receptacle ampere ratings must not be less than the feeder (or branch-circuit) overcurrent device ampere rating. Table 210.21(B)(2) does not apply »530.21(A)«.

Any ac single-pole portable cable connectors used must be listed and of the locking type. 400.10, 406.6, and 406.7 do not apply to listed single-pole separable connections, nor to single-conductor cable assemblies incorporating listed single-pole separable connectors. Paralleled sets of current-carrying single-pole separable connectors acting as input devices must carry prominent warning labels indicating the presence of internal parallel connections. Single-pole separable connectors must comply with at least one provision of 530.22(A)(1) through (3) »530.22(A)«.

Cellulose Nitrate Film Storage Vaults

A Cellulose nitrate film storage vault lamps must be installed in glass-enclosed, gasketed-type rigid luminaires (fixtures) »530.51«.

B Lamps must be controlled outside the vault by a switch having a pole in each ungrounded conductor and provided with a pilot light indicating "on" or "off." This switch must disconnect every ungrounded conductor terminating in any outlet within the vault from any supply source »530.51«.

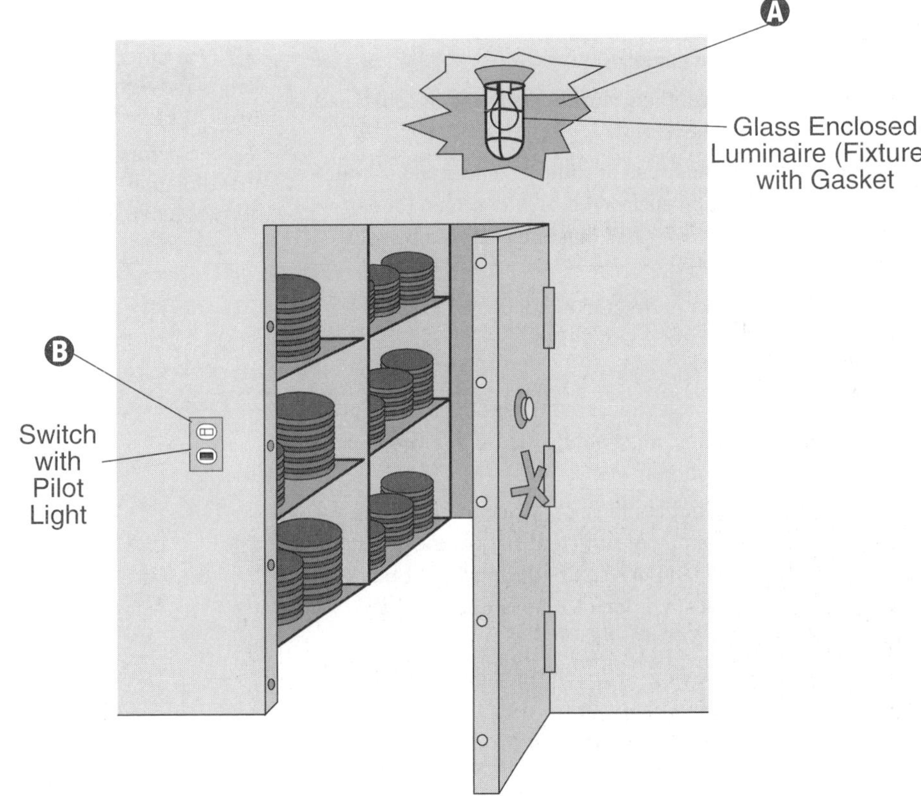

CAUTION *Unless otherwise permitted in 530.51, no receptacles, outlets, heaters, portable lights, or other portable electric equipment are allowed inside cellulose nitrate film storage vaults »530.52«.*

MOTION PICTURE PROJECTION ROOMS

General Provisions

Nonprofessional Projectors: Any type other than described in 540.2 »540.2 «. These projectors, including miniature types, employing cellulose acetate (safety) film, may be operated without a projection room »540.31 «.

Motor generator sets, transformers, rectifiers, rheostats, and similar equipment for the supply (or control) of current to projection (or spotlight equipment) using nitrate film must be located in a separate room. If inside the projection room, they must be located (or guarded) so that arcs/sparks cannot contact film, and the commutator end(s) of motor generator sets must comply with *one* of the following conditions:

(1) Be of the totally enclosed, enclosed fan-cooled, or enclosed pipe-ventilated type

(2) Be enclosed in separate rooms or housings built of noncombustible construction to exclude flyings or lint, and having proper clean air ventilation

(3) Have the motor-generator brush or sliding-contact end enclosed by solid metal covers

(4) Brushes or sliding contacts must be enclosed in substantial, tight metal housings

(5) The upper half of the brush or sliding-contact end of the motor-generator must be enclosed by a wire screen (or perforated metal) while the lower half must be enclosed by solid metal covers

(6) Have wire screens or perforated metal placed at the commutator of brush ends »540.11(A) «.

Extraneous equipment (switches, overcurrent devices, etc.), not normally a part of projection, sound reproduction, flood or other special effect lamps, must not be placed in projection rooms, unless an exception is met »540.11(B) «.

Conductors supplying outlets for arc and xenon projectors (professional-type) must be of sufficient size for the projector, but in no case smaller than 8 AWG. Conductors for incandescent-type projectors must conform to the normal wiring standards of 210.24 »540.13 «.

Insulated conductors having an operating temperature rating no less than 392°F (200°C) must be used on all lamps (or other equipment) where the ambient temperature at the installed conductors will exceed 122°F (50°C) »540.14 «.

Ⓐ Article 540 provisions apply to motion picture projection rooms, motion picture projectors, and associated equipment, both professional and nonprofessional types, which use incandescent, carbon arc, xenon, or other light source equipment capable of producing hazardous gases, dust, or radiation »540.1 «.

Ⓑ Any professional-type projectors must be located within a permanently constructed projection room approved for the type of building in which it is located. All projection ports, spotlight ports, viewing ports, and similar openings must be completely enclosed by use of glass or other approved material. Such rooms do not qualify as hazardous (classified) locations as defined in article 500 »540.10 «.

Ⓒ A minimum 30-in. (750 mm) wide working space must be provided on each side and at the rear of each motion picture projector, floodlight, spotlight, or similar equipment »540.12 «. Adjacent pieces of equipment can be served by one such space »540.12 *Exception* «.

Ⓓ **Professional Projector:** Either of two types of projectors: (1) using 35- or 70-mm film [a minimum of 1⅜ in. (35 mm) wide with 5.4 edge perforations per in. (212 perforations per meter)]; or (2) a type using carbon arc, xenon, or other light source equipment producing hazardous gases, dust, or radiation »540.2 «.

Ⓔ Projectors and enclosures for arc, xenon, and incandescent lamps and rectifiers, transformers, rheostats, and similar equipment must be listed »540.20 «.

Ⓕ Projectors and other equipment must be marked with the manufacturer's name (or trademark) as well as the voltage and current for which they are designed, per 110.21 »540.21 «.

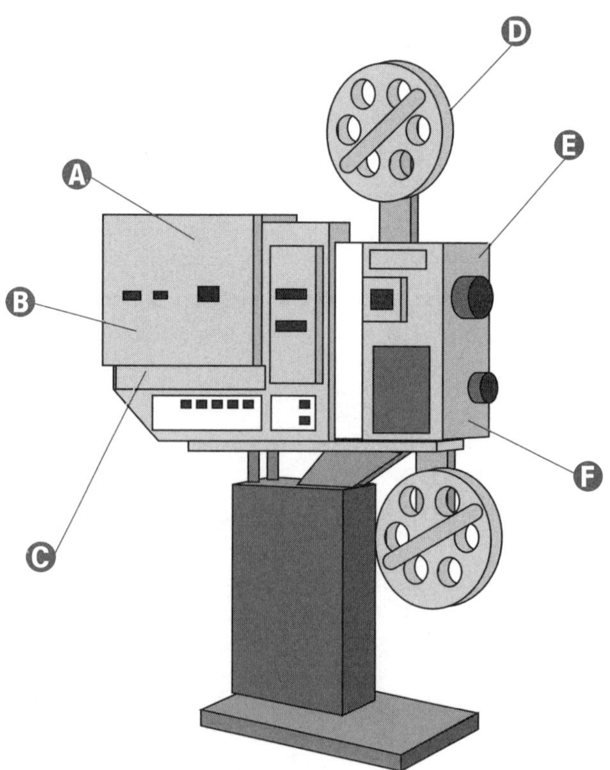

MANUFACTURED BUILDINGS

Manufactured Building Requirements

Wherever other *Code* article requirements differ from Article 545, the requirements of Article 545 apply »545.2«.

Fittings and connectors to be concealed at the time of on-site assembly (where tested, identified, and listed to applicable standards) are permitted for on-site module/component interconnection. Such fittings and connectors must equal the employed wiring method in insulation, temperature rise, and fault-current withstand. They must also be able to endure the vibration and minor relative motions occurring in the manufactured building components »545.13«.

In closed construction, cables can be secured only at cabinets, boxes, or fittings where using 10 AWG or smaller conductors and protection against physical damage is provided »545.4(B)«.

Service-entrance, service-lateral, feeder, or branch-circuit supply conductor routing provisions must be made to the service or building disconnecting means conductor »545.5«.

Install service-entrance conductors after the building is site-erected, unless the point of attachment location is known prior to manufacture »545.6«.

Exposed conductors and equipment must be protected during manufacturing, packaging, transit, and setup at the building site »545.8«.

Boxes of dimensions other than those required in Table 314.16(A) can be installed where tested, identified, and listed to applicable standards »545.9(A)«.

Any box no larger than 100 cubic in. (1650 cm^3), intended for closed-construction mounting, must be affixed with anchors (or clamps) to provide a rigid and secure installation »545.9(B)«.

A receptacle (or switch) with integral enclosure and mounting means can be installed where such is tested, identified, and listed to applicable standards »545.10«.

Prewired panels and building components must provide for the bonding, or bonding and grounding, of all exposed metals likely to be energized, per Article 250, Parts V, VI, and VII »545.11«.

Provisions must be made to route a grounding electrode conductor from the service, feeder or branch-circuit supply to the point of attachment to the grounding electrode »545.12«.

Closed Construction: Any building, building component, assembly, or system manufactured in such a manner that concealed parts of manufacture processes cannot be inspected prior to building site installation without disassembly, damage, or destruction »545.3».

Building Component: Any subsystem, subassembly, or other system designed for use in, integral with, or as part of a structure. A building component can be structural, electrical, mechanical, plumbing, fire protection, and other health and safety systems »545.3«.

Building System: Plans, specifications, and documentation for a manufactured building system, or for a type of system building component, such as structural, electrical, mechanical, plumbing, fire protection, and other systems affecting health and safety, including variations thereof specifically permitted by regulation, provided the variations are submitted as part of the building system or as an amendment thereto »545.3«.

(A) Article 545 covers manufactured building and building components requirements »545.1«.

(B) **Manufactured Building:** Any closed construction building made (or assembled) in manufacturing facilities, whether on or off the building site, for installation, or assembly and installation, on the building site, excluding manufactured homes, mobile homes, park trailers, or recreational vehicles »545.3«.

(C) All *Code* approved raceways and cable wiring methods, and such other wiring systems specifically intended and listed for use in manufactured buildings, are permitted with both listed fittings and fittings listed/identified for manufactured buildings »545.4(A)«.

NOTE

Service equipment must be installed in accordance with 230.70 »545.7«.

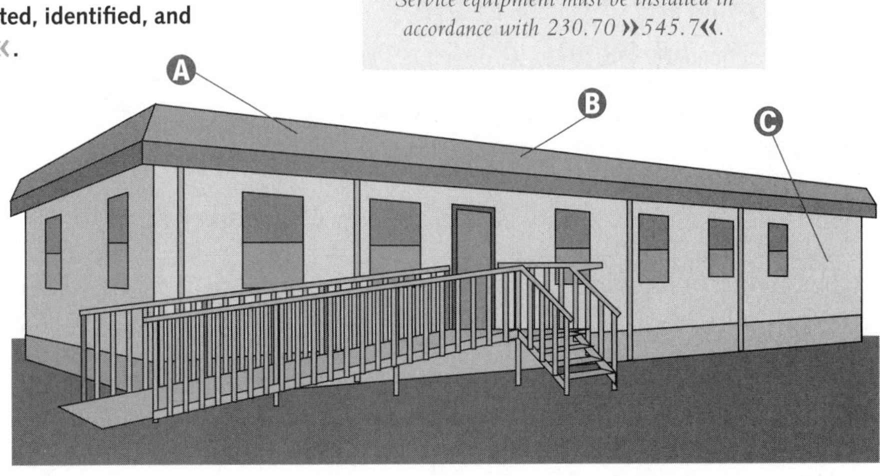

AGRICULTURAL BUILDINGS

Agricultural Building Provisions

Motors and other rotating electrical machinery must either be totally enclosed or designed to minimize the entrance of dust, moisture, or corrosive particles »547.7«.

Slatted floors, supported by structures that are part of an equipotential plane, do not require bonding »547.10(C)«.

All circuits providing electric power to equipment that is accessible to animals in dirt confinement areas must have GFCI protection »547.10(B)«.

For the purpose of 547.10, the term livestock does not include poultry »547.10«.

A This Article's provisions apply to agricultural buildings, part of buildings, or adjacent areas of similar nature as specified in (A) and (B).

(A) Agricultural buildings where excessive dust, and dust in combination with water, may accumulate. Included are all areas of poultry, livestock, and fish confinement systems, where litter dust or feed dust (including mineral feed particles) may collect.

(B) Agricultural buildings with a corrosive atmosphere. Such buildings include areas where:

(1) poultry and animal excrement can cause corrosive vapors

(2) corrosive particles may mix with water

(3) periodic washing for cleaning and sanitizing with water and cleansing agents creates a damp, even wet, environment.

(4) similar conditions exist »547.1«.

B The wiring method(s) employed must be types UF, NMC, copper SE cables, jacketed Type MC cable, RNC, LFNC, or other location-suitable cables/raceways having approved fittings. Article 320 and Article 502 wiring methods are permitted for areas described in 547.1(A) »547.5(A)«.

C An **Equipotential Plane** is an area where a wire mesh (or other conductive element) is: (1) embedded in or placed under concrete; (2) bonded to all metal structures and fixed nonelectrical equipment that may become energized; and (3) connected to the electrical grounding system, thus preventing a voltage difference from developing within the plane. »547.2«.

D All cables must be secured within 8 in. (200 mm) of each cabinet, box, or fitting. The ¼-in. (6-mm) airspace required for nonmetallic boxes, fittings, conduit, and cables in 300.6(C) does not apply in buildings covered by this article »547.5(B)«.

E Luminaires (lighting fixtures) must comply with all of the following:

(1) By installation, the entrance of dust, foreign matter, moisture, and corrosive material must be minimized.

(2) If subject to physical damage, must have a suitably protective guard.

(3) If exposed to moisture, whether from condensation, cleansing water, or solution, must be watertight »547.8«.

F Outdoor confinement areas such as feedlots, must have equipotential planes installed around metallic equipment accessible to animals and likely to become energized. The equipotential plane must completely enclose the area around the equipment where the animal will stand while accessing the equipment »547.10(A)«.

G Where grounding is required for noncurrent-carrying metal parts of equipment, raceways, and other enclosures, a copper equipment grounding conductor must be installed between the equipment and the building disconnecting means. If underground, the equipment grounding conductor must be insulated or covered »547.5(F)«.

CAUTION *Protect all electrical wiring and equipment subject to physical damage »547.5(E)«.*

Equipotential planes must be installed in concrete floors of livestock confinement buildings that contain metal equipment accessible to livestock that is likely to become energized »547.10(A)«.

NOTE

All 125-volt, single-phase, 15- and 20-ampere general purpose receptacles in equipotential plane areas, outdoors, and wet locations must have personnel GFCI protection »547.5(G)«.

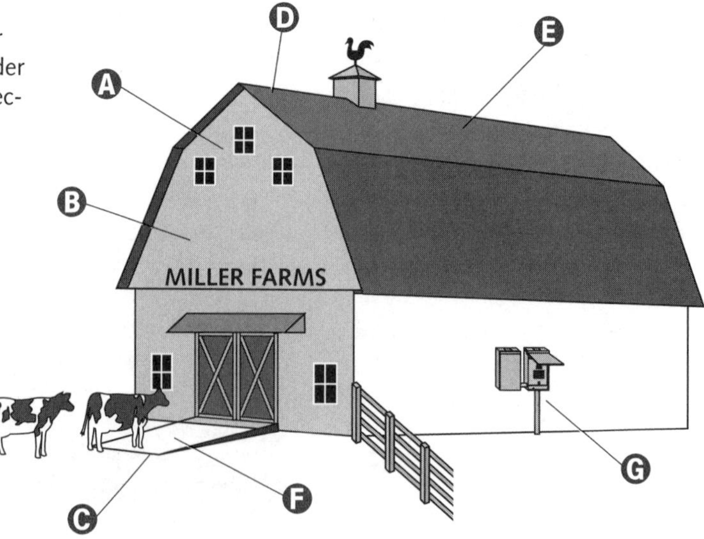

MOBILE HOMES, MANUFACTURED HOMES, AND MOBILE-HOME PARKS

General Requirements

Mobile Home: A factory-assembled structure transportable in one or more sections, built on a permanent chassis, designed for use as a dwelling having no permanent foundation where connected to the required utilities, and including the plumbing, heating, air-conditioning, and electric systems contained therein »550.2«.

Manufactured Home: A structure, transportable in one or more sections, that is 8 body-ft (2.5 m) or more in width, or 40 body ft (12 m) or more in length in the traveling mode, or when erected on site, is 320 ft^2 (30 m^2) or more, which is built on a chassis and designed for dwelling use (with or without a permanent foundation) where connected to the required utilities, inclusive of plumbing, heating, air-conditioning, and electrical systems contained therein »550.2«.

At least one receptacle outlet must be installed outdoors accessible at grade level, and not more than 6½ ft (2.0 m) above grade. A receptacle outlet located in an outside accessible compartment qualifies as an outdoor receptacle. These receptacle outlets must have GFCI protection for personnel »550.13(B) and (D)(8)«.

If a pipe heating cable outlet is installed, the outlet must be:

(1) Located within 2 ft (600 mm) of the cold water inlet.

(2) Connected to an interior branch-circuit, other than a small appliance branch-circuit. A bathroom receptacle circuit can be utilized for this purpose.

(3) On a circuit where all outlets are on the load side of the ground-fault circuit-interrupter protection for personnel.

(4) Mounted on the mobile home's underside, although this is not considered to be the outdoor receptacle outlet required in 550.13 (D)(8) »550.13(E)«.

CAUTION *All electrical materials, devices, appliances, fittings, and other equipment must be (1) listed or labeled by a qualified testing agency and (2) connected in an approved manner when installed »550.4(D)«.*

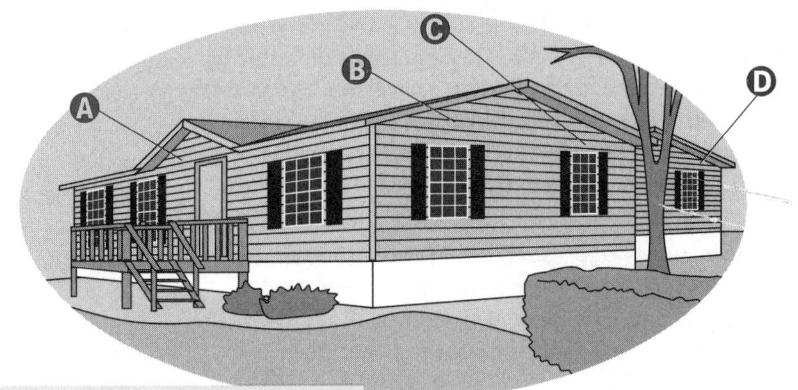

NOTE

Mobile homes, not installed in mobile-home parks, must comply with the provisions of this article »550.4(B)«.

Use only approved and listed fixed-type wiring methods to join portions of a circuit that must be electrically joined, and which are located in adjacent mobile-home sections after installation on its support foundation. The circuit's junction must be accessible for disassembly when preparing the home for relocation »550.19(A)«.

A mobile home not intended as a dwelling unit is not required to meet this Article's provisions pertaining to the number or capacity of circuits required. Included are those equipped for sleeping purposes only, contractor's on-site offices, construction job dormitories, mobile studio dressing rooms, banks, clinics, mobile stores, or intended for the display or demonstration of merchandise or machinery. Such non-dwelling units must, however, meet all other applicable requirements of this Article if provided with an electrical installation to be energized from a 120-volt or 120/240-volt ac power supply system. If design or available power supply systems dictate a different voltage, make adjustments in accordance with the appropriate articles/sections for the voltage used »550.4(A)«.

A Article 550 provisions cover the electrical conductors and equipment in/on mobile and manufactured homes, mobile and manufactured home electrical supply conductors, and the installation of electrical wiring, luminaires (fixtures), equipment, and appurtenances related to electrical installations within a mobile-home park up to the mobile-home service-entrance conductors or, if none, the service equipment »550.1«.

B For *Code* purposes (unless otherwise indicated), the term **mobile home** includes manufactured homes »550.2 Mobile Home«.

C This Article's provisions apply to mobile homes designed to connect to a wiring system rated 120/240 volts, nominal, three-wire ac, with grounded neutral »550.4(C)«.

D Outdoor or under-chassis line-voltage (120 volts, nominal, or higher) wiring, exposed to moisture/physical damage, must be protected by rigid or IMC. Conductors must be suitable for wet locations »550.15(H)«. EMT or RNC is permitted if closely routed against frames and equipment enclosures »550.15(H) *Exception*«.

Power Supply

Cords with adapters and pigtail ends, extension cords, and similar items must not be attached to, or shipped with, a mobile home »550.10(B)«.

A cord passing through walls or floors must be protected by means of conduits and bushings (or equivalent). The cord can be installed within the walls, provided a continuous raceway (maximum 1¼ in. [32 mm] in size) is installed from the branch-circuit panelboard to the underside of the mobile home floor »550.10(G)«.

The power-supply cord's attachment plug cap and any connector cord assembly or receptacle must be permanently protected against corrosion and mechanical damage, if such devices are externally located while the mobile home is in transit »550.10(H)«.

If the calculated load exceeds 50 amperes, or if a permanent feeder is used, the supply must be by means of:

(1) One mast weatherhead installation, installed per Article 230, containing four continuous, insulated, color-coded feeder conductors, one of which must be an equipment grounding conductor; or

(2) A metal raceway or RNC from the internal disconnecting means to the mobile home's underside, with provisions for attachment to the raceway on the mobile home's underside via a suitable junction box (or fitting) [with or without conductors as in 550.10(I)(1)] »550.10(I)«.

A The attachment plug cap must not only be a three-pole, four-wire, grounding type, rated 50 amperes, 125/250 volts with a configuration as shown in Figure 550.10(C) but also must be intended for use with a 50-ampere, 125/250-volt receptacle configuration as shown in Figure 550.10(C). It must be listed, individually or as part of a power-supply cord assembly, for such purpose, and molded to (or installed on) the flexible cord at the point where the cord enters the attachment plug cap »550.10(C)«.

B In a right-angle cap configuration the grounding member must be the farthest from the cord »550.10(C)«.

C The mobile home power supply must be a feeder assembly consisting of a single listed 50-ampere mobile home power-supply cord having an integrally molded (or securely attached) plug cap, or a permanently installed feeder »550.10(A)«.

D The power-supply cord must bear the marking: "FOR USE WITH MOBILE HOMES — 40 AMPERES" or "FOR USE WITH MOBILE HOMES — 50 AMPERES" »550.10(E)«.

E The cord must be a four-conductor listed type. One conductor must be identified by a continuous green color, or a continuous green color with one or more yellow stripes, for use as the grounding conductor »550.10(B)«.

F A suitable clamp (or equivalent) must be provided at the distribution panelboard knockout to afford cord strain relief and effectively prevent strain transmission to the terminals when the power-supply cord is handled as intended »550.10(B)«.

G A mobile home power-supply cord, if present, must be permanently attached to the distribution panelboard either directly or via a junction box also permanently connected thereto, with the free end terminating in an attachment plug cap »550.10(B)«.

H From the end of the power-supply cord (including bared leads) to the face of the attachment plug cap, the cord must not be less than 21 ft (6.4 m) nor more than 36½ ft (11 m) in length. From the face of the attachment plug cap to the mobile home entry point, the cord must be at least 20 ft (6.0 m) long »550.10(D)«.

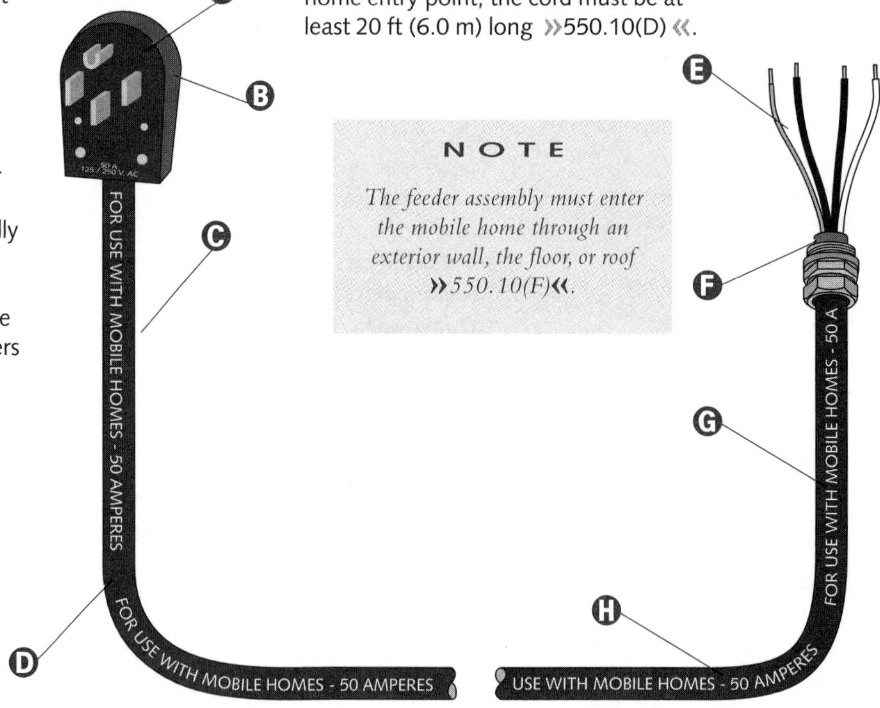

NOTE

The feeder assembly must enter the mobile home through an exterior wall, the floor, or roof »550.10(F)«.

Distribution Panelboards and Branch-Circuits

A distribution panelboard must be rated no less than 50 amperes and employ a two-pole circuit breaker rated 40 amperes for a 40-ampere supply cord, or 50 amperes for a 50-ampere supply cord. A distribution panelboard employing a disconnect switch and fuses must be rated 60 amperes and employ a single two-pole, 60-ampere fuseholder with 40- or 50-ampere main fuses for 40- or 50-ampere supply cords, respectively. The fuse size must be plainly marked on the outside of the distribution panelboard » 550.11(A)« .

Each mobile home must have branch-circuit distribution equipment that includes circuit breaker or fuse overcurrent protection for each branch-circuit. The branch-circuit overcurrent devices must be rated as follows: (1) not more than the circuit conductors, and (2) not more than 150% of a single appliance rated 13.3 amperes (or more) supplied by an individual branch circuit, but (3) not more than the overcurrent protection size and type marked on the air conditioner (or other motor-operated appliance)» 550.11(B)« .

A metal nameplate on the outside adjacent to the feeder assembly entrance must read:

THIS CONNECTION FOR 120/240-VOLT, THREE-POLE, FOUR-WIRE, 60-HERTZ, _____ -AMPERE SUPPLY

Fill in the blank with the correct ampere rating » 550.11(D)« .

Multiply 3 VA/ft^2 (33 VA/m^2) by the outside dimensions of the mobile home (coupler excluded) and divide by 120 volts to determine the number of 15- or 20-ampere lighting circuits » 550.12(A)« .

There must be one or more adequately rated circuits in accordance with the following:

(1) Ampere rating of fixed appliances not over 50% of circuit rating if lighting outlets (receptacles, other than kitchen, dining area, and laundry, considered as lighting outlets) exist on the same circuit.

(2) For fixed appliances on a circuit without lighting outlets, the sum of rated amperes must not exceed the branch-circuit rating. Motor loads or other continuous duty loads must not exceed 80% of the branch-circuit rating.

(3) The rating of a single cord- and plug-connected appliance on a circuit having no other outlets must not exceed 80% of the circuit rating.

(4) The rating of a range branch-circuit must be based on the range demand as specified in 550.18(B)(5) » 550.12(D)« .

A Each mobile home must have a single disconnecting means consisting of a circuit breaker, or a switch and fuses, and its accessories installed in a readily accessible location near the supply cord (or conductor's) entry point into the mobile home. The main circuit breakers (or fuses) must be plainly marked "Main." This equipment must contain a solderless-type grounding connector (or bar) with sufficient terminals for all grounding conductors. The neutral bar termination of the grounded circuit conductors must be insulated in accordance with 550.16(A) » 550.11(A)« .

B While the distribution panelboard must be in an accessible location, it cannot be located in a bathroom or clothes closet. A clear working space at least 30 in. (750 mm) wide and 30 in. (750 mm) in front of the panelboard must be provided. This space must extend from the floor to the top of the distribution panelboard » 550.11(A)« .

C The bottom of the distribution equipment, whether circuit breaker or fused type, must be at least 24 in. (600 mm) above the mobile home's floor level » 550.11(A)« .

D The distribution equipment must be suitably rated for the connected load » 550.11(A)« .

E Where circuit breakers provide branch-circuit protection, 240-volt circuits must be protected by a two-pole common or companion trip, or handle-tied paired circuit breakers » 550.11(C)« .

F Determine the required number of branch circuits in accordance with 550.12(A) through (E) » 550.12« .

G The branch-circuit equipment can be combined with the disconnecting means as a single assembly. Such a combination can be designated as a distribution panelboard. If a fused distribution panelboard is used, the maximum fuse size for the mains must be plainly marked with ¼-in. (6-mm), or taller, lettering visible during fuse replacements. If plug fuses and fuseholders are used, they must be tamper-resistant Type S, enclosed in dead-front fuse panelboards. Electrical distribution panelboards containing circuit breakers must also be dead-front type » 550.11« .

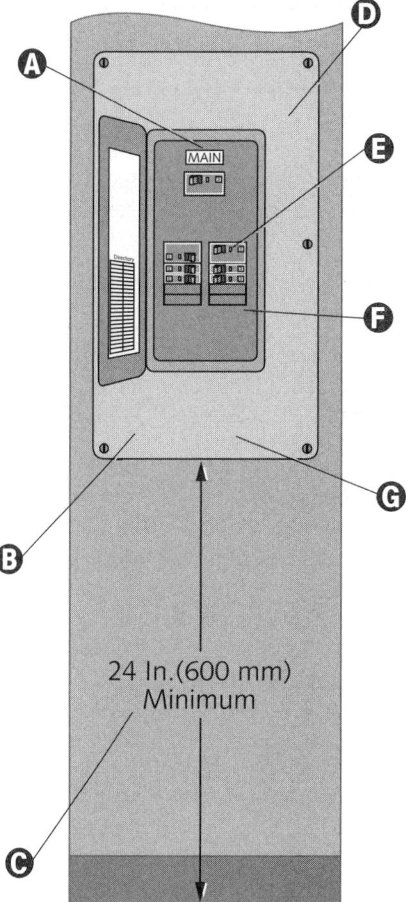

Grounding

Cord-connected appliances (such as washing machines, clothes dryers, refrigerators, and the electrical system of gas ranges, etc.) must be grounded via a cord having a grounding conductor and grounding-type attachment plug »550.16(B)(3)«.

Metallic pipes (gas, water, and waste) as well as metallic air-circulating ducts are considered bonded if connected to the chassis terminal by clamps, by solderless connectors, or by appropriate grounding-type straps »550.16(C)(3)«.

Any metallic exterior covering is considered bonded if (1) the metal panels overlap one another and are securely attached to the wood or metal frame parts by metallic fasteners, and (2) the lower panel of the metallic exterior covering is secured at opposite ends by metallic fasteners at a chassis cross member by two metal straps per mobile home unit (or section). The bonding strap must be at least 4 in. (100 mm) wide and of the same material as the skin, or of a material of equal or better electrical conductivity. The straps must be fastened with paint-penetrating fittings such as screws and starwashers or an equally appropriate substitute »550.16(C)(4)«.

In the electrical system, all exposed metal parts, enclosures, frames, lamp luminaires (fixtures) canopies, etc., must be effectively bonded to the grounding terminal or enclosure of the distribution panelboard »550.16(B)(2)«.

Ⓐ Grounding terminals must be of the solderless type, listed as pressure-terminal connectors, and recognized for the wire size employed »550.16(C)(2)«.

Ⓑ The supply cord or permanent feeder's green-colored insulated grounding wire must be connected to the grounding bus in the distribution panelboard or disconnecting means »550.16(B)(1)«.

Ⓒ The bonding conductor must be solid or stranded, insulated or bare, but must be 8 AWG copper minimum, or the equivalent. The bonding conductor routing must prevent physical damage exposure »550.16(C)(2)«.

Ⓓ All potentially energizable exposed, noncurrent-carrying metal parts must be effectively bonded to either the grounding terminal or the distribution panelboard's enclosure. A bonding conductor must be connected between the distribution panelboard and an accessible terminal on the chassis »550.16(C)(1)«.

CAUTION *Neither the mobile home's frame nor any appliance frame shall be connected to the grounded circuit conductor (neutral) in the mobile home »550.16«.*

Ⓔ Remove and discard bonding screws, straps, or buses in the distribution panelboard or in appliances »550.16(A)(1)«.

Ⓕ The grounded circuit conductor (neutral) must be insulated from the grounding conductors, from equipment enclosures, and other grounded parts. The grounded (neutral) circuit terminals in distribution panelboards, ranges, clothes dryers, counter-mounted cooking units, and wall-mounted ovens must be insulated from the equipment enclosure »550.16(A)(1)«.

Ⓖ In a mobile home, grounding of metal parts (both electrical and non-electrical) must be accomplished by connection to a distribution panelboard's grounding bus. The grounding bus must be grounded through the green colored insulated conductor in the supply cord (or the feeder wiring) to the service ground in the service-entrance equipment, located adjacent to the mobile home »550.16«.

NOTE

Connections of ranges and clothes dryers with 120/240-volt, three-wire ratings must be made with (a) four-conductor cord and three-pole, four-wire, grounding-type plugs or (b) by Type AC cable, Type MC cable, or (c) conductors enclosed in flexible metal conduit »550.16(A)(2)«.

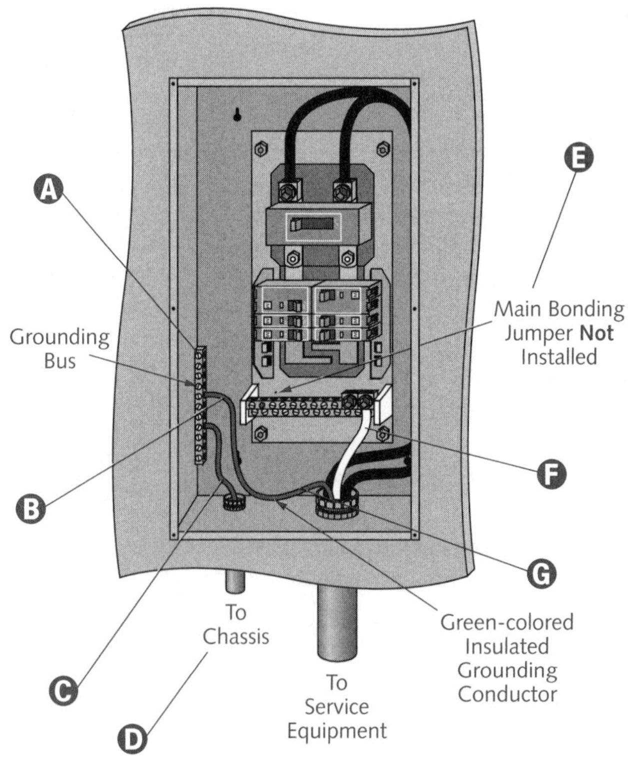

Services and Feeders

The mobile-home park secondary electrical distribution system supplying mobile home lots must be single-phase, 120/240 volts, nominal. For Part III purposes, where the park service exceeds 240 volts, nominal, transformers and secondary distribution panelboards must be treated as services ›› 550.30 ‹‹.

The manufactured home service equipment can be installed in, or on, a manufactured home, provided all of the conditions in 550.32(B) are met ›› 550.32(B) ‹‹.

Park electrical wiring systems must be calculated (at 120/240 volts) on the larger of (1) 16,000 volt-amperes for each mobile home lot, or (2) the load calculated per 550.18 for the largest typical mobile home that each lot accepts. The feeder or service load can be computed in accordance with Table 550.31. No demand factor is allowed for any other load, except as provided in this *Code*. Mobile home service and feeder conductors, in compliance with 310.15(B)(6), are permitted ›› 550.31 ‹‹.

Mobile home feeder conductors must consist of either a listed cord, factory-installed in accordance with 550.10(B), or a permanently installed feeder consisting of four insulated, color-coded conductors, each identified by the factory or field marking, in compliance with 310.12. Stripping the insulation is not an acceptable method of identifying equipment grounding conductors ›› 550.33(A) ‹‹.

If a mobile home feeder is installed between service equipment and a mobile home disconnecting means, as covered in 550.32(A), the equipment grounding conductor can be omitted where the grounded circuit conductor is grounded at the disconnecting means as required in 250.32(B) ›› 550.33(A) *Exception* ‹‹.

A Mobile home service equipment must be rated no less than 100 amperes at 120/240 volts. A mobile home feeder assembly must be connected by a permanent wiring method. Power outlets used as mobile-home service equipment can also contain receptacles rated up to 50 amperes with appropriate overcurrent protection. Fifty-ampere receptacles must conform to Figure 550.10(C)'s configuration ›› 550.32(C) ‹‹.

B Additional receptacles are allowed for connection of electrical equipment located outside the mobile home. All such 125-volt, single-phase, 15- and 20-ampere receptacles must be protected by a listed GFCI ›› 550.32(E) ‹‹.

CAUTION *Mobile home lot feeder circuit conductors must have adequate capacity for the loads supplied and must be rated no less than 100 amperes at 120/240 volts ›› 550.33(B) ‹‹.*

C Any mobile home service equipment utilizing a 125/250-volt receptacle must be marked as follows.

TURN DISCONNECTING SWITCH
OR CIRCUIT BREAKER OFF BEFORE INSERTING
OR REMOVING PLUG. PLUG MUST
BE FULLY INSERTED OR REMOVED.

The marking must be located adjacent to the service equipment's receptacle outlet ›› 550.32(G) ‹‹.

D The **mobile home** service equipment must be located adjacent to, not mounted in or on, the mobile home. It must be located in sight from and within 30 ft (9.0 m) from the exterior wall of the mobile home it serves. The service equipment can be located elsewhere on the premises, provided a disconnecting means suitable for service equipment is located as stated above for service equipment. Grounding at the disconnecting means must be compliant with 250.32 ›› 550.32(A) ‹‹.

E Mobile home service equipment or the local external disconnecting means permitted in 550.32(A) must provide a means for connecting (by a fixed wiring method) an accessory building, structure, or additional electrical equipment located outside the mobile home ›› 550.32(D) ‹‹.

F Outdoor mobile home disconnecting means must be installed so that the bottom of the disconnecting means enclosure is at least 2 ft (600 mm) above finished grade (or working platform). Installation must ensure that the center of the operating handle's grip, in the highest position, is no more than 6 ft 7 in. (2.0 m) above the finished grade (or working platform) ›› 550.32(F) ‹‹.

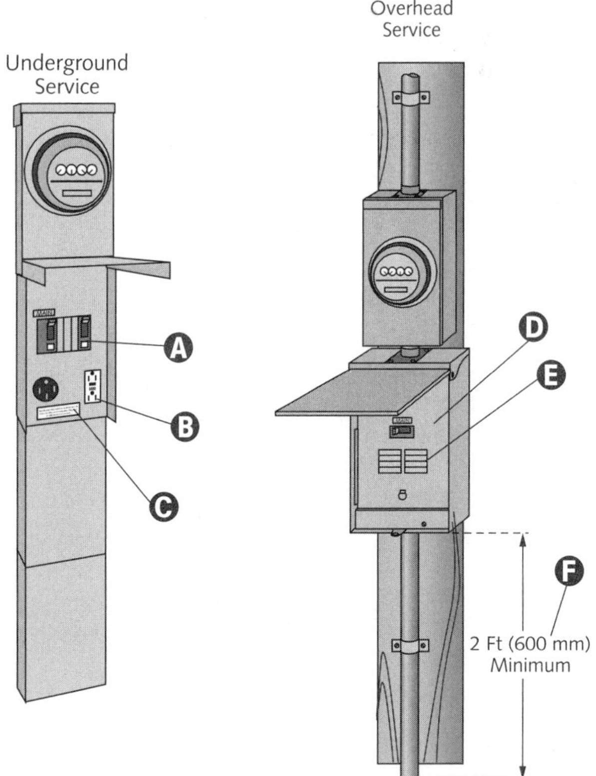

RECREATIONAL VEHICLES AND RECREATIONAL-VEHICLE PARKS

Recreational Vehicles

Recreational Vehicle Park: A plot of land on which multiple recreational vehicle sites are located, established, or maintained for general public recreational vehicle occupancy as temporary living quarters for recreation or vacation purposes »551.2«.

Recreational Vehicle Site: A plot of ground within a recreational vehicle park intended for the temporary accommodation of either a recreational vehicle, or a camping unit »551.2«.

Recreational Vehicle Stand: The area of a recreational vehicle site intended for the placement of a recreational vehicle »551.2«.

Low-voltage circuits furnished and installed by the recreational vehicle manufacturer, other than automotive vehicle circuits or extensions thereof, are subject to this *Code*. Circuits supplying lights subject to federal or state regulations must comply with applicable government regulations in addition to this *Code* »551.10(A)«.

Generator installations must comply with 551.30(A) through (E).

Recreational vehicle main power-supply assemblies must meet 551.44(A) through (D) requirements.

Recreational vehicle distribution panelboard provisions are found in 551.45(A) through (C).

The power-supply assembly (or assemblies) must be factory supplied/installed and be of a type specified in (1) or (2) »551.46(A)«.

Recreational vehicle power-supply attachment plugs must meet 551.46(C) requirements.

Recreational vehicle wiring methods must comply with 551.47(A) through (R).

A Article 551 provisions cover the electrical conductors and equipment installed within (or on) recreational vehicles, the conductors that connect recreational vehicles to an electrical supply, and the installation of electrical related equipment and devices within a recreational-vehicle park »551.1«.

B Recreational vehicle electrical equipment and material, intended for connection to a wiring system rated 120 volts, nominal, two-wire with ground, or a wiring system rated 120/240 volts, nominal, three-wire with ground, must be listed and installed in accordance with the requirements of Parts I, III, IV, V, and VI of Article 551 »551.40(A)«.

C **Recreational Vehicle:** A vehicular-type unit primarily designed as temporary living quarters for recreational, camping, or travel use, which either has its own motive power or is mounted on (or drawn) by another vehicle. Basically this includes: travel trailers, camping trailers, truck campers, and motor homes »551.2«. (See 551.2 for other word and phrase definitions relating to recreational vehicles and recreational vehicle parks.)

D Every appliance must be accessible for inspection, service, repair, and replacement without removal of permanent construction. Means must be provided to securely fasten appliances in place during travel »551.57«.

E **Motor Home:** A vehicular unit designed to provide temporary living quarters for recreational, camping, or travel use built on, or permanently attached to, a self-propelled motor vehicle chassis, or on a chassis cab/van that is an integral part of the total vehicle »551.2«.

NOTE

A recreational vehicle, not used for the purpose defined in 551.2, is not required to comply with Part I provisions relating to the number (or capacity) of circuits. The recreational vehicle must, however, meet all other applicable requirements of this Article if it has an electrical installation that can be energized from a 120- or 120/240-volt, nominal, ac power-supply system »551.4(A)«.

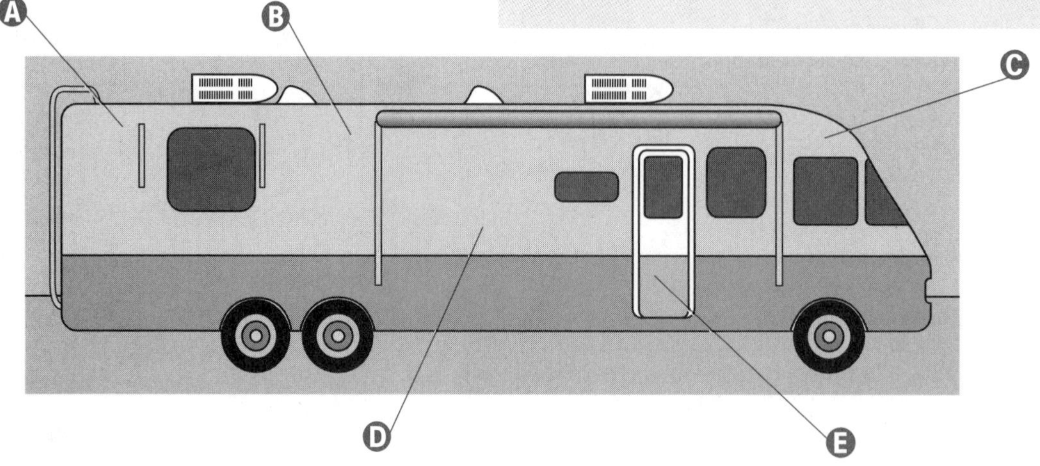

Recreational-Vehicle Parks

Where park service exceeds 240 volts, transformers and secondary distribution panelboards must be treated as services »551.73(B)«.

Recreational site feeder-circuit conductors must have adequate ampacity for the loads supplied and must be rated no less than 30 amperes. The grounded and ungrounded conductors must have the same ampacity »551.73(D)«.

No grounding electrode connection can be made to the neutral conductor on the load side of the service disconnecting means or to the transformer distribution panelboard »551.76(D)«.

If provided, the recreational vehicle site electrical equipment must be located on the parked vehicle's left (road) side, on a line that is 5 to 7 ft (1.5 to 2.1 m) from the left edge (driver's side of the parked RV), and from 16 ft (4.9 m) forward of the rear of the stand »551.77(A)«.

A disconnecting switch (or circuit breaker) must be provided in the site supply equipment for disconnecting the recreational vehicle power supply »551.77(B)«.

All site supply equipment must be accessible by an unobstructed entrance or passageway at least 2 ft (600 mm) wide and 6½ ft (2.0 m) high »551.77(C)«.

Sufficient space must be constantly available around all electrical equipment for ready and safe operation, in accordance with 110.26 »551.77(E)«.

A At least 70% of all electrically supplied recreational vehicle sites must be equipped with a 30-ampere, 125-volt receptacle conforming to Figure 551.46(C). Additional receptacle configurations conforming to 551.81 can be included. Dedicated tent sites having a 15- or 20-ampere electrical supply can be excluded when determining 30- or 50-ampere receptacles »551.71«.

B The recreational-vehicle park secondary electrical distribution system to 50-ampere recreational vehicle sites must be supplied from a branch-circuit of the voltage class and rating of the receptacle. Other recreational vehicle sites with 125-volt, 20- and 30-ampere receptacles can be derived from any grounded distribution system supplying 120-volt, single-phase power. The neutral conductors can be reduced in size below the minimum required ungrounded conductor size for 240-volt, line-to-line, permanently connected loads only »551.72«.

C Every electrical equipped recreational vehicle site must have at least one 20-ampere, 125-volt receptacle »551.71«.

D Five% (or more) of all electrically supplied recreational vehicle sites must be equipped with a 50-ampere, 125/250-volt receptacle conforming to the configuration identified in Figure 551.46(C). These electrical supplies can include additional receptacles configured per 551.81 »551.71«.

E Where the site supply equipment contains a 125/250-volt receptacle, the equipment must be marked as follows: "Turn disconnecting switch or circuit breaker off before inserting or removing plug. Plug must be fully inserted or removed." The marking must be located on the equipment adjacent to the receptacle outlet »551.77(F)«.

F Electrical service and feeders must be calculated on the basis of no less than 9600 volt-amperes per site equipped with 50-ampere, 120/240-volt supply facilities; 3600 volt-amperes per site equipped with both 20-ampere and 30-ampere supply facilities; 2400 volt-amperes per site equipped with only 20-ampere supply facilities; and 600 volt-amperes per dedicated tent site. Table 551.73 demand factors are the minimum allowable demand factors that can be used in calculating load for service and feeders. If the recreational vehicle site's electrical supply has more than one receptacle, the calculated load is computed for the highest rated receptacle only »551.73(A)«.

G Site supply equipment must be located no less than 2 ft (600 mm), and no more than 6½ ft (2.0 m), above the ground »551.77(D)«.

H Underground service, feeder, branch-circuit, and recreational vehicle site feeder-circuit conductors must be installed according to 551.80(A) and (B)

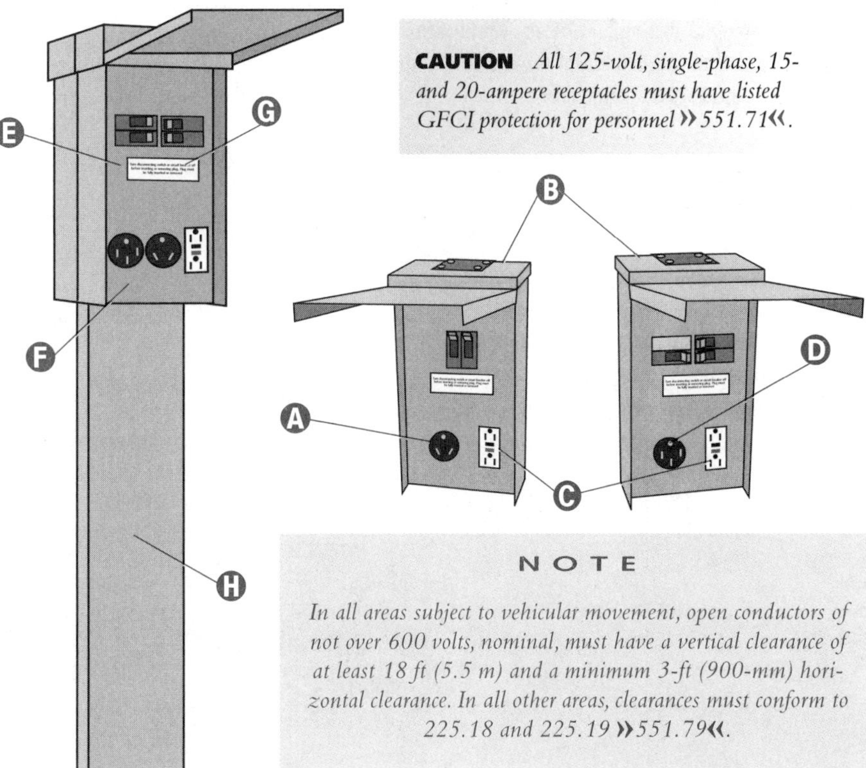

CAUTION *All 125-volt, single-phase, 15- and 20-ampere receptacles must have listed GFCI protection for personnel »551.71«.*

NOTE

In all areas subject to vehicular movement, open conductors of not over 600 volts, nominal, must have a vertical clearance of at least 18 ft (5.5 m) and a minimum 3-ft (900-mm) horizontal clearance. In all other areas, clearances must conform to 225.18 and 225.19 »551.79«.

FLOATING BUILDINGS

Floating Building Provisions

The service equipment for a floating building must be located adjacent to, but not in (or on), the building »553.4«.

One set of service conductors can serve multiple service equipment sets »553.5«.

Each floating building must be supplied by a single set of feeder conductors from its service equipment »553.6«. Where the floating building has multiple occupancy, each occupant can be supplied by a single set of feeder conductors extended from the occupant's service equipment to the corresponding panelboard »553.6 *Exception*«.

Wiring system flexibility must be maintained between the floating buildings and the supply conductors. All wiring must be installed so that water surface motion, and changes in water level, do not result in unsafe conditions »553.7(A)«.

LFMC, or LFNC with approved fittings, is allowed for feeders and wherever flexible connections are required. Extra-hard usage portable power cable, listed for both wet locations and sunlight resistance, are permitted for a feeder to a floating building where flexibility is required »553.7(B)«.

In a floating building, electrical and nonelectrical parts grounding must be provided via connection to a grounding bus in the building panelboard. The grounding bus must be grounded through a green-colored insulated equipment grounding conductor run with the feeder conductors and connected to a service equipment grounding terminal. The grounding terminal in the service equipment must be grounded by connection through an insulated grounding electrode conductor to an on-shore grounding electrode »553.8«.

The grounded circuit conductor (neutral) must be an insulated conductor identified in conformance with 200.6. The neutral conductor must be connected to the equipment grounding terminal in the service equipment, and, except for that connection, it must be insulated from the equipment grounding conductors, equipment enclosures, and all other grounded parts. The neutral circuit terminals in the panelboard and in ranges, clothes dryers, counter-mounted cooking units, and the like, must be insulated from the enclosures »553.9«.

Where cord-connected appliances require grounding, it must be accomplished by means of an equipment grounding conductor in the cord and a grounding-type attachment plug »553.10(B)«.

Wiring for floating buildings must comply with the applicable provisions of other Articles of this *Code*, except as modified by this Article »553.3«.

A Article 553 covers wiring, services, feeders, and grounding for floating buildings »553.1«.

B **Floating Building:** A building unit as defined in Article 100 that floats on water, is moored in a permanent location, and has a premises wiring system served through connection by permanent wiring to a remote electrical supply system »553.2«.

CAUTION *All metal parts in contact with the water, all metal piping, and all noncurrent-carrying metal parts that may become energized, must be bonded to the panelboard grounding bus »553.11«.*

NOTE

All electrical system enclosures and exposed metal parts must be bonded to the grounding bus »553.10(A)«.

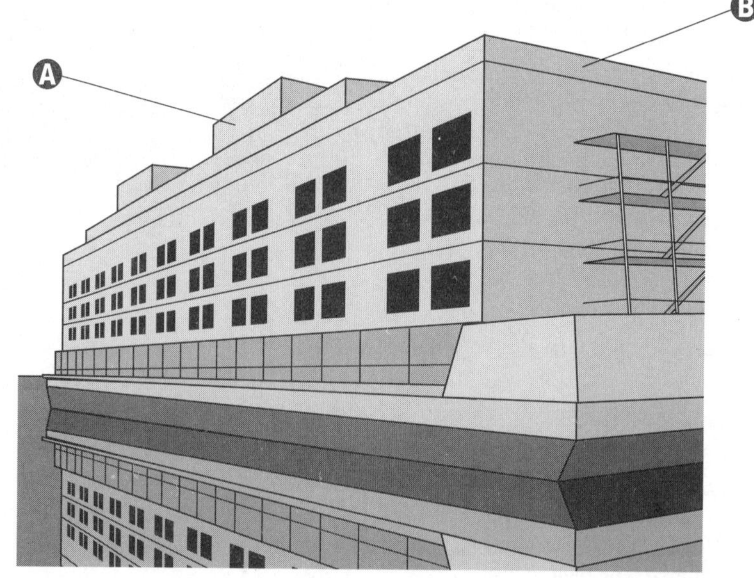

MARINAS AND BOATYARDS

Marina and Boatyard Requirements

The following items must be connected to an equipment grounding conductor run with the circuit conductors in the same raceway, cable, or trench:

(1) Metal boxes, metal cabinets, and all other metal enclosures

(2) Metal frames of utilization equipment

(3) Grounding terminals of grounding-type receptacles »555.15(A)«.

Branch-circuit insulated equipment grounding conductors must terminate either at a grounding terminal in a remote panelboard, or at the main service equipment grounding terminal »555.15(D)«.

A feeder suppling a remote panelboard must have an insulated equipment grounding conductor extending from a service equipment grounding terminal to a grounding terminal in the remote panelboard »555.15(E)«.

Floating dock or marina service equipment must be located adjacent to, rather than on or in, the floating structure.»555.7«.

The load calculation specifications for each feeder and/or service circuit supplying receptacles providing shore power for boats are located in 555.12.

Ⓐ Article 555 covers wiring and equipment installation in the areas comprising fixed (or floating) piers, wharfs, docks, and other areas in marinas, boatyards, boat basins, boathouses, and similar occupancies designed (or planned) for the purpose of repair, berthing, launching, storage, or fueling of small craft and the moorage of floating buildings. This Article does not cover a single, private, non-commercial docking facility for a one-family dwelling »555.1«.

Ⓑ Electrical equipment enclosures installed on piers above deck lever must be securely and substantially supported by structural members, independent of any conduit connected to them. If enclosures are not attached to mounting surfaces by means of external ears or lugs, the internal screw heads must be sealed to prevent seepage of water through mounting holes »555.10(A)«.

Ⓒ Receptacles intended to supply shore power to boats must be housed in marina power outlets listed for wet locations, listed enclosures protected from the weather, or listed weatherproof enclosures. The weatherproof integrity of the assembly must not be affected when the receptacles are in use with any of properly configured booted or non-booted attachment plug/cap inserted »555.19(A)(1)«.

Ⓓ Disconnecting means must be provided to isolate each boat from its supply connection(s) »555.17«. The disconnecting means can consist of a circuit breaker, switch, or both, and must be properly identified as to which receptacle it controls »555.17(A)«.

Ⓔ Receptacles providing shore power for boats must be rated at least 30 amperes and must be of the single outlet type »555.19(A)(4)«.

Ⓕ Electrical equipment enclosures on piers must be located so as not to interfere with mooring lines »555.10(B)«.

Ⓖ Receptacles rated between 30 and 50 amperes must be of the locking- and grounding-types »555.19(A)(4)(a)«. Receptacles rated for 60 or 100 amperes must be of the pin-and-sleeve type »555.19(A)(4)(b)«.

NOTE

Fifteen- and 20-ampere, single-phase, 125-volt receptacles installed outdoors, in boathouses, in buildings used for storage, maintenance, or repair where portable electrical hand tools, electrical diagnostic equipment, or portable lighting equipment are to be used must be provided with GFCI protection for personnel. Receptacles in other locations must be protected in accordance with 210.8(B) »555.19(B)(1)«.

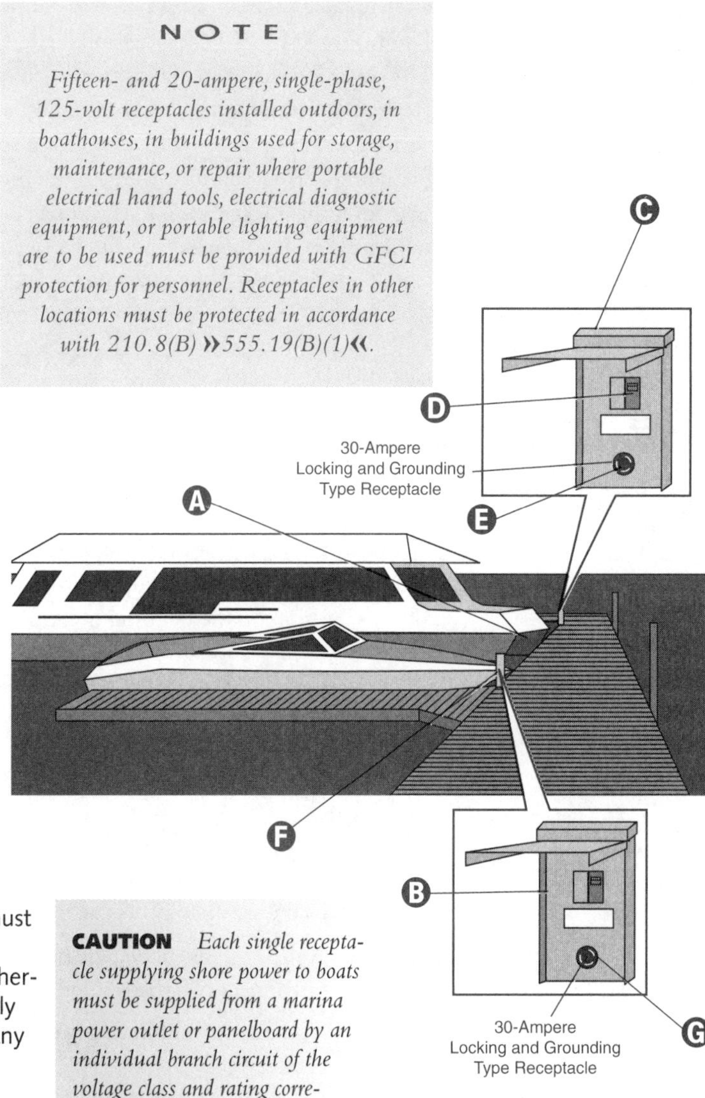

CAUTION *Each single receptacle supplying shore power to boats must be supplied from a marina power outlet or panelboard by an individual branch circuit of the voltage class and rating corresponding to the receptacle's rating »555.19(A)(3)«.*

Summary

- Article 518 covers all buildings, portions of buildings, or structures intentionally designed for the assembly of 100 or more persons.
- Article 520 provisions apply to motion picture (and television) studio audience areas, performance areas, theaters, and similar locations.
- While Article 520, Part II applies to fixed stage switchboards, Part IV applies to portable on-stage switchboards.
- Stage equipment, other than switchboards (fixed and portable), are covered in Article 520, Parts III and V, respectively.
- Theater dressing room provisions are located in Article 520, Part VI (520.71 through 73).
- Article 525 covers the installation of portable wiring and equipment for carnivals, circuses, exhibitions, fairs, traveling attractions, and similar functions, including wiring in (or on) all structures.
- Motion picture/television studios and similar locations are located in Article 530.
- The provisions of Article 540 apply to motion picture projector rooms, motion picture projectors, and associated equipment.
- Article 545 covers requirements for manufactured buildings and their components.
- The requirements of Article 547 apply to agricultural buildings (all or in part), or adjacent areas of like nature as specified in 547.1(A) and (B).
- Article 550 (mobile homes, manufactured homes, and mobile-home parks) is divided into three parts: Part I (general), Part II, (mobile and manufactured homes), and Part III (services and feeders).
- The provisions of Article 551 cover the electrical conductors and equipment installed within (or on) recreational vehicles, the conductors connecting recreational vehicles to an electrical supply, and the installation of electrical equipment and devices within a recreational vehicle park.
- Article 553 covers floating building wiring, services, feeders, and grounding.
- Article 555 covers wiring and equipment installation in the areas comprising fixed (or floating) piers, wharfs, docks, and other areas in marinas, boatyards, boat basins, boathouses, and similar occupancies designed (or planned) for repair, berthing, launching, storage, or fueling of small craft and the moorage of floating buildings.

Unit 18 Competency Test

***NEC*® Reference**	**Answer**	
__________	__________	1. A(n) _____ is any building that is of closed construction and is made or assembled in manufacturing facilities on or off the building site for installation, or assembly and installation on the building site, other than manufactured homes, mobile homes, park trailers, or recreational vehicles.
__________	__________	2. Every recreational vehicle site with an electrical supply shall be equipped with at least _____ 20-ampere, 125-volt receptacle. A minimum of _____ % of all recreational vehicle sites with an electrical supply shall each be equipped with a 30-ampere, 125-volt receptacle.
__________	__________	3. Outdoor mobile home disconnecting means shall be installed so the bottom of the enclosure containing the disconnecting means is not less than _____ feet above finished grade or the working platform.
__________	__________	4. A(n) _____ is an adapter cable containing one male and two female cord connectors used to connect two loads to one branch-circuit.
__________	__________	5. Agricultural building luminaires (lighting fixtures) shall be installed to minimize the entrance of _____.
__________	__________	6. All exposed incandescent lamps in theater dressing rooms, where less than _____ feet from the floor, shall be equipped with open-end guards riveted to the outlet box cover or otherwise sealed or locked in place.

NEC® Reference	Answer	
________	________	7. In exhibition halls used for display booths, as in trade shows, the temporary wiring shall be installed in accordance with Article _____.
________	________	8. Recreational vehicle site feeder-circuit conductors shall have adequate ampacity for the loads supplied and shall be rated at not less than _____.
________	________	9. A(n) _____ is a factory-assembled structure or structures transportable in one or more sections that is built on a permanent chassis and designed to be used as a dwelling without a permanent foundation where connected to the required utilities, and includes the plumbing, heating, air-conditioning, and electric systems contained therein.
________	________	10. The service equipment for floating docks or marinas shall be located _____ the floating structure.
________	________	11. Stage cables for motion picture studio stage set lighting shall be protected by means of overcurrent devices set at not more than _____ % of the ampacity given in the applicable tables of Articles 310 and 400.
________	________	12. A theater stage switchboard that is not completely enclosed dead-front and dead-rear or recessed into a wall shall be provided with a(n) _____ extending the full length of the board to protect all equipment on the board from falling objects.
________	________	13. Each carnival ride and concession shall be provided with a fused disconnecting switch or circuit breaker located within sight and within _____ feet of the operator's station.
________	________	14. Each motion picture projector, floodlight, spotlight, or similar equipment shall have clear working space not less than _____ in(es). wide on each side and at the rear thereof.
________	________	15. The overall length of a mobile home power-supply cord, measured from the end of the cord, including bared leads, to the face of the attachment plug cap shall not be less than _____ feet and shall not exceed _____ feet.
________	________	16. A travel trailer is a vehicular unit, mounted on wheels, designed to provide temporary living quarters for recreational, camping, or travel use, of such size or weight as not to require special highway movement permits when towed by a motorized vehicle, and of gross trailer area less than _____ square feet.
________	________	17. A 15-ampere duplex receptacle, installed in a television audience seating area, can be supplied by a _____ -ampere branch-circuit. I. 15 II. 20 III. 30 a) II only b) I only c) either I or II d) I, II, or III
________	________	18. Receptacles that provide shore power for boats shall be rated not less than _____ amperes, and shall be single outlet type.
________	________	19. Occupancy of any room or space for assembly purposes by less than _____ persons in a building of other occupancy, and incidental to such other occupancy, shall be classified as part of the other occupancy and subject to the provisions applicable thereto.
________	________	20. Open conductors of not over 600 volts, nominal, shall have a horizontal clearance of not less than _____ feet and a vertical clearance of not less than _____ feet in all areas subject to recreational vehicle movement.
________	________	21. Manufactured building service-entrance conductors must be installed after erection at the building site unless the _____ is known prior to manufacture.

NEC® Reference	Answer	
_______	_______	22. Circuits from portable switchboards (within theaters) directly supplying equipment containing incandescent lamps of not over _____ watts shall be protected by overcurrent protective devices having a rating or setting of not over 20 amperes.
_______	_______	23. The service equipment shall be located in sight from and not more than _____ feet from the exterior wall of the mobile home it serves.
_______	_______	24. Conductors supplying outlets for arc and xenon projectors of the professional type shall not be smaller than _____ and shall be of sufficient size for the projector employed.
_______	_______	25. Each recreational vehicle site service shall be calculated on the basis of not less than _____ volt-amperes per site equipped with 50-ampere, 120/240-volt supply facilities; and _____ volt-amperes per site equipped with only 20-ampere supply facilities that are not dedicated to tent sites.
_______	_______	26. Fixed stage switchboards (within theaters) shall be of the _____ type and shall comply with Part IV of Article 408 unless approved based on suitability as a stage switchboard as determined by a qualified testing laboratory and recognized test standards and principles.
_______	_______	27. A _____ is a building unit as defined in Article 100 that floats on water, is moored in a permanent location, and has a premises wiring system served through connection by permanent wiring to an electricity supply system not located on the premises.
_______	_______	28. The power supply to the mobile home shall be a feeder assembly consisting of not more than one listed _____ -ampere mobile power-supply cord with an integrally molded or securely attached plug cap, or a permanently installed feeder.
_______	_______	29. Amusement rides and amusement attractions shall be maintained not less than _____ feet in any direction from overhead conductors operating at 600 volts or less, except for the conductors supplying the amusement ride or attraction.
_______	_______	30. Lamps in cellulose nitrate film storage vaults shall be installed in rigid luminaires (fixtures) of the _____ type.
_______	_______	31. A professional projector is a type of projector using _____ film that has a minimum width of _____ in(es). and has on each edge _____ perforations per inch, or a type using carbon arc, xenon, or other light source equipment that develops hazardous gases, dust, or radiation.
_______	_______	32. In closed construction of manufactured buildings, cables shall be permitted to be secured only at cabinets, boxes, or fittings where _____ or smaller conductors are used and protection against physical damage is provided.
_______	_______	33. All cables installed in agricultural buildings shall be secured within _____ in(es). of each cabinet, box, or fitting.
_______	_______	34. A minimum of _____ % of all recreational vehicle sites, with electrical supply, shall each be equipped with a 50-ampere, 125/250-volt receptacle.
_______	_______	35. Footlights, border lights, and proscenium side lights shall be arranged so that no branch-circuit supplying such equipment will carry a load exceeding _____.

UNIT

19

SECTION FIVE: SPECIAL OCCUPANCIES, AREAS, AND EQUIPMENT

Specific Equipment

Objectives

After studying this unit, the student should:

- have a good understanding of electric sign and outline lighting system requirements, including neon and skeleton tubing.
- understand manufactured wiring system provisions.
- be able to properly install office partition wiring (power, communication, etc.)
- be familiar with electric vehicle charging system stipulations.
- thoroughly understand information technology (data processing) equipment requirements.
- be able to locate provisions relating to fire pumps.
- have a basic understanding of elevator, dumbwaiter, escalator, moving walk, wheelchair lift, and stairway chair lift requirements.
- be familiar with definitions pertinent to swimming pools, fountains, and similar installations.
- understand what constitutes a common bonding grid (for permanently installed pools) and what items must be bonded thereto.
- have a solid grasp of equipment, panelboard, receptacle, and lighting (including underwater lighting) provisions pertaining to permanently installed swimming pools.
- be able to accurately define the storable pool and know the applicable requirements.
- be familiar with spa and hot tub provisions.
- know where to find requirements for therapeutic pools and tubs.
- understand the fundamental hydromassage bathtub provisions.

Introduction

Bringing this text to a close, this unit addresses a wide variety of electrical equipment requirements. *NEC*® Chapter 6 is the basis for numerous equipment illustrations throughout this unit. While most of the articles of Chapter 6 are covered here, four additional articles are described under *Industrial Locations* (Unit 17). Those Articles are: 610 *(Cranes and Hoists)*, 630 *(Electric Welders)*, 669 *(Electroplating)*, and 670 *(Industrial Machinery)*. Some of the less commonly used articles are not referenced.

Beginning with illustrations for both *Electric Signs and Outline Lighting* and *Neon and Skeleton Tubing,* this unit recounts some of the many Article 600 requirements. Subsequently described in illustrative detail (with corresponding articles numbers in parentheses) are: *Manufactured Wiring Systems* (604); *Office Furnishings* (605); *Electric Vehicle Charging Systems* (625); *Audio Signal Processing, Amplification, and Reproduction Equipment* (640); *Information Technology Equipment* (645); *Sensitive Equipment* (647); *Pipe Organs* (650); and *Fire Pumps* (695). Next, this unit devotes several illustrations to Article 620's topics: *Elevators, Dumbwaiters, Escalators, Moving Walks, Wheelchair Lifts, and Stairway Chair Lifts.* Finally, the expansive subject of *Swimming Pools, Fountains, and Similar Locations* (Article 680) is introduced. Because Article 680 covers a broad range of equipment and installation applications, a significant number of illustrations has been dedicated to these requirements. It is imperative that the reader have a thorough understanding of the words and phrases relating specifically to swimming pools, fountains, and similar locations, found in 680.2. One particularly notable definition is that of **storable swimming or wading pool**. A misconception held by many is that a storable pool is, in simple terms, any above-ground swimming pool. Observe, however, that a storable pool has a maximum depth of only 42 in. While some of the smaller above-ground pools fall within this parameter, many, if not most, have a depth of at least 48 in. It must be concluded, therefore, that any above ground pool having a maximum depth greater than 42 in. qualifies as a permanently installed pool, and must be compliant with Article 680, Parts I and II. While space constraints prohibit an all-encompassing discussion of these provisions, the more predominant features are illustrated and discussed. Because of the potentially fatal nature of mistakes, study this unit carefully together with Article 680 when installing/wiring a pool, spa, hydromassage bathtub, or similar item.

As throughout this text, state and/or local jurisdictions may alter electrical requirements. Obtain a copy of any additional rules and regulations for your specific area.

EQUIPMENT

Pipe Organs

Electronic organs must comply with the appropriate Article 640 provisions »650.2«.

The power source must be a transformer-type rectifier, having a dc potential of 30 volts or less »650.3«.

The rectifier must be grounded according to 250.112(B) provisions »650.4«.

Use a minimum of 28 AWG conductors for electronic signal circuits, and at least 26 AWG for electromagnetic valve supply and the like. A main common-return conductor in the electromagnetic supply must be 14 AWG or larger »650.5(A)«.

Conductors must have thermoplastic or thermosetting insulation »650.5(B)«.

Except for the common-return conductor and conductors inside the organ proper, the organ sections and the console conductors must be cabled. An additional covering can enclose both cable and common-return conductors. Alternately, a separate common-return conductor is allowed to contact the cable »650.5(C)«.

Each cable must have an outer covering, either overall or on individual subassemblies of grouped conductors. Tape is an acceptable covering. If not housed in metal raceway, the covering must be flame retardant or the cable/cable subassembly must be wrapped by closely wound listed fireproof tape »650.5(D)«.

Securely fastened cables can attach directly to the organ structure without insulating supports. Cables must not be placed in contact with other conductors (except for the common return conductor) »650.6«.

Arrange circuits so that 26 and 28 AWG conductors are protected from overcurrent by an overcurrent device rated at no more than 6 amperes. Other conductor sizes must be protected in accordance with their ampacity. A common return conductor does not require overcurrent protection »650.7«.

Ⓐ Article 650 covers electrically operated pipe organ circuits and components employed for sounding apparatus and keyboard control »650.1«.

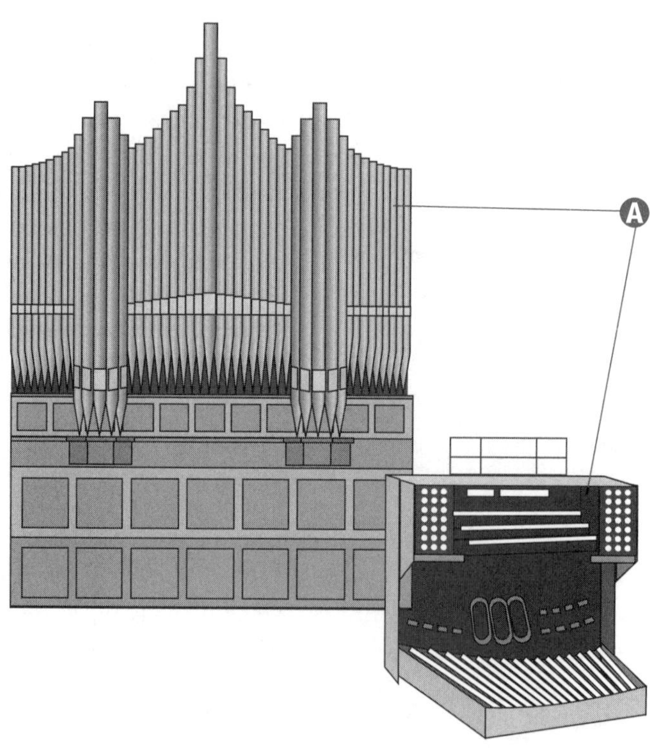

Electric Signs and Outline Lighting

Other than lamps and neon tubing, all live parts must be enclosed »600.8«.

Signs and outline lighting systems having incandescent lamp holders must be marked with the maximum allowable lamp wattage. The permanently installed markings must be in letters at least ¼ in. (6 mm) high, and must be visible during relamping »600.4(B)«.

Supplying branch-circuits of signs and outline lighting systems containing incandescent and fluorescent forms of illumination must not exceed a 20-ampere rating »600.5(B)(1)«.

The wiring method supplying signs and outline lighting systems must terminate within a sign, an outline lighting system enclosure, a suitable box, or a conduit body »600.5(C)(1)«.

Sign-supporting metal poles can enclose supply conductors, provided the stipulations of 410.15(B) are met »600.5(C)(3)«.

If operated by external electronic or electromechanical controllers, sign or outline lighting systems can have a disconnecting means located within sight of the controller or within the controller enclosure. The disconnecting means must disconnect both the sign (or outline lighting system) and the controller from all ungrounded supply conductors. The disconnecting means must be designed so that no pole can be operated independently, and must be capable of being locked in the open position »600.6(A)(2)«.

Ⓐ Each sign and outline lighting system, or its supplying feeder/branch, must be controlled by an externally operable switch (or circuit breaker) that opens all ungrounded conductors »600.6«. The disconnecting means must be within sight of the sign or outline lighting system that it controls. If out of the line of sight from any energizable section, it must be possible to lock the disconnecting means in the open position »600.6(A)(1)«.

Ⓑ Article 600 covers conductor and equipment installation for electric signs and outline lighting as defined in Article 100 »600.1«. By definition, electric signs and outline lighting includes all products and installations utilizing neon tubing, such as signs, decorative elements, skeleton tubing, or art forms »600.1 FPN«.

Ⓒ Signs and outline lighting systems must be marked with the manufacturer's name, trademark or other identification, input voltage, and current rating »600.4(A)«.

Ⓓ Signs and outline lighting system equipment in wet locations, unless of the listed watertight type, must be weatherproof and, if necessary, have drain holes in accordance with 600.9(D)(1) through (3) »600.9(D)«.

Ⓔ Signs and outline lighting system metal equipment must be grounded »600.7«. [See the additional grounding provisions in 600.7 (A) through (F)].

Ⓕ Unless protected from physical damage, sign or outline lighting system equipment must be at least 14 ft (4.3 m) above areas accessible to vehicles »600.9(A)«.

NOTE

Each commercial building and occupancy open to pedestrians must have at least one accessible outlet at every tenant space entrance for sign or outline lighting system use. The outlet(s) must be supplied by an exclusively dedicated branch-circuit rated 20 amperes (or more). Service halls or corridors are not considered pedestrian accessible »600.5(A)«.

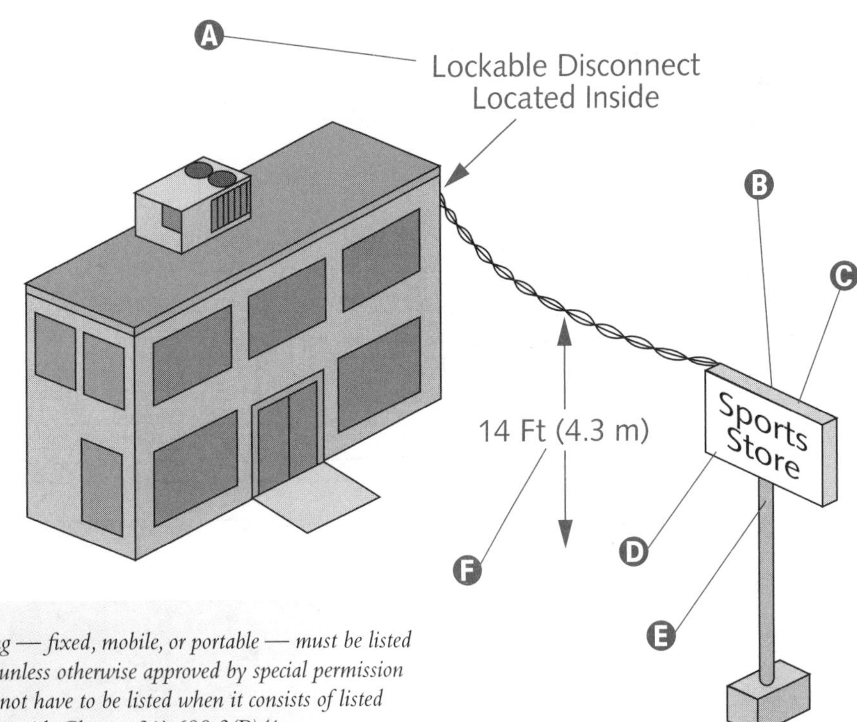

CAUTION *Electric signs and outline lighting—fixed, mobile, or portable—must be listed and installed in conformance with that listing, unless otherwise approved by special permission »600.3«. As a system, outline lighting does not have to be listed when it consists of listed luminaires (lighting fixtures) wired in accordance with Chapter 3 »600.3(B)«.*

Neon and Skeleton Tubing

Article 600, Part II provisions apply only to field-installed skeleton tubing, and are in addition to Part I requirements »600.30«.

Neon Tubing: **Electric-discharge tubing produced in shapes forming letters (or parts thereof), skeleton tubing, outline lighting, other decorative elements or art forms, and filled with various inert gases »600.2«.**

Skeleton Tubing: **Neon tubing that is, itself, the sign or outline lighting and not attached to an enclosure or sign body »600.2«.**

The supply branch-circuits of neon tubing installations must not be rated in excess of 30 amperes »600.5(B)(2)«.

Pedestrian accessible neon tubing, other than dry-location portable signs, must be protected from physical damage »600.9(B)«.

In damp or wet locations, all conductors' insulation must extend no less than 4 in. (100 mm) beyond the metal conduit or tubing »600.32(G)(1).«

A Sharp bends in insulated conductors must be avoided »600.32(D)«.

B Only one conductor must be installed per length of conduit or tubing »600.32(A)(2)«.

C Conductors must be installed on insulators, in RMC, IMC, RNC, LFNC, FMC, LFMC, EMT, metal enclosures, or other such listed equipment. Wiring methods must be installed per Chapter 3 requirements »600.32(A)(1)«.

CAUTION *Equipment having an open circuit voltage exceeding 1000 volts must not be installed in (or on) dwelling occupancies »600.32(I)«.*

D Ballasts, transformers, and electronic power supplies must meet the requirements in 600.21 through 600.23.

E Secondary conductors must be separated from each other and from all objects other than insulators or neon tubing by at least 1½ in. (38 mm) of space. GTO cable, installed in metal conduit or tubing, requires no spacing between the cable insulation and the conduit or tubing »600.32(E)«.

F In dry locations, all conductors' insulation must extend no less than 2½ in. (65 mm) beyond the metal conduit or tubing »600.32(G)(2)«.

G Neon tubing's length and design must not cause a continuous overcurrent beyond the capacity of the transformer or electronic power supply »600.41(A)«.

H The neon tubing and conductor must be supported within 6 in. (150 mm) of the electrode connection »600.42(C)«.

I Connections must be made by twisting the wires together, by use of a connection device, or by use of an electrode receptacle. Connections must be electrically and mechanically secure and must be within a specifically listed enclosure »600.42(B)«.

J Electrode terminals must not be accessible to unqualified persons »600.42(A)«.

NOTE

Conductor provisions pertaining to this illustration apply to neon secondary circuit conductors over 1000 volts, nominal (600.32). 600.31 covers neon secondary circuit conductors of lesser voltage.

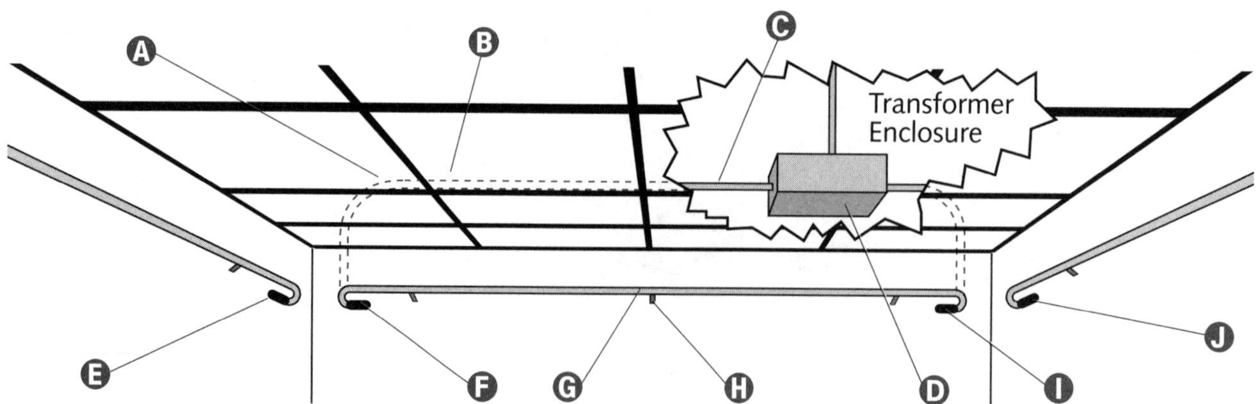

Manufactured Wiring Systems

Except as modified by this article's requirements, all other relevant *NEC*® Articles apply »604.3«.

Where listed for the purpose, manufactured wiring systems can be used in outdoor locations »604.4 *Exception No. 2*«.

Conduit must be listed FMC, or listed liquidtight flexible conduit, containing nominal 600-volt 10 or 12 AWG copper-insulated conductors, having a copper equipment grounding conductor (bare or insulated) equal in size to the ungrounded conductor »604.6(A)(2)«. (See *Exceptions No. 1 and 2*.)

Hard-usage flexible cord (minimum 12 AWG conductors) can be part of a listed factory-made assembly, not exceeding 6 ft (1.8 m) in length, as a transition method between manufactured wiring system components and utilization equipment, not permanently secured to the building structure. The cord must be visible along its entire length and must not be subject to strain or physical damage »604.6(A)(3)«.

A Article 604 provisions apply to field-installed wiring using off-site manufactured subassemblies for branch-circuits, remote control circuits, signaling circuits, and communications circuits in accessible areas »604.1«.

B The type of cable, flexible cord, or conduit must be clearly marked on each section »604.6(B)«.

C Receptacles and connectors must be (1) of the locking type, uniquely polarized and identified for the purpose, and (2) part of a listed assembly appropriate for the system »604.6(C)«.

D Manufactured wiring systems must be supported in accordance with the applicable cable or conduit article for the cable or conduit type employed »604.6(E)«.

E Ceiling grids cannot be used to support cables (and raceways). Secure support must be provided via wires and associated fittings *and* must be installed in addition to the ceiling grid support wires »300.11(A)«. (This illustration does not show cable supports.)

F **Manufactured Wiring System:** A system containing components assembled in the process of manufacture that after assembly cannot be inspected without damage or destruction »604.2«.

G Cable must be listed armored cable or metal-clad cable, containing nominal 600-volt 10 or 12 AWG copper-insulated conductors, having a bare or insulated copper equipment grounding conductor equal in size to the ungrounded conductor »604.6(A)(1)«.

H Manufactured wiring systems are permitted in accessible and dry locations, and in plenums and spaces used for environmental air, where listed for this application and installed per 300.22 »604.4«. In concealed spaces, one end of tapped cable can extend into hollow walls for direct switch/outlet termination »604.4 *Exception No. 1*«.

NOTE

Manufactured wiring systems cannot be used where limited by the applicable article in Chapter 3 for the wiring method used in its construction »604.5«.

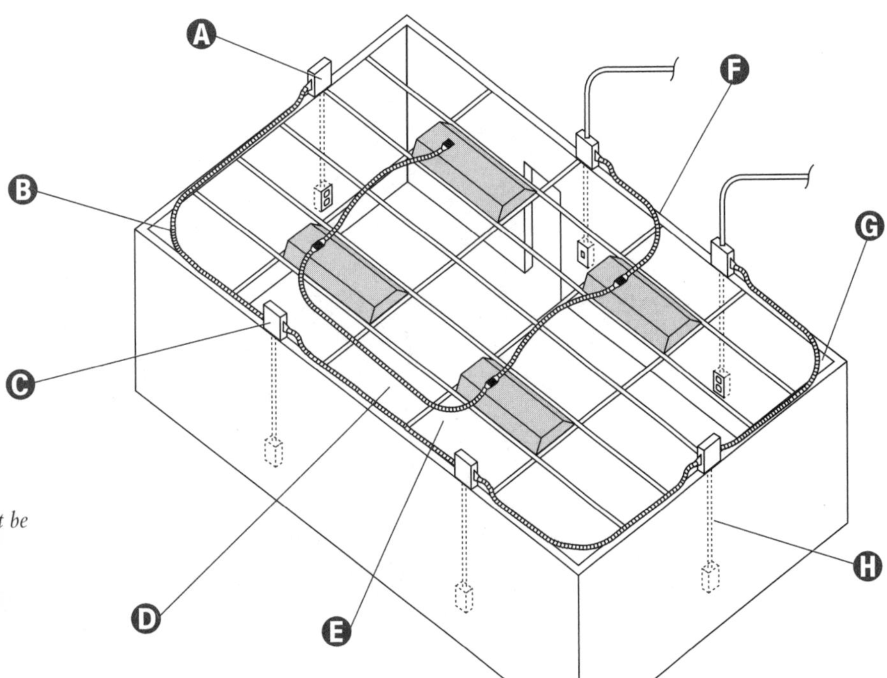

CAUTION *All unused outlets must be capped to effectively close connector openings »604.7«.*

Office Furnishings

Freestanding (not fixed) partitions can be permanently connected to the building's electrical system by one of Chapter 3's approved wiring methods »605.7«.

Individual freestanding partitions or groups of individual partitions that are electrically connected and mechanically contiguous, not exceeding 30 ft (9.0 m) when assembled, can connect to the building's electrical system by means of a single flexible cord and plug, provided all of the following are met:

(A) The flexible power-supply cord must be of the extra-hard usage type, with 12 AWG or larger conductors, with an insulated equipment grounding conductor, and not more than 2 ft (600 mm) in length.

(B) The receptacle(s) supplying power must be on a separate circuit serving only panels (no other loads), and must be located within 12 in. (300 mm) of the partition to which it is connected.

(C) Individual partitions, or groups of interconnected individual partitions, must contain no more than thirteen 15-ampere, 125-volt receptacle outlets.

(D) Individual partitions or interconnected groups must not contain multiwire circuits »605.8«.

A Article 605 covers electrical equipment, lighting accessories, and wiring systems that connect to, are contained within, or are installed on relocatable wired partitions »605.1«.

B Wiring systems must be identified as suitable power providers for lighting accessories and appliances in wired partitions. The partitions must not extend from floor to ceiling, unless the exception is met »605.2«.

C The electrical connection between partitions must be either a flexible assembly (identified for wired partition usage), or be of flexible cord, provided *all* of the following conditions are met: (1) the cord is extra-hard usage type with 12 AWG or larger conductors and must have an insulated grounding conductor; (2) the partitions are mechanically contiguous; (3) the cord, no longer than necessary for maximum partition positioning, is in no case to exceed 2 ft (600 mm); and (4) the cord terminates at an attachment plug and cord connector with strain relief »605.4«.

D Lighting equipment listed and identified for wired partition use must comply with all of the following:

(A A secure attachment or support means is provided.

(B) If cord and plug connection is provided, the cord length, while suitable for the intended application, must not exceed 9 ft (2.7 m) in length. The cord must be 18 AWG or larger, must contain an equipment grounding conductor, and must be of the hard usage type. Any other means of connection must be identified as suitable for the existing conditions.

(C) Convenience receptacles are not permitted in lighting accessories »605.5«.

E All conductors and connections must be contained within wiring channels (metal or other material) identified as suitable for the conditions of use. These channels must be free of projections or other conditions that may damage conductor insulation »605.3«.

F Except as modified by Article 605 requirements, all other *NEC*® Articles apply »605.2(B)«.

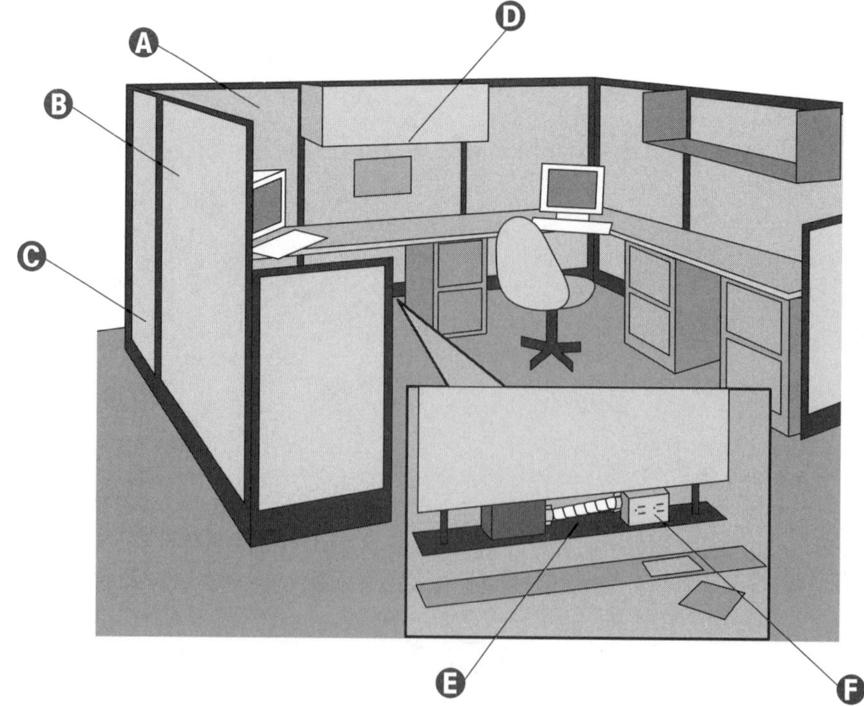

NOTE

Fixed wired partitions (secured to building surfaces) must be permanently connected to the building electrical system by one of Chapter 3's wiring methods »605.6«.

Electric Vehicle Charging Systems

Electric Vehicle Nonvented Storage Battery: A hermetically-sealed battery comprised of rechargeable electrochemical cell(s) without provisions for excessive gas pressure release, the addition of water (or electrolyte), or external electrolyte specific gravity measurements »625.2«.

Personnel Protection System: A system of personnel protection devices and structural features working together to provide personnel protection against electric shock »625.2«.

Feeders and branch-circuits supplying electric vehicle supply equipment require overcurrent protection, are sized for continuous duty, and must have a rating of no less than 125% of the maximum electric vehicle equipment load. Where the same feeder or branch-circuit supplies noncontinuous loads, the overcurrent device must have a rating at least equal to the sum of the noncontinuous loads plus 125% of the continuous loads »625.21«.

Mechanical ventilation is not required where electric vehicle nonvented storage batteries are used, or where the electric vehicle supply equipment is listed/labeled and marked as suitable for charging electric vehicles indoors without ventilation per 625.15(B) »625.29(C)«.

Ⓐ Unless specifically listed for the purpose and location, the electric vehicle supply equipment's coupling means must be situated no less than 18 in. (450 mm) and no more than 4 ft (1.2 m) above floor level »625.29(B)«.

Ⓑ Article 625 provisions cover the electrical conductors and equipment external to an electric vehicle that connect it to an electrical supply by conductive or inductive means, as well as the installation of equipment and devices related to electric vehicle charging »625.1«.

Ⓒ **Electric Vehicle:** Highway worthy automotive-type vehicle (such as a passenger automobile, bus, truck, van, and the like) primarily powered by an electric motor drawing current from a rechargeable storage battery, fuel cell, photovoltaic array, or other electric current source. Article 625 excludes electric motorcycles and similar type vehicles as well as off-road self-propelled electric vehicles (such as industrial trucks, hoists, lifts, transports, golf carts, airline ground support equipment, tractors, boats, etc.) »625.2«.

Ⓓ **Electric Vehicle Connector:** A device (part of the electric coupler) that, by insertion into an electric vehicle inlet, establishes an electrical connection for the purpose of charging and information exchange »625.2«.

Ⓔ **Electric Vehicle Coupler:** A mating electric vehicle inlet and electric vehicle connector set »625.2«.

Ⓕ **Electric Vehicle Inlet:** The electric vehicle device into which the electric vehicle connector is inserted for charging and information exchange. For *NEC*® purposes, the electrical vehicle inlet is considered part of the vehicle rather than part of the electric vehicle supply equipment »625.2«.

Ⓖ **Electric Vehicle Supply Equipment:** Includes the conductors (ungrounded, grounded, and equipment grounding conductors), electric vehicle connectors, attachment plugs, and all other fittings, devices, power outlets, or apparatus purposely installed to transfer energy from the premises wiring to the electric vehicle »625.2«.

NOTE

All electrical materials, devices, fittings, and associated equipment must be listed or labeled »625.5«.

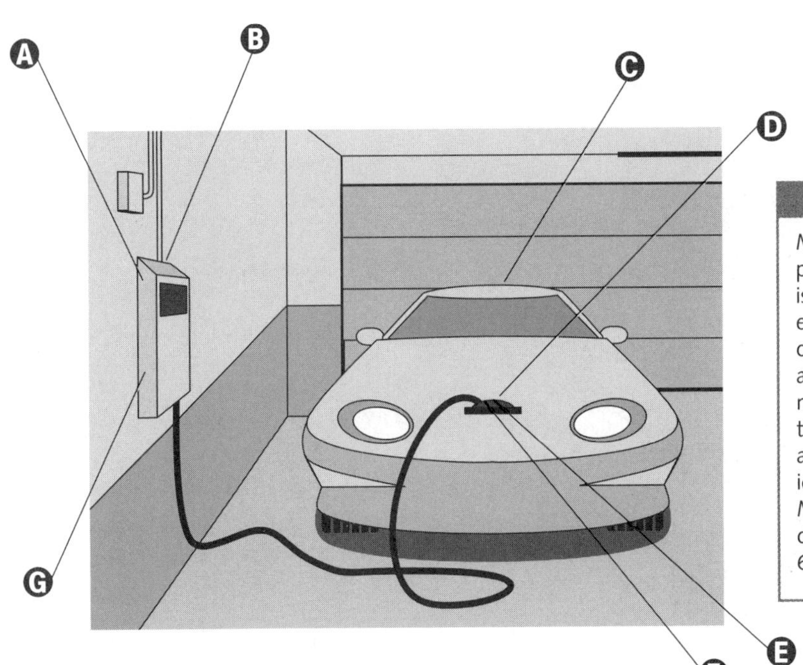

WARNING

Mechanical ventilation (such as a fan) must be provided where the electric vehicle supply equipment is listed/labeled and marked as suitable for charging electric vehicles that require ventilation for indoor charging, in accordance with 625.15(C). Both supply and exhaust ventilation equipment must be permanently installed to intake from, and vent directly to, the outdoors. Positive pressure ventilation systems are only permitted in locations that have been specifically designed and approved for that application. Mechanical ventilation requirements must be determined using one of the methods found in 625.29(D)(1) through (4) »625.29(D)«.

Audio Signal Processing, Amplification, and Reproduction Equipment

Grounding of separately derived systems with *60 volts to ground* must meet 647.6 requirements »640.7(B)«.

Equipment access must not be denied by accumulated wires and cables that prevent panel removal, including suspended ceiling panels »640.5«.

Wireways and auxiliary gutters must be grounded and bonded per Article 250. If the wireway (or auxiliary gutter) contains no power-supply wires, a 14 AWG copper equipment grounding conductor (or equivalent) is sufficient. If the wireway (or auxiliary gutter) contains power-supply wires, the equipment grounding conductor must meet 250.122 size specifications »640.7(A)«.

Isolated grounding-type receptacles are permitted as described in 250.146(D), as well as for the implementation of other Article 250 compliant technical power systems. For separately derived systems with *60 volts to ground*, the branch-circuit equipment grounding conductor must be terminated per 647.6(B) »640.7(C)«.

Portable Equipment: **Equipment fed with portable cords or cables and intended for movement from one place to another »640.2«.**

Temporary Equipment: **Portable wiring and equipment used for events of a transient or temporary nature where equipment removal is presumed at the event's conclusion »640.2«.**

Technical Power System **(often referred to as *tech power*): an electrical distribution system with 250.146(D) approved grounding, where the equipment grounding conductor is isolated from the premises grounded conductor except at a single grounded termination point within a branch-circuit panelboard or the origination (main breaker) branch-circuit panelboard, or at the premises grounding electrode »640.2«.**

Grouped (or bundled) insulated conductors of different systems, in close physical contact within the same raceway (or other enclosure), or in portable cords (or cables), must comply with 300.3(C)(1) »640.8«.

Ⓐ Article 640 covers equipment and wiring for audio signal generation, recording, processing, amplification and reproduction; distribution of sound; public address; speech input systems; temporary audio system installations; and electronic musical instruments (including organs). This encompasses audio systems subject to Article 517, Part VI, and Articles 518, 520, 525, and 530 »640.1«.

Ⓑ **Loudspeaker:** Equipment that converts an ac electric signal into an acoustic signal. The term **speaker** is commonly used to mean loudspeaker »640.2«.

Ⓒ **Mixer:** Equipment used to combine and level-match a multiplicity of electronic signals, as from microphones, electronic instruments, and recorded audio »640.2«.

Ⓓ **Audio Signal Processing Equipment:** Electrically operated equipment that produces and/or processes electronic signals that, when appropriately amplified and reproduced via loudspeaker, output an acoustic signal within the range of normal human hearing. Within Article 640, the terms **equipment** and **audio equipment** are equivalent to audio signal processing equipment »640.2«.

Ⓔ **Audio System:** The totality of equipment and interconnected wiring used to fabricate a fully functional audio signal processing, amplification, and reproduction system »640.2«.

Ⓕ **Equipment Rack:** A framework for the equipment support and/or enclosure, either portable or stationary »640.2«.

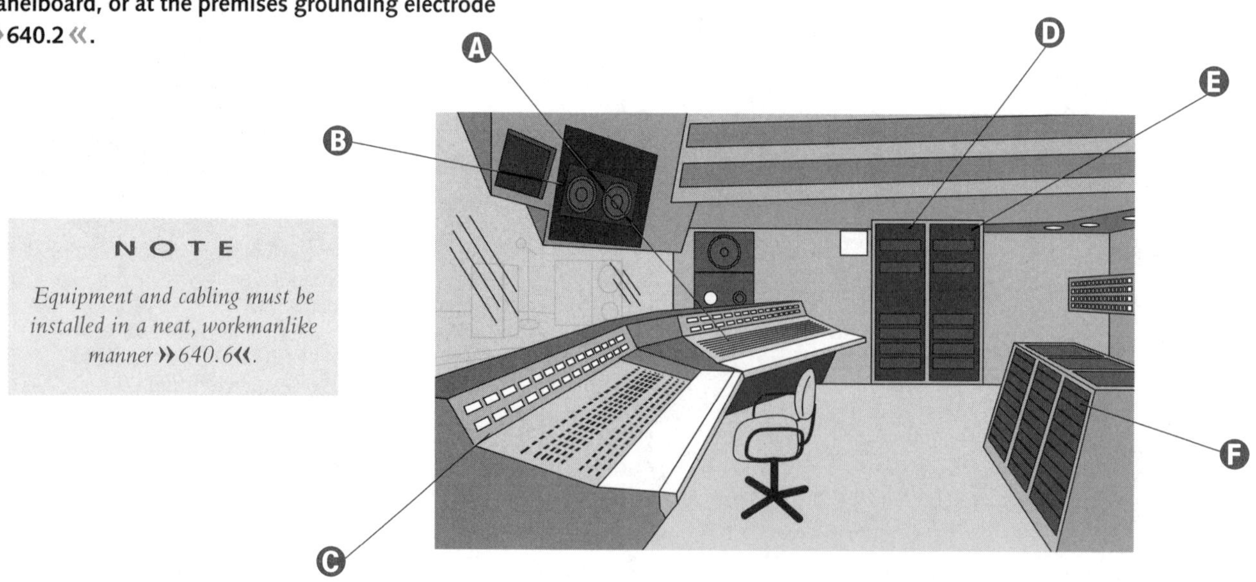

Information Technology Equipment

A A power disconnect means must be provided for all electronic equipment in the information technology equipment room. Likewise, a means to disconnect the power to all of the room's dedicated HVAC systems serving the room, closing all required fire/smoke dampers, must be provided. These disconnecting means' controls must be grouped, identified, and readily accessible at the principal exit doors. A single means to control both the electronic equipment and HVAC systems is acceptable. Where a push button is employed as a means to disconnect power, pushing the button in must disconnect the power »645.10«.

B Article 645 covers equipment, power-supply wiring, and equipment interconnected wiring, as well as grounding of information technology equipment (and systems), including information technology equipment room terminal units »645.1«.

C Article 645 applies, provided that listed information technology equipment is installed »645.2(3)«.

D The data processing system can be connected to a branch-circuit by any of the following listed means: (1) a flexible cord and attachment plug cap not to exceed 15 ft (4.5 m),or (2) a cord set assembly (protected from physical damage where run on the floor's surface) »645.5(B)«.

E Branch-circuit conductors supplying data processing system unit(s) must have an ampacity of at least 125% of the total connected load »645.5(A)«.

F Cables (power, communications, connecting, and interconnecting), and receptacles associated with the information technology equipment are permitted under a raised floor, provided 645.5(D)(1) through (6) requirements are met »645.5(D)«.

G Unless (1) or (2) permits otherwise, **uninterruptible power supply** (UPS) systems installed within the information technology room, together with their supply and output circuits, must comply with 645.10. The battery must also be disconnected from its load by the disconnecting means.

(1) Installations qualifying under Article 685 provisions.

(2) Power sources capable of supplying 750 volt-amperes (or less), derived either from UPS equipment or from battery circuits integral to electronic equipment »645.11«.

NOTE

*Article 645 applies, provided **all** of the conditions listed in 645.2(1) through (5) are met »645.2«.*

H Article 645 does apply when a dedicated heating/ventilating/air-conditioning (HVAC) system (separate from other occupancy areas) exclusively serves the information technology equipment. Any HVAC system serving other occupancies can also serve the information technology equipment room if fire/smoke dampers exist at room boundary penetration points. Such dampers must operate both upon activation of smoke detectors and by disconnecting means operation as required by 645.10 »645.2(2)«.

I All exposed noncurrent-carrying metal parts of an information technology system must either be grounded (per Article 250) or be double insulated. Power systems (derived within listed information technology equipment) supplying information technology systems through receptacles (or cable assemblies), which are part of this equipment, are not considered separately derived for the purpose of applying 250.20(D). Signal reference structures, where installed, must be bonded to the equipment grounding system provided for the information technology equipment »645.15«.

J It is not necessary to secure in place cables (power, communications, connecting, and interconnecting) and associated boxes, connectors, plugs, and receptacles that are listed as part of, or for, information technology equipment »645.5(E)«.

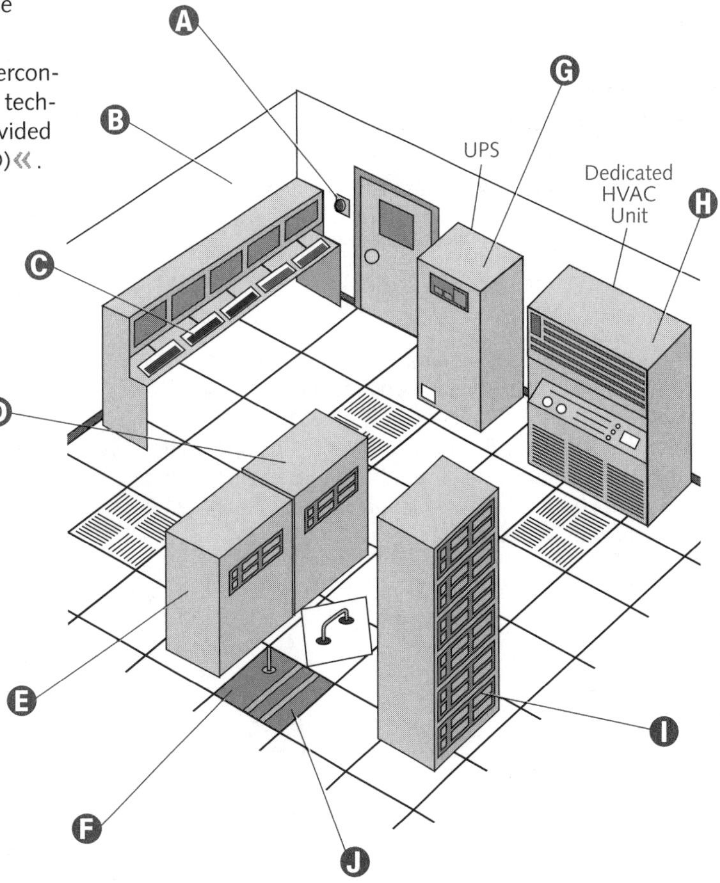

SENSITIVE ELECTRONIC EQUIPMENT

Technical Power-System Receptacles

A Article 647 covers the installation and wiring of sensitive electronic equipment that is connected to a separately derived system operating at 120 volts line-to-line and 60 volts to ground »647.1«.

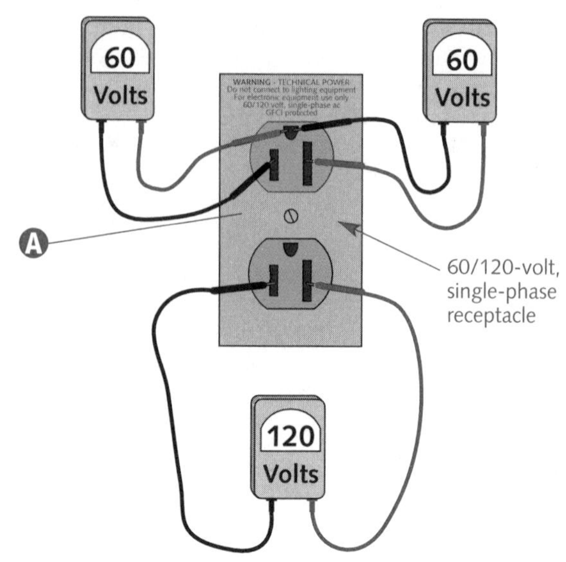

NOTE

While isolated ground receptacles are permitted as described in 250.146(D), the branch-circuit equipment grounding conductor must be terminated per 647.6(B) »647.7(B)«.

Receptacle Installation

A All outlet strips, adapters, receptacle covers, and faceplates must be marked as required by 647.7(A)(2)

WARNING — TECHNICAL POWER
Do not connect to
lighting equipment.
For electronic equipment use only.
60/120 V 1ø AC
GFCI protected.

B All 125-volt receptacles used for 60/120-volt technical power must be uniquely configured and identified for use with this class of system. All 125-volt, single-phase, 15- or 20-ampere rated receptacle outlets and attachment plugs, identified for use with grounded circuit conductors, are acceptable in machine rooms, control rooms, equipment rooms, equipment racks, and similar locations that are restricted to use by qualified personnel »647.7(A)(4)«.

C All 15- and 20-ampere receptacle outlets must be GFCI protected »647.7(A)(1)«.

D A 125-volt, single-phase, 15- or 20-ampere rated receptacle, having one of its current-carrying poles connected to a grounded circuit conductor, must be located within 6 ft (1.8 m) of all permanently installed 15- or 20-ampere rated 60/120-volt technical power-system receptacles »647.7(A)(3)«.

NOTE

Receptacles used for equipment connection must meet 647.7(A)(1) through (4) conditions »647.7(A)«.

CAUTION *Clear markings on all junction box covers must indicate the distribution panel and the system voltage »647.4(B)«.*

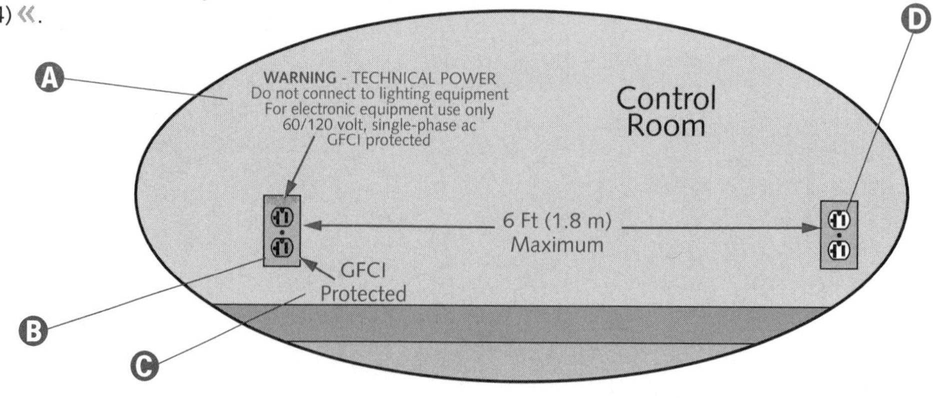

Separately Derived Systems with 60 Volts to Ground

A Standard single-phase panelboards and distribution equipment with higher voltage ratings can be used »647.4(A)«.

B The system must be grounded per 250.30 as a separately derived single-phase, three-wire system »647.6(A)«.

C A separately derived 120-volt, single-phase, three-wire system with 60 volts on each of two ungrounded conductors to a grounded neutral conductor can be used to reduce objectionable noise in sensitive electronic equipment locations provided: (1) the system is installed only in commercial or industrial occupancies, (2) the system's use is restricted to areas under close supervision by qualified personnel, and (3) all requirements in 647.4 through 647.8 are met »647.3«.

D Common-trip, 2-pole circuit breakers that are identified for operation at the system voltage must be provided for both ungrounded conductors in all feeders and branch-circuits »647.4(A)«.

E The system must be clearly marked on the panel's face or inside the panel's door »647.4(A)«.

F Permanently wired utilization equipment and receptacles must be grounded via an equipment grounding conductor, run with the circuit conductors, to an equipment grounding bus (prominently marked "Technical Equipment Ground") in the originating branch-circuit panelboard. The grounding bus must be connected to the grounded conductor on the line side of the separately derived system's disconnecting means. The grounding must be sized as specified in Table 250.122 and be run with the feeder conductors. The technical equipment grounding bus need not be bonded to the panelboard enclosure »647.6(B)«.

NOTE

All feeders and branch-circuit conductors installed under this section must be identified as to system at all splices and terminations by color, marking, tagging, or equally effective means. The means of identification must be posted at each branch-circuit panelboard and at the building's disconnecting means »647.4(C)«.

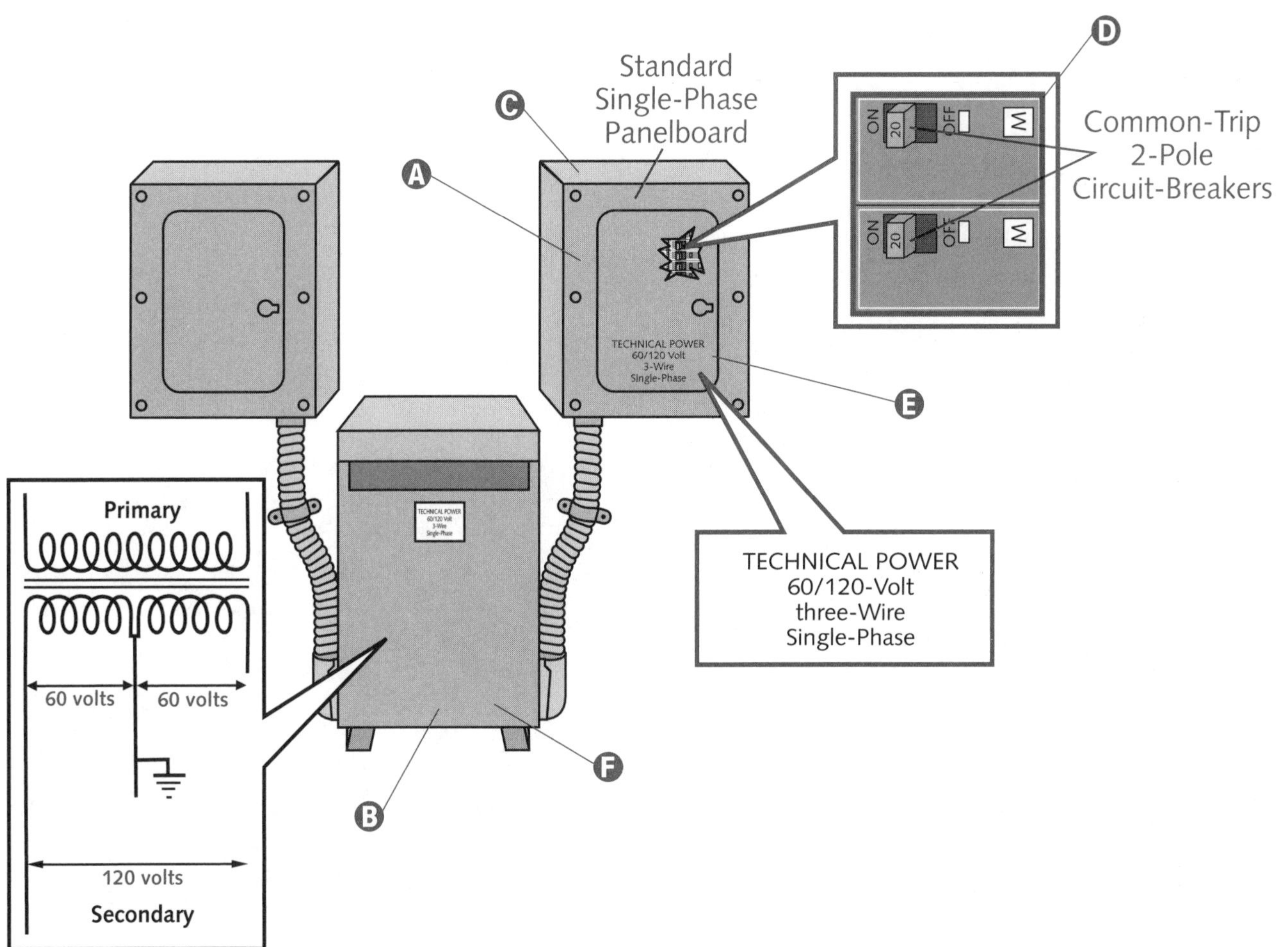

Fire Pumps

Electric motor-driven fire pumps must have a reliable power source »695.3«. See 695.3(A) for individual source requirements and 695.3(B) for multiple source requirements.

A fire pump can be supplied by a separate service »230.2(A) and 695.3(A)(1)«.

If service or system voltage differs from the fire pump motor's utilization voltage, one or more transformers (protected by disconnecting means and overcurrent protective devices) can be installed between the system supply and the fire pump controller in accordance with 695.5(A), (B), or (C). Only transformers covered in (C) can supply loads unrelated to the fire pump system »695.5«.

Power circuits and wiring methods must comply with the requirements in (A) through (G), and, as permitted in 230.90(A), *Exception No. 4;* 230.94, *Exception No. 4;* 230.95, *Exception No. 2;* 240.13; 230.208; 240.4(A); and 430.31 »695.6«.

The location of electric motor-driven fire pump controllers and power transfer switches must be within sight, and as close as practicable to, the motors they control »695.12(A)«. Engine-driven fire pump controllers must be located as close as practical to, and within sight of, the engines they control »695.12(B)«.

Fire pump controllers and power transfer switches, by location and/or protection, must remain undamaged by water escaping from pumps or pump connections »695.12(E)«.

The disconnecting means must be marked "Fire Pump Disconnecting Means." The minimum 1-in. (25-mm)-high letters must be visible without opening enclosure doors or covers »695.4(B)(3)«.

Fire pump supply conductors on the final disconnecting means load side, and overcurrent device(s) permitted by 695.4(B), must be entirely independent of other wiring. They must supply only those loads directly associated with the fire pump system, and must be amply protected from potential damage by fire, structural failure, or operational accident. These conductors can be routed through a building(s); (1) encased in 2 in. (50 mm) of concrete, or (2) within enclosed construction dedicated to the fire pump circuit(s) having a minimum of 1-hour fire resistance; or (3) be a listed electrical circuit protection system with a minimum 1-hour fire rating »695.6(B)«.

Automatic protection against overloads is not allowed on power circuits. Except as noted in 695.5(C)(2), branch-circuit and feeder conductors must be protected against short circuits only »695.6(D)«.

All engine controller/battery wiring must be protected against physical damage, being installed in accordance with the controller and engine manufacturer's instructions »695.6(G)«.

A All wiring from the controllers to the pump motors must be in RMC, IMC, LFMC, or LFNC Type LFNC-B, or Type MI cable »695.6(E)«.

B Conductors supplying fire pump motor(s), pressure maintenance pumps, and associated accessory equipment must have a minimum rating of 125% of the sum of the fire pump motor(s) and pressure maintenance motor(s) FLC(s), plus 100% of the associated fire pump accessory equipment »695.6(C)(1)«. Conductors supplying only fire pump motor(s) must have a minimum rating of 125% of the fire pump motor(s) full-load current(s) »695.6(C)(2)«.

C Article 695 covers the installation of: (1) electric power sources and interconnecting circuits, and (2) switching and control equipment dedicated to fire pump drivers »695.1(A)«.

Article 695 does not cover: (1) performance, maintenance, and acceptance testing of the fire pump system and the system components' internal wiring, and (2) pressure maintenance (jockey or makeup) pumps »695.1(B)«.

Taps used to supply fire pump equipment, if provided with service equipment and installed per service-entrance conductor requirements, can be connected to the supply side of the service disconnecting means »230.82(4)«.

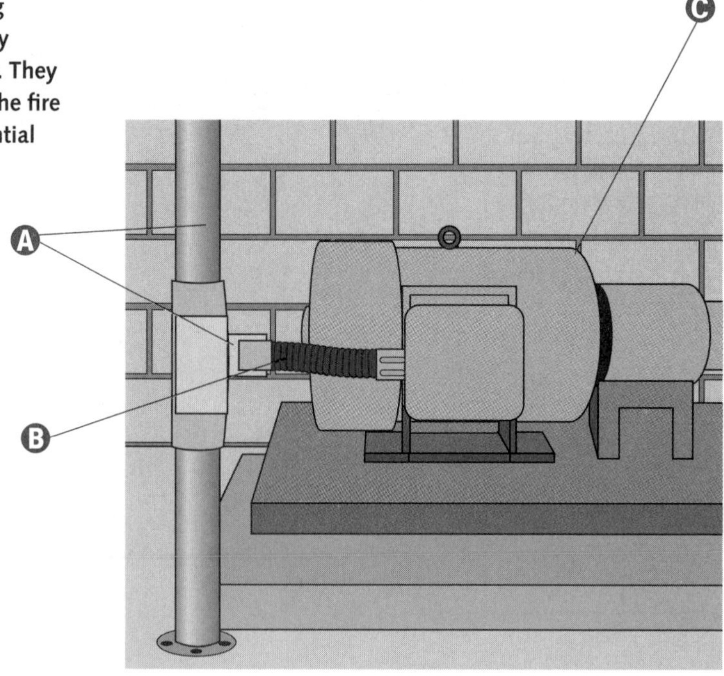

ELEVATORS, DUMBWAITERS, ESCALATORS, MOVING WALKS, WHEELCHAIR LIFTS, AND STAIRWAY CHAIR LIFTS

General

Control System: Overall system governing starting, stopping, direction of motion, acceleration, speed, and retardation of the moving member »620.2«.

Motion Controller: Electrical device(s) for that portion of the control system that governs acceleration, speed, retardation, and stoppage of the moving member »620.2«.

Motor Controller: The control system's operative units, i.e., the starter device(s) and power conversion equipment that drive an electric motor, or the pumping unit used to power hydraulic control equipment »620.2«.

Operation Controller: Electric device(s) for that portion of the control system that initiates starting, stopping, and direction of motion in response to an operating device's signal »620.2«.

Door operator controller and door motor branch-circuits, as well as feeders to motor controllers, driving machine motors, machine brakes, and motor-generator sets must have a circuit voltage of 600 volts or less »620.3(A)«.

Operating Device: Devices used to activate the operation controller, such as the car switch, push buttons, key or toggle switch(es) »620.2«.

Conductors and optical fibers within hoistways, escalator and moving walk wellways, wheelchair lifts, stairway chair lift runways, machinery spaces,control spaces, in/on cars, and machine/control rooms (exclusive of traveling cables connecting the car or counterweight and hoistway wiring), must be installed in RMC, IMC, EMT, RNC, or wireways; or must be Type MC, MI, or AC cable unless otherwise permitted in 620.21(A) through (C) »620.21«.

Ⓐ Working space must be provided about controllers, disconnecting means, and other electrical equipment. The minimum working space must meet 110.26(A) specifications. Where maintenance and supervision conditions ensure that only qualified persons will examine, adjust, service, and maintain the equipment, 110.26(A) clearance requirements can be waived per 620.5(A) through (D) »620.5«.

Ⓑ Circuit voltage of heating/air-conditioning equipment branch-circuits, located on the elevator car, must not exceed 600 volts »620.3(C)«.

Ⓒ Lighting circuits must comply with Article 410 requirements »620.3(B)«.

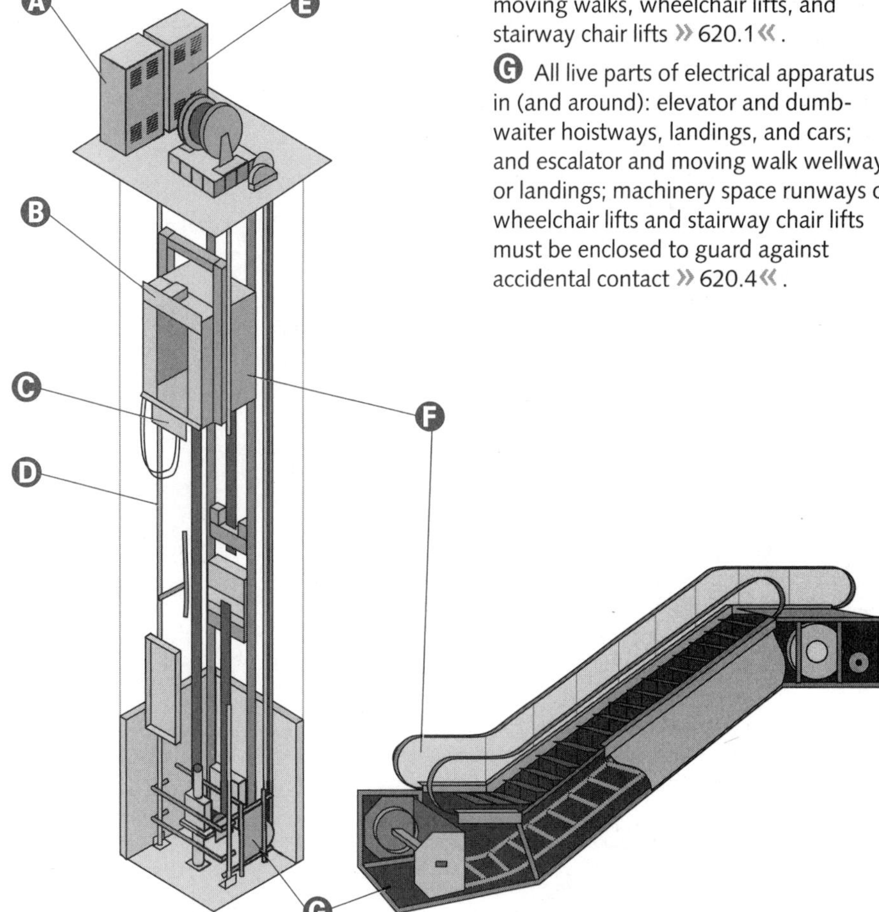

Ⓓ The minimum conductor size for traveling cables which feed lighting circuits is 14 AWG copper. 20 AWG (or larger) copper conductors are permitted in parallel, provided the ampacity equals or surpasses that of 14 AWG copper. For traveling cable circuits, other than lighting, 20 AWG copper is the minimum size »620.12(A)«.

Ⓔ The motor controller rating must comply with 430.83. The rating can be less than the nominal rating of the elevator motor, when the controller inherently limits the available power to the motor, being appropriately marked as power limited »620.15«.

Ⓕ Article 620 covers the installation of electric equipment and wiring relating to elevators, dumbwaiters, escalators, moving walks, wheelchair lifts, and stairway chair lifts »620.1«.

Ⓖ All live parts of electrical apparatus in (and around): elevator and dumbwaiter hoistways, landings, and cars; and escalator and moving walk wellways or landings; machinery space runways of wheelchair lifts and stairway chair lifts must be enclosed to guard against accidental contact »620.4«.

Elevator Wiring

FMC, LFMC, LFNC or flexible cords and cables, or conductors grouped together (taped or corded) that are part of listed equipment, a driving machine, or a driving machine brake can be installed on the car assembly, in lengths not to exceed 6 ft (1.8 m) without being installed in a raceway if of a jacketed flame-retardant type, and by location protected from physical damage »620.21(A)(2)(d)«.

Flexible cords and cables that are components of listed equipment, used in circuits operating at 30 volts rms (or less), or 42 volts dc (or less), are permitted in lengths not exceeding 6 ft (1.8 m) provided the cords and cables are (1) supported and protected from physical damage and (2) of a jacketed and flame-retardant type »620.21(A)(2)(c)«.

Motor-generators, machine motors, or pumping unit motors and valves, located adjacent to, or underneath, control equipment and provided with extra-length terminal leads, no longer than 6 ft (1.8 m), can be connected directly to controller terminal studs without regard to Articles 430 and 445 carrying-capacity requirements. Auxiliary gutters are permitted in machine/control rooms between controllers, starters, and similar apparatus »620.21(A)(3)(b)«.

Existing or listed equipment conductors can be grouped together (taped or corded) without being installed in a raceway. Such cable groups should by location be protected from physical damage, being supported at intervals of 3 ft (900 mm) or less »620.21(A)(3)(d)«.

A FMC, LFMC, or LFNC of ⅜ in. nominal trade size or larger, no longer than 6 ft (1.8 m), can be installed between control panels and machine motors, machine brakes, motor-generator sets, disconnecting means, and pumping unit motors and valves »620.21(A)(3)(a)«. LFNC, as defined in 356.2(2), can be installed in lengths exceeding 6 ft (1.8 m) »620.21(A)(3)(a) *Exception*«.

B FMC, LFMC, or LFNC of ⅜ inch nominal trade size (or larger), not exceeding 6 ft (1.8 m) in length, allowed for installation on cars, must be securely fastened in an oil free location »620.21(A)(2)(a)«. LFNC of ⅜ in. nominal trade size or larger, as defined by 356.2(2), is permitted in lengths greater than 6 ft (1.8 m) »620.21(A)(2)(a) *Exception*«.

C Hard-service cords and junior hard-service cords, conforming to Article 400 (Table 400.4) requirements, are permitted as flexible connections between the car's fixed wiring and the car door/gate devices. Only hard-service cords are permitted as flexible connections for top-of-car operating devices or car-top work lights. Devices, or luminaires (fixtures), must be grounded via an equipment grounding conductor run with the circuit conductors. Cables having smaller conductors as well as other insulation (jacket) types and thicknesses can serve as flexible connections between the car's fixed wiring and the car door/gate devices, if specifically listed for this use »620.21(A)(2)(b)«.

D FMC, LFMC, or LFNC can be installed within hoistways between risers and limit switch interlocks, operating buttons, and similar devices »620.21(A)(1)(a)«.

E Class 2 power-limited circuit cables installed in hoistways between risers, signal equipment, and operating devices must be of a jacketed flame-retardant type, securely supported and protected from physical damage »620.21(A)(1)(b)«.

F FMC, LFMC, LFNC or flexible cords and cables, or conductors grouped together (taped or corded) that are part of listed equipment, a driving machine, or a driving machine brake are permitted on the counterweight assembly, in lengths not to exceed 6 ft (1.8 m) without being installed in a raceway if of a jacketed flame-retardant type, and protected from physical damage »620.21(A)(4)«.

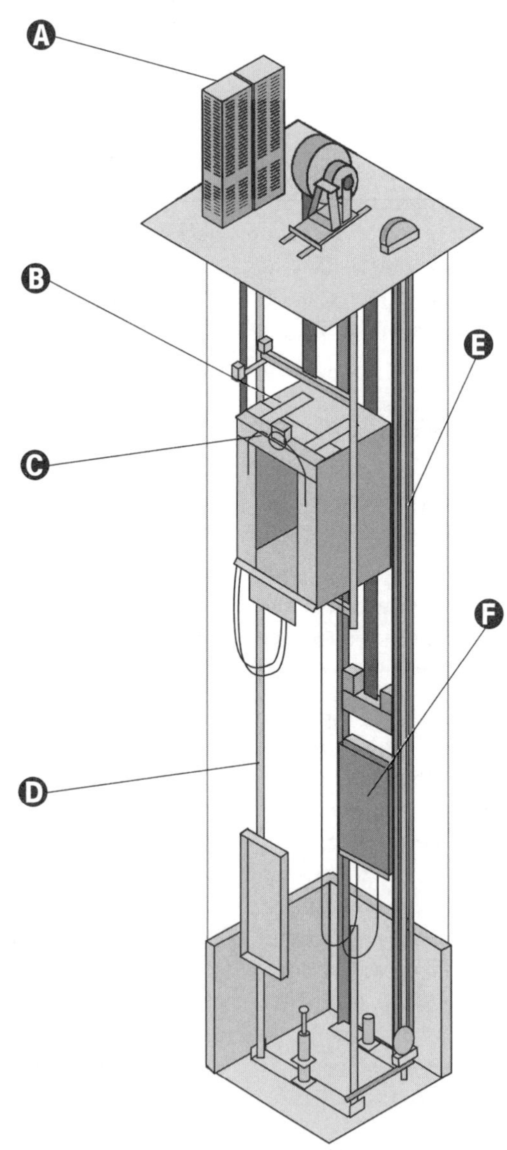

Escalator and Moving Walkway Wiring

NOTE *Wheelchair lift and stairway chair lift raceways must be installed according to 620.21(C) provisions.*

Escalators, moving walks, wheelchair lifts, and stairway chair lifts must comply with Article 250 »620.84«.

Ⓐ FMC, LFMC, or LFNC are acceptable in escalator and moving walk wellways. FMC or liquidtight flexible conduit of ⅜ inch nominal trade size is permitted in lengths of 6 ft (1.8 m) or less »620.21(B)(1)«. LFNC of ⅜ in. nominal trade size or larger, as defined in 356.2(2), can be installed in lengths greater than 6 ft (1.8 m) »620.21(B)(1) *Exception*«.

Ⓑ Class 2 power-limited circuit cables can be installed within escalators and moving walkways provided the cables are (1) supported and protected from physical damage and (2) of a jacketed and flame-retardant type »620.21(B)(2)«.

Ⓒ Hard-service cords, conforming to Article 400 (Table 400.4) requirements, are permitted as flexible connections on escalator and moving walk control panels and disconnecting means, where the entire control panel and disconnecting means can be removed from machine spaces per 620.5 »620.21(B)(3)«.

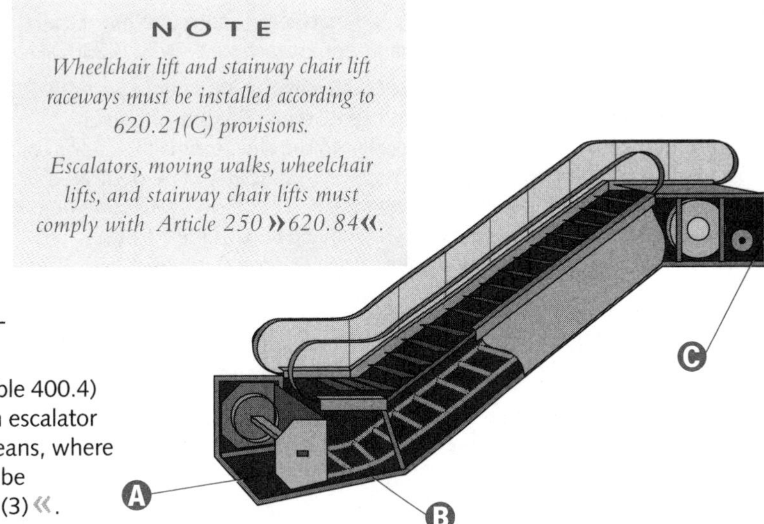

Conductor Installation

Auxiliary gutters are not subject to the length restrictions of 366.3, nor to the number of conductors per 366.6 »620.35«.

Ⓐ Only electric wiring, raceways, and cables used in direct connection with the elevator or dumbwaiter are permitted inside the hoistway, machine rooms, control rooms, machinery spaces, and control spaces. Wiring for signals; for communication with the car; for lighting, heating, air conditioning and ventilating the elevator car; for fire-detecting systems; for pit sump pumps; and for lighting and ventilating the hoistway are allowed »620.37(A)«.

Ⓑ Cables or raceway supports in a hoistway, escalator/moving walk wellway, or wheelchair lift and stairway chair lift runway, must be securely fastened to the guide rail; escalator/moving walk truss; or to the hoistway, wellway, or runway construction »620.34«.

Ⓒ Optical fiber cables and conductors for operating devices, operation and motion control, power, signaling, fire, alarm, lighting, heating, and air-conditioning circuits of 600 volts (or less) can run in the same traveling cable (or raceway) system if (1) all conductors are insulated for the maximum voltage applied to any conductor within the cable (or raceway) system; and (2) all live parts of the equipment are insulated from ground for the same maximum voltage. Such a traveling cable (or raceway) can also include shielded conductors and/or coaxial cable(s), if insulated for the maximum voltage applied to any conductor within the cable (or raceway) system. Conductors can be suitably shielded for telephone, audio, video, or higher frequency communications circuits »620.36«.

Ⓓ Bonding of elevator rails (car and/or counterweight) to a lightning protection system grounding-down conductor(s) is permitted. The lightning protection system must be located outside the hoistway. Elevator rails or other hoistway equipment cannot be used as the grounding-down conductor for lightning protection systems »620.37(B)«.

Ⓔ Main feeders supplying elevator and dumbwaiter power must be installed outside the hoistway unless: (1) by special permission, elevator feeders are permitted within an existing hoistway containing no spliced conductors; or (2) feeders are permitted inside the hoistway for elevators with driving machine motors located in the hoistway, on the car, or on the counterweight »620.37(C)«.

Ⓕ The cross-sectional area's sum of the individual conductors in a wireway must be no more than 50% of the wireway's interior cross-sectional area. Vertically run wireways must be securely supported at intervals of 15 ft (4.5 m) or less, and must have a maximum of one joint between supports. Adjoining wireway sections must be securely fastened together to form a rigid joint »620.32«.

CAUTION *Electrical equipment and wiring used for elevators, dumbwaiters, escalators, moving walks, and wheelchair lifts and stairway lifts in garages must comply with Article 511 requirements »620.38«.*

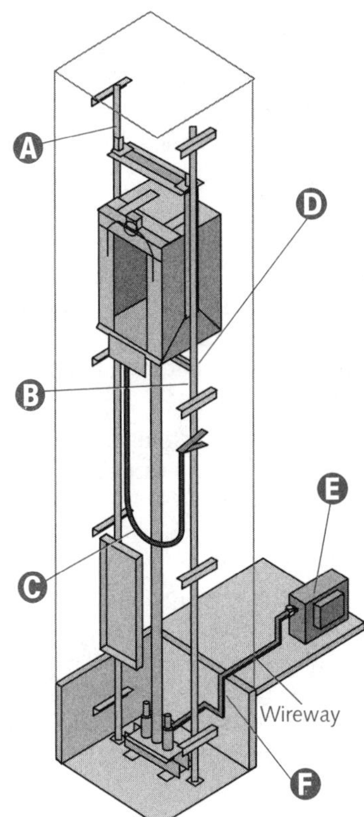

Branch-Circuits and Traveling Cables

A Each machine room or control room and machinery space or control space requires at least one 125-volt, single-phase, duplex receptacle »620.23(C)«.

B A separate branch-circuit must supply each elevator car's lights, receptacle(s), auxiliary lighting power source, and ventilation. The branch-circuit's overcurrent protective device must be located in the elevator machine room or control room/machinery space or control space»620.22(A)«. A dedicated branch-circuit is required for each elevator car's air-conditioning/heating unit. Locate this branch-circuit's overcurrent protective device in the elevator machine room or control room/machinery space or control space »620.22(B)«.

C Traveling cables must be suspended at both the car and hoistway ends (or counterweight end if applicable), thereby reducing the strain on individual copper conductors to a minimum. Support traveling cables by any one of the following means:

(1) Steel supporting member(s)

(2) Looping the cables around supports for unsupported lengths less than 100 ft (30 m)

(3) Suspending from the supports by a means that automatically tightens around the cable with increased tension, for unsupported lengths up to 200 ft (60 m) »620.41«.

D Careful location of traveling cable supports can reduce to a minimum the possibility of damage due to cables contacting hoistway construction/equipment. Provide suitable guards where necessary to protect the cables against damage »620.43«.

E Each hoistway pit must have at least one 125-volt, single-phase, duplex receptacle »620.24(C)«.

F A separate branch-circuit must supply the machine room or control room/machinery space or control space lighting and receptacle(s). Required lighting must not be connected to a GFCI's load side »620.23(A)«. Locate the lighting switch at the machine room or control room/machinery space or control space point of entry »620.23(B)«.

G Metal raceways and cables of Types MC, MI, or AC attached to an elevator car must be bonded to the car's grounded metal parts »620.81«.

H Traveling cable can be run without the use of a raceway for a distance of 6 ft (1.8 m) or less (measured from the first support point on the elevator car, hoistway wall, or counterweight [where applicable]), provided the conductors are grouped and taped (or corded) together, or are still in the original sheath. Traveling cables can continue, as fixed wiring, to elevator controller enclosures as well as elevator car and machine room, control room, machinery space, and control space connections, if suitable support and physical damage protection is provided »620.44«.

I Hoistway pit lighting and receptacle(s) must be supplied by a separate branch-circuit. In no case should required lighting be connected to the load side of a GFCI »620.24(A)«. The lighting switch's location must be readily

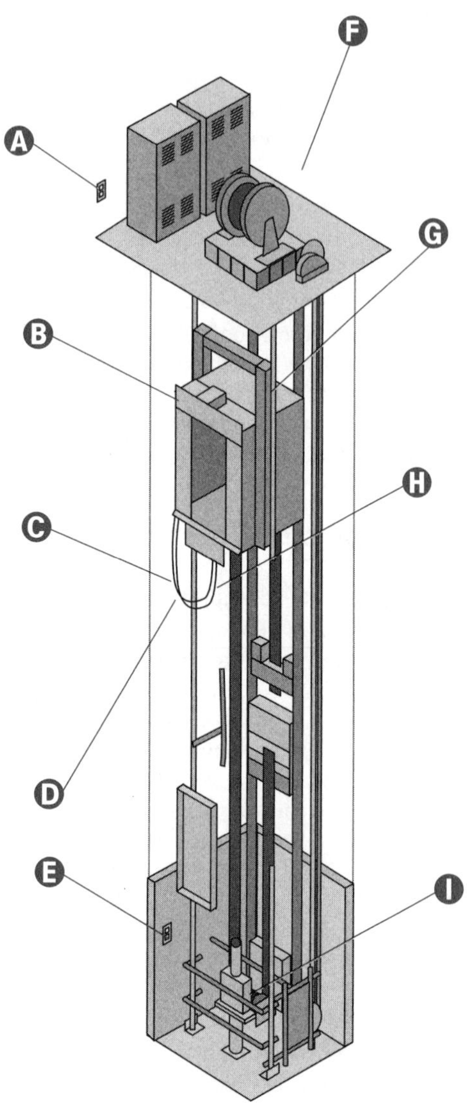

NOTE

For electric elevators, all motor frames, elevator machines, controllers, and metal electrical-equipment enclosures in (or on) the car or in the hoistway, must be grounded per Article 250 »620.82«.

CAUTION *Each 125-volt, single-phase, 15- and 20-ampere receptacle installed in pits, hoistways, on elevator car tops, and in escalator and moving walk wellways must be of the GFCI type. All 125-volt, single-phase, 15- and 20-ampere receptacles installed in machine rooms, and machinery spaces, must have GFCI protection for personnel. A single receptacle supplying a permanently installed sump pump does not require GFCI protection »620.85«.*

Protection and Control

Each unit must have a single means for disconnecting all ungrounded main power supply conductors that is designed so that no pole operates independently. Where multiple driving machines are connected to a single elevator, escalator, moving walk, or pumping unit, a single means must be provided to disconnect the motor(s) and control valve operating magnets »620.51«.

The disconnecting means' location must be readily accessible to qualified persons »620.51(C)«. Depending on the application, the disconnecting means' location must comply with 1, 2, 3, or 4 in 620.51(C).

Where there are multiple driving machines in a machine room, the disconnecting means must be numbered to correspond with the identifying number of the driving machine being controlled. The disconnecting means must have a sign identifying the location of the supply side overcurrent protective device »620.51(D)«.

Each elevator car must have a single means for disconnecting all ungrounded car light, receptacle(s), and ventilation power-supply conductors. The disconnecting means must be an enclosed externally operable fused motor-circuit switch or circuit breaker capable of locking in the open position and must be located in that elevator car's machine/control room. Each disconnecting means must be numbered to correspond with the identifying number of the car whose light source it controls. The disconnecting means must have a sign identifying the location of the supply side overcurrent protective device »620.53«.

Operating devices and control/signaling circuits must be protected against overcurrent according to 725.23 and 725.24. Class 2 power-limited circuits must be protected against overcurrent in accordance with the requirements of Chapter 9, Notes to Tables 11(A) and 11(B) »620.61(A)«.

Duty on elevator and dumbwaiter driving machine motors and driving motors of motor-generators (used with generator field control) is rated as *intermittent*. Such motors must be protected against overload per 430.33 »620.61(B)(1)«.

Duty on escalator and moving walk driving machine motors is rated as *continuous*. Protect such motors against overload in accordance with 430.32 »620.61(B)(2)«. Escalator and moving walk driving machine motors, and driving motors of motor-generator sets, must be protected against running overload as provided in Table 430.37 »620.61(B)(3)«.

Wheelchair lift and stairway chair lift driving machine motor duty is rated as *intermittent*. Such motors must be protected against overload according to 430.33 »620.61(B)(4)«.

CAUTION *Make no provision to open (or close) the disconnecting means from any remote part of the premises. If sprinklers are installed in hoistways, machine rooms, control rooms, machinery spaces or control spaces, the disconnecting means can automatically open the power supply to the affected elevator(s) prior to water release. Automatic closure of this disconnecting means is prohibited. Power must be restored only by manual intervention »620.51(B)«.*

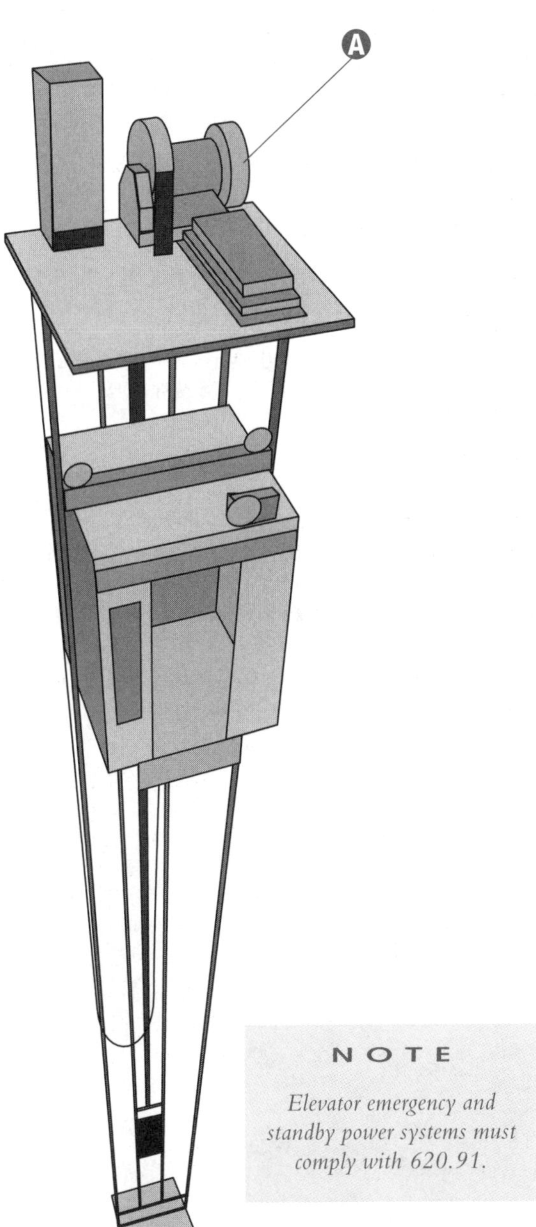

A Elevator, dumbwaiter, escalator, and moving walk driving machines; motor-generator sets; motor controllers; and disconnecting means must be installed in a room or space dedicated for that purpose, unless otherwise permitted in 620.71(A) or (B). The room, or enclosure, must be secured against unauthorized access »620.71«.

NOTE

Elevator emergency and standby power systems must comply with 620.91.

SWIMMING POOLS, FOUNTAINS, AND SIMILAR INSTALLATIONS

General

All electrical equipment installed in the water, walls, or decks of pools, fountains, and similar applications must comply with the provisions of this article »680.4«.

Except as modified by this section, wiring and equipment in, or adjacent to, pools and fountains, must comply with other applicable requirements of this *Code* »680.3«.

Storable pool provisions are located in Article 680, Part III (680.30 through 33).

The provisions of Article 680, Part 1 and Part VI shall apply to therapeutic pools and tubs in health care facilities, gymnasiums, athletic training rooms, and similar areas »680.60«.

Hydromassage bathtub requirements are located in Article 680, Part VII.

A Article 680 provisions apply to the electrical wiring construction (and installation) and for equipment in, or adjacent to, all swimming, wading, therapeutic, and decorative pools, fountains, hot tubs, spas, and hydromassage bathtubs, whether permanently installed or storable, as well as to similar metallic auxiliary equipment (such as pumps, filters, etc.) »680.1«.

B **Fountain:** Specific to this article, fountain includes fountains, ornamental pools, display pools, and reflection pools. Drinking fountains are not included »680.2«.

C Part I and V provisions apply to all permanently installed fountains as defined in 680.2. Self-contained, portable fountains [no larger than 5 ft (1.5 m) in any dimension] are not covered here »680.50«.

D **Permanently Installed Decorative Fountains and Reflection Pools:** Those that are constructed in, or on, the ground, or in a building so that the fountain cannot be readily disassembled, with or without electrical circuits of any nature. These units are constructed primarily for their aesthetic value, and are not intended for swimming or wading »680.2«.

E **Spa or Hot Tub:** A hydromassage pool, or tub for recreational or therapeutic use, not located in health care facilities, designed for immersion of users, and usually having a filter, heater, and motor-driven blower. It may be installed either indoors or out, on, or in, the ground or supporting structure. Generally, a spa or hot tub is not designed to be drained after each use »680.2«.

F Spa and hot tub requirements can be found in Article 680, Part IV (680.40 through 44).

G Bonding provisions, for permanently installed pools, are located in 680.26.

H **Pool:** Manufactured or field-constructed equipment designed to contain water on a permanent or semi-permanent basis, and used for swimming, wading, or other purposes »680.2«.

I Permanently installed pool underwater luminaire (lighting fixture) requirements are located in 680.23.

J **Permanently Installed Swimming, Wading, and Therapeutic Pools:** Those that are constructed totally or partially in the ground, all others with depth capacities greater than 42 in. (1.0 m), and all pools installed inside of a building, regardless of water depth, whether or not served by electrical circuits of any nature »680.2«.

K Permanently installed pools must comply with Article 680, Part II.

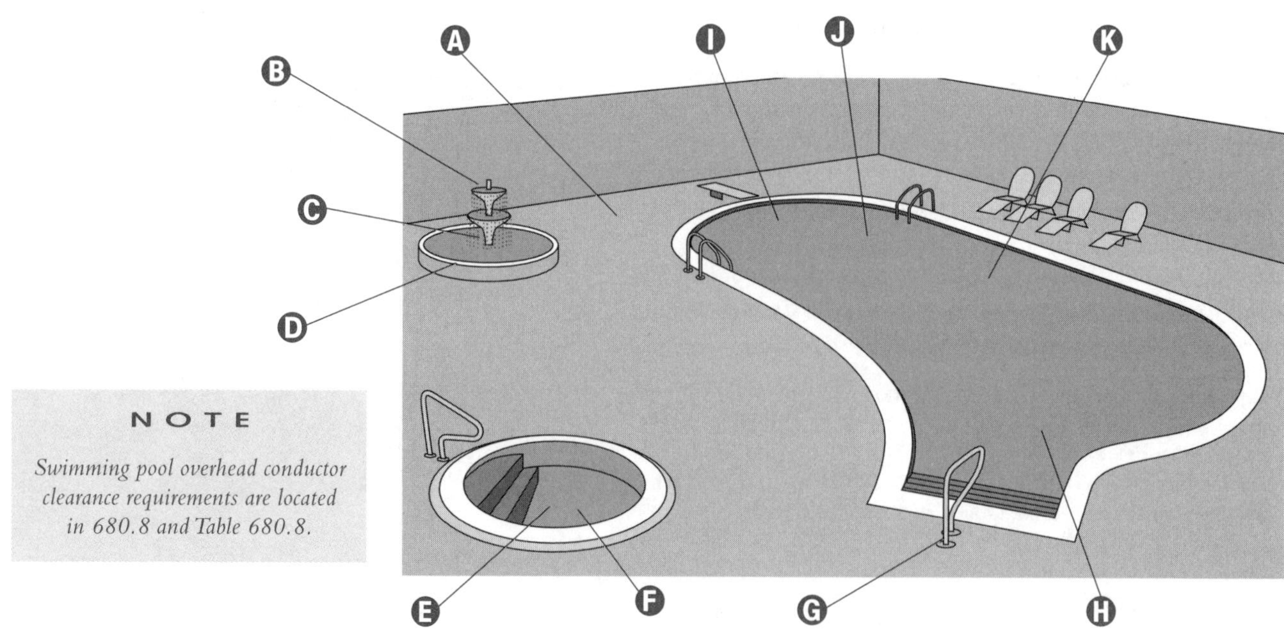

NOTE

Swimming pool overhead conductor clearance requirements are located in 680.8 and Table 680.8.

Definitions

A Self-Contained Spa or Hot Tub: Factory-fabricated unit consisting of a spa or hot tub vessel having integrated water-circulating, heating, and control equipment. Equipment may include pumps, air blowers, heaters, lights, controls, sanitizer generators, etc. »680.2«.

B Hydromassage Bathtub: A permanently installed bathtub with recirculation piping, pump, and associated equipment. It is designed to accept, circulate, and discharge water upon each use »680.2«.

C Wet-Niche Luminaire (Lighting Fixture): A luminaire (lighting fixture) intended for installation in a pool or fountain structure's forming shell, where completely surrounded by water »680.2«.

D Storable Swimming or Wading Pool: Those constructed on, or above, the ground having a maximum water depth capacity of 42 in. (1.0 m), or a pool with nonmetallic, molded polymeric walls (or inflatable fabric walls) regardless of dimension »680.2«.

E Pool Cover, Electrically Operated: Motor-driven equipment designed to cover and uncover the pool's water surface by means of a flexible sheet or rigid frame »680.2«.

F Cord- and Plug-Connected Lighting Assembly: A lighting assembly consisting of a luminaire (lighting fixture) intended for installation in the wall of a spa, hot tub, or storable pool, having a cord- and plug-connected transformer »680.2«.

G Forming Shell: A support structure designed for a wet-niche luminaire (lighting fixture) assembly and intended for pool or fountain structure mounting »680.2«.

H Packaged Spa or Hot Tub Equipment Assembly: A factory-fabricated unit consisting of water-circulating, heating, and control equipment mounted on a common base, which operates as a spa or hot tub. Equipment may include pumps, air blowers, heaters, lights, controls, sanitizer generators, etc. »680.2«.

I Self-Contained Therapeutic Tubs or Hydrotherapeutic Tanks: A factory-fabricated unit consisting of a therapeutic tub or hydrotherapeutic tank with integrated water-circulating, heating, and control equipment. Equipment may include pumps, air blowers, heaters, lights, controls, sanitizer generators, etc. »680.2«.

J No-Niche Luminaire (Lighting Fixture): A luminaire (lighting fixture) intended for above or below water installation without a niche »680.2«. Note: Not to be mistaken for a dry-niche luminaire (lighting fixture).

Dry-Niche Luminaire (Lighting Fixture): A luminaire (lighting fixture) intended for installation in the wall of a pool or fountain in a niche that is sealed against water entry »680.2«.

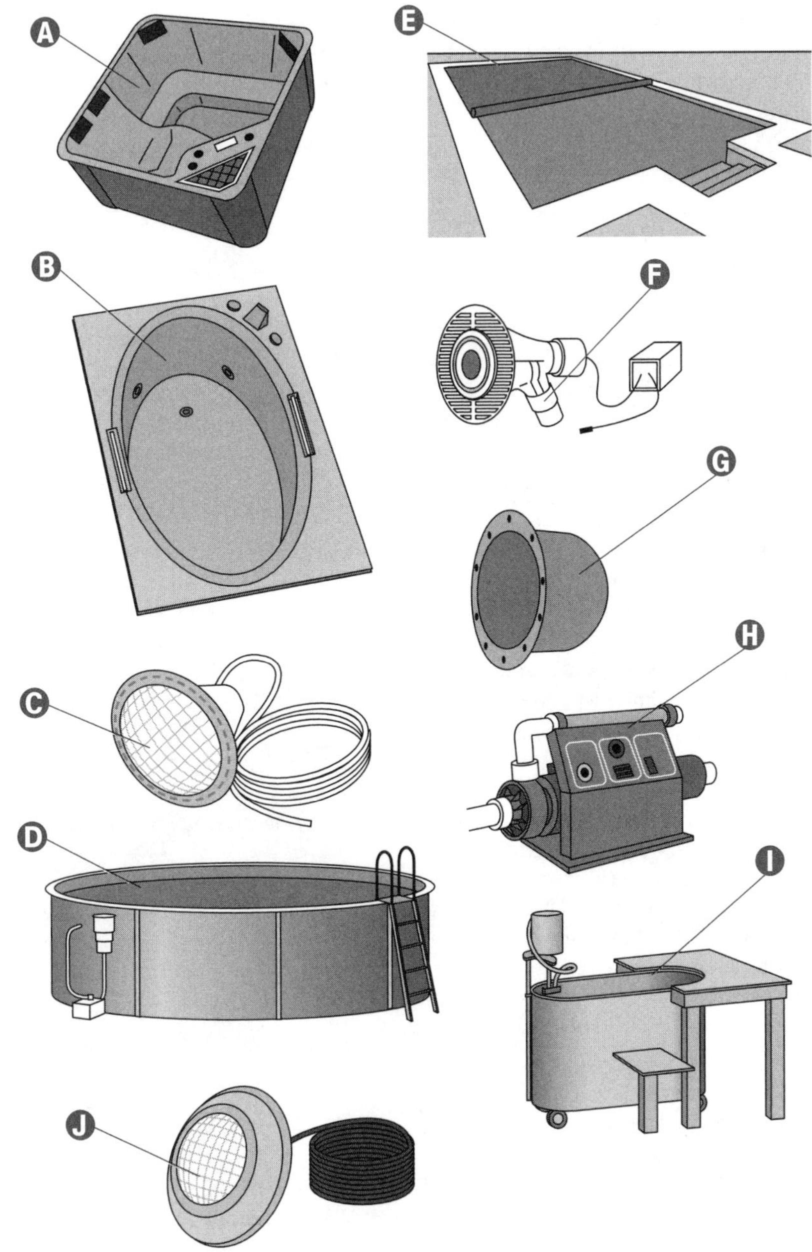

Common Bonding Grid

Structural reinforcing steel, or the walls of bolted or welded metal pool structures, can serve as a common bonding grid for nonelectrical parts where connections are made in accordance with 250.8 »680.26(D)«. Grounding conductors and bonding jumpers must be connected by exothermic welding, listed pressure connectors, listed clamps, or other listed means »250.8«.

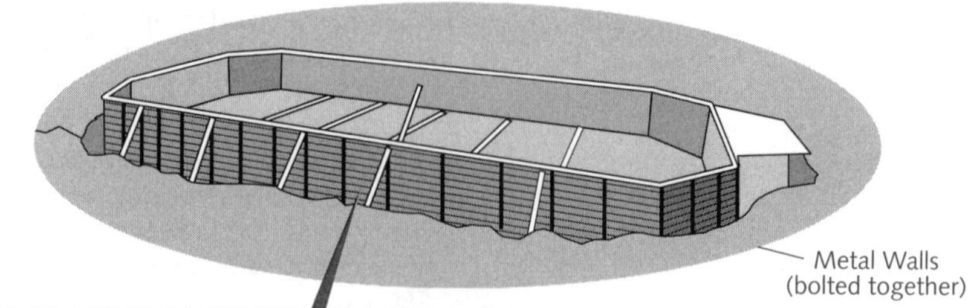

A common bonding grid can be any of the following:

(1) The structural reinforcing steel of a concrete pool where the rods are bonded together by the usual steel tie wires, or an equivalent.

(2) The wall of a bolted or welded metal pool.

(3) A solid copper conductor (insulated, covered, or bare), no smaller than 8 AWG.

(4) Brass RMC or IMC, or other identified corrosion-resistant metal conduit »680.26(C)«.

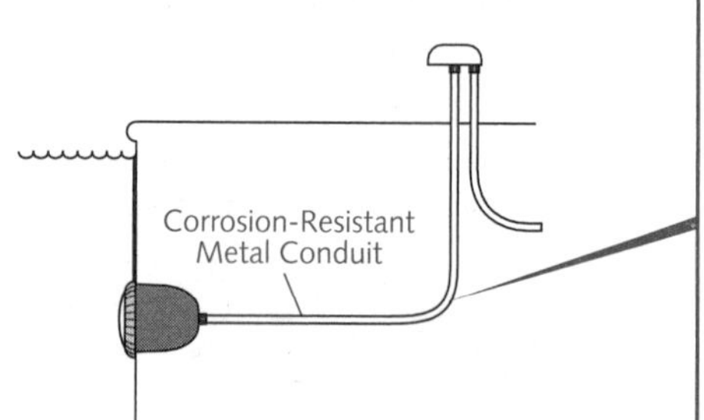

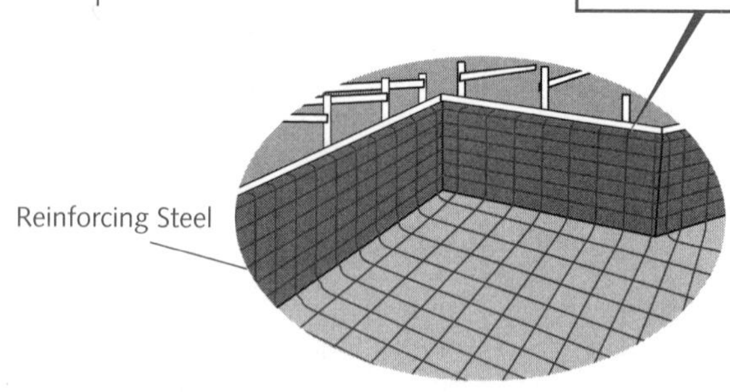

CAUTION *The parts specified in 680.26(B) must be connected to a common bonding grid with an 8 AWG or larger solid copper conductor (insulated, covered, or bare). Connections must be made by exothermic welding, or by pressure connectors/clamps* ***labeled*** *as suitable for the purpose and made of the following material: stainless steel, brass, copper, or copper alloy »680.26(C)«.*

Receptacles

Where a pool is within 10 ft (3.0 m) of a dwelling, and the dimensions of the lot prevent meeting the required clearances, not more than one receptacle outlet shall be permitted, if not less than 5 ft (1.5 m) measured horizontally from the pool's inside wall »680.22(A)(4)«.

A Receptacle(s) providing power for a permanently installed pool's water-pump motor(s), or other loads directly related to the circulation and sanitation system, can be located between 5 ft (1.5 m) and 10 ft (3.0 m) from the inside walls of the pool or fountain. If so located, the receptacle(s) must be **single**, of the **locking and grounding type**, protected by GFCI(s) »680.22(A)(1)«.

B A dwelling unit's permanently installed pool must have at least one 125-volt 15- or 20-ampere receptacle on a general-purpose branch-circuit that is located a minimum of 10 ft (3.0 m), but no more than 20 ft (6.0 m), from the pool's inside wall. This receptacle must not be more than 6 ft, 6 in. (2.0 m) above the floor, platform, or grade level serving the pool »680.22(A)(3)«.

NOTE

All 125-volt receptacles located within 20 ft (6.0 m) of the pool's inside walls must be protected by a GFCI »680.22(A)(5)«. This distance is determined by measuring the shortest path that the appliance supply cord (connected to the receptacle) would follow without piercing a floor, wall, ceiling, hinged or sliding panel doorway, window opening, or other effective permanent barrier »680.22(A)(6)«.

CAUTION *A receptacle installed in a wet location, unattended while the plugged-in product is in use, must have an enclosure that is weatherproof at all times, i.e., attachment plug cap inserted or removed »406.8(B)(2)(a)«.*

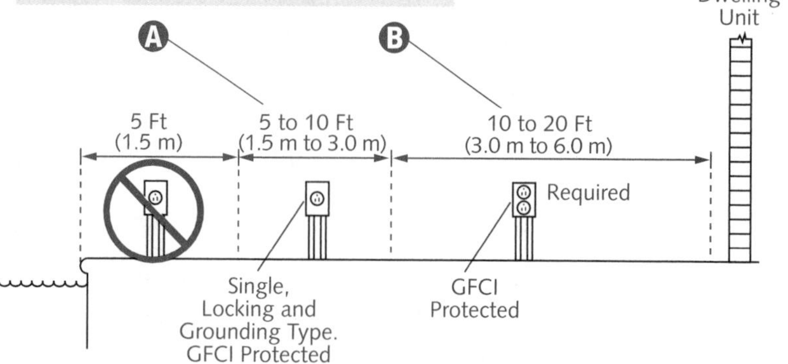

Bonding

In the case of pool water heaters (rated more than 50 amperes) having specific bonding and grounding instructions, only those parts designated to be bonded shall be bonded, and only those parts designated to be grounded shall be grounded »680.26(E)«.

A Bond all metal-sheathed cables/raceways, metal piping, and all fixed metal parts that are not separated from the pool by a permanent barrier within 5 ft (1.5 m) horizontally of the pool's inside wall, and within 12 ft (3.7 m) above the pool's maximum water level »680.26(B)(5)«. Examples of fixed metal parts include: fences/fence posts, gutter down-spouts, door/window frames, raceways, etc.

B Bond all forming shells and mounting brackets of no-niche luminaires (fixtures), unless a listed low-voltage lighting system is used which does not require bonding »680.26(B)(2)«.

> **NOTE**
>
> *This section's intention is not to require that 8 AWG or larger solid copper bonding conductor be extended (or attached) to any remote panelboard, service equipment, or any electrode, but rather that it must be employed to eliminate voltage gradients in the pool area as prescribed »680.26(A) and FPN«.*

C Bond any diving structure, observation stands, towers, or platforms not separated from the pool by a permanent barrier »680.26(B)(5)«.

D Bond all metal fittings within, or attached to, the pool structure »680.26(B)(3)«.

E Isolated parts, not more than 4 in. (100 mm) in any dimension, which do not penetrate the pool's structure more than 1 in. (25 mm), do not require bonding »680.26(B)(3)«.

F All parts listed in 680.26(B)(1) through (5) must be bonded together »680.26(B)«. Those parts must also be connected to a common bonding grid »680.26(C)«.

G Bond all metal fittings within, or attached to, the pool structure »680.26(B)(3)«. Note: Each ladder/hand rail anchor must be bonded unless not over 4 in. (100 mm) in any dimension.

H Bond all metal parts of equipment associated with pool covers, including electric motors »680.26(B)(4)«.

I Bond metal parts of electrical equipment associated with the pool water circulating system, including pump motors. (Metal parts of listed equipment incorporating an approved system of double insulation and providing a means for grounding internal inaccessible, noncurrent-carrying metal parts must not be bonded) »680.26(B)(4)«. Note: Pool equipment must be bonded to the common bonding grid regardless of the intervening distance and/or location to the pool.

J If reinforcing steel is effectively insulated at the time of manufacture and installation, by a listed encapsulating nonconductive compound, bonding is not required »680.26(B)(1)«.

K All of the pool structure's metallic parts (including the reinforcing metal of the pool shell, coping stones, and deck) must be bonded. The usual steel tie wires are suitable for bonding the reinforcing steel together; no welding or special clamping is required. These tie wires must be made tight »680.26(B)(1)«. Note: Steel tie wires are permitted to bond only reinforcing steel *together*. Steel tie wires must not be used to connect the 8 AWG copper bonding conductor to the reinforcing steel.

> **CAUTION** *The parts specified in 680.26(B) must be connected to a common bonding grid via a solid copper conductor (insulated, covered, or bare) no smaller than 8 AWG. Connections must be made by exothermic welding, or by pressure connectors/clamps* ***labeled*** *as suitable for that purpose and made of the following material: stainless steel, brass, copper, or copper alloy »680.26(C)«.*

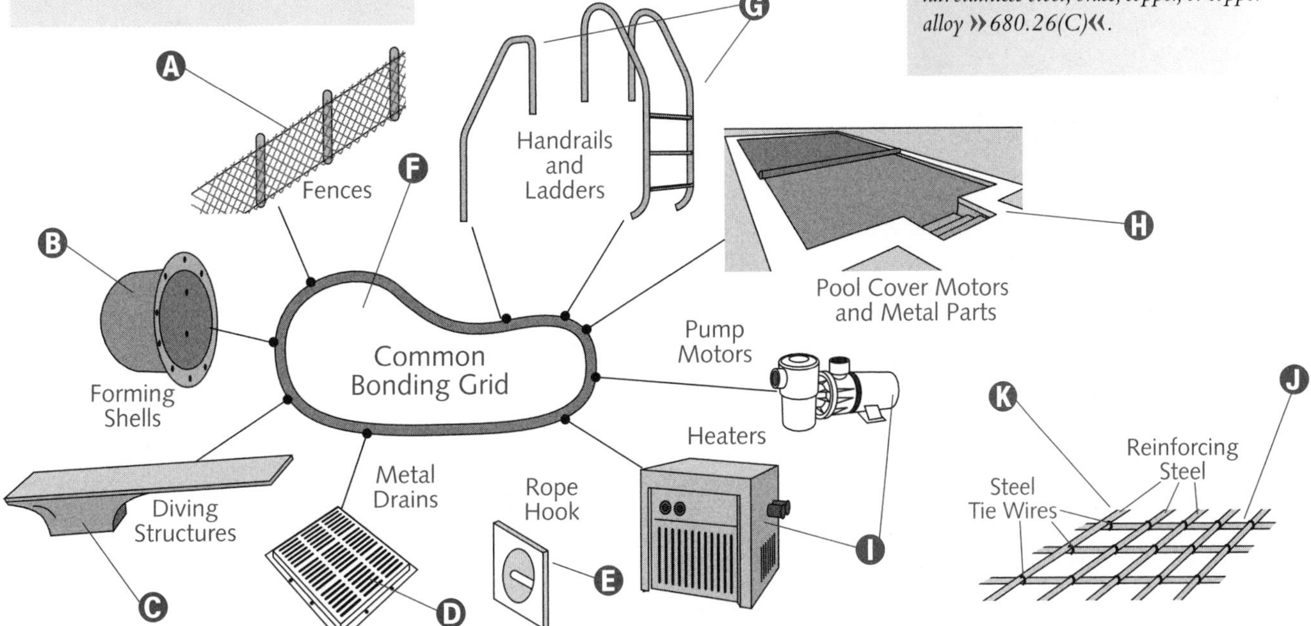

Luminaires (Lighting Fixtures), Lighting Outlets, and Ceiling-Suspended (Paddle) Fans

680.22(B)(1) limitations do not apply to indoor pool areas if all of the following conditions are satisfied: (1) luminaires (fixtures) are of a totally enclosed type; (2) ceiling-suspended (paddle) fans are identified for use beneath ceiling structures such as provided on porches or patios; (3) the branch-circuit supplying the equipment is GFCI protected; and (4) the bottom of the luminaire (fixture) or ceiling-suspended (paddle) fan is at least 7 ft, 6 in. (2.3 m) above the maximum water level »680.22(B)(2)«.

Cord- and plug-connected luminaires (lighting fixtures) must meet the same 680.7 specifications, where installed within 16 ft (4.9 m) of any point along the water's surface, measured radially »680.22(B)(5)«.

A In outdoor pool areas, luminaires (lighting fixtures), lighting outlets, and ceiling-suspended (paddle) fans must not be installed over the pool, or over the area extending 5 ft (1.5 m) horizontally from the pool's inside walls, unless located not less than 12 ft (3.7 m) above the maximum water level »680.22(B)(1)«.

B Existing luminaires (lighting fixtures) and lighting outlets located less than 5 ft (1.5 m), measured horizontally from the pool's inside walls, must be: (1) at least 5 ft (1.5 m) above the maximum water level surface; (2) rigidly attached to the existing structure; and (3) protected by a GFCI »680.22 (B)(3)«.

C Luminaires (lighting fixtures) and lighting outlets, and ceiling-suspended (paddle) fans, installed in the area extending between 5 ft (1.5 m) and 10 ft (3.0 m) horizontally from a pool's inside walls, must be protected by a GFCI, unless 5 ft (1.5 m) above the maximum water level and rigidly attached to the pool's adjacent (or enclosing) structure »680.22(B)(4)«.

NOTE

Locate all switching devices on the property at least 5 ft (1.5 m) horizontally from a pool's inside walls, unless separated from the pool by a solid fence, wall, or other permanent barrier »680.22 (C)«.

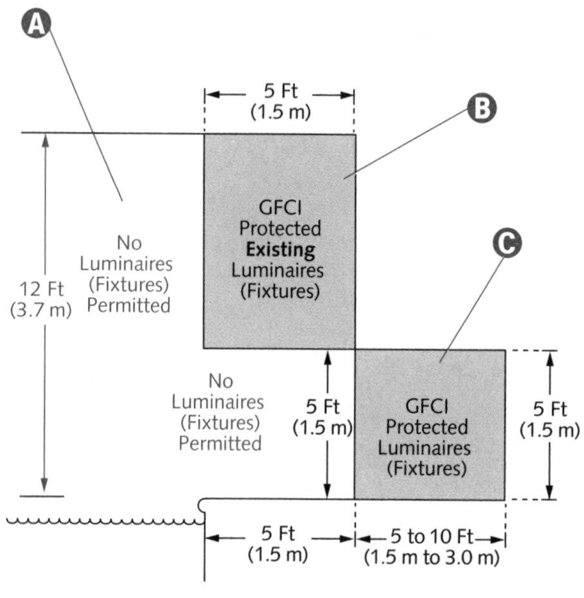

Hydromassage Bathtubs

A All bathroom receptacles must have GFCI protection for personnel »210.8(A)(1) and (B)(1)«.

B All 125-volt, single-phase receptacles not exceeding 30 amperes and located within 5 ft (1.5 m), measured horizontally, of the hydromassage tub's inside walls must be protected by GFCI »680.71«.

C Hydromassage bathtubs, and associated electrical components, must be protected by a GFCI »680.71«.

D All metal piping systems, electrical equipment metal parts, and pump motors associated with the hydromassage tub must be bonded together using an 8 AWG (or larger) solid copper bonding jumper (insulated, covered, or bare). Metal parts of listed equipment incorporating an approved double insulation system and providing a grounding means for internal nonaccessible, noncurrent-carrying metal parts must not be bonded »680.74«.

E Hydromassage bathtub electrical equipment must be accessible without damage to the building structure or finish »680.73«.

NOTE

Unrelated luminaires (lighting fixtures), switches, receptacles, and other electrical equipment located in the same room, must be installed in accordance with the requirements of Chapters 1 through 4 in the NEC® covering the installation of that equipment in bathrooms »680.72«.

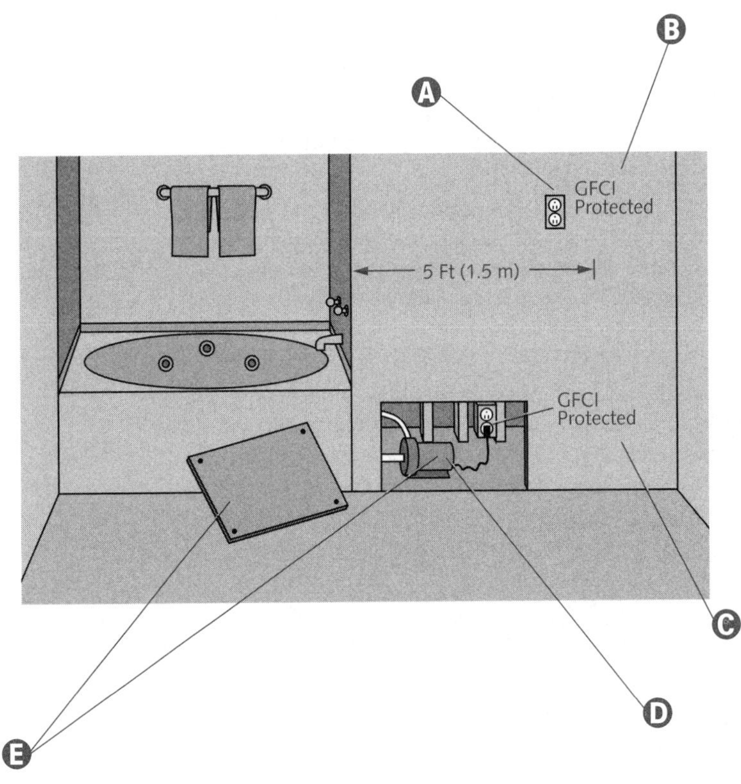

CAUTION *A receptacle must not be installed within a bathtub or shower space »406.8(C)«.*

Underwater Luminaires (Lighting Fixtures) (Permanently Installed Pools)

The junction box or transformer/other enclosure in the supply circuit to a wet-niche, or no-niche, luminaire (lighting fixture) and the field-wiring chamber of a dry-niche luminaire (lighting fixture), must be grounded to the panelboard's equipment grounding terminal. This terminal must be directly connected to the panelboard enclosure »680.24 (F)«.

A Luminaires (lighting fixtures) mounted in walls must be installed so that the top of the luminaire (fixture) lens is at least 18 in. (450 mm) below the pool's normal water level, unless the luminaire (fixture) is listed and identified for use at a depth of not less than 4 in. (100 mm) below the pool's normal water level »680.23(A)(5)«.

B The 8 AWG conductor termination in the forming shell must be covered or encapsulated by a listed potting compound in order to protect it from the possibly deteriorating effect of pool water »680.23(B)(2)(b)«.

C The luminaire (fixture) must be bonded to and secured to the forming shell by a positive locking device that ensures a low-resistance contact, and which can only be removed from the forming shell by use of a tool. Luminaires (fixtures) listed for the application, having no noncurrent carrying metal parts, do not require bonding »680.23(B)(5)«.

D Through-wall lighting assemblies,wet-niche, dry-niche, or no-niche luminaires (lighting fixtures) must be connected to an 12 AWG or larger equipment grounding conductor sized per Table 250.122 »680.23(F)(2)«.

E All forming shells must be bonded to a common bonding grid with a solid copper conductor (insulated, covered, or bare) no smaller than 8 AWG »680.26(B)(2) and (C)«.

F A junction box connected to a conduit extending directly to a forming shell, or mounting bracket of a no-niche luminaire (fixture), must comply with 680.24(A)(1) and (2).

G The equipment grounding conductor must be: (1) an insulated copper conductor, and (2) installed along with the circuit conductors »680.23(F)(2)«.

H A GFCI must be installed in the branch-circuit supplying luminaires (fixtures) operating above 15 volts »680.23(A)(3)«.

I Conduit must extend from the forming shell to a suitable junction box (or other enclosure) located as provided in 680.24. Conduit must be RMC, IMC, LFNC,or RNC. Metal conduit must be approved and must be made of brass or other approved corrosion-resistant metal »680.23(B)(2)«.

J When using nonmetallic conduit, an 8 AWG insulated copper equipment grounding conductor must be installed in the conduit with provisions for terminating in the forming shell, junction box/transformer enclosure, or GFCI enclosure, unless using a listed low-voltage lighting system that does not require grounding »680.23(B)(2)(b)«.

NOTE

Conductors on the load side of a GFCI or transformer, used to comply with 680.23(A)(8) provisions, must not occupy raceways, boxes, or enclosures containing other conductors, unless the other conductors are either protected by GFCIs or are grounding conductors. Feed-through type GFCI supply conductors can occupy the same enclosure. GFCIs are permitted in a panelboard containing circuits protected by means other than GFCIs »680.23(F)(3)«.

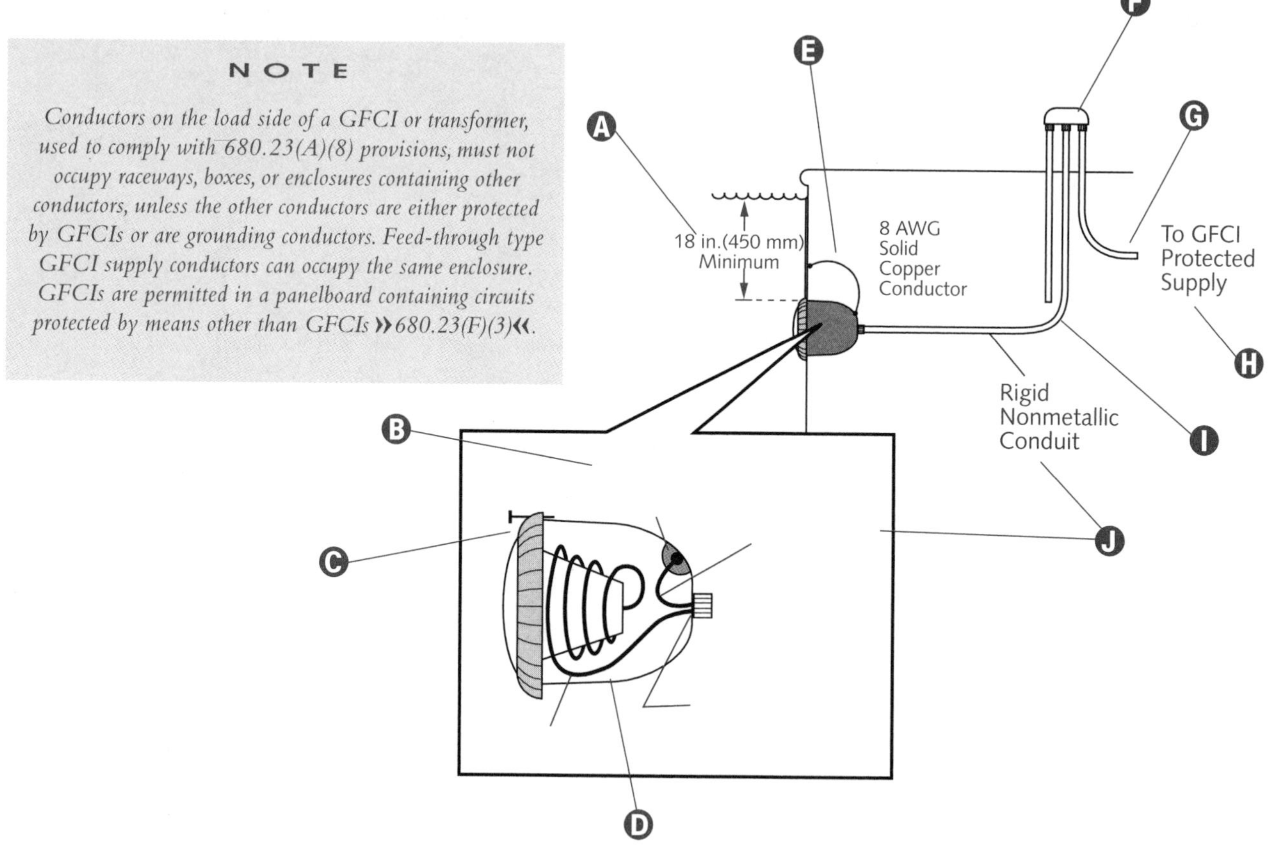

Junction Boxes for Underwater Lighting

Any enclosure for a transformer, GFCI, or a similar device connected to a conduit extending directly to a forming shell, or mounting bracket of a no-niche luminaire (fixture), must comply with 680.24 (B) requirements.

Junction boxes and enclosures, mounted above the grade of the finished walkway around the pool, must not be located in the walkway unless afforded additional protection, such as by location under diving boards, adjacent to fixed structures, and the like. »680.24(C)«.

Junction boxes, transformer enclosures, and GFCI enclosures connected to a conduit extending directly to a forming shell or mounting bracket of a no-niche luminaire (fixture) must have a number of grounding terminals that exceeds the number of conduit entries by at least one »680.24(D)«.

Strain relief must be provided for the termination of an underwater luminaire's (lighting fixture's) flexible cord within a junction box, transformer enclosure, GFCI, or other enclosure »680.24(E)«.

Ⓐ Locate the junction box at least 4 ft (1.2 m) from the pool's inside wall, unless separated from the pool by a solid fence, wall,or other permanent barrier »680.21(A)(2)(b)«.

Ⓑ A junction box connected to a conduit extending directly to a forming shell, or mounting bracket of a no-niche luminaire (fixture), must comply with 680.24(A)(1) and (2).

Ⓒ It must be listed and labeled for the purpose »680.24(A)(1)«.

Ⓓ The junction box must be made of copper, brass, suitable plastic, or other approved corrosion-resistant material »680.24(A)(1)(2)«.

Ⓔ It must have electrical continuity (between every connected metal conduit and the grounding terminals) by means of copper, brass, or other approved corrosion-resistant metal that is integral with the box »680.21(1)(3)«.

Ⓕ Measured from the inside of the bottom of the box, the junction box must be located at least 8 in. (200 mm) above the maximum pool water level »680.24(A)(2)(a)«.

Ⓖ The junction box must be equipped with threaded entries/hubs or a specifically listed nonmetallic hub »680.24(A)(1)(1)«.

Ⓗ Locate the junction box no less than 4 in. (100 mm) above the ground level, or pool deck (measured from the inside of the bottom of the box) »680.24(A)(2)(a)«.

Ⓘ A swimming pool junction box must comply with 314.23(E) support provisions. It must be supported by *two* or more conduits (rigid or intermediate metal conduit only) threaded wrenchtight into the enclosure or hubs. Secure each conduit within 18 in. (450 mm) of the enclosure since all entries are on the same side. Note: RNC is not permitted for swimming pool junction box support.

Ⓙ Maximum water level is defined as the highest level that water can reach before it spills out »680.2«.

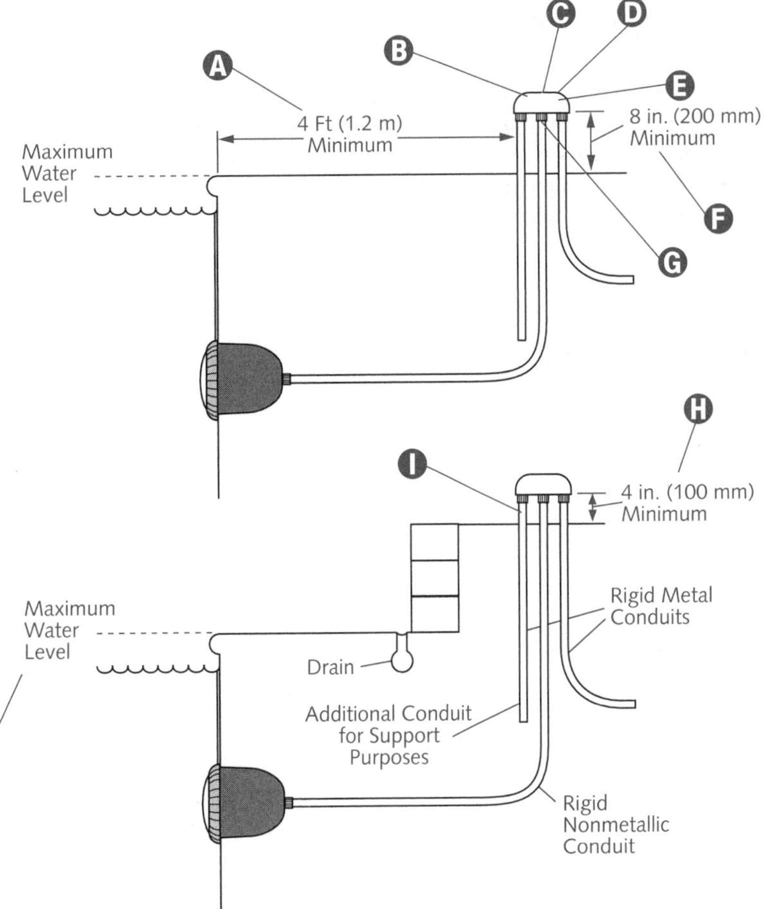

Equipment and Feeders

The following equipment must be grounded:

(1) Through-wall lighting assemblies and underwater luminaires (lighting fixtures), other than those low-voltage systems listed for the application without a grounding conductor

(2) All electrical equipment located within 5 ft (1.5 m) of the inside wall of the pool or fountain

(3) All electrical equipment associated with the pools or fountain's recirculating system

(4) Junction boxes

(5) Transformer enclosures

(6) GFCIs

(7) Panelboards (not part of the service equipment) that supply any pool associated electrical equipment »680.6«.

A separate building's panelboard can supply swimming pool equipment if the feeder meets 250.32 grounding requirements. If an equipment grounding conductor is installed, it must be an insulated conductor »680.25(B)(2)«.

Branch-circuits for pool-associated motors, in the interior of a one-family dwelling (or in another associated accessory building), can be installed in any of Chapter 3's wiring methods where meeting the requirements of this paragraph. Where run in a raceway, the equipment grounding conductor must be insulated. Where run in a cable assembly, the equipment grounding conductor can be uninsulated, but it must be enclosed within the cable assembly's outer sheath»680.21(A)(4)«.

The heating elements of electric pool water heaters must be subdivided into loads not exceeding 48 amperes, and protected at no more than 60 amperes. The branch-circuit conductor ampacity and the overcurrent protective device's rating/setting must be at least 125% of the nameplate rating's total load »680.9«.

Electric equipment must not be installed in rooms (or pits) without adequate drainage to prevent water accumulation during normal operation or filter maintenance »680.11«.

A permanently installed pool can have listed cord- and plug-connected pool pumps that incorporate an approved double insulation system and provide a grounding means for the pump's internal and inaccessible, noncurrent-carrying metal parts »680.21(B)«.

A The branch-circuits for pool-associated motors must be installed in RMC, IMC, RNC, or Type MC cable listed for location. Other wiring methods and materials are permitted in specific locations or applications as covered in this section. Any wiring method employed must contain a copper equipment grounding conductor sized per 250.122 but not smaller than 12 AWG »680.21(A)(1)«. Where installed on or within buildings, EMT is permitted »680.21(A)(2)«.

B One or more accessible disconnecting means from all ungrounded conductors must be provided within sight from all utilization equipment other than lighting »680.12«.

C 680.25 provisions apply to any feeder on the supply side of panelboards supplying branch-circuits for pool equipment covered in Part II and on the load side of the service equipment or the source of a separately derived system »680.25«. Feeders must be installed in RMC, IMC, LFNC, or RNC. EMC is permitted where installed on, or within, a building, and ENT is permitted where installed within a building »680.25(A)«.

D Where connections must be flexible at (or adjacent to) the motor, liquidtight flexible metal or nonmetallic conduit with approved fittings is permitted »680.21(A)(3)«.

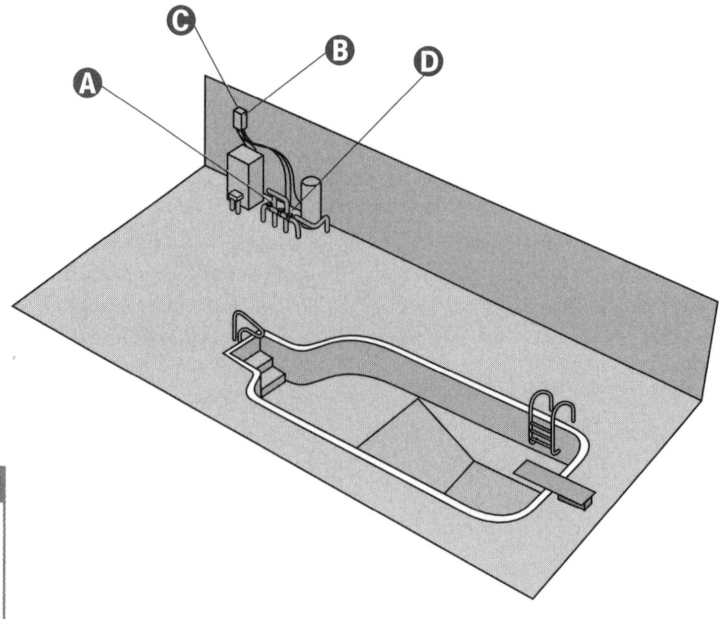

NOTE

The only underground wiring permitted under the pool, or within the area extending 5 ft (1.5 m) horizontally from the pool's inside wall, is the wiring necessary to supply pool equipment (permitted by this article). Where space limitations prevent wiring outside the 5-ft (1.5-m) restricted area, such wiring is permitted if installed in RMC, IMC, or a nonmetallic raceway system. All metal conduit must be corrosion resistant and suitable for the location »680.10«. Minimum burial depths are also provided in this section.

WARNING

Receptacles supplying pool pump motors rated 15 through 20 amperes, 120 through 240 volt, single phase, must have GFCI protection for personnel »680.22(A)(5)«.

Storage Pools

A luminaire (lighting fixture) installed in, or on, a storable pool's wall must be part of a cord- and plug-connected lighting assembly. This assembly must be listed as an assembly for the purpose, and have the following construction features:

(1) no exposed metal parts;

(2) a luminaire (fixture) lamp operating at no more than 15 volts;

(3) an impact-resistant polymeric lens, luminaire (fixture) body, and transformer enclosure; and

(4) a transformer meeting 680.23(A)(2) requirements with a primary rating not over 150 volts »680.33(A)«.

A lighting assembly (without a transformer) whose luminaire (fixture) lamp(s) operate at 150 volts or less, can be cord- and plug-connected where the assembly is listed as an assembly for the purpose. The installation must comply with 680.23(A)(5), and the assembly must have the following construction features:

(1) no exposed metal parts;

(2) an impact-resistant polymeric lens and luminaire (fixture) body;

(3) a GFCI with open neutral protection as an integral part of the assembly;

(4) the luminaire (fixture) lamp is permanently connected to the GFCI with open-neutral protection; and

(5) it complies with 680.23(A) requirements »680.33(B)«.

Ⓐ **Storable Swimming or Wading Pool:** Those constructed on or above ground with a maximum water depth capacity of 42 in. (1.0 m), or a pool with nonmetallic, molded polymeric walls, or inflatable fabric walls, of unlimited dimension »680.2«.

Ⓑ A wet location receptacle, into which is plugged a pool pump, requires a continuously weatherproof enclosure, i.e., whether the attachment plug is inserted or removed »406.8(B)(2)(a)«.

Ⓒ A cord-connected pool filter pump must incorporate an approved double insulation system, or its equivalent, and must have a means for grounding the appliance's internal and nonaccessible noncurrent-carrying metal parts only. The grounding means must consist of an equipment grounding conductor (run with the power-supply conductors) in the flexible cord that properly terminates in a grounding-type attachment plug with a fixed grounding contact member »680.31«.

CAUTION *Only GFCI protected electrical equipment (including power-supply cords) can be used with storable pools »680.32«. If flexible cords are used, see 400.4 »680.32 FPN«.*

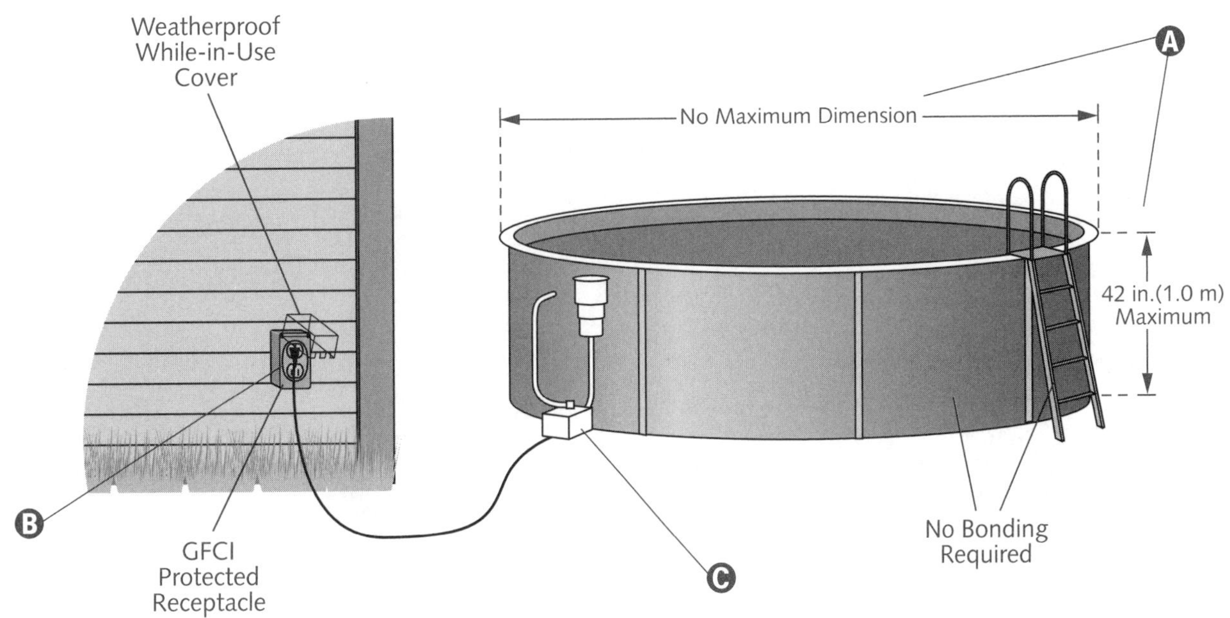

Spas and Hot Tubs

Receptacles providing spa or hot tub power must be GFCI protected »680.43(A)(3)«.

An indoor spa or hot tub must comply with Article 680 provisions (Parts I and II) unless modified by this section, and must be connected by a Chapter 3 approved wiring method. Listed spa and hot tub packaged units, rated 20 amperes or less, can be cord- and plug-connected to facilitate the removal or disconnection of the unit for maintenance and repair »680.43«.

All of the following must be grounded: (1) electric equipment located within 5 ft (1.5 m) of the spa or hot tub's inside wall; and (2) electric equipment associated with the spa or hot tub's circulating system »680.43(F)«.

A Luminaires (lighting fixtures), lighting outlets, and ceiling-suspended (paddle) fans located over the spa/hot tub, less than 5 ft (1.5 m) from the spa or hot tub's inside walls must not only be a minimum of 7 ft, 6 in. (2.3 m), above the maximum water level but must also be GFCI protected »680.43(B)(1)(b)«.

B Luminaires (lighting fixtures), meeting either (1) or (2) requirements, *and* protected by a GFCI, can be installed less than 7 ft, 6 in. (2.3 m), over a spa or hot tub:

(1) Recessed luminaires (fixtures) having a glass or plastic lens and nonmetallic or electrically isolated metal trim, suitable for damp location use

(2) Surface-mounted luminaires (fixtures), with a glass or plastic globe, and a nonmetallic body or a metallic body isolated from contact, and suitable for use in damp locations »680.43(B)(1)(c)«.

C Luminaires (lighting fixtures), lighting outlets, and ceiling-suspended (paddle) fans, located 12 ft (3.7 m) or more above the maximum water level, do not require GFCI protection »680.43(B)(1)(a)«.

D Switches must be located at least 5 ft (1.5 m), measured horizontally, from the spa or hot tub's inside walls »680.43(C)«.

E A readily accessible, clearly labeled emergency shutoff or control switch, for stopping the motor(s) that provide recirculating and jet system power, must be installed at least 5 ft (1.5 m) away, adjacent to, and within sight of the spa or hot tub. Note: This requirement does not apply to single-family dwellings »680.41«.

F Receptacles of 125 volts and 30 amperes or less, within 10 ft (3.0 m) of the spa or hot tub's inside walls, must be protected by a GFCI »680.43(A)(2)«. Determine this distance by measuring the shortest path the supply cord of an appliance connected to the receptacle would follow without piercing a building's floor, wall, ceiling, doorway with hinged or sliding door, window opening, or other effective permanent barrier »680.43(A)(4)«.

G At least one 125-volt, 15- or 20-ampere general-purpose branch-circuit receptacle must be located a minimum of 5 ft (1.5 m), and no more than 10 ft (3.0 m), from the inside wall of an indoor spa or hot tub »680.43(A)«.

H Locate property receptacles at least 5 ft (1.5 m), measured horizontally, from the spa or hot tub's inside walls »680.43(A)(1)«.

NOTE

A spa or hot tub installed ***outdoors*** *must comply with Article 680, Part I and II provisions unless otherwise permitted in the following: (A) listed packaged units utilizing a factory-installed assembly control panel or panelboard can be connected with 6 ft (1.8 m) or less of liquidtight flexible conduit, or can be cord- and plug-connected with a cord no longer than 15 ft (4.6 m), if protected by a GFCI. (B) bonding by metal-to-metal mounting on a common frame or base is permitted. The metal bands or hoops used to secure wooden staves do not require bonding as stipulated in 680.26 »680.42«.*

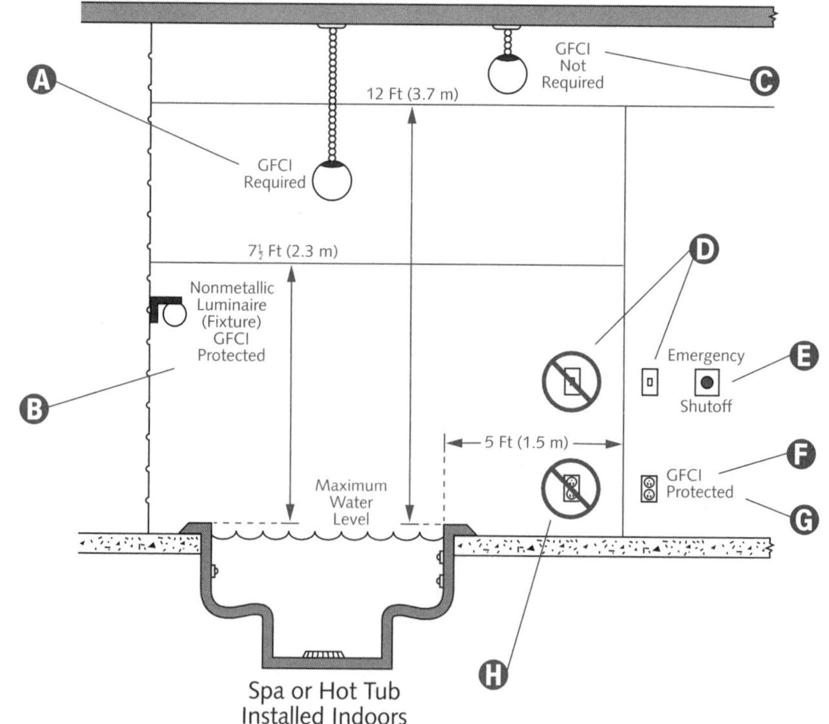

Pools and Tubs for Therapeutic Use

Outlet(s) supplying the following must be protected by a GFCI: (1) self-contained therapeutic tub or hydrotherapeutic tank; or (2) packaged therapeutic tub or hydrotherapeutic tank; or (3) field-assembled therapeutic tub or hydrotherapeutic tank »680.62(A)«.

A listed self-contained unit, or listed packaged equipment assembly, marked to indicate that integral GFCI protection is provided for all electrical parts within the unit or assembly (pumps, air-blowers, heaters, lights, controls, sanitizer generators, wiring, etc.), does not require that the outlet supply be GFCI protected »680.62(A)(1)«.

A therapeutic tub or hydrotherapeutic tank, rated greater than 250 volts or rated three-phase or with a heater load exceeding 50 amperes, does not require supply protection by means of a GFCI »680.62(A)(2)«.

All receptacles within 5 ft (1.5 m) of a therapeutic tub must be protected by a GFCI »680.62(E)«.

Ⓐ Article 680, Part I and Part VI provisions apply to therapeutically used pools and tubs in health care facilities, gymnasiums, athletic training rooms, and similar areas.Portable therapeutic appliances must comply with Parts II and III of Article 422 »680.60«.

Ⓑ All tub associated metal parts must be bonded by any of the following methods: (1) interconnection of threaded metal piping and fittings; (2) metal-to-metal mounting on a common frame or base; (3) suitable metal clamp connections; or (4) the provision of a solid copper bonding jumper (insulated, covered, or bare) no smaller than 8 AWG »680.62(C)«.

Ⓒ Tubs used for the therapeutic submersion and treatment of patients, which during normal use are not easily moved from place to place, or which are fastened or secured at a given location (including associated piping systems), must conform to 680.62(A) through (F) »680.62«.

Ⓓ All of the following must be **grounded**: (1) Electrical equipment located within 5 ft (1.5 m) of the tub's inside wall. (2) Electrical equipment associated with the tub's circulating system »680.62(D)(1)«.

Ⓔ The following parts must be bonded together:

(1) All metal fittings within, or attached to, the tub structure.

(2) Metal parts of electrical equipment associated with the tub's water circulating system, including pump motors.

(3) Metal-sheathed cables and raceways and metal piping within 5 ft (1.5 m) of the tub's inside walls and not separated from the tub by a permanent barrier.

(4) All metal surfaces within 5 ft (1.5 m) of the tub's inside walls not separated by a permanent barrier.

(5) Electrical devices and controls unrelated to the therapeutic tubs must either be located at least 5 ft (1.5 m) away from such units, or be bonded to the therapeutic tub system »680.62(B)«.

Ⓕ Therapeutic pools constructed in, or on, the ground, or within a building, in such a manner that the pool cannot be readily disassembled, must comply with Article 680, Parts I and II »680.61«. The limitations of 680.22(B)(1) through (B)(4) do not apply if all luminaires (lighting fixtures) are of the totally enclosed type »680.61 *Exception*«.

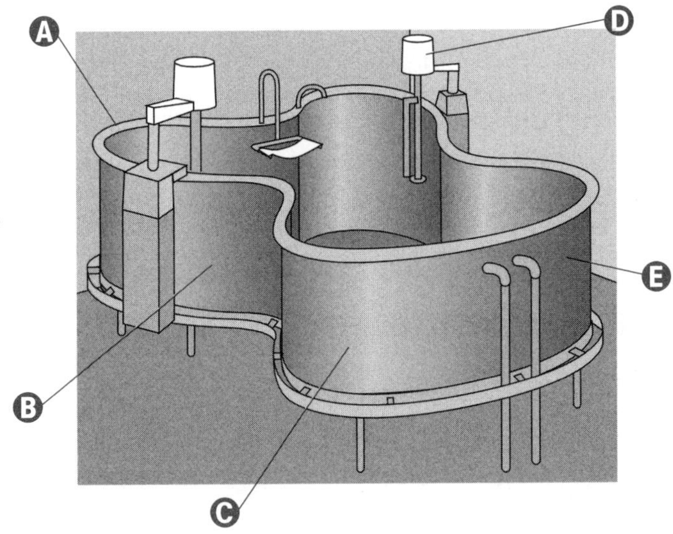

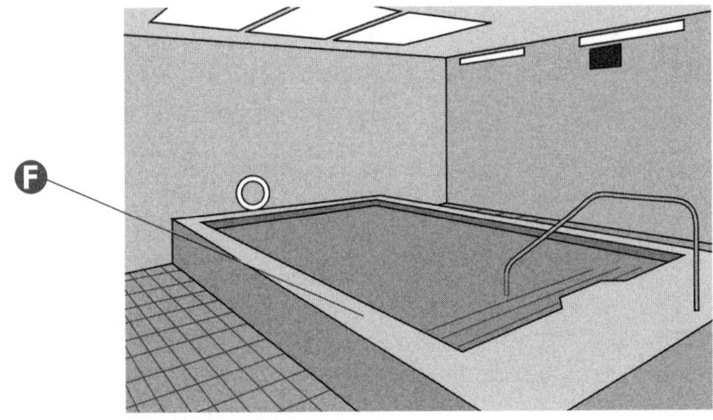

CAUTION *All luminaires (lighting fixtures) used in therapeutic tub areas must be of the totally enclosed type »680.62(F)«.*

Summary

- Article 600 covers the conductor and equipment installation for electric signs and outline lighting (including neon and skeleton tubing).
- Manufactured wiring system provisions are located in Article 604.
- Article 605 (Office Furnishings) covers electrical equipment, lighting accessories, and wiring systems used to connect, or contained within, wired partitions.
- Electric vehicle charging system provisions are located in Article 625.
- Audio signal processing, amplification, and reproduction equipment requirements can be found in Article 640.
- Article 645 covers information technology (data processing) equipment.
- Electrically operated pipe organ circuits and components employed for sounding apparatus and keyboard control are located in Article 650.
- Article 695 (Fire Pumps) covers the installation of electric power sources and interconnecting circuits as well as switching and control equipment dedicated to fire pump drivers.
- Elevator, dumbwaiter, escalator, moving walk, wheelchair lift, and stairway chair lift provisions are contained in Article 620.
- Article 680 provisions apply to the construction and installation of electrical wiring and equipment in, or near, all swimming, wading, therapeutic, and decorative pools, fountains, hot tubs, spas, and hydromassage bathtubs, whether permanently installed or storable, as well as to metallic auxiliary equipment (such as pumps, filters, etc.).
- Parts specified in 680.26(B) must be connected to a permanently installed pool's common bonding grid with an 8 AWG or larger solid conductor (insulated, covered, or bare).
- Very specific and detailed receptacle and lighting (including underwater) requirements are stipulated in Article 680.
- While Article 680, Part II covers permanently installed pools, storable pool requirements are found in Part III.
- Indoor spa and hot tub installations are covered in 680.43(A) through (G).
- Article 680, Part VI provisions apply to pools and tubs for therapeutic use in health care facilities, gymnasiums, athletic training rooms, and similar areas.
- Hydromassage bathtubs (and associated electrical components) are covered in Article 680, Part VII

Unit 19 Competency Test

***NEC*® Reference**	**Answer**	
________	________	1. Storable swimming pools are capable of holding water to a maximum depth of ____.in(es).
________	________	2. Overcurrent protection for feeders and branch-circuits supplying electric vehicle supply equipment shall be sized for _____ duty and shall have a rating of not less than _____ % of the maximum load of the electric vehicle supply equipment.
________	________	3. A clearly labeled emergency shutoff or control switch for the purpose of stopping the motor(s) that provide power to the recirculation system and jet system shall be installed at a point readily accessible to the users and at least _____ ft away, adjacent to, and within sight of the spa or hot tub.
________	________	4. Transformers used for the supply of underwater luminaires (fixtures) shall be of an isolated winding type with an ungrounded secondary that has a _____ between the primary and secondary windings.
________	________	5. Duty on elevator driving machine motors shall be rated as _____.
________	________	6. Not more than _____ ft of high-voltage cable shall be permitted in nonmetallic conduit from a high-voltage terminal of a neon transformer (over 1000 volts, nominal) supply to the first neon tube.

NEC® Reference	Answer	
________	________	7. All fixed metal parts that are within _____ ft horizontally of the inside walls of a permanently installed swimming pool, and within _____ ft above the pool's maximum water level, must be bonded together.
________	________	8. Electrically operated pipe organ circuits shall be arranged so that 26 and 28 AWG conductors shall be protected by an overcurrent device rated at not more than _____.
________	________	9. All live parts of electrical apparatus in the hoistways, at the landings, in or on the cars of elevators and dumbwaiters, in the wellways or the landings of escalators or moving walks, or in the runways and machinery spaces of wheelchair lifts and stairway lifts shall be _____.
________	________	10. Unless specifically listed for the purpose and location, the coupling means of electric vehicle supply equipment shall be stored or located at a height of not less than _____ in. and not more than _____ ft above an outdoor parking surface.
________	________	11. A(n) _____ is a structure designed to support a wet-niche luminaire (lighting fixture) assembly and intended for mounting in a pool or fountain structure.
________	________	12. Switches, flashers, and similar devices controlling transformers and electronic power supplies shall be rated for controlling inductive loads or have a current rating not less than _____ of the current rating of the transformer. a) 80% b) 100% c) 125% d) 200%
________	________	13. Audio system equipment supplied by branch-circuit power shall not be placed laterally within _____ ft of the inside wall of a pool, spa, hot tub, or fountain, nor within _____ ft of the prevailing or tidal high water mark.
________	________	14. Where a permanently installed pool is installed at a dwelling unit(s), at least one 125-volt, 15- or 20-ampere receptacle on a general-purpose branch-circuit shall be located a minimum of _____ ft from and not more than _____ ft from the inside wall of the pool.
________	________	15. The voltage at a fire pump controller's line terminals shall not drop more than _____ below normal (controller-rated voltage) under motor starting conditions.
________	________	16. Luminaires (lighting fixtures) located over a spa or within 5 ft (1.5 m) from the inside walls of the spa shall be permitted to be not less than _____ ft above the maximum water level where protected by a GFCI.
________	________	17. Traveling cable shall be permitted to be run without the use of a raceway for a distance not exceeding _____ ft in length as measured from the first point of support on the elevator car or hoistway wall, or counterweight where applicable, provided the conductors are grouped together and taped or corded, or in the original sheath.
________	________	18. Underground wiring shall not be permitted under the pool or within the area extending _____ ft horizontally from the inside wall of the pool unless this wiring is necessary to supply pool equipment permitted by this Article.
________	________	19. Hydromassage bathtub electrical equipment shall be _____ without damaging the building structure or building finish.
________	________	20. All 125-volt receptacles located within _____ ft of the inside wall of the pool or fountain shall be protected by a GFCI.
________	________	21. Manufactured wiring system cable shall be listed armored cable or metal-clad cable containing nominal 600-volt _____ AWG copper-insulated conductors with a bare or insulated copper equipment grounding conductor equivalent in size to the ungrounded conductor.

NEC® Reference	Answer	
________	________	22. Luminaires (lighting fixtures) mounted in permanently installed swimming pool walls shall be installed with the top of the luminaire (fixture) lens at least _____ in. below the normal water level of the pool.
________	________	23. Sign or outline lighting system equipment shall be at least _____ ft above areas accessible to vehicles unless protected from physical damage.
________	________	24. Existing luminaires (lighting fixtures) and lighting outlets located less than _____ ft measured horizontally from the inside walls of a pool shall be at least _____ ft above the surface of the maximum water level, shall be rigidly attached to the existing structure, and shall be protected by a GFCI.
________	________	25. A fixed or stationary electric sign installed inside a fountain shall be at least _____ ft inside the fountain measured from the outside edges of the fountain.
________	________	26. The conductors to the hoistway door interlocks from the hoistway riser shall be flame retardant and suitable for a temperature of not less than _____.
________	________	27. In outdoor pool areas, ceiling-suspended (paddle) fans shall not be installed above the pool or over the area extending _____ ft horizontally from the inside walls of the pool unless no part of the fan is less than _____ ft above the maximum water level.
________	________	28. The branch-circuit conductors supplying one or more units of a data processing system shall have an ampacity not less than _____ of the total connected load.
________	________	29. A junction box connected to a conduit that extends directly to a forming shell shall be provided with a number of grounding terminals that shall be _____.
________	________	30. Electric vehicle supply equipment shall be provided with a(n) _____ that de-energizes the electric vehicle connector and its cable whenever the electric connector is uncoupled from the electric vehicle.
________	________	31. Branch-circuits that supply neon tubing installations shall not be rated in excess of _____.
________	________	32. The termination of the 8 AWG equipment grounding conductor in a permanently installed swimming pool forming shell shall be covered with, or encapsulated in, a(n) _____ to protect the connection from the possible deteriorating effect of pool water.
________	________	33. Conductors supplying a fire pump motor(s), pressure maintenance pumps, and associated fire pump accessory equipment shall have a rating not less than _____ of the sum of the fire pump motor(s) and pressure maintenance motor(s) full-load current(s), and _____ of the associated fire pump accessory equipment.
________	________	34. Switching devices shall be located at least _____ ft horizontally from the inside walls of a pool unless separated from the pool by a solid fence, wall, or other permanent barrier.
________	________	35. Article _____ covers electrical equipment, lighting accessories, and wiring systems used to connect, contained within, or installed on relocatable wired partitions.
________	________	36. A junction box connected to a conduit that extends directly to a forming shell shall be located not less than _____in., measured from the inside of the bottom of the box, above the ground level, or pool deck, or not less than _____ above the maximum pool water level, whichever provides the greater elevation.

Index

M

N

O

S